Best regards

Jim Moore

&

Dorothy M. Moore

CAPSOL-T

Greentea + Guajillo red Pepper

[illegible]

ECTO-NOX Proteins

D. James Morré · Dorothy M. Morré

ECTO-NOX Proteins

Growth, Cancer, and Aging

D. James Morré
Mor-NuCo, LLC
Purdue Research Park
West Lafayette, IN, USA

Dorothy M. Morré
Mor-NuCo, LLC
Purdue Research Park
West Lafayette, IN, USA

ISBN 978-1-4614-3957-8 ISBN 978-1-4614-3958-5 (eBook)
DOI 10.1007/978-1-4614-3958-5
Springer New York Heidelberg Dordrecht London

Library of Congress Control Number: 2012941093

Printed on acid-free paper

Springer is part of Springer Science+Business Media (www.springer.com)

Preface

One does not discover new lands without consenting to lose sight of the shore for a very long time.

André Gide

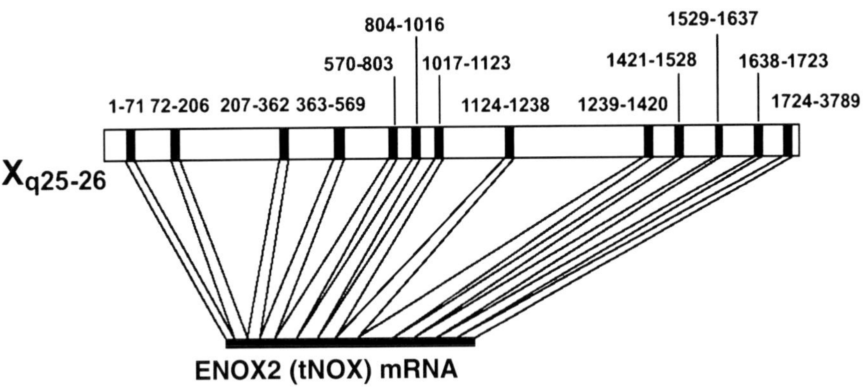

The discovery of the ENOX proteins was inexorably interwoven with the concept of a plasma membrane electron transport first indicated from the findings of Cathy Frantz, a masters student at Purdue University. Cathy found microsomal electron transport activities associated with highly purified plasma membranes from rat livers. The first report of these findings in 1973 at the American Society of Cell Biology meetings was greeted with indifferent disbelief. Clear exceptions were Prof. Fred Crane of Purdue University, a member of Cathy's examining committee and Fred's close friend and colleague, Prof. Hans Löw of the Karolinska Institute in Sweden. The Morrés continued to pursue these observations in their efforts to solidify the flow-differentiation model of Golgi apparatus functioning whereas Fred and Hans set upon a quest to discover the elusive plasma membrane electron transport system. That quest culminated with two co-edited volumes, one on animals in 1990 and one on plants in 1991 entitled *Oxidoreduction at the Plasma Membrane: Relation to Growth and Transport* published by CRC Press (Crane et al. 1990b, 1991). The existence of growth-related NADH oxidases of the plant and animal plasma membrane were first reported in 1986 from plants (Morré et al. 1986a) but it was not until 1990 (Morré and Crane 1990; Fig. 4.8; cover) that their role as the terminal oxidases of the plasma membrane electron transport chain was correctly formulated.

We thank our good friends and colleagues Frederick L. Crane and Hans Löw for championing the concept that the plasma membranes might contain redox active proteins with important functions in growth and disease even in the face of general non-acceptance of that notion by the overall scientific community and a universal lack of enthusiasm by extramural funding agencies for studies of plasma membrane electron transport no matter how well conceived.

It is a singular but sometimes lonely privilege to write a book on a potentially important new family of proteins virtually single handedly identified, cloned, characterized and clinically implemented with minimal independent outside confirmation at the time. While masquerading as intractable proteins, they have and continue to offer remarkable opportunities for research, commercial development and outside confirmation (Chap. 11; Table 11.1). The latter underscores the complexity of these proteins and the many unusual difficulties especially in their assay. We recall a visit from Dr. Warren McKellar of Eli Lilly to our laboratory in the very early days of anti-cancer sulfonylurea research to carry out a spectrophotometric assay on his own to validate the activity. He was pleased to see a rate only to remark a few moments later that it had stopped momentarily only to start up again a few minutes later. Several more years were required before the true meaning of that observation and its subsequent reproduction with other systems finally led to the still controversial conclusion that ENOX activities were oscillatory.

The ENOX proteins were discovered as a result of a search for a growth-related protein at the cell surface that was the target for immobilized forms of the anticancer drug doxorubicin (Adriamycin®). In the late 1970s several groups demonstrated that doxorubicin exhibited an enhanced anti-cancer activity if it was first conjugated to an impermeant support and not permitted to enter the cell (Chap. 11). As doxorubicin is a redox-active quinone site inhibitor, a redox protein was sought that was at the cell surface and cancer-specific. The search led eventually to discovery of the ECTO-NOX (ENOX) family of external hydroquinone oxidases, also capable of oxidizing external NADH. One subset of the ENOX proteins, the tumor-specific tNOX or ENOX2 proteins, was inhibited by doxorubicin and other quinone site-targeted anti-cancer drugs, was cancer-specific, was absent from the surface of non-cancer cells and tissues and was the first ENOX protein to be cloned (Chueh 1997; Chueh et al. 2002b). Three additional family members including the constitutive human ENOX1 in 2008 (Jiang et al. 2008) followed with the most recent, an age-related ENOX protein, cloned in 2010 and a constitutive ENOX1 from plants in 2011. Being proteins of the external cell surface and lacking *trans*-membrane helices to anchor the protein in the membrane, ENOX proteins were found to be shed and to appear in soluble form in patient sera and urine where they serve as early diagnostic markers for cancer presence and organ site to permit very early intervention strategies prior to advanced disease and metastatic spread (Chap. 12).

Findings of Chap. 8 suggest that the cancer- or tumor-related ENOX2 (tNOX) proteins are all splice variants from a single gene. More importantly, each major type of human cancer is characterized by characteristic transcript variants of unique molecular weight and isoelectric point (Chap. 12). Using a proteomics approach and a recombinant antibody specific for a common exon, it is possible not only to detect cancer but to diagnose it as well.

The purpose of this book is to document this unique family of cell surface proteins (the ECTO-NOX or ENOX protein family) involved in growth, biological time keeping, cancer, aging and viral infections and having properties of prions. The ENOX proteins are the exclusive discovery (subsequently confirmed by others) of the authors, Drs. D. James and Dorothy Morré, and their students and research associates at Purdue University. Roles in plasma membrane electron transport (Chap. 4), growth (Chap. 5), biological time keeping (Chap. 6), cancer (Chaps. 8, 11 and 12), prevention of viral infections (Chap. 7), crop production through control of plant growth (Chap. 10), and coronary artery disease and skin aging (Chap. 9), are among the many developing opportunities for new discovery and commercialization surrounding the ENOX proteins.

The book provides an ENOX-based mechanism for how cells become larger (increase in size) that is both unique and well documented with applications not only to cancer and cancer therapy but for production agriculture as well (Chap. 10) with increase of biomass for biofuel production as one exciting future prospect.

Finally, the concept of and the evidence for oscillations in the ratios of electron spin pairs defining *ortho* and *para* water as the basis for highly coordinated populations of coherent water that appear vital to water's biological and physical properties is completely new, of interest to the biological and physical sciences and now becoming widely accepted by the physical scientists involved with the study of the properties of water (Chap. 6).

Special mention is accorded to Michael Berridge, Frederick Crane, Iris Sun, Rita Barr and Hans Löw who have unwaveringly promoted plasma membrane redox and a functional role of ENOX proteins in the overall process, to the late Albert Overhauser for encouragement to seek an explanation of the oscillatory patterns of ENOX proteins at the atomic level (Chap. 6), to Michael Böttger for assistance with pivotal growth measurements (Chap. 5), to Ron Brightmore for his inspiring surveys of the literature and to Profs. Jacob Levitt and Hale Fletchall of the University of Missouri for planting the initial seeds of inquiry. Special thanks to Don Lee, Tom Shelton, Graham Kelly and Richard Greaves for recognizing the commercial potential of the ENOX protein family.

We express our appreciation to the many colleagues, postdoctorals, graduate students, undergraduate assistants, and technicians whose invaluable assistance made possible the experimental studies especially as graduate students, Andrew Brightman, P.-J. Chueh, Chinpal (James) Kim, Xiaoyu Tang and Ziying Jiang for ground breaking protein purification and molecular cloning efforts. Appreciation is extended as well to the even greater numbers who challenged and criticized the work to force us to work even more diligently to distinguish among possible interpretations of the findings. We thank Peggy Runck for manuscript preparation and Aya Ryuzoji for preparation of the figures. We are especially indebted to the unwavering support of the Morré children, Connie, Jeffrey and Suzanne, and grandchildren, Christopher, Eric and Katherine Chalko, Matthew, Timothy and Nicholas Miner, Suzanna Morré and Aren and Mariah Rudder. May our ENOX proteins always oscillate in synchrony.

West Lafayette, IN, USA

D. James Morré
Dorothy M. Morré

Contents

1 The ENOX Protein Family 1
1.1 The ENOX Protein Family Members 1
1.2 ENOX Proteins Are Associated with the External Cell Surface as Ecto Proteins and Are Shed into the Environment 4
1.3 Two Activities of ENOX Proteins Alternate 6
1.4 ENOX Proteins Participate Directly in the Enlargement Phase of Cell Growth 7
1.5 ENOX Proteins Are Resistant to Degradation and Tend to Form Insoluble Aggregates 7
1.6 ENOX Proteins Are Dicopper Proteins Lacking Both Iron and Flavin 8
1.7 The Oscillatory Behavior Complicates Assays of ENOX Activities 10
1.8 The Distinctive 2 + 3 Pattern of ENOX Oscillations Is a Unifying Characteristic of All Family Members 11
1.9 ENOX Proteins Differ Markedly from the NOX Proteins of Host Defense 13
1.10 ENOX Proteins Are of Low Specific Activity 14
1.10.1 Natural Electron Donors and Acceptors for Cell Surface-Associated ENOX Proteins 14
1.10.2 Hydroquinones as Natural Electron Donors 15
1.10.3 Reduced Pyridine Nucleotides as Artificial Electron Donors 15
1.10.4 Protein Thiols and Tyrosines as Electron Donors for arNOX Proteins and Generation of Superoxide 15
1.10.5 Aggregation and Formation of Amyloid 16
1.11 Why an External NADH Oxidase? 17
1.12 Summary 17

2 **Measurements of ECTO-NOX (ENOX) Activities** 19
2.1 Spectrophotometric Assay of NADH Oxidase 21
2.2 Statistical Analysis 22
2.3 Data Reduction Methods 23
2.3.1 Diode Array Instruments 30
2.4 Measurement of Hydroquinone Oxidase Activity with Reduced Coenzyme Q_{10} or Phylloquinone as Substrate 32
2.4.1 Enzyme Assay for Reduced Coenzyme Q_{10} Oxidase 32
2.4.2 Enzyme Assay for Reduced Phylloquinone Oxidase 33
2.5 Dissolved Oxygen Measurement 35
2.6 Estimation of Protein Disulfide-Thiol Interchange Activity 38
2.7 Preparation of Scrambled RNase Substrate 38
2.8 Estimates of Protein Disulfide-Thiol Interchange from Enzymatic Assay of Dipyridyl-Dithio Substrate Cleavage 41
2.9 Measurement of *Trans*-Plasma Membrane Redox by Reduction of Cell-Impermeable Dyes 42
2.9.1 CoQ_1 Can Function as an Intermediate Electron Carrier in WST-1 Reduction 42
2.9.2 Measurement of Plasma Membrane Electron Transport Based on WST-1 Reduction 42
2.10 Summary 44

3 **The Constitutive ENOX1 (CNOX)** 47
3.1 ENOX1 Function 47
3.2 ENOX1 Cloning 50
3.3 ENOX1 Characterization 53
3.4 ENOX1 Activity Requires the Presence of Copper 57
3.5 Copper Binding and Site-Directed Mutagenesis of Potential Copper-Binding Sites 58
3.6 Response to Nucleotides 60
3.7 Aggregation and Electron Microscopy 61
3.8 ENOX1 Fulfills Essential Roles in Cell Enlargement and Cellular Time-Keeping 62
3.9 ENOX1 of Human Platelets 62
3.10 Summary 62

4 **Role in Plasma Membrane Electron Transport** 65
4.1 Composition of the PMET 65
4.1.1 NADH Coenzyme Q Reductases 66
4.1.2 Hydroquinones 68
4.1.3 Terminal Oxidases 69
4.2 Electron Donors and Acceptors 73
4.3 Rates of PMET 73
4.4 Energetics of PMET 76
4.5 PMET Driven Outward Proton Pumping and Alkalinization of the Cytoplasm 78

4.6 PMET Function in Electron Import 80
4.7 PMET and Growth 81
4.7.1 Cell Cycle Check Point Control of Cell Enlargement 81
4.7.2 PMET Activity and Growth Are Correlated 82
4.8 Regulation of PMET 82
4.8.1 Feedback Regulation of PMET 85
4.8.2 ENOX Cell Surface Receptor Proteins 86
4.9 PMET and Glycolysis 86
4.9.1 PMET and Glycolysis of Cancer Cells 86
4.10 PMET Links to Major Signaling Pathways 88
4.10.1 Sirtuins 89
4.10.2 Sphingolipid Rheostat 91
4.10.3 NADH Modulation of PTEN Provides Link of PMET to Ras-Raf-Mek-Erk, PI3-AKT-mTOR and NF-κB 91
4.10.4 AMP-Activated Protein Kinase 93
4.10.5 Hypoxia 93
4.11 NAD^+ Homeostasis 95
4.12 Summary 95

5 Role in the Enlargement Phase of Cell Growth 97
5.1 Cell Enlargement Linked to ENOX Activities 97
5.2 ECTO NADH Oxidase of Liver Plasma Membranes Stimulated by Hormones and Growth Factors 98
5.3 ECTO NADH Oxidase of Rat Hepatoma Plasma Membrane Constitutively Activated and No Longer Growth Factor or Hormone-Responsive 98
5.3.1 Thiol Reagents 100
5.4 Relationship to Growth 101
5.4.1 Plants 101
5.4.2 Vertebrate Cells 113
5.5 Pathological Implications 123
5.5.1 Apoptosis 123
5.6 Physical Membrane Displacements 124
5.7 ATP- and p97 AAA-ATPase-Dependent and Drug-Inhibited Vesicle Enlargement Reconstituted Using Synthetic Lipids and Recombinant Proteins 124
5.8 Summary 138

6 Roles as Ultradian Oscillators of the Cells Biological Clock 141
6.1 Time Keeping Properties 141
6.2 Molecular Studies 142
6.3 Studies with Deuterium Oxide 148
6.4 The Role of Copper 149
6.5 The Copper Clock 150

6.6 EXAFS Investigations 152
6.7 Oscillations Inherent in the Structure of Water.............................. 154
6.8 Period Length Determined by Ionic Radius of Liganded Cation 158
6.9 Spectral Evidence for Disequilibrium of *ortho:para* Spin States in Liquid Water That Oscillate 161
6.10 Other Mechanisms Proposed for *ortho/para* Conversions and Departures from Their Equilibrium Ratio of 3:1 164
6.11 The 24-min Period Has Properties of a Carrier Wave Generated from the Basic Underlying *ortho–para* Water Oscillations? The Heart Rate Model 166
6.11.1 Growth Oscillations of Elongating Pollen Tubes 168
6.12 Phasing of the Rhythm.. 168
6.12.1 EMF Sets the Copper Clock 169
6.13 A Mechanism to Explain How Oscillations of Redox Potential of Aqueous Solutions Become Synchronous and Remain So 172
6.14 ENOX Clock and Cancer... 177
6.15 Are ENOX Oscillators Linked to the Drivers of the Circadian Clock and How Are They Linked?...................... 178
6.16 Why Oscillate? A Consequence of Active Sites in Metalloproteins? 184
6.17 Summary 185

7 Other Potential Functional Roles of ENOX Proteins 187
7.1 Cell Cycle Control 187
7.2 Gene Regulation.. 187
7.3 Endomembrane Function, Membrane Displacements, Vesicle Budding 188
7.3.1 Membrane Budding 188
7.3.2 Energy Requirements for Physical Membrane Displacement 190
7.4 Endocytosis and Autophagy... 192
7.5 Host Defense... 193
7.6 pH Control 193
7.7 Lipid Oxidation.. 195
7.7.1 arNOX Inhibitors and Prevention of Coronary Artery Disease 197
7.8 Life Extension and Calorie Restriction.. 197
7.9 Control of Apoptosis and Cell Survival... 198
7.10 Neurodegenerative Disorders... 200
7.11 Memory.. 201
7.12 Gametogenesis 202

7.13 Role in Viral Pathogenesis ... 203
7.13.1 ENOX2 Inhibitor (−)-Epigallocatechin-3-Gallate Blocks Virus Infections Alone and in Combination with Capsicum Vanilloids and Other Green Tea Catechins ... 205
7.13.2 Brefeldin A and Antitumor Quassinoids ... 207
7.14 Summary ... 209

8 ENOX2 (tNOX) and Cancer ... 211
8.1 ENOX2 Discovery ... 211
8.2 ENOX2 Activity ... 213
8.2.1 Biochemistry ... 217
8.3 Sequence ... 218
8.4 Structural Properties ... 222
8.4.1 ENOX2 Protein Phosphorylation ... 228
8.5 ENOX2 Presence and Cancer ... 228
8.5.1 ENOX2 Autoantibodies Generated in Cancer Patients ... 230
8.5.2 ENOX2 Gene Present in Genome as a Single Copy ... 230
8.5.3 ENOX2 Lacks Intrinsic Membrane-Binding Motifs ... 230
8.5.4 ENOX2 Has Properties of a Prion and Is Protease Resistant ... 231
8.6 ENOX2 Has Characteristics of an Oncofetal Protein ... 233
8.7 Transgenic Mouse Strain Overexpressing ENOX2 ... 235
8.8 Alternative Splicing as Basis for Specific ENOX2 Localization to the Cell Surface ... 242
8.8.1 Full-Length ENOX2 MRNA Identical to That of Cancer Cells Exists in Human Non-cancer Cells and Tissues ... 247
8.8.2 Full-Length 71 kDa ENOX2 Protein Not Translated ... 248
8.8.3 Cancer-Specific Expression of ENOX2 ... 250
8.8.4 Splice Variants of ENOX2 Were Found in Cancer Cells ... 250
8.8.5 Expression of Exon 4 Minus and Exon 5 Minus Forms of ENOX2 in COS Cells ... 250
8.8.6 Delivery of 34 kDa ENOX2 Protein to the Plasma Membrane ... 252
8.8.7 Mutation of Met 231 Blocked Expression of the Exon 4 Minus Splice Variant ... 252
8.8.8 Subcellular Localization of E4m ENOX2-EGFP and Full-Length ENOX2-EGFP Fusion Proteins ... 253
8.8.9 Regulation of ENOX2 Expression ... 256
8.9 hnRNP F Splicing Factor Directs Formation of the Exon 4 Minus Variant of ENOX2 ... 257
8.10 Summary ... 259

9 Age-Related ENOX Proteins (arNOX) 261
9.1 arNOX Discovery 264
9.2 Measurement of Superoxide Formation by arNOX 266
9.3 Characteristics 268
9.4 arNOX Cloning 272
9.5 Characterization of Recombinant arNOX Proteins 279
9.6 arNOX as a Biomarker of Aging 284
9.7 Role in Skin Aging 286
9.8 Role in Oxidation of Serum Lipoproteins 291
9.9 arNOX Activity Correlates with Life Span in Sea Urchins 297
9.10 arNOX in Plants 298
9.11 arNOX Inhibitors 299
9.11.1 Coenzyme Q 299
9.11.2 Botanical Sources of arNOX Inhibitors 305
9.12 Beneficial Biological Function Associated with Superoxide Production: Physiological Roles of Superoxide 307
9.13 NQO1 (Cytoplasmic NAD(P)H: Quinone Oxidoreductases, DT-Diaphorase EC 1.6.99.2) and Plasma Membrane Electron Transport 309
9.14 Summary 310

10 The Auxin-Stimulated ENOX and Auxin Stimulation of Plant Growth 313
10.1 Early Evidence for Auxin-Modulated Enzymes Involved in Plant Cell Enlargement 313
10.2 The Plasma Membrane as the Subcellular Location of the Auxin-Responsive Mechanism 317
10.3 Direct Effects of Auxin on Signaling Molecules Fail to Parallel Those of Mammalian Growth Factors 318
10.4 Evidence for a Redox-Related Plasma Membrane-Located Auxin Target 321
10.4.1 Separation of Auxin-Activated and Constitutive NADH Oxidase Activities 327
10.5 Auxin-Stimulated NADH Activity and Growth Oscillates with a Period Length of 24 min 327
10.6 Golgi Apparatus Transport Important to Sustained Cell Enlargement but not Specifically Required for Auxin-Induced Cell Enlargement 331
10.7 Response of ENOX of Isolated Plasma Membrane Vesicles to Osmotica 331
10.8 Inhibitors of the Auxin-Stimulated NADH Oxidase of Plants 332
10.9 Cell Elongation Oscillates with a Period of 24 min and Exhibits a Second set of Oscillations in Response to 2,4-D 335

10.10 The Auxin-Stimulated ENOX Has Properties of a Prion: How 2,4-D Kills Plants 337
10.11 Summary 347

11 Cancer Therapeutic Applications of ENOX2 Proteins 345
11.1 PMET as a Target for Anticancer Drug Development 348
11.1.1 Arsenicals as Unspecific Anticancer PMET Inhibitors 349
11.2 Inhibition of PMET and Induction of Apoptosis 350
11.2.1 Mechanism of Induction of Apoptosis When Plasma Membrane Electron Transport Is Inhibited 351
11.3 Mechanism of Growth Arrest When Plasma Membrane Electron Transport Is Inhibited 355
11.3.1 Elevation of Ceramide 355
11.3.2 Links for Elevated Ceramide and Cell Cycle Arrest 355
11.3.3 ENOX2 Inhibitors Slow the Growth of HeLa Cells and Induce Apoptosis in Cancer But Not in Noncancer Cells 356
11.3.4 ENOX Inhibitors Increase Cytosolic NADH Levels 358
11.3.5 Increased NADH Resulting from ENOX2 Cell Surface Inhibition Inhibits Plasma Membrane-Associated Sphingosine Kinase (SK) and Lowers Levels of Prosurvival Sphingosine-1-Phosphate 359
11.3.6 Sphingomyelinase 359
11.4 ENOX2 Inhibitors 362
11.4.1 Vanilloids (Capsaicinoids) as PMET Inhibitors 363
11.4.2 Anthracycline Antibiotics 368
11.4.3 Cisplatin Targets ENOX2 of the PMET 377
11.4.4 Antitumor Sulfonylureas 378
11.4.5 Antitumor Quassinoids Target ENOX2 387
11.4.6 Acetogenins 390
11.4.7 EGCg 391
11.4.8 Phenoxodiol Targets the PMET Through Inhibition of ENOX2 400
11.4.9 Sulforaphane 408
11.4.10 Suramin 409
11.4.11 Callipeltin 409
11.5 Nonsteroidal Anti-inflammatory Drugs 410
11.6 Retinoids and Calcitriol Agents of Differentiation 411
11.6.1 Retinoic Acid Inhibition of PMET 411
11.6.2 Retinoid Inhibition of ENOX2 411
11.7 ENOX2-Directed Therapeutic Antibodies 412

11.8 Antisense ... 414
11.9 ENOX Inhibitors Enhance the Response of Tumors to Radiation ... 415
11.10 ENOX2 as a Target for Cancer Prevention Through Early Intervention ... 415
11.11 Summary ... 416

12 Cancer Diagnostic Applications of ENOX2 Proteins ... 419
12.1 Cancer Cell Surface ENOX2 Shed into Sera as Biomarkers of Cancer Presence ... 420
12.1.1 Biomarker Discovery Based on ENOX2 Activity ... 421
12.1.2 Characteristics of ENOX2 as Cancer Biomarker Based on Activity ... 421
12.1.3 Transcript Variants Detected by Two-Dimensional Gel-Western Blot Analysis Are Cancer Site-Specific Biomarkers ... 424
12.1.4 Transcript Variants of ENOX2 ... 425
12.2 Two-Dimensional Gel-Western Blot Cancer Detection System ... 425
12.2.1 Two-Dimensional Gel-Western Blot Analysis of ENOX2 Transcript Variants Provide for Very Early Detection ... 428
12.3 Early Intervention ... 430
12.4 ENOX2 Autoantibodies May Preclude Conventional ELISA Tests for Early Appearance of ENOX2 in Serum or Plasma ... 430
12.5 Lipid-Associated Sialic Acid (LASA) Fractions from Sera of Cancer Patients Contain ENOX2 Fragments as Major Nonlipid Constituents ... 432
12.6 RTPCR Detection of Cancer Cells in Blood Based on Presence of ENOX2 Splice Variant mRNA ... 433
12.7 Summary ... 433

Epilogue—Remaining Challenges ... 435

Appendix Detailed Description of Two Dimensional Gel Electrophoresis-Western Blot Early Cancer Detection Protocol ... 441

References ... 443

Index ... 493

Chapter 1
The ENOX Protein Family

ENOX (ECTO-NOX = Ecto-Nicotinamide Dinucleotide Oxidase Disulfide Thiol Exchanger) denotes a family of cell surface proteins, exhibiting a cyanide-insensitive, time-keeping $CoQH_2$ (NAD(P)H) oxidase (NOX) activity and a protein disulfide-thiol interchange activity that alternate (Morré 1998c; Morré and Morré 2003a). The ECTO designation derives from their external location on the outer surface of the plasma membrane (Fig. 1.1) and to distinguish them from the *phox*Nox proteins of host defense (i.e., Lambeth et al. 2000). This external location and alternation of oxidative and protein disulfide interchange activities has been demonstrated for a wide range of animal and plant tissues and cell lines.

1.1 The ENOX Protein Family Members

Based on response to various effectors, four distinct groupings of ENOX proteins have been described (Table 1.1), all characterized by properties, unprecedented in the biochemical literature, of having two distinct biochemical activities, hydroquinone (NAD(P)H) oxidation and protein disulfide-thiol interchange, that alternate (Fig. 1.2). One (CNOX or ENOX1) is constitutive with an activity that is hormone-responsive but largely drug-resistant and is widely distributed among animals, plants, and yeasts (Chap. 3). A second ENOX activity, studied mostly in man, is tumor- or cancer-associated and designated tumor-associated ENOX (tNOX or ENOX2) (Chap. 8). ENOX2 proteins are inhibited by a series of quinone site inhibitors all with anticancer activity including the vanilloid, capsaicin (Morré et al. 1995b, 1996b), the antitumor sulfonylurea LY181984 (Morré et al. 1995f, g, h, i), doxorubicin (Adriamycin®) (Morré et al. 1997c), and the isoflavene phenoxodiol (Morré et al. 2007a). ENOX2 represents a largely unexploited potential cancer drug and therapeutic target (Chap. 11).

The constitutive NADH oxidase (ENOX1 or CNOX) is located on chromosome 13q14.11 (Jiang et al. 2008; Table 1.1). The tumor-associated NADH oxidase

D.J. Morré and D.M. Morré, *ECTO-NOX Proteins: Growth, Cancer, and Aging*,
DOI 10.1007/978-1-4614-3958-5_1,

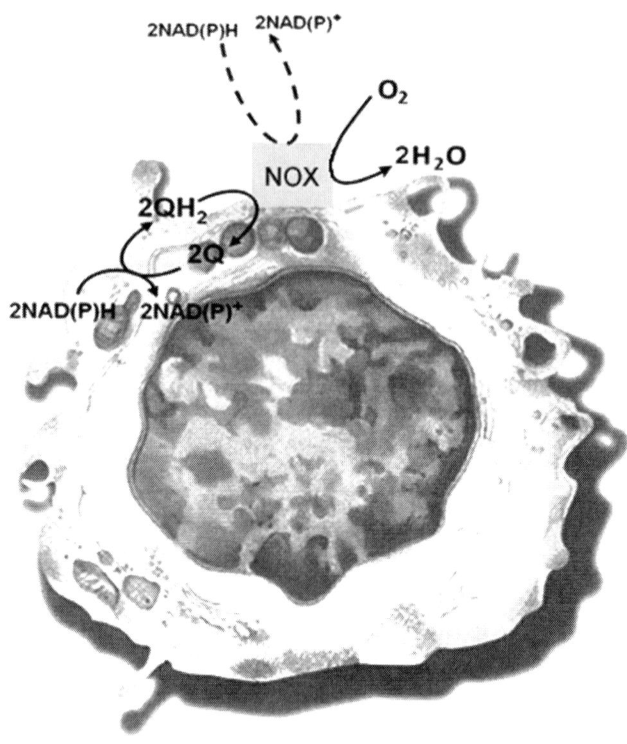

Fig. 1.1 Diagrammatic representation of the cell surface location of the ENOX proteins and their function in plasma membrane as terminal oxidases of a plasma membrane electron transport chain whereby cytosolic NAD(P)H is oxidized to reduce hydroquinone (QH_2) which serves as a trans-membrane electron carrier to deliver reducing equivalents to the ECTO-NOX (ENOX) proteins at the cells' exterior where they carry out four electron transfers to reduce molecular oxygen to $2H_2O$. External NAD(P)H also can serve as an artificial donor for the ENOX proteins to directly reduce molecular oxygen to $2H_2O$

Table 1.1 ENOX activity forms

	GenBank	Chromosome	Source	Inhibitors
ENOX1 (CNOX)[a]	EF432052	13q.14.11	Animal Plant Yeast/bacteria	Simalikalactone D
ENOX2 (tNOX)	AF207881	Xq25-q26.2	Cancer cells and tissues Cancer patient sera	Thiol reagents Antitumor drugs (Chap. 11)
TM9SF1a (arNOX)	NM-006405	14q11.2	Cells, tissues, and body fluids of aged individuals	Coenzyme Q_{8-10} Tyrosol Savory
TM9SF1b (arNOX)	NM-01014841	14q11.2		
TN9SF2 (arNOX)	NM-004800	13q32.3		
TM9SF3 (arNOX)	NM-020123	10q24.1		
TM9SF4 (arNOX)	NM-014742	20q11.21		

[a]Phased by melatonin

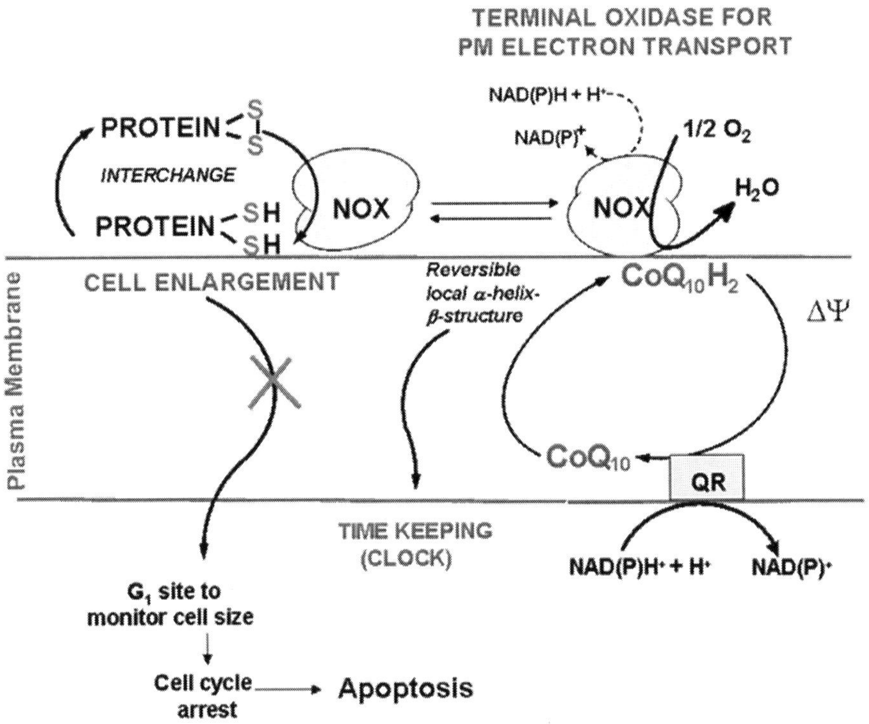

Fig. 1.2 In the oxidative portion of the ENOX cycle, the ENOX proteins function as terminal oxidases of plasma membrane electron transport (Chap. 4) to transfer reducing equivalents of NAD(P)H from the cytosol via two electron transfers mediated by membrane quinones (coenzyme Q in mammalian cells and vitamin K in plant cells). The protein disulfide interchange portion of the cycle is essential to the enlargement phase of cell growth (Chap. 5). The alternation of activities between the oxidative and protein disulfide-thiol interchange portions of the ENOX cycles lead to a periodicity which imparts a time-keeping function to members of the ENOX protein family (Chap. 6)

(also known as ENOX2 or tNOX) is present on the surface of cancer cells and located on chromosome Xq25-q26.2 (Chueh et al. 2002b). The age-related NADH oxidases (arNOX or ENOX3) that appear around age 30 and increase steadily thereafter have been identified as five members of the TM9 superfamily all with different chromosomal locations (Chap. 9). The TM9 arNOX proteins initially are membrane anchored. They are functionally similar to other ENOX proteins and are expressed with their N-termini exposed at the cell's exterior. These three ECTO-NOX forms have been cloned (GenBank Accession No. EF432052 for ENOX1 and AF207881 for ENOX2; see Chap. 9 for ENOX3). Human ENOX1 sequence shares 64 % identity and 80 % similarity with ENOX2. Except for critical functional motifs located near their N-termini (Table 1.2), ENOX3 proteins have sequences distinct from both ENOX1 and ENOX2 and have commonality only in their nine transmembrane spanning domains.

Table 1.2 Functional motifs comparing human ENOX1, ENOX2, and arNOX and plant dNOX

Motifs	ENOX1	ENOX2	arNOX (TM9SF2)	dNOX (ABP20)
Protein disulfide	C117KSC	C505XXXXC[a]	C120KLVC	C44KK
Drug binding [EEMTE]	Absent	E394EMTE	Absent	Absent[b]
Copper site I [H(Y)XH(Y)]	H260YSEH	H546VH	Y150QH	H106TH
Copper site II [H(Y)XH(Y)]	H579VH	Y560LH	H242TH	L150LH
Adenine nucleotide binding [GXGXXG]	G623VGATL	G590VGASL	G97QVLFG	G59LGIAG

[a]A C103KSC motif occurs in Exon 4 but is absent from splice variants with protein disulfide-thiol interchange activity. C510A and C569A mutants lack activity (Chueh et al. 2002a, b)
[b]Additionally dNOX (auxin binding protein 20) contains the auxin binding motif H106THP109GASEVLIVAQ which includes the copper site I motif

A fourth member of the ENOX protein family occurs in plants where it is associated with growth hormone (auxin)-stimulated rapid cell enlargement (Chap. 10). This ENOX form is normally inactive and becomes activated upon binding of the low molecular weight naturally occurring plant hormone indole-3-acetic acid or synthetic auxins such as the herbicide 2,4-dichlorophenoxyacetic acid (2,4-D). An auxin-regulated ENOX (dNOX) protein has been cloned and expressed (Sect. 10.4). A putative virus-related ENOX (vNOX) (Chap. 7) remains molecularly uncharacterized.

1.2 ENOX Proteins Are Associated with the External Cell Surface as Ecto Proteins and Are Shed into the Environment

ENOX proteins are not stringently bound to the plasma membrane but, instead, are released into extracellular fluids including interstitial fluids or free space in plants, sera, saliva, and urine in man or in culture medium conditioned by growth of animal or plant cells (Morré et al. 1996c; Yantiri et al. 1998). Sera of cancer patients and spent media from cancer cell lines grown in culture contained both ENOX1 and ENOX2 (Morré et al. 1996b; Wang et al. 2001). ENOX1 is found with noncancer cells and sera and urine of healthy volunteers or of patients with diseases other than cancer which lack ENOX2. ENOX3 proteins are shed through proteolytic cleavage of a ca. 30-kDa N-terminal fragment normally directed to the cells exterior and appear in body fluids primarily after age 30.

ENOX1 and ENOX2 lack transmembrane domains. They are readily released experimentally from plasma membranes by treatment with 0.1 M sodium acetate, pH 5.2 (del Castillo-Olivares et al. 1998). The age-related ENOXs are initially embedded in the plasma membrane by nine membrane-spanning helices but a ca. 30-kDa terminal fragment is cleaved and circulates.

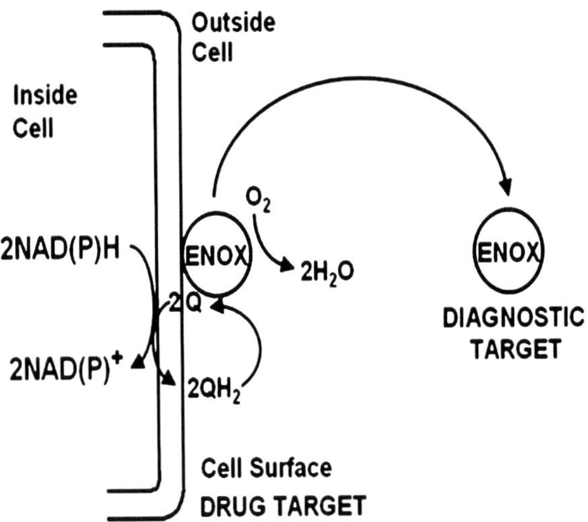

Fig. 1.3 Diagram illustrating the therapeutic and diagnostic cancer-specific ENOX2 (tNOX) utility of the ECTO-NOX (ENOX) proteins (Chaps. 11 and 12). Their extracellular location and involvement in the growth process makes them potential therapeutic targets especially for drug conjugates that need not enter cells to be effective as a strategy to improve efficacy and to reduce dose-limiting toxicities (Chap. 12). The ENOX proteins are shed into blood and other body fluids where they serve as diagnostic targets for early detection of cancer (Chap. 11) and the age-related NADH oxidases (arNOX) proteins monitor antiaging therapies (Chap. 9)

An external plasma membrane location of the ENOX proteins has been demonstrated rigorously from direct assay with whole cells and tissues (Pogue et al. 2000; Hicks and Morré 1998) (NADH is an impermeant substrate), assay of right-side-out (active) vs. inside-out (inactive) plasma membrane vesicles (Morré and Morré 2000), inhibition by antibodies (Cho et al. 2002), use of impermeant inhibitors (Kim et al. 1997), and immuno-, cyto-, and histochemistry (Cho et al. 2002). This unique subcellular location, consistent with the role of ENOX proteins as terminal oxidases of plasma membrane electron transport (Chap. 4) and in cell enlargement (Chap. 5), makes the ENOX proteins potential targets for cancer therapeutic agents (Fig. 1.3) and further serves to distinguish ENOX proteins from other cellular or plasma membrane-associated NAD(P)H oxidases.

A 33.5-kDa protease-resistant fragment with capsaicin-inhibited NADH oxidase activity was first purified from spent media of cultured HeLa cells (Wilkinson et al. 1996) and from sera of cancer patients (Chueh et al. 1997b). The NADH activities from sera of healthy volunteers or spent media of noncancer cells were unaffected by ENOX2 inhibitors (Morré et al. 1997a; Morré and Reust 1997). Both auxin-responsive and auxin-unresponsive plant ENOX proteins are shed from right-side-out plasma membrane vesicles prepared from dark-grown soybean hypocotyls.

Inhibition of shed ENOX3 provides a rational basis for antiaging interventions. Included are dietary or oral interventions to retard formation of aging-related arterial lesions through prevention of LDL oxidation and a reduction in the formation of foam cells and the use of topical arNOX inhibitors to prevent, reduce, or even reverse external manifestations of skin aging (Chap. 9). A relationship to neurological disorders may also be indicated (Chap. 9). Shed forms of the cancer-specific ENOX2 provided an opportunity for noninvasive early detection of cancer based on serum analysis (Fig. 1.3; Chap. 12).

1.3 Two Activities of ENOX Proteins Alternate

The distinguishing characteristic of ENOX proteins that not only permits their unequivocal identification but was suggestive of a time-keeping role is based on the observation that the ENOX proteins exhibit two distinct activities that alternate (Morré 1998c; Morré and Morré 2003a; Fig. 1.2). The first activity is that of a hydroquinone oxidase with NAD(P)H serving as an alternate nonphysiological substrate (Kishi et al. 1999). The second is that of a protein disulfide-thiol interchange measured either from the restoration of activity to inactive (scrambled) RNase (Morré et al. 1995d) or from the cleavage of dithiodipyridine substrates (Morré et al. 1999a; Chap. 2). Each activity generates a distinct oscillatory activity with a period length of 24 min for ENOX1 and the auxin-induced plant ENOX, a period length of 22 min for ENOX2, and a period length of 25–26 min for ENOX3 (Table 1.3). The strictly periodic nature of the NOX proteins distinguishes their activities from all other oxidase or protein disulfide isomerase (PDI) forms (Chueh et al. 2002b) and imparts to the ENOX1 protein a potential role as an ultradian (with period lengths of less than 24 h) oscillator of the cellular biological clock (Morré et al. 2002a; Chap. 6).

Table 1.3 ENOX period length in minutes generates circadian period in hours based on measurements of glyceraldehyde-3-phosphate dehydrogenase when transfected into CHO cells

	Period length		
	ENOX (min)	Circadian (h)[a]	Source
ENOX1	24	24	Wild type
	30	30	C120A
ENOX2	22	22	Wild type
	36	36	C558A
	42	42	C575A
Yeast	24	24	Wild type
	25	25	YDR005C

[a]Glyceraldehyde-3-phosphate dehydrogenase (GAPDH) activity is regulated on a daily basis by the circadian clock (Shinohara et al. 1998)

1.4 ENOX Proteins Participate Directly in the Enlargement Phase of Cell Growth

The ENOX1, auxin-activated plant ENOX, and ENOX2 proteins function in the enlargement phase of cell growth (Pogue et al. 2000; Morré et al. 2001a, 2002d; Chaps. 6 and 10). When their activity is inhibited, cells fail to enlarge (Morré and Grieco 1999; Morré et al. 1995a, b, f; Morré and Morré 2003a). Cancer cells in culture when inhibited either by capsaicin (Morré et al. 1995b) or other antitumor drugs and substances, being unable to enlarge to a size sufficient to allow division, undergo apoptosis (Morré and Morré 2003a; Morré et al. 1995b, f, 2000a). *Arabidopsis* seedlings treated with the quassinoid ENOX1 inhibitor simalikalactone D ceased growth until the inhibitor was metabolized but did not die and subsequently grew normally (Morré and Grieco 1999; Chap. 6). Other functional roles for ENOX proteins are summarized in Chap. 7.

1.5 ENOX Proteins Are Resistant to Degradation and Tend to Form Insoluble Aggregates

A second distinguishing characteristic shared by all ENOX proteins is resistance to proteases, cyanogen bromide, and other forms of digestion and an ability to irreversibly aggregate into amyloid rods (filaments) with concomitant loss of enzymatic activity (Kelker et al. 2001). Similar characteristics are exhibited by proteins associated with neurodegenerative and prion diseases. The ENOX proteins may impart protease resistance to normally protease susceptible proteins (i.e., glyceraldehyde-3-phosphate dehydrogenase (GAPDH) or recombinant ENOX proteins generated in bacteria), also a defining characteristic of the group of proteins designated as prions. This conversion process not only imparts protease resistance and resistance to chemical degradation but the proteins also become resistant to N-terminal sequencing. This latter property has long restricted rapid progress toward the characterization and analysis of ENOX proteins and their various transcription variants. The observations are consistent with the prion model where a certain subset of proteins become modified which then remember the modification as well as recruit other family members into a similarly modified state. In the normal situation, such phenomena may be beneficial and important to growth control and developmental processes. However, when carried to extremes as with 2,4-D treatment to kill plants (Chap. 10) or spongiform encephalopathies (prion diseases of animals and man), the result may be an incurable pathological state.

Sedlak et al. (2001) first purified a protease-resistant ENOX fragment with drug-resistant ENOX1 activity and a period length of 24 min from human serum. The protein fragment was blocked to direct sequencing and was resistant to further protease digestion. Polyclonal antisera raised to the fragment partially blocked total NOX activity of human sera and from the surface of human noncancer cells but did not react with recombinant ENOX2 or with molecular species identified as ENOX2 from sera of cancer patients or from cancer cell lines grown in culture. These antisera were shown subsequently to inhibit the activity of the recombinant human ENOX1 and, as such, has aided in its identification as the first candidate ENOX1 to be cloned and expressed as a recombinant protein (Jiang et al. 2008). The ENOX2 protein purified from sera of cancer patients (Chueh et al. 1997b) as the source of antigen to elicit an ENOX2-specific monoclonal antibody (Cho et al. 2002) and subsequent expression cloning of the ENOX2 protein was both heat and protease resistant as well as refractory to N-terminal sequencing (Chueh et al. 1997b, 2002b).

1.6 ENOX Proteins Are Dicopper Proteins Lacking Both Iron and Flavin

Since ENOX proteins lack iron or iron sulfur clusters and still reduce oxygen, a non-ferrous metal in a redox site was sought. ENOX2 contains a copper site conserved with superoxide dismutase (Schiniá et al. 1996; Chueh et al. 2002b). In early studies with ENOX2 purified from HeLa cells, the copper content was estimated to be about 1 mol bound copper/mol ENOX2 protein. Subsequently, purified ENOX1 and fully active processed 43 or 34 kDa ENOX2 proteins were shown to bind 2 mol of copper/mol of protein. Both of the sites responsible for copper binding have been identified and were required for activity (Chaps. 2 and 8; Jiang et al. 2008; Tang et al. 2010). Based on size exclusion chromatography, the soluble form of ECTO-NOX2 is predominantly a dimer. Therefore, the ECTO-NOXs may be viewed as dimeric proteins containing four coppers per dimer, thus consistent with an ability to carry out four electron transfers to molecular oxygen as required to form water (Fig. 1.4).

Molecular oxygen (Orczyk et al. 2005) and protein disulfide (Chueh et al. 1997a) both function as electron acceptors for ENOX protein-catalyzed reactions with a stoichiometry of 1 NAD(P)H or hydroquinone oxidized per 1/2 O_2 or disulfide pair reduced (Chueh et al. 1997a). Like NADH oxidation, oxygen consumption is periodic with a period length for ENOX1 of 24 min (Orczyk et al. 2005).

In addition to the transfer of electrons and proteins from NAD(P)H or quinols to oxygen to form water, ENOX3 proteins characteristically generate superoxide during one phase of the ENOX cycle (Morré et al. 2003a; Chap. 9). Substrates for the shed forms of arNOX appear to be proteins contacted by the body fluids which become oxidized in the process. The superoxide produced and its conversion to hydrogen peroxide would be only one part of the potentially destructive properties of circulating arNOX proteins. However, the amounts of hydrogen peroxide produced are substantial and contribute to lipid oxidation. Circulating lipoproteins and skin matrix proteins emerge as potentially important health-related targets. arNOX in the blood is structured as an integral component of the LDL particle through site-specific binding and is implicated as a major risk factor for cardiovascular disease. ENOX3 proteins contain a pyridine nucleotide-binding site, a disulfide-thiol interchange site, and two potential copper-binding sites in common with ENOX1

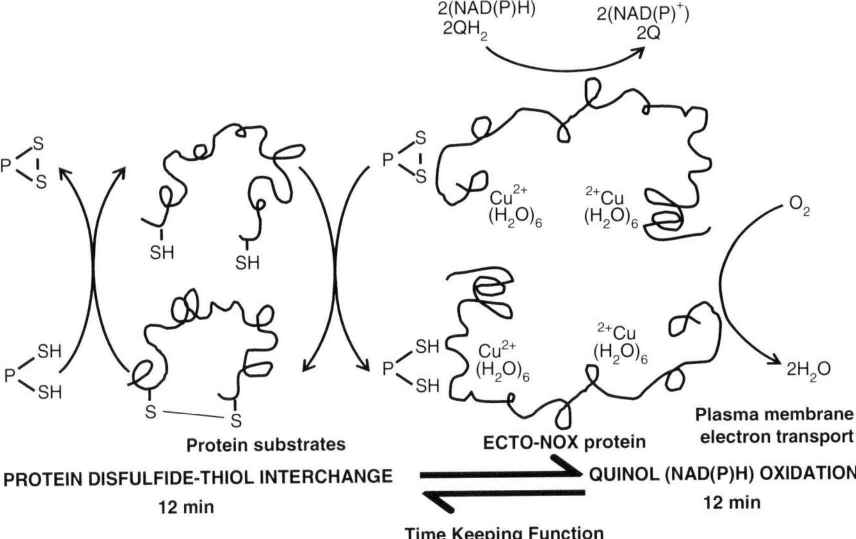

Fig. 1.4 Diagrammatic representation of the functional unit of ENOX proteins which is a dimer, each monomer of which contains two copper centers. During the oxidative portion of the ENOX cycle on the *right*, the net result is the transfer of 4 electrons plus 4 protons to molecular oxygen to form $2H_2O$. The *left portion* of the diagram illustrates the protein disulfide-thiol interchange activity portion of the cycle also shown in Fig. 1.2 where the result is an interchange of protons and electrons that results in the breakage and formation of disulfide bonds

and ENOX2 but otherwise lack significant sequence similarity or homology with ENOX1 or ENOX2 (Chap. 9). ENOX3 proteins and enzymatic activities (superoxide dismutase inhibited reduction of ferricytochrome c, a standard indicator of superoxide formation) are evident only on cell surfaces and body fluids of aged individuals and on plasma membrane of late passage cultured cells or senescent plant parts (Morré and Morré 2003a, b; Morré et al. 2003a).

Zinc is present in recombinant ENOX2 preparations in the ratio of 2 mol of bound zinc/mol of protein (Tang et al. 2010). Zinc is not redox active and its functional role in ENOX2 has not been investigated. ENOX proteins are not general thiol oxidases, lack thioredoxin reductase activity (Bosneaga et al. 2009), and do not normally function as peroxidases (Chap. 4).

The ENOX protein disulfide thiol interchange activity is similar to that catalyzed by protein disulfide isomerase (PDI). The interchange results in no net oxidation or reduction of protein thiols and does not require exogenous donors or acceptors. NAD(P)H, for example, is not required. With the dithiodipyridine substrates, reductive cleavage of the dithio bond is equivalent to the reductive cleavage of a protein disulfide and occurs at the expense of protein thiols. ENOX proteins, however, contain only one of the two characteristic –C–X–X–C– PDI motifs normally required for flavin binding (Table 1.2). However, such motifs are not necessarily required for PDI activity (Wovcehowsky and Raines 2003).

The protein disulfide-thiol interchange activity appears to be involved in controlling cell growth (Sect. 1.4). After division, the resultant cells must enlarge before dividing again. A variety of correlative and experimental data show that cell enlargement and ENOX activity are correlated. In particular, cell enlargement requires the formation of disulfide bonds in membrane proteins and, therefore, the disulfide-thiol interchange activity of ENOX proteins appears to serve an essential role in the growth process (Chap. 5).

1.7 The Oscillatory Behavior Complicates Assays of ENOX Activities

The functional diagram summarizing the two activities of the ECTO-NOX proteins emphasizes distinct oxidative and protein disulfide-thiol interchange events that alternate (Fig. 1.2). The continuous traces, even though illustrating the pattern of oscillatory activity (Fig. 1.5), lack resolution. To improve resolution, rates determined over 1 min at intervals of 1.5 min have been most commonly used (Chap. 2).

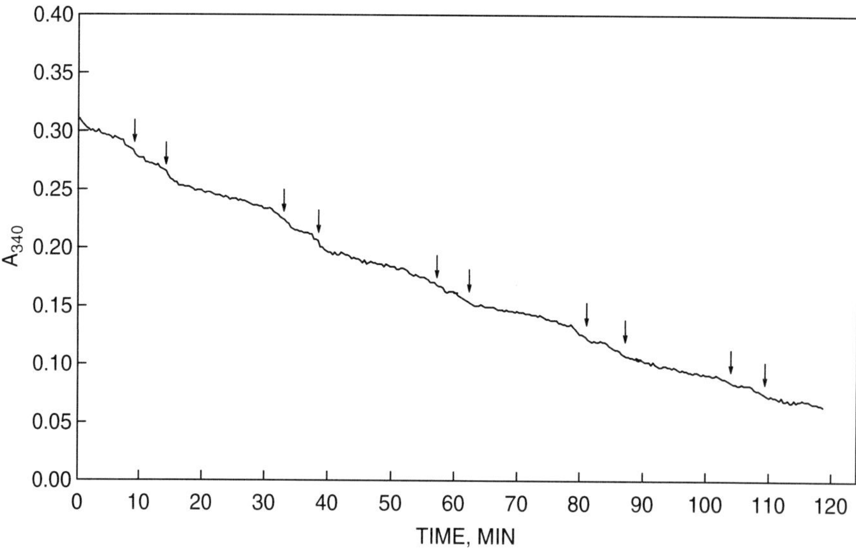

Fig. 1.5 Continuous traces of the activity of recombinant ENOX1 purified by isoelectric focusing. NADH oxidation (decreases in A_{340}) was assayed over 120 min with data collected automatically and stored using a SPECTRA max 340 PC microplate reader. Reproduced from Jiang et al. (2008) with permission from ACS Publications

The oxidative portion of the ECTO-NOX protein cycle of activity has been most often measured from the decrease in A_{340} from the oxidation of NADH (or NADPH) (Chap. 2). With turbid membrane preparations, use of a dual beam or grating instrument with the photo multiplier tube proximal to the sample is essential to reduce light scattering error in continuous measurements.

For the protein disulfide-thiol interchange activity, the restoration of activity to reduced, denatured, and oxidized (scrambled) yeast RNase through reduction, refolding under nondenaturing conditions, and reoxidation to form a correct secondary structure stabilized by internal disulfide bonds (Lyles and Gilbert 1991) is a reliable assay. Hydrolysis of cCMP catalyzed by RNase is the end point estimated from the increase in A_{490}. An alternative assay which requires subtraction of a substrate blank uses dithiodipyridine where cleavage produces pyridinethione which absorbs strongly at A_{340} (Morré et al. 1999a; Chap. 2).

The statistical evaluation of an oscillatory activity that is neither sinusoidal nor monotonic (Fig. 1.6; Sect. 1.8) is equally challenging (Chap. 2). The period length over successive cycles may be determined by Fourier analysis. The reproducibility of the complexities within the major period is amenable to time series (decomposition) analysis in which successive periods are superimposed, the statistical agreement with the average activity pattern is calculated, and an average "predicted" pattern is generated (Foster et al. 2003). These statistical methods have been borrowed from economic forecasting where time series analysis is a widely used approach to analysis and prediction of seasonal trends.

1.8 The Distinctive 2+3 Pattern of ENOX Oscillations Is a Unifying Characteristic of All Family Members

The typical pattern of oscillations in ENOX1 activity consists of five maxima that are neither sinusoidal nor monotonic (Fig. 1.6, see Fig. 6.1 for diagrammatic interpretation). Two of the five maxima labeled ① and ② are separated by intervals of 6 min. These intervals confer a characteristic asymmetry to the pattern of oscillations and for ENOX1 repeat every 24 min imparting a time-keeping aspect to the activity oscillations (Chap. 6). The remaining three maxima, labeled ③, ④, and ⑤, are separated by intervals of 4.5 min from each other and from the two maxima separated by 6 min. The result (6 min + 4 × 4.5 min) is the 24-min ENOX1 period.

The 6-min separation between maxima ① and ② is characteristic of all ENOX proteins. For ENOX2 with the shorter 22-min period and with ENOX3 with the longer period length of 25–26 min, the separation of maxima ① and ② remains at 6 min and the interval between maxima ③, ④, and ⑤ and their separation from the two maxima labeled ① and ② are correspondingly shortened to about 4 min for ENOX2 and lengthened to 4.8–5 min for ENOX3.

The statistical validity of this complex pattern of oscillations is provided by decomposition fits of the data where the patterns duplicated in successive cycles are evaluated for reproducibility (Sect. 1.7). In Fig. 1.6c, the mean average deviation

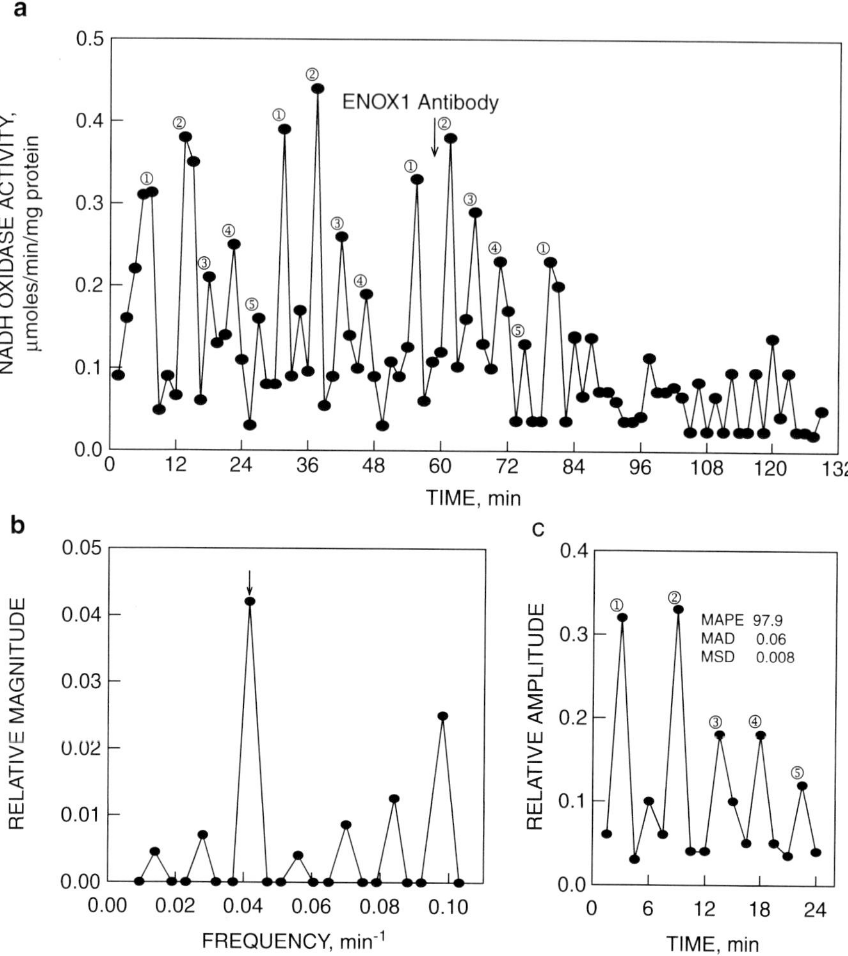

Fig. 1.6 The alternation of activities between the oxidative and protein disulfide-thiol interchange portion of the NOX cycle is not sinusoidal but in the form of a synchronous repetition pattern of five activity bursts (maxima) much in the manner of limit oscillators in biology. The oxidative portion consists of two maxima labeled ① and ② separated by 6 min and the protein disulfide-thiol interchange activity consists of three maxima separated by intervals of 4.5 min labeled ③, ④, and ⑤ to generate the characteristic 24-min period [6+(4×4.5) min] of ENOX1. The activity for ENOX1 is inhibited by ENOX1 antibody (**a**). Fourier analysis yields a period length of 24 min (frequency of 0.04/min) (**b**) and decomposition fits provide statistical confirmation of the 2+3 pattern of oscillations (**c**). Reproduced from Jiang et al. (2008) with permission from ACS Publications

of 0.06 is ca. 0.2 the mean activities of maxima ① and ② and ca. 0.3 that of the mean activities of maxima ③, ④, and ⑤. Maxima ① and ② are associated with hydroquinone or NADH oxidation whereas maxima ③, ④, and ⑤ correlate with maxima in protein disulfide thiol interchange (see Chap. 3). Cell enlargement is restricted to the protein disulfide-thiol interchange portion of the cycle (Fig. 1.2, see also Fig. 5.16) and oscillates with three maxima corresponding to maxima ③, ④, and ⑤ during each 24-min period (Chap. 5).

In keeping with their proposed clock function, the period lengths of all ECTO-NOX proteins are independent of temperature. Whereas NOX activities double with every 10° C rise in temperature ($Q_{10}=2$), period length is invariant with a Q_{10} of about 1 (Morré and Morré 1998; Pogue et al. 2000; Wang et al. 2001; Morré et al. 2002a, b), a hallmark of the biological clock (Edmunds 1988; Dunlap 1996). Estimates based on measurements made at liquid nitrogen temperatures yield a Q_{10} of 1.01 (Morré and Morré 2008b). Also implicit in the evaluation of any periodic or oscillatory phenomena is the need for a high level of synchrony (entrainment). ECTO-NOX proteins auto-synchronize (auto-entrain) in solution (Morré et al. 2002b). Red (plants) or blue (plant and animal) light entrains with organisms, tissue explants, and cells but not with membranes or molecules (ECTO-NOX proteins lack chromophores) (Morré et al. 1999b, 2002c; Morré and Morré 2003e). Also, small molecules such as melatonin will entrain the constitutive ENOX1 with a 24-min period of both plants (unpublished) and animals (Morré and Morré 2003e) but not of ENOX2 or arNOX. Entrainment is also achieved by exposure to low frequency electromagnetic fields (Morré et al. 2008b). Entrainment is a hallmark of the biological clock second in importance only to the temperature independence of period length (Edmunds 1988; Dunlap 1996). No time-keeping role for arNOX proteins has been indicated.

For the ENOX1 and ENOX2 proteins, cysteine replacement by alanines results in period lengths greater than 24 min (Table 1.3). In every instance for all examples studied, the length of the circadian day based on measurements of glyceraldehyde-3-phosphate dehydrogenase (GAPDH) activity in hours is equal to the ENOX period in minutes × 60 (Table 1.3).

1.9 ENOX Proteins Differ Markedly from the NOX Proteins of Host Defense

Examples of non-ENOX NOX proteins include the series designated as NOX1 through NOX4 which are part of the gp91*phox* protein family. These proteins function as part of a plasma membrane-associated flavo-hemoprotein complex containing one flavin-adenine dinucleotide (FAD) and two hemes that catalyze the NADPH-dependent reduction of O_2 to form superoxide and hydrogen peroxide. These reactive oxygen species contribute to host defense as well as more specific roles in signaling pathways (Lambeth et al. 2000). In resting cells (e.g., phagocytes), the enzymes are dormant but become activated by assembly with regulatory proteins p47*phox*, *p67phox,* and Rac. The *phoxNOX* proteins are located on the inner (cytosolic) surface of the plasma membrane and where, like ENOX3 proteins, they generate superoxide that dismutes to form H_2O_2 to support a postulated general function in peroxidative reactions at the cell surface/cytosol interface. ECTO-NOX (ENOX) proteins including ENOX3, by way of contrast, are located on the external

Table 1.4 Properties of ENOX proteins

Two activities alternate to generate a period length of 24 min for ENOX
Low turnover number: 200–500
Specific activity: 10–20 μmol/min/mg protein
No flavin, heme, or iron–sulfur centers
Ancillary proteins not required
Metals (Cu^{2+} and Zn^{2+}) present and required for activity
Have prion-like properties (protease resistance, aggregate to form insoluble amyloid rods)

(environmental) surface of the plasma membrane, reduce either NADH + H^+ or NADPH + H^+, lack both flavin and heme, and do not require ancillary proteins either for activation or for activity (Table 1.4).

The terminology suggested by the HUGO Human Gene Nomenclature Committee (HGNC) (http://www.gene.ucl.ac.uk/nomenclature/) and adopted for the ECTO-NOX (ENOX) proteins of the external plasma membrane surface was adopted to distinguish ENOX proteins from the *phoxNOX* proteins of host defense of the internal plasma membrane surface (eukaryotic homologs of *gp9/phox*) (Table 1.1).

1.10 ENOX Proteins Are of Low Specific Activity

NADH or NADPH are artificial donors for ENOX proteins. Their use has been predicated primarily on convenience of assay and the historical precedent of ECTO-NOX discovery based on NADH oxidation (Morré 1998c). Specific activities are low, in the range of 10–20 μmol/min/mg of protein purified to homogeneity giving rise to very small turnover numbers (the number of substrate molecules converted to product per minute with the enzyme fully saturated with substrate) of 200–500 for NADH oxidation (Table 1.4). Rates of oxidation of natural hydroquinone or other substrates contained within the plasma membrane or in biological fluids most often require measurement in the nmol/min/mg protein range. Yet, in well-synchronized NOX preparations, NADH oxidation rates oscillate between rapid and slow to create reproducible and statistically significant recurring patterns within each period. Spectroscopic approaches to rate analyses are described in Chap. 2 that validate the basic oscillatory phenomenon associated with ECTO-NOX proteins. Approaches to the measurement of low rates of NADH oxidation based on least squares slopes of spectrophotometric traces are illustrated along with statistical evaluations of the reproducibility of the oscillatory patterns in Chap. 2 as well.

1.10.1 Natural Electron Donors and Acceptors for Cell Surface-Associated ENOX Proteins

Among the natural electron donors for the oxidative activity of ECTO-NOX proteins are hydroquinones (coenzyme Q in animals (Kishi et al. 1999), vitamin K_1

hydroquinone in plants (Bridge et al. 2000)). Natural acceptors are molecular oxygen (Morré et al. 1998a; Orczyk et al. 2005) and protein disulfides (Chueh et al. 1997a).

1.10.2 Hydroquinones as Natural Electron Donors

Reduced quinones (coenzyme QH_2 and phylloquinone (vitamin K_1H_2)) are abundant in plasma membranes of animals and plants, respectively, and can function as lipophilic *trans*-plasma membrane shuttles ferrying reducing equivalents from cytosolic NAD(P)H to molecular oxygen with the ENOX protein functioning as the terminal oxidase (Morré et al. 1999c). This transplasma membrane electron transport chain is initiated at the cytosolic plasma membrane surface by quinone reduction catalyzed by NAD(P)H-quinone reductases (Kishi et al. 1999; Bridge et al. 2000). These relationships among plasma membrane electron transport constituents and ENOX proteins are shown in Fig. 1.2 and detailed in Chap. 4.

1.10.3 Reduced Pyridine Nucleotides as Artificial Electron Donors

All ENOX proteins thus far examined utilize both NADH and NADPH as the electron donor. The reduced pyridine nucleotides, NADH and NADPH, are regarded as nonphysiological substrates since reduced pyridine nucleotides in the concentrations required to sustain ENOX activities are encountered rarely, if at all, at the external cell surface. ENOX1 activities tend to utilize NADH preferentially whereas ENOX2 activity may be greater with NADPH. ENOX activity with NADPH as electron donor including recombinant ENOX2 is as a single burst inhibited by diphenyleneiodonium (DPI). ENOX activity with NADH as electron donor is largely unaffected by DPI (Morré 2002). DPI is regarded widely as a specific inhibitor of flavin-containing oxidases. ENOX2 lacks both flavin and flavin-binding domains and ENOX activities are unaffected by added flavin (FMN or FAD). Yet, ENOX activities are inhibited by DPI with NADPH as substrate. DPI inhibition of ENOX activities, therefore, appears to involve some aspect of NADPH binding not encountered with NADH but not related obligatorily to the presence of bound flavin. The possibility of the difference being due to two different enzymes was eliminated from studies with recombinant ENOX proteins.

1.10.4 Protein Thiols and Tyrosines as Electron Donors for arNOX Proteins and Generation of Superoxide

For arNOX proteins, protein thiols and tyrosines appear to serve as the primary electron donors. Unlike ENOX1 and ENOX2, superoxide anion is generated periodically during the characteristic ENOX cycle (Sect. 1.8) at or just prior to maximum ③.

Table 1.5 Common properties of ENOX proteins

Reduce NAD(P)H
Carry out disulfide interchange
Activities are synchronous and pulse
2 + 3 Periodic pattern
Period lengths 22–26 min
ENOX1 24 min
Resistant to cleavage
N-terminal sequencing
Proteases
Peptide bond cleavage
Primary amino acid sequence largely inferred from DNA sequence
Aggregate irreversibly when purified
Characteristic commonly associated with prions
Copper hexahydrate driven synchronous *ortho:para* departures from equilibrium of Cu-associated water functions and five limit oscillations to cause fluctuations in redox potential of bound copper

Table 1.6 Amyloid-forming proteins that exhibit periodic (copper-dependent) oscillations in NADH oxidation

ENOX1	24 min	ECTO-NOX proteins
ENOX2	22 min	
arNOX	26 min	
Mouse prion	24 min	Spongioform encephalopathies
Aβ peptides	24 min	Alzheimer's
α-Synuclein	54 min	Parkinson's

Modified from Morré and Morré (2003a)

1.10.5 Aggregation and Formation of Amyloid

Enzymatic assays with ENOX2 purified from the HeLa cell surface or purified recombinant ENOX1 or ENOX2 are especially difficult due to the propensity of the proteins to aggregate (Table 1.5). Greatest specific activities have been obtained with partially purified preparations of ENOX2 from HeLa cells still complexed with other proteins prior to final purification (del Castillo-Olivares et al. 1998). Aggregation becomes especially problematic with concentrated solutions of purified proteins, precluding, for example, meaningful solution NMR studies. Insoluble aggregates of amyloid which, once formed, are devoid of enzymatic activity and resistant to restoration of activity by disaggregation. The recombinant ENOX proteins have been successfully assayed most often with dilute solutions (ca. 10 ng/mL). Table 1.6 lists some amyloid forming proteins all of which exhibit periodic oscillation in NADH oxidase activity and require bound copper for activity. Included along with the ENOX proteins are mouse prions with copper-binding octarepeats (Garnett and Viles 2003; Wells et al. 2006), Alzheimers Aβ peptides (Miura et al. 2000; Curtain et al. 2001) with three histidines and an adjacent tyrosine available for binding of copper and α-synuclein of Parkinson's disease where conserved tyrosines in the acidic C-terminus are most likely involved in copper binding (Palk et al. 1999; Clayton and George 1998; see also Sect. 7.10).

1.11 Why an External NADH Oxidase?

The question has often arisen as to why cells should express one or more NADH oxidases on their cell surface capable not only of electron transfer from hydroquinones to molecular oxygen but also for the direct oxidation of externally supplied NAD(P)H. All the more puzzling is that NADH levels at the external surface of the cell are minuscule and several orders of magnitude below the K_ms of the external oxidases.

One explanation not previously considered was that the ability of the ENOX family of proteins to carry out very low specific activity oxidations of NADH arose from an early need for conserved adenine nucleotide binding sites such as those universally present in ENOX proteins at the external cell surface and perhaps unrelated to either hydroquinone binding or to hydroquinone oxidation. External adenine nucleotide binding proteins (adenine receptors) with various signaling functions are widely distributed among mammalian cells (and plants) (Burnstock 2007) that respond to low levels of purines, adenosine, and adenine nucleotides released by cells into the external milieu. ATP is considered the most important and ancient intracellular metabolite that might have given rise to extracellular adenosine through the action of cell surface 5′-nucleotidases (Burnstock 2007). The most well-studied example of external purinergic receptors that have evolved as a system of feedback regulators with signaling functions modulated by external levels of ATP have been with both the peripheral and central nervous systems (Burnstock 2007) where adenosine is considered one of the most important modulators of neurotransmission (Cunha 2008; Dunwiddie and Masino 2001; Fredholm and Hedqvist 1980; Latini and Pedata 2001; Sebastião and Ribeiro 2000; Stone 2005). Also part of the taste signaling cascade in the mouth, depolarization resulting from sodium influx and a rise in intracellular Ca^+ opens pannexin channels in the taste cell membrane to release ATP from the cell to activate purigenic receptors of innervating sensory nerve fibers (Huang and Roper 2010). Modulatory influences of adenine nucleotides other than NAD(P)H on NADH oxidase activity of ENOX proteins have been observed for both plants (Morré 1998d) and animals (Morré et al. 1994d, 1997e, Sect. 3.6).

1.12 Summary

ECTO-NOX or ENOX (because of their cell surface location) proteins comprise a family of dicopper NAD(P)H oxidases of plants, animals, yeasts, and bacteria that exhibit both oxidative and PDI-like activities. The two biochemical activities, hydroquinone [NAD(P)H] oxidation and protein disulfide-thiol interchange alternate, a property unprecedented in the biochemical literature. The constitutive ENOX (CNOX or ENOX1) is ubiquitous and refractory to drugs. A tumor-associated tNOX or ENOX2 is cancer-specific and cancer drug-inhibited. The physiological substrate for the oxidative activity appears to be hydroquinones of the plasma membrane such as reduced coenzyme Q_{10}. ENOX1 and ENOX2 proteins are growth-related and drive cell enlargement. Also indicated are roles in aging of an age-related or arNOX (ENOX3), and in neurodegenerative diseases. The regular pattern of oscillations appears to be related to transitions in secondary structure involving protein-bound copper hexahydrate. ENOX1 serves as a biochemical core oscillator of the cellular biological clock. Period length is independent of temperature and synchrony is achieved through entrainment.

Perhaps the final challenge to a complete understanding of the functional role of ENOX proteins will be to extend our understanding of plasma membrane electron transport and the role of ENOX1, ENOX2, and ENOX2-like (dNOX) proteins as drivers of cell enlargement (Chap. 5). How might physical membrane displacements both at the cell surface and at the level of the Golgi apparatus be mediated during the protein disulfide-thiol interchange portion of the ENOX cycle? How might the energy of NADH oxidation then be converted into an energy form such as ATP or ATP equivalents or a membrane potential during the oxidative portion cycle to drive cell enlargement? As such, physical membrane displacements driven by energy from NADH oxidation and mediated by ENOX proteins would offer a new paradigm in support of both plasma membrane redox and cell enlargement as among life's essential metabolic processes.

Chapter 2
Measurements of ECTO-NOX (ENOX) Activities

A major challenge in ENOX discovery and validation has been limitations imposed by methods of measurement of ENOX activities. Thus far, measurement opportunities for the oxidative activity have been restricted to NAD(P)H, the natural electron donors (hydroquinones such as coenzyme Q in animals (Kishi et al. 1999) and vitamin K_1 hydroquinone in plants (Bridge et al. 2000)) and electron acceptors for the oxidative activities of molecular oxygen (Morré et al. 1998a; Orczyk et al. 2005) and protein disulfides (Chueh et al. 1997a) and substrates that support protein disulfide-thiol interchange. Except for certain tetrazolium dyes and despite exhaustive studies, artificial dye donors and/or acceptors, frequently utilized with NADH dehydrogenases, have not been discovered for ENOX proteins. Measurements of protein disulfide-thiol interchange activity have utilized primarily activation of inactive ribonuclease A and cleavage of dithiodipyridine (DTDP) derivatives (Sects. 2.6, 2.7, and 2.8). Protein may be estimated by the method of Smith et al. (1985) with bovine serum albumin as the standard.

Despite their widespread use as ENOX substrates, NADH or NADPH are artificial donors. Their use has been predicated principally on convenience and the historical precedent of ENOX discovery based on NADH oxidation (Morré 1998c). Specific activities are low, in the 10–20 μmol/min/mg of recombinant protein purified to homogeneity giving rise to very small turnover numbers (the number of substrate molecules converted to product per minute with the enzyme fully saturated with substrate) of 200–500 for NADH oxidation (Table 1.4). Rates of oxidation of natural hydroquinone or other substrates contained within the plasma membrane or in biological fluids most often require measurement in the nmol/min/mg protein range. Yet in well-synchronized ENOX preparations, NADH oxidation rates oscillate between rapid and slow to create reproducible and statistically significant recurring patterns within each period. In this chapter, spectroscopic approaches to rate analyses are described that validate the basic oscillatory phenomenon associated with ENOX proteins. Our approaches to the measurement of low rates of NADH oxidation are based on least squares slopes of spectrophotometric traces along with statistical evaluations of the reproducibility of the oscillatory patterns.

D.J. Morré and D.M. Morré, *ECTO-NOX Proteins: Growth, Cancer, and Aging*,
DOI 10.1007/978-1-4614-3958-5_2, © Springer Science+Business Media New York 2013

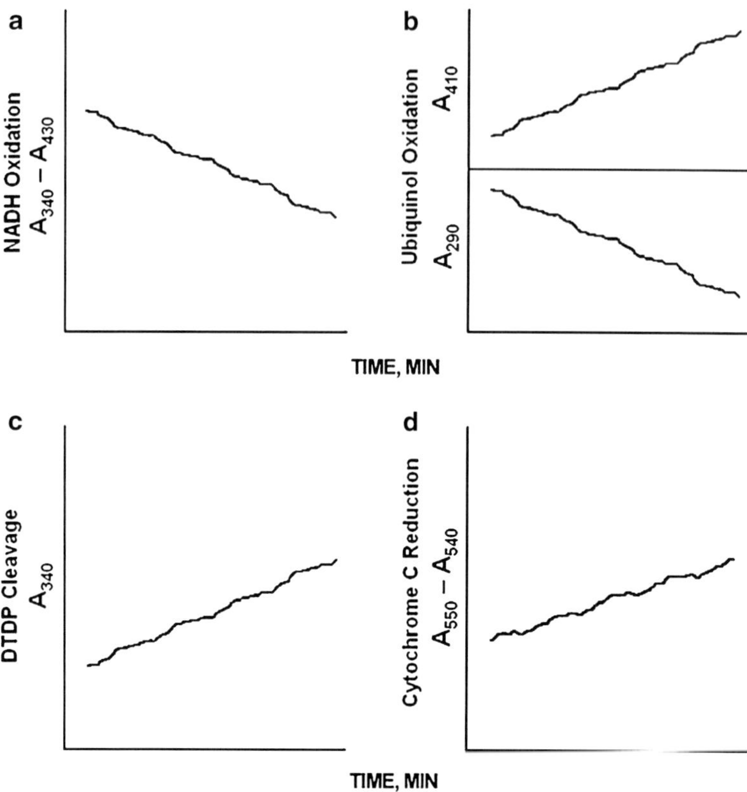

Fig. 2.1 Schematic representation of primary spectrophotometric assays used in the characterization of oscillatory activities. (**a**) The standard NADH oxidase activity that records the decrease in A_{340} resulting from NADH oxidation with reference at A_{430}. (**b**) Hydroquinone oxidation indicated by increasing absorbance at A_{410} or decreasing absorbance at A_{290}. (**c**) Cleavage of dithiodipyridine (DTDP) resulting in increased absorbance at A_{340} as a measure of protein disulfide-thiol interchange. (**d**) Reduction of cytochrome c (increase in A_{550} with reference at A_{540}) used as a measure of superoxide production by arNOX proteins (Chap. 9)

The four principal spectrophotometric assays to monitor periodicity of ENOX activities are illustrated diagrammatically in Fig. 2.1. The standard NOX assay (Fig. 2.1a) measures NADH oxidation (Sect. 2.1). The quinone oxidation assay (Fig. 2.1b) measures either an increase in absorbance of A_{290} or a decrease in absorbance at A_{410} (Sects. 2.4.1 and 2.4.2). Protein disulfide-thiol interchange activity (Sects. 2.6, 2.7, and 2.8) is most conveniently measured either as activation of scrambled and inactive RNase or from the increase in absorbance at 340 nm from the cleavage of DTDP (Fig. 2.1c). Production of superoxide as is characteristic of the age-related NOX proteins (Chap. 9) is estimated from the reduction of ferricytochrome c (Fig. 2.1d).

2.1 Spectrophotometric Assay of NADH Oxidase

NAD(P)H oxidase activity with NAD(P)H as substrate is determined from the disappearance of NAD(P)H measured at 340 nm in a reaction mixture of 25 mM Tris-Mes buffer (pH 7.2), 1 mM KCN, and 150 μM NADH at 37 °C and continuous stirring (Morré et al. 1995b) using Hitachi U3210 UV–visible (Hitachi Instruments, San Jose, CA) or SLM Aminco DW2000 (Milton Roy Company, Rochester, NY) spectrophotometers, the latter in the dual wavelength mode of operation with continuous recording or a comparable instrument. While these instruments are capable of recording data electronically, chart recorder output has been the preferred means of equipment use. The instruments were employed in double-beam (Hitachi U3210) or dual wavelength (SLM 2000 with reference at 430) modes with the reference path empty. Data were recorded continuously over intervals of 1 or 5 min each for periods of up to 90 min. For measurements of 1.5-min intervals, rates are monitored continuously over 1 min. After each 1-min measurement interval, the chart paper is returned to the starting position, the baseline is offset by a convenient increment, and recording restarted exactly 1.5 min after initiation of the previous trace. If timing is digital, uncertainty in restart interval is likely less than 1 s and there is no cumulative timing error. When paired instruments were employed, they were directly adjacent to each other and shared a common timer and thermostated water bath. Assays are normally initiated by addition of enzyme. ENOX activities can be measured conveniently with whole cells as NAD(P)H are impermeant substrates.

Circulating water bath temperatures are normally set to 37 °C for mammalian cells and 25 °C for plant cells and tissues. Cuvettes were typically at room temperature when loaded and temperature equilibrated up to 20 min before beginning the assays.

It is important that all reactants be equilibrated at the spectrophotometers' temperature prior to measurement of the oscillatory patterns. During temperature equilibration, sample warming contributes significantly to absorbance values. Near 34 °C, the mean temperature used, a 1° change in temperature results in a density change of 0.03 % (Lide 1992). For turbid samples ($A \cong 1$), this corresponds to an absorbance change of 3×10^{-4}.

For ENOX assays, a grating instrument with an end-on window photomultiplier tube along with the double beam stability offered by this type of instrument is preferred and, with turbid preparations, is essential. For example, the smooth signal generated by the Aminco SLM 2000 offers an unequivocal demonstration of oscillatory behavior with continuous tracings and an opportunity for unambiguous fitting of line slopes to discontinuous traces.

Diode array instruments are excellent for scanning. Full spectrum incident light is selectively absorbed by the sample and transmitted. However, for situations other than for clear aqueous samples of 1 cm path length, their utility is limited. With ENOX proteins it is rarely possible to provide a clear sample. Even highly purified ENOX preparations tend to aggregate and become turbid. With a turbid sample, polychromatic light is scattered. The transmitted light is no longer focused and may not even reach the detector. Within a conventional diode array instrument, the sample is placed too far from the detector to avoid serious interference from light scatter (see Sect. 2.3.1).

Unfortunately, spectrophotometers of the Aminco SLM 2000 design are no longer available commercially. A comparable instrument with an end-on photomultiplier and dual beam stability is the Spectronics UV 500 (Thermo Electron Scientific Instruments, Rochester, NY). This instrument, appropriately modified to achieve both stirring and temperature control, should fulfill the basic requirements for either continuous or discontinuous measurements of ENOX activities.

Unlike chemical oscillators (Scott 1994), ENOX oscillations are inherent in the protein itself and the period length is not dependent on the chemical environment or temperature per se (period length is independent of temperature). These factors, however, do influence absolute reaction rates and consequently, resolution. However, the greatest contributors to a lack of resolution are turbidity and particle settling. Stirring is essential to eliminate the latter. Turbidity is difficult or impossible to eliminate in most ENOX preparations. Improved resolution with the SLM DW2000 derives from minimization of turbidity influences by the end on photomultipliers. Double beam dual wavelength with A_{430} reference subtraction from the same cuvette to reduce the turbidity component is an aid with very turbid samples but most of the data generated have utilized only a single wavelength. A reference in the reference cell lacking NADH, for example, only exacerbates the turbidity problem so that the routine reference of choice is air. Turbidity may arise from vesicles (membrane preparations), lipid micelles (disrupted vesicles), and protein aggregates (solubilized and purified ENOX proteins). Purified ENOX proteins tend to aggregate to form insoluble (and inactive) amyloid rods under most standard conditions of assay (del Castillo-Olivares et al. 1998; Kelker et al. 2001). Addition of detergent is of limited benefit and may actually aggravate the situation since detergents enhance aggregation of ENOX proteins (Morré et al. 1998e), a phenomenon also encountered with prions (Prusiner et al. 1983).

2.2 Statistical Analysis

From the rate data, period lengths may be verified by fast Fourier analysis. One method for statistical validation of the reproducibility of rate data is time series analysis (Foster et al. 2003). In time series analyses, decomposition fits compare successive oscillatory patterns of period length determined by Fourier analysis (Fig. 1.6b). The decomposition fits (Fig. 1.6c) serve to evaluate the reproducibility of the oscillatory patterns by generating a predicted or forecasted oscillatory pattern and by comparison to the predicted pattern to yield mean average percentage error (MAPE), a measure of the periodic oscillation, mean average deviation (MAD), a measure of the absolute average deviations from the fitted values, and mean standard deviation (MSD), the measure of standard deviation from the fitted values (Foster et al. 2003). A trend line is first fitted to the data. The data are then smoothed by subtracting a centered moving average of length equal to the length of the period determined by Fourier analysis. Finally, the time series are decomposed into periodic and error components. The decomposition fits are used to validate the periodic oscillatory pattern and to demonstrate that minor intervening fluctuations also recurred within each period as part of a reproducible pattern. The decomposition fits used MINITAB®, a statistical package.

Implicit in the need to visualize periodic oscillations in ENOX activity is a high degree of synchrony. Unsynchronized ENOX preparations as a population would not generate discernable oscillations with a regular period. However, such populations, once synchronized, will generate oscillations under the same identical chemical and thermal conditions as for the unsynchronized preparations (Morré et al. 2002a, b).

2.3 Data Reduction Methods

From the SLM Aminco DW2000 and Hitachi U3210 spectrometers, reaction rate was determined from the least squares linear slope of the absorbance vs. time chart data. In practice this was accomplished by measuring rise vs. run of the linear absorbance vs. time line best fitting the data. Typically, slopes were >10 absorbance standard deviations/min for the SLM Amico DW2000 and normally >3 absorbance standard deviations/min for the Hitachi U3210. Despite the fact that some slopes were within the noise envelope of the traces especially with the Hitachi U3210, slopes were estimated to an accuracy of ±0.015 nmol/min compared to the actual rates recorded at 1–10 nmol/min.

Figure 2.2a is a continuous trace of a well-synchronized single culture of CHO cells determined over 84 min using the Aminco SLM 2000 spectrophotometer to illustrate the oscillatory activity characteristic of the ENOX enzymes from a continuous trace. Intervals of rapid activity were interspersed with intervals of lesser activity. The period length was about 24 min.

No oscillations were observed with NADH alone in the absence of cells (Fig. 2.2b) or with cells alone in the absence of NADH (Fig. 2.2c). In the absence of cells, a slow rate of non-enzyme-catalyzed NADH oxidation was observed. In the absence of NADH, the absorbance increase was due to light scattering as turbidity increased.

The continuous tracings provided by the Aminco SLM DW2000 and Hitachi U3210 recording spectrophotometers have been analyzed variously to elucidate details of the oscillatory patterns. One method developed over several years of experimentation is to estimate rates from line slopes fitted to continuous traces obtained over 1 min at 1.5-min intervals. The 0.5-min intervals between measurements are sufficient to allow for chart repositioning and instrument restart. The rise over run ratio allows for accurate measurement of slopes determined over 1 min with a precision of ±0.015 nmol, approximately equivalent to instrument and/or thermal variation (variation in trace width) for the Aminco SLM DW2000. The actual rates estimated were in the range of 1–10 nmol/min, 50–500 times the

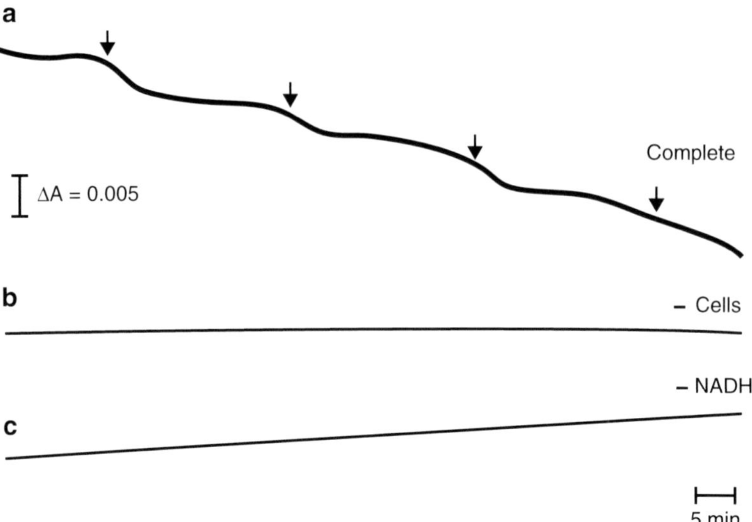

Fig. 2.2 A continuous trace of the decrease in A_{340} as recorded using an Aminco SLM 2000 spectrophotometer in the dual wavelength mode of operation as a measure of consumption of NADH over 96 min for a well-synchronized (by light exposure) culture of CHO cells. (**a**) Complete reaction mixture. Periods of rapid oxidation of NADH alternate with periods of much slower NADH oxidation. (**b**) Reaction mixture lacking cells. (**c**) Reaction mixture lacking NADH. Reproduced from Morré and Morré (2003c) with permission of International Hormesis Society

precision with which the line slopes were determined. This ratio represents the power of the line slope method of minimizing instrument or thermal variation as a significant source of measurement error. Maxima and minima are readily discerned and data are obtained that permit statistical evaluations not possible from continuous traces. For example, the maxima in NADH oxidation of well-synchronized preparations frequently may be resolved into two maxima rather than a single maximum. In addition, minor oscillations coinciding with maxima in protein disulfide-thiol interchange also may be discerned (Sun et al. 2000).

Oscillations in the rate of oxidation of NADH by CHO cells illustrated in Fig. 2.3 also with a period length of 24 min become more obvious when analyzed over 5 min (Fig. 2.3a, b) or over 1 min at intervals of 1.5 min (Fig. 2.3c, d) using the Aminco SLM 2000 in the dual wavelength of operation with reference at 430 nm. Both measuring and reference wavelengths pass simultaneously through the same cuvette so that errors due to electronic fluctuation and turbidity changes, for example, are minimized. Panel A illustrates rates determined over 5 min as would be done for an average determination of specific activity. Since the rates were not steady state, rates from a number of such traces (usually 2–5) normally were averaged where determination of an average rate is a primary consideration. With 5-min traces linked front to back, the oscillatory pattern shown in (b) was revealed which was similar to that observed with the continuous recording shown in Fig. 2.2a.

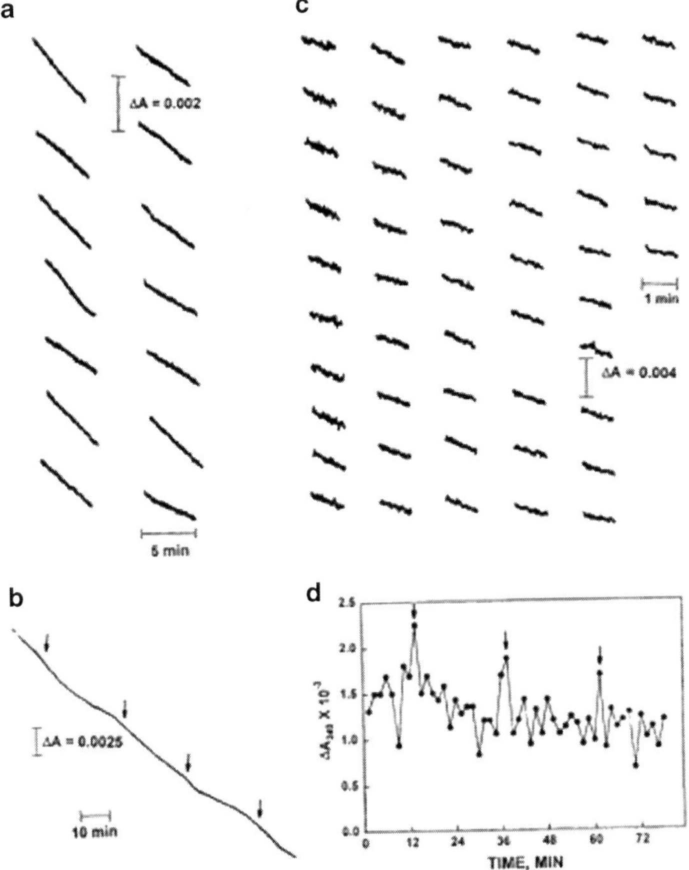

Fig. 2.3 Oscillatory oxidation of NADH by 10^6 CHO cells as in Fig. 2.2 (Aminco SLM 2000) illustrating three methods of spectroscopic analyses employed to demonstrate oscillations. (**a**) Activity analyzed for successive intervals of 5 min each. Traces are arranged from *top* to *bottom* and from *left* to *right* (see Fig. 2.3d). (**b**) The 5-min assays aligned as a continuous trace. (**c**) Activity analyzed over 1 min at intervals of 1.5 min. Traces are arranged from *top* to *bottom* and from *left* to *right* (see Fig. 2.3d). (**d**) Slopes calculated from (**c**). Assays were for 70 min. The period length was 24 min (*arrows*). HeLa cells contain a background activity approximately equal to that of the oscillatory activity that is proteinase K susceptible and does not oscillate. Reproduced from Morré and Morré (2003c) with permission of International Hormesis Society

For more detailed evaluations, determinations of rates over 1 min at intervals of 1.5 min are illustrated in Fig. 2.3c. The measured slopes when represented graphically in Fig. 2.3d illustrate maxima separated by intervals of 24 min. With the SLM 2000, the width of the trace was such that when slopes were estimated to the nearest 0.1 mm, measurement errors were ±0.015 nmol/min. Measurement errors greater than ±1 mm (±0.05 nmol/min) would have been virtually impossible to introduce. These assays all were for 70 min. HeLa cells contained a background activity as seen in

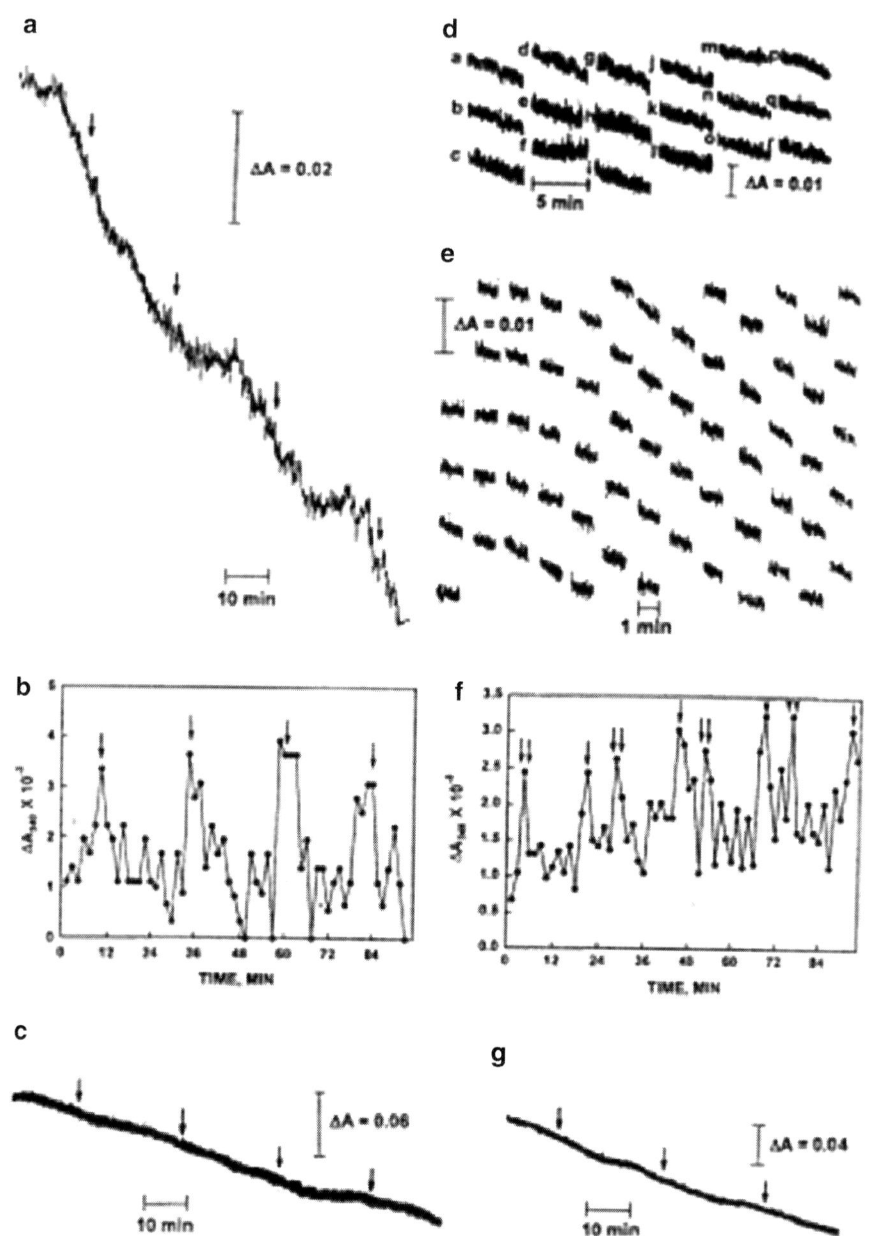

Fig. 2.4 Oxidation of NADH by a crude preparation of isolated, intact plasma membrane vesicles from soybean measured using paired Hitachi U3210 dual beam spectrophotometers with air as reference illustrating typical oscillatory patterns. The solutions were very turbid (initial absorbance of 1.4) and represent a worst-case scenario at the limits of the capacity of the photospectrometer. (**a**) Continuous trace over 90 min with alternating rapid and slow rates of NADH oxidation. (**b**) An identical aliquot of the same preparation as in (**a**) but analyzed in parallel using

Fig. 2.3d approximately equal to that of the oscillatory activity. The background activity was susceptible to digestion with proteinase K and did not oscillate (Morré and Morré 2000). The oscillatory ENOX activities are resistant to protease digestion, including digestion with proteinase K (del Castillo-Olivares et al. 1998).

A double peak oscillatory pattern is also typically observed as illustrated with plasma membranes from stem sections of dark-grown soybean seedlings analyzed using the SLM 2000 (Fig. 2.4). The plasma membranes were solubilized with 0.5 % Triton X-100 to minimize turbidity. Detergent treatment, however, introduced a non-oscillatory background NADH oxidase activity nearly equal to the oscillatory NADH activity. The background activity introduced by detergent treatment derives from the inner surface of the solubilized plasma membrane vesicles (Morré and Morré 2000) and is removed by digestion with proteinase K. These preparations were well synchronized by light treatment of the soybeans prior to harvest of the plant material and show two maxima in NADH oxidase activity with the second maximum (double arrow) following the first (single arrow) by about 6 min (Fig. 2.4f).

Much of the previous work, especially with plants, compared measurements using two side-by-side Hitachi U3210 dual beam spectrophotometers with air as reference. The most challenging measurements were with turbid untreated preparations of plasma membrane vesicles (initial absorbances between 1 and 2). Continuous traces exhibited more noise than those with the SLM 2000. Figure 2.5a, b compare a continuous trace of soybean plasma membranes with a starting absorbance of 1.4 generated in one of the two machines with data obtained from analyses of individual rates determined over 1 min at 1.5-min intervals in (b) as illustrated in (e). A regular pattern of rapid rates alternating with very slow rate was observed with both data sets. These intact plasma membranes have a very low background rate and the rate of NADH disappearance in between the maxima actually reached zero (Fig. 2.4b). Panel (c) illustrates a series of 18 5-min traces assembled end to end for the same plasma membrane preparation assayed in one of the two machines subsequent to the data of (a) and (b) but timed to be in phase. The individual traces are shown in (d). The sequence with which the traces were acquired is given by the letters a–r. The same sequence was followed for Figs. 2.3a, c and 2.4e (top to bottom, left to right). These types of traces are those most commonly used to measure effects of inhibitors and other parameters affecting ENOX activities. Examples of sequential 5-min assay segments where activity in m and f exhibits little or no activity are encountered as well (Fig. 2.3c). A minimum also is encountered at the end of i and the

Fig. 2.4 (continued) the second Hitachi instrument based on rates determined over 1 min at 1.5-min intervals. (**c**) An assembly of 18 assays of NADH oxidation each over 5 min determined on the same preparation sequentially but in phase with (**a**) and (**b**). (**d**) Examples of 5-min traces used in the generation of (**c**). These preparations, largely lacking background activity and relatively well synchronized by light, exhibit periods where activity slows or stops between periods of rapid activity. The period length is 24 min. The 5-min assay segments labeled f and m exhibit little or no activity. A minimum is also encountered at the end of i and the beginning of j. (**e**) Examples of rates determined over 1 min at intervals of 1.5 min. (**f**) Analyses of (**e**). (**g**) Sequential 5-min traces as in (**d**) except determined sequentially without interruption. All assays were for 90 min

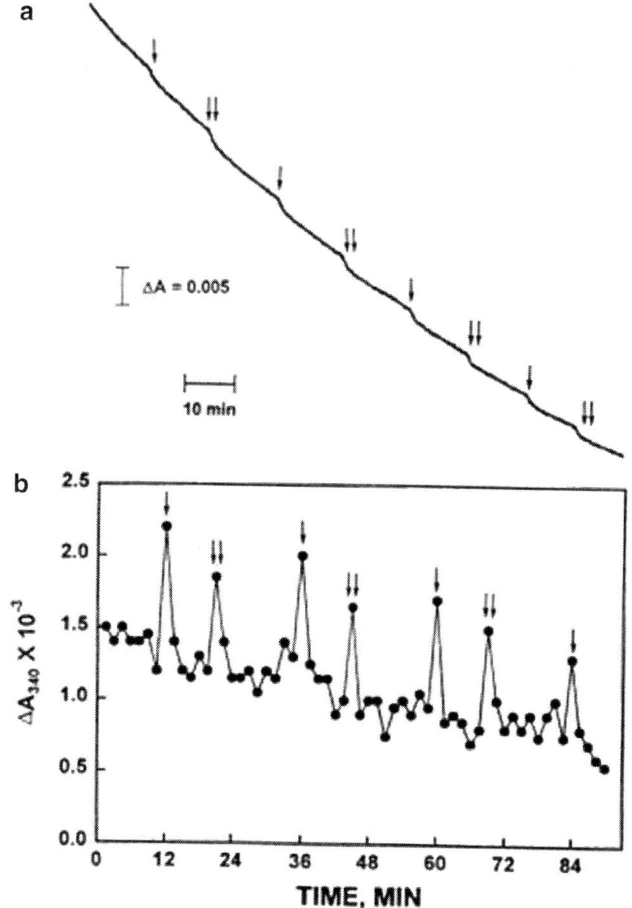

Fig. 2.5 Oxidation of NADH by a well-synchronized preparation of soybean plasma membranes solubilized in Triton X-100 to reduce turbidity also using the Aminco SLM 2000 in the dual wavelength mode of operation showing two distinct activity patterns (single and double arrows). (**a**) Continuous trace over 108 min. (**b**) Activity analyzed over 90 min from slopes determined over 1 min at 1.5-min intervals as illustrated for Fig. 2.3c, d. Since the vesicles were solubilized, a major contribution to background comes from a NADH oxidase of the internal plasma membrane surface that does not oscillate (Morré and Morré 2000). Reproduced from Morré and Morré (2003c) with permission of International Hormesis Society

beginning of j which would be detected with 1-min traces as a rate near zero. Figure 2.4e gives examples of sequential 1-min traces measured at intervals of 1.5 min as for Fig. 2.4b, f. Despite the width of the traces, slopes even with these relatively turbid samples were determined with an accuracy and reproducibility sufficient to reliably reveal details of the oscillatory pattern (Fig. 2.4b). A further illustration is provided in Fig. 2.4f, g where the two instruments again were used in parallel. Figure 2.4f shows the results of the 1-min analyses at intervals of 1.5 min whereas in (g), the same preparation was analyzed based on 18 consecutive 5-min

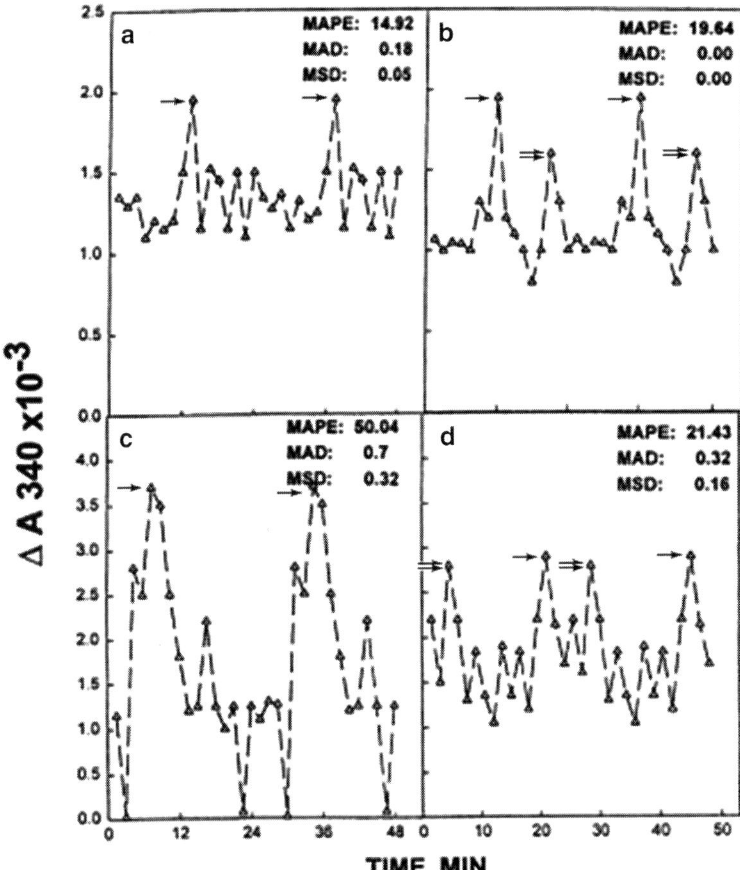

Fig. 2.6 Decomposition (time series) analyses of the oscillatory patterns of NADH oxidation of Figs. 2.3, 2.4, and 2.5. (**a**) Decomposition fit of data of Fig. 2.3d. The data consist of a pattern of repeating oscillations, all conforming to a 24-min period length (*arrows*). (**b**) Decomposition fits (*open symbols, dashed lines*) for two full cycles of data of Fig. 2.5b. (**c**) Decomposition fits for data of Fig. 2.4b. (**d**) Decomposition fits for data of Fig. 2.4f. Period lengths were determined by Fourier analysis. The decomposition fits show the reproducibility of the patterns of oscillations. Three measures of the accuracy of the statistically fitted values are provided in the *insets*. Reproduced from Morré and Morré (2003c) with permission of International Hormesis Society

scans as in (c) and (d) except that the rates were determined sequentially without interruption. Both methods revealed a similar pattern of oscillations.

To illustrate the reproducibility of the pattern of oscillations, decomposition fits for two full periods of each of the experiments of Figs. 2.3d, 2.4b, f, and 2.5b are shown in Fig. 2.6 along with three measures of the accuracy of the statistically fitted values. The MSD comparing all four data units analyzed averaged 13 %. Differences between maxima and minima were highly significant ($p<0.001$). Also evident from the decomposition fits were the double-peak pattern of Fig. 2.4f (Fig. 2.5b) compared to the single-peak patterns of Fig. 2.3d.

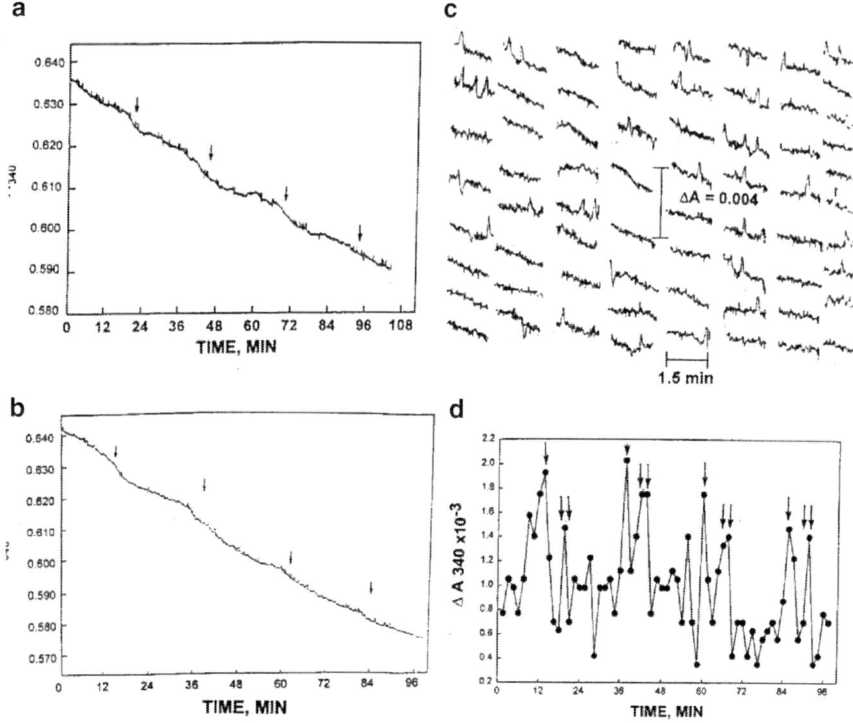

Fig. 2.7 Decrease in A_{340} from the oxidation of NADH measured using the Spectronics UV 500 spectrophotometer. (**a**) Continuous trace over 105 min of a preparation containing 10^6 intact CHO cells where synchrony was achieved by addition of 1 μM melatonin at $t=12$ min. *Arrows* spaced at 24 min coincide with intervals where rapid rates of NADH oxidation were observed. (**b**) As in (**a**) except for a preparation of plasma membranes of dark-grown soybean hypocotyls (150 mg total protein). Melatonin (1 μM) was added at $t=12$ min. (**c**) Data of (**b**) displayed as successive 1.5-min segments. (**d**) Analysis of (**c**) giving rates determined by least squares analyses of the 1.5-min segments shown in (**c**). Reproduced from Morré and Morré (2003c) with permission of International Hormesis Society

Absorbance values recorded at intervals of 10 sec using the Beckman DB200 spectrophotometer (Beckman Instruments, Palo Alto, CA) also provided useful information (Fig. 2.7). These data were analyzed to determine best fit slopes at intervals of 1.5 min. Oxidation was calculated from the formula $N-(N+1)$ for each 1.5-min interval to generate data comparable to that of Fig. 2.4f.

2.3.1 Diode Array Instruments

With assay of A_{340} from diode assay instruments, the variation in ratios normally fall within the envelope of machine variation, even with preparations treated with detergents such as Triton X-100 to reduce light scattering (Fig. 2.8). Thus, with the

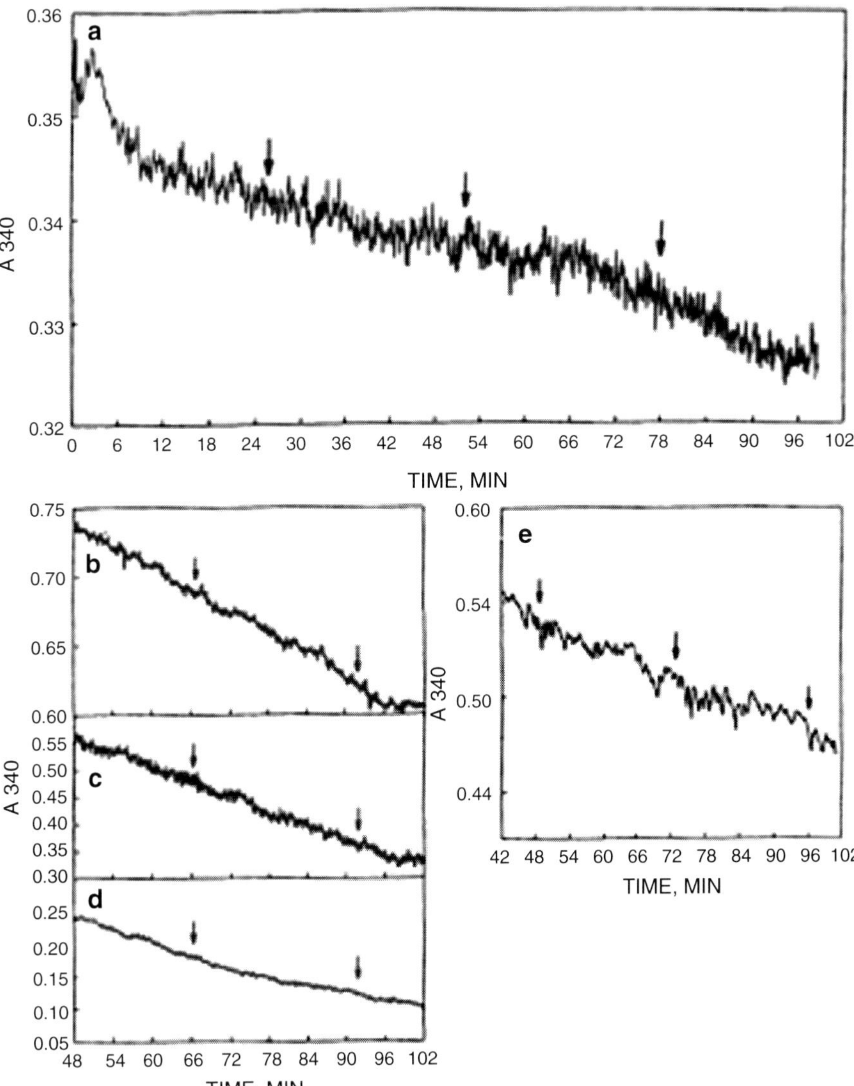

Fig. 2.8 Comparable assays of the decrease in A_{340} due to NADH oxidation using a diode array spectrophotometer (HP8452A). *Arrows* are spaced at 24 min to coincide with at least one interval where rapid rates are separated by a slow rate. (**a**) CHO microsomes dissolved in 0.5 % Triton X-100 to reduce turbidity. (**b–d**) Three different preparations of CHO membranes in which synchrony was induced by addition of 1 μM melatonin at $t = 0$. (**d**) With 5 % Triton X-100 added followed by centrifugation to remove floating lipids. A double peak with a 24-min period similar to that of Fig. 2.3 was seen at the *arrows* in all three repetitions on subsequent days with different preparations. (**e**) A skunk cabbage (*Symplocarpus foetidus*) spadix plasma membrane preparation comparable to the soybean plasma membrane preparation of Fig. 2.3a. Reproduced from Morré and Morré (2003c) with permission of International Hormesis Society

Hewlett Packard 8452 single beam diode array spectrophotometer, variations due to light scatter, even in well-synchronized preparations, were sufficient to prevent observation of statistically significant deviations in reaction rates of the continuous traces that would distinguish them from single experiential or linear decay (Fig. 2.8). When the diode array data were numerically differentiated using the Savitsky–Golay algorithm presuming local linearity and with averaging times ranging from 11 to 121 s (Savitsky and Golay 1964), no departures from linearity could be discerned above the variations due to light scatter. At least six different single beam diode array instruments have been evaluated and compared with outcomes similar to those observed with the Hewlett Packard 8452 instrument.

2.4 Measurement of Hydroquinone Oxidase Activity with Reduced Coenzyme Q_{10} or Phylloquinone as Substrate

Hydroquinone oxidase activity may be estimated as for NADH except from absorbance changes at 275 nm, 410 nm or both with reduced coenzyme Q_{10} (Tishcon, Westbury, NY) (Kishi et al. 1999) or reduced phylloquinone (Bridge et al. 2000) as the substrate (Fig. 2.9).

2.4.1 Enzyme Assay for Reduced Coenzyme Q_{10} Oxidase

The method for preparation of reduced coenzyme Qs (50 mM coenzyme Q_0 stock solution or 7.5 mM coenzyme Q_{10} stock solution in ethanol) involved addition of an equal volume of 0.25 % $NaBH_4$ under nitrogen followed after several min by 0.1 vol. of 0.1 N HCl to degrade the excess $NaBH_4$. The colorless solution held at room temperature must be prepared fresh for each experiment.

For oxidation of $Q_{10}H_2$ the reaction mixture contained the enzyme source, in 2.5 mL of 50 mM Tris-Mes buffer, pH 7.0. Triton X-100 (0.08 %) was used to solubilize the $Q_{10}H_2$ in the assay buffer. The reaction was started with the addition of 40 μL of 5 mM $Q_{10}H_2$. An extinction coefficient of 0.805/mM/cm was used to calculate the rate of $Q_{10}H_2$ oxidation (Sun et al. 1995). Rate measurements are illustrated for continuous traces (Fig. 2.10) as well as over 1 min at 1.5-min intervals (Fig. 2.11).

Q_{10} in ethanol exhibits a maximum at 275 nm, a broad band at about 410 nm, and a minimum at 236 nm (Crane et al. 1957; Fig. 2.9). Upon reduction of the quinone, the bands at 275 and 410 nm disappear and a new peak characteristic of $Q_{10}H_2$ appears at 290 nm (Hatefi 1963) shown schematically in Fig. 2.1.

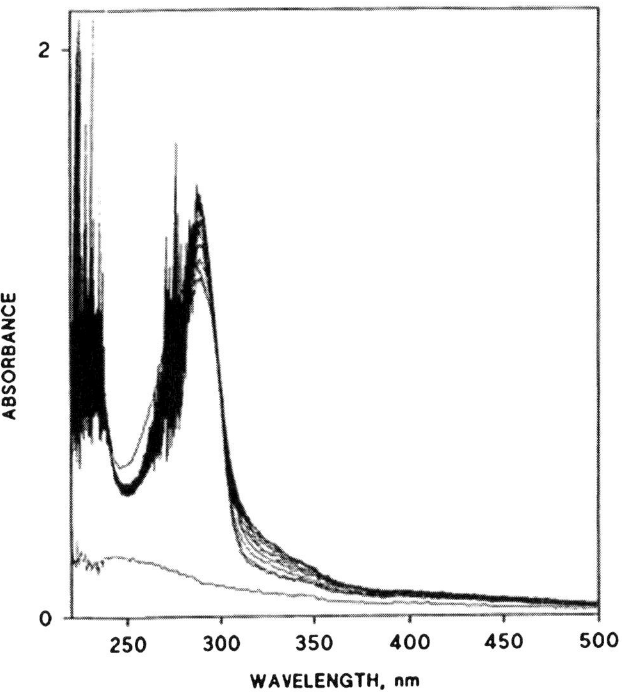

Fig. 2.9 Oxidation of $Q_{10}H_2$ (0.2 mM) in the presence of 0.2 mM Q_0. The preparation contained 0.6 mg protein, 1.2 mM EDTA, 0.08 % Triton X-100 and 50 mM Tris-Mes buffer, pH 7.0, and was scanned every 5 min for 50 min. Oxidation of $Q_{10}H_2$ is indicated by the increase in absorbance at 410 nm accompanied by a decrease in absorbance at 290 nm as illustrated diagrammatically in Fig. 2.1b. Reproduced from Kishi et al. (1999), Copyright 1999 with permission of Elsevier

2.4.2 *Enzyme Assay for Reduced Phylloquinone Oxidase*

For oxidation of reduced phylloquinone (Bridge et al. 2000), the reaction mixture contained 2.5 mL of 50 mM Tris-Mes buffer, pH 6.5. The reaction was started by the addition of 40 μL of the reduced 20 mM phylloquinone. The reduced phylloquinone oxidase activity was measured spectrophotometrically at a wavelength of 410 nm at 27 °C as described by Sun et al. (1995) for reduced coenzyme Q_{10} using a Hitachi U3210 spectrophotometer. A blank rate was subtracted in which the assay was carried out in the absence of added proteins. The extinction coefficient used for K_1H_2 oxidation was 0.74/mM/cm. The oscillatory pattern of K_1H_2 oxidation is illustrated in Fig. 2.12.

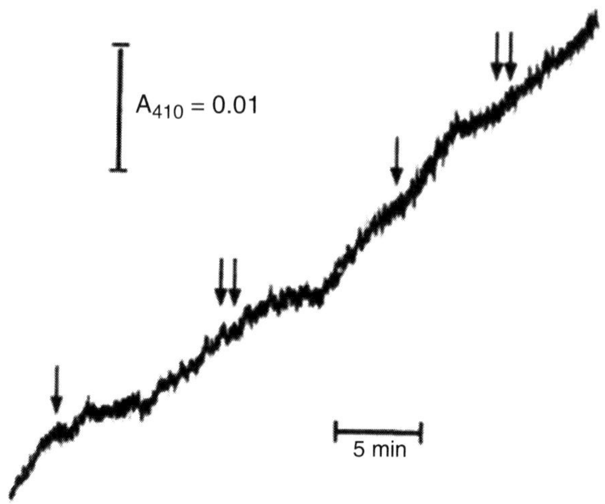

Fig. 2.10 Oxidation of $Q_{10}H_2$ as determined from the increase in absorbance at 410 nm monitored with an Aminco SLM 2000 in the dual wavelength mode of operation. The enzyme source was a solubilized and partially purified ENOX preparation from the surface of human cervical carcinoma (HeLa) cells that contained both ENOX1 (24-min period) and ENOX2 (22-min period). Periods of rapid oxidation (*arrows*) of $Q_{10}H_2$ (Tishcon, Westburg, NY) alternate with periods of much slower $Q_{10}H_2$ oxidation. The *single arrows* are separated by 24 min, whereas the *double arrows* are separated by 22 min. Results where $Q_{10}H_2$ oxidation was monitored from the decrease in A_{290} were similar. Reproduced from Morré (2004), Copyright 2004 with permission of Elsevier

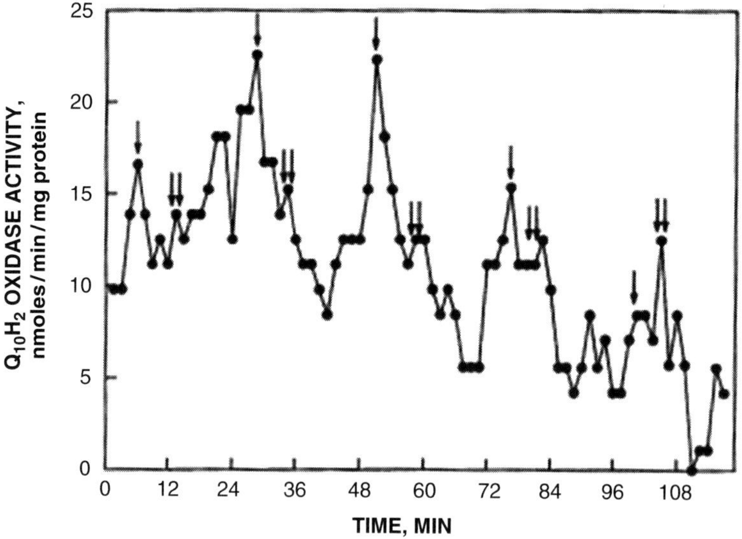

Fig. 2.11 Rate measurements over 1 min at 1.5-min intervals to illustrate the periodic variation in the rate of oxidation of $Q_{10}H_2$ as a function of time over 114 min. The enzyme source was a solubilized and partially purified ENOX preparation from the surface of human cervical carcinoma (HeLa) cells that contained both ENOX1 (24-min period length, *single arrows*) and ENOX2 (22-min period length, *double arrows*). Reproduced from Kishi et al. (1999), Copyright 1999 with permission of Elsevier

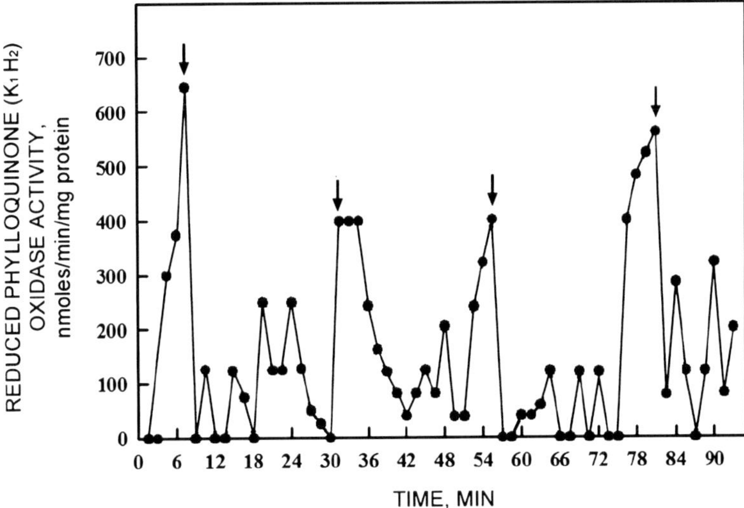

Fig. 2.12 Periodic variation in the rate of oxidation of reduced phylloquinone (K_1H_2) (0.32 mM) as a function of time over 93 min showing four maxima (*arrows* at 7.5, 31.5, 55.5, and 79.5 min) for an enzymatic preparation enriched in NADH oxidase activity solubilized from isolated plasma membrane vesicles. The average period length was about 24 min. The concentration of phylloquinone was 0.32 mM, the pH was 7, and the protein amount was 35 μg/assay. Values are from a single experiment. Reproduced from Bridge et al. (2000), Copyright 2000 with permission of Elsevier

2.5 Dissolved Oxygen Measurement

In our published studies, oxygen was measured using Yellow Springs Instruments model 53 Clark-type polarographic dissolved oxygen probe fitted with a Yellow Springs Instruments high-sensitivity membrane and electrolyte solution (5.25 g KCl and 63 μL Kodak Photo-Flo in 350 mL of distilled H_2O) connected to an amplifier and placed in a reaction chamber (Orczyk et al. 2005). A steady stream of 37 °C water was pumped through a water jacket surrounding the chamber. In addition, a Micro Stirring Bar, measuring 7 mm in length and 2 mm in diameter, was used to mix the solution. Measurements with a total sample volume of 2.5 mL were recorded with a Kipp and Zonen type BD112 chart recorder set at a rate of 2 mm/min and a scale of 10 mV (Figs. 2.13, 2.14, and 2.15). Assays were in the presence of phosphate-buffered (pH 7.4) saline and 2 mM KCN.

Calibration was based on 100 % air saturated water at 37 °C and 0 salinity (0.1869 mM) as compared to 0 % air saturated water obtained by anhydrous sodium sulfite titration (Clesceri et al. 1998). Instrument traces of oxygen consumption for CHO (Fig. 2.12) and HeLa cells (Fig. 2.13) show the characteristic oscillatory pattern of alternating fast and slow rates. Rates determined by numerical averaging show the oscillatory pattern more clearly (Fig. 2.14).

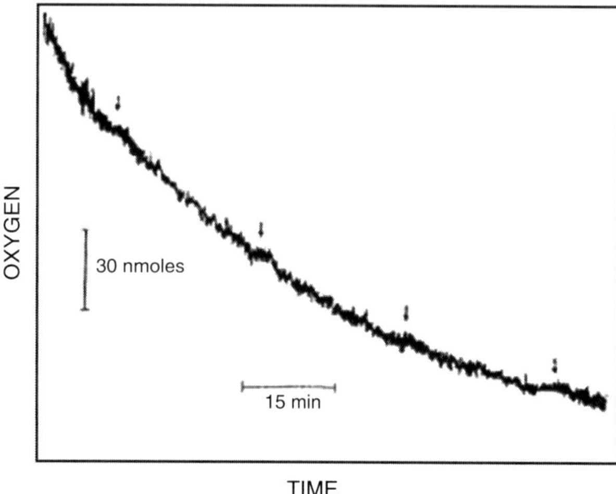

Fig. 2.13 Instrument trace of oxygen consumption by 3×10^7 CHO cells. The rates show a pattern of slow rates separated by rapid rates. *Single arrows* marking slow rates are separated by intervals of 24 min. The assay was in phosphate-buffered (pH 7.4) saline and 2 mM KCN. The initial rate of oxygen consumption was estimated to be 0.2 nmol molecular oxygen/min/10^6 cells. Reproduced from Orczyk et al. (2005) with permission of Springer-International

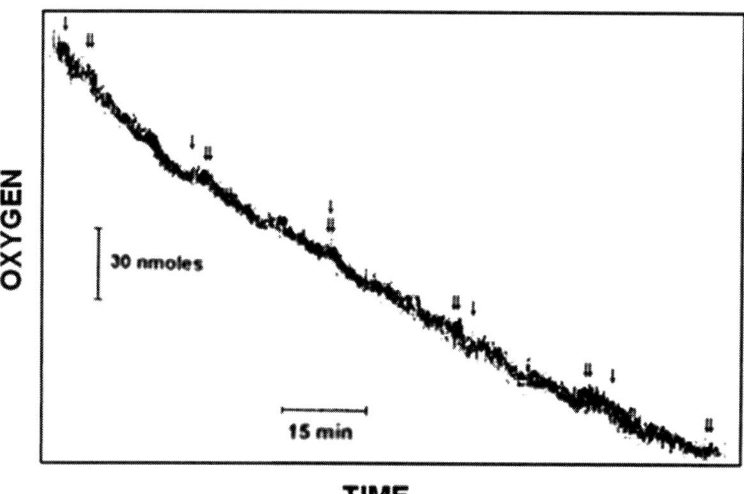

Fig. 2.14 Instrument trace of oxygen consumption by 4×10^7 HeLa cells. The rates show a pattern of alternating slow and fast rates. *Single arrows* marking slow rates are separated by intervals of 24 min characteristic of the constitutive ENOX1. *Double arrows* denote intervals of 22 min characteristic of the tumor-specific, ENOX2 form. Assay was in phosphate-buffered (pH 7.4) saline and 2 mM KCN. Rapid rates of oxygen consumption were estimated to be between 0.15 and 0.2 nmol molecular oxygen/min/10^6 cells. Reproduced from Orczyk et al. (2005) with permission of Springer-International

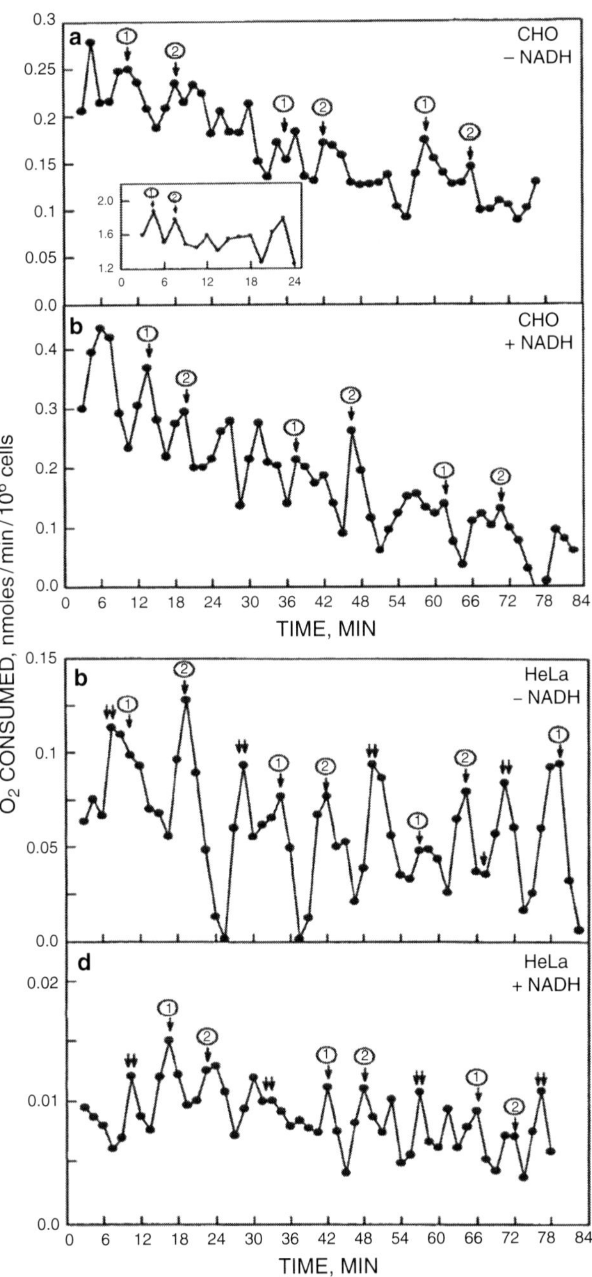

Fig. 2.15 Oxygen consumption measured over 1.5-min intervals by numerical averaging from the cumulative oxygen consumption as illustrated in Figs. 2.13 and 2.14. (**a**, **b**) 2×10^7 CHO cells in the absence (**a**) or presence (**b**) of 150 μM NADH. The *inset* in (**a**) is a decomposition fit of the type used to validate the reproducibility of the recurrent pattern and to provide the three measures of accuracy. The CHO cells exhibit maxima (*arrows* labeled ① and ②) that recur at regular intervals of 24 min as is characteristic of the constitutive ENOX1. (**c**, **d**) 4×10^7 HeLa cells in the absence (**c**) or presence (**d**) of 150-μM NADH. HeLa cells exhibit the maxima labeled ① and ② recur at regular intervals of 24 min as is characteristic of ENOX1 together with a second set of maxima (*double arrows*) separated by 22 min as is characteristic of the cancer-associated ENOX form, ENOX2. The assay was in phosphate-buffered (pH 7.4) saline containing 2 mM KCN. Reproduced from Orczyk et al. (2005) with permission of Springer-International

2.6 Estimation of Protein Disulfide-Thiol Interchange Activity

RNase A activity was assayed by two different methods. In the first method, a spectrophotometric assay based on cCMP as an RNase substrate was used (Lyles and Gilbert 1991) as diagramed in Fig. 2.16–2.18a. Inactive, oxidized, and scrambled (see below) RNase A was incubated together with cCMP in Tris-Mes buffer, pH 6.4, with or without GSH or GSSG. In the absence of a protein disulfide isomerase or protein disulfide-thiol interchange activity, the scrambled RNase was inactive (activity about 2 % that of native RNase). However, in the presence of a protein disulfide isomerase or protein disulfide-thiol interchange activity, the scrambled RNase became active as evidenced by an increase in A_{296} from the RNase-catalyzed hydrolysis of cCMP.

The assay mixture contained 50 mM Tris-Mes buffer (pH 6.0), 0.45 mM cCMP, 1 μM GSH or GSSG or GSH plus GSSG, and 400 μg of plasma membrane protein preincubated at 30 °C for 20 min. The assay was initiated by the addition of scrambled RNase A (see below). Hydrolysis of cCMP resulting from the activation of the initially randomly oxidized and inactive RNase was recorded continuously as an increase in A_{296}. The concentrations of cCMP were determined at a wavelength of 296 nm (extinction coefficient = 0.19/mM/cm). Spectrophotometric measurements were in parallel using two Hitachi (Tokyo, Japan) U3210 spectrophotometers with thermostatic cell compartments maintained at 30 °C with continuous stirring (Figs. 2.16 and 2.18a).

2.7 Preparation of Scrambled RNase Substrate

To prepare the scrambled and inactive RNase substrate, native RNase A (Sigma, type 1-AS from bovine pancreas) (30 mg/mL) was incubated for 1 h at 35 °C in 50 mM Tris-acetate, pH 8.6, containing 9 M urea and 130 mM DTT (Hillson et al. 1984) to denature and reduce the protein.

The denatured and reduced protein was then isolated by adjusting the pH to 4.0 with glacial acetic acid, followed by elution from a column of Sephadex G-25 with degassed 0.1 M acetic acid. The estimation of the protein concentration was followed by spectrophotometric measurement at 280 nm using native RNase A as the standard. The samples were diluted to approximately 0.5 mg/mL with 0.1 M acetic acid. Solid urea was added to the eluted RNase to a final concentration of 10 M, after which 0.1 M sarcosine hydrochloride was added and the pH adjusted to 8.5 with 1 M Tris. The mixture was then incubated in the dark for 2–3 days during which time the denatured protein was randomly oxidized (scrambled). The scrambled product was recovered by acidification of pH 4.0 with glacial acetic acid and elution from Sephadex G-25 in 0.1 M acetic acid. Fractions containing protein were pooled,

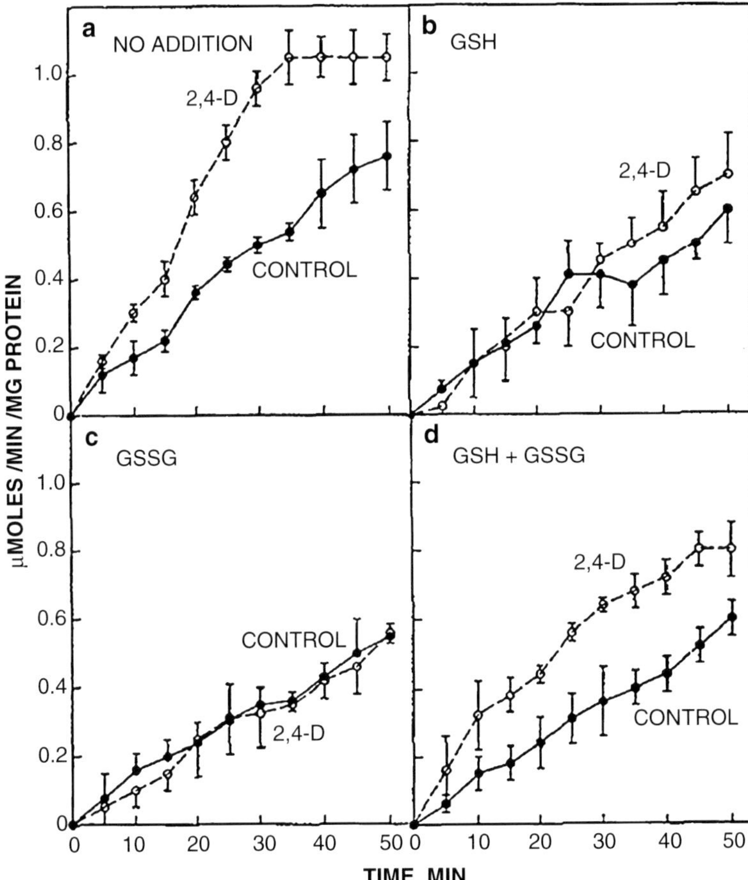

Fig. 2.16 Protein disulfide-thiol interchange activity estimated from the activation of scrambled RNase in the presence of cCMP substrate. Soybean plasma membranes (400 μg) were incubated together with scrambled (inactive) RNase and the cCMP substrate for the times indicated and the specific activity was determined over 5 min at successive 5-min intervals for each preparation. The GSH and GSSG (alone or in combination) solutions were freshly prepared and the final concentrations were 1 μM. The activity was stimulated by the synthetic plant growth hormone, 2,4-dichlorophenoxyacetic acid (2,4-D) at a final concentration of 1 μM. Results are means ± standard deviation of triplicate experiments. Reproduced from Morré et al. (1995d), Copyright 1995 with permission of Elsevier

adjusted to pH 8.0, and stored at 4 °C. Thiol groups, as determined using 5,5′-dithio(2-nitrobenzoic acid) (Ellman 1959) were 80–90 % oxidized.

The preparation of the scrambled RNase substrate is illustrated diagrammatically in Fig. 2.18.

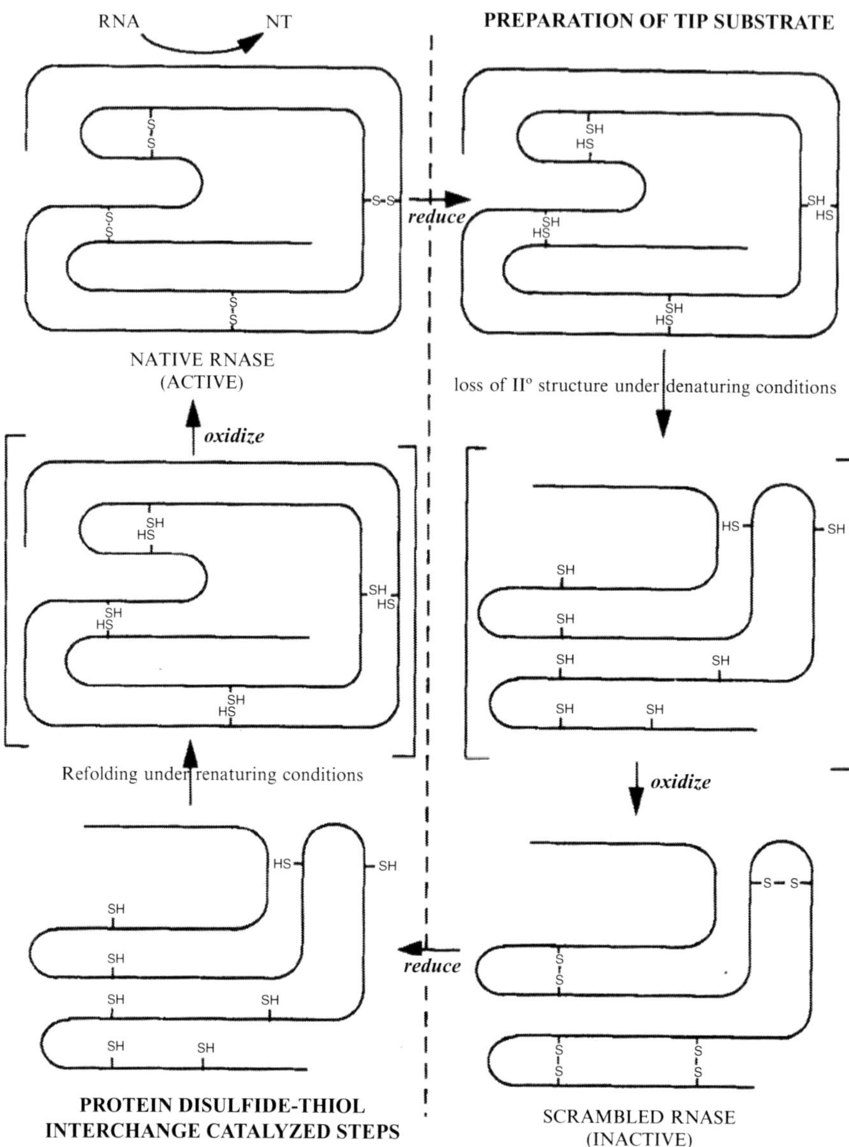

Fig. 2.17 Schematic representation of the procedure for preparation of scrambled and inactive RNase (*left*) and the restoration of activity through cleavage of disulfide bonds refolding under renaturating conditions and reformation of disulfide bonds as the protein disulfide-thiol interchange activity restores activity to the preparations

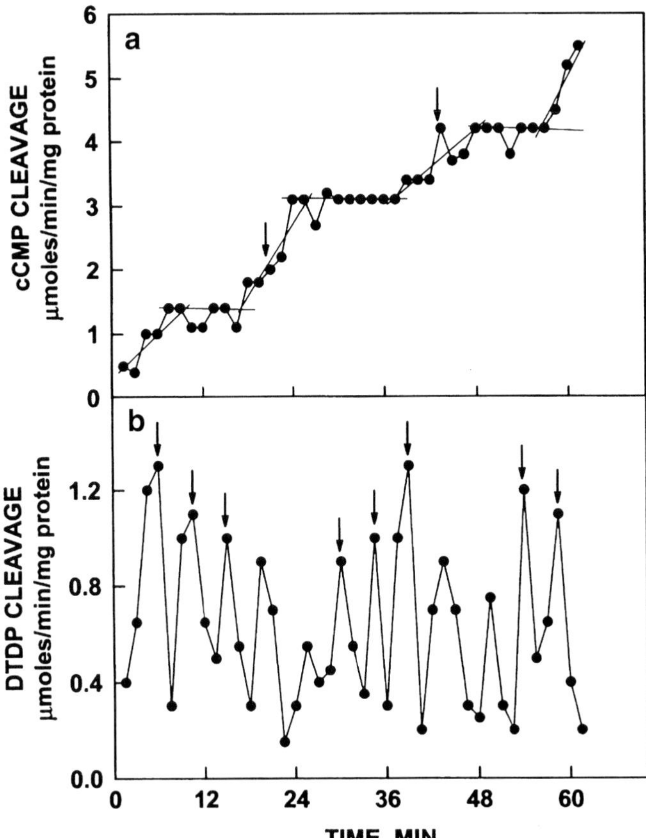

Fig. 2.18 Protein disulfide-thiol interchange activity of recombinant human ENOX1 measured from the activation of scrambled and inactive RNase to cleave cCMP measured spectrophotometrically (**a**) or from the cleavage of a DTDP substrate (**b**). Both activities exhibited an oscillatory activity. However, the activities were most strongly associated with the three maxima separated by 4.5 min rather than with the two maxima separated by 6 min dominated by oxidation of NADH. From Jiang et al. (2008) with permission from ACS publications

2.8 Estimates of Protein Disulfide-Thiol Interchange from Enzymatic Assay of Dipyridyl-Dithio Substrate Cleavage

The assay for the ability of 2,2′-dipyridyl-dithio substrates decompose to 2 mol of pyridinethione in buffered (50 mM Tris-Mes) aqueous solution at pH 7. The latter absorb strongly at 340 nM (Morré et al. 1999a). The assay was preincubated with 0.5 μmol 2,2′-DTDP or 6,6′-dithiodinicotinic acid (DTNA) in 5 μL of DMSO to react with endogenous reductants present with the plasma membranes. After 10 min, a further 3.5 μmol DTDP or DTNA were added in 35 μL DMSO to start the reaction. The final reaction volume was 2.5 mL.

The reaction was monitored from the increase in absorbance at 340 nm using a Hitachi Model U3210 spectrophotometer. The change in absorbance was recorded as a function of time by a chart recorder. Alternatively, absorbance changes were measured at 340 nm with reference at 430 nm using a SLM DW2000 spectrophotometer in the dual wavelength mode of operation. The specific activity was calculated using a millimolar absorption coefficient of 6.21/cm recognizing that 2 mol of product were generated for each mole of substrate cleaved (Fig. 2.17b).

The 2,2′-dipyridyl-dithio substrates were rapidly and completely hydrolyzed in the presence of reducing agents such as dithiothreitol and reduced glutathione (GSH). They were relatively more stable in the presence of the weak reducing agent cysteine and the weak oxidizing agent oxidized glutathione (GSSG). The substrates also spontaneously decompose such that it is necessary to subtract a blank rate with no enzyme present.

2.9 Measurement of *Trans*-Plasma Membrane Redox by Reduction of Cell-Impermeable Dyes

Artificial cell-impermeable dyes such as WST-1 (2-(4-iodophenyl)-3-(4-nitrophenyl)-5-(2,4-disulfophenyl)-2H-tetrazolium, monosodium salt) used in combination with mPMS (1-methoxy-5-methylphenazinium methyl sulfate) or coenzyme Q have been used to measure NADH oxidase driven plasma membrane electron transport (Tan and Berridge 2004; Figs. 2.19 and 2.20).

2.9.1 *CoQ_1 Can Function as an Intermediate Electron Carrier in WST-1 Reduction*

Cellular reduction of WST-1 involves an obligate intermediate electron carrier, namely, 1-methoxyPMS (Tan and Berridge 2004; Fig. 2.19). They also investigated whether CoQ_1, a homologue of CoQ_{10} with a short isoprenoid side chain and decreased lipophilicity could function as an alternative intermediate electron carrier. The results demonstrated that mPMS and CoQ_1 can directly access low potential electrons from the PMET system whereas WST-1 cannot. mPMS mediates electron transfer to WST-1 extracellularly, whereas CoQ_1 partitions in the plasma membrane.

2.9.2 *Measurement of Plasma Membrane Electron Transport Based on WST-1 Reduction*

Cells (2×10^4/0.1 mL for mPMS or 10^5/0.1 mL for CoQ_1) in HBSS, with or without mPMS or CoQ_1, were incubated in 96-well microtiter plates at 37 °C in a shaking

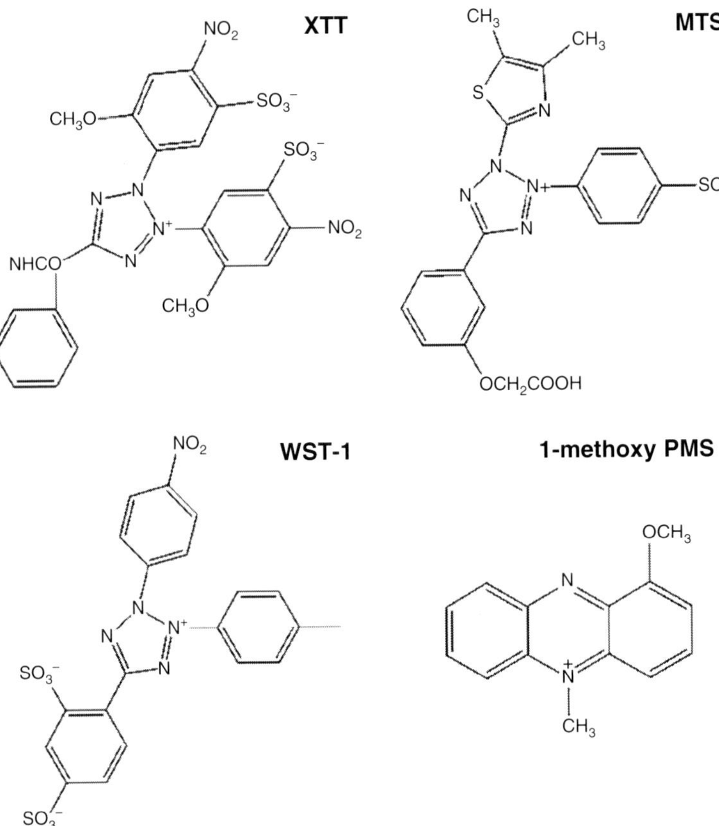

Fig. 2.19 Chemical structure of tetrazolium dye (XTT, MTF, and WST-1) and of the immediate electron acceptor 1-methoxy-5-methylphenazinium methyl sulfate (mPMS). Redrawn from Del Principe et al. (2011), Published with permission

BMG Fluostar microplate reader (Fig. 2.21). The reaction was initiated by adding 0.45 mM WST-1 followed by 25 μM mPMS or 50 μM CoQ_1 for 40 min. The absorbance was read in real time at 450 nm and the rate of WST-1 reduction as a measure of the activity of *trans*-plasma membrane electron transport is expressed as A_{450}/min (change in absorbance at 450 nm/min, multiplied by 1,000) (Tan and Berridge 2004).

For data of Fig. 2.21a, for example, when average rates were determined at 1.5-s intervals over 1 min at intervals of 1.5 min, oscillatory patterns characteristic of ENOX activities were observed (Fig. 2.22). Maxima corresponding to period lengths of 22 min (ENOX2), 24 min (ENOX1), or 26 min (arNOX) were evident from the analyses confirming the validity of using artificial cell-impermeable dyes in combination with mPMS to measure ENOX-driven plasma membrane electron transport.

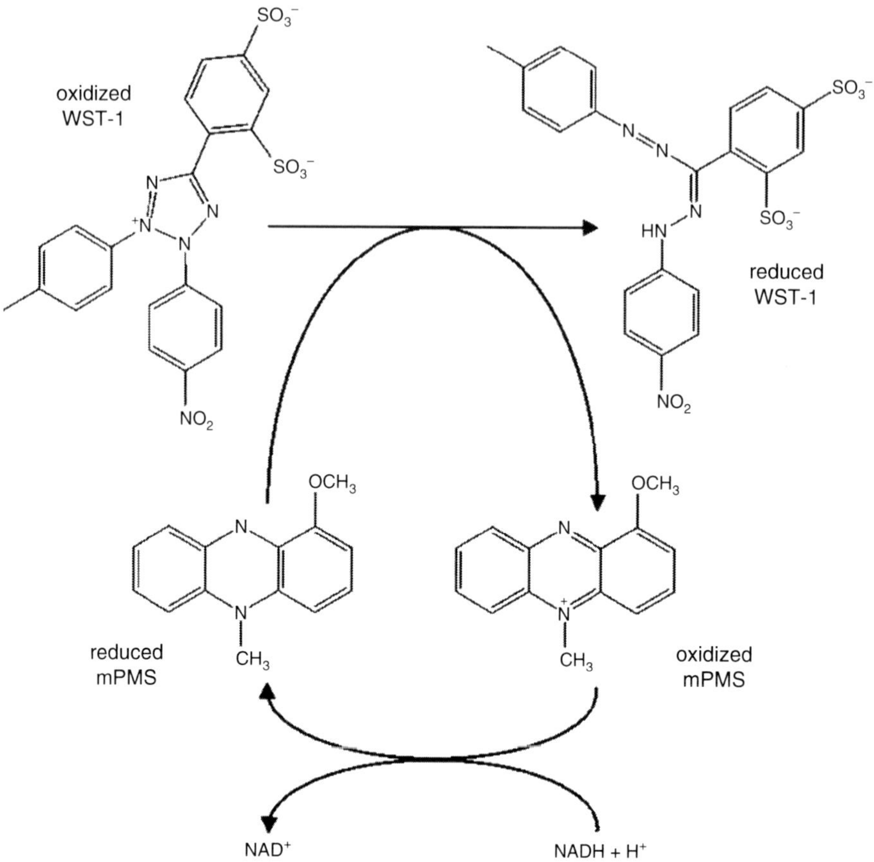

Fig. 2.20 Reduction of WST-1 (2-(4-iodophenyl)-3-(4-nitrophenyl)-5-(2,4-disulfophenyl)-2H-tetrazolium, monosodium salt) by the intermediate electron acceptor mPMS and formation of the reduced formazan. Redrawn from Del Principe et al. (2011), Published with permission

2.10 Summary

Spectroscopic strategies that substantiate periodic oscillations in low rates of NADH oxidation exhibited by ENOX proteins at the animal and plant cell surface are described. Both continuous display and discontinuous rate determinations exhibit the oscillations but continuous displays lack sufficient resolution to determine fine structural details. A procedure is documented where rates are determined by least squares analyses of traces recorded over 1 min at intervals of 1.5 min. These traces recapitulate the continuous displays but offer an opportunity to reliably estimate changes in reaction rates over short time intervals not afforded by the continuous traces. Turbidity is identified as the major contributor to losses in resolution. Even highly purified ENOX preparations tend to aggregate to form turbid suspensions.

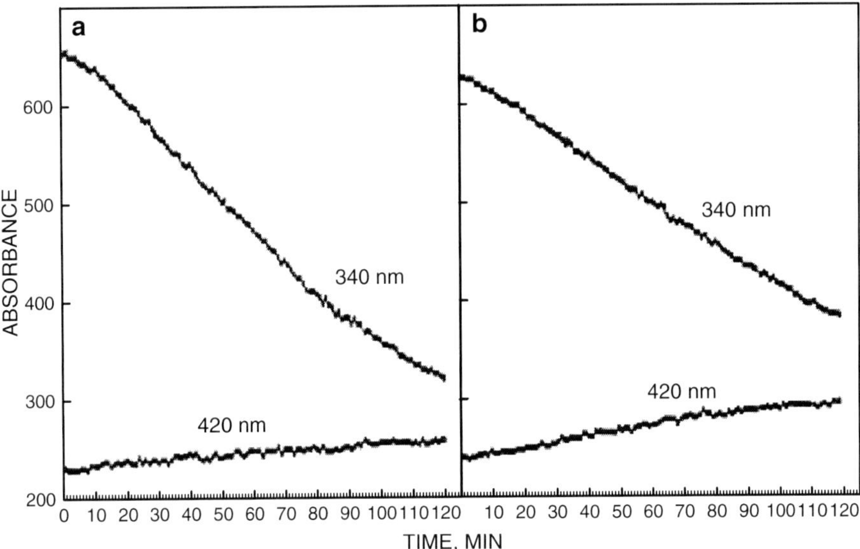

Fig. 2.21 Replicate measurements of NADH oxidase activity of intact 143B-osteocarcoma cells determined at 342 nm (with reference at 420 nm) based on the reduction of WST-1 in combination with mPMS (1-methoxy-5-methylphenazonium methyl-sulfate). Data courtesy of Dr. Michael Berridge, Malaghan Institute of Medical Research, Wellington, New Zealand, published with permission

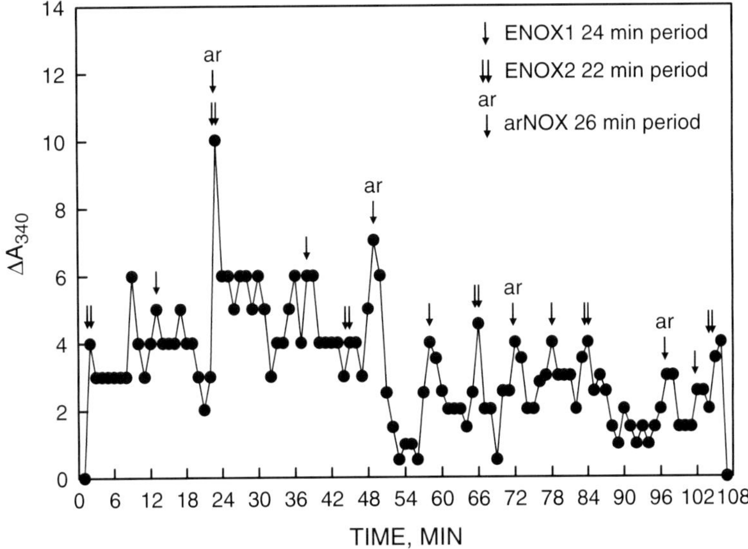

Fig. 2.22 Average rates from Fig. 2.21a determined at 1.5-s intervals averaged over 1 min and taken at intervals of 1.5 min reveal oscillatory patterns typical of established cultures of cancer cells expressing both ENOX1 and ENOX2 as well as the aging-related NADH oxidase (arNOX)

With turbid suspensions, double beam or dual wavelength instrumentation where the sample is placed immediately adjacent to the photomultiplier tube is required to reduce losses in resolution from turbidity. Also required are high levels of synchronous ENOX function. Blue or red (plants) light, small molecules (i.e., melatonin), electromagnetic fields, and autosynchrony alone or in combination may be used to synchronize the oscillations in asynchronous preparations. Special problems are posed by preparations containing more than one ENOX form that do not cross entrain, i.e., ENOX1 and ENOX2. Also described are assay methods based on hydroquinone (ubiquinol or phylloquinol) oxidation, and oxygen consumption for the oxidative activity and activation of scrambled RNase in the presence of cAMP substrate and from the cleavage of DTDP substrate measured spectrophotometrically for the protein disulfide-thiol interchange activity as well as the use of tetrazolium dyes.

Chapter 3
The Constitutive ENOX1 (CNOX)

3.1 ENOX1 Function

All animals and plants thus far investigated exhibit at least one hormone-responsive external plasma membrane hydroquinone oxidase capable of catalyzing protein disulfide interchange and that oxidizes NAD(P)H as an alternate substrate with a precise period length of 24 min (Morré 1998c; Morré and Morré 2003a). These proteins, known collectively as constitutive ECTO-NOX proteins (ENOX1 or CNOX), are widely distributed among animals, plants, and yeast, function as terminal oxidases of plasma membrane electron transport, are hormone responsive when assayed with plasma membranes from mammalian cells and tissues (Table 3.1), act as ultradian oscillators of the biological clock, and are essential to the enlargement phase of cellular growth (Fig. 3.1). They were first described in plants (Morré et al. 1986a; Brightman et al. 1988; Morré and Brightman 1991) as the result of a decade long search for a growth-related protein of the cell surface stimulated by plant growth regulators of the auxin type (Chap. 10). Shed forms of ENOX1 are found in culture media conditioned by the growth of mammalian or plant cells (Morré et al. 1996c) and in sera. Sera of cancer patients and plasma membranes of cancer cell lines grown in culture contain both ENOX1 and ENOX2 (Wang et al. 2001; Chap. 8). Yet, the ENOX activity of tumor cell plasma membranes, while inhibited by various anticancer drugs (Chap. 11), is no longer hormone stimulated (Table 3.1). Only ENOX1 is present in spent media from noncancer cells and in sera from healthy volunteers or in patients with diseases other than cancer which lack ENOX2.

The ENOX1 proteins function in the cell enlargement phase of cell growth (Pogue et al. 2000; Morré et al. 2001a, 2002d). When their activity is inhibited, cells are unable to enlarge (Morré and Grieco 1999; Morré and Morré 2003a). *Arabidopsis* seedlings treated with a quassinoid ENOX1 inhibitor simalikalactone D ceased growth until the drug was metabolized but did not die (Morré and Grieco 1999). Similarly, monoclonal antibodies raised to ENOX1 of soybean completely blocked enlargement of hybridoma cells producing the antibodies.

D.J. Morré and D.M. Morré, *ECTO-NOX Proteins:Growth, Cancer, and Aging*,
DOI 10.1007/978-1-4614-3958-5_3,

Table 3.1 NADH oxidase of plasma membranes of liver and hepatomas of the rat

		NADH oxidase activity[a]	
Addition	Concentration	Liver	Hepatoma
None		1.05±0.7	2.46±0.12
Insulin	0.1 nM	2.35±0.29	2.54±0.15
EGF	10 nM	2.23±0.21	2.52±0.13
Transferrin	12.5 μM	1.96±0.27	2.06±0.20
Lactoferrin	25 μM	1.62±0.32	2.27±0.14
Glucagon	3 nM	1.27±0.20	2.71±0.12

From Bruno et al. (1992)

[a]NADH oxidase activity was determined from the change in absorbance of NADH monitored at 340 nm with reference at 430 nm using an SLM-2000 (Aminco) dual-beam spectrophotometer in the dual-wavelength mode of operation. Unit is nmol/min/mg protein

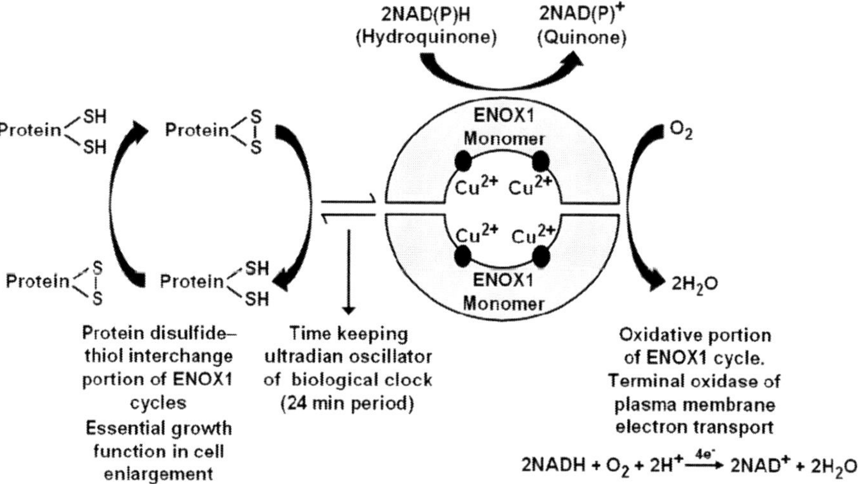

Fig. 3.1 Schematic indicating the several essential functions of the ENOX1 proteins

A distinguishing characteristic of ENOX1 and ENOX proteins in general that permits their unequivocal identification is that the proteins exhibit two activities that alternate with a period length of 24 min (Morré 1998c; Morré and Morré 2003a; Chap. 1). The first activity is that of a hydroquinone oxidase (NADH serves as an alternate substrate for this activity) (Kishi et al. 1999). The second is that of a protein disulfide-thiol interchange activity measured either from the restoration of activity to inactive (scrambled) RNase (Morré et al. 1995d; Chap. 2) or from the cleavage of dithiodipyridine substrates (Morré et al. 1999a; Chap. 2). Each activity generates a distinct oscillatory activity with a period length of 24 min for ENOX1. The strictly periodic nature of the ENOX proteins distinguishes them from all other oxidase or protein disulfide isomerase forms and imparts to the ENOX1 proteins a

```
ENOX2        1 --------MQRDFRWLWVYEIGYAADNSRTLNVDSTAMTLPMSDPTAWAT 42
cENOX1       1 MVDAGGVENITQLPQELPQMMAAAADGLGSIAIDTTQLNMSVTDPTAWAT 50
                          ::      :. ***.  :: :*:* :.:.::*******

ENOX2       43 AMNNLGMAPLGIAG------QPILPDFDPALGMMTGIPPITPMMPGLGIV 86
cENOX1      51 AMNNLGMVPVGLPGQQLVSDSICVPGFDPSLNMMTGITPINPMIPGLGLV 100
               *******.*:*:.*      .  :*.***:*.*****.**.**:****:*

ENOX2       87 PPPIPPDMPVVKEIIHCKSCTLFPPNPNLPPPATRERPPGCKTVFVGGLP 136
cENOX1     101 PPPPPTEVAVVKEIIHCKSCTLFPQNPNLPPPSTRERPPGCKTVFVGGLP 150
               *** *.::.*************** *******:*****************

ENOX2      137 ENGTEQIIVEVFEQCGEIIAIRKSKKNFCHIRFAEEYMVDKALYLSGYRI 186
cENOX1     151 ENATEEIIQEVFEQCGDITAIRKSKKNFCHIRFAEEFMVDKAIYLSGYRM 200
               **.**:** *******:* *****************:*****:******:

ENOX2      187 RLGSSTDKKDTGRLHVDFAQARDDLYEWECKQRMLAREERHRRRMEEERL 236
cENOX1     201 RLGSSTDKKDSGRLHVDFAQARDDFYEWECKQRMRAREERHRRKLEEDRL 250
               **********:*************:********* ********::**:**

ENOX2      237 RPPSPPPVVHYSDHECSIVAEKLKDDSKFSEAVQTLLTWIERGEVNRRSA 286
cENOX1     251 RPPSPPAIMHYSEHEAALLAEKLKDDSKFSEAITVLLSWIERGEVNRRSA 300
               ******.::***:**.:::*************: .**:************

ENOX2      287 NNFYSMIQSANSHVRRLVNEKAAHEKDMEEAKEKFKQALSGILIQFEQIV 336
cENOX1     301 NQFYSMVQSANSHVRRLMNEKATHEQEMEEAKENFKNALTGILTQFEQIV 350
               *:****:**********:****:**::******:**:**:*** ******

ENOX2      337 AVYHSASKQKAWDHFTKAQRKNISVWCKQAEEIRNIHNDELMGIRREEEM 386
cENOX1     351 AVFNASTRQKAWDHFSKAQRKNIDIWRKHSEELRNAQSEQLMGIRREEEM 400
               **::::::*******:*******.:* *::**:** :.::**********

ENOX2      387 EMSDDE-IEEMTETKETEESALVSQAEALKEENDSLRWQLDAYRNEVELL 435
cENOX1     401 EMSDDENCDSPTKKMRVDESALAAQAYALKEENDSLRWQLDAYRNEVELL 450
               ******  :. *:. ..:****.:** ***********************

ENOX2      436 KQEQGKVHREDD-PNKEQQLKLLQQALQGMQQHLLKVQEEYKKKEAELEK 484
cENOX1     451 KQEKEQLFRTEENLTKDQQLQFLQQTMQGMQQQLLTIQEELNNKKSELEQ 500
               ***: ::.* ::  .*:***::***::*****:**.:*** ::*::***:

ENOX2      485 LKDDKLQVEKMLENLKEKESCASRLCAS-------------------NQD 515
cENOX1     501 AKEEQSHTQALLKVLQEQLKGTKELVETNGHSHEDSNEINVLTVALVNQD 550
                *::: :.: :*: *:*: . :..*   :                  ***

ENOX2      516 SEYPLEKTMNSSPIKSEREALLVGIISTFLHVHPFGASIEYICSYLHRLD 565
cENOX1     551 RENNIEK--RSQGLKSEKEALLIGIISTFLHVHPFGANIEYLWSYMQQLD 598
                *  +**  .*. :***:****:**************.***: **:::**

ENOX2      566 NKICTSDVECLMGRLQHTFKQEMTGVGASLEKRWKFCGFEGLKLT     610
cENOX1     599 SKISANEIEMLLMRLPRMFKQEFTGVGATLEKRWKLCAFEGIKTT     643
               .**.:.::* *: ** : ****:*****:******:*.***:* *
```

Fig. 3.2 ENOX1 sequence and comparison to ENOX2 (Chap. 8). The drug-binding site *underlined* is present in ENOX2 but is absent in ENOX1. The NADH-binding sequences denoted by a *dashed underline* and the copper-binding sites indicated by *dotted underlines* are common to both ENOX2 and ENOX1. The C117XXC putative half of the protein disulfide interchange motif is indicated by *closely spaced dots*. *Asterisks* identical sequence; *colons* conserved substitutions; *dots* semiconserved substitutions. The nucleotide sequence of ENOX1 is available from GenBank under Accession No. EF432052. Reproduced from Jiang et al. (2008) with permission from ACS Publications potential role as ultradian (with period lengths of less than 24 h) oscillators of the cellular biological clock (Morré et al. 2002a; Chap. 6).

Purification of a protease-resistant protein fragment with drug-resistant ENOX1 activity and a period length of 24 min from human sera was described (Sedlak et al. 2001). The protein fragment was blocked to direct sequencing and resistant to further protease digestion. Polyclonal antisera raised to the protease-resistant fragment partially blocked total ENOX activity of human sera and from the surface of human noncancer cells. The antisera blocked completely the activity of human recombinant ENOX1 (Fig. 3.4f) but did not react with recombinant ENOX2 or with molecular species identified as ENOX2 from sera of cancer patients or from cancer cell lines grown in culture (Sedlak et al. 2001). The antisera inhibit the growth of HeLa cells by about 50 % presumably through some interaction with the ENOX1 protein also present on the HeLa cell surface.

ENOX1 proteins lack iron or iron sulfur clusters but still reduce oxygen. Human ENOX1 cloned and sequenced contains a copper site conserved with the copper site of the enzyme Mg^{2+} superoxide dismutase (Jiang et al. 2008) as does ENOX2 (Chueh et al. 2002b). Site-directed mutagenesis of this copper site resulted in the loss of enzymatic activity when the mutant protein was expressed in *Escherichia coli* (Chueh et al. 2002a). The discovery of a second copper site also required for activity (Jiang et al. 2008) raised the possibility that the concerted four electron transfers required to reduce molecular oxygen to water are carried out by dimers of the dicopper-containing ENOX monomers (see also Fig. 1.1).

3.2 ENOX1 Cloning

ENOX1 was cloned from a HeLa cDNA library (Accession number EF432052) and inserted into the vector pET43.1a for expression of a fused protein with NusA and His tags (Jiang et al. 2008). The sequencing of the construct yielded an insert of 1,929 bp that encoded an open reading frame of 643 amino acids with 64 % identity and 80 % similarity to ENOX2 (Fig. 3.2). Sequencing of two clones of pET43.1a-ENOX1 revealed a new isoform of proliferation-inducing gene 38 as the candidate CNOX having a semiconserved mutation of E16 to D. A homolog of proliferation-inducing gene 38 protein in *Dania rerio* also has D in this position. Homologs in *Equus caballus* and *Canis lupus familiaris* have another semiconserved substitution (E to Q) in this position, while in other species the E is conserved. A gene coding for the proliferation-inducing protein is present in genomes of all so far sequenced vertebrate and insect species and is highly conserved. No similar gene was found in plants, yeast, or prokaryotes.

The ENOX1 gene has been identified and cloned from the yeast *Saccharomyces cerevisiae* (S. S. Dick, A. Ryuzoji, S. M. Morré and D. J. Morré, unpublished) as well as the plant homolog from *Arabadopsis lyrata* (X. Tang, L. Ades, D. M. Morré and D. J. Morré, unpublished). The yeast and *Arabadopsis* ENOX1 genes have significant homology but lack significant homology with the corresponding vertebrate and insect ENOX1 genes. The vertebrate and yeast and plant ENOX1

Exon Map of ENOX1

ENOX1 protein [Homo sapiens]

```
AAH24178                    643 aa          linear   PRI 11-SEP-2007
DEFINITION   ENOX1 protein [Homo sapiens].
ACCESSION    AAH24178
VERSION      AAH24178.1  GI:18848204
```

Homo sapiens ecto-NOX disulfide-thiol exchanger 1, mRNA

```
LOCUS        BC024178                   2466 bp    mRNA

(cDNA clone  MGC:26180 IMAGE:4826032), complete cds.
ACCESSION    BC024178
VERSION      BC024178.1  GI:18848203
```

Exon 1 (nt 15-nt 55)

Exon 2 (nt 56-nt 92)

Exon 3 (nt 93-nt 234)

Exon 4 (nt 235-nt 378)

Exon 5 (nt 379-nt 518)

Exon 6 (nt 519-nt 692)

Exon 7 (nt 693-nt 901)

Exon 8 (nt 902-nt 1131)

Exon 9(nt 1132-nt 1346)

Exon 10 (nt 1347-nt 1451)

Exon 11 (nt 1452-nt 1569)

Exon 12 (nt 1570-nt 1754)

Exon 13 (nt 1755-nt 1866)

Exon 14 (nt 1867-nt 1922)

Exon 15 (nt 1923-nt 2022)

Exon 16 (nt 2023-nt 2108)

Exon 17 (nt 2109-nt 2451)

Fig. 3.3 Located on chromosome 13q14.11, the human ENOX1 gene is composed of 17 exons

recombinant proteins have virtually identical functional properties when expressed in bacteria and comparable functional motifs yet appear to have evolved independently to fulfill similar, if not identical functions, with vastly different genetic origins.

The human ENOX1 gene is located on 13q14.11 and composed of 17 exons (Fig. 3.3). In mammalia with the XY system of sex determination, the gene has

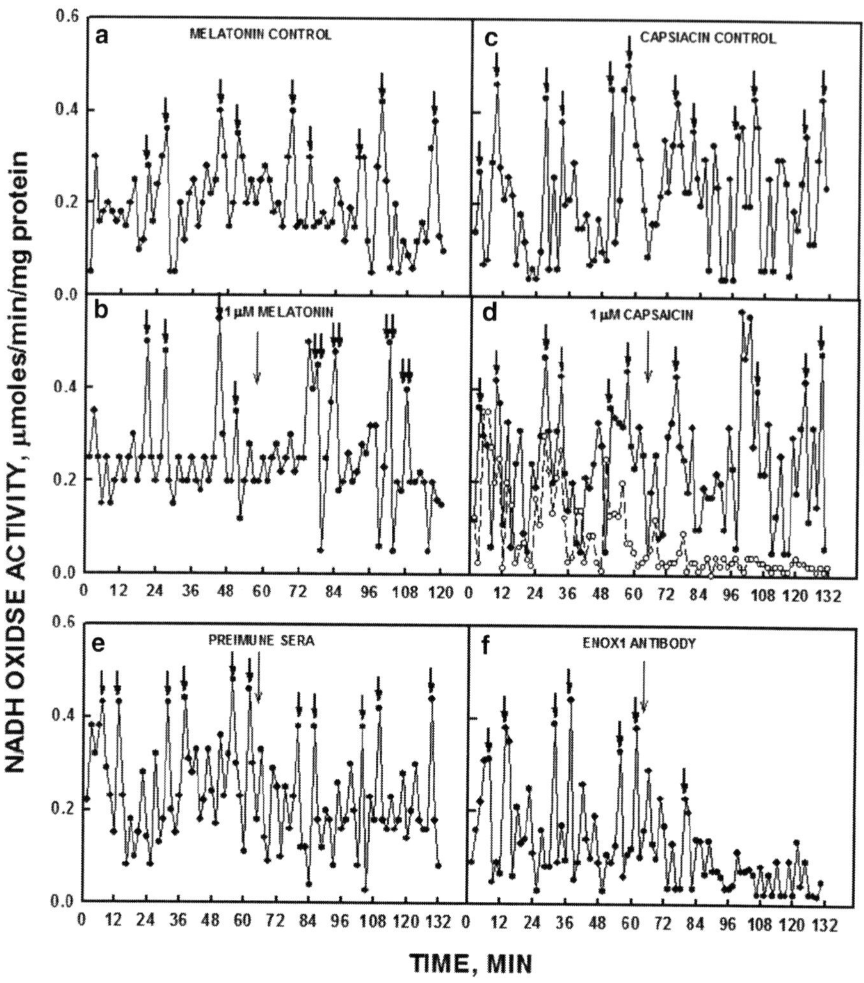

Fig. 3.4 NADH oxidase activity of recombinant human ENOX1 response to 1 μM melatonin (**a**, **b**), 1 μM capsaicin (**c**, **d**), and antihuman ENOX1 antibody (1:1,000) (**e**, **f**) added were indicated by the *large arrows* in (**b**), (**d**), and (**f**), respectively. In (**d**), the response of recombinant ENOX2 to capsaicin (*open circles, dotted lines*) is included for comparison. For each treatment, two assays were carried out simultaneously in parallel using paired Hitachi spectrophotometers with aliquots of the same ENOX1 preparation. Illustrated is the oscillatory pattern of five maxima. The major maxima separated by 6 min were indicated by *single arrows*. The three minor maxima that follow are separated from the major maxima and each other by 4.5 min creating the 24-min period [6+(4.5×4)=24]. Maxima appear at the same times comparing (**a**) and (**b**) prior to melatonin addition, (**c**) and (**d**) prior to capsaicin addition, and (**e**) and (**f**) prior to antibody addition illustrating the reproducibility of the oscillatory patterns. After addition of melatonin in (**b**), new maxima appear 24 min following melatonin addition (*double arrows*), an ENOX1 characteristic. Capsaicin, a standard ENOX2 inhibitor, had no effect in (**d**), and the activity was inhibited after one cycle in (**f**) by the addition of antibody to native ENOX1 from human sera. Reproduced from Jiang et al. (2008) with permission from ACS Publications

Table 3.2 Functional motifs of human ENOX1

Activity	Motif
Protein disulfide	C117KSC
Copper sites [H(Y)XH(Y)]	H260YSEH
	H579VH
Drug binding [EEMTE]	Absent
Copper site II [H(Y)X(HY)]	H579VH
Adenine nucleotide binding [GXGXXG]	G623VGATL

autosomal localization in contrast to the ENOX2 coding gene which is located on the X chromosome in these species. For both proteins, UniGene database contains multiple data confirming their expression (presence of mRNA) in different tissues and at different developmental stages of different species.

The identified functional motifs of human ENOX1 are listed in Table 3.2. The putative NADH-binding site, a disulfide-thiol interchange site, and copper-binding sites of ENOX2 are conserved in ENOX1 while a putative anticancer drug-binding site is not (Table 3.2).

Additional ENOX1 proteins have been discovered with sequence substantially different from the ENOX1 reported by Jiang et al. (2008) (Sect. 3.2). Thus far, ENOX1 proteins purified to homogeneity from natural sources all have failed to yield N-terminal amino acid sequence due to being blocked to N-terminal sequencing and being resistant to proteolytic degradation including treatment with proteinase K.

3.3 ENOX1 Characterization

Recombinant cENOX1 expressed in bacteria exhibited an oscillatory pattern of oxidation of exogenously supplied NADH characteristic of ENOX proteins (Fig. 3.4a). The repeating pattern was that of five maxima, two of which were separated by 6 min (arrows) and the remainder separated by 4.5 min [6 min + (4 × 4.5)min = 24 min]. As is characteristic of ENOX1 proteins from other sources, the oscillatory pattern could be phased by the addition of 1 μM melatonin (Fig. 3.4b). A new maximum was observed exactly 24 min after melatonin addition.

That the activity was not that of ENOX2 was shown by resistance of the activity to inhibition by the active pungent principle of chili peppers, capsaicin (8-methyl-*N*-vanillyl-6-noneamide). In the presence or absence of 1 μM capsaicin, a concentration sufficient to inhibit ENOX2 completely, the activity was unaffected (Fig. 3.4c, d).

A third criterion to identify the expressed protein as an ENOX1 was inhibition of the activity by the ENOX1-specific antibody (Sedlak et al. 2001) raised against the circulating form of ENOX1 from human sera (Fig. 3.4e, f). As is typical of antibody inhibition of ENOX activity, inhibition was delayed through approximately one full cycle of activity to permit antibody binding after which the activity ceased (Fig. 3.4f). In the absence of antibody, the activity continued (Fig. 3.4e).

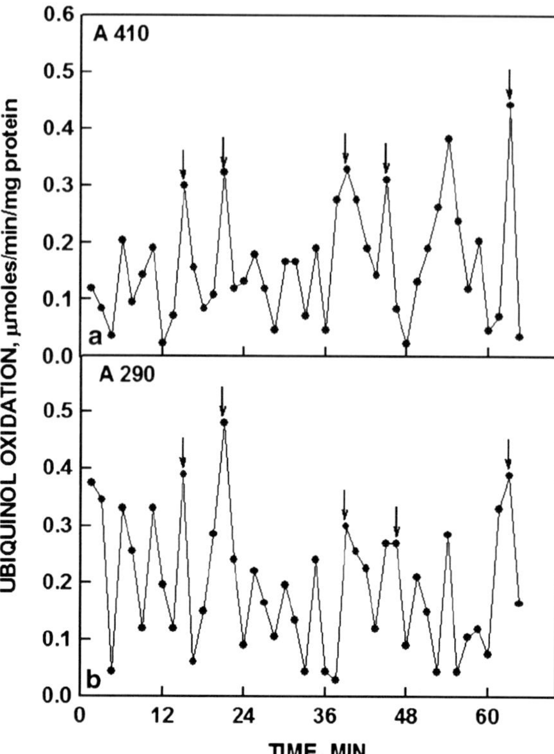

Fig. 3.5 Ability of recombinant human ENOX1 to oxidize reduced coenzyme Q (hydroquinone) measured either by an increase in A_{410} (**a**) or by a decrease in A_{290} (**b**). As with NADH oxidation of Fig. 3.4a, the activity oscillates with prominent maxima separated by 6 min (*arrows*) to create a 24-min period containing three additional maxima separated by 4.5 min (total of five maxima). Reproduced from Jiang et al. (2008) with permission from ACS Publications

All ENOX proteins not only oxidize reduced pyridine nucleotides and hydroquinones but also carry out protein disulfide-thiol interchange (Fig. 2.18). Yet, they contain only one of the two –CXXC– motifs characteristic of the flavin containing protein disulfide isomerases. The protein disulfide-thiol interchange (protein disulfide isomerase) activity is evident from the time-dependent restoration of activity to inactive, scrambled ribonuclease in a standard protein disulfide isomerase assay (see Fig. 2.18). The two activities, NADH oxidation and protein disulfide interchange, alternate such that the activation of inactive, scrambled ribonuclease also oscillates with a period length of 24 min (Morré and Morré 2003a) as does the cleavage of a dithiodipyridyl protein disulfide-thiol interchange substrate (see Fig. 2.18).

Reduced coenzyme Q is oxidized by ENOX1 in a standard assay (Fig. 3.5; Chap. 2) with activity measured either as an increase at A_{410} (Fig. 3.5a) or as a decrease at A_{290} (Fig. 3.5b). The pattern of oscillations with a 24-min period (arrows) characteristic of NADH oxidation (Fig. 3.4a) and dithiodipyridine cleavage (see Fig. 2.18b) was seen as well for hydroquinone oxidation (Fig. 3.5). Hydroquinones of the plasma membrane (reduced coenzyme Q for animals/reduced coenzyme Q or phylloquinone for plants) are the physiological substrates for ENOX1 proteins.

The ENOX1 activities of the recombinant proteins described above were restricted to proteins at the calculated molecular weight for NusA-tagged ENOX1 (Fig. 3.6)

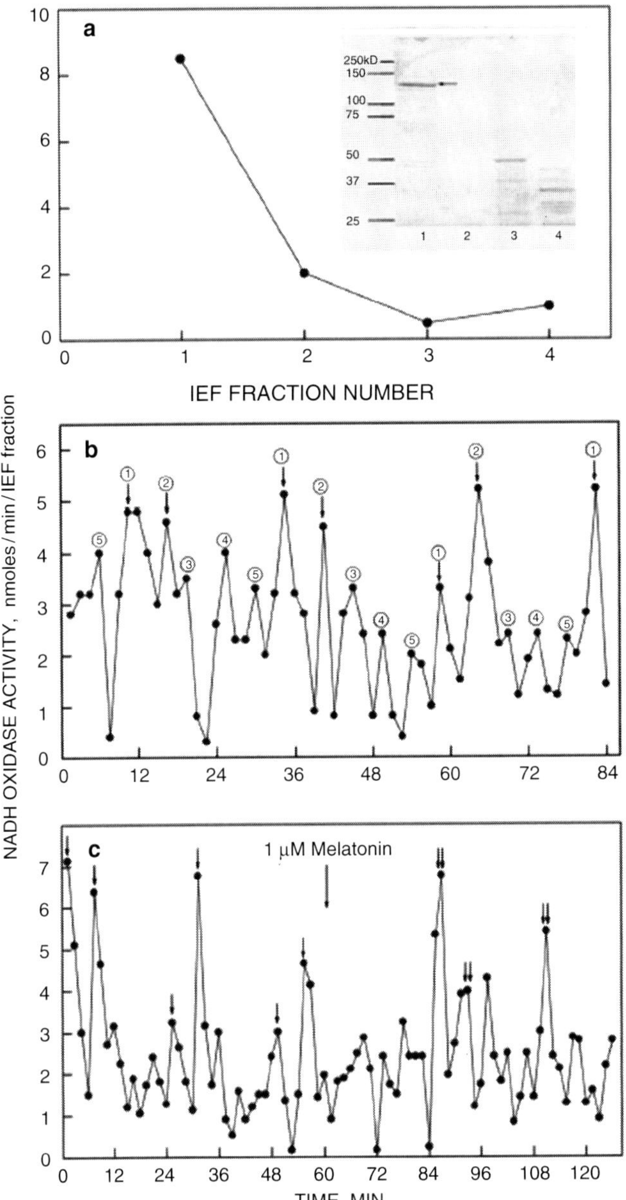

Fig. 3.6 Purification of human recombinant ENOX1 protein by isoelectric focusing. (**a**) Activity assay of recombinant ENOX1 eluted from slices of isoelectric focusing gel in which the activity correlated with the position of the protein on the gel. Silver-stained SDS-PAGE gel protein fractions eluted from the isoelectric focusing gel of the ENOX1. *Lane 1*: NusA-ENOX1 protein fraction eluted from p*I* 7 to 10 gel slice (IEF fraction 1). *Lane 2*: Protein fraction eluted from p*I* 6 to 7 gel slice (IEF fraction 2). *Lane 3*: Protein fraction eluted from p*I* 4.5 to 6 gel slice (IEF fraction 3). *Lane 4*: Protein fraction eluted from p*I* 3 to 4.5 gel slice (IEF fraction 4). (**b**, **c**) NADH oxidase activity of gel slices corresponding to the position of 138 kDa ENOX1 (*arrow* shown for IEF fraction 1 of (**a**) *inset*). Other gel regions were without activity. The characteristic oscillatory pattern in (**b**) was observed. As shown in (**c**), the phase of the activity oscillations was phased by addition of 1 μM melatonin (*double arrows*) as is characteristic of ENOX1 proteins in general. Reproduced from Jiang et al. (2008) with permission from ACS Publications

Table 3.3 Response of recombinant human ENOX1 to ENOX2 inhibitors, the arNOX inhibitor tyrosol, and the ENOX1 inhibitor simalikalactone D[a]

Inhibitor	μmol/min/mg
None	0.43±0.07
cis-Platinum (100 μM)	0.42±0.06
Phenoxodiol (10 μM)	0.41±0.06
EGCg (500 μM)	0.46±0.07
Capsaicin (1 p1VI)	0.43±0.06
Tyrosol	0.43±0.06
Simalikalactone D	0.08±0.01

[a]Averages of three determinations±standard deviations. The concentrations of *cis*-platinum, phenoxodiol, EGCg, and capsaicin used result in 90 % inhibition of recombinant NusA-tagged ENOX2 assayed in parallel. From Jiang et al. (2008)

by isoelectric focusing. All other proteins present in any of the preparations lacked these activities (Fig. 3.6a). Additionally, the material eluted from the isoelectric focusing gel exhibited the characteristic oscillatory pattern of ENOX proteins (Fig. 3.6b) with a 24-min period (arrows), and the period was shifted by the addition of melatonin based on phase at time of addition (Fig. 3.6c) as is characteristic of ENOX1 (but not ENOX2) proteins. The new maxima occurred exactly 24 min after melatonin addition and continued thereafter as phased by the melatonin addition. ENOX1 is also phased by blue (red or bluc in plants) light and in response to low frequency electromagnetic fields (Morré et al. 2008b) as well as auto-entrained. The response to light appears to be somehow receptor mediated as ENOX proteins have no chromophores, whereas the response to melatonin does not require association of the ENOX1 with a membrane and is presumed to be a direct response of the ENOX1 protein to melatonin. The period length of 24 min appears to be related directly to the normally 24-h circadian daily cycle. A C120A replacement in human ENOX1 yields a 30-min period length (Jiang et al. 2008) and a 30-h circadian cycle for COS cells transfected with the C12A replacement cDNA based on measurements of glyceraldehyde-3-phosphate dehydrogenase activity. See also inhibitor experiments with *Chlamydomonas reinhardtii* below (Forbes-Stovall et al. 2008).

The ENOX activity eluted from the IEF gel was further identified as ENOX1 rather than ENOX2 by its resistance to various ENOX2 inhibitors including *cis*-platinum, phenoxodiol, (−)-epigallocatechin-3-gallate (EGCg), and capsaicin all tested at concentrations sufficient to inhibit ENOX2 activity completely (Table 3.3). With the eluted ENOX from the IEF gels, no inhibition was observed under conditions where the activity was inhibited by the ENOX1-specific quassinoid inhibitor simalikalactone D (Morré and Grieco 1999). Properties of ENOX1 proteins (human, yeast, and plant) are summarized in Table 3.4.

Geng et al. (2009) have reported that the indolyl-quinuclidinol, (*Z*)-(+)-2-(1-benzylindol-3-ylmethylene)-1-azabicyclo[2.2.2]octan-3-ol, inhibited ENOX1 activity

Table 3.4 Properties of ENOX1 proteins

Activities are synchronous and pulse
Two activities alternate to generate a 2 + 3 periodic pattern with a period length of 24 min
Low turnover number: 200–500
Specific activity: 1–10 μmol/min/mg protein
No flavin, heme, or iron–sulfur centers
Ancillary proteins not required
Metals (Cu^{2+} and Zn^{2+}) present and Cu^{2+} required for activity
Have prion-like properties (protease resistance, propensity to irreversibly aggregate resulting in formation of insoluble amyloid rods)
Resistant to inhibition by ENOX2 and arNOX inhibitors (Table 3.2)
Inhibited by simalikalactone D
Periodicity is phased by melatonin and low frequency electromagnetic field and is auto-entrained (Chap. 6)

of the endothelial cell surface (EC_{50} = 10 μM). Inhibited in parallel was retroviral-mediated endothelial cell migration, the ability to form capillary-like structures in Matrigel, and neoangiogenesis by 70 % or more. Retroviral-mediated shRNA suppression of endothelial ENOX1 expression inhibited cell migration and tubule formation with effects similar to those observed with the small molecule analog. Genetic or chemical suppression of ENOX1 significantly increased radiation-mediated caspase3-activated apoptosis to suggest that targeting ENOX1 of tumor microvasculature might provide a novel therapeutic strategy to enhance the response of tumors to radiation (Geng et al. 2009).

In experiments with *C. reinhardtii*, the above ENOX1-specific indolyl-quinuclidinol blocked the circadian rhythm of phototaxis without interfering with the organism's ability to swim (Forbes-Stovall et al. 2008). The observations support both the ENOX1-directed specificity of the drug as well as the indicated role of ENOX1 as the ultradian driver of the circadian clock. In experiments with other organisms, including the yeast *S. cerevisiae*, the indolyl-quinuclidinol blocked ENOX1 activity in a manner similar to that observed with the quassinoid simalikalactone D (Sect. 3.3).

3.4 ENOX1 Activity Requires the Presence of Copper

Copper presence was necessary for ENOX1 activity (Fig. 3.7). The IEF-purified maltose binding protein (MBP)-tagged ENOX1 when unfolded in the presence of trifluoroacetic acid retained activity after dialysis and at physiological pH (Fig. 3.6a). However, if the ENOX1 was unfolded in the presence of the copper chelator bathocuproine, activity was lost (Fig. 3.7b). Activity was subsequently restored by refolding in the presence of copper at physiological pH.

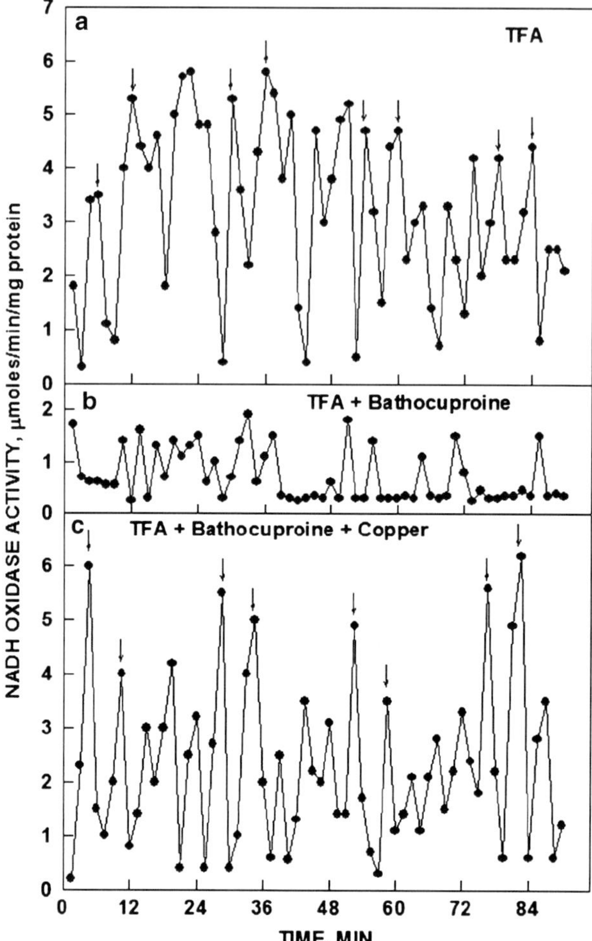

Fig. 3.7 Recombinant human ENOX1 eluted from an IEF gel was unfolded by treatment with trifluoroacetic acid (TFA) in the absence (**a**) or presence (**b**) of 0.3 mM bathocuproine to remove copper. The bathocuproine treated preparation was no longer enzymatically active following dialysis to remove the bathocuproine and the TFA. However, activity to the preparation of (**b**) was restored by refolding in the presence of 100 μM copper chloride at physiological pH (**c**). Reproduced from Jiang et al. (2008) with permission from ACS Publications

3.5 Copper Binding and Site-Directed Mutagenesis of Potential Copper-Binding Sites

Several motifs were identified as potential copper-binding sites. They were HCKSC120, H260YSEH, Y304SM, H531SH, H579VH, and M608LLM. C120, H260, Y304, H531, and H579 were individually replaced by alanines. The mutant MBP-tagged proteins were assayed for activities of NADH oxidation (Table 3.5).

Table 3.5 Calculated specific activities of mutated recombinant ENOX1 to identify copper-binding sites[a]

Protein	nmol/min/mg
WT	0.40±0.03
C120A	0.36±0.07
H260A	0.05±0.05[b]
Y304A	0.37±0.09
H531A	0.50±0.03[b]
H579A	0.07±0.09[b]
M608A	0.03±0.03
H260A+I579A	0.05±0.05[b]

[a]The activity of vector protein was subtracted as background. From Jiang et al. (2008)
[b]Significantly different from wild type (WT), $p<0.001$

Table 3.6 Bound copper comparing MBP-tagged ENOX1, mutant MBP-ENOX1, and pMAL-c2E vector protein

	Protein (μM)	Copper (μM)	Ratio[a]
pMAL-c2E vector	25.49±2.02	5.80±1.76	0.22±0.06[b]
MBP-ENOX1	10.60±1.17	21.07±3.85	1.98±0.14[c]
MBP-ENOX H260A	9.38±1.92	12.38±1.62	1.33±0.10[d]
MBP-ENOX1 H579A	9.44±1.31	10.60±1.42	1.14±0.27[e]
MBP-ENOX1 H260A+H579A	8.75±1.61	4.76±0.92	0.56±0.17[f]

[a]By comparing five means of ratios using statistical adjustment by analysis of variance (ANOVA), Ratios designated d and c are not significantly different from each other at $\alpha=0.05$. However, groups of b, c, and f were different from groups d and e at $\alpha=0.05$. From Jiang et al. (2008)

The wild type MBP-tagged ENOX1 was the control. Mutations in H260YSEH or H579VH resulted in loss of NADH oxidase activities. Both H260YSEH and H579VH when mutated in the same protein also produced a product with no activity.

For measurement of copper binding, the requirement for purified protein lacking an interfering tag was met by expressing ENOX1 carrying a MBP tag by ligation of the ENOX1 DNA into the pMAL-c2E vector. After the pMAL-c2E vector protein, MBP-tagged ENOX1 protein and mutant MBP-tagged NOX1 proteins were purified by amylose resin column chromatography, and the copper amounts bound by each were measured by colorimetric assay with vector protein as control. The ENOX1 protein bound 2 mol of copper/mol of protein (Table 3.6). After either of the two copper-binding motifs was mutated, ca. 1 mol of copper was bound per mol of protein. When both copper-binding motifs were mutated, <1 mol copper/mol protein was bound (Table 3.6).

An important consideration is that the ENOX1 protein bound approximately 2 mol of copper/mol of protein. As the prevailing soluble form of ENOX2 proteins may be a dimer based on size exclusion chromatography, the presence of two coppers per monomer permits construction of a model depicting ENOX1 as a dimeric protein containing four coppers/dimer capable of carrying out four electron transfers from NADH or reduced coenzyme Q directly to molecular oxygen as required to form water (Fig. 3.1; see also Fig. 1.1). ENOX proteins do reduce molecular oxygen to water (see Figs. 2.13, 2.14, and 2.15; Orczyk et al. 2005) despite the fact that they lack flavin and/or electron carriers other than copper.

3.6 Response to Nucleotides

Inhibition of plasma membrane NADH oxidase activity (ENOX1) by submicromolar concentrations of ATP have been observed with pig (Morré et al. 1994d) and rat liver (Morré et al. 1997e) plasma membranes and with dark-grown soybean seedlings (Morré et al. 1993a; Morré 1998d). For pig liver plasma membranes (Morré et al. 1994d) and also for rat liver plasma membranes (Morré et al. 1997e), the ATP response was accelerated by the coaddition of cyclic AMP at concentrations about 0.1 those of the ATP. The plant membranes did not respond to cyclic AMP.

Literature reports suggest that extracellular ATP in the submillimolar range inhibited growth of transformed mammalian cells in culture and to a lesser extent the growth of nontransformed cells (Friedberg and Kübler 1990). Externally supplied ATP convincingly restored auxin (IAA)-induced growth to auxin-starved hypocotyls of *Helianthus annus* under anaerobic conditions (Hager et al. 1971). The response was to millimolar concentrations of nucleotide and was given as well by ITP, GTP, UTP, and CTP. A source of external ATP, at least for mammalian cells, would be the corelease of ATP during secretion from the contents of Golgi apparatus secretory vesicles fusing with the plasma membrane. ATP is costored with catecholamines within chromaffin granules in the ratio of 4 mol of catecholamine:1 mol of ATP (Winkler and Carmichael 1982). In the vas deferens of the guinea pig, both ATP and norepinephrine were not only cosecreted but also appeared to function as cotransmitters based on a variety of pharmacological and electrophysiological experimentation (White 1988). Even in plants, low concentrations of secreted ATP might function as part of a feedback signaling system to coordinate secretion and cell enlargement (Morré 1998).

Phosphorylation of ecto-proteins by externally supplied ATP has been demonstrated with aortic endothelial cells (Pirotton et al. 1992), brain neurons (Hogan et al. 1995), and in hippocampal slices (Chen et al. 1996) presumably through the action of protein kinases located at the extracellular surface of the plasma membrane. Also indicated is a role for guanine nucleotides in the regulation of plasma membrane NADH oxidase, but with properties that differ from those of either trimeric or the low-molecular-mass G proteins thus far described (Morré et al. 1993a). The results further demonstrate regulation of NADH oxidase activity of rat liver plasma membranes through cyclic AMP-mediated phosphorylation by membrane-located protein kinase activities (Morré et al. 1997c).

3.7 Aggregation and Electron Microscopy

The MBP-tagged ENOX1 and pMAL-c2E vector proteins were purified by using an amylose resin column. The MBP tag was then removed by digesting the MBP-tagged ENOX1 (lane 1) with TEV protease (Fig. 3.8, inset). Upon treatment with TEV protease, both the native ENOX1 (71 kDa) and the MBP tag (43 kDa) were recovered (lane 2).

After removal of the MBP tag, both the NusA-NOX1 (MW 138 kDa) and the MBP ENOX1 formed insoluble and enzymatically inactive aggregates which were collected by centrifugation (Fig. 3.8). Electron microscopic examination after negative staining with uranyl acetate revealed the presence of 10 nm diameter linear fibrils 100–200 nm long resembling amyloid in both preparations and similar to those characteristic of prions (Griffith 1967; Prusiner 1994), Alzheimer's Aβ protein (Multhaup 1997), and aggregated ENOX2 proteins (del Castillo-Olivares et al. 1998). The aggregated preparations not only lacked enzymatic activity but were refractory to resolubilization. Thus, like ENOX2 (Kelker et al. 2001; Kim and Morré 2004), the ENOX1 proteins share characteristics in common with prions especially those properties of resistance to protease degradation and the ability to aggregate with the formation of amyloid fibers.

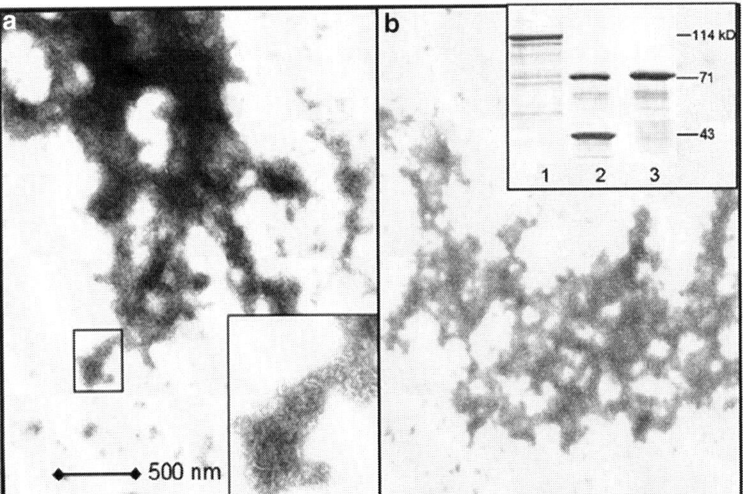

Fig. 3.8 Aggregated and precipitated MBP-tagged recombinant ENOX1 analyzed by electron microscopy before (**a**) and (**b**) after removal of the MBP tag. The *inset* in (**b**) shows a silver-stained SDS-PAGE gel of undigested and digested MBP-tagged ENOX1 proteins. *Lane 1*: MBP-tagged ENOX1 (113 kDa). *Lane 2*: Digested recombinant protein solution containing MBP (42 kDa) and ENOX1 (71 kDa). *Lane 3*: Precipitated ENOX1 after ultracentrifugation. Reproduced from Jiang et al. (2008) with permission from ACS Publications

For reasons possibly related to dimerization rather than multimerization, ENOX1 proteins prepared by isoelectric focusing or eluted from isoelectric focusing gels did not aggregate, retained full enzymatic activity, and were stable in solution for several weeks or months stored frozen or at 4 °C.

3.8 ENOX1 Fulfills Essential Roles in Cell Enlargement and Cellular Time-Keeping

The protein disulfide-thiol interchange activity of the ENOX protein has been implicated as obligatory to the cell enlargement phase of cell growth for both animal (Pogue et al. 2000; Wang et al. 2001) and plant (Morré et al. 2001a, 2002d) cells (Chap. 5) as well as in cellular time-keeping (Chap. 6). However, the mechanism whereby disulfide bonds are formed and broken through the action of ENOX proteins may differ substantially from that of the classical flavoprotein disulfide isomerases. In any event, a source of protons and electrons is not required for the interchange activity as exemplified by the ability of ENOX1 alone to catalyze the restoration of activity to inactive and denatured ribonuclease A. Furthermore, ENOX proteins are not thioredoxin reductases and neither glutathione nor dithiothreitol support the protein disulfide-thiol interchange activity. The activity appears to be protein thiol and protein disulfide specific. Moreover, the mutation of HCKSC120 in the ENOX1 protein changed the oscillatory period length of the protein from 24 to 30 min (Jiang et al. 2008). This finding is consistent with the results with ENOX2 proteins in which cysteine mutations caused similar changes in the period length of the activity oscillations (Chueh et al. 2002b; Morré et al. 2002a).

3.9 ENOX1 of Human Platelets

An ENOX1 of human platelets has been described where expression is modulated by capsaicin (Savini et al. 2010) whereas ENOX1 per se is capsaicin-insensitive (Jiang et al. 2008). Capsaicin triggers a signaling cascade in platelets whereby ENOX1 expression is upregulated via activation to TRPV1 (The transient receptor potential cation channel subfamily V member 1, also known as the capsaicin receptor and the vanilloid receptor 1), the natural receptor for capsaicin. Binding of capsaicin to its receptor triggers ROS production which, in turn, enhances expression and activity of the platelet ENOX1.

3.10 Summary

ENOX1 (CNOX) proteins are constitutive growth-related ENOX proteins of the cell surface that catalyze both hydroquinone and NAD(P)H oxidation as well as protein disulfide-thiol interchange and exhibit both prion-like and time-keeping (clock) properties. The two enzymatic activities they catalyze alternate to generate a regular period of 24 min in length. Human ENOX1, when cloned, yielded a sequence of 1,929 bp and an open reading frame encoding 643 amino acids. The gene is located on chromosome 13 (13q14.11). Functional motifs include a binding and adenine nucleotide copper-binding motifs along with essential cysteines. The drug-binding motif (EEMTE) sequence of ENOX2 is absent. The activities of ENOX1 proteins, in general, and the recombinant protein expressed in *E. coli* were not affected by capsaicin, EGCg, and other ENOX2-inhibiting substances. The purified recombinant protein bound ca. 2 mol of copper/mol of protein. Bound copper was necessary for activity. H260 and H579 were required for copper binding as confirmed by site-directed mutagenesis, loss of copper-binding capacity, and resultant loss of enzymatic activity. Addition of melatonin phases the 24-min period such that the next complete period begins exactly 24 min after the melatonin addition as appears to be characteristic of ENOX1 activities in general. Oxidative activity was exhibited both with NAD(P)H and reduced coenzyme Q as substrate. Concentrated solutions of the purified ENOX1 protein irreversibly formed insoluble aggregates resembling amyloid and devoid of enzymatic activity as is characteristic of ENOX proteins in general.

Chapter 4
Role in Plasma Membrane Electron Transport

A central role common to all ENOX proteins is to function as terminal oxidases of plasma membrane electron transport (PMET). PMET is an essential component of the enlargement phase of cell growth on the one hand and functions both to maintain the cytoplasmic $NAD^+/NADH$ ratio and to generate NAD^+ required to support glycolytic ATP production on the other. PMET provides a ubiquitous cell surface redox system that occurs in all eukaryotic cells and organisms (bacteria, yeast, plants, and animals) thus far investigated (Crane et al. 1985)) with the potential to fulfill multiple roles in regulating cellular redox status (Crane et al. 1990a, b). Changes in the ratios of oxidized and reduced pyridine coenzymes in response to metabolic changes are important not only to regulate intracellular energy levels but also as a source of signals derived from changes in the redox state of the micro environment at the interface of the cytosol and the plasma membrane (Lin and Guarente 2003).

4.1 Composition of the PMET

A transplasma membrane electron transport chain comprised of four components as diagrammed in Fig. 4.1 is the principal route for flow of electrons through the PMET (Morré et al. 1999c; Crane et al. 1994).

- One or more NADH-CoQ reductase enzymes located on the cytosolic side of the plasma membrane, which regulate the cytosolic $NADH^+/NADH$ ratio and ascorbate reduction (Morré 2004)
- Coenzyme Q (COQ) within the membrane
- Hydroquinone oxidases (ECTO-NOX or ENOX proteins) on the outer surface of the plasma membrane that function as terminal oxidases

In addition, cells are able to transfer electrons to artificial electron impermeant acceptors such as ferricyanide (Fig. 4.2) and diferric transferrin. Such transfers involve other carriers such as porin isoform 1 (VDAC) of the plasma membrane

D.J. Morré and D.M. Morré, *ECTO-NOX Proteins: Growth, Cancer, and Aging*,
DOI 10.1007/978-1-4614-3958-5_4,

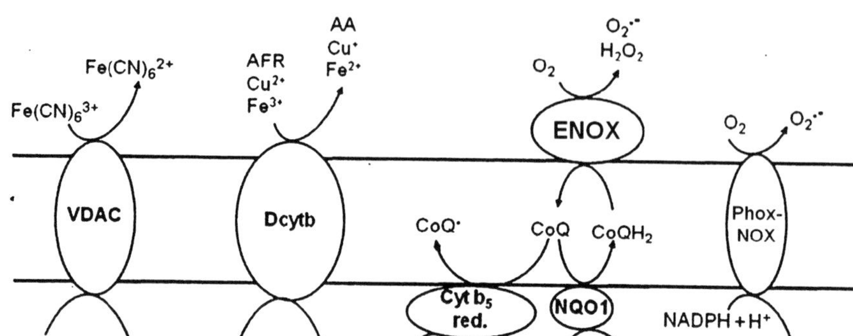

Fig. 4.1 Key components of the PMR system. Intracellular electron donors, provided by NAD(P)H or vitamin C (AA), generate outward electron flow to reduce extracellular acceptors [oxygen, ascorbyl free radical (AFR), ferric or cupric ions]. Enzymes contributing to the PMR system located at the cytosolic surface of the plasma membrane include NAD(P)H:quinone oxidoreductase 1 (NQO1) and cytochrome b_5 reductase (cyt b_5 red). Others, such as VDAC, Dcytb, and members of the PhoxNox (NOX) family, are transmembrane ENOX proteins located at the external surface. Coenzyme Q (CoQ), a mobile component of the PMR system, is located in the hydrophobic phospholipid bilayer and acts as an electron shuttle. Redrawn from Del Principe et al. 2011. Published with permission

(Sect. 4.1.3.2) that bypass the cell surface ENOX proteins along with a 57-kDa doxorubicin-inhibited NADH-quinone (NADH-ferricyanide reductase) (Sect. 4.1.3.3).

4.1.1 NADH Coenzyme Q Reductases

For both animals and plants, isoforms of NADH-cytochrome b_5 reductase (EC. 1.6.2.2) (Navarro et al. 1995; Villalba et al. 1997, 1998) and/or NADH-NAD(P)H dehydrogenases (1.6.99.3) (Navarro et al. 1995; Villalba et al. 1995, 1997) located on the cytoplasmic surface of the plasma membrane are required to reduce the ubiquinone within the plasma membrane (phylloquinone in plants). Cytoplasmic NAD(P)H:quinone oxidoreductase 1 [NQO_1, DT-diaphorase (EC. 1.6.5.2)] of the cytosol is unlikely to be responsible for the reduction of plasma membrane coenzyme Q based on substrate specificity and its inhibition by dicumarol

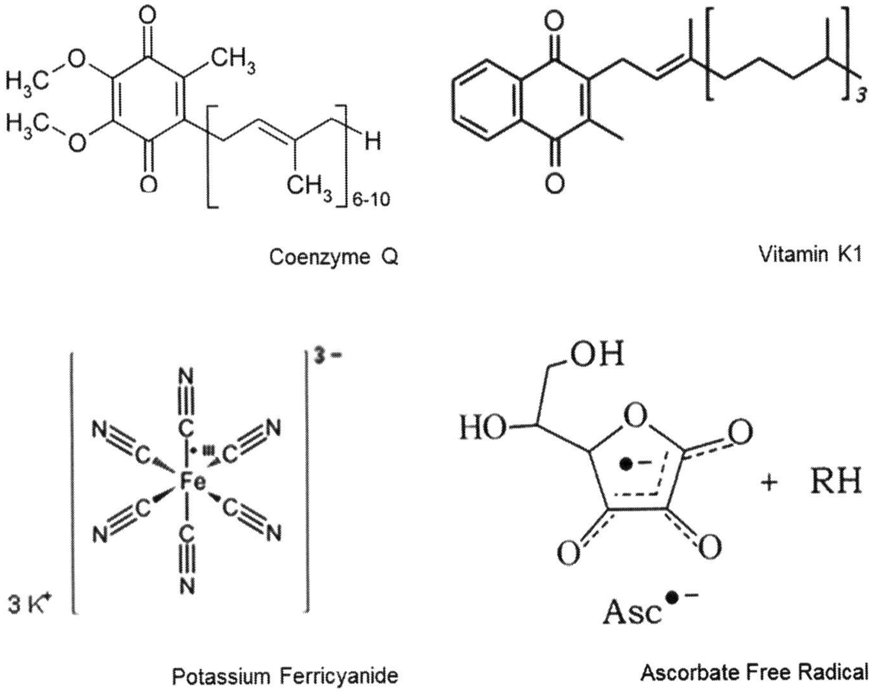

Coenzyme Q

Vitamin K1

Potassium Ferricyanide

Ascorbate Free Radical

Fig. 4.2 Chemical structures of coenzyme Q_{10}, phylloquinone, transferrin, and ascorbate free radical electron acceptors of plasma membrane electron transport

[3,3′-methylene-bis(4-hydroxycoumarin)] and dicumarol analogs, which compete with NAD(P)H for binding to the enzyme. A concentration of 5 μM dicumarol that did not inhibit cytosolic NAD(P)H-dependent ubiquinone reductases in vitro also did not inhibit reduction of endogenous CoQ_9 or CoQ_{10} added to cultured rat hepatocytes, suggesting that NQO_1 is not involved in maintaining CoQ in the regeneration of cellular $CoQH_2$ or in its antioxidant actions (Kishi et al. 2002). Furthermore, CoQ homologs with long isoprenoid chains were not good substrates for NQO_1 compared with low molecular weight quinones such as menadione and 2,3-dimethyl-6-methyl-1,4-benzoquionone.

For phylloquinone in plants, reduction would be accomplished by a NAD(P)H-quinone reductase or diaphorase also at the cytosolic surface of the plasma membrane (Sparla et al. 1996; Van Gestelen et al. 1996) the "standard system" in Fig. 4.3. Direct coupling of phylloquinol of the vitamin K pool to the external NADH oxidase would complete the electron transport chain with reduction of molecular oxygen to form water. Analyses of plasma membrane fractions from all plant species tested

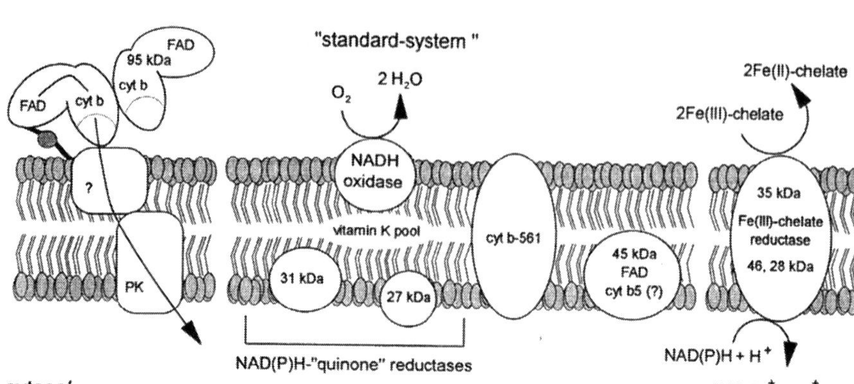

Fig. 4.3 Plasma membrane-associated NAD(P)H oxidoreductase activities and redox constituents of plants. The identified proteins use NAD(P)H as electron donor and, except the for the NADH oxidase, reduce HCF III. It was assumed that cytochrome *b*-561 and the Fe(III)-chelate reductase are membrane spanning. A nitrate reductase activity isolated from plasma membrane of algae may be involved in signal transduction for nitrate uptake. Fe^{3+}-chelate reductase activities have been identified in the plasma membrane of all plants investigated. In the "standard system," phylloquinone would be reduced by a NAD(P)H-quinone reductase or diaphorase at the cytosolic plasma membrane surface. Coupling of the phylloquinol of the vitamin K pool to the external NADH oxidase with reduction of molecular oxygen to form water would complete the electron transport chain. Redrawn from Lüthje et al. 1997. Published with permission

thus far suggest the presence of both b-type cytochromes (Asard et al. 1998), flavins (Bérczi et al. 1998) and proteins potentically capable of cytochrome b_5 reduction.

4.1.2 Hydroquinones

Hydroquinones, present in the membrane lipid bilayer and required for activity (Table 4.1), are the transmembrane component of PMET. Approximately 30 % of the membrane located coenzyme Q of rat liver, for example, is associated with extra mitochondrial membranes (10 % in plasma membranes) (Table 4.2; Morré and Morré 2011). In animals, the hydroquinones are primarily coenzyme Q_{10} (Fig. 4.2) (coenzyme Q_8 in some species) (Åberg et al. 1992). The hydroquinone phylloquinone (Döring and Lüthje 1996; Lüthje et al. 1997, 1998; Fig. 4.3) serves as an electron donor for the plasma membrane ECTO-NOX proteins of plants (Bridge et al. 2000) with a specific activity comparable to that for ubiquinol in animals (Kishi et al. 1999). Yeast (*Saccharomyces cerevisiae*) harboring a CoQ3 gene deletion (coq 3Δ) that do not synthesize CoQ (Clarke et al. 1991) provides genetic evidence for a CoQ requirement for plasma membrane redox activities (Santos-Ocaña et al. cited by Villalba et al. 1998).

Table 4.1 Effect of coenzyme Q and inhibitory analogs of coenzyme Q and NADH oxidase NADH oxidase activity of rat liver plasma membranes

Addition	Percent stimulation (+) or inhibition (−)
Coenzyme Q_{10}, 10 μM	+185
Chloroquin, 500 μM	−92
Piericidin, 0.1 μM	−85
Piericidin + CoQ	−25
Capsiacin, 150 μM	−88
Capsiacin + CoQ	−15
Retinoic acid, 10 μM	−57

Data from Alcain et al. (1994), Brightman et al. (1992) and Sun et al. (1987c). From Crane and Löw (2008)

Table 4.2 Distribution of coenzyme Q among endomembranes of rat liver compared to mitochondria

Cell component	CoQ, μg/mg protein	Protein, mg/g liver	CoQ, μg/g liver
Mitochondria	1.40	33	46.2
Nuclear envelope	0.15	6	0.9
Endoplasmic reticulum	0.15	51	7.7
Golgi apparatus	2.62	2	5.2
Plasma membrane	0.74	7	5.1
Total			65.1

Data from Kalén et al. (1987) and Morré and Morré (1989). From Morré and Morré (2011)

4.1.3 *Terminal Oxidases*

As terminal oxidases, ubiquitous NADH oxidases (now ENOX1) (Morré and Brightman 1991; Morré and Morré 2003a; Chap. 3) use electrons from hydroquinone to reduce oxygen, forming water (Kishi et al. 1999; Orczyk et al. 2005) or a 57-kDa ferricyanide ($FeCN_6^{=}$) (Fig. 4.4) reductase (Kim et al. 2002) or Porin 1 (VDAC) (Baker et al. 2004a, b) that oxidize NADH directly using ferricyanide as acceptor (Fig. 4.1).

Highly purified preparations of plasma membranes of plants contain at least two oxidase forms designated NADH oxidase I and NADH oxidase II (Pupillo et al. 1986; Brightman et al. 1988). That the final product of the type I or plant ENOX1 is water is based on the stoichiometry of 2 NADH molecules oxidized to 1 oxygen molecule reduced. The second type of activity has characteristics of a membrane-bound peroxidase that is stimulated by phenolics such as SHAM, coniferyl alcohol or ferrulic acid, and manganese and ferrous ions and inhibited by cyanide and catalase (Møller and Bérczi 1985, Møller and Bérczi 1986; Asard et al. 1987; Vianello and Macri 1989). More typically, plant peroxidases are soluble or cell wall-bound rather than associated with the plasma membrane. An apparent $S_{0.5}$ (Saturation difference) for O_2 of 30 μM (pH 7.0) was calculated from the deflection of O_2 uptake

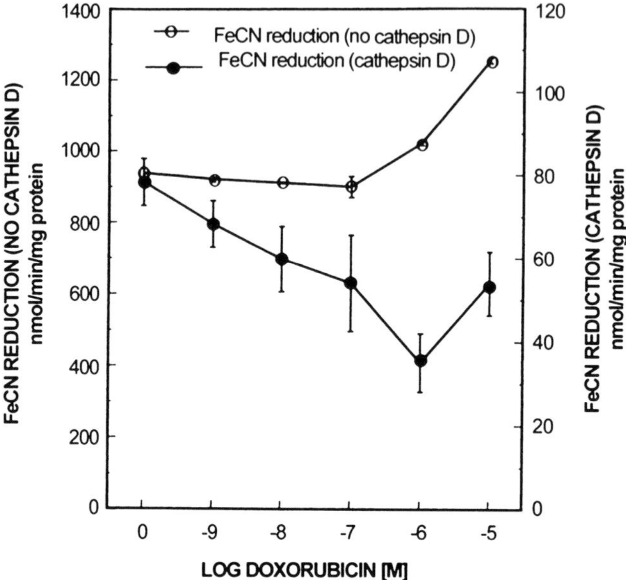

Fig. 4.4 Effect of doxorubicin on NADH-ferricyanide reductase activity. The doxorubicin-inhibited activity normally is marked by a dominating doxorubicin-resistant NADH-cytochrome b_5 reductase. Upon its removal by treatment with cathepsin D (Choury et al. 1981), the activity that remained and attributed to a ferricyanide reductase of the plasma membrane was strongly inhibited by doxorubicin. Reproduced from Kim et al. 2002 with permission Copyright 2002 by the American Society for Biochemistry and Molecular Biology, Inc

curves at saturating concentrations of pyridine nucleotides for duroquinone-dependent electron transport with *Curcubita* microsomes (Pupillo et al. 1986).

Also previously described are vanadate-stimulated NADH oxidases from red cell membranes that generate hydrogen peroxide (Vijaya et al. 1984) or vanadate-stimulated NADH oxidase that generates superoxide (Darr and Fridovich 1984). Liochev and Fridovich (1988) present evidence from rat liver microsomes that superoxide is responsible for the vanadate stimulation of NAD(P)H oxidation by biological membranes. NADH oxidation by the superoxide-generating ENOX (age-related NOX or arNOX, Chap. 10) present on red cell membranes and, associated with plasma membrane fragments in rat liver microsomes, is unaffected by either decavanadate or metavanadate alone or in combination over the concentration range 1–100 μM and does not appear to be responsible for the reported stimulations by vanadates of NADH oxidation with microsomes and red cell membranes.

4.1.3.1 ENOX Proteins as Terminal Oxidases of PMET

On the outer surface of the plasma membrane the polyfunctional ENOX proteins, hydroquinone oxidases, act as the principal physiological terminal oxidases to complete the transmembrane NADH oxidation electron transport chain (Fig. 4.1). The roles of NADH ferricyanide reductases in the absence of external ferricyanide or other sources of Fe^{3+} remain unresolved.

The ENOX proteins of PMET are low affinity NADH oxidases as well as hydroquinone (quinol) oxidases described initially as KCN-insensitive NADH oxidases present on the external surface of animal as well as plant cells (Brightman et al. 1988; Morré 1998c). Under normal conditions, high levels of NADH are not found outside cells such that the primary function of ENOX proteins of PMET is hydroquinone oxidation (Chaps. 1 and 3). ENOX proteins also have protein disulfide isomerase activity important to cell enlargement (Chap. 5). Among the potential terminal oxidases, the ENOX proteins uniquely are oxidases carrying out four electron transfers to oxygen to form water as expected for a physiologically functional PMET.

4.1.3.2 Porin Isoform 1 or VDAC

Porin isoform 1 or VDAC (voltage-dependent anion selective channel) 1, a predominant protein in the outer mitochondrial membrane, is expressed as well in the plasma membrane and may also function there as a NADH:ferricyanide reductase (Baker et al. 2004a, b; Lawen et al. 2005).

The two distinct functions of VDAC1, anion channel activity in the mitochondrion and NADH:ferricyanide reductase activity in the plasma membrane, are based on subcellular locations of two different forms of the VDAC1 protein generated by the use of alternative first exons (Buettner et al. 2000; Baker et al. 2004a, b). One isoform contains a leader peptide in its N-terminus. It enters the secretory pathway and reaches the plasma membrane via the Golgi apparatus. The second isoform lacks this sequence and is directed into the mitochondrial membrane (Buettner et al. 2000). VDAC1 from both plasma membrane and mitochondria possess NADH-ferricyanide reductase activity in vitro.

Autistic children show elevated autoantibodies to several proteins essential for normal brain function including VDAC (Gonzalez-Gronow et al. 2010). Antibodies to VDAC inhibit transplasma membrane electron transport and impair growth and induce apoptosis in human neuroblastoma cells in culture suggesting a possible causal role for neurological disorders seen in autistic children (Gonzalez-Gronow et al. 2010).

4.1.3.3 57-kDa Doxorubicin-Inhibited NADH-Quinone (NADH-Ferricyanide) Reductase

A 57-kDa doxorubicin (adriamycin)-inhibited NADH-quinone (NADH-ferricyanide) reductase was purified from rat liver plasma membrane (Kim et al. 2002). Peptide sequence homology search showed no sequence homology with known proteins, but a segment of the peptide sequence had homology with aldehyde dehydrogenase, pyruvate:ferredoxin oxidoreductase, and trans-2-enoyl-acyl carrier protein reductase II. To demonstrate the activity, it was first necessary to remove the predominant NADH-cytochrome b_5 reductase that is not inhibited by doxorubicin from the membranes. When the plasma membranes were treated with cathepsin D, the enzyme activity was reduced to 7 % of the initial rate. The activity that remained, however, was strongly inhibited by doxorubicin with maximum inhibition at 1 μM (Fig. 4.4). The doxorubicin-inhibited NADH quinone reductase may provide a target for the anthracycline antitumor agents (Chap. 12) and remains a candidate protein for the elusive ferricyanide reductase of PMET but is not considered as a candidate terminal oxidase in the absence of Fe^{3+} oxidants.

4.1.3.4 Other Potential PMET Terminal Oxidases

In addition to the ENOX hydroquinone oxidases that also oxidize external NADH, the VDAC 1, and the 57-kDa ferricyanide reductases, other less well-characterized plasma membrane redox activities have been described. In rat synaptosomes, more than one plasma membrane oxidoreductase activity appears to be differentially expressed during CNS development (Dreyer 1990; Zurbriggen and Dreyer 1994; Yong and Dreyer 1995). A synaptosomal 2,6-dichloroindophenol (DCIP) reductase of unknown function was purified and sequenced that showed no homology to the ubiquinone-linked oxidases (Bulliard et al. 1997). The DCIP reductase activity was upregulated by oxidative stress and varied with the cell cycle (Zurbriggen and Dreyer 1996; Bulliard et al. 1997).

Based on the differential responses of the DCIP and ferricyanide reductase activities of chick forebrain neurons to CoQ_1 and DPI, Wright and Kuhn (2002) concluded that DCIP and ferricyanide were likely reduced via distinct, substrate-specific redox pathways. A cell surface localization was indicated from inhibition of both reductase activities by the membrane impermeant sulfhydryl reagent muchloric acid (MCA). Only a small portion of total transplasma membrane electron flow associated with DCIP and ferricyanide reduction was inhibited by exogenous superoxide dismutase suggesting a relatively minor contribution from superoxide production to distinguish their activity from the neutrophil-like NADPH oxidase capable of producing superoxide expressed in mouse cortical (Noh and Koh 2000) and sympathetic neurons (Tammariello et al. 2000) and in neurons of the chick (Murakami et al. 1986; Cross 1987) or the arNOX proteins (X. Tang, D.Parisi, D.M. Morré and D.J. Morré, unpublished) none of which utilize molecular oxygen as acceptor and are unlikely to account for the direct reduction of either DCIP or ferricyanide.

Duodenal cytochrome b (Dcytb, also known as Cybrd1) that belongs to the cytochrome b_{561} family is induced in response to hypoxia and iron deficiency (McKie et al. 2001). Dcytb was first identified as a ferrireductase involved in dietary iron uptake in duodenal enterocytes (McKie et al. 2001). It was subsequently identified in the plasma membrane of other cell types (e.g., Turi et al. 2006) where one of its biological functions may be to reduce extracellular ascorbate radical (AFR) (Su et al. 2006). AFR reduction could be important to maintain ascorbic acid in a reduced state which may be important especially for biological fluids (e.g., plasma, interstitial, and cerebrospinal fluid).

4.2 Electron Donors and Acceptors

Principally, PMET transports electrons from intracellular NAD(P)H to physiological extracellular electron acceptors including molecular oxygen (Møller and Bérczi 1985; Morré et al. 1998a; Orczyk et al. 2005), disulfides (Chueh et al. 1997a; Morré et al. 1998a), ascorbate free-radical (Alcaín et al. 1991; Navas et al. 1992; Gomez-Diaz et al. 1997), and diferric transferrin (Sun et al. 1987a, b) or to artificial external electron acceptors such as ferricyanide or DCIP (Cherry et al. 1981; Crane et al. 1985; Sun et al. 1984a).

PMOR activity with intact cells may be measured using NADH (Chap. 2; Wang et al. 2001) or other membrane impermeant electron acceptors such as ferricyanide ($FeCN_6^{□}$), DCIP (Crane et al. 1985), or ascorbate free radical (Alcaín et al. 1991; Navas et al. 1992; Gomez-Diaz et al. 1997). Except for NADH, the plasma membrane proteins catalyzing the electron transfer to the artificial electron donors, especially as anticipated membrane-spanning proteins, remain to be identified in part due to the complexity of the multicomponent redox pathway. Plants, likewise, carry out transmembrane electron transport (Sijmons et al. 1984; Döring et al. 1990) (Fig. 4.3). However, a membrane-spanning protein capable of direct transfer of electrons from the cytosol to external acceptors also has yet to be demonstrated for the plasma membrane of plants (Lüthje et al. 1997).

4.3 Rates of PMET

Although cellular ATP is generally considered to be generated by oxidative phosphorylation, compensatory mechanisms (such as the pyruvate/lactate couple and enhanced PMET activity) are sufficient to allow cell survival even in the presence of mitochondrial dysfunction. The potential for PMET to drive glycolytic ATP production can be seen in mitochondrial knock-out ρ° cells (Scarlett et al. 2004 and ref. cit.) as well as in lymphocytes derived from diabetes mellitus patients, which lack fully functional mitochondria (Lenaz et al. 2002). Loss of mitochondrial function is frequently accompanied by upregulation of PMET (Malik et al. 2004). The ratio of free NAD^+ to NADH is lower in transformed cells than in nontransformed cells (Schwartz et al. 1974) possibly due in part to increased PMET activity (Sect. 4.9.1).

The ENOX proteins are not inhibited by mitochondrial inhibitors such as cyanide or rotenone. Diphenyleneiodonium (Fig. 4.5) has been used to distinguish ENOX activity from the activity of flavoenzymes but it also inhibits ENOX activities with NADPH as substrate (Morré 2002). Crane and Löw (2008) reported that diphenyleneiodonium inhibited H_2O_2 formation in SK-30 melanoma cells by 80 %. These cells have no cytochrome b_{358}, which is the catalytic component of the phox-NOX NADPH oxidases (Lambeth et al. 2000). Since mitochondria and other cytosolic oxidases can be inhibited without effects on H_2O_2 production, it may be one of

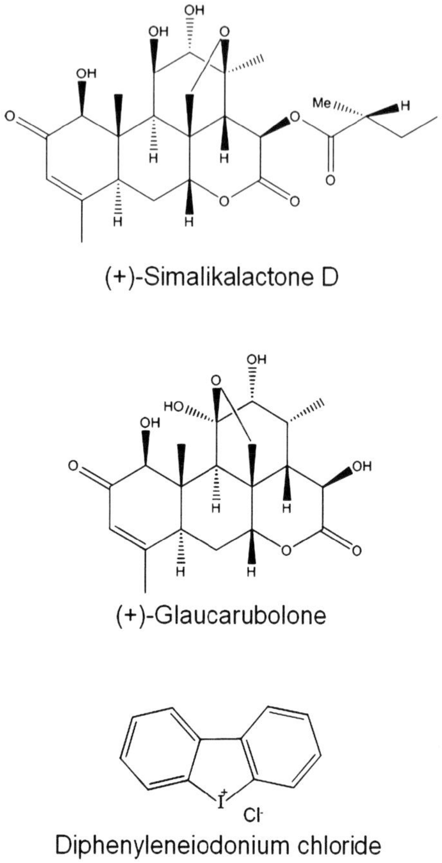

Fig. 4.5 The chemical structure of the specific quassinoid ENOX1 inhibitor simalikalactone D, the specific quassinoid ENOX2 inhibitor glaucarubolone, and the inhibitor of NAD(P)H oxidases diphenyleneiodonium chloride

(+)-Simalikalactone D

(+)-Glaucarubolone

Diphenyleneiodonium chloride

the ENOX proteins, most likely arNOX (Chap. 9), that is inhibited. As an example, superoxide generation by a megakaryotic cell line was inhibited 32 % by apocyanin [a naturally occurring and selective flavonoid inhibitor for the NADPH phoxNOX oxidases (Choi et al. 2007; Prata et al. 2006)] that also inhibits arNOX. When lactate is supplied to increase cytosolic NADH, the plasma membrane oxidase generates more superoxide (Bassenge et al. 2000). If pyruvate is added, the oxidation is decreased (Crane and Löw 2008). A similar dependency on NADH is found with transplasma membrane electron transport in pulmonary arterial endothelial cells (Merker et al. 2002).

The ENOX proteins of the external cell surface are subject to different controls and involves redox carriers different from the family of superoxide-generating NADPH oxidases, also named NOX or phoxNOX (phagocyte oxidase) (Navarro et al. 1995) which includes at least seven eukaryotic homologs of gp91 phox including Nox 1 to Nox 5, Duox 1, and Duox 2 (Lambeth et al. 2000) (Fig. 4.6). The primary control is exerted by the concentration of NADPH in the cytosol. Adriamycin

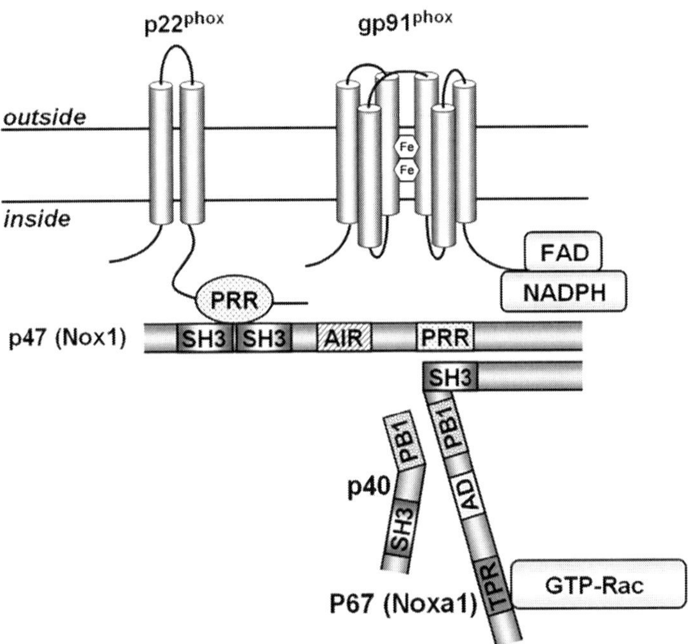

Fig. 4.6 Schematic model of the structure of phoxNOX (phagocyte oxidse) 1-4 enzymes. Cylinders represent transmembrane domains. *PRR* proline-rich region, *SH3* Src homology three domain, *AIR* auto-inhibitory region, *PB1* Phox/Bem1 domain, *AD* activation domain, *TPR* tetratricopeptide repeat motif, *Fe* heme group. Redrawn from Del Principe et al. 2011. Published with permission

and other antitumor drugs also inhibit the NADH diferric transferrin reductase and have not been reported to affect the NADPH oxidase of host defense (Sun and Crane 1990; Sun et al. 1992a).

The prevailing dogma is that the antitumor effects of the anthracycline antibiotics such as doxorubicin and the bleomycins are due to their effects on DNA (Remers 1988). However, there is also evidence that there are target sites of these drugs at the plasma membrane (Murphee et al. 1976). Similarly, actinomycin D is well known for the inhibition of cell growth and mitosis (Hoober and Cohen 1967) but a cell membrane target also may exist. The potent nonantibiotic antitumor drug, cisplatin, also has a definite effect on the cell surface (Macquet and Butour 1983) whereas transplatin has little or no effect (Table 4.3). These observations are further expanded in Chap. 11.

In rapidly dividing cell lines, the PMOR is capable of transferring electrons at rates exceeding activated neutrophils undergoing the respiratory burst (Berridge and Tan 2000). The importance of the plasma membrane redox system is exemplified by observations that, during generation of human Namalwa ρ° cells (a Burkitt lymphoma line which lacks mitochondria with a functional respiratory chain), the rate of whole cell ferricyanide reduction is increased about fourfold. In these cells, the

Table 4.3 Inhibition of NADH ferricyanide reductase activity of isolated plasma membranes by platinum drugs

Addition	NADH ferricyanide reductase (nmol/min mg protein)
None	394
10^{-7} M cisplatin	225
10^{-7} M transplatin	390

Data from Sun and Crane (1984) for pig erythcrocyte plasma membranes

PMOR can compensate for the total lack of mitochondrial oxidative phosphorylation by oxidizing excess NADH necessary for survival through glycolic generation of ATP (Lawen et al. 1994). Upregulation of the plasma membrane NADH-ferricyanide reductase activity is a prerequisite for the viability and growth of the Namalwa ρ° cells which are devoid of mitochondrial DNA (Lawen et al. 1994). Conversely, inhibition of the PMOR in ρ° cells (Lawen et al. 1994), and other cell lines (Wolvetang et al. 1996), causes cell death. The PMOR also modulates cellular processes some of which are redox sensitive, including cell growth, intracellular signaling, and apoptosis (reviewed in Baker and Lawen 2000). By maintaining the cell's antioxidant systems, PMOR has been suggested to retard cell death by reducing oxidative stress (reviewed in Villalba and Navas 2000).

4.4 Energetics of PMET

As introduced in the previous sections, the most obvious energetic considerations involving PMET are to regenerate NAD^+ from NADH in order to drive the glycolytic production of ATP through substrate level phosphorylations (Fig. 4.7). However, as NADH is oxidized at the plasma membrane, approximately 10 kcal of energy are released. An equally important question concerns the fate of an energy release sufficient to generate approximately 3 moles of ATP per mole of NADH oxidized.

Assuming that the energy resulting from NADH oxidation at the plasma membrane is simply not dissipated as heat, two possibilities for its conservation and subsequent utilization in the growth process have been considered.

The first possibility is that the energy released from NADH oxidation is conserved as an increase in membrane potential either to maintain or to create an inside-out proton gradient at the plasma membrane and to account for alkalinization of the cytoplasm or to fulfill the energy requirements of cell growth. The second possibility is that the energy is conserved either as ATP or as some yet unknown ATP equivalent also to fulfill the requirements for ATP in the cell enlargement phase of the growth process (Morré et al. 2006a; Hicks-Berger et al. 2006; Chap. 5).

In support of the latter possibility, Chi and Pizzo (2006) report that ATP synthase is present and active on the surface of endothelial cells. Angiostatin binds and inhibits the f_1f_0 ATP synthase and an antibody directed against the catalytic-β-subunit of

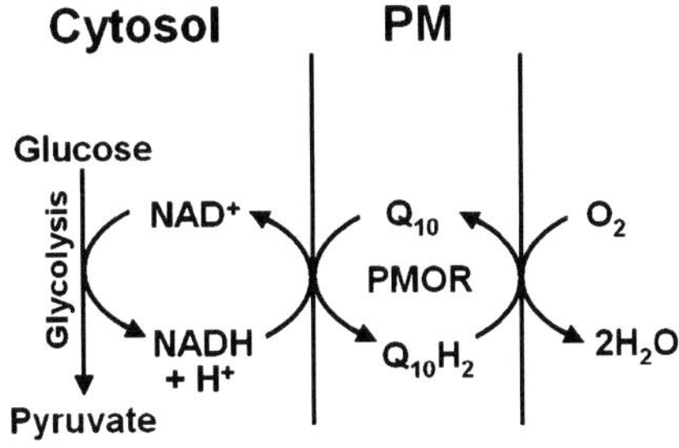

Fig. 4.7 Enhanced PMOR activity accelerates growth (Chap. 5) and oxidizes NADH to support glycolysis

Table 4.4 Isolated plasma membrane vesicles from dark-grown soybean seedlings oriented inside out generate ATP in response to NADH

	ATP formed (nmoles/10 min)
Complete system[a]	0.22
– Plasma membrane	0.055
– NADH	0.02
– Reduced coenzyme Q	0.02
– NADH–vanadate	0.03
– Reduced coenzyme Q–vanadate	0.015
– ADP	0.02
– Inorganic phosphate	0.02
– Reduced coenzyme Q + oxidized coenzyme Q	0.135
+ KCN + Antimycin A + Rotenone	0.13
+ f_1/f_o antibody	0.11
+ Simalikalactone D	0.055

Right-side-out vesicles were unresponsive to NADH. Unpublished results of Zoran Kvrgic, Valparaiso University, Valparaiso, IN

[a]The complete system contained (final concentrations) in a final volume of 100 μL, 160 μg plasma membrane protein, 30 μM NADH, 120 μM reduced or oxidized coenzyme Q, 100 μM ADP, 2 mM inorganic phosphate, and 60 μM vanadate. The concentrations of KCN, rotenone, antimycin A, and simalikalactone D were 1 μM. The f_1/f_o antibody (1 μL) was from rabbit, polyclonal, Agrisera Antibodies (Vännäs, Sweden). The luminescence assay for ATP was according to Burwick et al. (2005). Determinations were with a GloMax®-Multi+ Microplate Reader (Promega, Fitchburg, WI)

ATP synthase inhibited the activity and reduced endothelial cell proliferation. This raised the interesting possibility of a cell surface ATP synthase (Chap. 5). Based on a luciferase luminescence assay, NADH/reduced coenzyme Q-dependent ATP formation requiring the presence of ADP and inorganic phosphate was catalyzed by inside-out vesicles of soybean plasma membranes (Table 4.4). Involvement of ENOX1 proteins was indicated from inhibition by the ENOX1-specific inhibitor simalikalactone D. The bulk of the ATP formed was resistant to mitochondrial electron transport inhibitors. Right-side out plasma membrane vesicles were ineffective.

With the assumption that ATP formation would be restricted to the oxidative portion of the NADH cycle, the molar ratio of ATP formed to NADH oxidized was determined to be 1.5. Further functional implications of these observations are discussed in Chap. 5.

4.5 PMET Driven Outward Proton Pumping and Alkalinization of the Cytoplasm

Transplasma membrane electron transport is correlated with modulation of internal pH and redox homeostasis, as it is able to activate proton release, a biological function which strictly requires the presence of coenzyme Q in the plasma membrane (Sun et al. 1992b). Proton efflux in mammalian cells and the consequent alkalinization of the cytoplasm is most often attributed to the Na^+/H^+ antiport targeted by amiloride (Grinstein and Rothstein 1986; Rozengurt 1986; Moolenaar et al. 1986; L'Allemain et al. 1984) (Fig. 4.8). However, much, but not all, of PMET-related

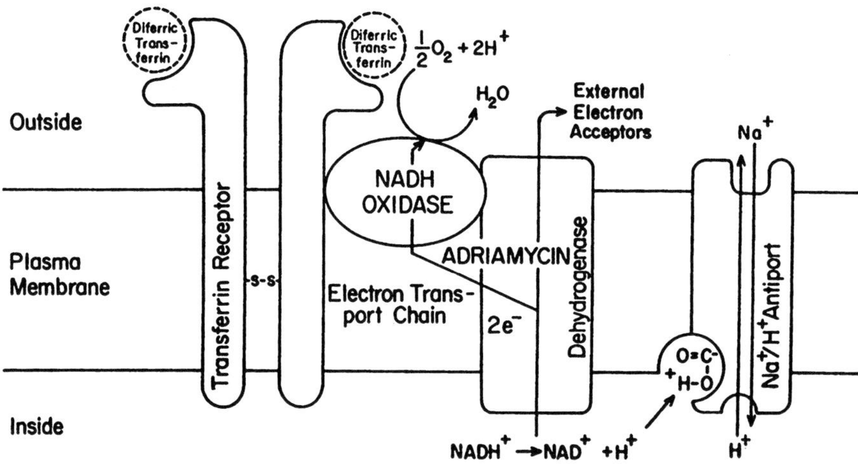

Fig. 4.8 Diagram illustrating the hypothetical relationship of constituents within the mammalian plasma membrane to proton transport. The flow of electrons from cytoplasmic NADH is depicted as being coupled to molecular oxygen via the NADH oxidase. The protons generated from NADH oxidation are exported via the Na^+/H^+ antiport in exchange for imported Na^+ ions. From Morré and Crane (1990)

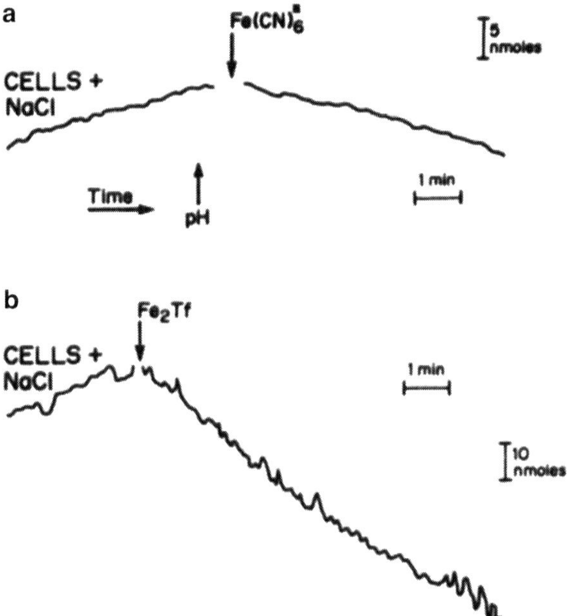

Fig. 4.9 Proton release from pineal cells stimulated by external oxidants. (**a**) Ferricyanide. (**b**) Diferric transferrin. (**a**) Ferricyanide-stimulated change in pineal cells (10 mg) in 1.5 mM tris buffer with indicated salts at 150 mM were incubated in circulating air used to equilibrate CO_2. The pH was measured with a glass electrode and proton change titrated with 0.01 M HCl at end of the experiment. Potassium ferricyanide (1.1 mM) or diferric transferrin (17 μM) was used to initiate proton release. Modified and redrawn from Crane et al. 1990a. Published with permission

proton efflux is amiloride insensitive (Sun et al. 1995). A link of superstoichiometric protein efflux (protein pumping) to PMET was first indicated from observations that proton efflux could be induced by addition of either diferric transferrin (Grinstein and Rothstein 1986), a growth factor, or the external oxidant ferricyanide (Sun et al. 1987a) (Fig. 4.9) [See also Barr et al. (1990) for evidence from cultured plant cells]. The proton release by HeLa cells was a mixture of amiloride-sensitive (antiport) and -insensitive (blocked by ENOX inhibitors) proton efflux mechanisms. Both Sun et al. (1984a) and Garcia-Cañero and Guerra (1988) showed previously that proton release by HeLa cells in response to ferricyanide was stimulated by sodium ions but was only partly amiloride sensitive. Perhaps no more than 50 % of the proton efflux of HeLa cells induced by either diferric transferrin or ferricyanide is inhibited by amiloride (Sun et al. 1995). Proton export is another major difference between PMET and the phagocyte NADH oxidases.

A second line of evidence in addition to amiloride resistance for a role of PMET in proton release from HeLa cells in response to diferric transferrin or ferricyanide was the demonstration of a requirement for quinones in the proton release (Sun et al. 1992a). As shown in the Fig. 4.10, the internal pH increases with diferric transferrin addition. With intact cells, the coenzyme Q analogs,

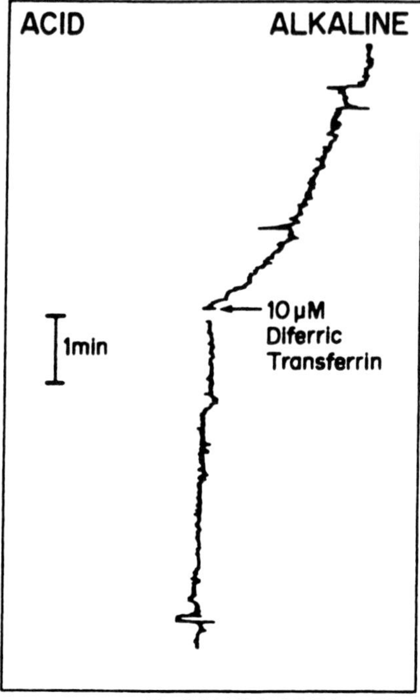

Fig. 4.10 Change of 0.2 units in internal pH of HeLa cells measured with the fluorescent indicator BCECF in response to 10 μM diferric transferrin. From Crane et al. 1990a. Published with permission

DCIQ (2,3-dimethoxy-5-chloro-6-naphthylmercapto-1,4-benzoquinone) or ETHOXQ (2-methoxyl-3-ethoxy-5-methyl-6-hexadecylmerapto-1,4-benzoquinone) inhibited both ferricyanide- and diferric transferrin-stimulated proton release in HeLa cells, and this release could be reversed by the addition of coenzyme Q (Sun et al. 1992b).

The increase in membrane potential accompanying activation of Na^+/H^+ exchange has been attributed to an increase in activity of the Na^+/H^+-ATPase to remove the extra Na^+ taken up by the antiport (Garcia-Cañero et al. 1987). Both ferricyanide and diferric transferrin produce an increase in membrane potential and the increase has been reported to be ouabain sensitive (Crane et al. 1990a) which would be consistent with some relationship to activation of the Na^+/H^+-ATPase. A key feature of transmembrane proton pumping is that it generates a transmembrane proton gradient which is an alternative means to conserve the large free energy of oxidation of NADH (de Grey 2003).

4.6 PMET Function in Electron Import

A possibility that electron transport constituents of the plasma membrane might function as electron import pathways has been reviewed by Kennett and Kuchel (2003). Extracellular GSH apparently is unable to act as an electron donor for GSSG reduction inside the cell via transmembrane thiol-disulfide interchange. However, there are reports that NADH when placed in the extracellular medium produces effects that are consistent with the concept of electron import (Zemkova et al. 1984; Birkmayer et al. 1993; Birkmayer 1996; Forsyth et al. 1999; Slade et al. 1999; Nadlinger et al. 2002).

Extracellular NADH exerts a hyperpolarizing effect on mouse skeletal muscle plasma membranes (Zemková et al. 1984). The hyperpolarizing effect was inhibited by ouabain, an inhibitor of Na^{+}/K^{+} ATPase and plasma membrane oxidoreductase inhibitors including adriamycin, pCMBS, and atebrin. NADH is oxidized in the process and K^{+} membrane permeability is increased (Zemková et al. 1984). Such actions of extracellular NADH may indicate that plasma membrane oxidoreductase components could potentially participate in electron import.

NADH, given orally in a stabilized absorbable form, has been reported to have positive therapeutic effects on patients with Parkinson's disease (Birkmayer et al. 1993), Alzheimer's disease (Birkmayer 1996), and chronic fatigue syndrome (Forsyth et al. 1999). The NADH is thought to alleviate motor and cognitive dysfunction by stimulating endogenous dopamine biosynthesis and by triggering energy production through ATP generation (Birkmayer 1996). NADH also exerts an antiproliferative effect on some but not all carcinoma cell lines (Slade et al. 1999) and reduces oxygen consumption in mitochondrial gene knockout cells (Herst et al. 2004).

4.7 PMET and Growth

A role for PMET in growth has been implicit from the onset (Crane et al. 1985, 1990a, b; Morré et al. 1988a, b, c) and, in part, was instrumental in the discovery of the ENOX proteins (Morré et al. 1986a, 1988b). Subsequently, the protein disulfide-thiol interchange portion of the ENOX cycle was shown to correlate with the enlargement phase of cell growth (Chap. 5). As cells must reach some minimal size following division in order to divide again, a response of progression through the cell cycle to PMET activity might be anticipated as well.

Ferricyanide, differic transferrin, and other external oxidants are capable of promoting growth of mammalian cells in culture (Faulk et al. 1991; Martinus et al. 1993) especially in serum-deficient media (Ellem and Kay 1983; Sun et al. 1984b).

4.7.1 Cell Cycle Check Point Control of Cell Enlargement

Cell enlargement and the cell cycle are coordinated but separable processes (Jorgenson and Tyers 2004). The observation that continuously proliferating cells precisely double their size during each cycle to maintain constant volumes has resulted in the suggestion of the existence of an active cell size control mechanism in eukaryotic cells. Such checkpoint control would function to prevent delayed or premature cell division at inadequate size or mass. This mechanism seems well defined in yeasts but in some respects remains an open issue with mammalian cells. Eukaryotic cells coordinate cell size with cell division by regulating the length of the G_1 and G_2 phase of the cell cycle (Sweiczer et al. 1996).

4.7.2 PMET Activity and Growth Are Correlated

Diferric transferrin can replace serum for growth of many cells (Sect. 4.8) through stimulation of PMET, an observation that initiated a series of studies focused on the correlation of PMET activity and growth. Interest in the ENOX1 proteins in this regard is predicated on nearly three decades of research indicative of a vital and essential role for ENOX1 to drive cell enlargement in both plant and animal cells. The work began with investigations in the early 1960s relating to a general understanding of the enlargement phase of cell growth followed by studies on the involvement of the ENOX proteins in the mid-1970s to mid-1980s and culminating with the cloning of the human ENOX1 in 2008 (Jiang et al. 2008).

That the growth stimulations related to PMET are largely correlated with effects on ENOX proteins is supported by:

1. A strong correlation between rate of cell enlargement and ENOX1 activity (Chap. 5)
2. Inhibition of cell enlargement (and growth) by relatively specific inhibitors of both ENOX1 and cell enlargement (Morré and Grieco 1999)
3. Overexpression of cloned ENOX1 in a mammalian cell line (HEK) that resulted in increased rates of cell enlargement and increased cell volume (Bosneaga and Tang, unpublished) as well as from overexpression of ENOX2 in both cultured mammalian cells (Chueh et al. 2004) and in transgenic mice (Yagiz et al. 2006, 2007)

4.8 Regulation of PMET

Regulation of plasma membrane redox can be achieved by activation or upregulation of rate-limiting components or by changes in substrate availability. External acceptors such as ferricyanide (Sun and Crane 1984; Fig. 4.11) or ascorbate free radical (Alcaín et al. 1991) have been used to drive PMET with rates being proportional to the levels of external acceptor. With molecular oxygen as the physiological acceptor, the general assumption would be that the ENOX proteins as terminal oxidases of the electron transport chain appear to be rate limiting especially in plants (Chap. 5). Unfortunately, little is known about changes in ENOX protein or mRNA levels during normal development and growth or in response to effectors. Clear relationships of cell enlargement and ENOX levels in the membrane are indicated from overexpression studies with cultured cells (Chueh et al. 2004) and transgenic

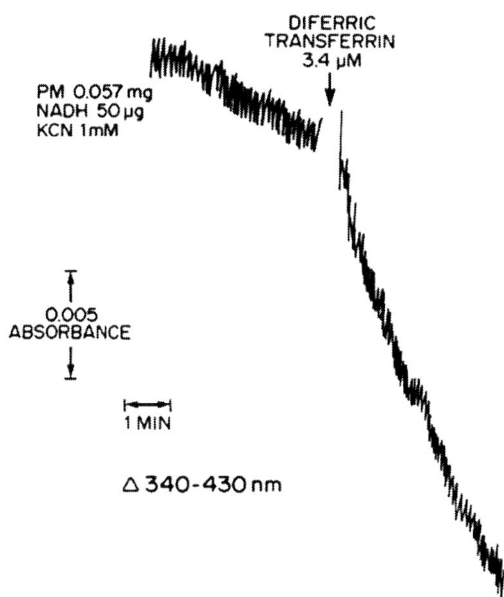

Fig. 4.11 Spectrophotometric assay for NADH-diferric transferrin reduction by isolated rat liver plasma membranes. Absorbance was measured at 340–430 nm. From Löw et al. 1990 with permission

mice (Yagiz et al. 2006, 2007). Overexpression resulted in both enhanced rates of cell enlargement and, with ENOX2, an enhanced level of drug inhibition. In stomach cancer cells, capsaicin-induced cytotoxicity resulted in apoptosis and a concomitant downregulation of ENOX2 (Wang et al. 2009). Comparable studies to determine if PMET is rate limited by ENOX levels are currently missing from our information.

On the other hand, there are several clear examples of where ENOX activities of cells and tissues are elevated in response to growth factors in both animals (Morré et al. 1988c; Bruno et al. 1992; Brightman et al. 1992) and plants (Morré et al. 1986a; Brightman et al. 1988; Morré and Brightman 1991; Chaps. 5 and 10). The oxidase is also strongly stimulated by diferric transferrin (Sun et al. 1987b; Bérczi et al. 1991) and ceruloplasmin (Alcain et al. 1992). A chelatable iron site on the outer surface of the plasma membrane has been reported to be necessary for NADH oxidation (Alcaín et al. 1994). Especially well studied is the response of plants and excised plant parts to the auxin-type regulators of plant cell enlargement such as 2,4-D where the enhanced rate of cell enlargement correlates in direct proportion to the activation of an auxin-stimulated cell surface ENOX activity (Chap. 10). In contrast to mammalian and plant ENOX1 activities which are hormone and growth factor stimulated, the tumor-associated ENOX2 proteins (Chap. 8) and age-related ENOX proteins (Chap. 9) are refractory to hormone and growth factor stimulation. Similarly cancer cells are, for the most part, unresponsive to hormones and growth factors despite the presence on their surface of the normally hormone and growth factor-responsive ENOX1 in addition to the hormone and growth factor-unresponsive ENOX2. The ENOX2 proteins appear to be constitutively activated to account for the unregulated growth which is universally associated with the cancer phenotype (Chap. 8).

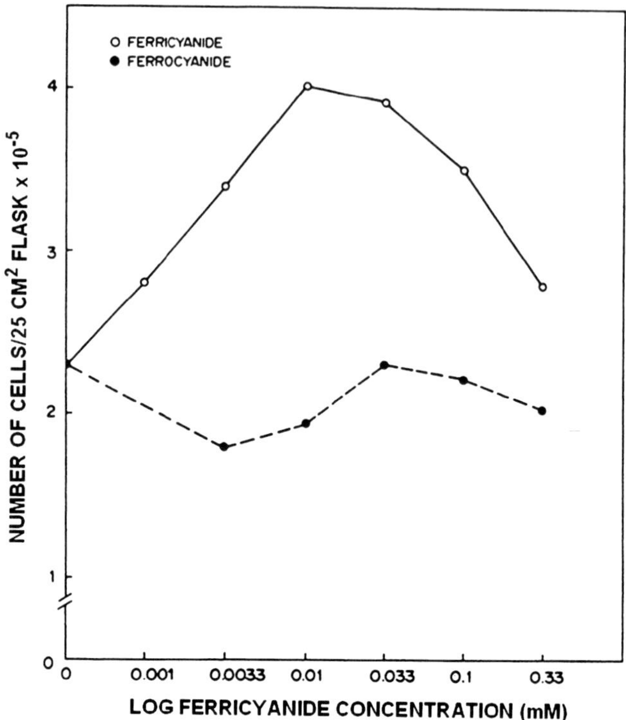

Fig. 4.12 Stimulation of HeLa cell growth in serum-free media supplemented with ferricyanide and lack of stimulation by low concentrations of ferrocyanide. Redrawn from Crane et al. 1990a. Published with permission

The oxidation of cytosolic NADH is observed as well when the oxidase is stimulated by diferric transferrin (Faulk et al. 1991) an external oxidant capable of promoting growth in serum-deficient media in human melanoma (Ellem and Kay 1983) and HeLa (Sun et al. 1984b) cells. Differic transferrin can replace serum for the growth of many cells (Ellem and Kay 1983) through stimulation of the oxidase (Fig. 4.12).

A frequent approach to experimental regulation of PMET is through the use of inhibitors such as anticancer drugs targeted to the ENOX2 proteins of cancer cells (Morré 1998c) (Chap. 11) or resiniferatoxin for noncancer cells (Wolvetang et al. 1996). The relevance of PMOR activity to cell growth and viability has been well documented in proliferating cell lines (Chueh 2000; Baker and Lawen 2000; Berridge and Tan 2000). Several anticancer drugs selectively toxic to transformed cells are all inhibitors of the terminal oxidase, which exhibits atypical characteristics in transformed cells (Morré 1998c, Chap. 8). Yet, studies of the PMOR are constrained by a lack of specific inhibitors especially for nontransformed cells (e.g., diphenyliodonium (DPI) inhibits many flavoenzymes with similar redox mechanisms) (Morré 2002; O'Donnell et al. 1994). Membrane impermeant thiol reagents have been used in this capacity (Morré et al. 1995a; Morré and Morré 1995a). However, thiol reagents are not specific inhibitors per se. Their inactivation of different proteins may be highly concentration-dependent, involving factors like thiol accessibility, protein tertiary structure, and neighboring functional groups (van Iwaarden et al. 1992). Yet, HL60 ρ° cells, where NADH ferricyanide reductase activity is upregulated, are much more sensitive to the membrane impermeant sulfhydryl reagent PCMBS than their parental cells (Larm et al. 1994). The parental cells tolerated continuous exposure to 100 μM PCMBS over several days without loss of viability while the ρ° cells rapidly died. Quassinoids have been used for this purpose as well including studies with plants (Morré and Grieco 1999) and yeast (S. Dick, A. Ryuzoji, D.M. Morré and D.J. Morré unpublished). Simalikalactone D specifically inhibits ENOX1 (Chap. 3) and glaucarubolone specifically inhibits ENOX2 (Chap. 8).

PMET is dependent upon the presence of quinones in the membrane (Morré and Morré 2011). Removal by solvent extraction results in cessation of activity which can be restored by quinone addition (Sun et al. 1992b). However, there are no studies to suggest that plasma membrane quinone levels are rate limiting to PMET under normal conditions. Studies of patients and organisms deficient in coenzyme Q_{10} have focused on mitochondrial coenzyme Q requirements such that the status of PMET in coenzyme Q deficiency has been little studied.

Cytosolic NADH levels at the plasma membrane undoubtedly profoundly affect the rate of PMET (Crane and Löw 2008; Crane et al. 1985; Baker and Lawen 2000; Herst and Berridge 2006). Under anaerobic conditions, NADH will be elevated as a result of anaerobic glycolysis but oxygen will be in short supply. Electrons apparently can be transferred to protein disulfides to replace oxygen under anaerobic conditions (Chueh et al. 1997a). Calorie restriction upregulates the plasma membrane redox system and a role in protection of brain cells against age-related increases in oxidative and metabolic stresses has been suggested (Hyun et al. 2006).

4.8.1 Feedback Regulation of PMET

Feedback regulation of PMET based on intracellular energy demands is implicated in several studies including those of Nadlinger et al. (2002) who reported an in vitro assay using extracellular NADH as a measure of the intracellular ATP content of human RBCs. RBCs withdrawn from healthy individuals immediately before and after exercise were incubated with 0.5 mM NADH. The rate of extracellular NADH oxidation was faster in RBCs isolated after exercise than those isolated before. For example, 80–100 % of the NADH was oxidized in RBCs isolated after exercise compared to 5 % oxidized in RBCs isolated before exercise. After aerobic activity, the ATP-to-ADP ratio decreased to 80 % of the baseline level and correlated with the extent of NADH oxidation. That the rate of NADH oxidation apparently reflects the oxidation state of the intracellular medium, which presumably would be more oxidized in ATP depleted cells, is of interest in regard to a mechanism whereby the rate of glycolytic ATP production and growth rate might be coupled. Since the PMET pathways may sustain glycolytic ATP production (Berridge and Tan 2000), it is possible that ENOX expression and activity may be regulated by a feedback mechanism that responds to intracellular energy metabolism (Scarlett et al. 2005).

4.8.2 ENOX Cell Surface Receptor Proteins

Although ENOX1 and ENOX2 proteins are known to transmit and receive information across the cell membrane, they do not contain a transmembrane hydrophobic sequence and are not membrane spanning (Morré et al. 2001c). Thus, ENOX1 and ENOX2 receptor proteins at the surface of the cell membrane may be required in order to send in and receive signals from the cell. Studies using idiotypic antibodies are consistent with a 52-kDa surface protein as the ENOX2 receptor (X. Tang, results unpublished).

A requirement for transmembrane exchange of information between ENOX proteins at the cell surface and proteins within the cells interior derives from the response of ENOX proteins to signals such as light that require communication with intracellular photoreceptors and the ENOX proteins located at the cell surface. ENOX receptors might bridge the gap between the cytosolic photoreceptor proteins and the surface ENOX molecules to account for light phasing of the ENOX1 period for example. Alternatively the receptor would allow signals such as those generated in response to microwaves where the ENOX proteins are regarded as the primary sensor (Morré et al. 2008b) to be transformed into a transplasma membrane signaling cascade to modulate intracellular responses.

4.9 PMET and Glycolysis

A major impact of PMET, especially in mitochondrial-deficient cells, is to oxidize NADH to support the glycolytic production of ATP (Fig. 4.7). High intracellular NADH/NAD^+ ratios generally inhibit major metabolic pathways such as glycolysis and fatty acid oxidation (Williamson et al. 1993) as do high levels of ATP. High levels of ATP repress glycolysis via allosteric inhibition of phosphofructokinase, a key enzyme of glycolytic regulation. This is a possible basis of the Pasteur effect in that the presence of O_2 allows ATP synthesis through oxidative phosphorylation which causes an allosteric inhibition of phosphofructokinase resulting in the inhibition of glycolysis. Glycolysis appears not to be directly inhibited by O_2 but by ATP such that the presence of O_2 will cause the inhibition of glycolysis when O_2 is not being used to generate ATP (Lopez-Lazaro 2008).

4.9.1 PMET and Glycolysis of Cancer Cells

Among the first identified biochemical alterations of cancer cells was a shift in glucose metabolism from oxidative phosphorylation to aerobic glycolysis (Warburg 1929, 1930; Warburg et al. 1924). Much of this metabolic conversion has been shown subsequently to be controlled by specific transcriptional events.

Tumor cells that are actively proliferating have been observed to undergo a switch from oxidative to glycolytic metabolism (Warburg 1929, 1956), named the

Crabtree effect, where the energy requirements of the rapidly dividing cells are provided by ATP from glycolysis (Sener et al. 1988; Cavalli and Liang 1998; Hofhaus et al. 1996). Tumors characteristically metabolize the majority of the glucose they take up through glycolysis even in the presence of an adequate oxygen supply. Most have a substantial reserve capacity to produce ATP by oxidative phosphorylation when glycolysis is suppressed such that the high rate of glycolysis is required to support cell growth rather than to compensate for defects in mitochondrial function (Fantin et al. 2006).

The conventional view is that in glycolytic cells, most of the reoxidation of NADH occurs via lactate dehydrogenase which reduces pyruvate to lactate. Glycolytic cancers have been known to produce large amounts of lactate with cellular growth rates directly related to rates of lactate production in fast growing hepatomas (Warburg 1921, 1956; Pedersen 1978).

In glycolytic tumor cells, there is an increase of all enzymes and glucose transporters with respect to normal cells (Martinez-Sánchez and Giuliana 2007; Moreno-Sanchez et al. 2007). Hexokinase II (HK-II) is over-expressed, increasing both the activity and its binding to the outer mitochondrial membrane, which in turn increases the HK-II access to newly synthesized ATP by oxidative phosphorylation. The over-expressed HK is strongly inhibited by its product, glucose-6-phosphate, whereas phosphofructokinase type 1 (PFK-1) activation by fructose-2,6-bisphosphate overcomes the citrate and ATP inhibition.

An increased flux towards ribose-5-phosphate (and nucleotide) synthesis is documented for several tumor cell lines. Tumor PFK-1 (C and L subunits) and the PFK-2FB3 isoform are over-expressed. In some tumors, the amount of α-glycerol-3-phosphate dehydrogenase (αGPDH) decreases. In addition, pyruvate may be oxidized by mitochondria to generate alanines and, in tumor cells, synthesize malate in a reaction catalyzed by an over-expressed cytosolic malic enzyme.

Low NAD^+ availability increases intracellular levels of glycolytic triose phosphates (glyceraldehyde-3-phosphate and dihydroxyacetone phosphate) which, if not metabolized further, decompose spontaneously into methylglyoxal, a glycating agent and source of mitochondrial dysfunction and reactive oxygen species (Hipkiss 2010). Methylglyoxal has anticancer activity and is inhibitory to ENOX2 (results unpublished). Many glycolytic intermediates can glycate proteins and other macromolecules (Hipkiss 2006).

NAD^+:NADH ratios in the normal cells range from 3.1 to 3.6 and increase three to fourfold as they grow from the logarithmic to the stationary phase (Schwartz et al. 1974). In contrast, the NAD^+:NADH ratio of transformed cells was low (1.2–3.2) and did not rise when the cells reached a heavy density. The authors suggest that the inability of cells to elevate their NAD^+:NADH ratio at confluency is a characteristic of transformed cells and related to their defective growth control. More importantly, the low NAD^+ to NADH ratio of cancer cells reflects the over production of NADH through enhanced glycolysis and the increased reliance on the PMET to generate NADH in support of glycolytic ATP formation.

Using a noninvasive assay based on two-photon autofluoresence lifetime imaging, the average NADH concentration in human Hs578T breast cancer cells is $168 \pm 49\ \mu M$,

1.8-fold higher than Hs578Bst breast normal cells (99±37 μM) (Yu and Heikal 2009). Also revealed was a statistically larger fraction of free NADH in the Hs578T cancer cells compared to the Hs578Bst normal cells.

Poor vascularization, leading to hypoxia and acidification of the extracellular matrix is characteristic of early tumor development potentially related to the switch to glycolysis and dependency on PMET (Gatenby and Gillies 2004). The often steep oxygen gradients in the tissue adjacent to blood vessels (Dewhurst et al. 1994; Helmlinger et al. 1997) may present one of the first growth-limiting challenges encountered by a developing tumor since pO_2 levels decline more rapidly with distance from blood vessels than glucose levels (Gatenby and Gillies 2004). Cells that can switch at an early stage from a mitochondrial to a glycolytic energy metabolism might be expected to gain a significant proliferative advantage under hypoxic conditions (Herst and Berridge 2006). Guppy et al. (2005) have argued that the OXPHOS to glycolysis switch is both temporary and reversible (Guppy 2002; Guppy et al. 2002; Zu and Guppy 2004; Yasuda et al. 2004). However, if cancer cells switch quickly and efficiently to favor glycolysis under hypoxic conditions, this may give them a selective advantage. Under normoxic conditions some cancer cells may be no more glycolytic than normal cells although solid tumors often maintain their glycolytic status even after being cultivated under normoxic conditions in vitro. Tumor aggressiveness has been correlated with level of glycolytic metabolism (Weber 1983; Kunkel et al. 2003; Schomack and Gillies 2003; Simonnet et al. 2003).

A common consequence of a wide variety of mutations underlying human cancer is activation of the hypoxia-inducible factor (HIF) (Shaw 2006). HIF stimulates expression of glycolytic enzymes and decreases reliance on mitochondrial oxidative phosphorylation in tumor cells, even under aerobic conditions. In addition, the master metabolic regulator AMP-activated protein kinase (AMPK) has been connected to several human tumor suppressors (Alessai et al. 2006). Therapeutic strategies based on modulation of AMPK, HIF, and other metabolic targets related to increased glucose uptake and glycolysis by tumor cells have been proposed (Garber 2006).

4.10 PMET Links to Major Signaling Pathways

The most obvious link of PMET to major signaling pathways derives from observations that NAD^+/NADH not only act as oxidoreductase cofactors but also exhibit profound effects in a signaling capacity (Fig. 4.13). NADH was reported early to inhibit adenylate cyclase of adipocyte plasma membranes (Löw and Werner 1976). In most instances it is not the ratio of NAD^+/NADH but the absolute amount of one or the other. Some pathways respond to NAD^+ levels and not to NADH. Other pathways respond to NADH levels but not to NAD^+. The sections which follow summarize potential links of NADH levels to signaling pathways.

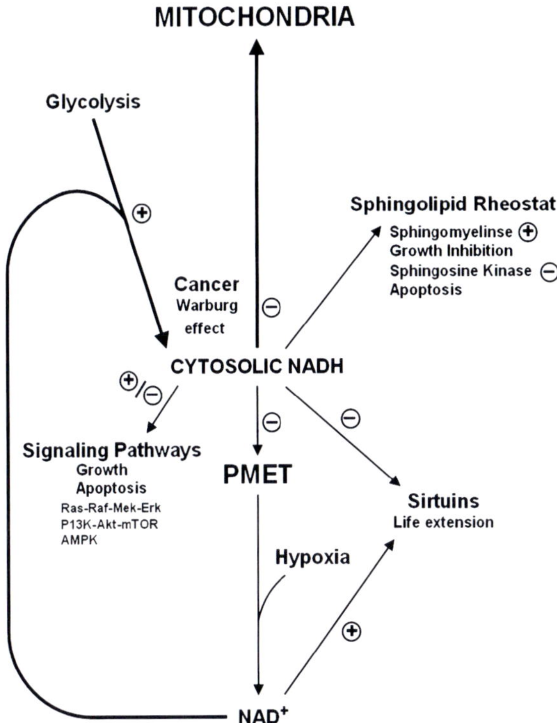

Fig. 4.13 Regulatory roles for NADH and relationship of NADH oxidation by PMET to cytosolic NADH levels and maintenance of glycolytic ATP production. NAD^+ is carried into the mitochondrion by a specific membrane transport protein

4.10.1 Sirtuins

Since the discovery of NAD-dependent deacetylases, sirtuins, it has been recognized that maintaining intracellular levels of NAD^+ is crucial to the activation of SIRT1. Originally linked to longevity in yeast, sirtuins and more specifically, SIRT1s have been implicated in numerous biological processes especially in the operation of the biological clock.

Rhythmic oscillations in NADH and NAD^+ levels resulting from the oscillatory activity of the ENOX terminal oxidases (Chap. 6) are reflected in the periodic expression and activity of clock proteins (Fig. 4.14).

Experiments using cultured cells have suggested that the cellular redox state is capable of influencing rhythms. CLOCK and its homolog NPAS2 can bind efficiently to BMAL1 and consequently to E-box sequences in the presence of reduced nicotinamide adenine dinucleotides (NADH and NADPH) (Fig. 4.14a). On the other hand, the oxidized forms of the nicotinamide adenine dinucleotides (NAD^+ and $NADP^+$) inhibit DNA binding of CLOCK:BMAL1 or NPAS2:BMAL1. As the

a

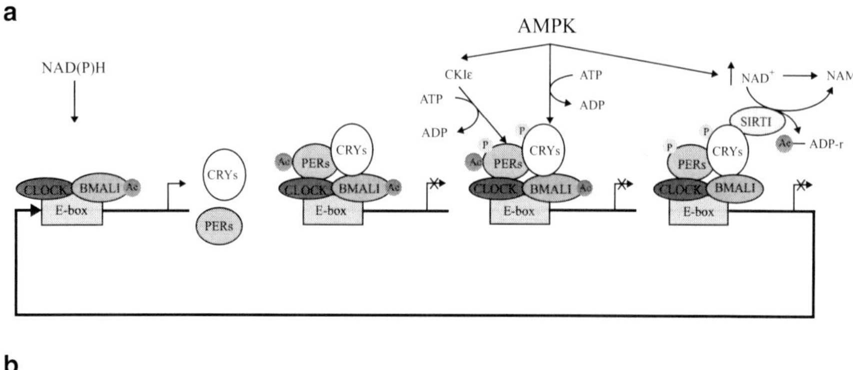

b

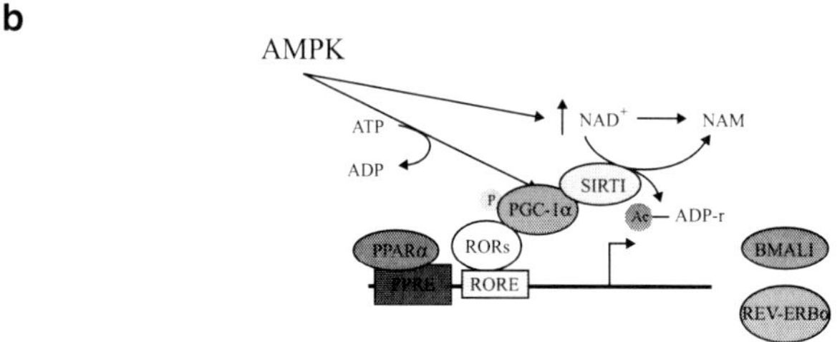

Fig. 4.14 Role of NADH in regulation of sirtuin 1 activity as related to clock function and line to energy metabolism. (**a**) High NAD(P)H levels promote CLOCK:BMAL1 binding to E-box sequences leading to the acetylation of BMAL1 and expression of Pers, Crys, and other clock-controlled genes. The negative feedback loop, PERs:CRYs, binds to CLOCK:BMAL1 and consequently PERs are acetylated. Activated AMPK leads to a rise in NAD+ levels, phosphorylation of CRYs, and phosphorylation of CKLε, which then phosphorylates the PERs. As a result of increased NAD+ levels, SIRT1 deacetylates PERs and BMAL1. This and the destabilization of phosphorylated PERs and CRYs relieves PERs:CRYs repression and another cycle starts. (**b**) Expression of BMAL1 and Rev-erbα genes are controlled by PPARα and binding of RORs to RORE sequences. RORs need a co-activator, PGC-1α, which is phosphorylated by activated AMPK. In parallel, AMPK activation leads to an increase in NAD+ levels, which in turn activate SIRT1. SIRT1 activation leads to PGC-1α deacetylation and activation. Acetyl adenosine diphosphate ribose (Ac-ADP-r) and nicotinamide (NAM) are released after deacetylation by SIRT1. Modified from Froy and Miskin 2010. Published with permission

NAD(P)+/NAD(P)H redox equilibrium depends on the metabolic state of the cell, this ratio could dictate the binding of CLOCK/NPAS2:BMAL1 to E-boxes and result in phase shifting of cyclic gene expression.

Hypoxia-inducible factor 1 (HIF-1) is a key regulator of O_2 homeostasis. Increased production of H_2O_2 and the accumulation of glycolytic metabolites are known to activate HIF-1. During hypoxia, however, SIRT-1 is downregulated due to deceased NAD+ levels. When glycolysis is blocked, SIRT1 is upregulated, leading to deacylation at Lys 674 and inactivation of HIF-1α. Recruitment of p^{300} is blocked and HIF-1α target genes are repressed (Lim et al. 2010) with attendant negative effects on tumor growth and angiogenesis.

4.10.2 *Sphingolipid Rheostat*

One important and well-studied link of ENOX2 inhibitors to cytosolic signaling pathways is the activation of acid sphingomyelinase and the resultant formation of ceramide-enriched membrane domains which lead to apoptosis (Fig. 4.13) demonstrated initially by De Luca et al. (2005) and reiterated by Dumitru and Gulbins (2006).

The stimulation of sphingomyelinase and inhibition of sphingosine kinase in response to ENOX2 inhibitors is predicated upon an increase in cytosolic NADH as a result of inhibited PMET. Based on fluorescence measurements, increases of 35–45 % were measured experimentally during the first 30 min following exposure to inhibitor (Wu et al. 2011). The source of ceramide as sphingomyelin hydrolysis rather than de novo biosynthesis is supported by observations that a similar response to the ENOX2 inhibitor capsaicin was not prevented by ceramide biosynthesis inhibitors (Sanchez et al. 2007).

A mechanism whereby these increased NADH levels might be linked to signaling pathways leading to apoptosis is provided by the reciprocal responses of plasma membrane-associated sphingomyelinases and sphingosine kinase to NADH levels (De Luca et al. 2005, 2010; Chap. 8). For sphingomyelinases, NADH stimulated the activity whereas NAD^+ was either inhibitory (neutral sphingomyelinase) or without effect (human acid sphingomyelinase). In contrast, the activity of sphingosine kinase was inhibited by NADH whereas NAD^+ was without effect.

Consistent with the simulation of sphingomyelinase by NADH and the inhibition of sphingosine kinase by NADH were significant changes in the level of ceramide (stimulated) and sphingosine-1 phosphate (inhibited) in response to treatment with HeLa cells with the ENOX2 inhibitors EGCG [(-)-epigallocatechin-3-gallate] and phenoxodiol (Wu et al. 2011).

External addition of 100 μM ferricyanide, known to deplete cytosolic NADH levels through marked stimulation of PMET, strikingly reduced the ability of ENOX2 inhibitors to block the growth of HeLa cells and induce apoptosis due to restoration of elevated NADH levels (Wu et al. 2011).

4.10.3 *NADH Modulation of PTEN Provides Link of PMET to Ras-Raf-Mek-Erk, PI3-AKT-mTOR and NF-κB*

NADH modulation of phosphatase and tensin homolog (PTEN) potentially links PMET to the Ras-Raf-Mek-Erk, P13-AKT-mTOR, and INF-κB pathways (Fig. 4.15). The link is provided by reports that the PTEN gene product is redox sensitive and dependent on NADPH/thioredoxin to maintain its enzymatic activity (Lee et al. 2002). Activation of Akt caused by the accumulation of NADH, which indirectly inactivates PTEN by blocking thioredoxin NADP(H)-dependent activation of PTEN, whose redox-sensitive phosphatase activity is essential for opposing

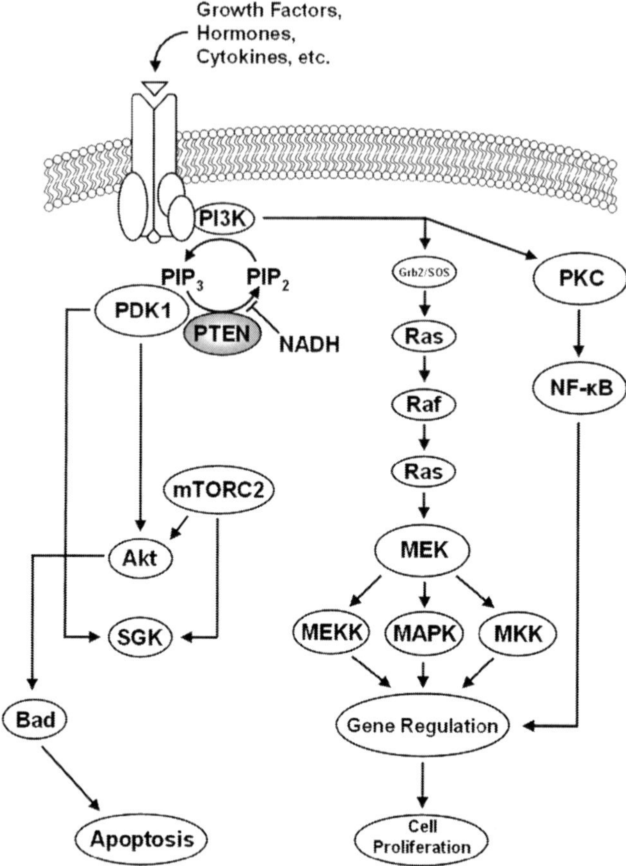

Fig. 4.15 Potential regulatory links to PMET effects on cytosolic NAD+/NADH ratios directly affecting PTEN as a key early event under NADH regulation potentially affecting a broad spectrum of regulatory activities

the activation of Akt. Because NADH competitively inhibits thioredoxin NADP(H)-dependent reactivation of PTEN, increased NADH in cancer cells caused by impaired mitochondrial function in respiration-deficient cells leads to inactivation of PTEN. The PTEN protein is primarily a phosphatidyl-inositol-3,4,5-triphosphate 3-phosphatase (Tamguney and Stokoe 2007). Unlike most of the protein tyrosine phosphatases, the PTEN protein preferentially dephosphorylates phosphoinositide substrates. It negatively regulates intracellular levels of phosphatidyl-3,4-5-triphosphate in cells and by negatively regulating Akt/PKB signaling pathways (Chu and Tarnawaski 2004) functions as a tumor suppressor. In the study by Pelicano et al. (2006), NADH caused a significant decrease of the phosphoprotein signal detectable under nonreducing conditions suggesting that PTEN was likely in an oxidized state, which masked the epitope for antibody binding by disulfide formation. One explanation would be that NADH might compete with NADTrx to affect the redox state of PTEN thereby indirectly enhancing via PDK1, the phosphorylation of Akt and of SGK in parallel to the similar response to elevated mTORC2 activity (Fig. 4.15).

4.10.4 AMP-Activated Protein Kinase

AMPK is a heterotrimeric complex of a catalytic subunit and two regulatory subunits represented by two or three isoforms each that provides an important link that integrates signaling with metabolism. AMPK is a sensor of the energy status within cells activated by lowered energy levels, which upon activation acts to restore energy balance. This is done in part by modulating NAD^+ levels and SIRT1 activity (Canto and Auwerx 2009; Canto et al. 2009).

AMPK is activated by NAD^+ and inhibited by NADH creating the possibility of a regulatory modulation by the cytosolic NAD^+/NADH ratio in addition to changes in the AMP/ATP ratio (Rafeloff-Phail et al. 2004). A network of metabolic pathways is stimulated by activated AMPK to restore energy balance by inhibiting ATP-consuming processes and stimulating ATP-generating pathways. Increased substrate oxidation, generation of reduced pyridine nucleotide cofactors, and production of ATP result. A high AMP/ATP ratio activates AMPK through phosphorylation on threonine 172 catalyzed by an upstream kinase. Activation of AMPK stimulates fatty acid oxidation, hexokinase activity, and uptake of glucose all of which repress hepatic gluconeogenesis and fatty acid synthesis enzymes (Kemp et al. 2003).

4.10.5 Hypoxia

Hypoxia is yet another opportunity of where PMET can maintain a high level of NAD^+ in the cytosol as a potential second messenger to activate vital processes such as those linked to the sirtuins (F. L. Crane, personal communication) (Fig. 4.13). Hypoxia has been defined as "oxygen availability not coping with aerobic ATP requirements" (Connett et al. 1990).

In tumors, hypoxia is often a prognostic indicator of poor patient survival in that hypoxic tumors emerge as being more aggressive (Denko et al. 2000). The rapid growth and poor vascularization of solid tumors exposed to hypoxia appears to promote the metastatic phenotype by reduced intercellular adhesion and increased cell motility and invasiveness (Zhang et al. 2006). Recruitment of CtBP (C-terminal binding protein), a NAD(H)-dependent transcriptional corepresser, to the E-cadherin promoter was increased in hypoxia as a result of increased NADH levels. Increased tumor cell migration was blocked by CtBP knockdowns suggesting that CtBP may reduce expression of cell adhesion genes under hypoxic conditions in response to levels of free NADH (Zhang et al. 2006).

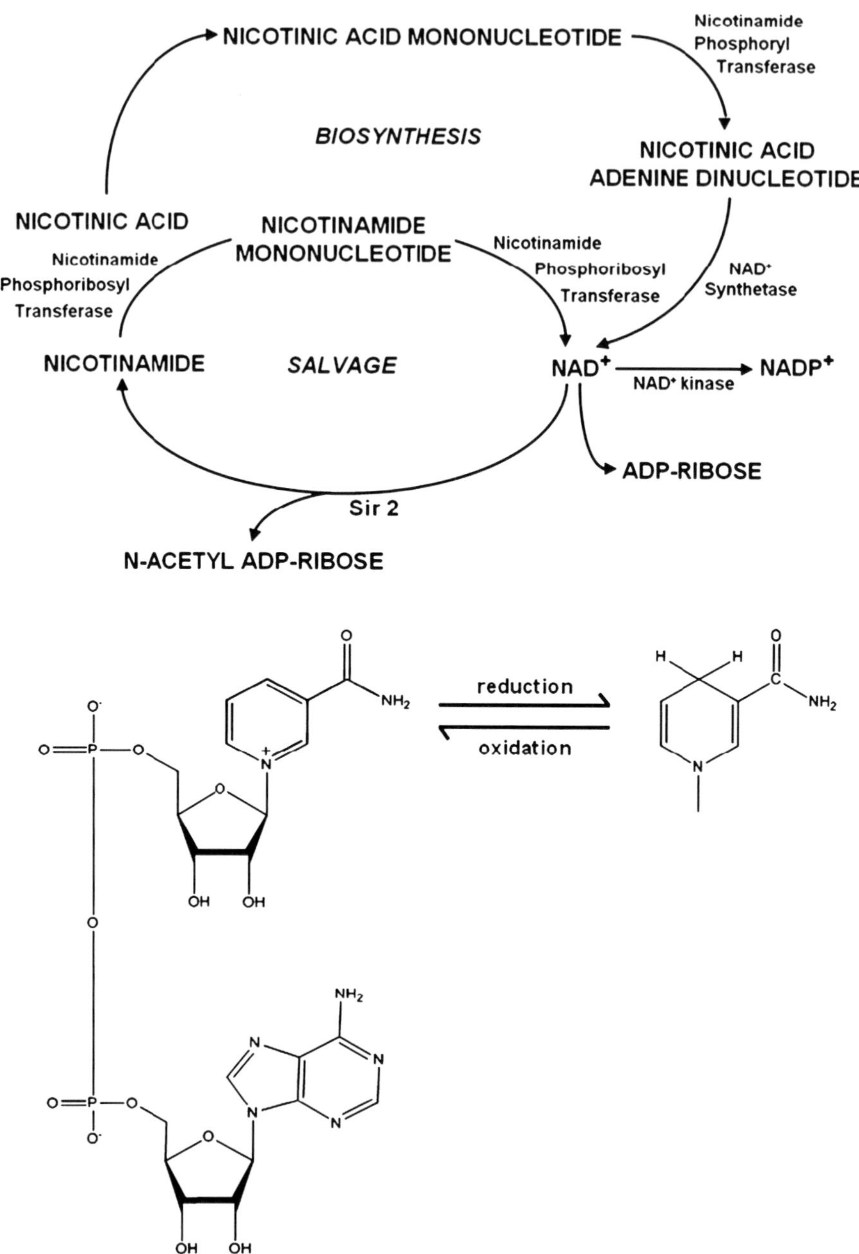

Fig. 4.16 NAD⁺ homeostasis

4.11 NAD⁺ Homeostasis

The principal effect of PMET is to alter the balance between oxidized and reduced forms of nicotinamide adenine dinucleotide. NAD^+ molecules are not consumed during oxidation reactions. Yet their half-life is relatively short per se (about 90 min in *Escherichia coli*) so that continuous renewal is required. The pyrimidine moiety of degraded NAD^+ can also be reutilized via a salvage pathway (Fig. 4.16).

In rat liver, the total amount of NAD^+ and NADH is approximately 1 μM per gram of wet weight, about ten times the concentration of $NADP^+$ and NADPH (Reiss et al. 1984). Estimates of the actual concentration of NAD^+ in cell cytosol is harder to measure and estimated to be around 0.3 mM (Yamada et al. 2006; Yang et al. 2007). Nicotinamide mononucleotide synthesis by Nampt is inhibited by both NAD^+ (K_i=0.14 μM) and NADH (K_i=0.22 μM), through an apparent regulatory feedback mechanism (Burgos and Schramm 2008).

4.12 Summary

All eukaryotic cells and organisms (bacteria, yeast, plants, and animals) possess a transplasma membrane electron transport chain. The chain is comprised of one or more NADH-coenzyme Q reductase enzymes located on the cytosolic side of the plasma membrane, coenzyme Q within the membrane, and ENOX proteins on the outer surface of the plasma membrane that serve as hydroquinone oxidases. In addition, cells are able to transfer electrons to artificial electron impermeant acceptors such as ferricyanide and diferric transferrin. Such transfers bypass the cell surface ENOX proteins. The transplasma membrane electron carriers for the latter remain to be identified although porin isoform 1 or VDAC of the plasma membrane and a 57-kDa doxorubicin-inhibited NADH quinone oxidoreductase have been proposed as candidates. Transplasma membrane electron transport is accompanied by proton pumping (acidification of the medium; alkalinization of the cytosol), of which, much is amiloride insensitive. A key feature of transmembrane proton pumping associated with PMET is that it generates a transmembrane potential as one means to conserve the large change in free energy from NADH oxidation.

Because ENOX proteins respond to external signals such as light that require communication with intracellular photoreceptors, ENOX receptor proteins located at the cell surface are indicated. NADH levels provide links of PMET to major signaling pathways including sirtuin-catalyzed reactions, regulation of clock function, modulation of the sphingosine rheostat, and the PDK-1 regulatory protein PTEN.

As many aggressive and invasive cancers rely on glycolysis for their energy requirements, the increased glycolytic rates involve increased PMET activity in order to ameliorate reductive stress caused by a build-up of intracellular NADH. Recycling NADH is a major function. The importance of PMET in cancer cell survival is demonstrated by its increased activity in cancer cells, as well as in glycolytic ρ° cancer cell lines that lack functional mitochondria. Inhibition of PMET in cancer cells is associated with cell death. Therefore, specific targeting of PMET with anticancer drugs remains a potentially useful but largely unexploited anticancer strategy.

Chapter 5
Role in the Enlargement Phase of Cell Growth

5.1 Cell Enlargement Linked to ENOX Activities

Cell enlargement is a necessary requisite for sustained growth of cells of all higher organisms (Taiz 1984; Baserga 1985, 2007). However, a prevailing view, especially for walled cells of plants established early (Lockhart 1965), was that cell enlargement was the result of a passive yielding of the cell walls in response to turgor pressure. For animal cells, enlargement is generally considered to follow cell division primarily as a result of biosynthesis of cytosolic and membrane proteins, lipids and nucleic acids (Baserga 1985; De Berardinis et al. 2008). However, the requirement for doubling of biomass which accompanies each passage through the cell cycle does not necessarily dictate that biomass doubling drives cell enlargement.

In this chapter, evidence is presented for a central role of ECTO-NOX (ENOX) proteins linked to the enlargement phase of cell growth through protein disulfide–thiol interchange (Fig. 5.1). This potential major physiological function of ENOX1 and ENOX2 proteins would parallel that of their role as terminal oxidases of plasma membrane electron transport (Chap. 4). A role of plasma membrane electron transport in growth was implicit in the discovery of the cell surface NADH oxidase proteins and has culminated in observations that the response of the different NADH oxidase activity forms to respond to hormones, growth factors, thiol reagents and antitumor or potential antitumor drug parallels similar effects on plasma membrane electron transport specifically and on growth in general (Crane et al. 1990b, 1991). However, the significance of the correlation was not fully appreciated until the discovery that ENOX proteins exhibited a second enzymatic activity, that of protein disulfide–thiol interchange (Morré 1998a; Chap. 2). This early information has been reviewed (Morré 1994a; Döring and Lüthje 1996; Lüthje et al. 1997; Medina 1998).

D.J. Morré and D.M. Morré, *ECTO-NOX Proteins: Growth, Cancer, and Aging*,
DOI 10.1007/978-1-4614-3958-5_5, © Springer Science+Business Media New York 2013

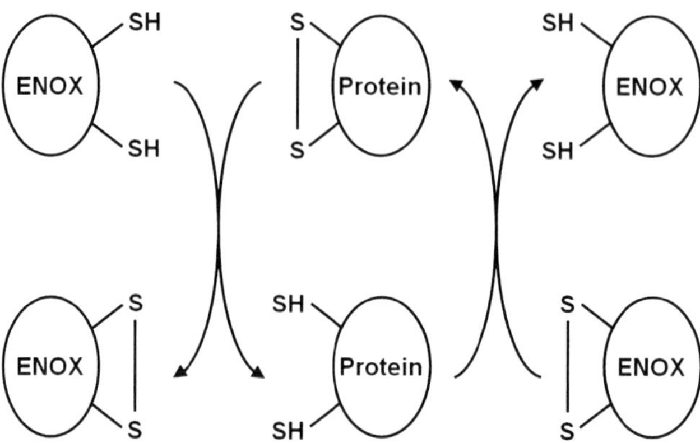

Fig. 5.1 Protein disulfide–thiol interchange portion of the ENOX cycle. Proton and electrons from thiol group of one protein are utilized to reduce a disulfide group of another protein

5.2 ECTO NADH Oxidase of Liver Plasma Membranes Stimulated by Hormones and Growth Factors

ENOX1 proteins of plasma membranes isolated from rat livers respond to a number of growth factors and hormones for which rat liver plasma membranes have receptors. These included epidermal growth factor (EGF), diferric transferrin, insulin and lactoferrin (Brightman et al. 1992; Bruno et al. 1992; Fig. 5.2). These findings have been extended to cells and membranes of other animal and plant sources with focus primarily on EGF and differic transferrin and plant growth hormones of the auxin type (Chap. 10). With differic transferrin the response is direct, whereas with EGF the response is indirect through coupling with EGF receptors. Stimulations of between 30 and 100 % were observed for each of the different growth factors and hormones. Previously, Gayda et al. (1977) and Morré et al. (1988a, b, c) reported a stimulatory effect of the thyroid hormone, T_3, on NADH oxidase activity of rat liver plasma membranes.

5.3 ECTO NADH Oxidase of Rat Hepatoma Plasma Membrane Constitutively Activated and No Longer Growth Factor or Hormone-Responsive

The first indications of altered ENOX activities in cancer (Chap. 8) came from studies of plasma membrane vesicles isolated from liver nodules induced by the hepatocarcinogen 2-acetylaminofluorene fed to rats (Morré et al. 1991a). In contrast to plasma membranes from liver, the plasma membranes of the preneoplastic liver nodules contained an ENOX oxidase activity that was no longer responsive to diferric transferrin despite an increased number of transferrin receptors. When extended to

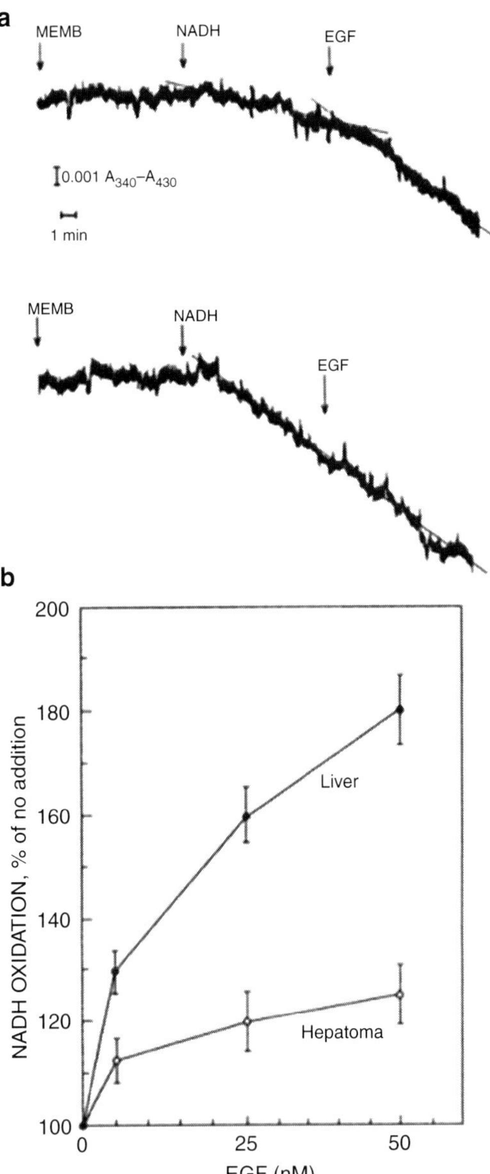

Fig. 5.2 (**a**) Spectrophotometric traces of NADH oxidation by plasma membranes. The reaction was initiated by the addition of NADH. After 5 min, epidermal growth factor (EGF) was added. Absorbance was measured at 340 nm with a reference at 430 nm. *Top*. Rat liver plasma membranes. Upon EGF addition, an accelerated rate of NADH oxidation was observed within the time of temperature equilibration and mixing of the system (less than 1 min). *Bottom*. Hepatoma (RLT-N) plasma membranes. No response to added EGF was discernible. Each reaction contained 0.1 mg of membrane protein. (**b**) Oxidation of NADH by rat liver (*filled circles*) and rat hepatoma RLT-N (*open circles*) plasma membranes in response to increasing concentrations of EGF. With hepatoma plasma membranes, the initial specific activity was about 2.5 nmol/min/mg of protein, 2.5 that of the rat liver plasma membrane. Modified from Bruno et al. (1992)

rat hepatomas, similar results were found. The specific NADH oxidase activity of the hepatoma plasma membrane in the absence of growth factors or hormones was at a level nearly that of the rat liver plasma membrane when fully hormone- or growth factor-stimulated (Bruno et al. 1992). Furthermore, the addition of growth factors and hormones to the hepatoma plasma membranes resulted in no further stimulations of activity. These findings were interpreted to indicate that the hepatoma plasma membrane contained an ENOX activity that was constitutively activated and growth factor unresponsive to parallel the unregulated growth that is characteristic of cancer cells in general. The unregulated and constitutively activated ENOX is the ENOX2 form (Chap. 8) that is distinct from the constitutive and regulated ENOX1.

5.3.1 Thiol Reagents

In addition to the response to anticancer drugs, an important criterion to distinguish ENOX1 and ENOX2 of mammalian cells is their response to thiol reagents (Morré and Morré 1995a; Fig. 5.3). A similar response was seen with plants where the auxin-induced ENOX activity was more susceptible to inhibition by thiol reagents than was the constitutive activity (Morré et al. 1995a). This activity now largely explains the exquisite sensitivity of cell enlargement to inhibition by thiol reagents.

Reagents capable of covalent reaction with protein thiols such as *N*-ethylmaleimide (NEM), *p*-chloromercuribenzoate (PCMP), and DTNB inhibited nearly quantitatively cell enlargement of elongating segments of soybean induced by auxin. Constitutive growth was unaffected or much less affected by these reagents (Chap. 10). Similarly, the increment of NADH oxidase activity stimulated by auxin growth regulators was inhibited by thiol reagents in parallel with cell enlargement whereas the basal NADH oxidase levels in the absence of auxin were resistant to thiol reagents (Morré et al. 1995a). A similar situation was encountered for the NADH oxidase activity of liver plasma membranes (Morré and Morré 1995a) which was relatively resistant to thiol reagents at concentrations where the NADH oxidase activities of rat hepatoma and HeLa (human cervical carcinoma) were inhibited.

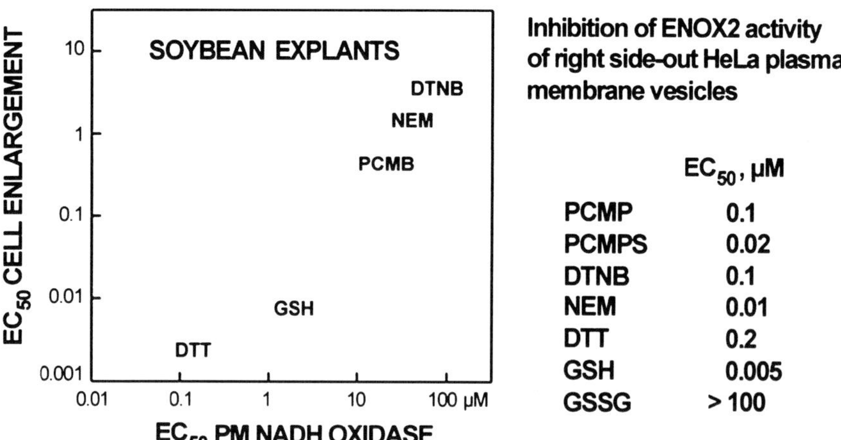

	EC_{50}, μM
PCMP	0.1
PCMPS	0.02
DTNB	0.1
NEM	0.01
DTT	0.2
GSH	0.005
GSSG	> 100

Fig. 5.3 Growth and ENOX activity blocked by thiol reagents

The basis for the differential sensitivity to thiol reagents has not been studied. The soluble serum forms of the activity also tend to be more responsive to thiol reagents with sera of cancer patients than that of sera of normal volunteers (Morré et al. 1997a). There may be some relationship to the number of thiol groups present as the expressed cancer-specific form of the protein (ENOX2) has eight cysteines (Chueh et al. 2002b) whereas the mammalian constitutive ENOX1 has only two (Jiang et al. 2008).

5.4 Relationship to Growth

A relationship of ENOX proteins to growth is indicated, as well, from inhibitor studies. Thus far, without exception, compounds that stimulate the NADH oxidase in plants stimulate cell enlargement in plants (Fig. 5.4a) and compounds that inhibit the NADH oxidase of plants also inhibit cell enlargement of plants (Fig. 5.4b; 2,4-D, Chap. 10). Mammalian cell and animal data are mostly restricted to inhibitors but the correlation is very much the same (Fig. 5.5).

5.4.1 Plants

Cell enlargement is especially relevant for plant cells where increases in size of the plant or of specific plant parts is sometimes due almost entirely to cell elongation rather than the result of an increase in the number of cells (Fig. 5.6). Using excised plant stems floated on solution, the rate of elongation can often be accelerated several-fold by the addition of cell elongation promoters known collectively as auxins (Morré and Key 1967). This auxin-induced growth is primarily the result of cell elongation in which displacement of the cell surface membranes normally is a significant end result.

Figure 5.4 is a correlation between stimulation of NADH oxidase activity of isolated plasma membrane vesicles vs. inhibition of cell enlargement of 1 cm long hypocotyl sections of soybeans measured after 18 h in darkness. Among the stimulatory compounds in addition to auxins are phenylacetate, triacontanol (a long-chain water-insoluble alcohol, Morré et al. 1991c), low concentrations of organic solvents and alcohols (Morré 1998a) and ascorbate free radical (onion roots) (Hidalgo et al. 1991).

In contrast, treatment of isolated plasma membrane vesicles with various osmotica was inhibitory to ENOX activity (Morré and Brightman 1997). These treatments preferentially inhibited auxin-stimulated ENOX and cell enlargement in parallel and required intact vesicles with semipermeable membranes. The findings were interpreted to indicate that maximum rates of NADH oxidase activity and of cell enlargement resulted only with plasma membranes at full turgor. A nearly complete cessation of auxin-induced cell enlargement and NADH oxidase activity could be detected at incipient plasmolysis (zero turgor). A variety of osmotica were examined and all showed similar NADH oxidase activity–plant cell enlargement relationships.

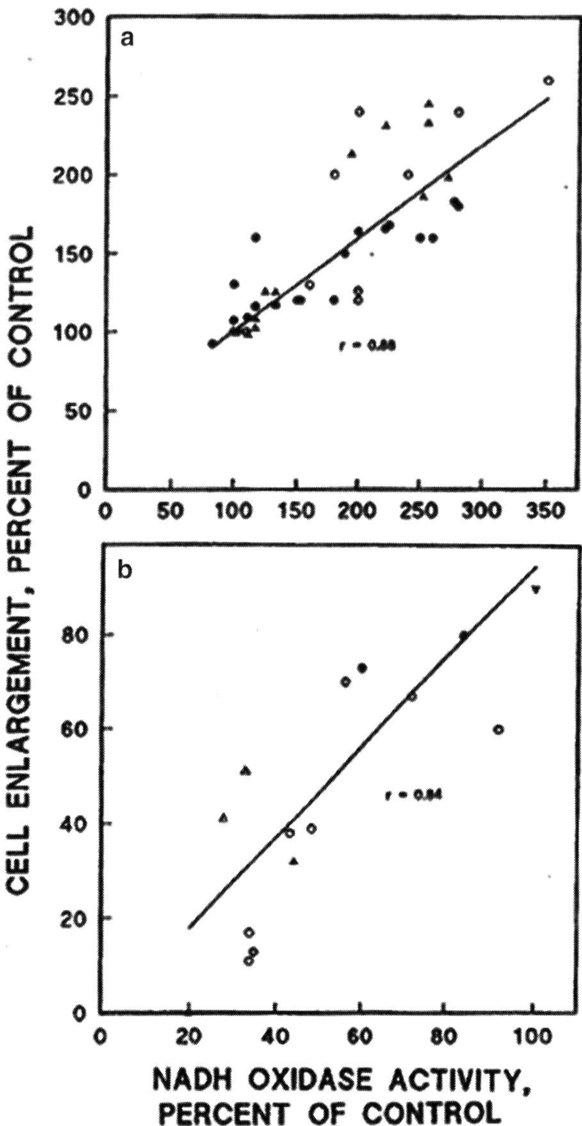

Fig. 5.4 Correlation of cell enlargement of 1 cm segments of cut hypocotyls of dark-grown seedlings of soybean and NADH oxidase activity of isolated plasma membrane vesicles. (**a**) Stimulation. (*Filled circles*) Auxins (Morré et al. 1988c; Morré and Brightman 1991). (*Open circles*) Lysolipids, fatty acids and fatty acid esters (Brightman et al. 1991). (*Filled triangles*) The auxin herbicide, 2,4-D. (**b**) Inhibition. (*Filled circles*) 2,4-D (Morré et al. 1988c; Morré and Brightman 1991). (*Open circles*) Osmotica with and without 2,4-D (Morré and Brightman 1997). (*Filled triangles*) Chlorsulfuron (Morré et al. 1995e). (*Inverted filled triangles*) Reduced glutathione. (*Open triangle*) Glaucarubolone. Reproduced from Morré (1998c) with permission from Springer Science+Business Media

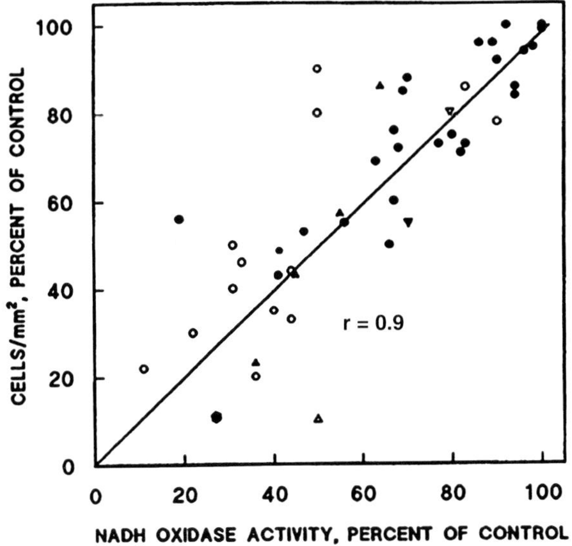

Fig. 5.5 Correlation between NADH oxidase activity of isolated plasma membrane vesicles and cells/mm^2 as percent of control comparing different drugs and drug-responsive mammalian cell lines. (*Filled circles*) HeLa cells treated with retinoids (Dai et al. 1997). (*Open circles*) HeLa cells treated with the antitumor sulfonylurea LY181984 (Morré et al. 1997f). (*Filled triangles*) BT-20 human adenocarcinoma cells treated with capsaicin (Morré et al. 1995b). (*Open triangles*) HL-60 cells treated with capsaicin (Morré et al. 1995b). (*Inverted open triangles*) K562 and NRK cells treated with LY181984 (Morré et al. 1995f). (*Solid hexagonal*) HeLa cells treated with doxorubicin (Morré et al. 1997c). Reproduced from Morré (1998c) with permission from Springer Science + Business Media

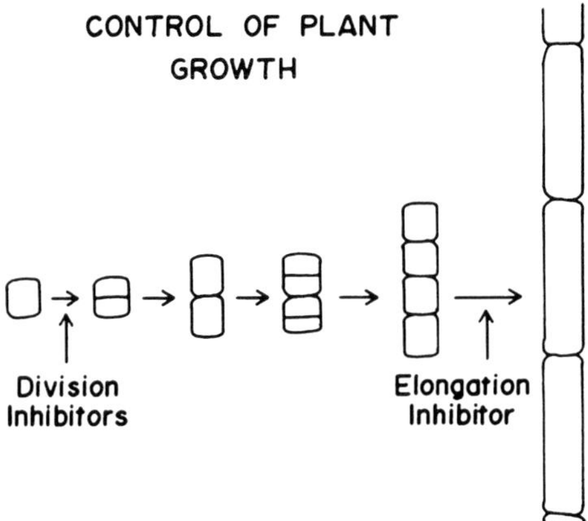

Fig. 5.6 Diagrammatic representation of plant growth as the summation of two cellular events—division and elongation. Stem growth in plants consists principally of cell enlargement. *Arrows* indicate sites of action of inhibitors which function as growth retardants

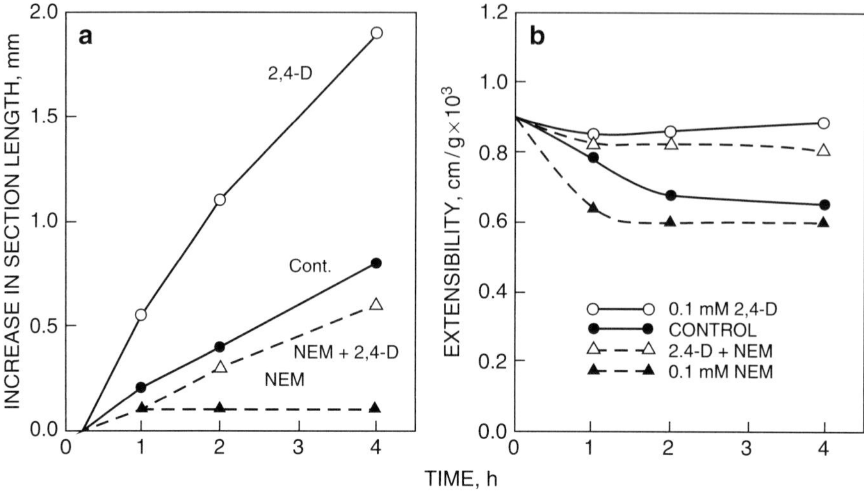

Fig. 5.7 Time course of cell elongation (**a**) and extensibility (**b**) of 1 cm etiolated pea third internode sections in the presence and absence of 0.1 mM 2,4-D with and without 0.1 mM *N*-ethylmaleimide (NEM). Redrawn from Eisinger and Morré (1968)

5.4.1.1 Turgor Pressure Is Not the Driving Force of Cell Enlargement in Plants

Despite the findings summarized in Sect. 5.4.1 above, turgor pressure is not the dominant driving force of cell enlargement. In early experiments, plant stem sections were treated with the auxin plant growth regulators such as 2,4-dichlorophenoxyacetic acid (2,4-D) or indole-3-acetic acid (IAA) to accelerate growth and to loosen cell walls. If the sections were then treated with NEM or other thiol reagents, wall extensibility and turgor remained high, at the control levels, whereas cell enlargement had ceased (Fig. 5.7).

Solute leakage measured by conductivity changes in the presence of NEM accounted for only a 5–6 % reduction in internal osmotic pressure (Eisinger and Morré 1968). Moreover, when sections were first incubated in NEM and then transferred to the sulfhydryl protectant dithiothreitol (DTT), growth was restored with no effect on the rate of solute leakage. These experiments demonstrate that plant cell enlargement, at least, is not the result of turgor-driven expansion of auxin-loosened cell walls but occurs by a sulfhydryl reagent-blocked, active mechanism.

In parallel studies using a variety of methods, no obligatory requirement for delivery of new membrane materials to the plasma membrane could be demonstrated for cell enlargement to occur (Morré 1994b; Morré and Mollenhauer 2009). For example, at temperatures of 18 °C or less, elongating segments of soybean showed extensive accumulations of membranes at the *trans* Golgi network. These accumulations were reminiscent of temperature blocks seen in other plant and

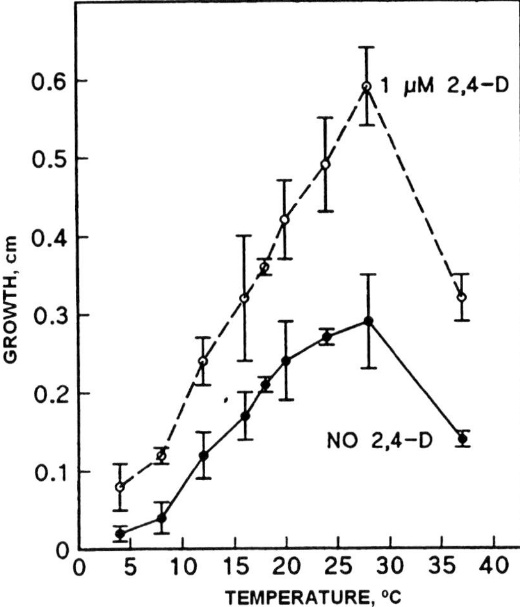

Fig. 5.8 Auxin-induced elongation of 1 cm segments of soybean hypocotyls exposed at different temperatures. The response to temperature shows no abrupt transition as would be expected for a typical low temperature block. Reproduced from Morré (1994b) with permission from Springer-Wien

animal cells (Tartakoff 1986; Morré and Mollenhauer 2009). Despite the low temperature blockage of delivery of new membrane to the cell surface, auxin-induced growth showed no sharp transition in response to temperature over the range of 4–25 °C (Fig. 5.8; Morré 1994b). This would argue that elongation growth in plants induced by auxin occurred in a manner independent of the vesicular transport pathway of delivery of new membrane to the plasma membrane.

Similar conclusions were reached earlier based on results with narrow setae of the moss *Pellia* treated with the sodium ionophore monensin (Morré et al. 1986c). The setae elongate rapidly and their enlargement is stimulated by auxin. Yet the setae are thin enough that monensin is able to penetrate all of the cells. The ability of monensin to penetrate the setae was shown by electron microscope observations of swollen *trans* Golgi apparatus elements following fixation with glutaraldehyde (Morré et al. 1986c). Despite the nearly complete inhibition by monensin of normal *trans* Golgi apparatus functioning, treated setae responded in a relatively normal fashion to auxin for times of 4 h or more after the onset of monensin inhibition. Beyond 4 h, auxin-induced elongation ceased abruptly, due to rupture of the excessively thinned plasma membrane which resulted from sustained cell elongation in the absence of delivery of essential plasma membrane precursors required for sustained cell expansion.

5.4.1.2 ECTO-NOX Proteins as Drivers of Cell Enlargement

To establish ENOX proteins as drivers of the cell enlargement, a number of correlative studies were carried out. Among those, the comparison of oscillatory patterns of ENOX activity and growth are among the most compelling (Pogue et al. 2000; Morré et al. 2001a, 2002d; Wang et al. 2001).

5.4.1.3 Cell Enlargement of Plant Tissue Explants Oscillates with a Temperature-Independent Period Length of ca. 24 min

If ENOX activity is correlated with cell enlargement, then the rate of cell enlargement also should be periodic with a major period of about 24 min. This expectation is borne out with both elongation of hypocotyl sections of dark-grown soybeans (Figs. 5.9 and 5.10a–c; Morré et al. 2001a, 2002d) and coleoptile sections of dark-grown maize (Claussen et al. 1997). Steady-state cell elongation in these tissues is strikingly periodic in parallel with cell surface NADH oxidation. The major period is about 24 min. There is also a report (Millet and Badot 1996) where enlargement of single epidermal cells of garden bean was followed at closely spaced intervals to

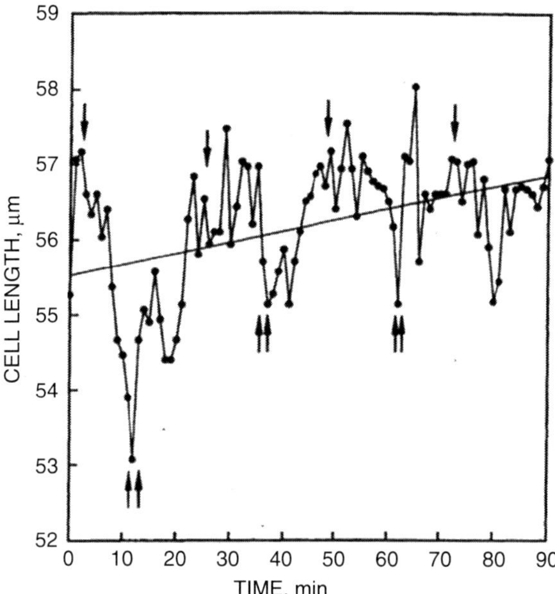

Fig. 5.9 Periodicity of enlargement of epidermal cells of soybean as a function of time over 90 min at 24 °C as determined by video-enhanced microscopy. *Single arrows* indicate maximum cell lengths within each 24-min period based on the decomposition fit. *Double arrows* indicate minima. Reproduced from Morré et al. (2001a) with permission from Springer Science+Business Media

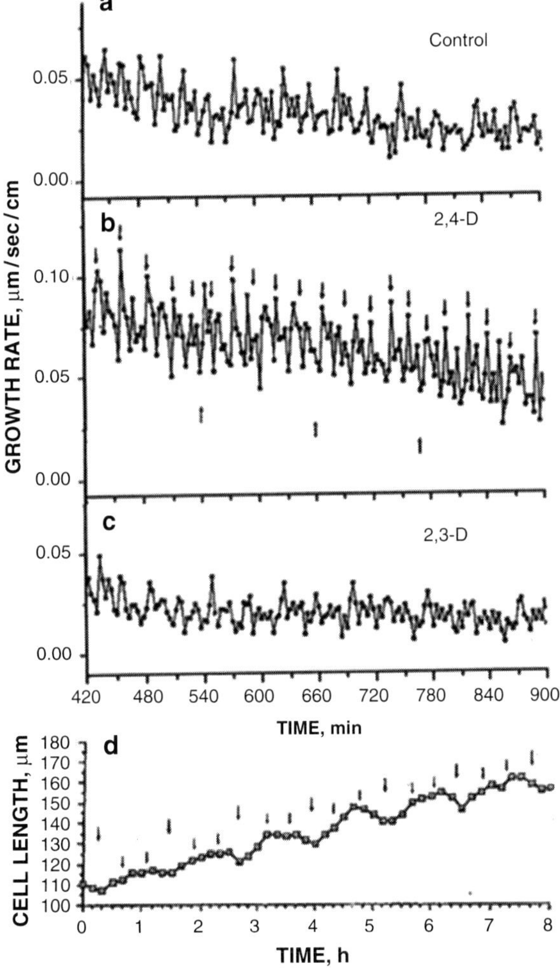

Fig. 5.10 Periodicity of cell enlargement. Elongation growth of 0.5 cm sections of dark-grown seedlings of soybean measured at 2 min intervals as described by Lüthen and Böttger (1992). Each result is for four 0.5 cm sections. Steady-state elongation rates between 420 and 900 min are shown. Every five periods (120 min) there was a discontinuity in the pattern in which the relative positions of the major and minor peaks become inverted for one 24 min period (**b**, *upward pointing arrows*). Triplicate experiments gave similar results. (**a**) No addition; (**b**) 1 µM 2,4-D; (**c**) 1 µM 2,3-D; (**d**) Time course of length variations for an epidermal cell of garden bean with an initial length of 100 µM. The major period is 72 min (*large upper arrows*) with evidence of minor periods of 24 min (*small lower arrows*) (from Millet and Badot 1996). Reproduced from Morré (1998c) with permission from Springer Science + Business Media

reveal a ca. 72 min major period (3 × 24 min). A 24 min minor period can be seen as well (Fig. 5.10d). Cell enlargement in HeLa cells monitored by light microscopy also is periodic (Wang et al. 2001). The frequencies correspond to those of the NADH oxidases present at the HeLa cell plasma membrane (Wang et al. 2001).

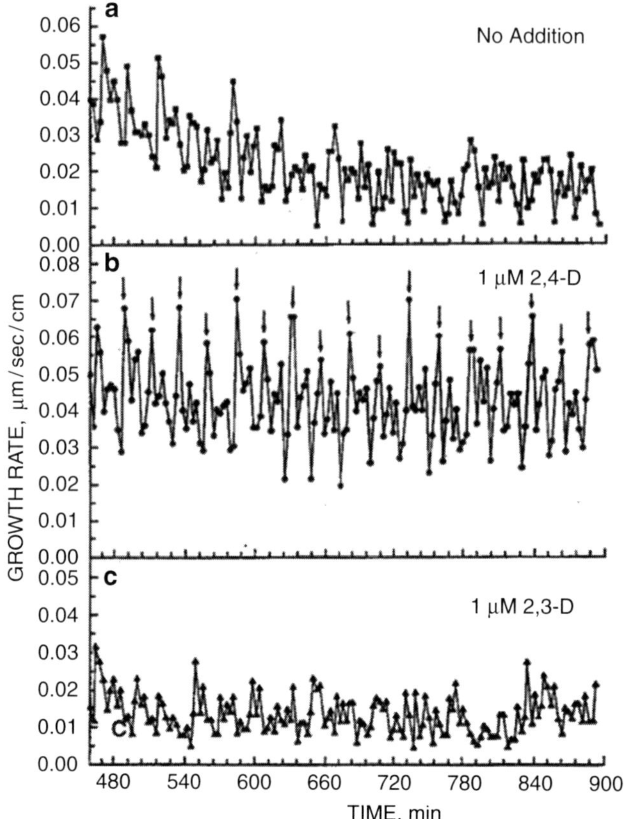

Fig. 5.11 Regular oscillations in the rate of elongation growth of 0.5 cm sections from hypocotyls of dark-grown seedlings of soybean measured at 3 min intervals as for Fig. 5.10. Steady-state elongation rates between 420 and 900 min were from a single experiment with rates representing four 0.5 cm sections. (**a**) No addition; (**b**) 1 µM 2,4-D; (**c**) 1 µM 2,3-D. Each experimental condition with four 0.5 cm sections per determination was repeated at least four times with similar results. In (**b**), every other maximum was accentuated, generating a 48 min super period. Reproduced from Morré et al. (2002d) with permission from Springer Science + Business Media

Also oscillating with a defined period were elongation rates enhanced by the auxin herbicide 2,4-D (Fig. 5.11b) or the natural auxin IAA (Fig. 5.12a) and diminished by the inactive auxin herbicide analog 2,3-D (Fig. 5.11c). While the length of the sections increased at an average steady-state rate of 0.03–0.05 µm/s/cm and was increased to 0.05–0.08 µm/s/cm in the presence of 1 µM 2,4-D, the rate of enlargement fluctuated by more than a factor of 2 around these means. The fluctuations were not random but exhibited a regular pattern of oscillations with a period of about 24 min as shown by a portion of the data of Fig. 5.11b expanded in Fig. 5.13a to show detail. Fourier analysis (Fig. 5.13b) gave major periods of 24 (frequency 0.042) and about 12 (frequency 0.08) min. The 24 min period length is seen in the

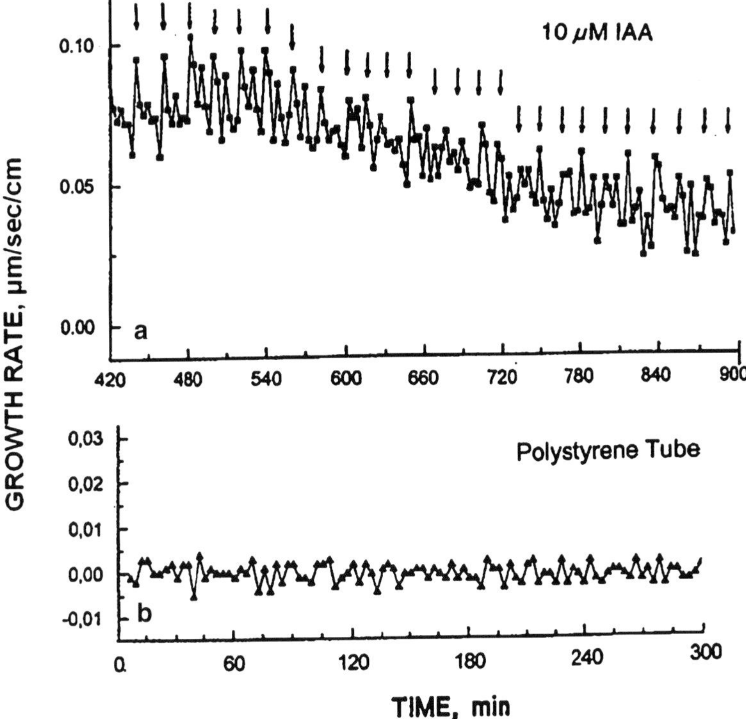

Fig. 5.12 In (**a**) are shown regular oscillations in the rate of elongation growth of 0.5 cm sections from hypocotyls of dark-grown seedlings of soybean measured at 3 min intervals as in Fig. 5.10 but in the presence of 10 μM indole-3-acetic acid (IAA). In (**b**), the tissue sections were replaced by polystyrene tubing of approximately the same diameter as the tissue sections. Reproduced from Morré et al. (2002d) with permission from Springer Science+Business Media

growth rate maxima shown at the arrows in Fig. 5.11b whereas the 12 min period defined the alternation of major and minor deviations also seen in the decomposition fits of Fig. 5.13c. The oscillations observed were not inherent in the instrumentation. When tissue was replaced by polystyrene tubing, a level of background variation was observed but without periodicity (Fig. 5.12b).

The 2,4-D-stimulated cell elongation actually is derived from the induction of a new auxin-induced set of oscillations not evident prior to 2,4-D addition. The new increased rate of elongation and oscillatory activity is observed within a few minute after 2,4-D addition. A complex new pattern of major and minor oscillations observed following 2,4-D addition is illustrated in Figs. 5.11b and 5.14. Steady-state is achieved eventually including restoration of the original 24 min period but at the 2,4-D-stimulated growth rate (see Chap. 10 for explanation). A striking feature of such growth oscillations with 2,4-D was that every other maximum was accentuated yielding a 48 min super period (Fig. 5.11b).

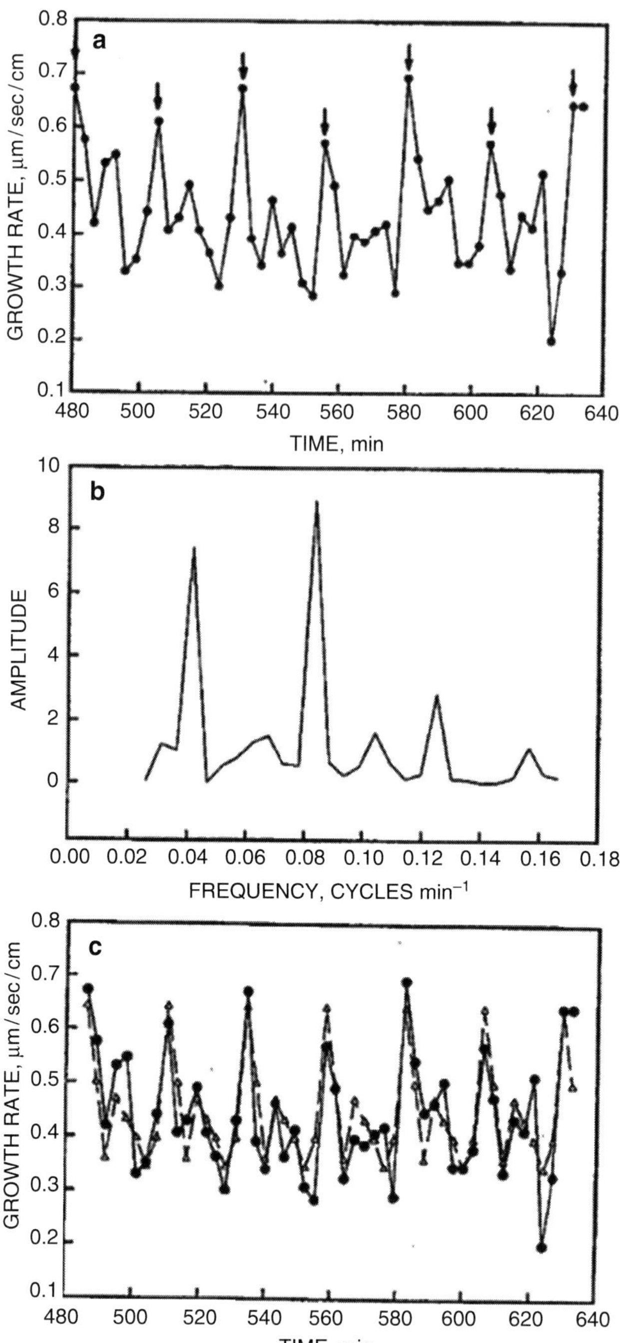

Fig. 5.13 Data of Fig. 5.11b expanded to illustrate detail. (**a**) Cell enlargement; (**b**) Fourier analysis verifies a major period (*arrows* in **a**) of 24 min; (**c**) time series (decomposition) analyses (*dashed lines*) verify a pattern of alternating major and minor maxima to account for the ca. 12 min period (frequency of 0.08) also observed in the Fourier analysis. Reproduced from Morré et al. (2002d) with permission from Springer Science + Business Media

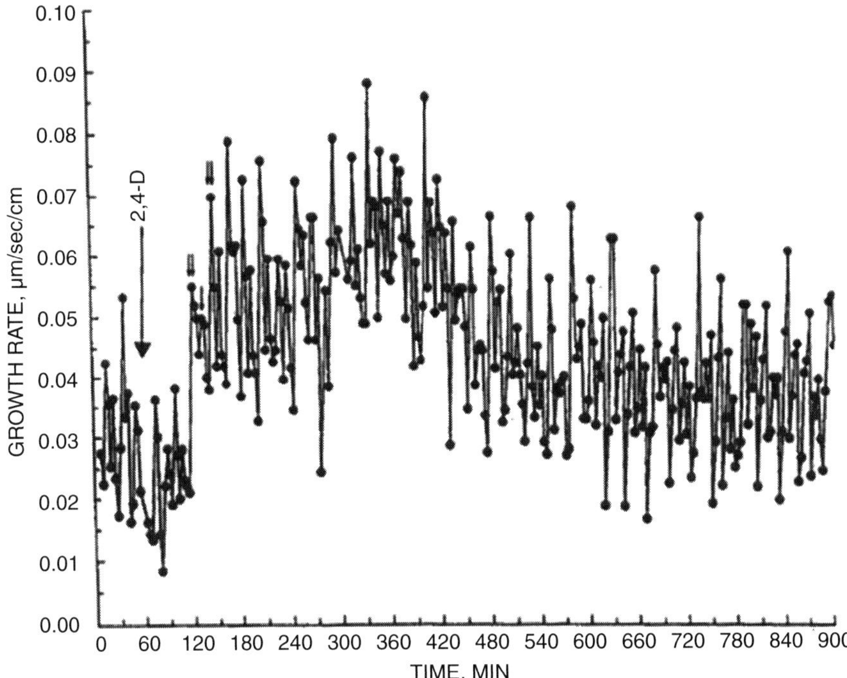

Fig. 5.14 As in Fig. 5.11 except that 1 μM 2,4-D was added after 60 min (*a large arrow*) and the elongation rate was determined for an additional 840 min. Immediately following the 2,4-D addition a new period (*small double arrows*) was observed alternating with the period observed in the absence of 2,4-D (*a small single arrow*). After about five periods, a single major period was again established, corresponding to the new period of 24 min. However, enhanced maxima every other 24 min period (48 min super period) persisted as in Fig. 5.11b. Reproduced from Morré et al. (2002d) with permission from Springer Science + Business Media

5.4.1.4 The Period Length of Cell Enlargement Is Independent of Temperature

Rate of enlargement of soybean sections increased with increasing temperature between 15 and 35 °C (Table 5.1). Below 8–10 °C, significant elongation was not observed. Above 35 °C, growth rates were not increased further in response to temperature. The average increase in growth rate for a 10 °C rise in temperature between 15 and 35 °C was by a factor of 2.15. However, over this same range of temperatures, the period length of these oscillations in the steady-state rate of elongation remained constant at 24 min (Fig. 5.15; Table 5.2). Additionally, the length of the major period was similar both in the presence and absence of 1 μM 2,4-D (Table 5.2).

Table 5.1 Elongation rate of 0.5 cm sections of soybean hypocotyls as a function of temperature

	Elongation rate (μm/s/cm)	
Temperature (°C)	No 2,4-D	1 μM 2,4-D
15	0.01 ± 0.002	0.02 ± 0.004
25	0.032 ± 0.003	0.06 ± 0.007
35	0.048 ± 0.004	0.1 ± 0.008

Results are averages of three experiments with duplicate determinations per experiment ± standard deviations among experiments. From Morré et al. (2002d)

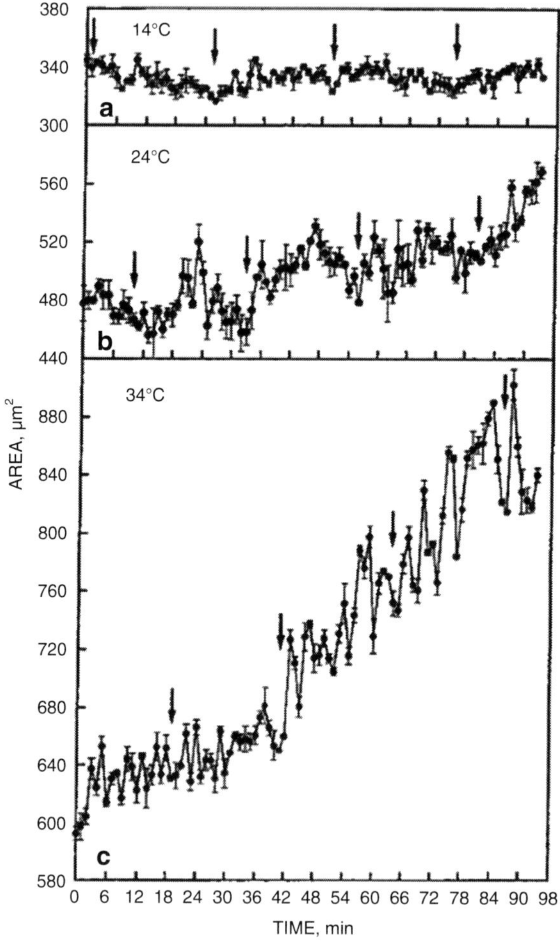

Fig. 5.15 Growth of CHO cells at 14 °C (**a**), 24 °C (**b**) and 34 °C (**c**) determined by video-enhanced microscopy. Cell area was measured every minute over 90 min. *Arrows* spaced 24 min apart coincide with the positions of statistically significant minima determined from detrended decomposition fits. Reproduced from Pogue et al. (2000) with permission from Elsevier

Table 5.2 Period length of elongation of 0.5 cm sections of soybean hypocotyls as a function of temperature determined from Fourier and time series (decomposition) analyses

	Period length (min)	
Temperature (°C)	No 2,4-D	+1 μM 2,4-D
15	24.1 ± 0.8	23.7 ± 1.5
25	24.3 ± 0.2	24.4 ± 1.0
35	24.3 ± 0.2	24.9 ± 0.5
Mean	24.2 ± 0.1	24.3 ± 0.6

Results are averages from three experiments ± standard deviations among experiments. From Morré et al. (2002d)

Values for plus 2,4-D are based on steady-state rates after about 300 min following 2,4-D addition

5.4.1.5 Cell Enlargement Is Restricted to the Protein Disulfide–Thiol Interchange Portion of the ENOX Cycle

In studies where cell enlargement and rates of NADH oxidation of epidermal strips of soybean were assayed simultaneously, in parallel, the two activities were verified to alternate (Fig. 5.16) as expected if the cell enlargement was restricted to the protein disulfide–thiol interchange portion of the ENOX cycle. Similar observations were made with CHO cells in culture where both cell enlargement and protein disulfide–thiol interchange activities exhibited the characteristic pattern of these activity maxima separated by intervals of 4.5 min whereas during the oxidative portion of the cycle (two maxima were separated by 6 min) both protein disulfide–thiol interchange and cell enlargement were reduced or not evident.

5.4.2 *Vertebrate Cells*

Growth of vertebrate cells has been measured as increase in cell number rather than cell enlargement as in plants. However, there may be a relationship between the two. Proliferating mammalian cells in culture tend to double their size and their mass before each division (Mitchison 1971). A dividing cell, if spherical, in order to double in volume, must increase the surface covered by the plasma membrane by a factor of 1.6 (Graham et al. 1973). Parallel inhibitions of both NADH oxidase activity of isolated plasma membrane vesicles and of growth have been noted for both the antitumor agents of Table 5.3 and the differentiation-inducing agents of Table 5.4. With the antitumor LY181984, inhibitor of growth, NADH oxidation, and protein disulfide–thiol interchange were closely correlated (Fig. 5.17). Growth was slowed after about 72 h when the drug became cytostatic. The resultant cells were smaller, approximately 50 % of the volume of untreated cells (Morré and Morré 1995b). Also cytostatic to cancer cells (HeLa, mammary adenocarcinoma) were ENOX2-specific monoclonal antibodies. Again, after about 72 h, the results were cytostatic populations of smaller cells unable to divide.

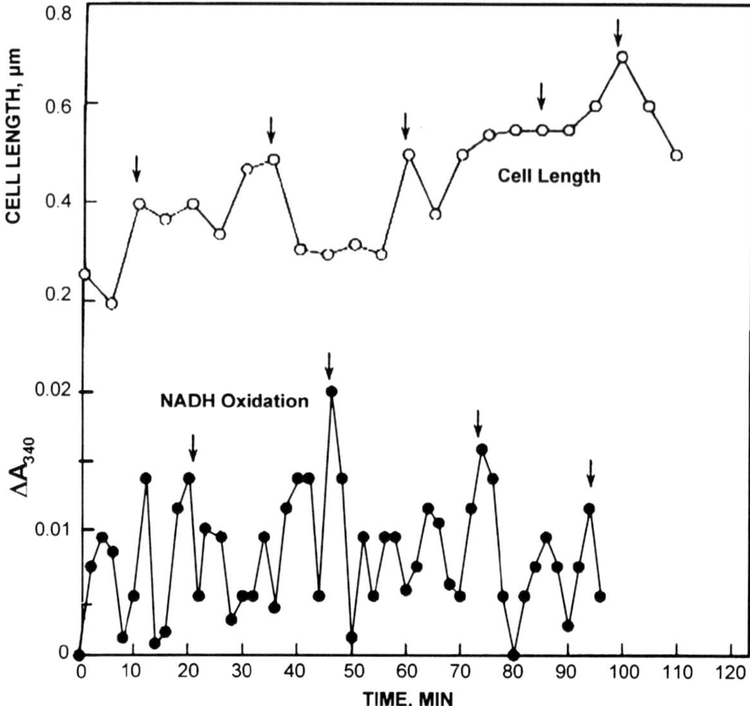

Fig. 5.16 Simultaneous measurements of cell enlargement and NADH oxidation by synchronized epidermal strips of soybean with maxima in both indicated at the arrows. Most rapid enlargement occurred during minima in NADH oxidation during the ENOX cycle when protein disulfide–thiol interchange occurs

One interpretation of these observations is that the constitutive activation of ENOX2 in cancer cells leads to unregulated cell enlargement. Vertebrate cells, as with yeast cells (Baserga 1985), once having divided must reach some minimum size before they are able to divide again. If by inhibiting ENOX2 in cancer cells, unregulated cell enlargement is also blocked, perhaps this is sufficient to restrict unregulated cell cycle entry characteristic of uncontrolled growth of cancer cells (Fig. 5.18).

5.4.2.1 Role of ENOX Proteins in Vertebrate Cell Enlargement

Increase in cell area, like ENOX activity, fluctuated with alternating rapid and slow rates for vertebrate cells in much the same manner as for plants. Data of Fig. 5.19a for CHO cells which lack ENOX2 show a single 24 min period of alternating maxima in rates of cell enlargement followed by relaxation. As observed previously, cell enlargement of CHO cells is not a simple sine function but each maximum, based on decomposition fits, is seen to consist of three separate oscillations, on average, labeled 1, 2 and 3 (Fig. 5.19a) separated by about 4.5 min. These oscillations

Table 5.3 Agents with antitumor activity inhibitory to NADH oxidase activity of isolated plasma membrane vesicles and growth (cell number after 72 h) of cancer cell lines and of the auxin-induced component of NADH oxidase activity of isolated soybean plasma membrane vesicles and of cell enlargement over 18 h of excised 1 cm hypocotyl segments from etiolated seedlings floated on solutions containing 1 μM 2,4-D with or without agent

	EC_{50}			
Antitumor drug	Growth	Plasma membrane NOX	Tissue or cell type	References
LY181984[a]	30 nM[b]	30 nM	HeLa	Morré et al. (1994f, 1997f)
Chlorsulfuron	1 μM[c]	0.1 μM[c]	Soybean	Morré et al. (1995e)
Capsaicin	10 nM[b]	5 M	HeLa	Morré et al. (1995b)
	1 μM	1 μM	BT-20 mammary adenocarcinoma	Morré et al. (1995b)
	No effect	No effect	MCF-10A mammary epithelia	Morré et al. (1995b)
	10 μM[d]	10 μM[d]	HL-60	Morré et al. (1995b)
	50 μM[c]	50 μM[c]	Soybean	Results unpublished
Doxorubicin	n.d.	0.7 nM	Rat hepatoma	Morré et al. (1997c)
	n.d.	0.1 μM	HL-6	Morré et al. (1997c)
	100 μM[c]	0.1 nM	Soybean	Morré et al. (1988c)
cis-Platinum[a]	10 μM[c]	0.1 μM[c]	Soybean	Morré et al. (1988b) Results unpublished
Acetogenins				
Annonacin A	n.d.	1 μM	HeLa	Morré et al. (1995c)
Asimicin	n.d.	5 nM	HeLa	Morré et al. (1995c)
Bullatacin	n.d.	5–10 nM	HeLa	Morré et al. (1995c)
	n.d.	0.1 μM	HL-60	Morré et al. (1995c)
Bullatacinome	n.d.	0.1–1 μM	HeLa	Morré et al. (1995c)

From Morré (1998c)
n.d. Not determined
[a]The inactive antitumor sulfonylurea LY181985, and *trans*-platinum were ineffective or required several orders of magnitude more substance to inhibit
[b]Plus 10 μM EGF
[c]2,4-D-induced component of growth and NADH oxidase
[d]Constitutively activated

correspond to the three oscillations in the 2+3 pattern of ENOX activity that correlated with maximum protein disulfide–thiol interchange activity and minima in oxidative activity (Chap. 1).

Maxima in one set of oscillations are separated from comparable maxima in subsequent sets of oscillations by 24 min. A similar set of oscillations marked by single arrows is discernable with HeLa cells (Fig. 5.19b). Cancer cells which have both ENOX1 (24 min period) and ENOX2 (22 min period) at their surfaces exhibit a complex pattern of oscillations of two different period lengths (Wang et al. 2001). However, with HeLa cells the pattern of growth oscillations is more complex reflecting the complexity of the oscillation patterns found with the oxidation of NADH. Single arrows indicate oscillations separated by 24 min characteristic of

Table 5.4 Differentiating agents that inhibit NADH oxidase activity of isolated plasma membrane vesicles and growth measured as in Table 5.3

Agent	EC_{50} Growth	EC_{50} NOX	Tissue or cell type	References
Retinoic acid	1 μM	1 μM	HeLa	Dai et al. (1997)
	10 μM[a]	1 μM[a]	Soybean	Results unpublished
	1 μM	0.1 μM	HKc	Morré et al. (1992a)
	10 nM	1 nM	HKc/HPV16	Morré et al. (1992a)
Retinol	~10 μM	10 μM	HeLa	Dai et al. (1997)
	100 μM[a]	1 nM[a]	Soybean	Results unpublished
Calcitriol	10 nM	10 μM	Hkc	Morré et al. (1992a)
	10 nM	0.1 μM	HKc/HPV16	Morré et al. (1992a)
Sodium phenylacetate[b]	1 nM	70 nM	HeLa	Results unpublished
Sodium nitrophenylacetate	100 μM[a]	n.d.	Soybean	Morré et al. (1988c)

From Morré (1998c)
n.d. Not determined
[a]2,4-D-induced component of growth and NADH oxidase
[b]With soybean, NADH oxidase activity of plasma membrane vesicles and growth of hypocotyl sections was stimulated

ENOX1 whereas the double arrows denote a second set of oscillations with a period length of 22 min derived from ENOX2 (Fig. 5.19b).

When increased cell area of CHO cells was measured, alternations of rapid and slow rates were observed. Maxima and minima were separated by ca. 12 min giving rise to a period length of 24 min (Fig. 5.19a). The cell enlargement data were most amenable to time series analysis (Foster et al 2003). With the trend component subtracted, the periodic fluctuations were resolved clearly in the decomposition fits. The small cell of Fig. 5.19b nearly doubled in size whereas the large cell of Fig. 5.17b enlarged more slowly. Yet, cell areas of both cells oscillated to approximately the same extent.

The oscillatory increases in cell area (cell enlargement) at 24 °C represented by maxima and minima (arrows) occurred reproducibly at intervals of 24 min (Fig. 5.15b). When measured at 14 °C, the cells failed to enlarge appreciably but cell area (volume) apparently continued to fluctuate with a period of ca. 24 min (Fig. 5.15a; Pogue et al. 2001). The fluctuations evidenced by statistically significant minima (arrows) were spaced at 24 min intervals. At 34 °C, the rate of cell enlargement more than doubled compared to 24 °C (Fig. 5.15c). However, during portions of the cycle where slowing of active growth occurs, the cells actually shrink suggesting that the portions of the ENOX cycle resulting in enlargement involve active stretching. Moreover, the periods of increase were separated by statistically significant minima (arrows). When subject to time series analysis, a constant period length of 24 min at all three temperatures, 14, 24 and 34 °C, was observed (Table 5.5). The portion of the ENOX cycle where the cells actually shrink coincides with the portion of maximum rates of NADH or hydroquinone oxidation (Pogue et al. 2000; Morré et al. 2002a). Net cell enlargement is linear but the periodic rates oscillate

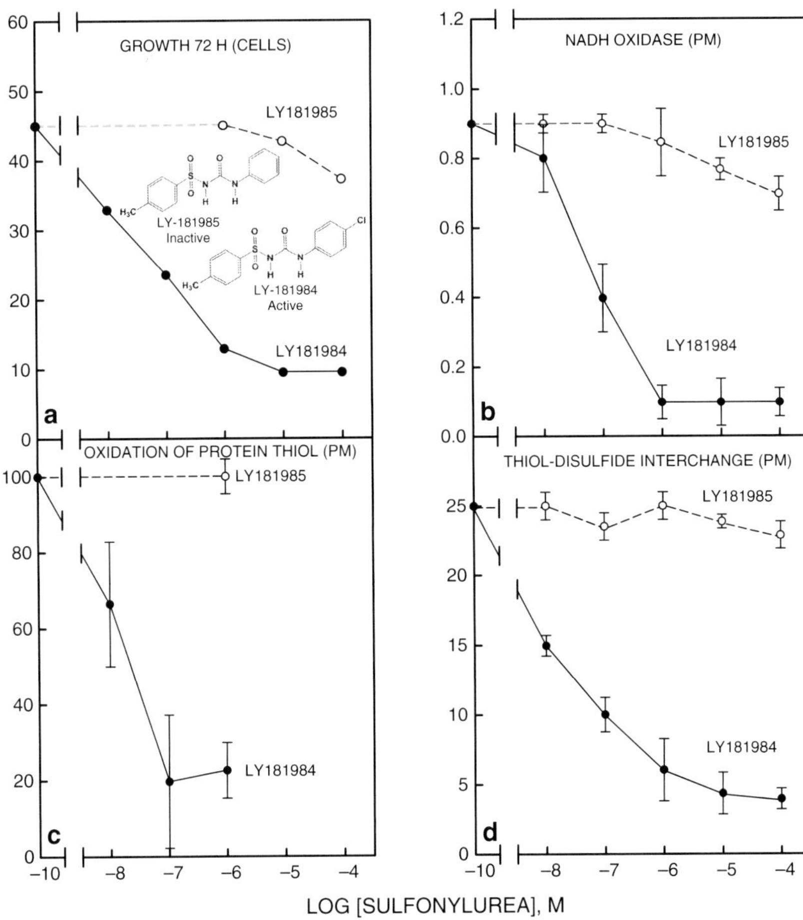

Fig. 5.17 Correlation of growth, NADH oxidation, oxidation of thiols and protein disulfide–thiol interchange activity of HeLa cells in culture in response to the active antitumor sulfonylurea LY181984 and lack of effect of the antitumor-inactive sulfonylurea, LY181985. (**a**) Growth of HeLa cells in culture. (**b**) Oxidation of NADH (NADH oxidase) of isolated vesicles of HeLa plasma membrane. (**c**) Oxidation of thiols measured by plasma membrane vesicles in the presence of 0.1 mM NADH measured directly by reaction with DTNB and a protein disulfide–thiol interchange activity estimated from the restoration of activity to scrambled and inactive bovine ribonuclease A. (**d**) Protein disulfide–thiol interchange activity measured from cleavage of dithiodipyridine substrate. The antitumor-inactive LY181985 (one chlorine difference on the B ring) was largely without effect. Growth (**a**) was measured in the presence of 10 nM EGF as described (Morré et al. 1997d) (from Morré et al. 1995i, 1997f). Reproduced from Morré (1998c) with permission from Springer Science + Business Media

well above and well below the average trend line at regular intervals with the 24 min period. In cells fully enlarged prior to division, cell size continues to oscillate with a 24 min period but the trend line is flat and the net rate of cell enlargement is zero (Pogue et al. 2001).

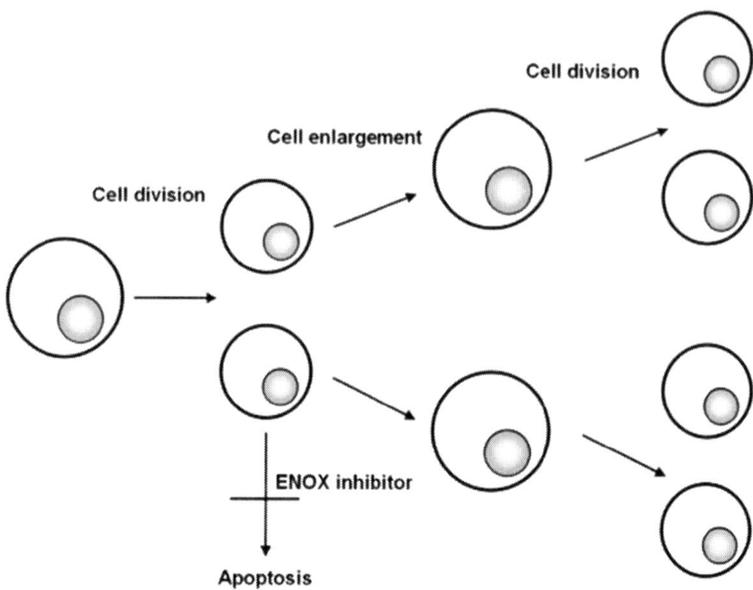

Fig. 5.18 Essential role of cell enlargement in growth. Following division, the resultant small cells must enlarge to some minimum size before they are able to divide again. If enlargement is prevented, the cells will be unable to divide again with programmed cell death (apoptosis) as the usual consequence

While correlations do not prove causality, our studies have consistently demonstrated that ENOX activity of isolated vesicles of plasma membrane or with intact cells and rate of cell enlargement are, without exception, correlated (Morré 1998a). The correlations hold both for growth stimulation and for growth inhibitors. Conditions that stimulate the activity of the NOX proteins stimulate cell enlargement, while conditions that inhibit the activity of the NOX proteins inhibit cell enlargement.

Why cells should pulse as they enlarge is apparently determined by the alternation of activities which is the basis for the periodic activity. Hydroquinone or NAD(P)H oxidation alternates with a protein disulfide–thiol interchange activity (Morré 1998c; Chap. 6). As developed in the next section, cell enlargement occurs specifically in response to protein disulfide–thiol interchange and rests during hydroquinone NAD(P)H oxidation.

5.4.2.2 Cell Enlargement and Cell Cycle Control

Continuously dividing cells double their size during each cell cycle to maintain constant volumes. This fact raises the possibility of control mechanisms whereby cell size is actively monitored to prevent delayed or premature cell division at an

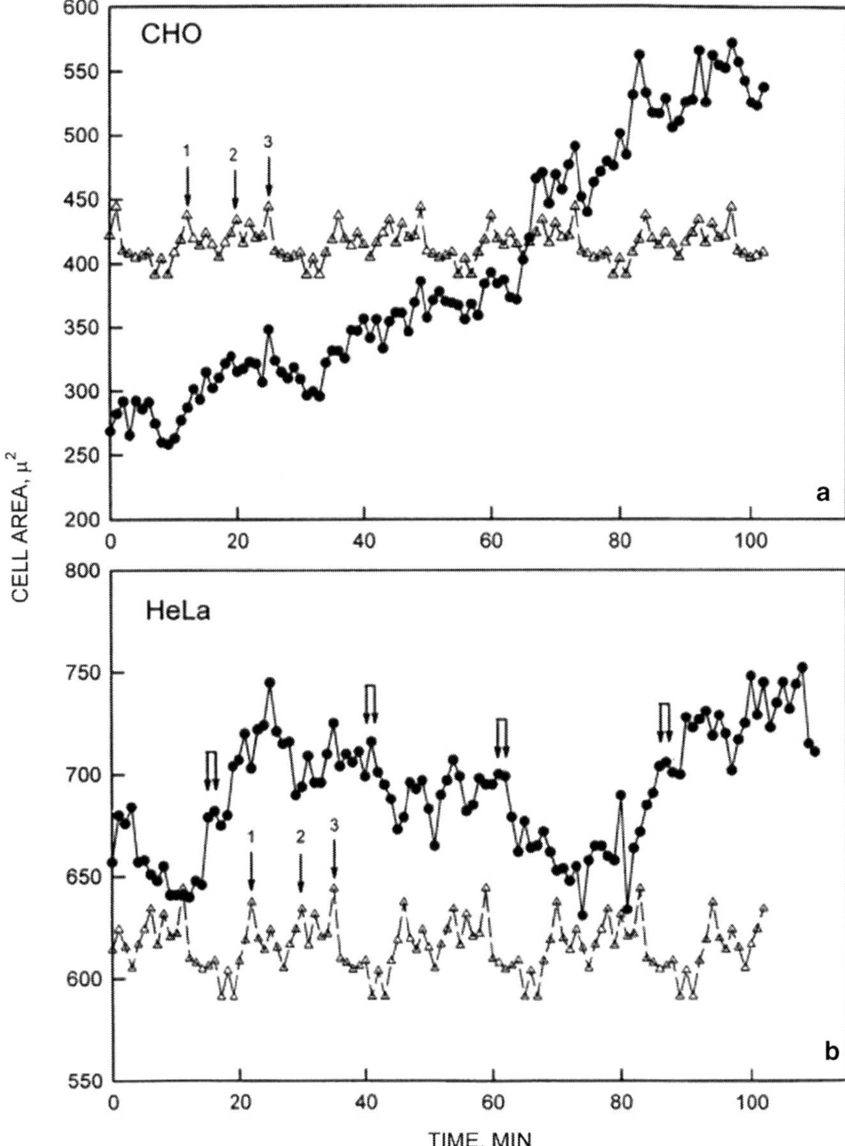

Fig. 5.19 Periodicity of growth (increase in area) of cells in culture at 37 °C determined using video-enhanced light microscopy. Measurements were at 1 min intervals over 110 min. Periods of rapid cell enlargement alternated with periods of relaxation where cell areas actually decreased. (**a**) CHO cells. The enlargement phase consists of three maxima (labeled 1, 2 and 3) separated by about 5 min. *Single arrows* coincide with the positions of maximum rates of enlargement spaced 24 min apart. Recurrent positions of maxima were determined from the detrended decomposition fits (*open triangles, dotted lines*). (**b**) HeLa cells. A more complex pattern of oscillatory components consisted of (*single arrows*) the oscillatory components spaced 24 min apart that correspond to those observed with CHO cells as determined from the detrended decomposition fits (*open triangles, doted lines*) with CHO cells plus a second set of oscillations (*double arrows*) spaced 22 min apart. Reproduced from Wang et al. (2001) with permission from Elsevier

Table 5.5 Period length of enlargement of CHO cells

Temperature (°C)	Period length (min)[a]
14	24.2 ± 1.5
24	24.9 ± 1.7
34	23.8 ± 2.1
Mean	24.0 ± 0.2

Results are means from three different cells at each of the three temperatures. From Pogue et al. (2000)

[a]Period length was determined by decomposition analysis from differential contrast images recorded at 1 min intervals

inappropriate cell mass (Grebien et al. 2005). Cell cycle check points that monitor cell size are known for yeasts but not for metazoan cells. Yet, even though cell growth and the cell cycle are separable processes, they are somehow coordinated (Jorgensen and Tyers 2004; Tzur et al. 2009) in whatever terms cell growth is measured (volume, length, mass or protein content). To a certain extent, metazoan cells coordinate cell size with cell division by regulating the length of the G_1 and G_2 phase of the cell cycle (Sweiczer et al. 1996).

The metabolic activity that distinguishes cell growth (i.e., increase in cell biomass) from proliferation is duplication of the genome and the ensuring massive commitment to nucleic acid biosynthesis (De Berardinis et al. 2008). Also growth is related to ribosome synthesis (Bernstein et al. 2007) which may provide an explanation for why ribonucleic acid synthesis is required for sustained cell elongation in plants (Key 1964) whereas transcription of auxin-regulated genes and synthesis of new auxin-specific proteins are not.

Investigators have looked downstream of TOR to see how exactly it stimulates cell growth, in terms of how big cells get and in terms of how fast cells divide. It seems as if TOR's main purpose is to regulate how the ribosome operates. TOR signaling turns on the expression of ribosome proteins and ribosomal RNA. Since the ribosome is the main factory in the cell, it is no surprise that a regulator of cell size would also regulate ribosome production. Active TOR actually sits on ribosomes as they get ready to translate an mRNA into protein. This active TOR adds phosphates directly to ribosomal proteins and even activates other kinases, such as S6 kinase, an enzyme that adds even more phosphates to the ribosome (S6 is a small ribosomal subunit).

5.4.2.3 Results with ENOX2 Overexpression

Results with ENOX2 overexpression in noncancer cell lines and results with antisense in cancer cell lines are consistent with the hypothesis that functional cell surface expression of ENOX2 is both necessary and sufficient for the cancer-specific inhibitions of cell growth attributed to various ENOX2 inhibitors such as the anticancer green tea catechins exemplified by (−)-epigallocatechin-3-gallate (EGCg)

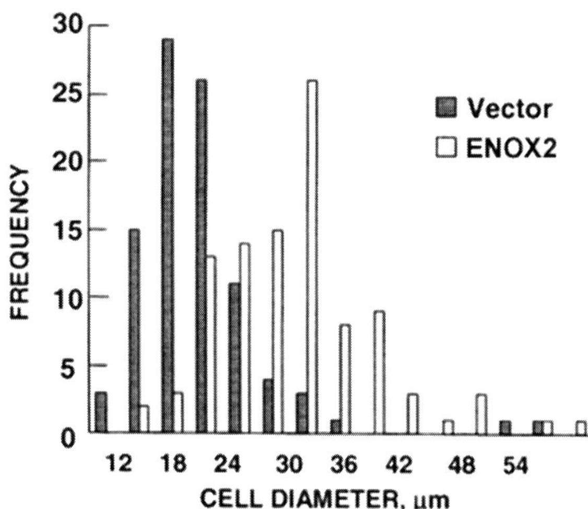

Fig. 5.20 Longitudinal diameters of ENOX2-transfected COS cells (*open bars*) measured directly from light microscope (×40) images. ENOX2-transfected COS cells were approximately two times larger than those of untransfected cells (*solid bars*). Reproduced from Chueh et al. (2004) with permission from Wiley

(Chueh et al. 2004). As an extension of these observations, transgenic mice overexpressing ENOX2 were found to exhibit the same level of unregulated cell enlargement and sensitivity to EGCg as observed with cancer cells (Yagiz et al. 2006).

When increases in cell area were measured, overexpression of ENOX2 cDNA in COS cells led to a more rapid rate of cell enlargement and several-fold increase in cell volume compared to nontransfected COS cells (Chueh et al. 2004; Figs. 5.20 and 5.21). Similar results were obtained with mouse embryonic fibroblast (MEF) cells isolated from ENOX2 transgenic mouse embryos compared to wild-type FVB mouse embryos as controls. After 24 h of growth, wild-type cells exhibited an area postfixation of $2{,}160 \pm 940\ \mu m^2$ whereas transgenic MEF cells had an area of $3{,}120 \pm 690\ \mu m^2$ assuming a proportionate increase in cell thickness corresponding to ca. 10 % of the leading edge. These measurements suggest that the average volume of the transgenic MEF cells overexpressing ENOX2 was approximately twice that of the wild-type MEF cells (Yagiz et al. 2007).

The enhanced rate of cell enlargement comparing wild-type and transgenic MEF cells overexpressing ENOX2 may account, as well for accelerated rate of division and for differences in cell number. As the MEF cells from transgenic cells enlarge more rapidly, they achieve a critical size sooner than wild-type cells after which they are able to divide. The combination of enhanced rate of cell enlargement and enhanced cell division then would combine to account for the doubling of cell mass observed as a result of ENOX2 overexpression in the mouse embryo fibroblast cells (Yagiz et al. 2007).

The tissue expressions of ENOX2 mRNA were greatest in heart, lung and liver of transgenic mice overexpressing ENOX2 (Yagiz et al. 2006; Fig. 5.22). When these tissues were analyzed for cell size, the cells of transgenic animals were, on average, 20 % larger in surface area than cells from corresponding wild-type tissues

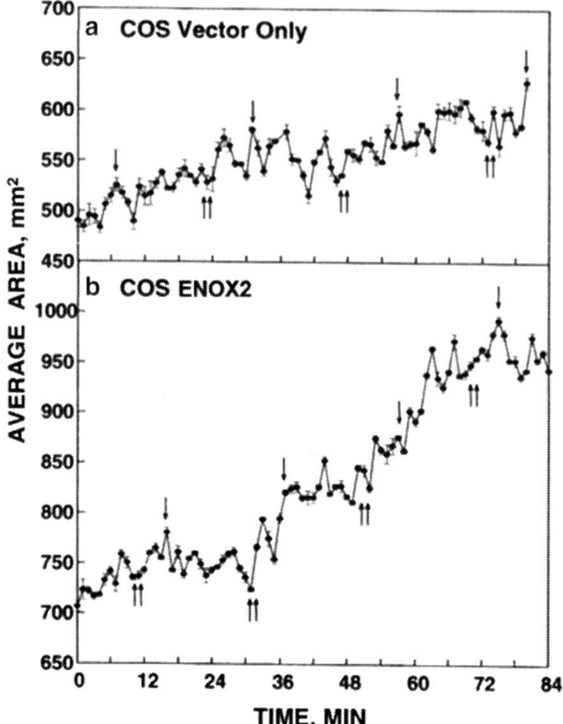

Fig. 5.21 Increase in area (enlargement growth) of COS cells transfected with the vector alone (**a**) or with ENOX2 cDNA (**b**) as determined by image-enhanced light microscopy. The growth rates fluctuated with a complex pattern of periodicity but were approximately twofold greater with the ENOX2-transfected cells. In (**a**), *single arrows* separated by intervals of 24 min indicate periods of rapid enlargement alternating with resting periods (*double arrows*). In (**b**), the *single arrows* indicating periods of rapid enlargement and intervening rest periods (*double arrows*) are separated by intervals of 22 min. Reproduced from Chueh et al. (2004) with permission from Wiley

(Yagiz et al. 2008). In intestine, spleen and kidney, where ENOX2 mRNA overexpression was lower, cell size increases were less evident. Carcass weights of the transgenic animals which had grown faster than wild type throughout the study (Yagiz et al. 2006) remained greater at the end of the study (Yagiz et al. 2008). The increase in carcass weight was reflected not only in muscle mass but also in femur weight and thickness but not length (Yagiz et al. 2008). Overall, the findings with transgenic animals were consistent with the property of ENOX2 observed in studies with cultured cells contributing to the enlargement phase of cell growth (Yagiz et al. 2008).

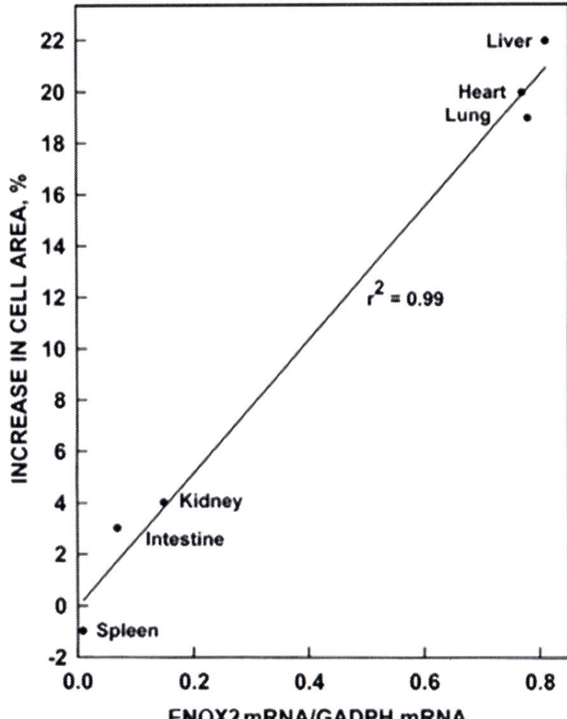

Fig. 5.22 Correlation between tissue expression of ENOX mRNA and increase in cell size in ENOX2 overexpressing transgenic mice. Tissue samples were analyzed for ENOX2 mRNA by PCR. Quantitation was by densitometry and expressed as the ratio to glyceraldehyde-3-phosphate dehydrogenase (GAPDH) mRNA as an internal control. Reproduced from Yagiz et al. (2008) with permission from Wiley

5.5 Pathological Implications

With the several functional indications of the NADH oxidase protein as a terminal acceptor of plasma membrane electron transport and in growth control, possible applications of significance emerge including cell death, oxidative stress, aging, and cancer.

5.5.1 Apoptosis

When growth of cancer cells is prevented by ENOX2 inhibitors, a major proportion of the cells in the cytostatic populations undergo apoptosis as indicated by nuclear changes detected by DAPI fluorescence, for example (Wolvetang et al. 1994). DAPI-detected chromatin clumping has been observed with HeLa cells treated with cytostatic concentrations of capsaicin (Morré et al. 1995b), antitumor sulfonylurea LY181984 (Morré and Morré 1995b), retinoic acid (Morré et al. 1996a) or phenoxodiol (De Luca et al. 2010; Wu et al. 2011). Development of apoptosis in response to cytostatic concentrations of ENOX2 inhibitors usually requires 72–96 h or sometimes longer as a response of small cells to being no longer able to divide.

In ρ° cells lacking mitochondria, survival appears to depend upon plasma membrane redox as a means to regenerate NAD^+ for glycolytic ATP production (Larm et al. 1994; Lawen et al. 1994; Wolvetang et al. 1996). When inhibited by capsaicin or resiniferatoxin in these cells, apoptosis is observed. The proposal that Bcl-2 induction results from sensing of $NAD^+/NADH^+ + H^+$ ratios seems especially intriguing in this regard (Wolvetang et al. 1996). Evidence for this suggestion comes from experiments where apoptosis was induced in ρ° cells by plasma membrane NADH oxidase inhibitors such as the vanilloid capsaicin and where it was then ameliorated by activators of the plasma membrane NADH-oxidoreductase systems such as potassium ferricyanide (Wolvetang et al. 1996; Wu et al. 2011). The findings point to potentially significant interactions among plasma membrane redox, cytosolic redox homeostasis and imbalance and oxidative stress.

5.6 Physical Membrane Displacements

According to a model published in 1994 (Morré 1994a; Fig. 5.23), cell enlargement is envisioned as a specific example of a more general physiological process referred to collectively as physical membrane displacement (Morré and Mollenhauer 2009). Included in these phenomena, in addition to cell enlargement, would be vesicle budding, protuberance formation and certain dynamic events associated with cell migration and motility (Ambrose and Rose 1975). Data are limited to two examples.

The first example is the response of the NADH oxidase and protein disulfide–thiol interchange activity (Jacobs et al. 1996) of transitional endoplasmic reticulum to retinol. Retinol stimulates NADH oxidase activity and promotes both vesicle budding (Morré and Morré 1987; Nowack et al. 1990) and Golgi apparatus dynamics determined by video-enhanced light microscopy (Morré et al. 1992b). The second example is found in the response of the *trans* Golgi apparatus membranes of cancer cells to the antitumor sulfonylurea LY181984 (Morré et al. 1994f). The latter findings suggest a subcellular location of a LY181984 target at the *trans* Golgi apparatus face presumably involving Golgi apparatus-located ENOX1 proteins. Related to the latter is a report of parallel inhibition of the ENOX1 activity of transitional endoplasmic reticulum by brefeldin A (Morré et al. 1994e).

5.7 ATP- and p97 AAA-ATPase-Dependent and Drug-Inhibited Vesicle Enlargement Reconstituted Using Synthetic Lipids and Recombinant Proteins

Based on prevailing models of cell enlargement (Sect. 5.1), reconstitution of cell enlargement in a cell-free environment should not be possible due to the absence of cell turgor (plants) and/or biosynthesis (animals and plants). However, if cell enlargement is as shown in this chapter the result of ENOX1- and ENOX2-mediated

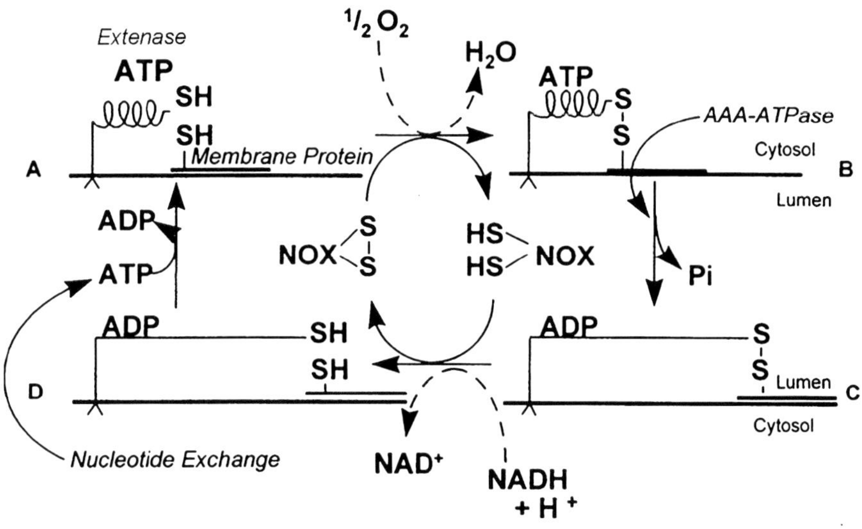

Fig. 5.23 Early physical membrane displacement model (Morré 1998c) to explain ECTO-NOX (ENOX) functions in growth. (**a**) Hypothetical membrane protein (now ATPase Associated with Diverse Cellular Activities (AAA-ATPase)) with a dynamic coiled region (now a ratchet mechanism, see Fig. 5.) with ATP bound. (**b**) One end of the protein is anchored to prevent slippage. A sulfhydryl link between the protein with ATP bound and a protein in the membrane is formed through the interchange activity of the ENOX protein. (**c**) The ATP is hydrolyzed and the coiled region extends, displacing the bound protein within the membrane. (**d**) The disulfide is reduced with return of the protons and electrons to the ENOX. As a result, the bound protein is released. The coiled region then binds another ATP which is exchanged for ADP. The coil reforms, and the system is ready to displace another membrane protein. Repeated cycles would result in membrane displacement or, with proper orientation, in the formation of membrane vesicles or blebs. Reproduced from Morré (1994a) with permission from Spring Science+Business Media

force extension of plasma membranes at the cell surface, cell-free cell enlargement becomes a demonstrable phenomenon.

Growth hormone-responsive and nucleoside triphosphate-dependent enlargement of membrane vesicles in a cell-free system was first achieved with inside-out vesicles of plasma membranes from soybeans prepared by aqueous two-phase partition and everted by freezing and thawing (Auderset and Morré 2006; Figs. 5.24 and 5.25). In the presence of 100 μM ATP in 40 mM HEPES buffer, pH 7, enlargement of isolated plasma membrane vesicles was accelerated by the synthetic plant growth factor, 2,4-D, compared to ATP alone, 2,4-D alone or no additions. After 20 min with 1 μM 2,4-D, vesicles increased in diameter on average by 20 %. Although vesicle diameters in the presence or absence of 2,4-D overlapped, the means were clearly separated. The 20 % increases in diameter corresponded to a doubling of vesicle volume. Measurements were on highly purified preparation of sealed vesicles (Fig. 5.26) identified as plasma membranes by the characteristic staining reaction with phosphotungstic acid at low pH (Roland et al. 1972). Both 100 μM ATP

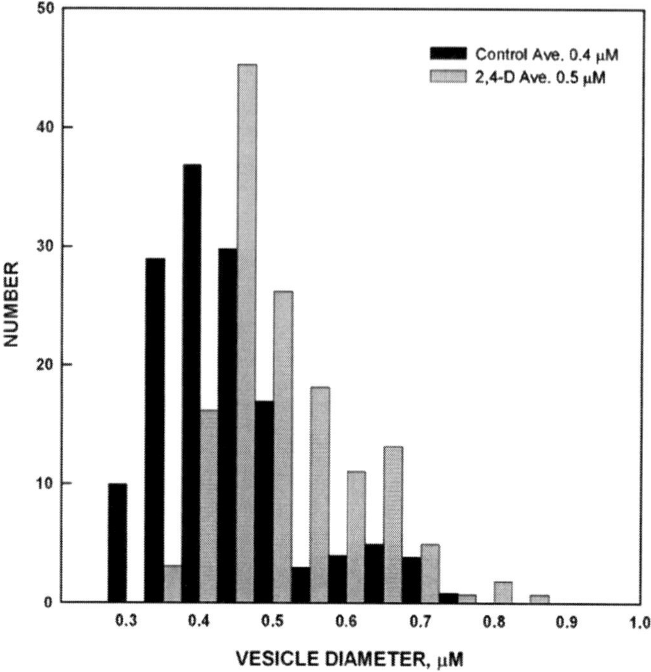

Fig. 5.24 Quantitation of the response of inside-out plasma membrane vesicles from soybean incubated for 20 min in the absence (*open bars*) or the presence (*hatched bars*) of buffered 100 μM ATP and 1 μM 2,4-D. Vesicle diameters were shifted in distribution toward larger diameters by the treatment compared to buffer alone. Membrane suspensions (500 μL) were mixed with 2 % glutaraldehyde fixative in 0.2 M sodium phosphate, pH 7.2 (50 μL). After 10 min, the membranes were collected by centrifugation (microfuge) and fresh 2 % buffered glutaraldehyde was added. Post fixation was in 2 % osmium tetroxide in the same buffer with embedment in Epon. Thin sections were post-stained with lead citrate and examined and photographed at a primary magnification of 14,000 and enlarged photographically two to fivefold. Diameters of all vesicles were measured. A total of 140 vesicles were measured from three electron micrographs photographed at random for each treatment. Reproduced from Auderset and Morré (2006) with permission from Wiley

and 1 μM 2,4-D were necessary to stimulate vesicle enlargement in the cell-free environment. In the presence of 1 μM 2,4-D, enlargement observed with 100 μM ATP was greater than with either 10 μM ATP or 500 μM ATP alone. In the presence of 100 μM ATP, vesicle enlargement was proportional to the logarithm of 2,4-D concentration. With the growth-inactive 2,4-D analog, 2,3-D, no vesicle enlargement was observed either alone or in the presence of 100 μM ATP. Right side-out vesicles did not enlarge in response to either ATP, 2,4-D or the two in combination suggesting that the responsible ATP binding site was on the side of the plasma membrane facing the interior of the cell.

The ATP requirement for cell enlargement adds a functional implication to the possibility raised in Chap. 4 that some or all of the 10 kcal of energy released by oxidation of NADH at the plasma membrane, based on studies with inside-out

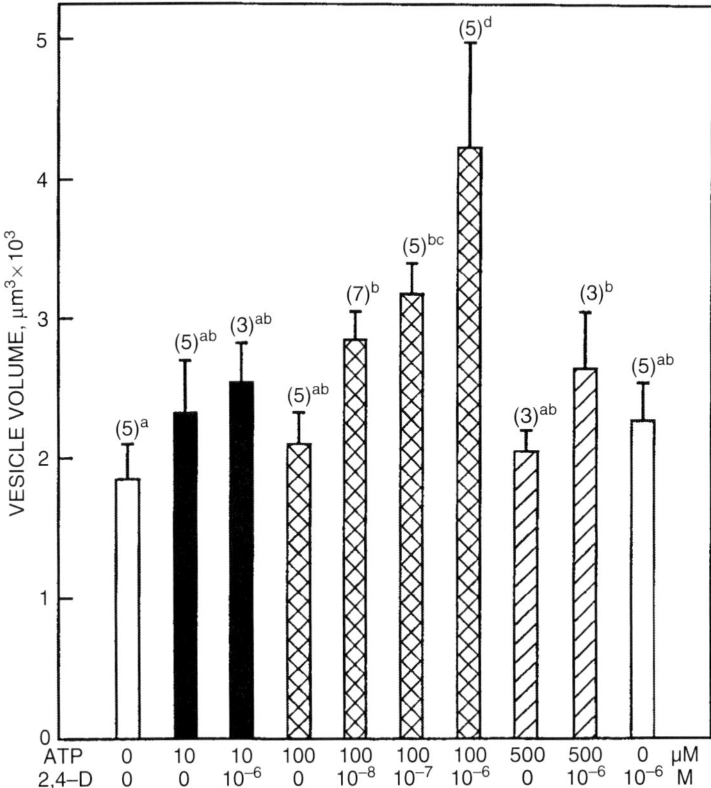

Fig. 5.25 Response of inside-out vesicles of plasma membranes from soybean to 1 μM 2,4-D or to varying concentrations of ATP alone or in combination with 1 μM 2,4-D. Incubations were for 30 min. Determinations were based on measurements from 140 to 200 vesicles for each of three electron micrographs for each treatment. The numbers of determinations are given in *parentheses*. Numbers not followed by the same letter are significantly different as determined by Student's two-tailed *t*-test. Reproduced from Auderset and Morré (2006) with permission from Wiley

plasma membrane vesicles from hypocotyl sections of soybean, is conserved as ATP (Sect. 4.4).

Near optimal enlargement of inside-out vesicles (to expose the ATP binding site) occurred at 100 μM ATP + 1 μM 2,4-D (Table 5.6). Right side-out vesicles did not respond to ATP as the ATP binding site was not exposed (Table 5.7). The cell-free enlargement was refractory to inhibition by the plasma membrane proton pumping inhibitor vanadate (Table 5.8) but was inhibited by the sulfonylurea herbicide chlorsulfuron which inhibits the auxin-stimulated NADH oxidase of the plasma membrane (Morré et al. 1995e). Antisera to *A*TPase *A*ssociated with Diverse Cellular *A*ctivities (AAA-ATPase) motif B inhibited, whereas preimmune sera did not.

Using light scattering, the increase in A_{520} (Morré et al. 1995e) was a measure of vesicle enlargement, the increase seen with 100 μM ATP and 1 μM 2,4-D was fourfold that with either 2,4-D or ATP alone (Table 5.9). Azide and oligomycin,

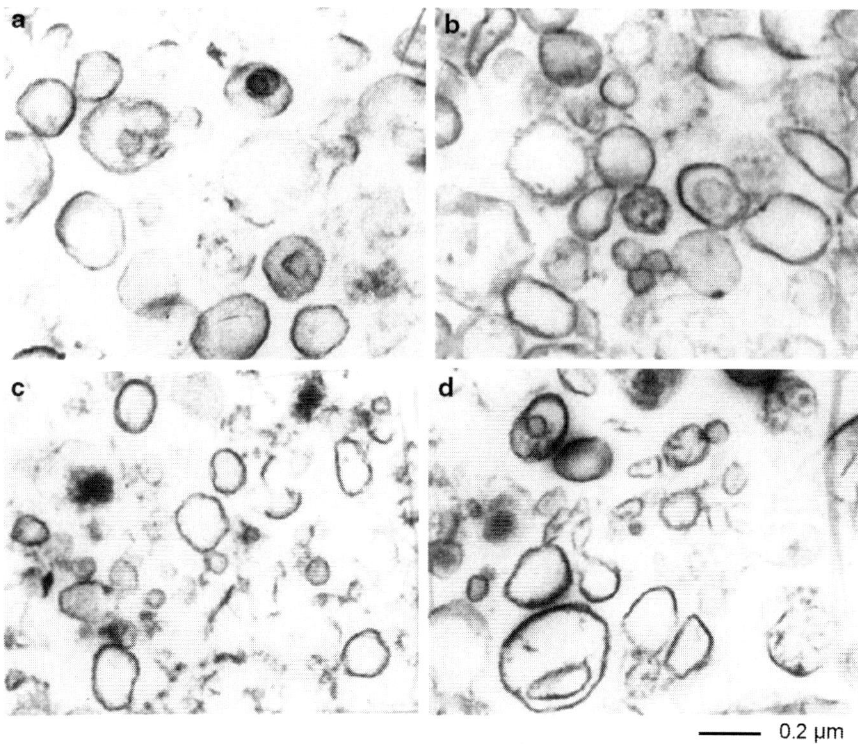

Fig. 5.26 Electron micrographs of sealed highly purified isolated soybean plasma membrane vesicles reactive with phosphotungstic acid at low pH to verify a plasma membrane origin (Roland et al. 1972). Vesicles appeared spherical and ranged in size from about 100 nm to about 350 nm diameter. (**a**, **c**). Inside-out vesicles incubated in HEPES buffer for 30 min. (**b**, **d**) Right side-out vesicles incubated in HEPES buffer for 30 min. (**a**, **b**) Stained with lead acetate to contrast all membranes present. (**c**, **d**) Stained with phosphotungstic acid at low pH to specifically stain plasma membranes. There were no indications for increase in size due to vesicle fusion, i.e., vesicle fusion profiles were no more abundant after incubation than before incubation. Reproduced from Hicks-Berger and Morré (2006) with permission from Wiley

mitochondrial respiratory inhibitors; ouabain, an inhibitor of the plasma membrane sodium–potassium ATPase; nitrate, an inhibitor of the v-type vacuolar ATPases and vanadate, an inhibitor of the plasma membrane proton-pumping ATPases were without effect on the increase in A_{520} inducted by 100 μM ATP+1 μM 2,4-D (Table 5.9). Only NEM, an inhibitor of both the auxin-stimulated plasma membrane NADH oxidase and of vesicle budding and cobalt chloride, an inhibitor of the AAA-ATPases, reduced the changes in A_{520} to levels seen with ATP or 2,4-D alone (Table 5.9).

As with cell-free vesicle enlargement monitored by electron microscopy (Table 5.8), two different antisera raised to AAA-ATPase motif B peptide sequence of a peptide antibody raised to a conserved peptide plasma membrane-associated

Table 5.6 Response of right side-out vesicles of plasma membranes prepared from hypocotyls of dark grown soybeans to 100 μM ATP or 1 μM 2,4-D alone or in combination

Treatment	Vesicle volume ($\mu m^3 \times 10^3 \pm$ standard deviation)
Initial (not incubated)	6.1 ± 1.8
Buffer	6.8 ± 1.4
+100 μM ATP	5.9 ± 1.1
+1 μM 2,4-D	5.6 ± 2.1
+100 μM ATP + 1 μM 2,4-D	4.1 ± 1.35

Incubations were for 30 min. Results are averages of three determinations ± standard deviations. Values were not significantly different ($p < 0.01$). From Hicks-Berger and Morré (2006)

Table 5.7 Response of inside-out vesicles of plasma membranes prepared from hypocotyls of dark-grown soybeans to 100 μM ATP or 10^{-6} M 2,4-D alone or in combination

Treatment	Vesicle volume ($\mu m^3 \times 10^3 \pm$ standard deviation)
Initial (not incubated) (6)	2.9 ± 0.6[a]
Buffer (6)	2.6 ± 0.4[a]
+100 μM ATP (6)	2.9 ± 0.3[a]
+1 μM 2,4-D (7)	3.2 ± 0.4[a]
+100 μM ATP + 1 μM 2,4-D (8)	4.3 ± 0.4[b]
+1 μM 2,3-D (3)	2.2 ± 0.45[a]
+100 μM ATP + 1 μM 2,3-D (3)	2.6 ± 1.1[a]

Incubations were for 30 min. The number of determinations is given in *parentheses*. Values not followed by the same letter were significantly different ($p < 0.001$). From Hicks-Berger and Morré (2006)

Table 5.8 Volumes of inside-out plasma membrane vesicles prepared from hypocotyls of dark-grown soybeans after incubation with ATP + auxins and the effect of potential inhibitors on vesicle enlargement

Treatment (*n*)	Vesicle volume ($\mu m^3 \times 10^3 +$ standard deviation)
100 μM ATP + 1 μM 2,4-D (8)	4.3 ± 0.4[b]
100 μM ATP + 1 μM IAA (3)	4.3 ± 0.4[b]
50 μM Vanadate + 100 μM ATP + 1 μM 2,4-D (3)	5.0 ± 1.1[b]
100 μM ATP + 1 μM 2,4-D + 1 μM chlorsulfuron (3)	2.4 ± 1.5[a]
100 μM ATP + 1 μM 2,4-D + 1 μg AAA-ATPase motif B antibody (3)	2.5 ± 0.6[a]
100 μM ATP + 1 μM 2,4-D + 1 μg pre-immune sera (3)	5.0 ± 1.0[b]

The vesicles were treated with the compounds as indicated in the table for 30 min at room temperature. The number of times (*n*) individual experiments were repeated is given in *parentheses*. Standard deviations are among repetitions. Values not followed by the same letter were significantly different ($p < 0.001$). From Hicks-Berger and Morré (2006)

Table 5.9 Enlargement rates of inside-out plasma membrane vesicles prepared from dark-grown hypocotyls of soybeans in response to metabolic effectors based on light scattering

Experiment (*n*)	Slope (A_{520}/min × 10^{-5})
HEPES buffer (3)	0 ± 4[d]
100 μM ATP (3)	6 ± 1[c]
1 μM 2,4-D (3)	9 ± 2[c]
100 μM ATP + 1 μM 2,4-D (8)	31 ± 7[a]
100 μM ATP + 1 μM 2,4-D + 1 mM NaN_3 (2)	35 ± 4[a]
100 μM ATP + 1 μM 2,4-D + 10 μg/mL oligomycin (3)	30 ± 2[a]
100 μM ATP + 1 μM 2,4-D + 1 mM ouabain (3)	31 ± 4[a]
100 μM ATP + 1 μM 2,4-D + 50 mM KNO_3 (2)	30 ± 3[b]
100 μM ATP + 1 μM 2,4-D + 50 mM vanadate (5)	28 ± 6[a]
100 μM ATP + 1 μM 2,4-D + 1 μM NEM (2)	6 ± 1[b]
100 μM ATP + 1 μM 2,4-D + 10 mM $CuCl_2$ (2)	10 ± 1[c]

The slopes are reported as average values relative to the HEPES buffer control. The number of times the experiments were repeated (*n*) is given in *parentheses*. Standard deviations are among repetitions. Numbers not followed by the same letter were found to be statistically different using a Student's *t*-test, $p < 0.05$. From Hicks-Berger and Morré (2006)

Table 5.10 Motif B and membrane AAA-ATPase antibodies prevent the absorbance increase, at 520 nm, of inside-out vesicles prepared from hypocotyls of dark-grown soybeans in response to ATP + 2,4-D

Experiment (*n*)	Slope (A_{520}/min × 10^{-5})
HEPES buffer (3)	0 ± 4[c]
100 μM ATP + 1 μM 2,4-D (8)	31 ± 7[a]
100 μM ATP + 1 μM 2,4-D + motif B AAA-ATPase antibody (3)	9 ± 2[b]
100 μM ATP + 1 μM 2,4-D + membrane-ATPase antibody (3)	7 ± 2[b]
100 μM ATP + 1 μM 2,4-D + 1 μg pre-immune sera (3)	39 ± 3[b]

The slopes are reported as average values relative to the HEPES buffer control. The number of times the experiments were repeated (*n*) is given in *parentheses*. Standard deviations are among repetitions. Numbers not followed by the same letter were found to be statistically different using a Student's *t*-test, $p < 0.05$. From Hicks-Berger and Morré (2006)

plant AAA-ATPase inhibited cell-free vesicle enlargement based on an increase in A_{520}, whereas preimmune sera did not (Table 5.10).

The vesicle enlargement observed both by changes in vesicle diameter and by increase in A_{520} was accompanied by membrane thinning. The lipid bilayers of inside-out plasma membrane vesicles incubated with ATP + 2,4-D for 30 min were 30 % thinner than vesicles incubated in HEPES buffer alone (Table 5.11).

With the availability of recombinant ENOX2 (Cho et al. 2002) and the identification, characterization, cloning and expression in bacteria of a plasma membrane-associated AAA-ATPase from soybean (Hicks-Berger et al. 2006), cell-free analyses were extended to completely define cell-free systems using the recombinant proteins and artificial lipid vesicles of defined compositions (Morré et al. 2006a).

Table 5.11 Bilayer thinning as determined from inside-out plasma membrane vesicles prepared from hypocotyls of dark-grown soybeans and enlarged ×100,000 and ×360,000

Treatment of ISO vesicles (*n*)	Membrane width (nm)
PTA stained vesicles enlarged ×100,000	
HEPES buffer (20)	7.5 ± 2.5
ATP + 2,4-D (25)	5.2 ± 1.0
HEPES buffer (36)	8.0 ± 2.0
ATP + 2,4-D (23)	5.6 ± 0.9
Lead stained vesicles enlarged ×360,000	
HEPES buffer (25)	7.2 ± 2.0
ATP + 2,4-D (39)	5.0 ± 2.0

Inside-out vesicles were stained with PTA and enlarged ×100,000 or stained with lead and enlarged ×360,000 in order to measure the width of the membrane bilayer. The lipid bilayer of inside-out vesicles thinned only when treated with ATP + 2,4-D. Incubations with HEPES buffer alone did not induce vesicle enlargement and the lipid bilayer of those vesicles did not thin with time. *n* the number of vesicles measured to determine membrane bilayer width. Standard deviations are among individual vesicles, From Hicks-Berger and Morré (2006)

Results from such systems using well synchronized preparations of recombinant ENOX2 and a spectrophotometric assay remarkably recapitulated the results with inside-out plasma membrane vesicles (Fig. 5.27). A net rate of enlargement required the presence in the vesicle of a thiol-containing protein in these experiments supplied as serum albumin (Fig. 5.28). Fully oxidized serum albumin was ineffective. There have been indications thus far of any high degree of specificity for the source of thiol-containing protein utilized as substrates for the protein disulfide–thiol interchange activities of the ENOX proteins involved in cell-free cell enlargement.

Recombinant AAA-ATPase, recombinant ENOX, a protein thiol source and ATP all were required (Hicks-Berger et al. 2006) to achieve enlargement of the synthetic lipid vesicles. Not only did the vesicles enlarge but the rate of enlargement oscillated precisely according to the pattern described for ENOX protein disulfide–thiol interchange (Fig. 2.18) and with a period length characteristic of the recombinant ENOX source (22 min for ENOX2, 24 min for ENOX1). In control experiments, no ATP-dependent vesicle enlargement or periodic oscillations were observed with the AAA-ATPase or the ENOX alone (Fig. 5.29). Also addition of ATP was without effect when only the single proteins were incorporated into the membrane vesicles. ADP and AMP were used as specificity controls for ATP and did not support cell-free vesicle enlargement.

Apparently there may be sufficient conservation of structure both among species and between the plant and animal kingdoms to allow for interchange of protein components. In the homologous system, recombinant soybean plasma membrane ATPase and purified 2,4-D-activated soybean NOX (dNOX) were combined in stoichiometric proportions into the liposomes together with albumin (necessary for sustained enlargement). In the presence of ATP and 2,4-D, these vesicles enlarged and the enlargement rate oscillated with a period length of 24 min. In the heterologous system, the lipid vesicles containing albumin were reconstituted with the

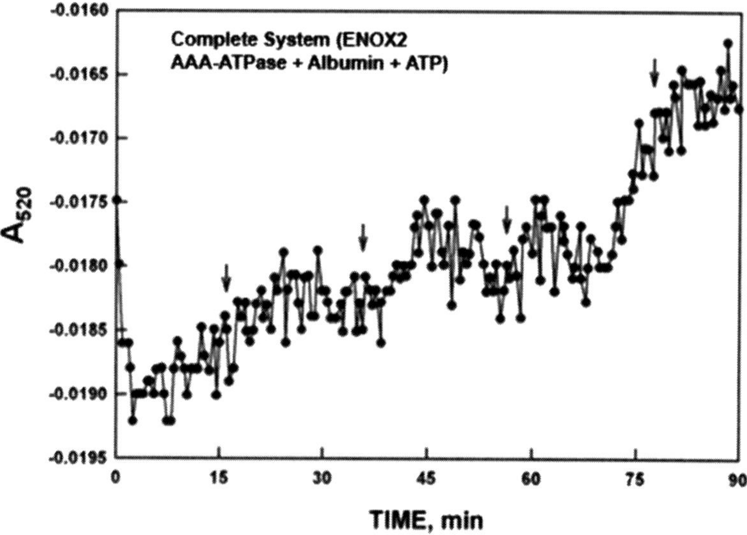

Fig. 5.27 Enlargement of lipid vesicles with a completely defined cell-free system reconstituted using recombinant proteins. Enlargement of large lipid vesicles of defined composition reconstituted with equimolar amounts of a synchronous recombinant ECTO-NOX (truncated ENOX2) with a 22-min period length and recombinant plasma membrane AAA-ATPase and serum albumin in the presence of 100 μM ATP. The lipid vesicles enlarge with intervals of rapid enlargement (*arrows*) separated by intervals of slow or negative rates of enlargement that parallel the pattern of enlargement of intact cells and tissues and of isolated inside-out plasma membrane vesicles. As recombinant ENOX2 with a period length of 22 min was used, the period length of the oscillations was 22 min. Large lamellar vesicles were formed from phosphatidylcholine, cholesterol and dicetylphosphate in a molar ratio of 50:45:5 where the phosphatidylcholine was a 2:1 mixture of synthetic dimyristoyl and dipalmitoyl phosphatidylcholines. The lipids were dried to a film and reconstituted into vesicles by resuspension in buffer containing the recombinant proteins in equimolar ratios of 0.04 nmol recombinant protein/mg lipid. Reproduced from Hicks-Berger and Morré (2006) with permission from Wiley

recombinant AAA-ATPase from soybean and recombinant human ENOX2. Again the vesicles enlarged in an ATP-dependent manner. The rate of enlargement was not influenced by the presence of 2,4-D but was inhibited by capsaicin, a quinone site inhibitor specific for ENOX2 (Morré et al. 2007a). The rate of enlargement was oscillatory with a period length of 22 min rather than of 24 min reflective of the shorter period length of the tumor-associated ENOX form, ENOX2 (Fig. 5.27). Only with the complete system did the vesicles respond with an enlargement rate significantly greater than with just buffer alone.

The ATP responsive component of the cell-free vesicle enlargement is a p97 AAA-ATPase similar to the p97 AAA-ATPase required for membrane budding and physical membrane displacements within the endomembrane system (Morré and Mollenhauer 2009, pp. 228–237). Based on cell-free systems, budding of vesicles from Golgi apparatus exhibited an absolute requirement for ATP (Morré 1998b). But even in the presence of ATP, vesicle formation and release was markedly accelerated by NADH and completely blocked by a variety of thiol reagents

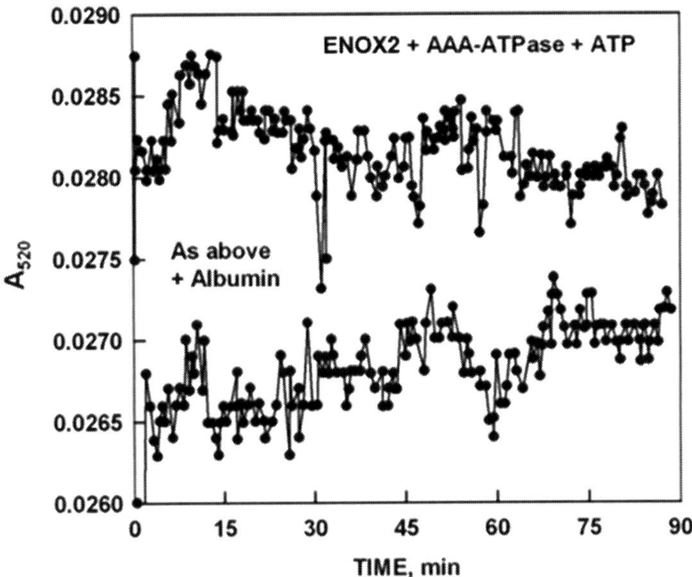

Fig. 5.28 As in Fig. 5.25 except ENOX2 plus AAA-ATPase+ATP with or without serum albumin. Both preparations exhibit an oscillatory pattern of lipid vesicle enlargement. However, only with the addition of serum albumin, does vesicle enlargement occur. Reproduced from Hicks-Berger and Morré (2006) with permission from Wiley

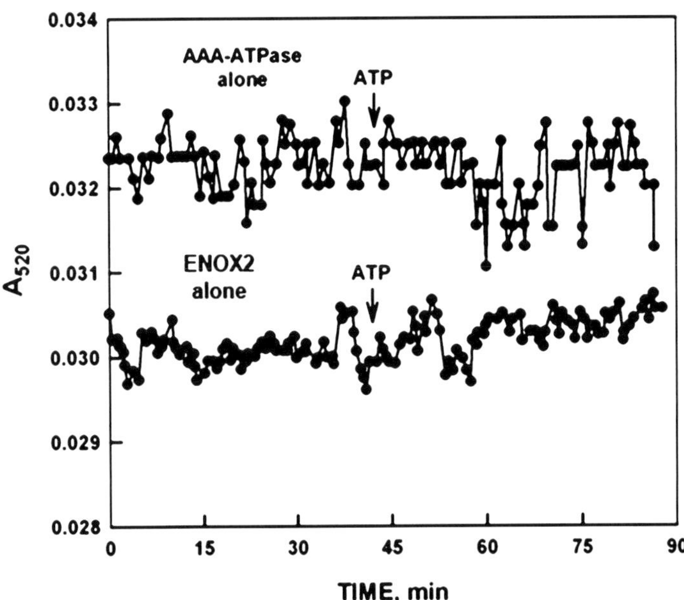

Fig. 5.29 Control preparations with AAA-ATPase alone (*upper curve*) or ECTO-NOX (ENOX2) alone (*lower curve*). Neither addition alone supported vesicle enlargement either before or after (100 μM ATP added at the *arrows*) addition of ATP. Lipid vesicles reconstituted with the serum albumin alone did not enlarge either before or after addition of ATP. Reproduced from Hicks-Berger and Morré (2006) with permission from Wiley

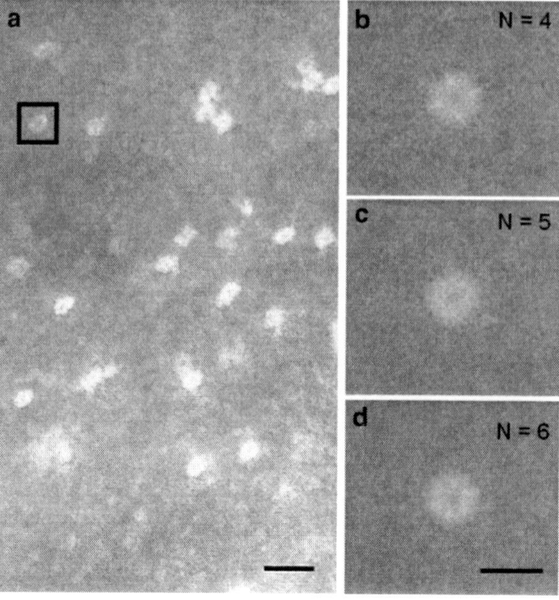

Fig. 5.30 Electron microscopy of negatively stained (uranyl acetate) recombinant AAA-ATPase expressed in *Escherichia coli* and purified as described (REF). Rotational analyses (Markham et al. 1963) of a single particle ($n=4$, 5 and 6) of average diameter of 9–10 nm revealed the presence of hexamers as is characteristic of purified AAA-ATPases have sixfold symmetry (REF). Scale marker=0.02 μm in (**a**) and 0.01 μm for (**b–d**). Reproduced from Hicks-Berger and Morré (2006) with permission from Wiley

(Rodriguez et al. 1992). These observations, together with evidence from studies of cell enlargement in plants, led us to postulate the central involvement of NADH- and ATP-requiring driving enzymes both for Golgi apparatus vesicle budding and cell enlargement according to the model of Fig. 5.23 (Morré 1994b, 1995b). An extensive evaluation of a series of different donor and acceptor combinations involving reduced pyridine nucleotide eventually yielded a NADH oxidase (ENOX) activity that was subsequently shown to be both cell surface (plasma membrane) (Morré 1998c) and Golgi apparatus (Morré 1998b) localized and exquisitely tied to the enlargement phase of cell growth (Morré 1998c).

The purified p97 AAA-ATPases typically form hexamers ca. 9–10 nm in diameter seen as ring-like structures when viewed by electron microscopy in negatively stained preparations (Zhang et al. 1994; Ogura and Wilinson 2001; Vale 2002). The purified AAA-ATPase from plants utilized in these studies similarly form ca. 9 nm hexamers (Fig. 5.30). Similar appearing hexamers were identified in the reconstituted lipid vesicles by transmission electron microscopy from their characteristic appearance (Fig. 5.31). The synthetic lipid vesicles were multilaminate (Fig. 5.31a). At higher magnification, structures aligned between the lamina had the appearance of the AAA-ATPase hexamers (Fig. 5.31b). Their sixfold symmetry was confirmed by rotational analysis ($n=6$) (Fig. 5.31d) (negatively stained particles of Fig. 5.30b–d). The artificial lipid membranes of Fig. 5.31 were unfixed but stained with osmium vapor and dehydrated to reveal proteins within the lipid layer in negative contrast. Figure 5.31c illustrates two apparent AAA-ATPase hexamers observed in tangential

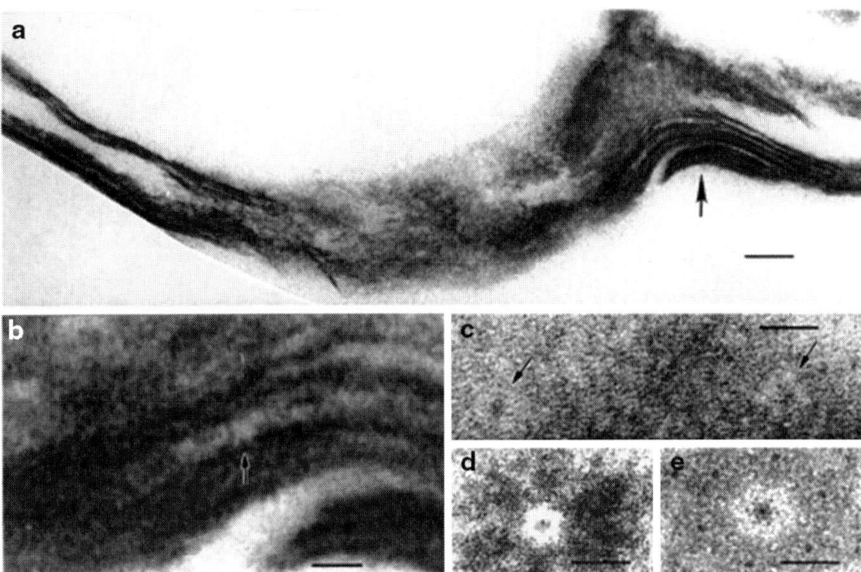

Fig. 5.31 Synthetic lipid vesicles reconstituted with recombinant ENOX and recombinant plasma membrane AAA-ATPase shown in cross section. The membranes were unfixed and stained with osmium vapor to show proteins in the membrane in negative relief. Scale marker = 0.1 μm. (**a**) Portion of the multilamellate membrane of a synthetic lipid vesicle. The *arrow* indicates the portion enlarged in (**b**). (**b**) Enlarged portions of (**a**) showing apparent AAA-ATPase hexamers in cross section oriented between the lipid layers possibly flanked by structures that may represent associated ENOX proteins. (**c**) Tangential cut through the synthetic membrane showing two putative AAA-ATPase hexamers (*arrows*) as seen in pure preparations (Fig. 5.4). (**d**) Rotational analysis ($n=6$) of one of the apparent AAA-ATPase hexamers of (**b**) (second to the *left* of *arrow*) showing sixfold symmetry. (**e**) Rotational analysis ($n=6$) of the apparent AAA-ATPase on the *right* in (**c**). (**c**–**e**) Scale marker = 0.01 μm. Reproduced from Hicks-Berger and Morré (2006) with permission from Wiley

section within the artificial membrane bilayer. Rotational analysis ($n=6$) confirmed a hexameric organization.

The basis for the ATP requirement for cell enlargement may be understood in the context of known characteristics of the p97 AAA-ATPase. The ATPase associated with the transitional endoplasmic reticulum involved in the budding of COP-II-coated vesicles (the TER-ATPase) was the first example of a membrane-associated p97 AAA-ATPase to be isolated, characterized and cloned (Zhang et al. 1994). A member of the unique AAA-ATPase family, the TER-ATPase was subsequently shown to fulfill the ATP requirement of both cell enlargement and vesicle budding. An AAA-ATPase equivalent of the TER-ATPase was isolated from plasma membranes of plants and cloned (Hicks-Berger and Morré 2006) for use in the development of the defined cell free systems exhibiting ATP-dependent enlargement of lipid vesicles.

Because of the requirement for inside-out plasma membrane vesicles and to achieve access to cytosolic ATP, the ATP-binding site of the plasma membrane-located AAA-ATPase must be located on the cytosolic surface of the plasma membrane. In order for ATP to reach this site with isolated vesicles, it was necessary to utilize plasma membrane vesicles that had been turned inside out. The small molecules such as 2,4-D for plants and ENOX2 inhibitors such as capsaicin are freely permeable and can reach their target. ENOX2 normally located on the outer plasma membrane surfaces can be reached by small molecules from either direction.

The response of cell-free cell enlargement to various ATPase inhibitors strongly implicates the AAA-ATPases not those of proton pumping at the plasma membrane or at vacuoles or within mitochondria. The cell-free enlargements of the inside-out plasma membrane vesicles from soybean were refractory to a series of known ATPase inhibitors except for the sulfhydryl inhibitor NEM and $CoCl_2$ a specific inhibitor of the AAA-ATPase of transitional endoplasmic reticulum (TER AAA-ATPase) (Zhang et al. 1994) and other AAA-ATPases (Hicks-Berger and Morré 2006). $BaCl_2$ also inhibited, as did antisera to AAA-ATPase motif B and antisera to the TER AAA-ATPase investigated by Zhang et al. (1994).

The members of the ATPase family of the AAA-type are defined by the presence and high degree of conservation of one or two sequences of 200–250 amino acids that include Walker A and Walker B motifs (Walker et al. 1982) as well as an AAA-specific region of homology to comprise a diverse and ubiquitous group of molecular chaperones (Ogura and Wilinson 2001; Pye et al. 2006; Vale 2002; Vale and Milligan 2000). Multiple sequence analysis suggests that these proteins share distinct structural and mechanistic features that distinguish them from other nucleoside 5′-triphosphatases (Neuwald et al. 1999; Woodman 2003). AAA proteins are considered to be ATPases that use ATP hydrolysis to fulfill a diversity of cellular functions (Patel and Latterich 1998; Ogura and Wilinson 2001; Pye et al. 2006; Vale 2002; Vale and Milligan 2000) in membrane trafficking and fusion, ubiquitin-dependent processes, protein folding, organelle biogenesis, and cell division and growth (Dalal and Hanson 2001; Feiler et al. 1995; Lupas and Martin 2002; Whiteheart et al. 1994; Woodman 2003). Family members use their ATPase activity to catalyze processes as diverse as proteolysis (six components of the 26S proteasome are AAA-ATPases), the disassembly of protein complexes (e.g., the NEM-sensitive factor which disassembles SNARE pairs during membrane fusion events) and protein unfolding (Dreveny et al. 2004; Neuwald et al. 1999). The implication of CDC48/p97, a widely conserved AAA-ATPase, in spindle disassembly (Cao et al. 2003), links AAA-ATPases and cell division (Cheeseman and Desai 2004).

The crystal structure of full-length mouse p97 complexed with a mixture of ADP and ADP-AlFx obtained at a resolution of 4.7 Å (DeLaBarre and Brunger 2003) or complexed with AMP-PNP or 2-Br AMP-PNP at 3.6 Å (Davies et al. 2005) together with small angle X-ray scattering measurements during nucleotide (ATP) hydrolysis (Huyton et al. 2003), reveal opportunities for large conformational changes in the helical subdomains during the hydrolysis cycle (DeLaBarre and Brunger 2005).

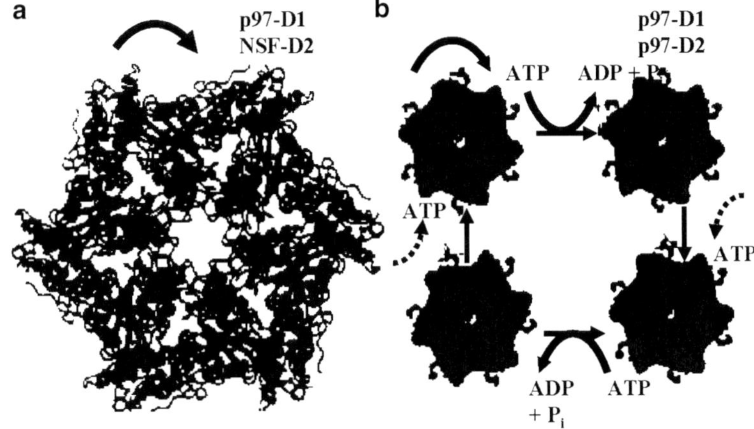

Fig. 5.32 Speculative model of p97 acting as a molecular ratchet (see text) (REF). (**a**) Superposition of p97 D1 in an ADP state with NSF D1 in an ATP state on their respective P loops. A small clockwise rotation around the hexamer axis is observed between the monomers. (**b**) Possible ratchet mechanism for p97 with D1 and D2 rings acting as interdependent and counter-balanced parts of the ATP hydrolysis cycle. *Top left*, the D1 and D2 rings are shown as ATP- and ADP-bound states, respectively. ATP hydrolysis of D1 would cause a clockwise rotation of D1. *Top right*, the rotation of D1 causes a conformational change in D2 to allow ADP release, resulting in an ATP-ready or empty state while D1 returns to an ADP-bound state. *Bottom right*, ATP hydrolysis in D2 would cause a counter clockwise motion. *Bottom left*, the rotation caused by D2 would allow the release of ADP in D1, resulting in an ATP-ready state, while D2 returns to an ADP-bound state. The binding of ATP by D1 would initiate another hydrolysis cycle (*top left*). Modified from Rouiller et al. (2002). Used by permission. Reproduced from Hicks-Berger and Morré (2006) with permission from Wiley

The AAA-ATPase proteins typically assemble into oligomers and seem to have conserved functions in a folding or unfolding step catalyzed in an ATP-dependent manner or to convert the energy from ATP binding and hydrolysis into mechanical force (Dreveny et al. 2004). Electron microscopy of p97 hexamers shows that the proteins form a barrel-like structure with the two Walker domains arranged on top of each other to form two hexameric rings (Rouiller et al. 2000, 2002). A conceptual model for p97 acting as an ATP-driven molecular ratchet, with D_1 and D_2 hexameric rings forming two interconnected but negatively cooperating parts, was developed (Rouiller et al. 2000, 2002) and serves as the basis for proposed involvements of AAA-ATPases in mediating physical displacement of membranes as, for example, in membrane budding (Zhang et al. 2000) and cell enlargement (Fig. 5.32). As summarized in Fig. 5.33, the ratchet model is only one of several different models proposed for translation of conformational changes in AAA-ATPases during the ATPase cycle into force extensions (Pye et al. 2006), any of which could be accommodated into the diagram of Fig. 5.20 including roles for yet unidentified adaptor proteins such as ENOX1 and ENOX2.

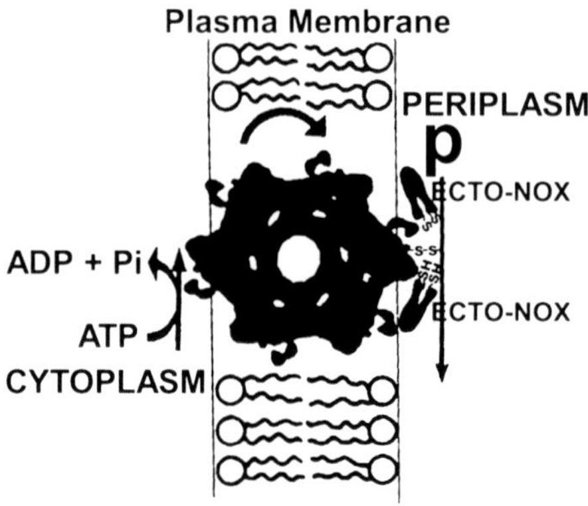

Fig. 5.33 Schematic interpretation of the organization of the plasma membrane AAA-ATPase hexamer within the bilayer as observed by transmission electron microscopy. Binding of ATP and subsequent hydrolysis and release of ADP would advance the AAA-ATPase hexamer ratchet linked through a disulfide bond to a structural membrane protein (p) on the opposite side of the membrane. As the AAA-ATPase hexamer rotates, the membrane protein (p) would be extended in the direction of the *arrow*. The function of the ENOX protein would be to periodically break and reform the disulfide linkages between the AAA-ATPase hexamer and the membrane protein (p) as depicted in the model of Fig. 5.20. Reproduced from Hicks-Berger and Morré (2006) with permission from Wiley

5.8 Summary

A close coupling of PMET and growth was indicated beginning with the earliest investigations of Frederick Crane, Hans Löwe and their colleagues. However, the significance of the correlation was not fully appreciated until the discovery that ENOX proteins exhibited a second enzymatic activity, that of protein disulfide–thiol interchange. A central role of the ENOX1 and ENOX terminal oxidases of PMET in the enlargement phase of cell growth is well established while effect on cell proliferation, i.e., cell cycle control may be indirect. The rate of cell enlargement and the rate of ENOX1 and/or ENOX2 at the cell surface, through use of activators, inhibitors, antitumor drugs or overexpression, are highly correlated. Both are blocked by thiol reagents. Additionally, the rate of cell enlargement is periodic as ENOX1 and ENOX2 activities with a major period of 24 min for ENOX1 and about 22 min for ENOX2. These observations initially were with cell elongation in plants but were subsequently extended to cell enlargement in HeLa and other mammalian cell lines. Enlargement is restricted to the protein disulfide–thiol interchange part of the ENOX cycle that normally alternates with hydroquinone (NAD(P)H) oxidation to generate the 24 min (22 min) ENOX1 (ENOX2) ultradian growth cycle.

Recombinant ENOX proteins when incorporated into synthetic lipid vesicles with other required constituents reconstitute physical membrane displacements (i.e., vesicle budding or cell enlargement) in completely cell-free systems as an approach to elucidations of functions in growth. Isolated inside-out plasma membrane vesicles from plants, when supplied both with ATP and auxin to stimulate ENOX1 activity, undergo cell-free vesicle enlargements equivalent to a doubling in volume over 20 min.

Implicit in the findings with ENOX proteins is the realization that the mechanism of cell enlargement in both plants and animals is similar and not exclusively a passive process resulting from osmotic stretching or even due to an active process obligatorily driven by biosynthesis. It is the result of ENOX1- or ENOX2-driven physical membrane displacement. Additionally, cell enlargement emerges as having an energy requirement. An energy requirement is universal among membrane displacement models and is met at the cell surface through coupling with a plasma membrane-associated AAA-ATPase.

Chapter 6
Roles as Ultradian Oscillators of the Cells Biological Clock

6.1 Time Keeping Properties

The considerable body of evidence reviewed in this chapter supports the hypothesis that ENOX proteins function as core ultradian oscillators of the cells' biological clock (Morré et al. 2002a). This hypothesis followed from the observed alternation of the two enzymatic activities they catalyze, hydroquinone (NADH) oxidation and disulfide-thiol interchange, to generate a regular period length of 24 min for ENOX1 (Morré and Morré 2003a; Chap. 1) and of 22 min for ENOX2 (Chap. 8). ENOX1 carries out hydroquinone oxidation or oxidation of external NAD(P)H for 12 min (two bursts separated by 6 min) and then that activity rests. While the hydroquinone (NADH) oxidative activity rests, the ENOX1 proteins engage in disulfide-thiol interchange activity for 12 min (three bursts separated by 4.5 min). That activity then rests as the 24-min cycle repeats. The overall pattern, however, is more complex than just the alternation of two different enzymatic activities. As alluded to above, within the hydroquinone oxidation part of the cycle, there are two maxima separated by 6 min. Within the protein disulfide-thiol interchange portion of the cycle, there are three maxima separated by intervals of 4.5 min (Fig. 6.1). The periodic behavior is exhibited by pure recombinant ENOX1 and ENOX2 proteins (Fig. 6.2) and is accompanied by a recurring pattern of amide I/amide II FTIR and CD spectral changes suggestive of α-helix–β-sheet transformations (Morré and Morré 2003a; Kim et al. 2005; Fig. 6.3). The periodic oscillations associated with the time-keeping activities of ENOX proteins require copper (Fig. 6.3b) and appear to be inherent in the molecular structure of the Cu^{II} aqua ion itself (Jiang et al. 2006). Ultradian rhythms are quite common in living organisms and most relate in some manner to the classical 24-h circadian rhythm (Lloyd and Stupfel 1991). Additional examples of ENOX-catalyzed oscillations and methods of measurement are given in Chap. 2.

D.J. Morré and D.M. Morré, *ECTO-NOX Proteins: Growth, Cancer, and Aging*,

DOI 10.1007/978-1-4614-3958-5_6,

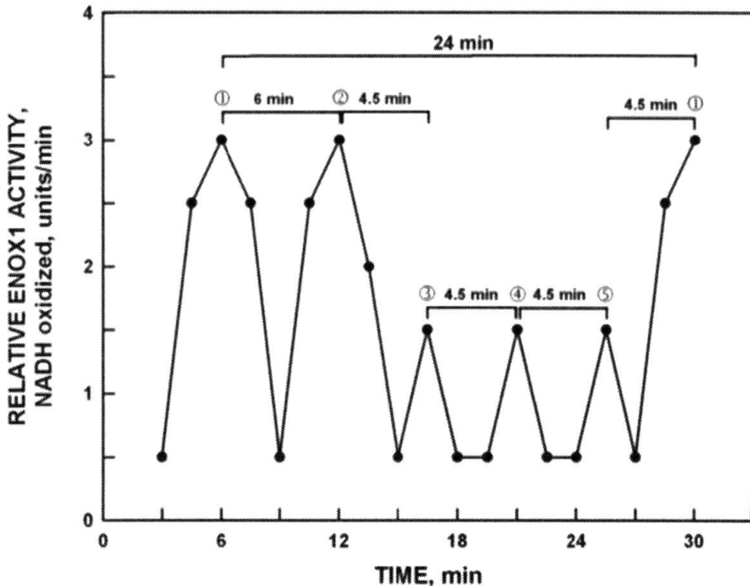

Fig. 6.1 Diagrammatic representation of the signature ENOX cycle with measurements averaged over 1 min at intervals of 1.5 min. Typically there are five maxima, two of which ① and ② are separated by 6 min and three of which ③, ④, and ⑤ are separated from each other and from ① and ② by 4.5 min. The asymmetry defines a 24-min period [6 min + (4 × 4.5 min)]

6.2 Molecular Studies

Investigation of the time-keeping functions of ENOX proteins has been facilitated by the availability of both ENOX1 (Chap. 3) and ENOX2 (Chap. 8) as recombinant proteins expressed in bacteria (Chueh et al. 2002b; Jiang et al. 2008). The expressed proteins have the same characteristics and properties as the membrane-associated proteins demonstrating that ancillary proteins are not involved in their activity.

The ENOX proteins exist in solution principally as dimers. Two moles of copper are bound per ENOX monomer (Jiang et al. 2008; Tang et al. 2010). The net result is a dimeric sharing of four copper atoms (two copper atoms/monomer) to allow for a concerted transfer of the required four electrons from NADH to reduce molecular oxygen to water (Chap. 1; Fig. 1.4).

To implicate ENOX proteins as the ultradian time keepers (pacemakers) of the biological clock, COS cells were transfected with cDNAs encoding ENOX2 proteins having a period length of 22 min or with C575A or C558A cysteine to alanine replacements having period lengths of 36 or 42 min (Morré et al. 2002a; Fig. 6.4). This differed from replacements of cysteines of ENOX2 numbered 505 and 569 which were required for activity as part of the disulfide-thiol exchange activity site (Chueh et al. 2002b). ENOX2 transfectants carrying the C575A or C558A to alanine replacements exhibited 36 or 40–42 h circadian patterns in the

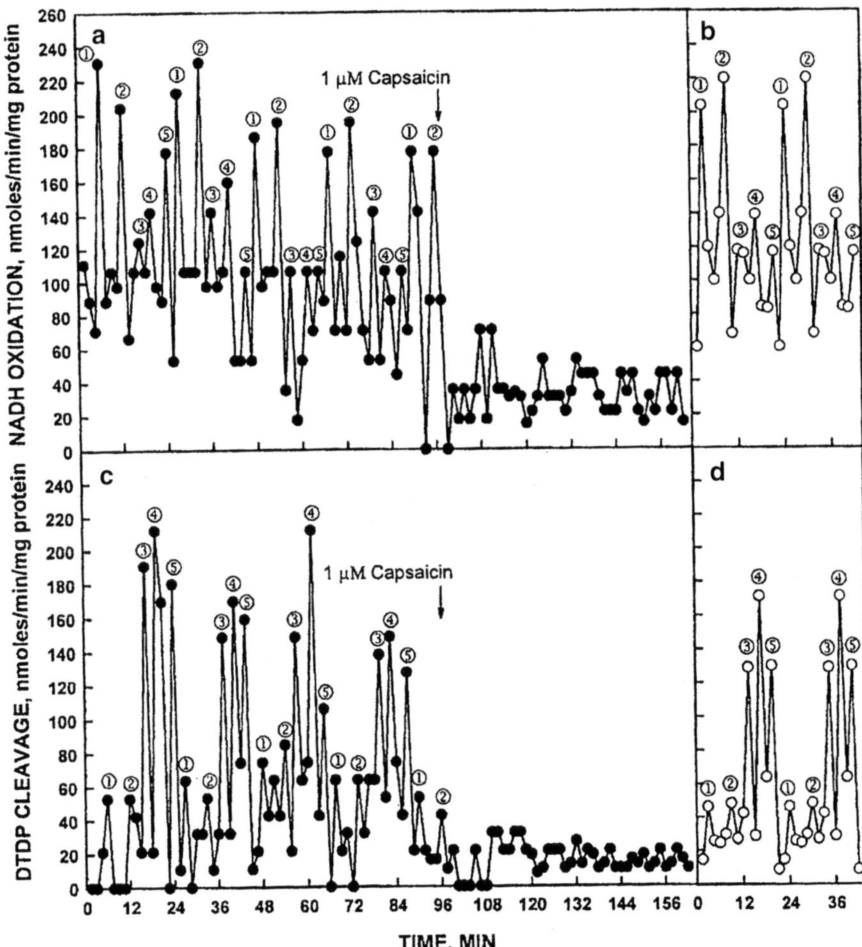

Fig. 6.2 Activity measurements of cellulose binding domain (CBD)-tagged recombinant ENOX2 (truncated ENOX2 (CBD)) at 37 °C as a function of time over 100 min at 35 °C. The truncated ENOX form has a shorter period length of 21 min compared to 22 min for full length ENOX2. Capsaicin (1 μM) was then added and the assays were continued for an additional 60 min. (**a**) Rate of oxidation of NADH. A repeating pattern of five maxima at intervals of 21 min (*arrows*). Two maxima separated by 5 min, ① and ②, dominate. (**b**) Decomposition fit of 0–100 min of (**a**) to show the reproducibility of the oscillatory pattern conforming to a period length of 21 min. The accuracy measures obtained were mean average percent error (MAPE) 21.43, mean average deviation (MAD) 10.7, and mean standard deviation (MSD) 3.1. (**c**) Protein disulfide-thiol interchange. The three maxima, ③, ④, and ⑤, separated by 4–4.5 min alternate as a triad with the two maxima in rate of NADH oxidation. (**d**) Decomposition fit of 0–100 min of (**c**) to show the reproducibility of the oscillatory pattern over two full cycles of 21 min each. The accuracy measures obtained were MAPE 50.0, MAD 19.0, and MSD 2.1. The estimation of NADH oxidation was based on the decrease in A_{340} as described (Morré et al. 1995b). Protein disulfide-thiol interchange was determined from the cleavage of an artificial dithiodipyridine (DTDP) substrate indicated in (**c**) (Morré et al. 1999a). Both activities were inhibited completely by 1 μM capsaicin added after 100 min (*arrows*). Decomposition fits for two full periods show the reproducibility of the oscillations and a period length of 21 min. Reproduced from Kim et al. (2005) with permission from The International Hormesis Society

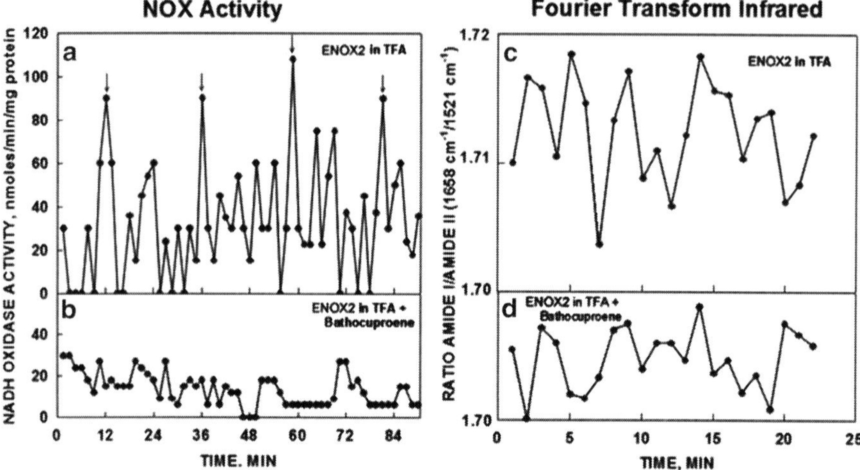

Fig. 6.3 Activity of recombinant ENOX2 protein unfolded in the presence of 0.1 % trifluoroacetic acid (TFA) alone (**a**, **c**) or in the presence of 25 μM bathocuproine to remove bound copper (**b**, **d**). Upon removal of the TFA by dialysis, activity in the absence of bathocuproine was retained (**a**, **c**). Upon removal of TFA from the bathocuproine-treated sample no activity was observed (**b**, **d**) but activity was restored by addition of 100 μM $Cu^{II}Cl_2$ (not shown). Similar results were obtained for recombinant human ENOX1 (Jiang et al. 2008) and for a plant ENOX1 isolated from plasma membranes of soybean (Jiang et al. 2006). Changes in circular dichroism and infrared absorbance suggestive of α-helix–β-sheet transitions accompany the activity oscillations (**c**) and also require the presence of copper (**d**). Reproduced from Morré and Morré (2008b) with permission from Springer Science + Business Media

activity of glyceraldehyde-3-phosphate dehydrogenase (GAPDH), a common clock-regulated protein, in addition to the endogenous 24-h circadian period length (Fig. 6.5). The fact that the expression of a single oscillatory ENOX protein determined the period length of a circadian biochemical marker (60 × the ENOX period length) was interpreted as compelling evidence that ENOX proteins were the biochemical ultradian drivers of the cellular biological clock.

When assayed for GAPDH activity at 4-h intervals over 76 or 88 h, a major 24-h circadian rhythm of alternating maxima and minima seen with nontransfected (wild type) COS cells was augmented in the GAPDH activity of COS cells transfected with ENOX2 by a second set of maxima with the period length of 22 h reflective of the 22-min period length of ENOX2 (Morré et al. 2002a).

To demonstrate the periodicities resulting from transfection, the activities of wild-type cells measured in parallel were subtracted from the activities of the transfected cells. As illustrated in Fig. 6.5c, d, with transfectants carrying the C575A replacement with a 36-min period. A new set of major oscillations spaced at intervals of 36 h (ENOX period in minutes × 60) was revealed within the circadian pattern of GAPDH activity (Fig. 6.5c). Similarly, with the C558A replacement with a 42-min ENOX2 period, a 42-h circadian rhythm based on analyses of GAPDH was revealed (Fig. 6.5d). A CKSC220A replacement in human ENOX1 did not result in

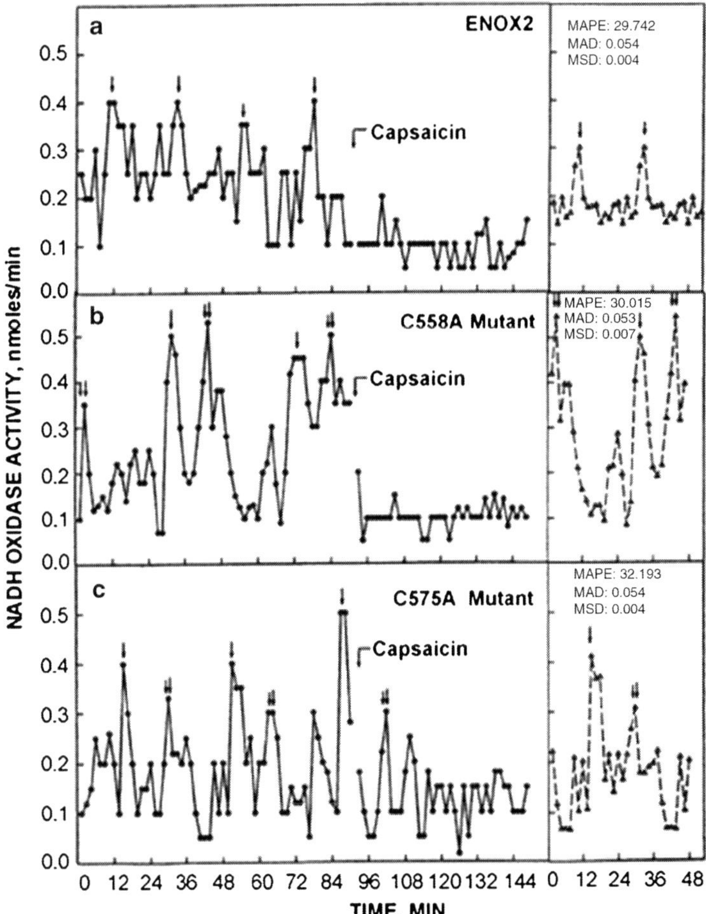

Fig. 6.4 Effect of cysteine to alanine replacements on the rate of NADH oxidation and period length of recombinant ENOX2 proteins. (**a**) Wild-type recombinant ENOX2. The period length is 22 min (*single arrows*) and the activity is inhibited by capsaicin. (**b**) Recombinant C556A replacement ENOX2. The maxima are a doublet (*single* and *double arrows*) with a period length of 42 min. Both members of the doublet were inhibited by capsaicin. (**c**) Recombinant C575A replacement ENOX2. The maxima are a doublet (*single* and *double arrows*) with a period length of 36 min. Both members of the doublet were inhibited by capsaicin. Decomposition fits (*filled triangles* and *dashed lines*) for two (**a**) or one (**b**, **c**) full periods presented in the panels on the *right* show the reproducibility of the patterns of oscillations. The two maxima patterns were noted previously as a feature of ENOX function. Reproduced from Morré et al. (2002a) with permission from American Chemical Society Publications

loss of activity but did change the length of the oscillatory period from 24 to 32 min (Jiang et al. 2008). Similarly yeast cells transfected with the CKSC120A replacement cDNA exhibited both the normal 24-h circadian cycle in GAPDH activity plus a 32-h circadian cycle corresponding to the CKSC120A replacement (S. Dick, unpublished results). The results are summarized in Tables 1.3 and 6.1.

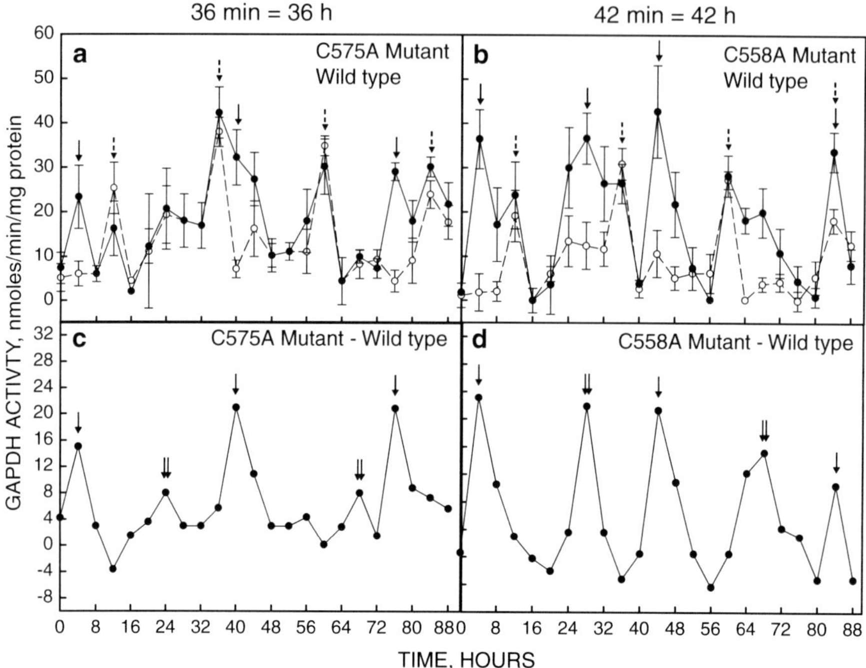

Fig. 6.5 Effect of cysteine replacements with alanine on the corresponding length of the circadian period of COS cells as determined from measurements of glyceraldehyde-3-phosphate dehydrogenase (GAPDH) activity in response to transfection. (**a**) The C575A replacement with an ENOX period length of 36 min exhibited maxima in GADH activity (*solid arrows*) absent from wild type (*open symbols*, *dashed lines*, and *broken arrows*). Maxima and minima were significantly different ($p<0.005$). (**b**) As in (**a**) except for the C558A replacement. (**c**) GAPDH activities of wild-type COS cells subtracted from those carrying the C575A replacement both showed a major (*single arrows*) and a minor (*double arrows*) circadian period length of 36 h (ENOX period in minutes × 60). (**d**) As in (**c**) but for the C558A replacement having a period length of 42 min and a circadian period length of 42 h ($p<0.005$). Reproduced from Morré et al. (2002a) with permission from American Chemical Society Publications

The period lengths of the activity oscillations of the NOX proteins are independent of temperature rather than temperature compensated (Sect. 6.5; Morré and Morré 1998; Pogue et al. 2000; Wang et al. 2001; Morré et al. 2002a; Table 6.2), and their phases are entrainable (Morré et al. 1999b, 2002b, c; Morré and Morré 2003e). These two characteristics, temperature independence (compensation) and entrainment (coupling the intrinsic clock to environmental cues), are two defining hallmarks of the biological clock (Lloyd et al. 1982; Dunlap 1996; Edmunds 1998). ENOX synchrony is achieved through autosynchrony in solution (Morré et al. 2002b), by coupling to red (plants) (Morré et al. 1999b) and blue (plants and animals) (Morré et al. 2002c; Morré and Morré 2003e) light photoreceptors and in direct response to

Table 6.1 Effect of site-directed mutagenesis of a recombinant truncated ENOX2 on NADH oxidase enzymatic activity, period length, and inhibition by capsaicin

Replacement[a]	Enzymatic activity[b]	Period length (min)	Inhibition by 1 μM capsaicin
None	+	22	+
C505A	–		
C510A	+	36	+
C558A	+	42	+
C569A	–		
C575A	+	36	+
C602A	+	36	+
M396A	+	22	–
H546A	–		
H562A	–		
G592V	–		

[a]Resequencing confirmed the expected DNA sequences for each of the indicated amino acid replacements

[b]Absolute specific activities were determined from detailed kinetic analyses of both NADH oxidation and protein disulfide-thiol interchange for each mutant. For example, with C505A and C569A replacements both NADH oxidase and protein disulfide-thiol interchange activities were absent. From Chueh et al. (2002b)

Table 6.2 ENOX period length is independent of temperature

	Soybean plasma membrane (Morré and Morré 1998)	Milk-fat globule membrane (Morré et al. 2002b)	HeLa plasma membrane (Wang et al. 2001)
Temperature (°C)	Period length (min)		
17	24.4 ± 1.3	23.8	24.5 ± 0.0
27	24.1 ± 1.9	23.8	24.7 ± 1.1
37	23.8 ± 2.4	23.5	24.0 ± 0.9
Average	24.1 ± 0.3	23.7 ± 0.2	24.4 ± 0.4

Period length is also temperature independent for:

- ENOX1 activity and enlargement of CHO cells and ENOX1 activity of CHO cell plasma membranes (Pogue et al. 2000)
- ENOX2 activity and enlargement of HeLa cells (Wang et al. 2001)
- Enlargement of soybean cells and elongation of soybean stem sections (Morré et al. 2001a)

melatonin (Morré and Morré 2003e) or to low frequency electromagnetic fields (LFEMF) (Morré et al. 2008b). As such, a clock function would be a secondary, albeit important, function of synchronizing the activities of living organisms to the 24-h day/night earth's rotational cycle.

In plants, the circadian clock is especially important to appropriately time physiological processes, ranging from drought resistance to hybrid vigor (Harmer 2010).

6.3 Studies with Deuterium Oxide

Deuterium oxide (^2H_2O, D_2O, heavy water) is an almost singular example of a substance that significantly and reproducibly alters the length of the circadian day in a wide range of organisms. That deuterium oxide lengthens the ENOX1 period in proportion to its effects on the circadian day (Fig. 6.6) is among the strongest evidence to support a role for ENOX1 as an ultradian oscillator of the cells' biological clock. Deuterium oxide reversibly lengthens both the period and the phase of the period of the circadian day in a variety of organisms (Bruce and Pittendrigh 1960; Suter and Rawson 1968). *Euglena gracilis* when adapted for long periods of several months to D_2O exhibits a period of the running circadian rhythm of phototaxis lengthened from its normal value (close to 23 h) to 28 or 29 h (Bruce and Pittendrigh 1960). These and subsequent studies by Suter and Rawson (1968) and Enright (1997) indicated that the effect of D_2O to lengthen the circadian day is of widespread occurrence. D_2O lengthens the circadian period in green plants (Bunning and Baltes 1963), in isopods (Enright 1997), birds (Palmer and Dowse 1969), mice (Suter and Rawson 1968; Palmer and Dowse 1969; Dowse and Palmer 1972; Richter 1977), hamsters (Richter 1977; Pichard and Zucker 1986; Lesauter and Silver 1993), insects (Pittendrigh et al. 1973), and unicellular organisms (Bruce and Pittendrigh 1960; McDaniel et al. 1974). Pittendrigh et al. (1973) suggest that the effect may be general since no exceptions were found in the 12 or more different species examined.

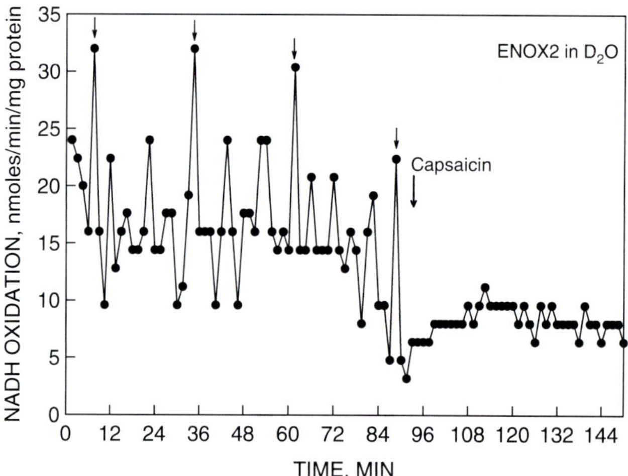

Fig. 6.6 NADH oxidation and inhibition by capsaicin of nus-tagged recombinant ENOX2 at 37 °C measured in the presence of a reaction mixture in which water was replaced by D_2O. The final NADH concentration was 150 μM and the entire system was buffered at pH 6.5 with 50 mM Tris–HCl. The period length was proportionately increased by 25 % from 22 to 25 min in D_2O compared to water. With the constitutive ENOX1, the period length in D_2O was increased from 24 to 30 min in keeping with the response of organisms to D_2O where the length of the circadian day is increased from 24 to about 30 h (see text). Reproduced from Morré and Morré (2008b) with permission from Springer Science + Business Media

6.4 The Role of Copper

Protein bound copper is required for ENOX activity and the characteristic 2+3 oscillatory pattern (Chap. 1). In copper depleting experiments, the material removed by chelation with bathocuproine was checked and exhibited a 2+3 pattern of NADH oxidation similar to that of the ENOX proteins except that rates of NADH oxidation were similar for all five oscillatory maxima instead of the ①, ② pattern of the oxidative portion of the ENOX cycle of the copper-containing ENOX1 protein (Fig. 6.3b; Jiang et al. 2006). Unexpectedly, the copper solutions of themselves exhibited the characteristic 2+3 oscillatory pattern characteristics of the ENOX proteins based on rates of NADH oxidation (Fig. 6.7). The major difference in the

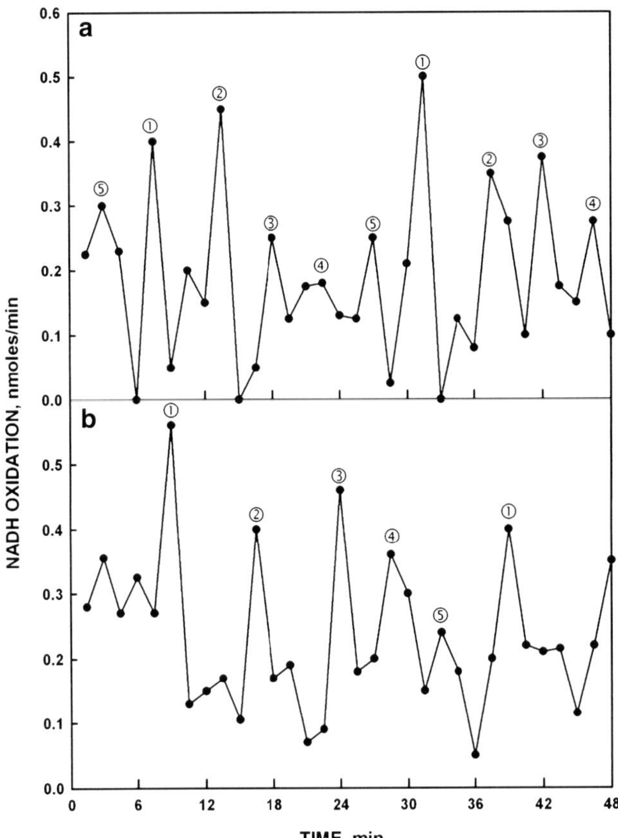

Fig. 6.7 Periodic catalysis of the oxidation of NADH by 25 μM aqueous Cu^{II}Cl$_2$ (**a**) with a period length of 24 and of 30 min with Cu^{II}Cl$_2$ solvated in D$_2$O (**b**). The final NADH concentration was 150 μM. The solution was buffered at pH 6.0 using 50 mM Tris–HCl. Within each period of 24 min two maxima are separated by 6 min (7.5 min for D$_2$O) and the remaining maxima are separated by 4.5 min (5.5 min for D$_2$O). The first maximum of the doublet separated by 6 min (7.5 min for D$_2$O) is designated as ① and the pattern repeats after the fifth maximum to generate the 24-min period (30 min for D$_2$O). Reproduced from Morré et al. (2007b) with permission from Elsevier

absence of protein was that the oscillations of the 2+3 pattern were approximately equal in magnitude. Asymmetry was still maintained with the maxima ① and ② being separated by 6 min whereas the maxima ③, ④, and ⑤ being separated from each other and from ① and ② by 4.5 min [6+(4×4.5)=24]. Apparently, the principal difference with the protein present is that redox potential of oscillations ③, ④, and ⑤ is somehow diverted into protein disulfide-thiol interchange. As for copper-containing ENOX proteins, the period length of the copper oscillations is both independent of temperature and independent of pH, copper concentration, and concentration of NADH (Jiang et al. 2006). Subsequently, a pattern of distortion in the four equatorial oxygens of the Cu^{II} aqua ion at a close distance relative to the two axial oxygen atoms at a longer distance indicated from extended X-ray absorption fine structure (EXAFS) studies was observed to underlie the changes in redox potential of copper in solution which are sufficient to catalyze NADH oxidation. These findings have been related to a metastable equilibrium condition in the ratio of energetically different *ortho* to *para* nuclear spin orientation of the hydrogen or deuterium atoms associated with the coordinated water molecules of the Cu^{II} aqua ion (Sect. 6.6).

6.5 The Copper Clock

Based on the findings that solvated Cu^{II} as the chloride or in combination with other anions alone was sufficient to catalyze NAD(P)H oxidation with the characteristic 2+3 periodicity, a copper "clock" was devised as illustrated in Fig. 6.8. The clock accurately predicted the precise times of initiation of new 24-min cycles. Aqueous copper solutions also exhibited periodic oscillations in redox potential sufficient to oxidize NAD(P)H (Fig. 6.9) with the same 2+3 oscillatory pattern as that of the copper clock. The temperature independence of the biological clock could now be understood as the consequence of a physical rather than a chemical basis for the timing events rather than some elaborate system of temperature compensation of a series of chemical reactions.

A striking feature of the copper clock was its high degree of synchrony. Copper solutions stored for weeks or months exhibited the same level of synchrony as those freshly prepared, i.e., they did not become less well synchronized with time. With unshielded samples, the solutions remained synchronous but were observed to occasionally depart abruptly from regular repetition by being displaced forward or backwards by a few minutes as if the copper clock had been phased (synchronized) by some external force, most probably fluctuations in the earth's electromagnetic field (Sect. 6.12.1; Morré et al. 2008b).

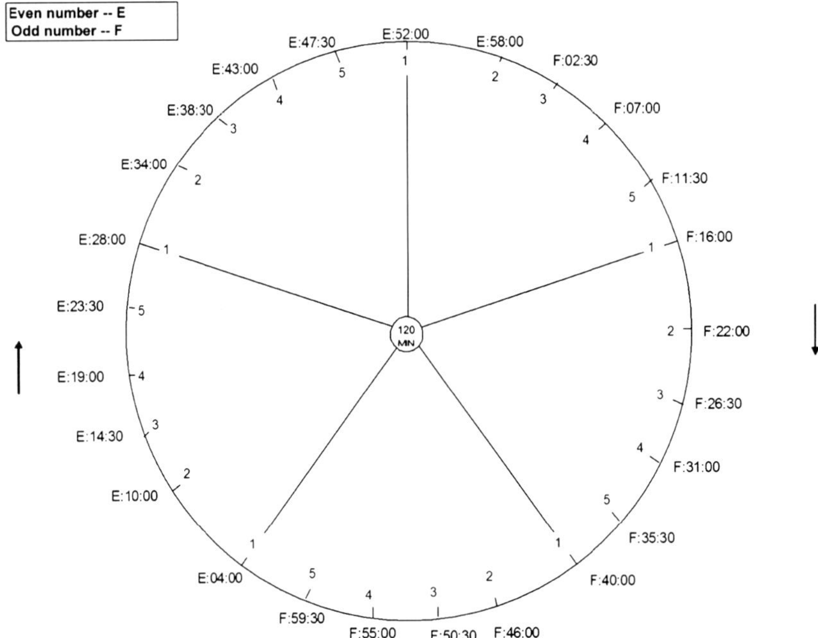

Fig. 6.8 The copper clock. Reproduced from Morré and Morré (2008b) with permission from Springer Science+Business Media

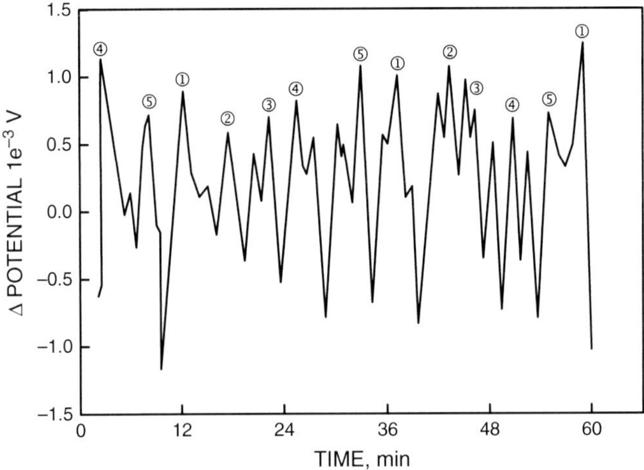

Fig. 6.9 Oscillations in redox potential of an unbuffered 10 mM solution of $Cu^{II}Cl_2$ as determined using a platinum electrode and a CH Instruments (Austin, TX), Model 600 B Series Electrochemical Analyzer/Workstation system as a function of time under oxygen-depleted conditions. The five maxima pattern is reproduced except that several of the maxima appear to bifurcate and are represented as doublets which may be reflective of the enhanced resolution of the method. Redrawn from Morré et al. (2007b)

6.6 EXAFS Investigations

In order to probe the physical basis for phased, highly synchronous oscillations in redox potential and rates of NAD(P)H oxidation, the local environment of the Cu^{2+} aqua ion in copper chloride solutions (Fig. 6.10) was investigated by X-ray absorption spectroscopy (Morré et al. 2007b). Detailed EXAFS analyses revealed a pattern of oscillations resembling closely those of the copper-catalyzed oxidation of NADH.

When individual values for aqueous $Cu^{II}Cl_2$ were plotted as a function of time for a k (Å^{-1}) value of 9, an oscillatory pattern was observed (Fig. 6.11) similar to that observed for rates of NAD(P)H oxidation. In the pattern of alternating maxima and minima, two of the maxima were separated by an interval of 6 min and the remaining maxima were separated by intervals of 4.5 min just as with measurements of NADH oxidation. A 24-min periodic spacing was confirmed by Fourier analysis (Fig. 6.12a). Statistical analysis of the changes in amplitude utilized decomposition fits to create the overall pattern of alternating maxima and minima based on the total time period of data collection of 92 min (Fig. 6.12b). Several such decomposition fits superpositioned and averaged for 8.9–9.1 k (Å^{-1}) to allow for calculation of standard errors revealed statistically different maxima and minima within the pattern of oscillations that was characteristic of both Cu^{II}-catalyzed oxidation of NAD(P)H and of changes in redox potential of the aqueous $Cu^{II}Cl_2$ solutions (Fig. 6.13).

When carried out in D_2O, the Fourier analysis now revealed extension of the period length from 24 to 30 min (Fig. 6.14a). The decomposition fits yielded patterns of oscillations similar to those observed for $Cu^{II}Cl_2$ in water except that the intervals between maxima and minima were increased by about 25 % in proportion to the overall lengthening of the period from 24 to 30 min (Fig. 6.14b).

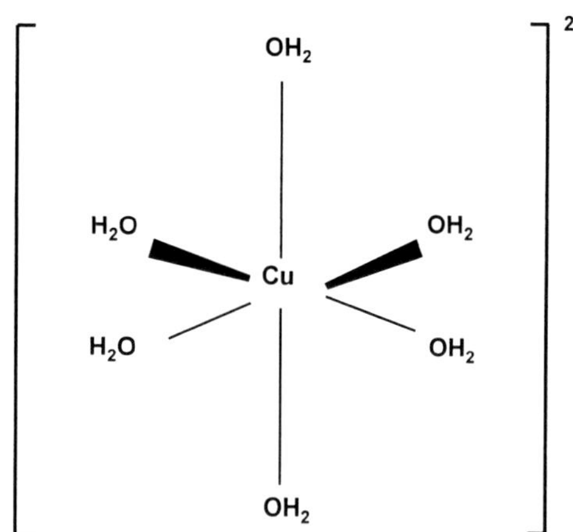

Fig. 6.10 The Cu^{II} hexaaqua ion

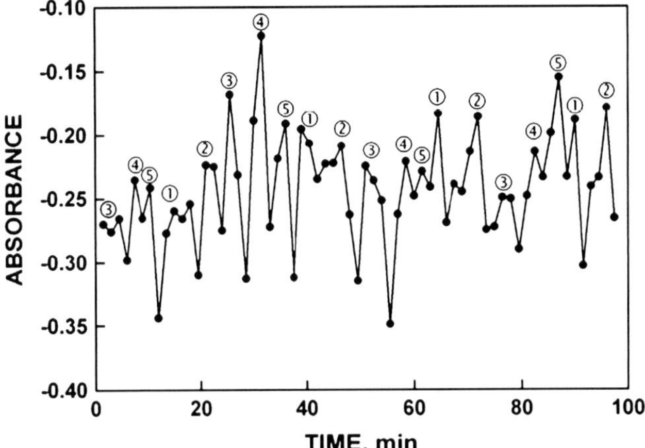

Fig. 6.11 Extended X-ray absorption fine structure (EXAFS) absorbance data of k^2 weighted $\chi^{(k)}$ spectra for a 0.1 M aqueous solution of $Cu^{II}Cl_2$ at 25 °C and 170 bar plotted as a function of time plotted at a k (Å^{-1}) of 8.9. Measurements were collected at ca. 1.5-min intervals (±0.1 min) over 90 min. Plotted values are real times with the plotted times corresponding to the times at the end of the measurements. Maxima ① and ② are separated by 6 min whereas maxima ③, ④, and ⑤ are separated from each other and from ① and ② by 4.5 min [6+(4×4.5) min=24 min]. Reproduced from Morré et al. (2007b) with permission from Elsevier

Superposition of several such decomposition fits revealed significant differences between maxima and minima within the repeating 2+3 five-peak pattern of oscillations (Fig. 6.15).

While the EXAFS studies did suggest involvement of distortion of the sixfold coordination in the axial and/or equatorial oxygen atoms of the coordinated water molecules with four short equatorial bonds and two longer axial bonds for $Cu^{II}Cl_2$ solution derived from previous EXAFS studies (Filipponi et al. 1994; Fulton et al. 2000; Korshin et al. 1998), the physical basis for the oscillations remained elusive. As the longer axial bonds have a large vibrational amplitude, they only contribute to the EXAFS at low k. The calculated distance changes between the copper and the water oxygens of the copper hexahydrate being relatively small, and the location of changes being at relatively high k, might indicate a greater involvement of the four short bonds of the copper aqua ion.

With copper solutions or with ENOX proteins, solvation in D_2O increases the period length from 24 to 30 min (Fig. 6.6) and for pure water from 20 to 24 min (Fig. 6.16). Heavy water is the only perturbation known to alter the period length of the circadian day and it is significant that both the NOX and copper clocks are similarly altered. Thus, focus turned toward the six water molecules of the Cu^{II} hexahydrate as the source of the oscillatory activity.

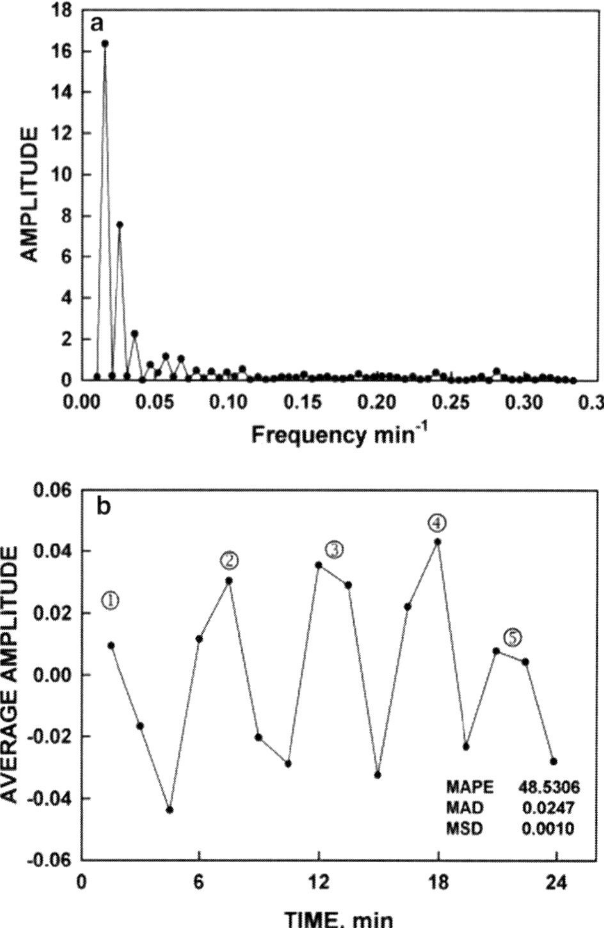

Fig. 6.12 Fast Fourier and decomposition analysis of data of Fig. 6.9. (**a**) Fast Fourier analysis revealed period lengths corresponding of 24, 36, and 60 min corresponding to measured frequencies of 0.017, 0.108, and 0.042/min. (**b**) Decomposition fit of the data utilizing a period length of 24 min determined by Fast Fourier analysis. The decomposition fits clearly show a pattern at k (Å^{-1}) of nine with recurrent maxima, two of which, labeled ① and ②, are separated in time by 6 min and three additional maxima separated in time by 4.5 min labeled ③, ④, and ⑤. The accuracy measures, MAPE, MAD, and MSD are indicative of a close fit between the original and the fitted data. Reproduced from Morré et al. (2007b) with permission from Elsevier

6.7 Oscillations Inherent in the Structure of Water

The oscillations with maxima separated by intervals of 4.5 or 6 min are slow by comparison even to Jahn–Teller equilibria (Bersuker 1984). The only atomic phenomenon that occurs on a similar time scale is the equilibration of the two alternative orientations of the nuclear spins (*ortho* and *para*) of the hydrogen atoms

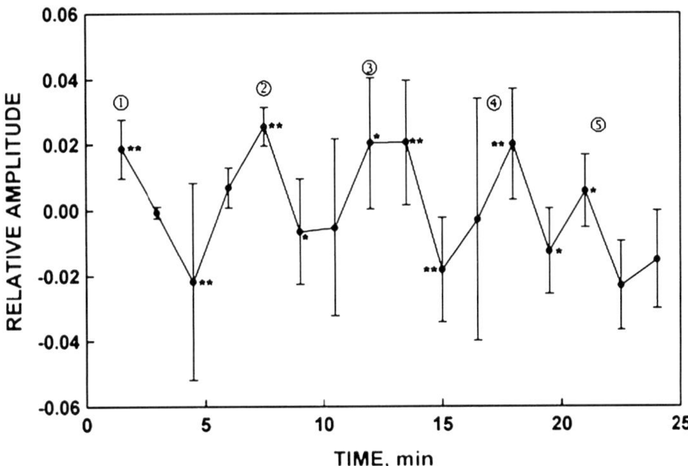

Fig. 6.13 Decomposition fits averaged from data collected at intervals of 0.1 k (Å^{-1}) in the range 8.9–9.1 ± standard deviations. Maxima and minima are significantly different (*$p<0.006$; **$p<0.003$) as determined by two-tailed t test comparisons. Values averaged over the range 8.8–9.4 k (Å^{-1}) were similar except for variations around the maximum labeled ④. Reproduced from Morré et al.(2007b) with permission from Elsevier

of water (Tikhonov and Volkov 2002). The hydrogen atom nuclear spins may either be parallel (total spin is 1) in *ortho* molecules or antiparallel (total spin is 0) in *para* molecules. Each spin isomer has its own system of rotational levels. The relaxation time to convert spin-modified ice to the 3:1 *ortho*:*para* equilibrium ratio has been estimated to require several months. Liquid lifetimes were measured as 26 ± 5 min for *para* water and 55 ± 5 min for *ortho* water. Metals and other impurities can result in *ortho*:*para* imbalance and conceivably an oscillatory nonequilibrium condition that would translate into a regular periodic pattern of energy states leading to redox changes ultimately affecting rates of NAD(P)H oxidation (Morré et al. 2007b).

Pure water shows fluctuations in redox potential similar to those seen with aqueous solutions of copper chloride and evidenced by periodic alterations in the rate of NADH oxidation (Fig. 6.16a; Morré et al. 2008d). The period length is about 18 min, however, compared to 24 min for copper chloride. Pure D_2O yielded a longer period length of about 24 min (Fig. 6.16b) following the pattern of ENOX oscillations in the presence of copper^{II} where the ENOX2 period in the absence of D_2O was increased from 24 to 30 min. (Fig. 6.17b).

Oscillations were observed as well from EXAFS measurements with pure water (Fig. 6.17; Morré et al. 2008d). Fast Fourier analysis and decomposition fits revealed a period length of 18 min comparable to that observed with redox potential amplified through changes in NADH oxidation (see Figs. 2 and 3 of Morré et al. 2008d). Also oscillating with an 18-min period were changes in redox potential determined using a platinum electrode (Fig. 5A of Morré et al. 2008d). Similar measurements for D_2O yielded a period length of about 24 min (Fig. 5B of Morré et al. 2008d).

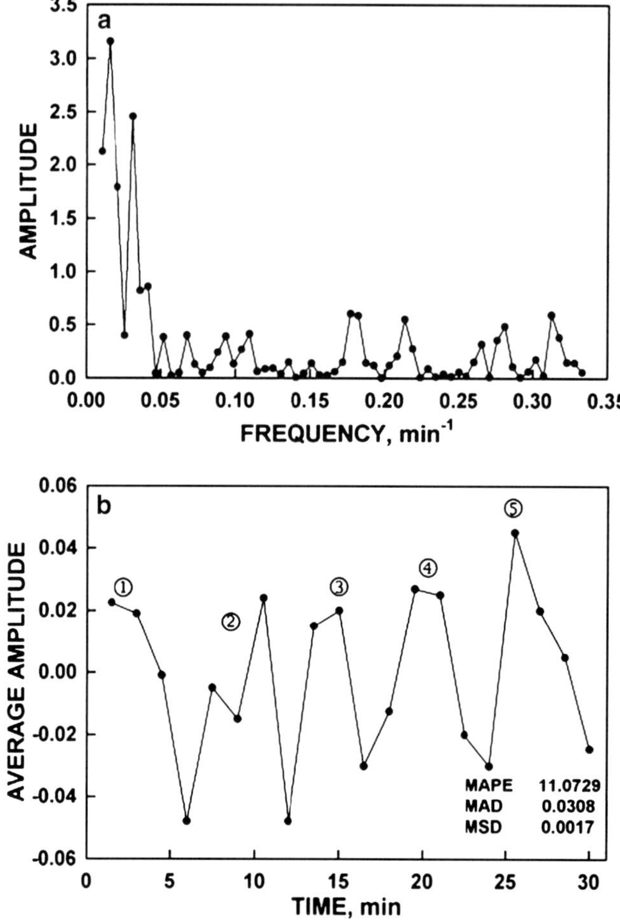

Fig. 6.14 As in Fig. 6.12 except that the $Cu^{II}Cl_2$ was solvated in D_2O and the determinations were for 60 min rather than 90 min. (**a**) Fast Fourier analysis at k (Å^{-1}) in the range of $10.0 < k < 10.4$ yielded period lengths of 30 and 60 min corresponding to frequencies of 0.0 17 and 0.033/min. (**b**) Decomposition fits at k (Å^{-1}) of ten utilizing a period length of 30 min revealed a five peak pattern within the 30-min period with maxima labeled ① and ② separated by 7.5 min and maxima labeled ③, ④, and ⑤ separated by intervals of 5.5 min. Maximum ② was resolved into two maxima not observed in averaged data (e.g., Fig. 6.9). The accuracy measures, MAPE, MAD, and MSD are indicative of a close fit between the original and the fitted data. Reproduced from Morré et al. (2007b) with permission from Elsevier

In order to monitor the *para*-H_2O/*ortho*-H_2O interconversion of water, the middle infrared spectral region suggested by Binhi and Stepanov (2000) and Mumma et al. (1987) as appropriate to such measurements was probed (Morré et al. 2008d). Infrared spectroscopic measurement of the *para*-H_2O/*ortho*-H_2O ratio above the water sample surface derived from the fact that the vibrational infrared spectra of water depend on its nuclear spin state. For a narrow spectral range encompassing

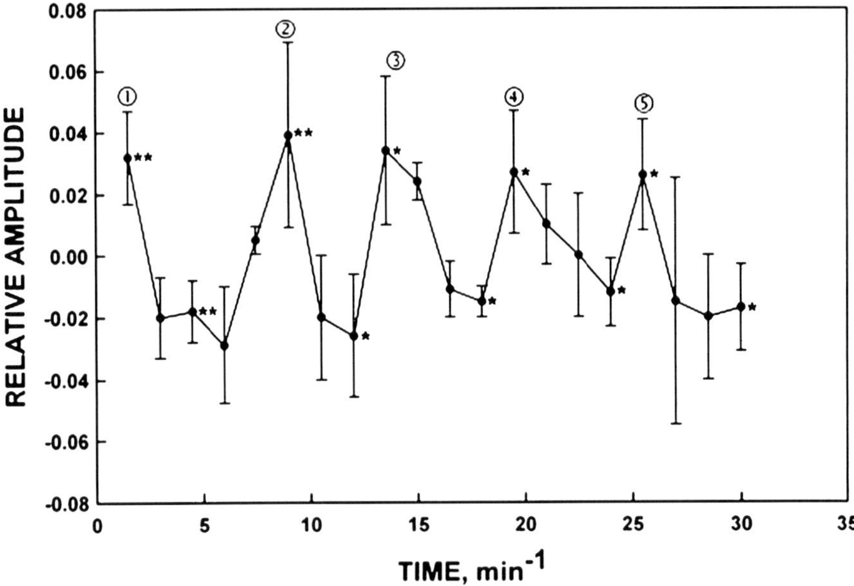

Fig. 6.15 As in Fig. 6.14 for data for $Cu^{II}Cl_2$ solvated in D_2O at intervals of 0.1 *k* ($Å^{-1}$) over the range of 9.8–10.1. Maxima and minima were significantly different ($*p<0.03$; $**p<0.009$) as determined by two-tailed *t* test comparisons. Values averaged over the range of 8.8–9.2 as in Fig. 6.13 were similar except for a less pronounced resolution in the regions of maxima labeled ④ and ⑤. Reproduced from Morré et al. (2007b) with permission from Elsevier

lines belonging to the different spin states of water-associated protons, the relative heights of lines will, on average, equal the natural *ortho* to *para* composition ratio in water of 3:1 (Binhi and Stepanov 2000).

Based on the infrared measurement, oscillatory patterns similar to those observed with aqueous copper solutions or with the ENOX proteins were seen to be inherent in the properties of pure water or of D_2O. Oscillations in the *ortho*:*para* hydrogen equilibrium of water and D_2O were recorded at 3,801 and 3,779/cm for water (Fig. 6.18) and in the range of 2,600–2,650/cm or 2,425–2,475/cm, for D_2O (not shown). For both, the oscillatory changes above baseline differed approximately threefold (Fig. 6.18a, b). Repeating patterns of five maxima at approximately 18-min intervals were observed. When analyzed by decomposition (time series) analyses (Fig. 6.18c–e), 2 + 3 patterns of oscillations were resolved similar to those observed previously for NAD(P)H oxidation and redox potential in aqueous copper solutions (Jiang et al. 2006; Morré et al. 2007b).

The time dependence of equilibration of the ratio of *ortho* to *para* spin isomers of water determined from far infrared spectra starting with *ortho*-enriched water vapor based on original data of Tikhonov and Volkov (2002) is shown in Fig. 6.19. Spins of the protons of water molecules satisfy the requirements necessary for the onset of the long-period oscillations. Experimentally, the relaxation process to reach equilibrium of 2:1 *ortho* to *para* ratio takes minutes to hours and even months, depending on temperature and some other conditions.

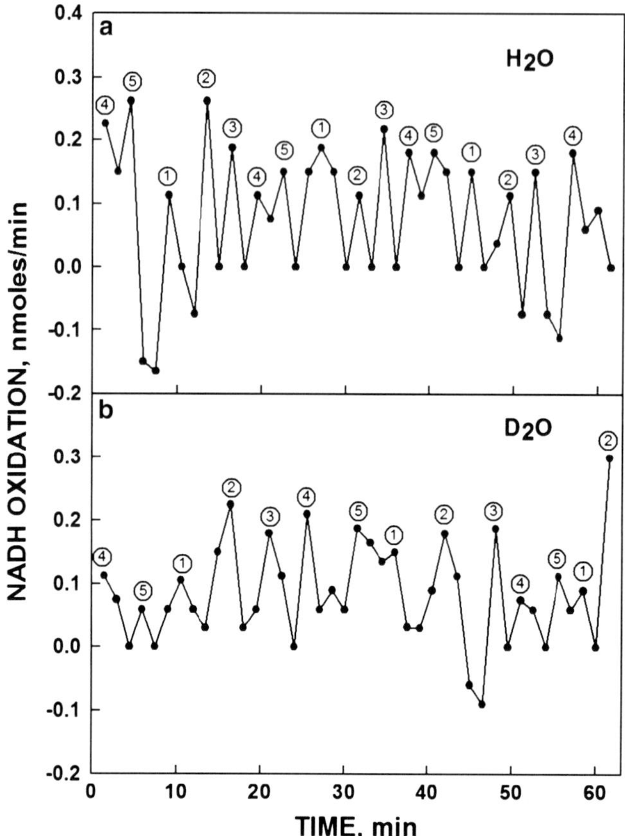

Fig. 6.16 Periodic catalysis of the oxidation of NADH by deionized, distilled water (**a**) with a period length of 18 min and D_2O (**b**) with a period length of 24 min. The final NADH concentration was 150 μM. The solution was unbuffered. Within each period of 18 min maxima labeled ① and ② are separated by 4.5 min for H_2O and by 5.8 min for D_2O. The remaining maxima designated ③, ④, and ⑤ are separated by 3.4 min for H_2O and by 4.4 min for D_2O. The pattern repeats after the fifth maximum to generate the 18-min period for H_2O or 24-min period for D_2O. Reproduced from Morré et al. (2008d) with permission from Elsevier

6.8 Period Length Determined by Ionic Radius of Liganded Cation

Chlorides in solution of other members of Group II, the so-called coinage metals, including silver and gold, also exhibited periodic oscillations but the period lengths were longer than those of copper. Some property of these different metals was then sought that might correlate with the period length of the oscillations. Ionic radii were found to do so (Fig. 6.20). Period length of the oscillations of an aqueous

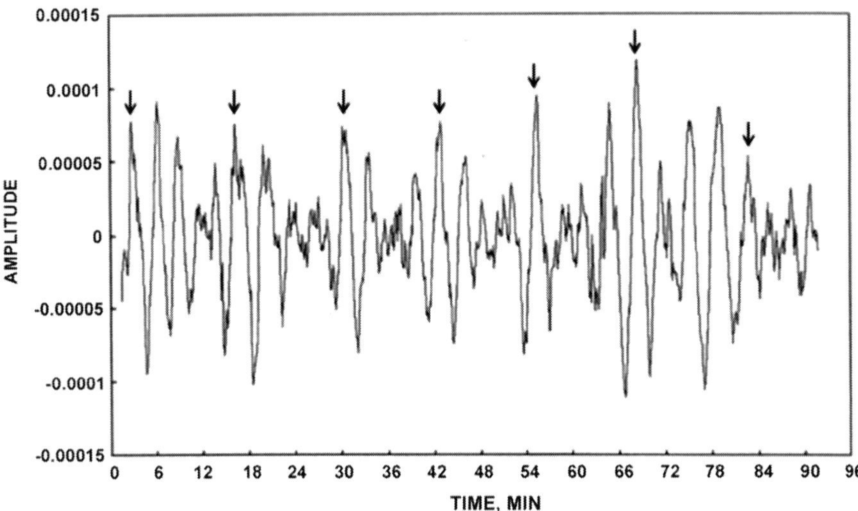

Fig. 6.17 Moving average of EXAFS measurements recorded at intervals of 10 s at a constant k ($Å^{-1}$) of 9.3 for deionized, distilled water. An oscillatory pattern was observed with a recurring pattern of five maxima (*arrows*). Fast Fourier analyses and decomposition fits revealed a period length of 18 min. Similar measurements for D_2O yielded a period length of about 24 min. Reproduced from Morré et al. (2008d) with permission from Elsevier

solution was directly proportional to the ionic radius of the cation present and independent of cation concentration. In the data of Fig. 6.20, all solutions were tested as the chloride at a final concentration of 10 μM. Only with Cu^{II} (replaceable by Ni^{II}) was the asymmetric period length of 5.8 min × 5 = 24 min observed. The 24-min ultradian period is apparently unique to copper and possibly nickel. The 22-min ENOX period may be modulated by zinc but not as a redox metal since the valence of zinc does not change. To what extent other metals might drive ultradian rhythms when appropriately linked to catalytic centers of proteins provide intriguing possibilities for future study.

As to why the period length of the *ortho–para* transition would increase from 18 to 24 min in the presence of Cu^{II} can be explained in the context of the previously observed response of *ortho*:*para* water to the ionic environment (Tikhonov and Volkov 2002). Period lengths of the oscillatory patterns of NAD(P)H oxidation for a series of aqueous ionic solutions were proportional to ionic radius with Cu^{II} uniquely giving rise to a 24-min period (Fig. 6.20). These observations serve to illustrate that period length of the oscillation in rates of NADH oxidation is very much dependent on the associated metal with Cu^{II} giving rise to the 24-min period for ENOX1 and perhaps a modulating effect of ENOX2-bound Zn^{II} for ENOX2 (Morré et al. 2002a).

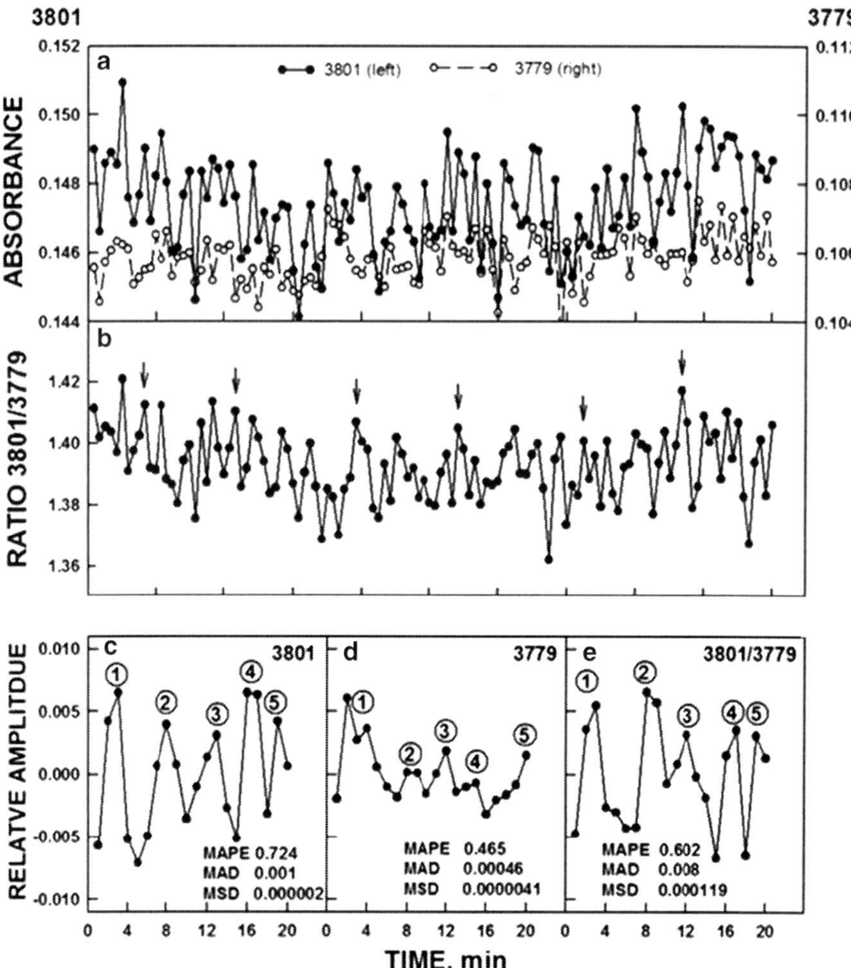

Fig. 6.18 FTIR spectroscopic measurement of the ratio of *para*-H_2O/*ortho*-H_2O above a water sample surface determined at 3,801 and 3,779/cm, respectively (**a**). The ratio of the two wavelengths exhibited a repeating pattern of oscillations of five maxima at intervals of about 18 min (*arrows*) (**b**). Decomposition fits using an imposed period length of 18 min of data collected at 3,801/cm (**c**), and at 3,779/cm (**d**) as well as the ratio of the two (**e**) revealed the oscillatory pattern typical of water with five recurrent maxima, two of which, labeled ① and ②, were separated in time by about 4.5 min and three additional maxima separated in time by 3.4 min labeled ③, ④, and ⑤. The accuracy measures, MAPE, MAD, and MSD are indicative of a close fit between the original and the fitted data. Reproduced from Morré et al. (2008d) with permission from Elsevier

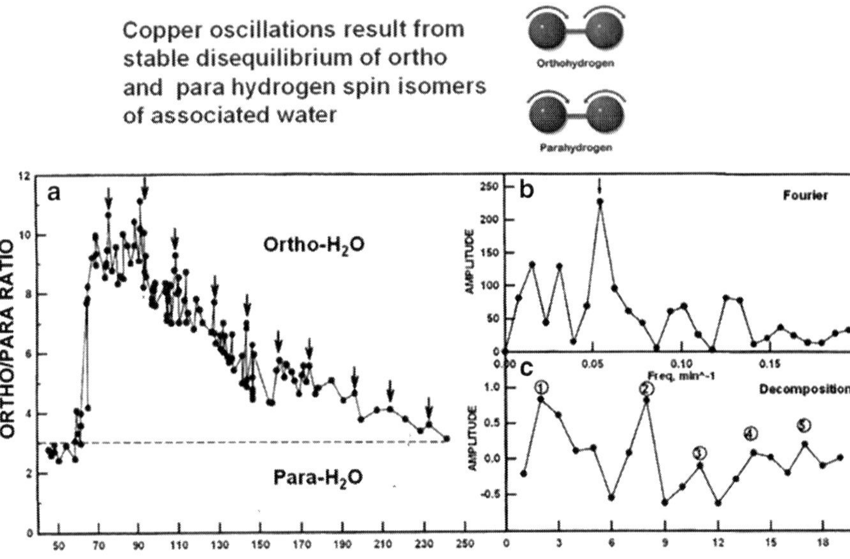

Fig. 6.19 (**a**) Time dependence of equilibrium of the ratio of the *ortho–para* spin isomers determined from far infrared spectra starting with *ortho* enriched (1:10) water vapor derived from the original data of Tikhonov and Volkov (2002). The lifetime for the water was estimated to be 55 ± 5 min. During equilibration, the *ortho–para* ratio oscillated with a ca. 19-min period length (*arrows*). (**b**) Period length determined by Fast Fourier Analysis (frequency arrow = 18.6 min). The amplitude of the effect exceeded the instrumental error (±5 %) by an order of magnitude. (**c**) Decomposition fit of the data of (**a**) for a 19-min period to show the five-peak (2 + 3) pattern characteristic of the ENOX-related time keeping

6.9 Spectral Evidence for Disequilibrium of *ortho:para* Spin States in Liquid Water That Oscillate

A spin state in liquid water that oscillates with a natural oscillatory cycle inherent to liquid water is proposed as the basis for the biological clock. The evidence for which has evolved experimentally from a clock-related protein-associated copper hexahydrate. A likely source may be instabilities in the water molecule itself, whose two hydrogen nuclei undergo ordered periodic shifts between *ortho* and *para* states. However, there is no clear mechanism as to how the ratio of *ortho* to *para* water is maintained in a steady oscillation. Why not simply reach the equilibrium ratio and remain there? Such phenomena are not without precedent in biology, however. An appropriate model might be that of a limit oscillator such as that which controls heart beat (Sect. 6.10).

Many materials preferentially absorb *para* water due to its nonrotation ground state. One possibility is that populations of water molecules spontaneously form some type of collective order whose structure subtly favors formation of *ortho* water. As *ortho* water forms, it might locally add to the field generated by the collective

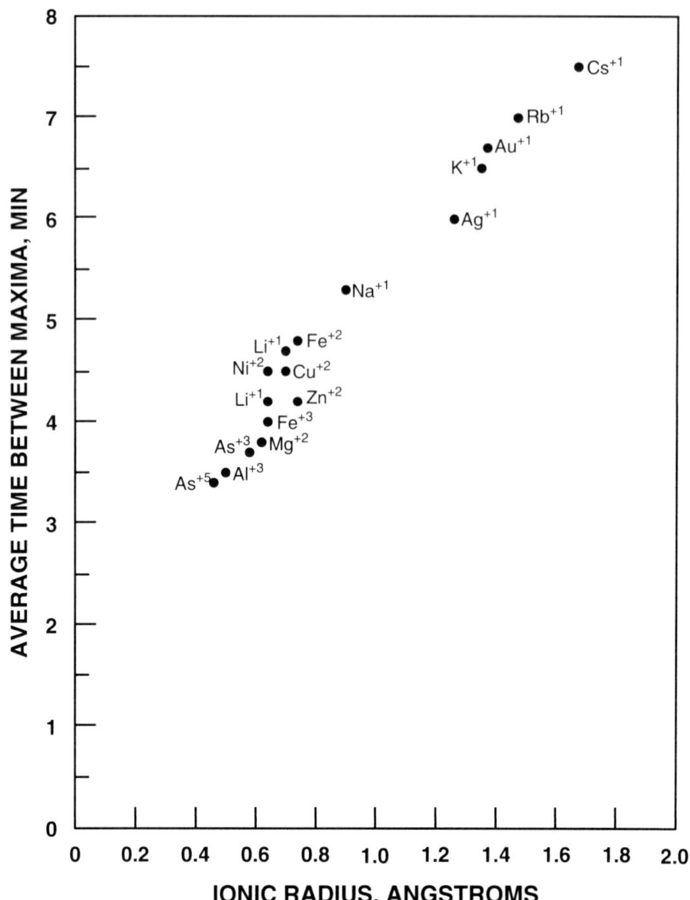

Fig. 6.20 Period length of the oscillations of an aqueous solution is directly proportional to the ionic radius of the cation present and independent of cation concentration. All solutions were tested as the chloride at a final concentration of 10 μM. Only with Cu^{2+} (replaceable by Ni^{2+}) was the asymmetric period length of 5.8 min × 5 = 24 min observed. Reproduced from Morré et al. (2008b) with permission from Elsevier

order which would further favor the accumulation of *ortho* water for some period of time. But then perhaps the initial condition favoring *ortho* water formation produces an increased potential which can accrue but cannot be released or discharged until it achieves some minimum threshold of activation energy. Once the minimum threshold is achieved there is a discharge of potential and the conditions favorable to *para* water result. The process now runs for a time in the opposite direction until some threshold level of *para* water is achieved and the cycle reverses (Fig. 6.21).

Spectral evidence for *ortho–para* spin isomer disequilibrium in liquid water has been provided by Pershin (2005, 2006) (Fig. 6.22). The spectrometric measurements were of backscattered Raman signal of water excited by the second-harmonic radiation of a Nd:YAG laser (a single 10-ns pulse or a train of such pulses with a repetition period of 1 s). The laser radiation is focused in a cell with room-temperature

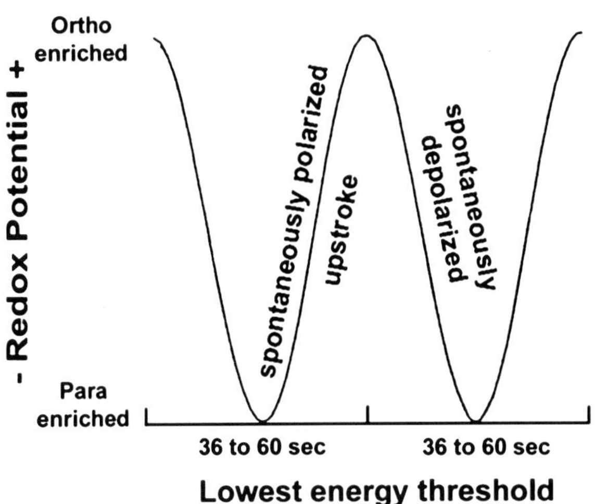

Fig. 6.21 Model based on the concept of limit oscillations to explain *ortho–para* oscillations and their failure to reach a steady-state equilibrium. *Ortho* water reaches some high energy threshold and then spontaneously converts to *para* water to reach some low energy threshold at which time the conversion reverses back to conversion to *ortho* water. Reproduced from Morré and Morré (2012a, b) with permission from WIT Press

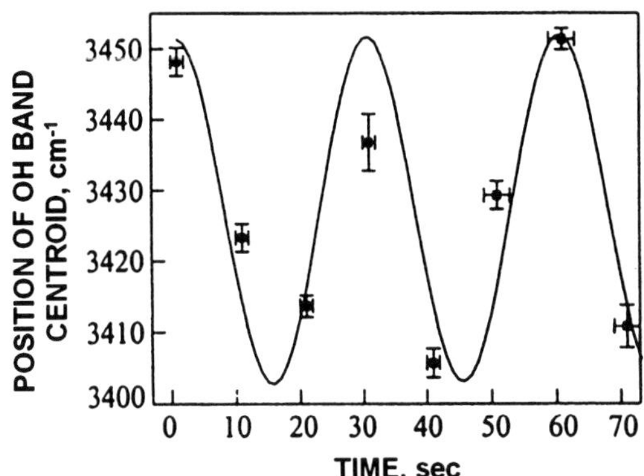

Fig. 6.22 Harmonic oscillations of the OH centroid band determined from Raman scattering as a measure of exchange between two water states related to *ortho/para* ratios. The period length was determined to be 35 ± 13 s. Reproduced from Pershin (2005) with permission from Springer Science + Business Media

water. The measurements were at laser intensities of 2, 15, 35, and 350 MW/cm^2. The Raman signal was focused on the entrance slit of a polychromator interfaced with an image intensifier and a diode array. Low-intensity measurements were carried out with a cooled diode array.

Spectral manifestation of a relatively high mobility of "hot" molecules was observed to result in the formation of a high-frequency wing of the OH band and a shift of the center of the band with a coefficient of about 1/cm/grad upon heating. Such a shift was interpreted as a decrease in the concentration of strong H-bonds as a result of disordering through destruction of polymer water and an increase in the number of complexes whose molecules can rotate (Pershin 2005, 2006).

The result is the experimental validation of the hypothesis of the independent existence of two liquids in water. Both liquids consist of hydrogen-bonded complexes of H_2O molecules, which exhibit characteristic frequency distributions of OH oscillators and form an integrated envelope of the OH-stretching vibration band.

These findings plus evidence that water vapor is a mixture of independent fractions of *ortho* and *para* modifications led Pershin (2005) to assume the existence of two independent states of liquid water with long lifetimes and two hydrogen bond types differing by energy. Specifically "the energy of the hydrogen bond between *ortho* isomers of molecules which always rotate seems to be lower than between *para* molecules, a part of which cannot rotate at room temperature."

The conclusion was reached that this overheating–overcooling process is a fundamental property of water, which manifests itself at any temperature and not the result of perturbation of overcooled water by an optical pulse. His findings show "that such an evolution of the band center is steadily observed in water and at room temperature. Moreover, this overheating-overcooling process in approximated by a harmonic function with a period of 35 ± 13 s without noticeable damping for ~16 min" (Fig. 6.22).

6.10 Other Mechanisms Proposed for *ortho/para* Conversions and Departures from Their Equilibrium Ratio of 3:1

Why the *ortho/para* hydrogen compositions should exhibit a regular pattern of oscillation from their equilibrium ratio of 3:1 remains unexplained but such variation would account for the oscillatory behavior of the redox potential and catalytic activity of aqueous solutions in relation to the proposed time-keeping model (Morré et al. 2002a).

Several mechanisms for the *ortho/para* conversion have been discussed in the literature. A paramagnetic mechanism for dihydrogen has been developed where the spin conversion is caused by the magnetic interaction of the hydrogen spins with paramagnetic centers such as the unpaired electron spins themselves (Wigner 1933). Normally, the *ortho/para* conversion is a slow process but paramagnetic impurities, i.e., unpaired electron spins or electrons with orbital momentum such as provided by metallic cations including copper would serve as catalytic centers for the interconversion process (Kummer 1962). In a recent model (Buntkowsky et al. 2006), *ortho/para*-H_2 molecules in one particular spin state were suggested to react with a suitable catalyst to create an X-H_2 complex. Within the complex a postulated evolution of the initial density matrix of the dihydrogen would initiate a partial conversion of one rotational state to the other. Upon decay of the bound state, the final density matrix would then be transferred to the free dihydrogen state.

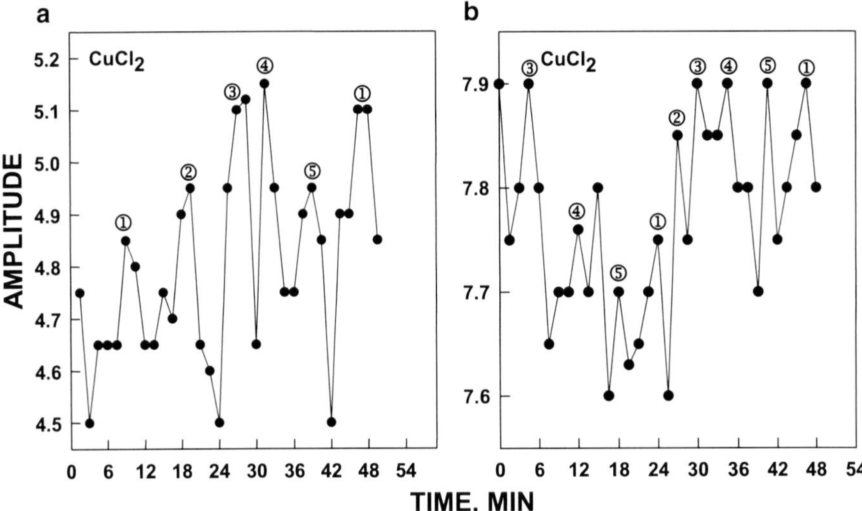

Fig. 6.23 EPR amplitude of unbuffered 10 mM solutions of $Cu^{II}Cl_2$ determined at liquid nitrogen temperature (110 K) oscillated with a period length of 22.8 min. (**a**) and (**b**) are from two different experiments. Reproduced from Morré et al. (2008b) with permission from Elsevier

Water molecules take on the two different spin states in a manner analogous to hydrogen molecules and the net interconversion of the two forms might occur by a similar mechanism even within the Cu^{II} hexahydrate. However, the oscillations we observe with water are apparently not due to the net final equilibrium position determining *ortho*/*para* conversions. Net *ortho*/*para* conversions are temperature dependent (Milenkl et al. 1997) and reduced in ice (Tikhonov and Volkov 2002). While amplitudes of the oscillatory phenomena reported here are temperature dependent, the period length by contrast is temperature independent (Jiang et al. 2006) and unchanged even at liquid nitrogen temperatures (Fig. 6.23).

The differences in energies of the different spin isomers of water are small (less than 10^{-24} erg) (Emsley et al. 1965) and are much lower than the energy of thermal motion. Therefore, a spin-only interaction would not be expected to affect intermolecular interactions (Emsley et al. 1965). On the other hand, the absorption rates of *ortho* and *para* water from water vapor to various organic and inorganic sorbents have been observed to differ markedly. The binding of the *para* isomer with such preparations is distinctly faster than the binding of the *ortho* isomer. Binding avidity is most likely determined from differences in quantum statistics for the two different spin isomers as noted by Potekhin and Khusainova (2005). Estimates of the energy barriers that determine rates of absorption suggested by these authors that the difference in free energy barriers may exceed the energy of spin–spin and spin–orbit interaction by many orders (Potekhin and Khusainova 2005). This raises the possibility that the spin state of water may substantially influence physical, chemical, and biological phenomena including redox potential.

6.11 The 24-min Period Has Properties of a Carrier Wave Generated from the Basic Underlying *ortho–para* Water Oscillations? The Heart Rate Model

Heart beat rate is an example of a limit-cycle oscillation in that the frequency of beats varies widely, while each individual beat continues to pump about the same amount of blood. Control of heart beat has its origins in a group of specialized cardiomyocytes (heart muscle cells) called the sinoatrial node or primary pacemaker. The action potentials that propagate along the plasma membranes of these excitable cells are generated by a sequence of changes in electrical potential between the interior of the cell and the surrounding extracellular space. The key to the rhythmic firing of pacemaker cells that makes it an attractive model of *ortho–para* water oscillations is that, unlike other muscle cells and neurons, the pacemaker cells slowly depolarize by themselves.

As *ortho* water forms, it might locally add to the field generated by the collective order (Sect. 6.14) which would further favor the accumulation of *ortho* water for some period of time (Fig. 6.21). But then it seems that the initial condition favoring *ortho* water formation as a result of this process produces an increased potential which can accrue but cannot be released or discharged until it achieves some minimum threshold of activation energy. Once the minimum threshold is achieved, there is a discharge of potential and the conditions favorable to *para* water result and the process now runs for a time in the opposite direction until some threshold level of *para* water is achieved and the cycle reverses.

Depolarization, a reduction in the degree of electron negativity in resting cells (the cells become more positive) occurs when positively charged sodium and calcium ions enter the cytosol to generate inward ionic currents. Depolarization continues until the threshold potential is reached between −40 and −50 mV. When threshold is reached, the cells enter phase 0 or upstroke. Following upstroke, the cardiomyocytes repolarize. Repolarization, the return to resting potential, occurs when outward currents restore the membrane potential to its resting level of electronegativity. In the heart, repolarizing currents are generated largely by potassium efflux, possibly supplemented by the inward movement of chloride ions.

Continuous recordings of heart rate data in both male and female subjects were analyzed by fast Fourier and time series analyses (decomposition fits) to determine if the periodicity yielded an asymmetric pattern of five maxima with two of the maxima separated by 6 min. Analyzed were both the continuous recording and the same data averaged over 1 min at 1.5-min intervals (Fig. 6.24).

Heart beat oscillates with a ca. 24-min period characteristic of *ortho/para* oscillations of water with five maxima two of which are separated by 6 min and three of which are separated by 4.5 min recapitulating the asymmetry of previous oscillatory measurements (Fig. 6.24).

Since the primary oscillations on average have a period length of only 6 s, the 2+3 pattern must represent a form of carrier wave generated by interaction of one to several parallel sets of primary oscillations of much greater frequency (Fig. 6.25) such as the *ortho/para* water oscillations with period lengths of 30–50 s reported by Pershin (2005) (Fig. 6.22).

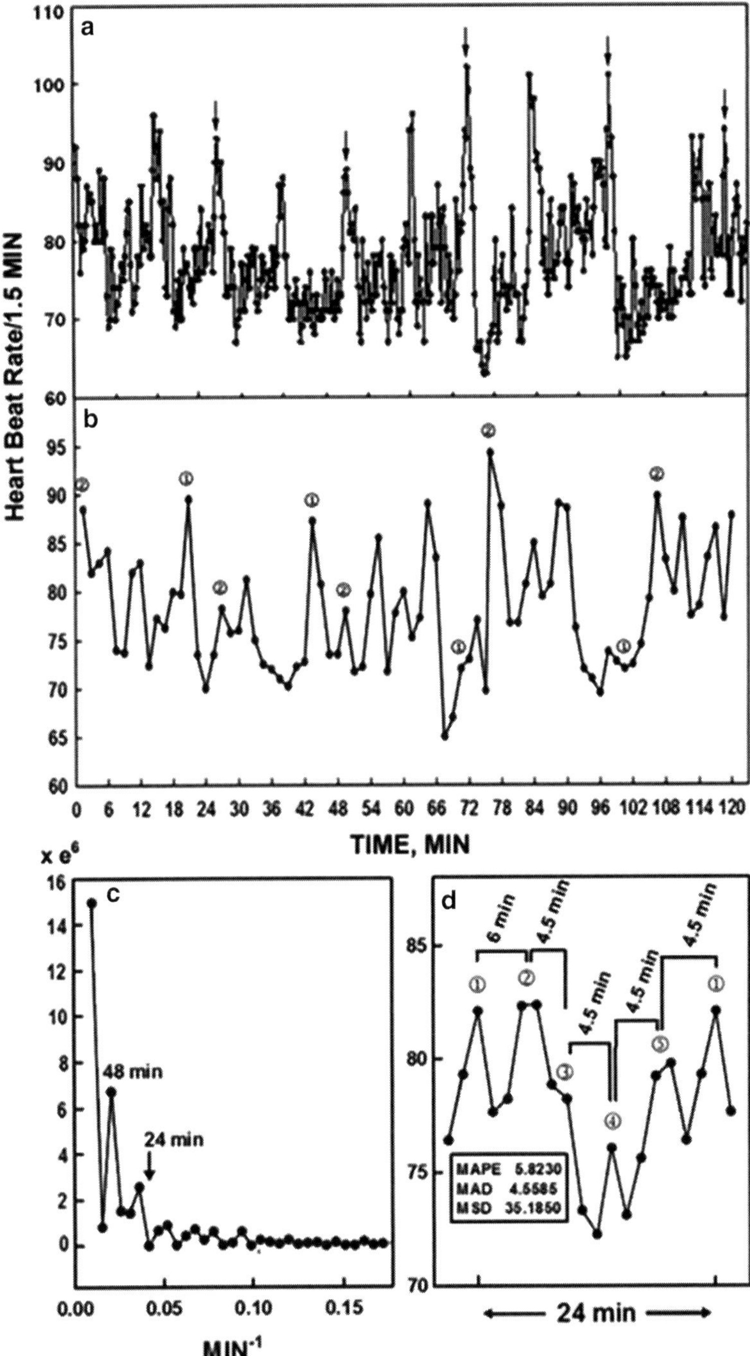

Fig. 6.24 Heart beat rate is the classic example of a limit oscillator. (**a**) Heart beat data from a 30-year-old male measured over 15 s every 15 s for 120 min. *Arrows* indicate a 24-min period length. (**b**) Data of (**a**) averaged over 1 min every 1.5 min to generate the 2+3 signature pattern illustrated schematically in Fig. 6.1. (**c**) Fourier analysis of data of Fig. 6.24b. (**d**) Time series analysis (decomposition fits) of data of Fig. 6.24b using a period length of 24 min determined from (**c**) to more clearly illustrate the 2+3 five maximum pattern

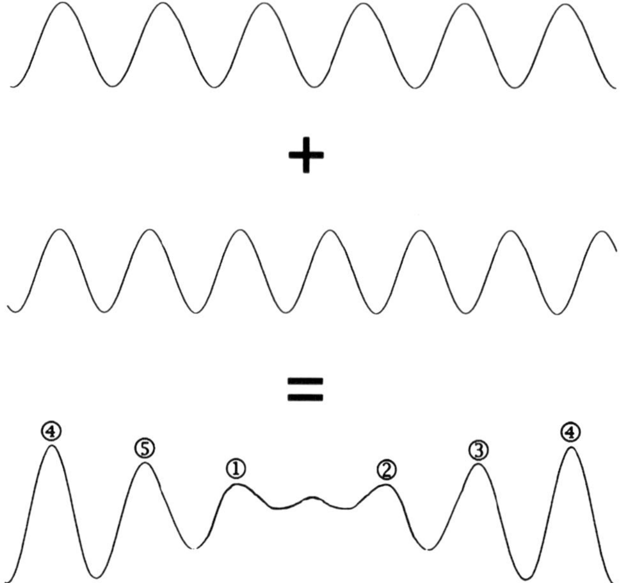

Fig. 6.25 Illustration of the generation of an asymmetric lower frequency carrier wave through algebraic summation of harmonics generated by primary oscillations of higher frequency (see Fig. 6.31c)

6.11.1 Growth Oscillations of Elongating Pollen Tubes

An oscillatory pattern of tip growth in lily pollen tubes based on measurements at 1–3 s intervals exhibits a fundamental period length of about 30 s (67 s for tobacco) (McKenna et al. 2009). This value is very close to the period length of the fundamental *ortho–para* oscillations of water. Also, oscillations with a similar period length were seen in changes in NAD(P)H fluorescence (Cárdenas et al. 2006). Growth oscillations and to a lesser extent, the NADH oscillations, are independent of aerobic energy metabolism (Rounds et al. 2010). Although period length increases following inhibition of mitochondrial electron transport, they are inherent in the cytosolic milieu in much the same manner as has been observed for water driven redox changes. The differences in period length observed are indicative of summation of more than one limit cycle (Rounds et al. 2010).

6.12 Phasing of the Rhythm

Since the operation of the "copper clock" demands a high level of synchrony in solution, there must be external stimuli that phase the oscillations. The oscillations were initially shown to be phased by light exposure following a period of dark adaptation in experiments with brine shrimp (Chalko et al. 2000). In cells and tissues, red

(Morré et al. 1999b) and blue (Morré et al. 2002c; Morré and Morré 2003b) light following a dark period initiate the ENOX1 rhythm so that at day break the ENOX1 rhythm of all of the cells in a sun-exposed plant are in synchrony. Blue light also synchronizes mammalian cells which offers potential basis for the successful use of blue light therapy to treat pathological skin conditions such as psoriasis. There is, however, no direct effect of light on the oscillatory activity of either copper solutions or ENOX proteins.

Phasing of ENOX rhythms by light is restricted to organisms, tissue explants, and cells but not with membranes and molecules that lack chromophores. Therefore, some form of communication between cell surface ENOX1 proteins and the blue (cryptochrome) and red (phytochrome in plants) photoreceptors that are light responsive are either cytosolic or associated with cytoplasmic endomembranes such as endoplasmic reticulum (i.e., Williamson et al. 1975) such that signals must be transmitted across the plasma membrane in order to achieve synchrony. Interestingly, following exposure of dark-grown hypocotyl segments, laboratory fluorescent lighting greatly reduced the incidence of close endoplasmic reticulum–plasma membrane associations present in segments exposed only to dim green light (Auderset et al. 1980).

Melatonin (Morré and Morré 2003b), lithium (Kromkowski et al. 2008), and caffeine (J. Kromkowski, Purdue University, results unpublished) do directly affect ENOX1 by phasing the rhythm. Presumably this occurs as a result of specific binding of these substances by ENOX1 proteins. Autoentrainment of ENOX1 proteins in solution also occurs (Morré et al. 2002b). If two out of phase preparations of ENOX1 proteins are mixed and incubated together for several hours to overnight, the ENOX1 proteins contained in the combined preparations will adjust their oscillatory cycles to generate a single new set of oscillations with the first maximum located midway between the start and finish of the two original periods. However, melatonin, lithium, or caffeine, all of which entrain the oscillatory activities of ENOX1 proteins, for example, has no effect on entrainment of copper in solution. Some other means of achieving synchrony must operate to ensure synchrony of the copper clock as will be developed in Sect. 6.12.1.

6.12.1 EMF Sets the Copper Clock

Findings from several laboratories suggest that electromagnetic fields affect a number of processes that appear to be clock related. Morré et al. (2008b) provide results from $Cu^{II}Cl_2$ solutions that suggest proton nuclear spin-dependent oscillations of water may be phased by exposure to low frequency electromagnetic fields (LFEMF) in the same manner as those responsible for ENOX-mediated activity oscillations (Fig. 6.26). The pattern of EMF phasing, however, is complex and highly dependent upon where in the ENOX or copper cycle the EMF is applied. LFEMF applied when the copper cycle was at maxima ②, ③, or ④, for the most part, delayed the activity by one cycle and then resumed at the same point in the cycle as when irradiated (total delay of 9 ± 5 min (2×4.5 min)). When EMF was applied when the copper cycle was at maxima ① or ⑤, the activity was delayed and shifted back one or two maxima to ④ or ⑤ for

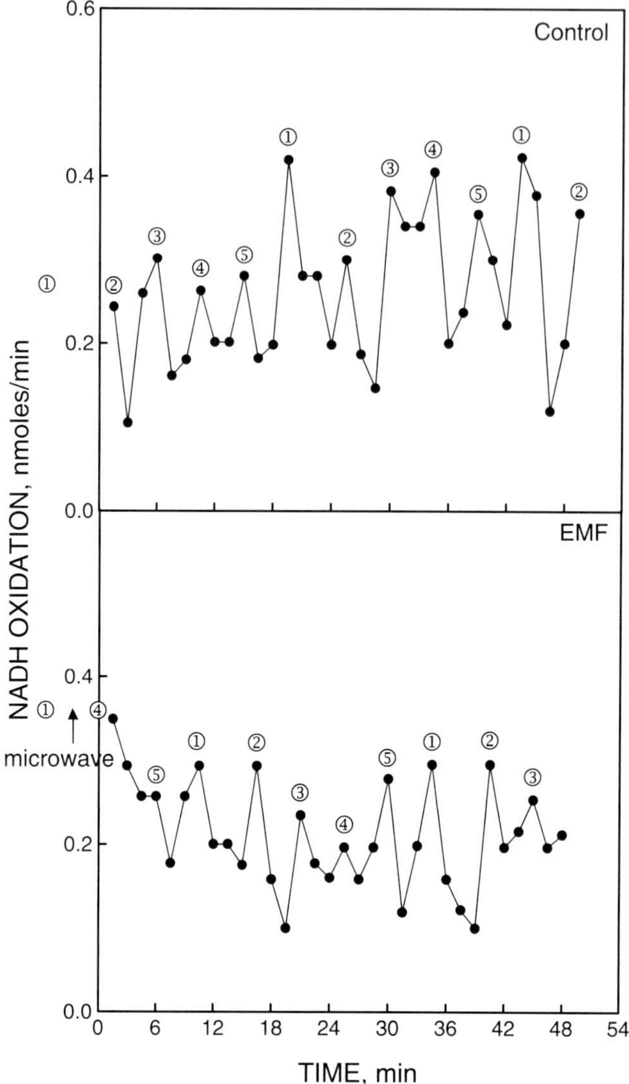

Fig. 6.26 The copper clock is phased (set) by EMF. Oscillatory pattern of NADH (120 μM) oxidation catalyzed by buffered (50 mM Tris-Mes) 100 μM copper at pH 6.0 in the absence of protein altered by EMF. Activity maxima labeled ① and ② are separated by 6 min and are followed by activity maxima labeled ③, ④, and ⑤ which are separated by 4.5 min. The two activity maxima separated by 6 min recur every 24 min to establish the 24-min periodic activity. The two assays were carried out in parallel with the same copper solution using two paired spectrophotometers with the only difference being the EMF exposure as indicated for b which displaced activity peak c ahead by about 15 min or back by about 9 min to coincide with peak ④ in the control preparation. The actual shift depended where in the cycle the EMF was initiated (Morré and Morré 2008b). Similar results were observed with both ENOX1 and ENOX2 preparations both as soluble proteins and associated with cells and membranes. Reproduced from Morré and Morré (2008b) with permission from Springer Science + Business Media

maxima ① or to ③ or ④ for maximum ⑤ (total delay of 13.5–16 min (2×4.5+4.5 or 6)). This pattern was confirmed in more than 33 repetitions (Morré et al. 2008b).

Melatonin is normally considered to act through a series of G-protein coupled receptors, MT1 and MT2, and a third binding site, MT3, recently identified as the enzyme NRH:quinone oxidoreductase 2 (QR2) (Mailliet et al. 2005). It has been suggested that melatonin binds to a co-substrate binding site of QR2 and donates an electron to the enzyme co-factor, flavin adenine dinucleotide (FAD) (Tan et al. 2007). FAD could be reduced to either FADH or FADH2 while melatonin would be converted to N^1-acetyl-N^2-formyl-5-methoxykynuramine and/or cyclic 3-hydroxymelatonin. The relationship between these several functional binding sites and binding to ENOX1 those results in phasing of the ENOX1 cycle are not known. ENOX1 lacks flavin and melatonin does not serve as a substrate.

The copper clock (Fig. 6.8) will keep very accurate time when shielded from environmental electromagnetic fields. However, abrupt phase shifts of 5–15 min were observed on occasion in unshielded preparations. These observations may bear on the central problem of how the copper-based oscillations become synchronized in nature even in the absence of light and outside a living organism. Synchrony seems to occur at irregular intervals, perhaps every 3–5 days. Since synchrony is achieved by exposure to LFEMF, it may occur in nature as a result of changes in the earth's electromagnetic field. Changes (ca. 800 nT) in the earth's magnetic field sufficient to phase the copper clock occur globally (http://www.anarctica.gov.au/science/antarctic-observatories/observatory-functions-at-davis) at irregular intervals similar to those we have observed for the natural resetting of the copper clock with unshielded preparations in the laboratory.

Minorsky (2007) reported that infradian rhythms in bean seeds correlate with extremely small oscillations in the vertical vector (Bz) of the interplanetary magnetic field. He concluded that the positive correlations between the lunar phase and seed imbibition reported previously (Brown and Chow 1973; Brown 1977; Spruyt et al. 1987) were coincidental. The hypothesis proposed by Minorsky (2007) is that the 7- and 14-day rhythms in bean seed imbibition are related to subharmonics of the 27.3-day solar rotation cycle.

Phase shifts in response to EMF exposure similar to those observed with copper^{II} solutions also occur in the oscillatory behavior of ENOX activities of HeLa cells grown in culture, ENOX proteins released from the surface of HeLa cells by treatment with 0.1 M sodium acetate, pH 4.5, or recombinant ENOX2 proteins. All showed a phase shift in response to 2 min of 50-Hz EMF exposure (Morré et al. 2008b). Measurements were in parallel on two aliquots of the same preparation, one exposed to EMF and the other not. Analyses were in parallel using duplicate Hitachi U3210 spectrophotometers such that EMF exposure was the only variable. Similarly, vesicles of plasma membrane isolated from a plant source (soybean) also exhibited an oscillatory pattern of NADH oxidation with a period length of 24 min that was phase shifted by EMF exposure. As with the other ENOX sources, the period length was unaltered by EMF exposure. Only the phase of the pattern of oscillation was shifted by 6 and 12 min depending on the phase of the ENOX cycle at the time of the EMF exposure.

6.13 A Mechanism to Explain How Oscillations of Redox Potential of Aqueous Solutions Become Synchronous and Remain So

Irrespective of the mechanism, the *ortho–para* oscillations must occur in a highly synchronized manner. Moreover, attendant oscillatory changes in redox potential offer an opportunity to monitor, as well, synchrony of oscillations in populations of water molecules. The periodicity of the ENOX1/copper/water clock can be phased by brief 10–20 s exposure to very LFEMF. In so doing, the synchronized populations of oscillating water molecules create a collectively coherent synchronous system. Two asynchronous but contiguous water samples separated by a nonmetal barrier become fully synchronized within times even as short as 1 min such that water molecules, being dipoles with strong polar interactions and a high degree of spatial orientation, may not only respond to LFEMF but may also generate such fields so that water molecules are able to communicate via the oscillating electromagnetic field and, therefore, remain highly synchronous perhaps over relatively long distances.

As a test for water synchrony, oscillatory changes correlated with fluctuations in redox potential as reflected in rates of NADH oxidation were monitored (Fig. 6.27). For example, 120-μM NADH prepared in HPLC grade water was placed in two plastic spectrometer cuvettes. One of the cuvettes was exposed to LFEMF (30 s, 40 μT) to be out of phase with the water in the other cuvette. When placed side by side for 1 h, subsequent measurements revealed identical oscillatory patterns for the two cuvettes when analyzed in parallel using two paired spectrophotometers (Fig. 6.28). When a thin copper or aluminum foil sheet was placed between the cuvettes, the communication between the two cuvettes was blocked (Fig. 6.28). The shortest exposure time sufficient to initiate synchrony thus far tested is 5 min.

Water sampled from two points in a shallow pond with sampling points separated by 100 ft (36 m) also were synchronous as was water collected from opposite ends of a 20-acre lake (Fig. 6.29). Flowing water also was synchronous (Fig. 6.30). Thus far the largest body of water sampled have been from Lake Ontario and from the Niagara River in New York with sampling points separated by approximately 20 miles for each. Samples from both were found to be synchronous. There have been previous reports supporting the concept of coherent water (Pollack and Clegg 2008; Del Giudice et al. 2010). Our findings support that concept but water coherence translates into highly correlated water populations to an extent much greater than may have been previously anticipated.

An alternative source of energy absorbed and emitted by water and correlated with *ortho–para* oscillation of *ortho–para* spin pairs on water is auto-oscillations in luminescence following infrared radiation (Gudkov et al. 2011; Fig. 6.31). The emissions oscillate with 300 and 1,150 s rhythms (Fig. 6.31c) that agree with our previously found period of oscillation of 18 min reflective of *ortho* to *para* spin isomers and redox potential of pure water (Morré et al. 2008d).

Metastable structural features in the organization of liquid water have been discussed extensively in both experimental and theoretical works as the target for electromagnetic fields (Binhi and Stepanov 2000; Binhi 2002). Low frequency

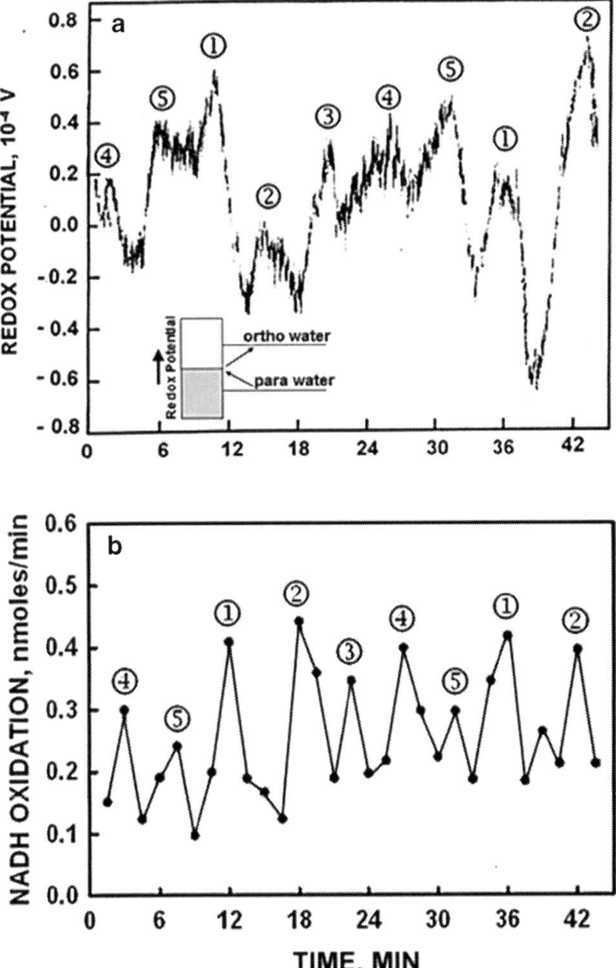

Fig. 6.27 The redox potential (**a**) of an aqueous solution of copper chloride measured continuously showing a 2+3 pattern of oscillation with NADH oxidation measured over 1 min at intervals of 1.5 min (**b**) in parallel. The period length of both is 24 min. Results with pure water are similar except that the period length is now 18 min (Morré et al. 2008d). The use of NADH to monitor redox potential of water or aqueous solutions is illustrated in Figs. 6.27b, 6.29, 6.30, 6.31, and 6.32. Reproduced from Morré and Morré (2012a, b) with permission from WIT Press

spectra of the electric conductivity of liquid water under electromagnetic field exposure were ascribed to formation of clusters of water molecules (Fresenko and Gluvstein 1995) or in UV luminescence spectra were ascribed to derive from defects of water structure at the microscopic level (Lobyshev et al. 1999). Within the water molecule, nuclear spins of protons of water were considered as a primary target for the external magnetic field (Binhi 2002). Proton spins were suggested to take part in spin–orbit interactions that would modify proton motion and influence the rate of formation and breakup of water clusters.

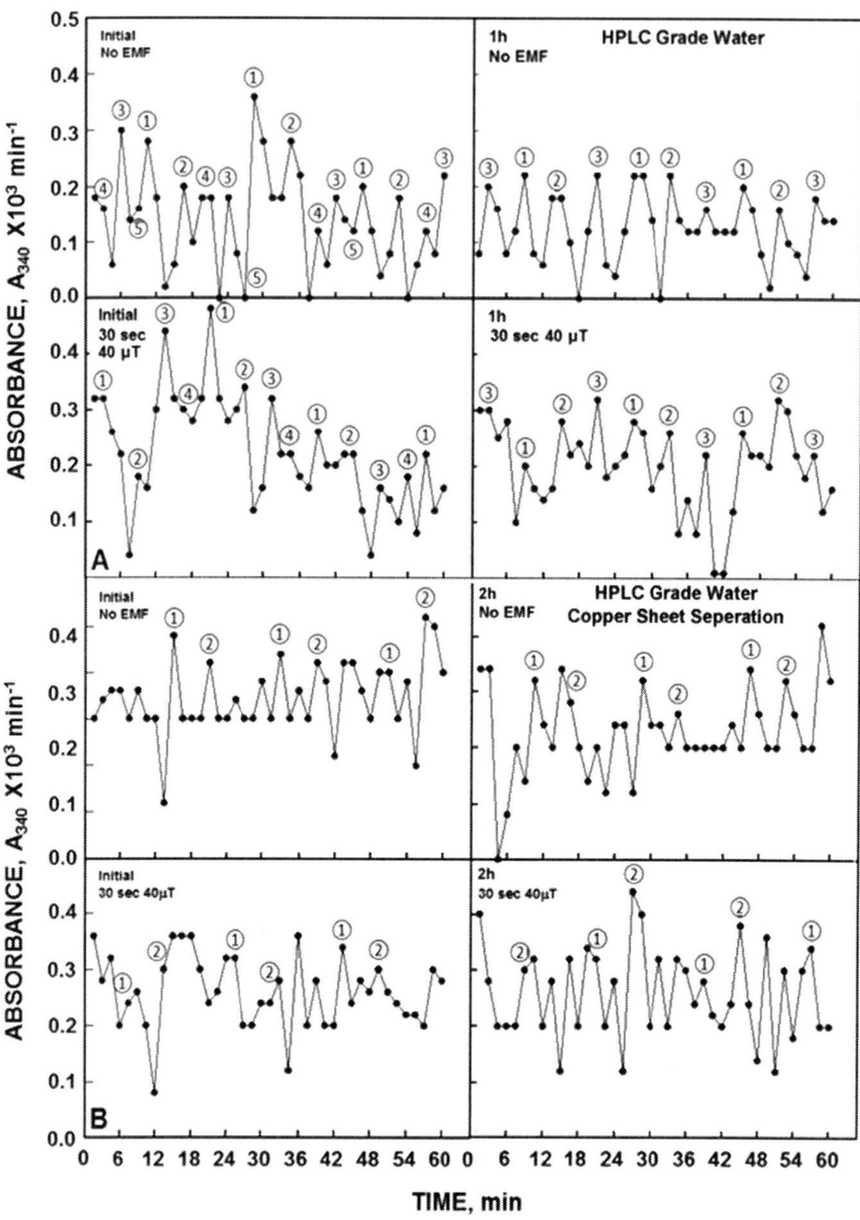

Fig. 6.28 Results demonstrating low frequency EMF synchronization of water samples through a thin plastic barrier. In (**a**), a sample of HPLC grade water was phased using low frequency EMF (30 s, 40 μT) as in Fig. 6.29 and shown now to be out of phase with the original water sample. However, when the two samples were placed adjacent to each other for 15 min, both samples then oscillated synchronously. (**b**) Experiment as in (**a**) except that the two samples were separated by a thin barrier of sheet copper which blocked the low frequency EMF coming from the adjacent water samples and prevented the phasing observed in (**a**). Reproduced from Morré and Morré (2012a, b) with permission from WIT Press

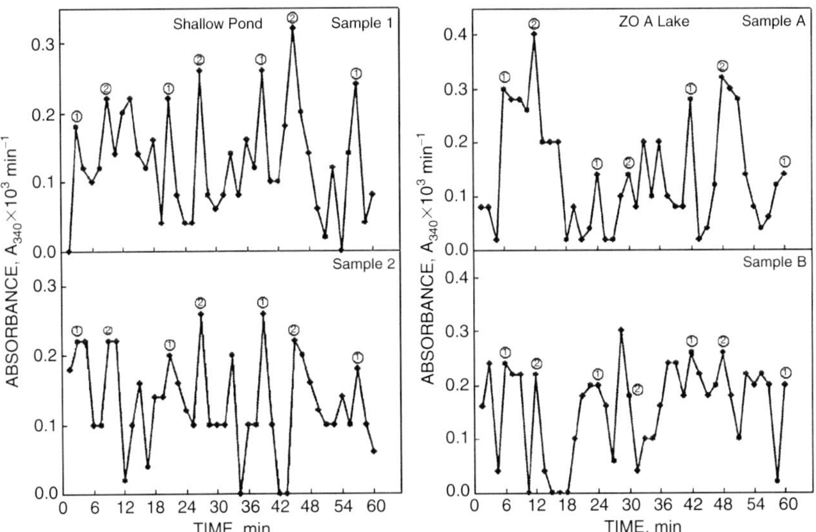

Fig. 6.29 Test of the prediction that contiguous water oscillates in phase through low frequency EMF communication. (**a**) Water samples from a shallow pond separated by a distance of 100 ft. (**b**) Water sampled at opposite ends of a 20-acre lake. In both instances, the samples were synchronous. Reproduced from Morré and Morré (2012a, b) with permission from WIT Press

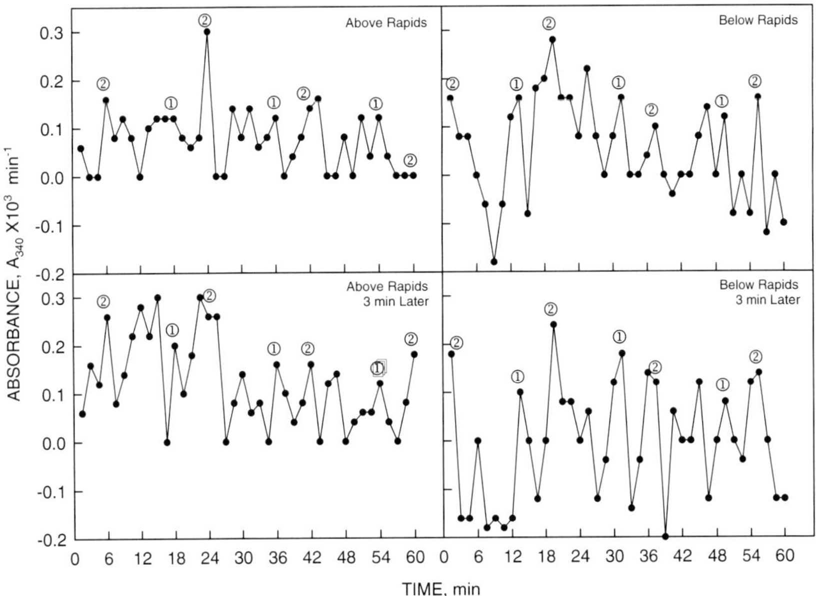

Fig. 6.30 As in Fig. 6.29 except water sampled from a flowing stream at $t=0$ and 15 min later. In between the two measurements, it was estimated that at least 50,000 gallons of water had passed the test site. Both measurements indicated that the water in the stream was synchronous. Reproduced from Morré and Morré (2012a, b) with permission from WIT Press

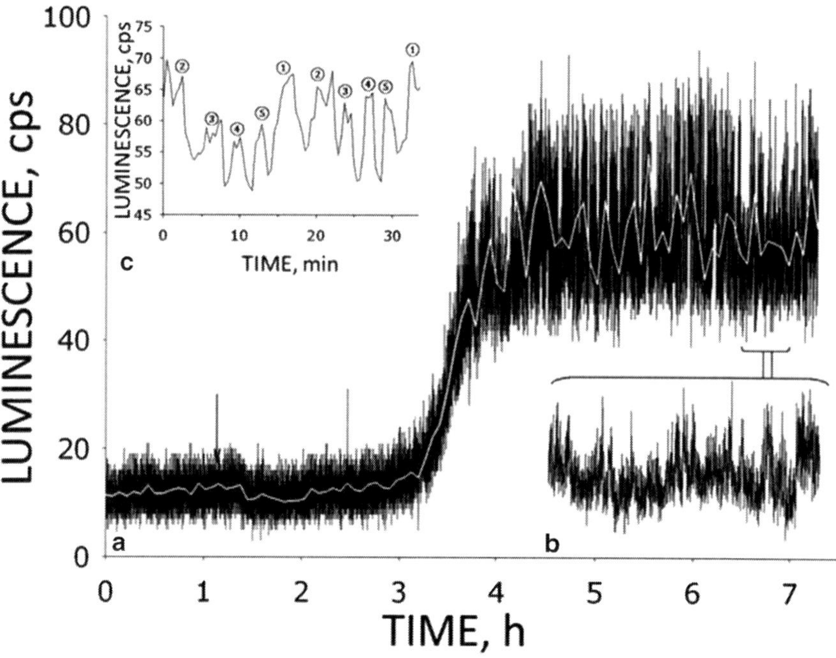

Fig. 6.31 Effect of laser radiation on water luminescence. (**a**) Data of a representative experiment. Water was exposed to an infrared laser ($A = 1{,}264$ nm, power 5 mW) for 5 min. The time of exposure is shown by a *vertical arrow*. To the *left* of the *arrow* is a record of the background water luminescence before the onset of laser irradiation. The *white line* on the basic plot represents the macrostructure of a SigmaPlot revealing the characteristic 2 + 3 pattern of oscillations with a period length of 18 min in a form of "carrier wave" generated from *ortho*/*para* water oscillations with period lengths of 30–50 s as shown in (**c**). (**b**) The microstructure of changes in water luminescence. (**c**) The integral intensity of water luminescence corresponding to that bracketed in (**b**). Maxima labeled ①–⑤ denote an 18-min period length. Modified from Gudkov et al. (2011)

Fast Fourier analysis of the luminescence data of Fig. 6.31 (data collected at 1 sec intervals) revealed multiple sinusoidal periodicities with period lengths in multiples of 1.2 sec (Results unpublished). Significant maxima were observed at period lengths of 2.4, 3.6, 6, 12, 24, 36, 48, 72 and 96 sec. When relative contributions of each were summed algabraically to generate a hypothetical carrier wave, the resultant summation was periodic but no longer sinusoidal. Analyses of areas under the generated wave when determined over intervals of 1 min every 1.5 min in the manner of collection of NADH oxidase activity and other data, revealed the characteristic 2 + 3 oscillatory pattern with an 18 min period typical of water and with two of the maxima separated by an interval of 6 min. A similar pattern was obtained when the primary data of Fig. 6.31 were averaged over 1 min at intervals of 1.5 min giving credence to the suggestion based on the heart beat rate model (Sect. 6.11) that the 18 to 28 min period 1engths of ENOX-related phenomena have their origins as carrier waves generated from the basic underlying ortho-para water oscillations which may have a fundamental period length in the range of 1.2 sec or shorter.

6.14 ENOX Clock and Cancer

Short-term disturbances of biological 24-h rhythms (chronodisruptions) that result in jet-lag and shift-lag symptoms in man are well known. However, more serious effects of chronic disruptions of circadian rhythms such as shift work, including long-term development of cancer, are a relatively new concept.

Beginning in the 1960s (see Fu and Lee 2003 for literature) several carefully designed studies were published to document that tumor development is affected by disruption of circadian rhythms. Breast cancer is accelerated by constant light exposure (as in people that work predominantly at night) (Schemhammer et al. 2001, 2006). Pinealectomy which accelerates proliferation of breast epithelial stem cells, induces mammary gland development, increases the formation of mammary tumors in rodent models, and is correlated with increased incidence and development of mammary tumors. Even the robustness of diurnal rhythmicity in patients with breast cancer may provide a significant predictor for survival (Sephton et al. 2000). With metastatic colorectal cancer, patients with clearly defined rest and activity rhythms had a fivefold higher survival rate and experienced less fatigue than patients with diminished rest/activity rhythms in a 2-year study (Mormont et al. 2000). Increased cancer risk to pilots, flight attendants, and airline cabin crews subjected to repeated circadian disruptions involving jet lag may also be an area of concern (e.g., Wartenberg and Stapleton 1998).

The dominant ENOX of cancer cells (ENOX2 or tNOX) oscillates with a period length of 21–22 min rather than 24 min. A 21- or 22-min ENOX period might be expected to generate a 21- or 22-h circadian day as observed in cultured cells over-expressing ENOX2 (Chueh et al. 2004) and in activity patterns of some cancer patients (Mormont and Levi 1997; Levi 2000). ENOX2 at the surface of cancer cells responds to several potent anticancer drugs (Chaps. 8 and 11). Consequently to increase the margin of safety, it should be possible to time chemotherapy to where the cancer cells and tissues are maximally sensitive to the drug compared to normal cells and tissues. Chronotherapy in oncology takes advantage of the asynchronies in cell proliferation and drug metabolic rhythms that distinguish normal and malignant tissues as a means to select the most efficacious time of day to administer. This has been a particular therapy accomplished on an empirical basis with patients and offers promise to minimize chemotherapeutic damage to host tissue and maximize toxicity to malignancies (Levi 2002; Levi and Schibler 2007).

There is extensive literature documenting the ability of weak electromagnetic fields to affect human cells and human health including generation of reactive oxygen species, immunological responses, inflammation, cell proliferation, wound healing, developmental processes, nerve regeneration, and cancer. All may be affected by electromagnetic fields (Byus et al. 1988; Goodman et al. 1993; Lyle et al. 1988; Matanoski 1995; Savitz 1995; Sisken et al. 1993). As a result of the wide range effects of EMF on biological progresses, concern has been expressed that exposure to low frequency magnetic fields, particularly those generated by electrical transmission lines, distribution lines, and electrical appliances, may increase the incidence of cancer (Fedrowitz et al. 2002; Thun-Battersby et al. 1999; Wolf et al. 2005; Yoshizawa et al. 2002; Simko et al. 2001; Lange et al. 2002; Henshaw 2002) including increased risk of childhood leukemia (Henshaw 2002). In the studies of Thun-Battersby et al. (1999), 50-Hz magnetic fields of low (100 microTesla = 100 μT) flux density enhanced mammary gland tumor development and growth in the 7,12-dimethylbenz(a)anthracene model of breast cancer in female Sprague–Dawley rats.

In other examples, a sinusoidal 50-Hz magnetic field with a magnetic flux density of 0.5 mT affected expression of cell adhesion molecules including VLA-2, the integrin receptor for collagen, and VLA-5, the integrin receptor for fibronectin in two human osteosarcoma cell lines (Santini et al. 2003). Exposure to a 0.2 T static magnetic field blocked the angiogenic response induced by treatment with prostaglandin E1 and fetal calf serum in the chick embryo chorioallantoic membrane assay (Ruggiero et al. 2004). A relatively low dose (0.1 mg/mL) of doxorubicin combined with a magnetic field inhibited proliferation of human osteosarcoma cells in culture by 82 %, whereas the magnetic field or adriamycin alone resulted in only 19 % and 44 % inhibition, respectively (Chakkalakal et al. 1999), thus offering the possibility of lowering the therapeutic dose of drugs by combination with EMF.

6.15 Are ENOX Oscillators Linked to the Drivers of the Circadian Clock and How Are They Linked?

That ENOX oscillations are linked to the drivers of the circadian 24-h clock is shown by two sources of evidence:

1. Transfection experiments with ENOX2 cysteine replacements (Sect. 6.2; Table 6.1).
2. Correlative experiments with heavy water (Sect. 6.3) and to a lesser extent with lithium and caffeine (this section). These substances are among the few that consistently alter the period length of the circadian day.

The transfection experiments have been primarily with COS cells transfected with ENOX2 cDNAs in which codons for specific cysteines were replaced by alanine codons (Chueh et al. 2002a, b). The result was ENOX2 proteins having periodic oscillations of 22 (normal ENOX2), 36, and 42 min (Sect. 6.2). The circadian period of COS cells was synchronized by exposure of dark-grown cells to light. To evaluate changes in circadian day length, the activity of a common glycolytic housekeeping protein GAPDH was monitored. GAPDH activity normally exhibits a 24 h circadian period length (Shinohara et al. 1998). In the transfected cells, the 24-h circadian day (corresponding to the 24-min period of the constitutive ENOX1) was retained by the transfected cells. In addition, however, 22, 34, and 42 h circadian daily rhythms were observed corresponding, respectively, to the 22, 34, and 42 min ultradian period lengths exhibited by the ENOX2 proteins introduced as a result of the transfection.

Also indicative of a linkage between the ultradian ENOX rhythm and the length of the circadian day are responses to substances known to consistently alter the period length of the circadian day in living organisms. Most compelling are the studies with heavy water (D_2O) summarized in Sect. 6.3 where both the ENOX period and the copper clock period lengths are increased by 30 % from 24 min to about 30 min in keeping with the response of the circadian day to D_2O which is to increase the period length from about 24 h to about 30 h (Pittendrigh et al. 1973). Among the many models proposed for the circadian clock is that of it being a counter of higher ultradian oscillations (Edmunds 1988) as proposed here (see also Sect. 6.16).

Lithium, used in the treatment of bipolar disorder, lengthens the circadian day by about 3 h and increases the period length of the ENOX and copper clock cycles from 24 to 27 min (Kromkowski et al. 2008). Some rapidly cycling manic-depressive patients exhibit free-running circadian rhythms which cycle faster than once per 24 h as well (Abe et al. 2000; Kripke et al. 1978). A therapeutic effect of lithium may be to slow or delay these rhythms (Kripke et al. 1978; Welsh and Moore-Ede 1990).

The use of lithium in the treatment of bipolar disorder is based in part on the hypothesis that individuals with bipolar disorder suffer from a disturbance of the master clock in the brain. Since sleep cycles, body temperature, and metabolism may not be properly synchronized (Abe et al. 2000; Yin et al. 2006), lithium is used to restore these daily rhythms (Atkinson et al. 1975; Kripke et al. 1978; Wehr and Goodwin 1983; Yin et al. 2006). Lithium has been used to delay the sleep–wake rhythm in healthy normal subjects (Kripke et al. 1979) and since lithium may be changing the circadian clock, bipolar, and other mood disorders such as major depression and seasonal affective disorder could, in fact, be the result of abnormalities in circadian rhythms (McClung 2007). Similarly sleep phase advance may sustain the effects of total sleep deprivation both with or without a combined antidepressant drug treatment (Benedetti et al. 2001). Improved clinical outcome observed in lithium-treated patients was attributed to the phase delaying effect of lithium on biological rhythms, leading to a better synchronization of biological rhythms within the sleep–wake cycle (Benedetti et al. 2001). Lithium delays circadian rhythms in the isolated *Aplysia* eye (Engelmann 1973) and slows or delays circadian rhythms in *Kalanchoe* (Engelmann 1972), a plant, in cockroaches (Engelmann 1973; Hofmann et al. 1978) and in rodents (Kripke and Wyborney 1980). With *Kalanchoe*, a concentration of 1 mM LiCl was demonstrated to lengthen the length of the circadian period by more than 1 h from 22.3 h in the control to 23.6 h (Engelmann 1972). With rats, lithium increased the free-running circadian wheel-running rhythm in the laboratory from 24 to 24.4 h (Kripke and Wyborney 1980).

The hypothesis that constitutive ENOX proteins are the drivers of the cells circadian clock is supported by results with lithium where the length of the circadian day is increased by about 4 % (Morré et al. 2002a). A phasing effect of lithium on ENOX where a new set of oscillations with a period length slightly shorter than 24 min was observed (Kromkowski et al. 2008) and might have relevance to bipolar disorder and lithium use in its treatment.

Elevated anxiety symptoms among consumers of large amounts of caffeine and cases of caffeine-induced manic depression with improvements related to discontinuation of caffeine in the diet (Iancu et al. 2007) may also be clock related. In unpublished work, a lengthening of the ENOX cycle of about 1 min/period was observed in the presence of 0.4-μM caffeine (J. Kromkowski, Purdue University, results unpublished).

Phasing by light also supports the ENOX protein being the ultradian driver of the circadian clock (Chalko et al. 2000). Phasing of both the ENOX activity and the circadian day is seen in response to red and blue light in plants and to blue light with mouse skin (Morré and Morré 2003b) and in clinical patients with blue light responsive skin disorders (J. H. Wilkins, D. J. Morré and D. M. Morré, results unpublished) as well as from a response to melatonin as well as valerian with both plant and animal CNOX where a new maximum ① appears exactly 24 min after melatonin addition (Morré and Morré 2003e). With blue light, the new maximum appears after $12+24=36$ min following light exposure. Neither purified ENOX proteins nor the copper clock respond to red or blue light. How then are the ultradian ENOX oscillators and the cellular elements responsive to light linked? The putative light receptors (phytochrome for red light in plants; cryptochrome for blue light) are largely cytosolic, or associated with cytosolic endomembranes (Sect. 6.12), such that a transmembrane receptor system should be required to link ENOX proteins to the cells interior. An anti-idiotypic antibody to a peptide complementary to the selected regions of ENOX2 which might be expected to recognize a cell surface receptor appropriately linked to the cytosol (X. Tang, results unpublished). Several candidate proteins have been located using this antibody including a promising candidate at 80 kDa that binds ENOX2.

The pathway nevertheless remains obscure as does the relationship, if any, between the ENOX1 24-min cycle and the eight or more core circadian genes that have been identified so far along with the attendant control models of transcription-translation feedback loops. Presumably the mechanism would be somewhat analogous to a traditional mechanical clock that counts pendulum swings (the 24-min ultradian rhythm) to drive a 2 h rhythm or gear, for example, which would engage 12 times in sequences that would eventually generate the 24-h circadian day (see Sect. 6.16 for example of the molecular segmentation clock of somite formation).

An opportunity to link the 120 min super period to downstream activation of clock genes is provided by a growing awareness of a general response of transcription factors to levels of NADH and/or NAD^+ and especially to influence their dimerization and migration to the nucleus (Ueda et al. 2002; Lin and Guarente 2003; Zhao et al. 2006). Rutter et al. (2001) first reported that binding of NPAS2/BMAL1 (NPAS2 is a homolog of CLOCK) is dependent on the redox state of nicotinamide cofactors in the cell. Rutter et al. (2001) examined the ability of the heterodimer to bind to oligonucleotides containing the E-box sequence in the presence of different concentrations and ratios of $NAD(P)^+$ and NAD(P)H in vitro. They found that increased NAD(P)H levels resulted in increased DNA binding of NPAS2/BMAL1, while increased $NAD(P)^+$ levels inhibited the binding. In addition, a low ratio of $NAD(P)^+$ to NAD(P)H increased DNA binding. Nicotinamide

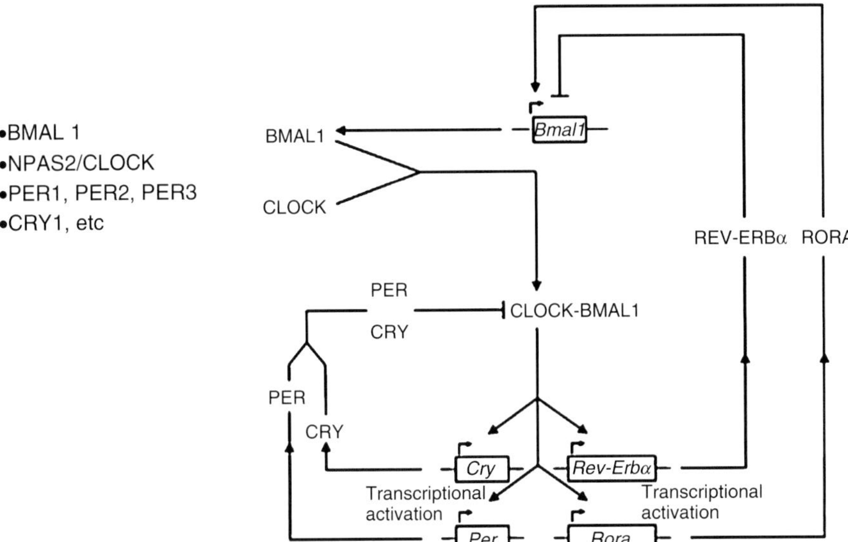

Fig. 6.32 Core circadian genes. Modified from Bell-Pedersen et al. (2005)

cofactors emerge as important effectors of both normal metabolism (Canelas et al. 2008) and for switching between survival and apoptosis in cancer cells (De Luca et al. 2005).

The molecular mechanisms that underlie the circadian clock (Fig. 6.32; For reviews, see Bell-Pedersen et al. 2005; Schibler and Naef 2005 and Ramsey et al. 2007) are based on a system of interconnected transcriptional-translational feedback loops in which specific clock proteins repress transcription of their own genes (Dunlap 1999; Young and Kay 2001; Hirayama and Sassone-Corsi 2005; Fig. 6.32). The various proteins of the mammalian clock include three period proteins (PER1, PER2, and PER3), two cryptochromes (CRY1 and CRY2), CLOCK, and BMAL1. Genes that are expressed in a circadian manner contain a binding site (CACGTG) in their promoters known as an E-box. This binding site is the target for a heterodimer made up of NPAS2 or CLOCK and BMAL1 whose formation is dependent upon reduced pyrimidine nucleotide levels. When the heterodimers bind to the E-box, transcription from that gene is activated. Genes that are activated in this manner include the clock genes Per1, Per2, and Per3, Cry1 and Cry2, Rev-Erbα, and Rora, as well as several circadian output genes involved in normal metabolism (Lowrey and Takshashi 2004). The Per and Cry gene products interact in the cytoplasm to form heterodimers as their concentration slowly builds. These heterodimers translocate to the nucleus to inhibit CLOCK/BMAL1-mediated transcription through direct protein–protein interactions. Inhibiting CLOCK/BMAL1 results in a drop in transcription of the Per and Cry genes and a decrease in the level of their gene products. This eventually leads to a release of the inhibition of CLOCK/BMAL1 to enable the start of another cycle of gene transcription. The complete cycle from CLOCK/BMAL1 binding to DNA to the release of inhibition as diagrammed in Fig. 6.32 requires approximately 24 h to complete.

The binding of the NPAS2/BMAL1 in response to increased NADH levels would be expected to occur during the minima of oxygen consumption when NADH concentration is greatest. Periodic elevations in NADH (every 2 h) may indicate downstreams signal counting events involving NADH and or NAD^+ levels remaining to be elucidated. Clearly, circadian clock functions are not simply the domain of suprachiasmatic nucleus neurons as thought for decades, but instead are common features of most, if not all cells (Nagoshi et al. 2004).

There are seven mammalian enzymes constituting class II histone deacetylases (HDAC) (Yang and Seto 2008). HDAC-mediated deacetylation of histones correlates with gene silencing (Grunstein 1997; Wade and Wolffe 1997; Struhl 1998; Workman and Kingston 1998; Suka et al. 2002). The class II enzymes are homologs of yeast Sir2 (silencing information regulator) and are known as SIRT1 to SIRT7. These proteins are uniquely different from other HDACs and potentially vital to the present proposal in that they have the property of dynamically sensing energy metabolism (Bordone and Guarente 2005). Unlike other HDACs, SIRT proteins catalyze a unique reaction that requires NAD^+ as a cofactor. In this reaction, nicotinamide is liberated from NAD^+ and the acetyl group of the substrate is transferred to cleaved NAD^+ generating the metabolite *O*-acetyl-ADP ribose (Sauve et al. 2006). Due to the NAD^+ dependency, SIRTs constitute one of the functional links between metabolic activity and genome stability. Because of the NAD^+ requirement for Sir2 deacetylase activity, silencing may play a critical role in the stringency of circadian gene expression and, in the oscillatory silencing that periodically follows a transcriptional peak.

In mammals, it has been proposed that SIRT1 functions as an enzymatic rheostat of CLOCK function (Nakahata et al. 2008). SIRT1 regulates the amplitude and the duration of circadian gene expression through the interaction and the deacetylation of key circadian clock regulators, such as BMAL1 and PER2. Both NAD^+ levels (Nakahata et al. 2009) and the rate-limiting enzyme in mammalian NAD^+ biosynthesis, nicotinamide phosphoribosyltransferase (NAMPT) (Nakahata et al. 2009; Ramsey et al. 2009) display circadian oscillations and modulate CLOCK:BMAL1-mediated circadian transcriptional regulation through SIRT1 that underlie a regulatory network regulated by NAMPT-mediated NAD^+ biosynthesis and SIRT1 (Imai 2010). Our hypothesis is that, as such, SIRT1 conveys the transducing signals coming from synchronized fluctuations in NAD^+ levels originating from the coincidence of the 24-min oscillator and the 40-min metabolic oscillator to the core circadian machinery. SIRT1 physically associates with CLOCK and contributes to the acetylated state of CLOCK targets. CLOCK-BMAL1 and SIRT1 co-localize in a chromatin-associated regulatory complex as promoters of clock-controlled genes.

The CLOCK protein is one of the few core circadian regulators whose levels do not oscillate under most conditions (Lee et al. 2001). However, its histone acetyl transferases (HAT) activity does oscillate leading to an oscillatory remodeling of chromatin and a regulated HDAC whose activity may function as rheostats of the HAT's function of CLOCK.

Histone acetylation is recognized as one of the most prominent epigenetic features leading to activation of gene expression (Strahl and Allis 2000). Acetylation of the ∈-amino groups of specific lysine residues in the N-terminus of core histones is generally associated with transcription activity. It is thought to result in an open chromatin conformation that allows access of the transcription machinery to promoters (Struhl 1998; Cheung et al. 2000; Li et al. 2008). Acetylation tips the balance toward relaxing chromatin. Deacetylation, on the other hand, shifts the balance back toward condensed chromatin and the silencing of gene expression. The enzymes responsible for these transitions are HATs and HDACs as follows:

1. SIRT1 deacetylase is circadian and independent of SIRT1 levels.
2. SIRT1 controls circadian histone acetylation.
3. CLOCK is a HAT.
4. SIRT1 is in a chromatin complex with CLOCK-BMAL1 on the dbp promoter. Findings of Nakahata et al. (2007) indicate that the CLOCK-BMAL1 dimer and SIRT1 coexist in a chromatin regulatory complex that operates on circadian promoters in a rhythmic pattern reflective of cytosolic NAD^+ levels (Fig. 4.13).
5. BMAL1 is rhythmically acetylated by CLOCK and this event is essential for control of circadian function (Hirayama et al. 2007).
6. BMAL1 deacetylation by SIRT1 responds to NAD^+.

The NAD(H)-dependent energy pathways in the cell could influence the HAT function of CLOCK:BMAL1. CLOCK-mediated acetylation and attendant transcriptional activation could be counterbalanced by transcriptional silencing induced by NAD^+-dependent HDACs such as SIRT1, a NAD^+-dependent HDAC2.

The scheme of Fig. 6.33 would provide for oscillations in circadian gene transcription resulting from high levels of NADH alternating with high levels of NAD^+. Such periodic alternations, which do occur in yeast, would be expected to be intensified at intervals of 2 h at the coincidence of the 24-min ENOX1 with the 40-min metabolic cycle. Thus, levels of reduced vs. oxidized pyridine nucleotides emerge as a mechanism whereby the oscillations might be linked to transcriptional control. It remains to be determined precisely how this level of control is translated into a 24 h day. However, downstream feedback loops exemplified by PER1 and PER2 in human cells and their equivalents in yeast and other organisms may be sufficient as described in detail by others (Dunlap 1999; King and Takahashi 2000; Bell-Pedersen et al. 2005; Langmesser et al. 2008).

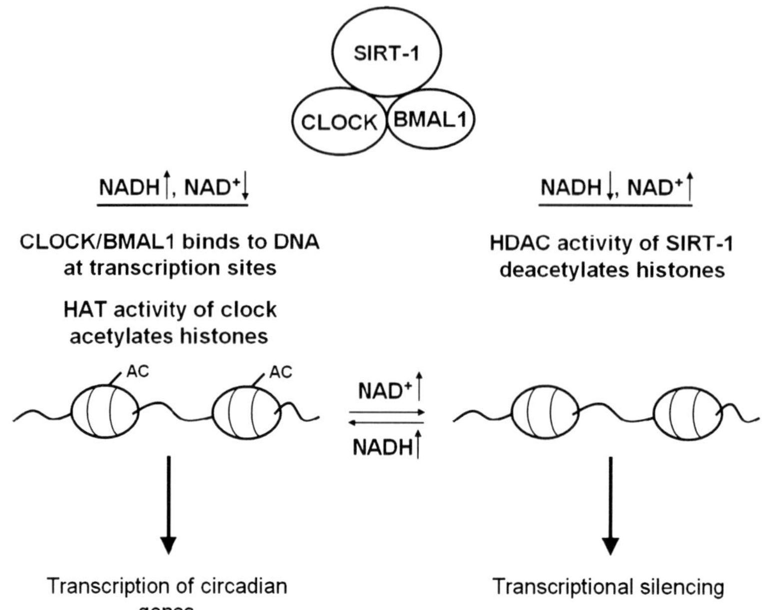

Fig. 6.33 Hypothesis to explain circadian oscillations in transcription of circadian genes based NADH/NAD^+ ratios. Under conditions of high NADH, CLOCK-BMAL1 dimer-SIRT1 complexes bind tightly to DNA at transcription sites. Histone acetyl transferases (HAT) activity of CLOCK acetylates histone to indicate transcription of circadian genes. In contrast, high NAD^+ activates histone deacetylases (HDAC) activity of SIRT1 to deacetylate the histone and result in transcriptional silencing

6.16 Why Oscillate? A Consequence of Active Sites in Metalloproteins?

There are no clear explanations as to why ultradian rhythmic activities should have adaptive advantages over continuous manifestations of the same activities. One suggestion is that periodic functions do allow for sequential alternations of activities within a single protein molecule. The ENOX proteins are examples of this. Another example is that of the molecular segmentation clock of somite formation. Regions of presomatic mesoderm that give rise to the precursors of the vertebral column and segmented muscles (Andrade et al. 2007; Shifley and Cole 2007) appear with an ultradian period of one every 2 h (120 min), i.e., every five ENOX1 24-min oscillations. If the 2 + 3 pattern of periodic oscillations persists through successive additional rounds of signal transduction, a 2-h ultradian period (5 × 24 min) might be anticipated to be one result.

As such, oscillatory alternation of functional activities of ENOX proteins in the growth process may be the primary functional outcome of the oscillatory pattern at the cellular level with a time keeping role being an important but perhaps secondary outcome. During the protein dithiol interchange portion of the cycle, cells undergo three consecutive cycles of cell enlargement of about 4 min each. This phase then rests as the oxidative portion of the cycle continues as plasma membrane electron transport possibly to fulfill some metabolic requirement essential to the process of cell enlargement.

6.17 Summary

A homodimeric and growth-related ENOX1 protein of the mammalian and plant cell surface with a binuclear copper center and protein disulfide-thiol interchange activity has characteristics of an ultradian driver of the biological 24-h circadian clock. The recurring complex 2+3 set of oscillations in secondary structure, enzymatic activity, and redox potential generates a period length of 24 min (repeats 60 times over 24 h). The period length is temperature independent and entrained by light, melatonin, and low frequency EMF. COS cells transformed with ENOX variants where specific cysteine codons were replaced by alanine codons yielded circadian periods of 22, 34, or 42 h, respectively, based on activity of GAPDH in response to ENOX oscillations with period lengths of 22, 36, or 42 min. The period length of the ENOX oscillator is phased by light, melatonin, lithium, and caffeine. The oscillations require bound copper and are recapitulated in solution by copper salts. The period length of both the ENOX and Cu^{II} oscillations in D_2O is increased to 30 min. Organisms grown in deuterium oxide exhibit a 30-h circadian day. The period length of the ENOX and Cu^{II} oscillations as well as those of water were phased by brief exposure to LFEMF. The oscillatory pattern appears to be determined by periodic variations in the ratios of *ortho* and *para* nuclear spins of the paired hydrogen or deuterium atoms of the elongated octahedral structure of the protein bound Cu^{II} hexahydrate as determined by spectroscopic analyses. The oscillations result from physical rather than from chemical events to account, for the first time, for the temperature independence of the period length of clock-related phenomena. Based on a heartbeat model of limit cycle oscillations, these oscillations also appear to give rise by algebraic summation to the generation of the observed 24-min periodicity as a form of carrier wave. Attendant changes in redox potential offer an opportunity to monitor oscillations in synchronous populations of water molecules through redox-catalyzed oscillation in the rate of NADH oxidation. The findings demonstrate that water molecules communicate with each other via very LFEMF most likely generated by the energetics of the synchronous *ortho–para* interconversions of the nuclear spins of the paired water hydrogens. The level of synchrony achieved is ultimately translated into highly coordinated populations of coherent water extending over large distances. An opportunity to link the ultradian ENOX oscillations to downstream activation of clock genes may be provided by responses of transcription factors to levels of NADH and especially to NAD^+ resulting from oscillatory ENOX activity. For example, the amplitude and duration of expression of circadian genes such as BMAL1 and PER2 is regulated through the interaction and deacetylation interaction and deacetylation by SIRT1. NAD^+ levels and NAMPT, the rate-limiting enzyme in mammalian NAD^+ synthesis, both exhibit circadian oscillations and modulate CLOCK:BMAL1-mediated circadian transcriptional regulation through SIRT1. Evidence supportive of a regulatory network regulated by NAMPT-mediated NAD^+ biosynthesis and SIRT1 which, in turn, may be regulated by ENOX-mediated fluctuations in NADH/NAD^+ levels, is discussed in Chap. 4. Chronic disturbances or disruptions of circadian rhythms may have adverse effects on human health and well-being including increased risk to cancer.

Chapter 7
Other Potential Functional Roles of ENOX Proteins

The principal functional roles of ENOX proteins are in plasma membrane electron transport (Chapter 4), in growth (Chapter 5) and in cellular time keeping (Chapter 6). In this chapter indications for other possible functional roles of ENOX proteins in diverse cellular processes are considered.

7.1 Cell Cycle Control

Both cell division and cell enlargement are fundamental to the growth of organisms. By serving as drivers of cell enlargement, ENOX proteins may at least indirectly exert some degree of control on the cell cycle. Eukaryotic cells coordinate cell division with cell size by regulating the length of the G_1 and G_2 phases of the cell cycle (Sweiczer et al. 1996). In order to divide again, a newly divided cell must achieve some minimum size in order to pass the G_1 checkpoint that monitors cell size (Morgan 1997; Fig. 5.18). If cell enlargement is prevented as with treatment of cancer cells with ENOX2 inhibitors (Chap. 8), cell enlargement is blocked, the cells fail to divide, and after 48–72 h undergo apoptotic cell death.

7.2 Gene Regulation

There is no evidence for a transcription factor function for ENOX proteins despite the fact that both ENOX1 and ENOX2 sequences are suggestive of such a function (Chueh et al. 2002b).

An indirect example is by exerting a significant impact on cytosolic NADH levels through their function as terminal oxidases of plasma membrane electron transport, a major impact on NADH/NAD^+-regulated transcription factors would be expected. For example, the mammalian SIR2 ortholog, SIRT1 encodes an NAD-requiring NAD-dependent histone deacetylase required for chromatin silencing (Guarente and Picard 2005). NADH is a competitive inhibitor of SIR2 (Bordone and Guarente 2005).

D.J. Morré and D.M. Morré, *ECTO-NOX Proteins:Growth, Cancer, and Aging*, DOI 10.1007/978-1-4614-3958-5_7,

Thus any overall reduction in NADH as for example by inhibition of ENOX1 in normal cells or of ENOX2 in cancer cells will activate SIR2 (see Chap. 6 for an extension of this concept to the control of circadian rhythms and Sect. 7.7 for applications to calorie restriction).

7.3 Endomembrane Function, Membrane Displacements, Vesicle Budding

Vesicle formation and membrane budding are well-established manifestations of the vectorial displacement of membranes that occur at the transitional endoplasmic reticulum to deliver membranes to the *cis* Golgi apparatus, at the *trans* Golgi apparatus to deliver materials to the plasma membrane and the cell exterior along an exocytic pathway (Farquhar 1985), and at the plasma membrane to internalize both membrane and substances from the external milieu along an endocytic pathway (Silverstein et al. 1977). Displacements occur as part of normal cellular activities and cellular movements especially in cultured cells. These include ruffling of the cell surface, and the formation and retraction of pseudopods along with pleomorphic changes in cell shape (Hynes 1979). Three manifestations of physical membrane displacement (budding, pleomorphic shape changes and lateral displacement of membrane sheets) are illustrated in Fig. 7.1.

7.3.1 Membrane Budding

Budding phenomena (Fig. 7.1b) are central to both exocytic and endocytic processes (Farquhar 1985). Use of specific inhibitors has contributed much to our understanding of membrane budding and as a process essential to membrane

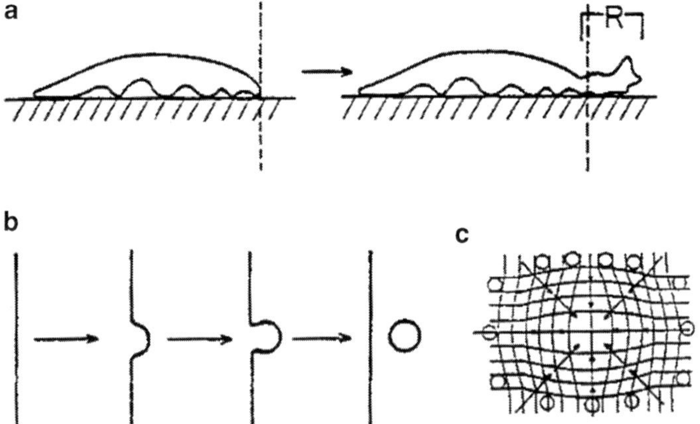

Fig. 7.1 Diagram depicting three manifestations of physical membrane displacements. (**a**) Pleomorphic shape changes associated with mammalian cells during movement. (**b**) Membrane budding. (**c**) Planar physical membrane displacement as would occur in membrane sheets during cell enlargement. Reproduced from Morré (1994b) with permission from Springer-Wien

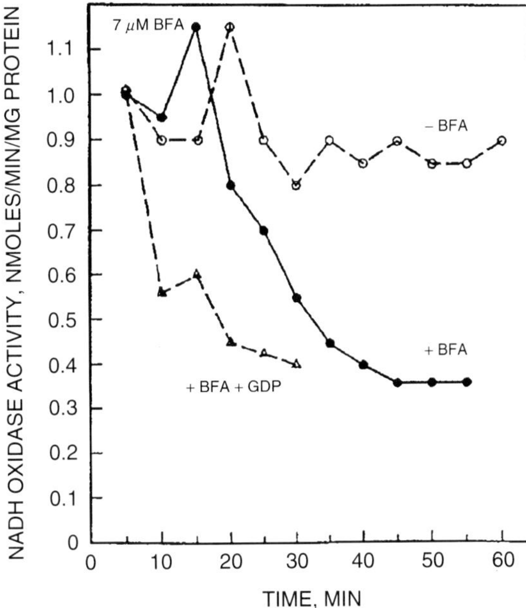

Fig. 7.2 Inhibition of the NADH oxidase of rat liver Golgi apparatus by 7 μM brefeldin A (BFA) is time dependent and accelerated by GDP. Reproduced from Morré et al. (1994e) with permission from FEBS

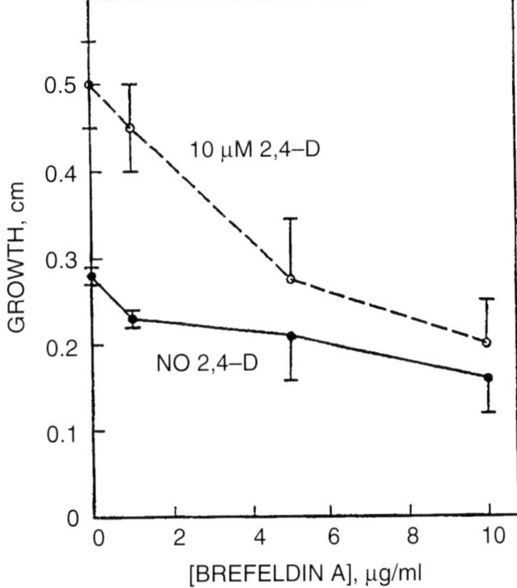

Fig. 7.3 Brefeldin A inhibits elongation of 1 cm sections of soybean hypocotyls. The growth induced by plant auxin growth hormones such as indole-3-acetic acid or 2,4-dichlorophenoxyacetic acid (2,4-D) is preferentially inhibited by brefeldin A. Reproduced from Morré (1994b) with permission from Springer-Wien

trafficking. One such inhibitor is the fungal macrolide antibiotic, brefeldin A. NADH oxidation is inhibited by brefeldin A and the brefeldin A inhibition is augmented by GDP (Morré et al. 1994e; Fig. 7.2). Not only is trafficking from the Golgi apparatus to the plasma membrane blocked by brefeldin A (Miller et al. 1992) but brefeldin A also specifically inhibits auxin-induced cell elongation in plants (Schindler et al. 1994; Fig. 7.3). As brefeldin A is normally thought to act exclusively at the level of

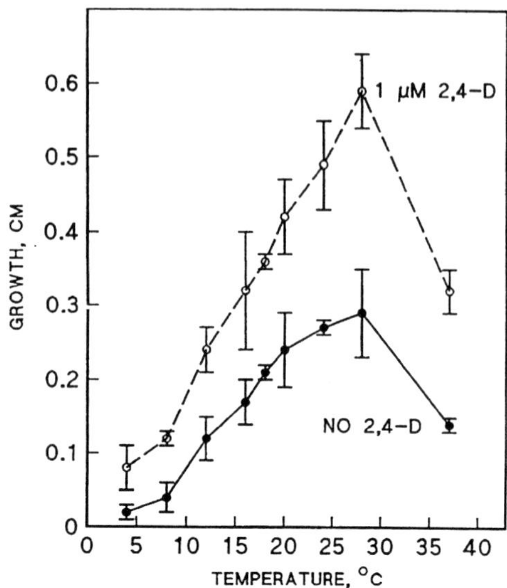

Fig. 7.4 Auxin-induced elongation of 1 cm segments of soybean hypocotyls exposed at different temperatures. The response to temperature shows no abrupt transition as would be expected for a typical low-temperature block of membrane traffic to the cell surface. Reproduced from Morré (1994b) with permission from Springer-Wien

the Golgi apparatus, such findings have been regarded as evidence for an obligatory role of secretory vesicles derived from the Golgi apparatus in cell enlargement especially in plants where cell enlargement in the absence of cell division plays a major role in the growth process.

There are, however, several lines of evidence that argue against this interpretation. One of these is the response to temperature where vesicular traffic is blocked at temperatures of 18 °C or less (Tartakoff 1986). Yet cell enlargement induced by auxin shows no sharp transition in response to temperature over the entire range of 4–25 °C (Fig. 7.4). These observations suggest that elongation growth in plants may continue for a time in a manner independent of the vesicular transport pathway at least in plants.

Experiments where vesicle delivery to the cell surface in setae of the moss *Pellia* was blocked by monensin led to similar conclusions. Here, cell enlargement could occur for a time independently of the delivery of new membrane to the cell surface. Despite the nearly complete inhibition by monensin of vesicular transport from the Golgi apparatus, cell enlargement continued at a normal rate for up to 4 h or more after the onset of monensin inhibition (Schnepf et al. 1979; Morré et al. 1986c).

7.3.2 *Energy Requirements for Physical Membrane Displacement*

An energy source emerges as a universal requirement for physical membrane displacement. Budding of endoplasmic reticulum vesicles requires ATP whereas transfer between *trans* Golgi apparatus membranes and the plasma membrane appears to

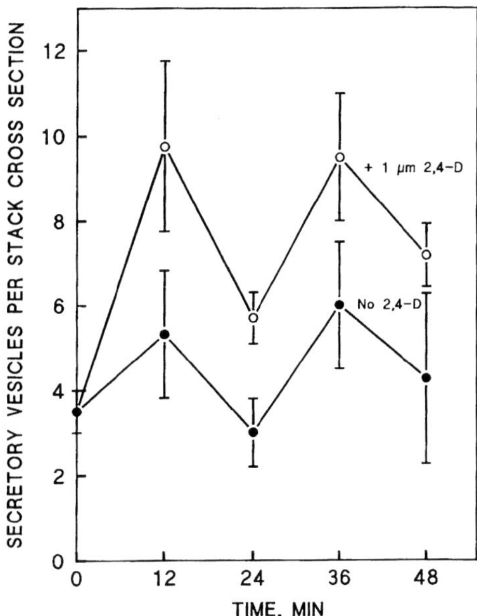

Fig. 7.5 Secretory vesicle formation by Golgi apparatus of excised soybean stem segments exhibit oscillations with a period length of 24 min suggesting involvement of ENOX1 in their formation. Unpublished data of G. Auderset and D.J. Morré

be supported by reduced pyridine nucleotides. Rodriguez et al. (1992) found that transfer was promoted by reduced pyridine nucleotide (NADH) or reduced pyridine nucleotide plus ascorbate free radical using a reconstituted Golgi apparatus to plasma membrane transfer system; ATP largely was without effect. Using the cell-free system described by Rodriguez et al. (1992), transfer of radiolabeled lipids or proteins from *trans* Golgi apparatus elements to inside-out plasma membrane vesicles immobilized on nitrocellulose was found to be promoted by NADH. In support of a role of ENOX proteins in membrane displacements leading to endomembrane vesicle formation is a 24-min periodicity of vesicle formation as is characteristic of the time-keeping properties of ENOX1. Figure 7.5 shows data from experiments with plant Golgi apparatus from etiolated hypocotyls of soybean (*Glycine max*) synchronized by exposure to white light at $t=0$ where vesicle production by the Golgi apparatus stacks oscillated with a 24-min period and was stimulated by the synthetic plant growth substance, 2,4-dichlorophenoxyacetic acid (2,4-D), at a near optimum concentration for growth of 1 μM. Oscillations in Golgi apparatus-derived secretory vesicles were reported previously for dark grown coleoptiles of oat (*Avena sativa*) in response to gravity stimulation (Shen-Miller and McNitt 1978; Shen-Miller and Miller 1972). Plant ENOX proteins are gravity responsive (Garcia et al. 1999) and their presence in the gravity-responsive oat cells might play a role in the Golgi apparatus response to geostimulation.

Previously, *trans* elements of Golgi apparatus of rat liver (Morré et al. 1978) were shown to oxidize NADH by cytochemistry as determined by reaction in the presence of a complex of copper and ferricyanide. When reduced, electron-dense deposits of insoluble copper ferrocyanide (Hatchett's brown) are visible in the electron microscope. Highly purified fractions of Golgi apparatus from rat liver oxidize NADH.

Vesicle budding from endoplasmic reticulum to the Golgi apparatus includes a requirement for ATP, an involvement of low molecular weight G proteins and other cytosolic factors and inhibition by thiol reagents such as *N*-ethylmaleimide (Wattenberg 1991). Coat proteins are involved (Donaldson et al. 1990) and the macrolide antibiotic brefeldin A inhibits (Donaldson et al. 1990; Miller et al. 1992). Polyclonal antisera to the p^{97} ATPase block transfer in a cell-free system where the inhibition is primarily at the level of vesicle budding. An absolute requirement for ATP has been established in vesicle budding from transitional endoplasmic reticulum using permeabilized cells (Balch and Keller 1986; Simons and Virta 1987; Beckers et al. 1987, 1990) and cell-free systems (Morré et al. 1986b; Balch et al. 1987; Paulik et al. 1988; Warren et al. 1988; Wattenberg 1991). The p^{97} ATPase of transitional endoplasmic reticulum is distinct from standard ATPases from other membrane fractions such as the plasma membrane, mitochondria or vacuolar apparatus in its response to ATPase inhibitors. This characteristic has proven useful in linking the p^{97} to physical membrane displacement phenomena such as vesicle budding and cell enlargement which show a similar profile of inhibition as p^{97} (Zhang et al. 1994). The ATPase of the transitional endoplasmic reticulum is not inhibited by ouabain, oligomycin, nitrate, or vanadate; however, cobalt chloride inhibits. The activity corresponding to a monomeric molecular weight of approximately 97 kDa isolated from rat liver has a native molecular weight of approximately 600 kDa. Highly purified preparations by electron microscopy are seen as ring-like structures comprised of six, apparently identical, subunits (Zhang et al. 1994; Chap. 5). Partial amino acid sequence revealed homology with valosin, CDC58 and the *N*-ethylmaleimide-sensitive factor (NSF). As a unique ATPase family, these proteins are recognized by antisera raised against the 97-kDa monomer. These same antisera inhibit both cell-free budding and transfer between transitional endoplasmic reticulum and *cis* Golgi apparatus with liver fractions (Morré and Mollenhauer 2009).

7.4 Endocytosis and Autophagy

A role for a group of ENOX proteins in endocytosis has emerged recently from the structural studies of the age-related NOX (arNOX) shed from the cell surface into the circulation. A family of 9-membered transmembrane proteins, the C-Terminal ca. 24 kDa fragment, is shed into the circulation as a soluble protein and, as originally described, also appears in the content of endosomes (Singer-Krüger et al. 1993; Schimmöller et al. 1998).

The catalytic activity of the arNOX proteins is retained by the C-terminal fragment which is capable of generating superoxide (Chap. 9). Superoxide generation by arNOX family members shed into the circulation correlates with various "action at a distance" phenomenon such as oxidation of serum low-density lipoproteins associated with cardiovascular disease (see Sects. 7.6 and 7.7) and oxidation of supporting skin collagen and elastin associated with skin aging (C. Meadows, D. J. Morré, D. M. Morré, Z. D. Draelos and D. Kern, unpublished results).

7.5 Host Defense

The oxidative burst of the phox-NOX family of cell surface oxidases in host defense is well established (Lambeth et al. 2000). This protective function is based on the ability of these enzymes to generate superoxide and cause oxidative damage to invading organisms. A similar role might be envisioned as well for the superoxide generating arNOX proteins especially as they occur on the internal surface and content of endosomes (Sect. 7.4).

7.6 pH Control

While there are many cell surface transporters contributing to pH control, evidence from effector and inhibitor response implicates ENOX proteins in pH control, especially in proton extrusion and the attendant alkalinization of the cytosol. Evidence for a link between plasma membrane electron transport and proton extrusion is summarized in Chap. 4. The pharmacological sulfonylurea inhibitor of ENOX2, LY181984, reduced proton transport in HeLa cells (Morré and Morré 1995b; Fig. 7.6) but the change in internal pH was only about 0.5 units at pH 7.4 even after treatment with 100 μM LY181984 compared to the inactive LY181985.

A cell surface-associated electrogenic proton pump driven by ATP is considered to be responsible in part for the generation of a membrane potential and to establish a pH gradient across the plasma membrane of plant cells (Poole 1978; Spanswick 1981). While the generally accepted mechanism for creating an electrochemical proton gradient is that of a plasma membrane H^+-ATPase, H^+ transport may be driven by a plasma-membrane redox system (Crane et al. 1985; Hassidim et al. 1987) as well. Proton pumping across the plasma membrane is stimulated by the auxin plant growth hormones (Rayle 1973; Cleland 1975; Kurkdjian et al. 1979; Kutschera and Schopfer 1985a; Lüthen et al. 1990) as well as by the fungal toxin fusicoccin (Marré et al. 1973, 1974; Cleland 1976; Kurkdjian et al. 1979; Marré 1979; Kutschera and Schopfer 1985b) and inhibited by vanadate (Rubinstein and Stern 1986). The significance of outward proton pumping has generally been considered to achieve cell wall acidification (loosening) (Brummer and Parish 1983). However, the necessity and the sufficiency of cell wall acidification for cell elongation to occur has been challenged (Kutschera and Schopfer 1985a; Lüthen et al. 1990). There is a lag, for example, of 5–7 min between the addition of the auxin and the onset of external proton secretion with plant stem or coleoptile segments (Cleland 1976; Cleland and Rayle 1978).

The explanation for the lag in growth and cell wall acidification may reside in the work of Felle (1987) (Fig. 7.7), where both membrane potential and cytoplasmic pH

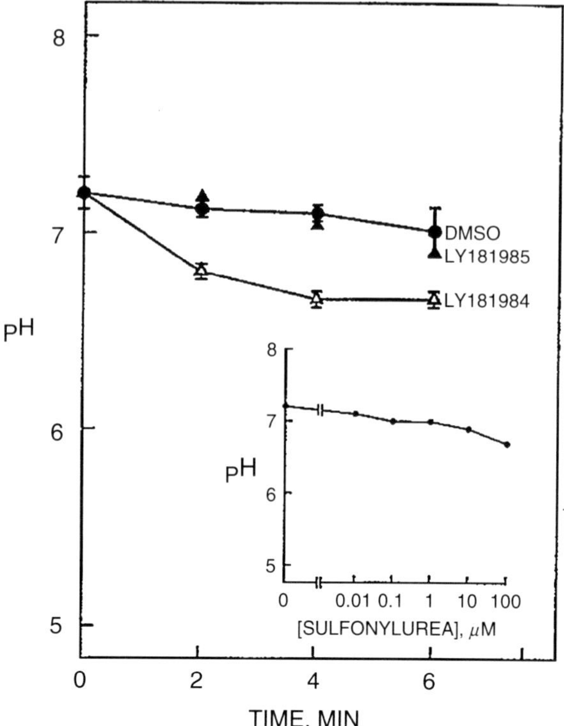

Fig. 7.6 Cytoplasmic pH as a function of time of treatment of HeLa cells with either LY181984 (active antitumor sulfonylurea) or LY181985 (inactive analog). *Inset*: Dose–response of cytoplasmic pH as a function of the logarithm of sulfonylurea (LY181984) concentration. Reproduced from Morré and Morré (1995b) with permission from Springer-Wein

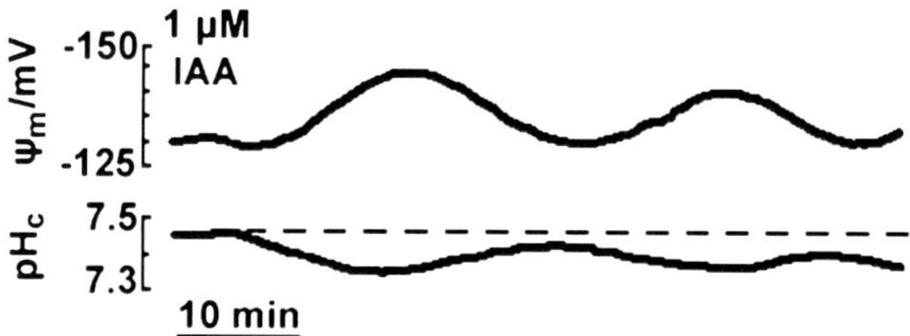

Fig. 7.7 Oscillations of cytosolic pH (pH_c) and membrane potential (Ψ_m) following treatment of maize coleoptiles with 1 μM indole-3-acetic acid (IAA) added to the medium, $pH_o = 6$, measured simultaneously with double-barreled pH microelectrodes. Dashed line indicates control-pH_c before IAA addition. Reproduced from Felle (1988) with permission from Springer Scientific + Business Media

were measured. The lag in the rise in membrane potential (Fig. 7.7, upper curve) parallels the lag in proton extrusion such that it may be assumed that the two are related. At the same time, there is a parallel acidification of the cytoplasm (Fig. 7.7, lower curve).

The activation of proton release by plant plasma membranes is paralleled in HeLa cells where cytoplasmic acidification is related to the activation of an H^+/Na^+ antiport (Sun et al. 1987a, 1988). As mentioned above, the source of protons for cytoplasmic acidification might very well be NADH oxidation at the inner surface of the plasma membrane (Ivankina and Novak 1980). The induction of proton extrusion in animal cells by external impermeant electron acceptors such as ferricyanide and diferric transferrin has been explained accordingly. A link between proton pumping and ENOX activity in plants is provided from the observation that both the proton pumping and membrane potential responses are sinusoidal (Fig. 7.7). They oscillate with the same period of about 48 min but are displaced from each other by 1/2, i.e., 24 min.

Unfortunately, interest in ENOX proteins and pH control diminished in the late 1980s without a clear resolution. This area remains largely unexplored and represents an area of opportunity for future investigation.

7.7 Lipid Oxidation

One subset of the ENOX terminal oxidase proteins with unique properties is that of the age-related ENOX protein or arNOX (Morré and Morré 2003b, 2006d, 2008a; Morré et al. 2003a, 2009c) (Chap. 9). Unlike the ENOX1 and ENOX2 proteins which carry out 4-electron transfers to molecular oxygen to form water, the arNOX proteins result in the generation of superoxide at the cell surface (Chap. 9). The superoxide generated affords an opportunity to form H_2O_2 and other reactive oxygen species (ROS) for propagation to adjacent cells and tissues. In keeping with ectoproteins in general, ENOX proteins, including arNOX, are shed to the circulation. The shed forms of the protein appear in body fluids (serum, saliva, urine and perspiration and interstitial fluids that percolate through the basement membrane). Protein thiols have been demonstrated to serve as the electron donors for the shed forms of arNOX. The ROS generated from the superoxide can become accessible to lipoproteins in the circulation resulting in their oxidation and increased atherogenic risk as well as damage to adjacent cells and extracellular supporting matrices that are important to skin health.

Age and oxidative stress are major risk factors for heart disease (Schmuck et al. 1995). A large body of evidence supports the notion that ROS provide a causal link in the appearance of oxidized circulating lipoproteins such as oxidized low-density proteins (LDLs) and their subsequent clearance by macrophages and delivery to the arterial wall. It now appears likely that oxidized LDL is a major contributor to progressive atherogenesis by enhancing endothelial injury, by inducing foam cell (lipoprotein engorged macrophages) generation and associated smooth muscle proliferation (Holvoet 1999). Macrophages clear the circulation of oxidized lipoprotein particles by internalizing them and in so doing they are transformed into foam cells. The foam cells deliver their cargo of oxidized fats and cholesterol where

Table 7.1 arNOX-mediated oxidation of serum lipoproteins

Donor	MDA-LM (μmol/L)		
	Acceptor – SOD	Lipoprotein + SOD	Change due to arNOX
Plasma membrane overexpressing arNOX	0.93±0.03	0.57±0.03	0.40±0.06
Plasma membrane lacking arNOX	0.43±0.06	0.43±0.06	0.03±9.95

MDA-LM formed 2 h of incubation with donor plasma membranes overexpressing high levels of arNOX compared to plasma membranes not overexpressing arNOX
arNOX aging-related cell surface NADH oxidase; *MDA-LM* malondialdehyde-like material; *SOD* superoxide dismutase. Data of Morré and Morré (2006d)

they are deposited beneath the arterial wall. Such progressive delivery of oxidatively damaged lipoprotein particles eventually leads to atherosclerotic plaques and advanced heart disease.

However, the basis for LDL oxidation has been little studied. The levels of common antioxidants including α-tocopherol, β-carotene, and ascorbate decline with age, but there is no apparent correlation between ingestion of these common antioxidants and amelioration of the aging process or decreased mortality (Bjelakovic et al. 2007). The implication is that the oxidative damage leading to aging and increased atherogenic risk is the result of a more specific causation. Why does LDL oxidation increase in the elderly, and why is it greater in some individuals than in others? Our findings suggest that LDL oxidation in the elderly and in individuals at high risk for heart disease correlates with the levels of circulating aging-related cell surface NADH oxidase (arNOX) (Table 7.1). The arNOX proteins are shed into the milieu surrounding the cells. In aged individuals, the amount of superoxide generated by the shed arNOX proteins has been measured to be quite substantial reaching a maximum at age 65–75 in males and age 55–65 in females. Of those who die of a heart attack, 85 % are 65 or older (American Heart Association 2008). Women surviving beyond age 65 usually have diminished arNOX levels compared to men and a lower risk of cardiovascular disease compared to men (Levine and Kannel 2003) further suggesting some causal relationship between arNOX levels and atherogenic risk.

Progressive development of age-associated systematic disease has long been associated with the production of ROS (Linnane et al. 2007). Our work has focused on age-related oxidase (arNOX) as an important source of ROS especially in the circulation. Because arNOX proteins are shed from the cell surface and circulate, they coexist with LDLs in the blood and appear to be responsible for their oxidation. Oxidized lipoproteins are endocytosed by macrophages to form the foam cells which are ultimately responsible for the deposition of cholesterol and oxidized lipids in the arterial walls as the basis for formation of fatty streaks leading to atherosclerotic plaques and coronary artery disease (Holvoet et al. 1998, 2001).

7.7.1 arNOX Inhibitors and Prevention of Coronary Artery Disease

If the hypothesis that arNOX is responsible for LDL oxidation and atherogenesis (Morré et al. 2010c) is correct, the expectation is that arNOX inhibitors would ameliorate and possibly prevent progression of coronary artery disease. On this basis, a concept of anti-arNOX nutritional supplements based on the arNOX inhibitory properties of various French seasonings known collectively as Herbs de Provence was advanced (Morré et al. 2010c).

Ingestion of such preparations results in a decrease in the generation of ROS by arNOX and provides a possible explanation for the French Paradox where the French diet or lifestyle leads to reduced atherogenic risk despite a cholesterol-rich diet high in cheese and butter. Previous studies attributed the reduction in risk to the consumption of red wine as a natural source of the polyphenol resveratrol (Teissedre and Waterhouse 2000). However, the active herbs, which are staples of the French diet, offer a more compelling explanation. Certain of these herbs and phenolics may have benefit as well by inhibiting platelet adhesion and aggregation (Yazdanparast and Shahriyary 2008).

7.8 Life Extension and Calorie Restriction

Calorie restriction is the only experimental manipulation known that consistently appears to extend the life span of living organisms (Lin et al. 2002; Weindruch and Walford 1998; Roth et al. 2001). A major cause of aging is thought to result from the cumulative effects of cell loss over time (Cohen et al. 2004). The proposed mechanism underlying the extension of life span involves a shift from a state of growth and proliferation to one of maintenance and repair (Walford et al. 1987; Weindruch et al. 1988; Yu et al. 1985).

Initially, calorie restriction was believed to extend life span by decreasing metabolic rate, decreasing mitochondrial oxygen consumption, and, therefore, attenuating oxidative stress. However, the exact opposite appears to be true as evidenced by increases in mitochondrial content and oxygen consumption in response to calorie restriction (Nisoli et al. 2005). Observations that calorie restriction does not increase life span when the gene encoding cytochrome c is deleted (Lin et al. 2002) or in the presence of electron transport inhibitors (Bishop and Guarente 2007; Panowski et al. 2007) suggest that mitochondria are critical factors in the phenomenon of life span extension by calorie restriction.

There are several mechanisms by which mitochondria may be responsible for the life-enhancing effects of calorie restriction (Guarente 2008). However, the simplest and most likely explanation is that enhanced mitochondrial activity results in the depletion of reduced pyridine nucleotides (NADH) of the cytosol and the increased production of NAD^+ and that lowering of the level of NADH results in life extension.

Among the several determinants of life span, the one common thread that has emerged is a variety of species from yeast to rodents is a correlation with lowered levels of NADH.

The basis for life extension by lowered NADH and increased NAD^+ is due in part to inhibition of SIR2 by NADH, a key regulator of life span in a number of organisms including yeast, worms, flies, rodents, and non-human primates (Bordone and Guarente 2005). It is well known that the silent information regulator (Sir) proteins regulate life span (Haigis and Guarente 2006). In yeast, an extra copy of the SIR2 gene extends replicative life span by 50 %, while deleting SIR2 shortens life span (Kaeberiein et al. 1999). SIR2 encodes an NADP-dependent deacetylase which emerges as a critical mediator of calorie restriction (Guarente and Picard 2005).

The mammalian SIR2 ortholog SIRT1 encodes an NAD-requiring NAD-dependent histone deacetylase required for chromatin silencing and life-span extension. NADH is a competitive inhibitor of SIR2. Therefore, an overall reduction in NADH will activate SIR2 and result in life extension. An extension of this concept is that of calorie restriction mimics (mimetics) that will modify metabolism to reduce NADH levels and increase NAD^+ comparable to that achieved under calorie restriction conditions without the usual reduction in calorie content. With extended use, calorie restriction mimics would up- and down-regulate gene expression and cellular proteins that recapitulate calorie-restricted profiles as well as decrease insulin resistance (lower fasting blood glucose levels) and increase glucose uptake similar to changes seen in calorie restriction.

Longevity regulatory genes in addition to the NAD-dependent histone deacetylase silent information regulator (SIR2) also include the forkhead transcription factor FOXO (Giannakou and Partridge 2004).

7.9 Control of Apoptosis and Cell Survival

The ENOX2 inhibitor phenoxodiol is a potent inducer of apoptosis (Alvero et al. 2006; Kamsteeg et al. 2003). Phenoxodiol treatment of ovarian cancer cells has provided evidence that the apoptotic effect is due to the regulation of the Akt-FLIP-XIAP pathway through modulation of upstream signaling cascades (Kamsteeg et al. 2003). Inactivation of FLIP and XIAP in response to phenoxodiol restores the sensitivity of ovarian cancer cells to Fas-mediated apoptosis through the Akt signal transduction pathway. Akt translocates to the nucleus and may regulate the transcription of genes mediating cell survival. The induction of FLIP expression and blocking the extrinsic apoptotic pathway is one possible mechanism by which Akt functions as a promoter of survival (Kamsteeg et al. 2003). Ovarian cancer cell lines investigated exhibited high levels of FLIP. Inhibition of Akt expression followed by a decrease in FLIP expression and reversal of the sensitivity to Fas-mediated apoptosis was consistently observed in response to phenoxodiol. As an alternative, phosphor-Akt might target XIAP for proteosomal degradation through phosphorylation (Kroesen et al. 2003).

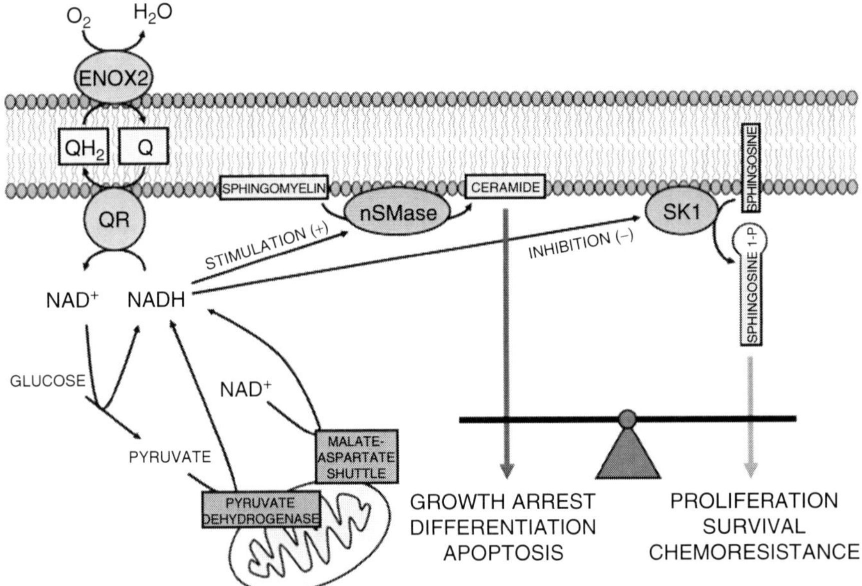

Fig. 7.8 ECTO-NOX proteins at the cell surface keep the plasma membrane pool of coenzyme Q_{10} (CoQ) largely oxidized through the oxidation of $CoQH_2$. ENOX2 is a cancer-specific ECTO-NOX form absent from non-cancer cells. When ENOX2 is active, CoQ of the plasma membrane is oxidized and NADH is oxidized at the cytosolic surface of the plasma membrane. However, when ENOX2 is inhibited and plasma membrane electron transport is diminished, both $CoQH_2$ and NADH accumulate. CoQ and NAD^+ block sphingomyelinase activity, preventing ceramide accumulation and the ensuing G_1 arrest. Inhibition of ENOX2 leads to the replacement of CoQ by $CoQH_2$ and the accumulation of cytosolic NADH. The result is sphingomyelinase release from inhibition, ceramide accumulation and G_1 arrest. Sphingosine kinase is inhibited either by $CoQH_2$, NADH or both resulting in decreased sphingosine-1-phosphate (S1P) and diminished phosphorylation of protein substrates by Akt along with activation of the FAS pathway leading to apoptosis as FLIP and XIAP would no longer be available to serve as alternative substrates for caspase 8

ENOX2 is the primary drug target of phenoxodiol (Herst et al. 2007; Morré et al. 2007a). When inhibited, the plasma membrane content of ubiquinol is increased, which in turn results in cytosolic accumulation of NADH and the resultant decoupling of the S1P pro-survival signal transduction cascade which then appears to be linked to the inhibition of Akt, XIAP, and FLIP and initiation of the Akt-FLIP-XIAP apoptotic cascade (De Luca et al. 2005, 2009; Morré et al. 2007a; Fig. 7.8). ENOX2 inhibition also results in an accumulation of ceramide through NADH-mediated activation of plasma membrane sphingomyelinase (De Luca et al. 2005). Both events (concurrent reduction in S1P and accumulation of ceramide) appear to initiate caspase-3 dependent programmed cell death (Fig. 7.9). The cancer specificity of the ENOX2 protein provides a mechanistic explanation for the apparent targeted toxicity of phenoxodiol for cancer cells and the lack of a phenoxodiol response with most normal cells (Morré et al. 2007a). The role of the ENOX2 proteins is to drive

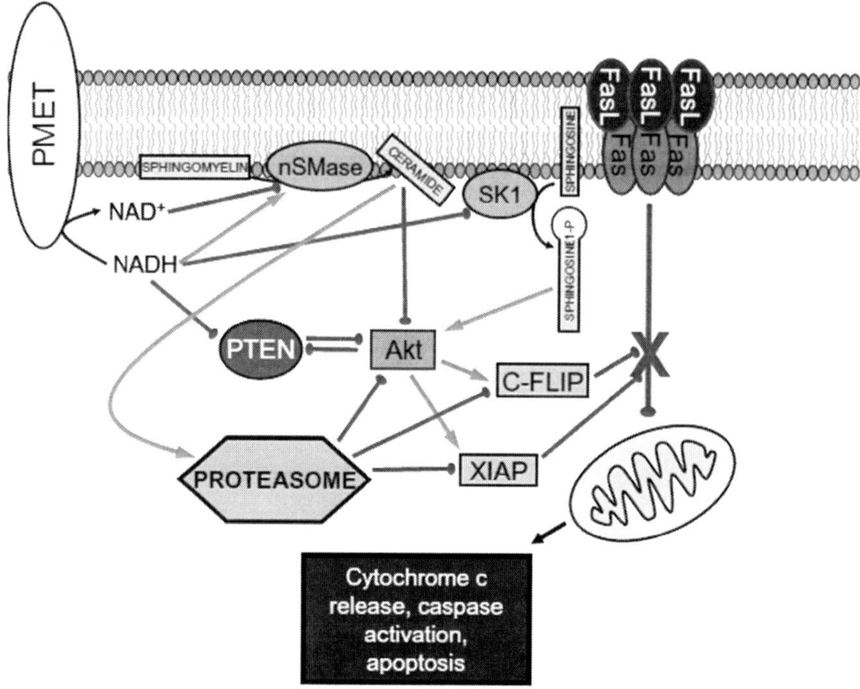

Fig. 7.9 Phenoxodiol, a ENOX2-specific inhibitor, alters cytosolic NADH levels to initiate a regulatory cascade linking sphingolipid metabolism and the PT3K/Akt pathway to apoptosis. Evidence is consistent with Akt function as a promoter of survival through the induction of FLIP expression and blockage of the extrinsic apoptotic pathways

cell enlargement (Morré and Morré 2003a) and when ENOX2 is inhibited, the cells fail to enlarge to a size sufficient to pass the G_1 checkpoint that monitors cell size and cannot divide. This may explain how phenoxodiol disrupts the cell cycle in the majority of cancer cell lines thus far investigated. Cycling of A2780 ovarian cancer cells is strongly blocked at S phase by phenoxodiol with a more moderate block at G_2M (Brown et al. 2005). The latter was elicited by inhibiting cdk2 activity through ρ53-independent increased $ρ21^{waf1}$ expression (Aguero et al. 2005).

7.10 Neurodegenerative Disorders

It was summarized in Chap. 1 that ENOX2 and prions share a set of similar properties. Recombinant mouse prion protein when examined for ECTO-NOX activities exhibited a copper-dependent pattern of oscillating and alternating NADH oxidase and disulfide–thiol interchange with a 24-min period indistinguishable from that of ENOX2 (Kim and Morré 2004) (Chap. 8). Evidence supports a role for thiol–disulfide exchange in the conversion of PrPc to the infectious PrP^{Sc} (Welker et al. 2001).

Structures virtually indistinguishable from aggregates associated with neurodegenerative amyloid-forming proteins were observed by high-resolution electron microscopy for enzymatically inactive amyloid rods and open cylinders (rings) of ENOX2.

When extended to Alzheimer's Aβ 1–43 peptide and α-synuclein of Parkinson's disease, copper-dependent oscillations in NADH oxidase activity similar to that of ENOX2 and the mouse prion with a 24-min period length have been observed (Table 1.6). The Aβ peptide lacks a cysteine and does not carry out disulfide–thiol interchange. α-Synuclein also has a copper-dependent and oscillating NADH oxidase activity but the period length is 54 min.

Porin isoform 1 or VDAC, a plasma membrane NADH-ferricyanide reductase, essential for normal brain functioning has also been implicated in playing a role in neurological disorders such as autism (Sect. 4.1.3.2).

7.11 Memory

While ENOX proteins per se have not been implicated in brain function related to memory, they, like prions, participate in a primitive form of learning and teaching that might serve as a model for more complex learning processes. Entrainment (Chap. 6) is a form of learning and teaching in which all ENOX proteins participate without which their characteristic periodic functions would be largely obscured by lack of synchrony.

One such proposed learning and teaching model is that of the scrapie and related prions (Tompa and Friedrich 1998). Here the infective form of the prion protein, also protease resistant, is capable of eliciting conformational changes in other non-infective prion proteins to render them infective. What emerges is the potential for a family of amyloid-forming proteins all capable of binding copper and exhibiting oscillatory oxidation of NADH of low specific activity (Table 1.6). The oscillations are copper dependent as is their oxidative capacity. Complete amino acid sequences are known for each of the proteins listed. There is virtually no sequence similarity. Even though all bind copper, the copper-binding motifs differ. For ENOX2, the putative copper sites are HVH 546 together with His 562 conserved with superoxide dismutase, one copper site and HS 82 as a second copper site as determined from site-directed mutagenesis (Tang et al. 2010). The mouse prion utilizes the octarepeats (Garnett and Viles 2003; Wells et al. 2006). For α-synuclein and Alzheimer's Aβ, different combinations of histidines or tyrosines are utilized. The copper binding motifs of human amyloid beta include three histidines (His, His^{13} and His^{14}) along with an adjacent tyrosine (Tyr^{10}) (Miura et al. 2000; Curtain et al. 2001). For α-synuclein, copper binding appears to be dependent on the acidic C terminus (Palk et al. 1999) most likely involving conserved tyrosines Y^{125}, Y^{133} and Y^{136} (Clayton and George 1998). Yet all examples in Table 1.6 bind adenine nucleotides (NADH) although the amino acid sequences within the putative binding regions differ. Probably the best source of a new set of tools to understand ENOX periodicity and oxidative function may be provided by the Alzheimer's Aβ peptides of 43 ca. amino acids in length (Markert et al. 2004). The copper site is known and a putative NADH site is found near the C terminus as in ENOX2 in the vicinity of M-35.

An important emerging function beyond the oxidative response contributing to neurodegenerative pathology is the ability of ENOX proteins, prions and other amyloid-forming proteins to "remember." Entrainment of ENOX proteins, both in solution and as receptor-mediated events, involves memory (learning) in the form of a particular conformation within a protein molecule and which can then be imparted to other protein molecules (teaching) of the same or similar primary structure to achieve a synchronous population. In a similar manner, prions correctly folded to be proteinase K resistant impart this property to other prion molecules initially protease susceptible to amplify and propagate the protein species associated with neurodegeneration. Thus, ENOX proteins and prions and possibly all amyloid forming proteins share the property of protein memory and transmissible alterations passed from one protein molecule to another ultimately imparting protease resistance, for example, to an entire population of protein molecules (learning and teaching). Even plant ENOX proteins may possess such properties (Chap. 10). A process of learning and teaching has been applied to plants where a class of herbicides mimics a native growth factor (hormone) to alter a plant-specific ENOX protein to become unregulated and to recruit other related ENOX proteins to become unregulated with the death of the plant as the eventual result. The basic strategy, once again, is that of the prion model involving modification of a certain subset of ENOX proteins which then remember and recruit other family members into their modified aberrant state. In the normal situation, these phenomena may be beneficial and important, for example, to growth control and developmental processes. When carried to extremes, as with the herbicide treatment or with scrapie or mad cow disease, the result is a pathological state leading to the demise of the affected organism.

7.12 Gametogenesis

There have been reports of protein disulfide isomerase activities associated with plasma membranes of several cell types (Kroning et al. 1994) with multiple functions attributed to the protein (Zai et al. 1999). By expressing both an isomerase and a chaperone-like activity (Wang 1998; Chen et al. 1999), protein folding is assisted. A protein disulfide isomerase-like activity associated with the plasma membranes of sperm (Bohring et al. 2001) has been suggested to be important for capacitation and the acrosome reaction as has a similar activity associated with the acrosome directly (Ohtani et al. 1993).

Mouse sperm contains a protein disulfide isomerase-like activity (protein disulfide thiol interchange) based on assays using cell impermeant substrates (Morré 2006). The activity has characteristics of a constitutive ENOX (ENOX1) activity as is associated with plasma membranes of a variety of cells and tissues (Chap. 1) rather than that of a classical protein disulfide isomerase (e.g., ER 60) typically associated with the endoplasmic reticulum (Essex et al. 2001). The activity is resistant to both capsaicin and bacitracin (Morré 2006). The latter distinguishes it from the more classical protein disulfide isomerases (Essex et al. 2001) which are inhibited by bacitracin.

7.13 Role in Viral Pathogenesis

Protein disulfide isomerase activities associated with plasma membrane have been implicated in viral infections (Fig. 7.10). Enveloped mammalian viruses generally enter cells via fusion of viral and cellular membranes (Battini et al.

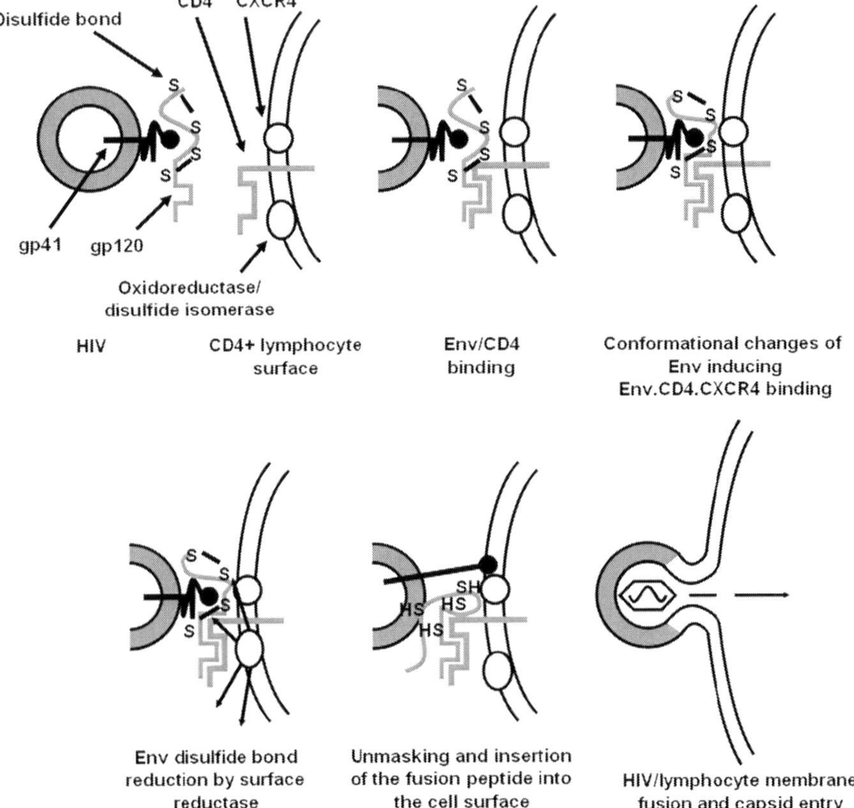

Fig. 7.10 Env reduction and HIV/lymphocyte fusion. Viral infection (enveloped viruses) appears to require protein disulfide–thiol interchange. Conformational changes required for viral entry are triggered. Modified from Barbouche et al. (2003)

1995) frequently in a process involving endocytosis (Marsh and Peichen-Matthews 2000). Even during the early stages of viral replication, virus-infected cells transport and express viral glycoprotein antigens at their cell surface (early literature reviewed by Morré and Ovtracht 1977). Ryser et al. (1994) and Abell and Brown (1993) were among the first to publish data suggesting that thiol–disulfide interchange reactions that occurred during the interaction of viruses with cells were necessary for fusion of viral and cellular membranes (reviewed by Sanders 2000). Examples include protein disulfide isomerase-like thiol-disulfide exchange with human immunodeficiency virus type 1 (HIV-1) envelope proteins that may trigger changes in conformation required for HIV-1 entry (Fenouillet et al. 2001; Matthias and Hogg 2003; Tichopad et al. 2005; Ou and Silver 2006; Ho and Douglas 1992), disulfide bond restructuring post-receptor binding conformational changes within ENV that induce fusion competence (Barbouche et al. 2003) and the observations of Ciriolot et al. (1997) and Garci et al. (1992) with Madin Darby canine cells, of Abell and Brown (1993) with Sindbis virus and of Battini et al. (1995) with murine leukemia virus. Infection of lymphoid cells by HIV-1 was inhibited by membrane impermeant sulfhydryl blocking reagents and by inhibitors of cell surface protein disulfide isomerase (Ryser et al. 1994). Implicit in the findings of Ryser et al. (1994) was the interpretation that the PDI-like activity mediated a thiol–disulfide exchange with HIV-1 envelope proteins, triggering changes in conformation required for HIV-1 entry. While the target protein resembled that of a classic protein disulfide isomerase (Mandel et al. 1993; Ryser et al. 1991; Gallina et al. 2002), it may be due to an activity of a different protein (Nakano and Ono 1990).

Based on drug response and other characteristics, the activity might be a member of the ENOX family. ENOX proteins carry out protein disulfide–thiol interchange in addition to hydroquinone (NADH) oxidation (Morré 1998c). The cancer-specific ENOX2 forms which originates as splice variants of a single gene, for example, are inhibited by several important classes of quinone site or potential quinone site inhibitors (Chap. 8) many of which exhibit both antiviral and anticancer activity (Morré et al. 1998h, 2000a; Paulik et al. 1999; Morré 1998c). Another group of ENOX2 inhibitors, the callipeltins, also exhibit both anticancer and anti-HIV activity (Freedman 1989). The antimalarial Artesunate is also a potent inhibitor of cancer cell growth (Efferth et al. 2003). The antitumor and anti-trypanosomal naphthylsulfonylurea, suramin, is yet another example (Stein et al. 1989; Jentsch et al. 1987). Suramin is an effective inhibitor of ENOX2 with an EC_{50} of about 0.1 μM with fully oxidized HeLa cell plasma membranes (Fig. 7.11).

While it is most likely that the protein disulfide isomerase activity associated with viral entry is due to a protein not restricted to cancer cells, it is of interest that the majority of the known ENOX2 inhibitors were discovered initially as anti-trypanosomal or as antiviral agents. The anti-trypanosomal activity is now understood since the parasites express on their cell surface an ENOX form similar or identical to that of cancer cells (results unpublished). The antiviral activities remain enigmatic. Yet, the antitumor sulfonylureas (Paulik et al. 1999), the antitumor quassinoids (Morré et al. 1998h) and brefeldin A (Morré et al. 1994e), all known

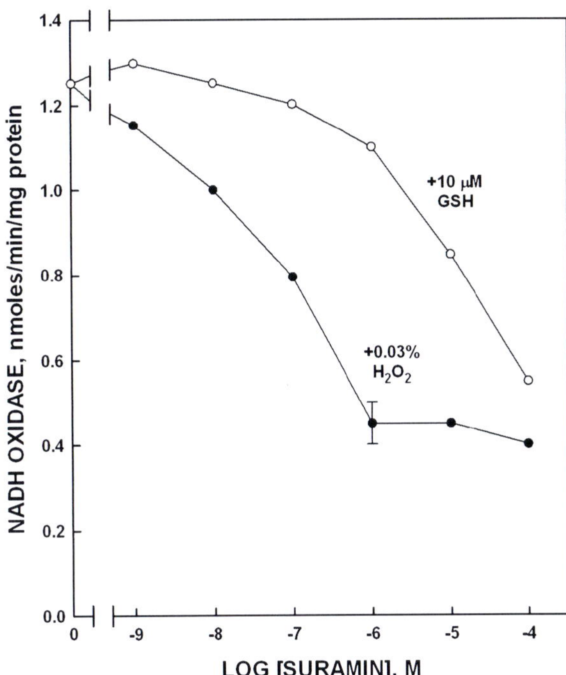

Fig. 7.11 Inhibition of NADH oxidase activity of isolated plasma membrane vesicles of HeLa cells by increasing concentrations of suramin in the presence of reduced glutathione (GSH) or hydrogen peroxide

ENOX2 inhibitors, inhibited entry of HIV. One explanation may be that the protein disulfide–thiol interchange activity necessary for virus entry is specifically induced by virus infection at the surface of the infected cells (Morré et al. 1998h; Paulik et al. 1999) as well as the major catechin with anticancer activity from green tea, (−)-epigallocatechin-3-gallate (EGCg) (Nakano and Ono 1990; Mukoyama et al. 1991) as well as (−)-epigallocatechin-3-gallate sulfate (Mizuno et al. 1992).

7.13.1 ENOX2 Inhibitor (−)-Epigallocatechin-3-Gallate Blocks Virus Infections Alone and in Combination with Capsicum Vanilloids and Other Green Tea Catechins

Non-nutritional polyphenolic compounds such as the principal green tea catechin, (−)-epigallocatechin gallate (EGCg), have been reported as candidate HIV agents (Freedman 1989; Fassina et al. 2002) based on laboratory tests. EGCg is impressively effective in reducing infectivity of HIV in human MT-2 cells grown in culture or peripheral blood monocytes (PBMCs) from healthy donors (Maxeiner et al. unpublished; Morré and Morré 2009; Fig. 7.12). An EC_{50} of near 1 μM with a control virus production of 14,000 infective particles/ml and near 10 μM was observed

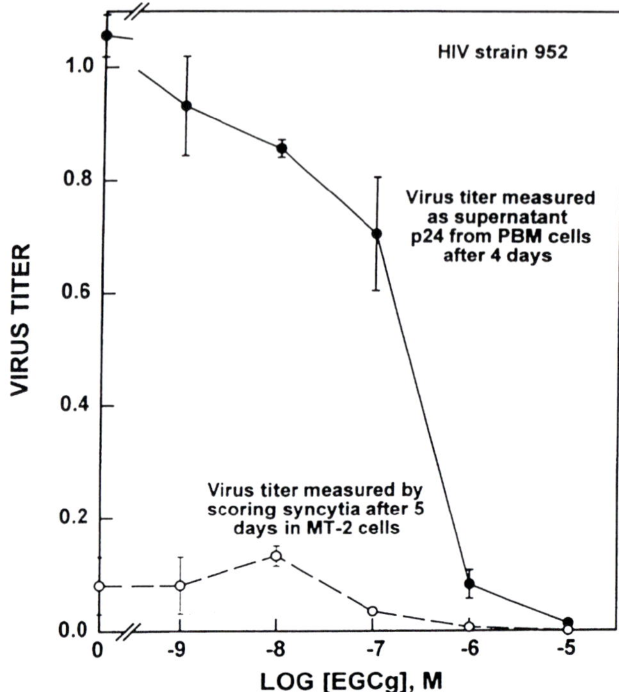

Fig. 7.12 ELISA of supernatant p24 (*solid line*; *solid symbols*) or virus titer based on number of syncytia observed by light microscopy on day 5 by titration of the day 4 supernatants in previously uninfected MT-2 cells (*dashed line*; *open symbols*) as a measure of mature virus production after 4 days with HIV strain 952 (City Laboratory, Hamburg) in peripheral blood monocytes (PBMCs) prepared from healthy volunteers. Assays were in triplicate ± standard deviations. Unpublished data of Maxeiner et al.

with a control virus production of 40,000 infective particles/mL. When the supernatants on day 4 from the experiment were reassayed with a second group of MT-2 cells based on syncytia formation after 5 days (a biological measure of infective virus particles), the EC_{50} for prevention of infection was near 0.1 μM. At 1 μM, the infectivity was almost entirely prevented. The EGCg was not toxic and did not affect the viability of the human MT-2 cells as based on MTT assay at the 1 μM concentration of EGCg where virus production was prevented. Similarly, the EGCg was without effect on the viability of the human peripheral blood lymphocytes.

While it is not known if the catechins affect virus replication per se, the findings suggest that they block at some novel cell surface NOX target. They appear to exert antiviral activity by interfering with either reinfection or budding and release of mature virus or both.

7.13.1.1 HIV and FIV

In parallel, combinations of green tea catechins (Morré et al. 2003c) and green tea catechins + vanilloids (Morré and Morré 2003d) were compared to EGCg for prevention of HIV infection (Morré and Morré 2009). Green tea catechins inhibited infectivity by 98 % at a concentration based on EGCg content of 1,000 nM. Inhibition of virus infectivity by the green tea catechins reached an IC_{50} at about 100 nM. In contrast, the catechin–vanilloid mixture inhibited virus infection almost 100 % at a concentration of 100 nM and reached an IC_{50} between 0.1 and 1 nM with concentrations based on EGCg content. These results demonstrated that the catechin–vanilloid mixture was 100 times more effective on an EGCg basis than the catechins alone.

Comparatively the potent antiviral activity of EGCg and epicatechin gallate against HIV reverse transcriptase at 10–20 ng/mL and against RNA polymerase has been claimed as well (Nakano and Ono 1990; see also Tichopad et al. 2005). The effect seemed to correlate with competition for the template-primer rather than to an antioxidant action of the EGCg.

7.13.1.2 Other Viruses

Antiviral action was demonstrated in protecting cultured rhesus monkey kidney MA104 cells against rotaviruses and enteroviruses by EGCg (Mukoyama et al. 1991). The effect, however, was most pronounced when the virus was treated with the agent before infection of the cells. Both EGCg and theaflavin digallate inhibited the infectivity of both influenza A and B virus in cultured Madin-Darby canine kidney cells (Mukoyama et al. 1991). Electron microscopic inspection revealed virus agglutination and inhibition of virus adsorption to the target cells. Pre- and post-treatment of the cells themselves with EGCg produced significantly weaker effects. Inhibition of the infectivity of influenza virus by tea polyphenols has also been reported (Nakayama et al. 1993; Yamaguchi et al. 2002) as has the inhibition of the Epstein-Barr lytic cycle (Chang et al. 2003) and a reduction of herpes simplex virus titers (Isaacs et al. 2008) all in cultured cells.

7.13.2 *Brefeldin A and Antitumor Quassinoids*

Inhibitors of cell surface ENOX proteins with protein disulfide isomerase activity, brefeldin A (Morré et al. 1994e) and glaucarubolone (Morré et al. 1998h; Fig. 4.5), were shown to selectively inhibit the growth of Crandall Feline Kidney (CFK) cells permanently infected with FIV (Fig. 7.13) as well as human MOLT-4 cells permanently infected with HIV-1 (Morré et al. 1994e, 1998h; Fig. 7.14) whereas the growth of uninfected cells was not inhibited. The results suggest that cell lines permanently infected with either the feline or the human lentivirus exhibit growth

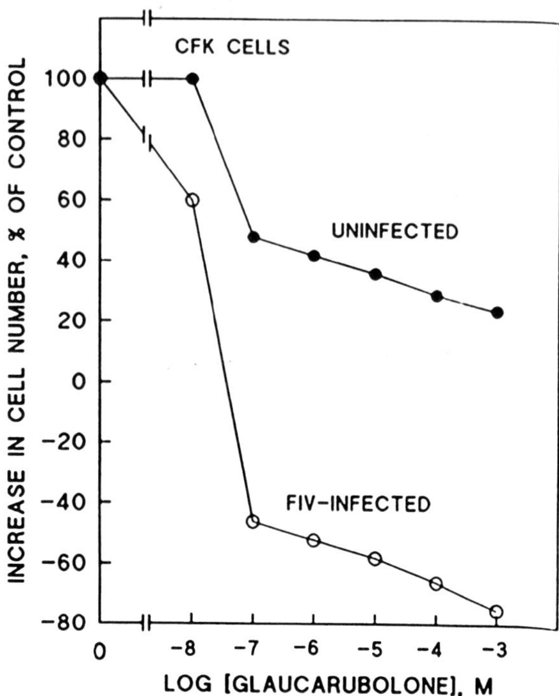

Fig. 7.13 Dose–response after 48 h of uninfected and HIV-infected Crandall feline kidney (CFK) cells to glaucarubolone. Growth of uninfected cells was slowed but cells were not killed. All FIV-infected cells treated with glaucarubolone eventually died except at the two lowest concentrations tested. Reproduced from Morré et al. (1998h) with permission from Elsevier

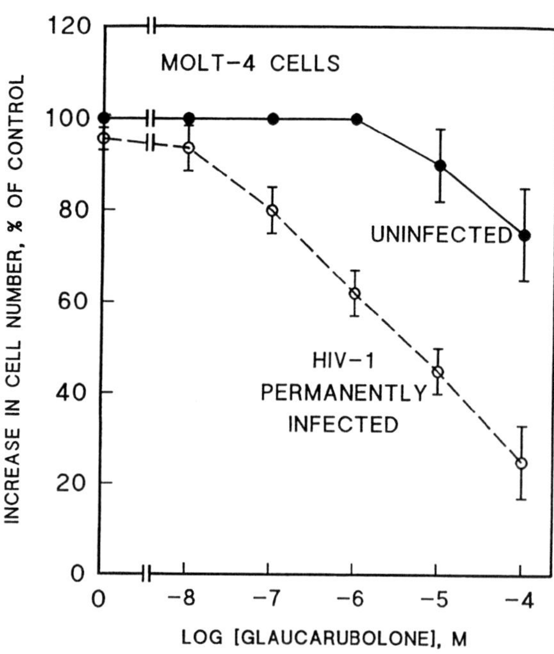

Fig. 7.14 Dose–response after 48 h of uninfected and HIV-1-infected human MOLT-4 cells to glaucarubolone. Reproduced from Morré et al. (1998h) with permission from Elsevier

response characteristics to the quassinoids in common with other malignantly transformed cell lines. In addition, the quassinoids may delay viral infection suggesting some commonality between the mechanism responsible for inhibition of the growth of the transformed phenotype and viral infection. Previously, the quassinoids glaucarubolone and isobrucine at concentrations of 0.5 μM were found to inhibit virus production in chick embryo fibroblasts infected with the Schmidt-Ruppin strain of Rous sarcoma virus, type D (Pierré et al. 1980). Even during the early stages of viral replication, virus-infected cells transport and express viral glycoprotein antigens at their cell surface (Owens and Compans 1989; Stephens and Compans 1988; Saraste and Kruismanen 1984; Tooze et al. 1984; early literature reviewed in Morré and Ovtracht 1977; Appendix Table 4, page 264 of Morré and Mollenhauer 2009). The expression of these glycoprotein antigens provides opportunities (a) for drug targeting to virus-infected cells and (b) to immobilize the drugs to restrict their action to cell surface targets. The drug conjugates were even more effective than the free drug (Paulik et al. 1999).

The drug-inhibited NADH oxidation site of the cell surface-located drug-inhibited protein disulfide–thiol interchange protein is at the external cell surface (Morré 1995a) as is the activity postulated to be essential to virus entry. Thus, it is tempting to speculate that an ENOX protein exhibiting a protein disulfide–thiol interchange protein of the mammalian cell surface responsive to virus presence is the target responsible for the antiviral activities, especially of the conjugated antiviral drugs. This ENOX or ENOX-like protein might represent a novel new target for drug prevention of viral infection as well as an opportunity to expand the understanding of how enveloped viruses gain entry into cells during the infective process.

7.14 Summary

This chapter addresses a series of processes where ENOX proteins have been consistently implicated but such involvement is not generally recognized, incompletely understood, or overshadowed by other more familiar mechanisms. Among these is cell cycle control where resolution primarily awaits a better understanding of how cell enlargement and cell division are coupled and the validity of postulated cell cycle check points in higher plant and mammalian cells that monitor cell size. Similarly, that ENOX proteins have molecular characteristics in common with known transcription factors has been noted, but evidence for their involvement in this role is lacking. Somewhat more well substantiated by experimental observations is a role in endomembrane (i.e., endoplasmic reticulum and/or Golgi apparatus) function, membrane displacements and vesicle budding. Most intriguing in this regard is evidence for oscillations having a period length of 24 min associated with Golgi apparatus activity in plants. An important area remaining to be investigated is the role of ENOX proteins and plasma membrane electron transport in general as a means by which the energy requirements for physical membrane displacements and, especially, cell enlargement are met. A change in free energy of ca −220 kJ/mol upon oxidation of NADH to form water is available but the mechanism whereby this energy release is conserved (ATP?/membrane potential?) and coupled to membrane displacements has been little investigated. A role of the arNOX (ENOX3) proteins in endocytosis may be indicated from their presence in endosomes. Similarly, since the ENOX3 proteins generate superoxide dismutases to form hydrogen peroxide, there is a potential role for these enzymes in host defense. Adverse roles for arNOX in lipid oxidation (primarily lipids associated with membranes and/or low-density lipoprotein particles) in skin health and coronary artery disease are clearly implicated as well. Roles for ENOX proteins and plasma membrane electron transport in pH control were topics of extensive research in the mid to early 1980s but without clear resolution. Resolution of this topic may tie to the role of ENOX proteins and plasma membrane electron transport in the utilization of energy derived from NADH oxidation at the plasma membrane. NAD^+ and/or NADH levels which may be affected by levels of ENOX activity and/or expression are determining factors in life extension and calorie restriction as well as control of apoptosis and cell survival. The many parallels between the physical properties of prions and ENOX proteins offer implications in neurodegenerative disorders and primitive forms of learning and memory. Finally, the protein disulfide–thiol interchange (protein disulfide isomerase) activities of ENOX or ENOX-like proteins in gametogenesis and in viral infections have been reported as having potential roles in these processes as well.

Chapter 8
ENOX2 (tNOX) and Cancer

Unregulated NADH oxidases associated with the cell surface, designated ENOX2, that are responsive to anticancer drugs and differentiating agents are characteristic of human cancer cells (Medina et al. 1997; Morré 1998c). Their association with the cancer cell surface helps explain the well-known cancer cell characteristic of uncontrolled growth. In contrast to the ENOX1 proteins, which are activated by growth factors, the cancer-associated ENOX2 proteins are unresponsive to growth factors and constitutively activated (Bruno et al. 1992; Table 3.1). ENOX2 is not the result of an oncogenic mutation (Chueh et al. 2002b). Rather, it appears to be similar to a form of NAD(P)H oxidase important to maintenance of unregulated cell enlargement in very early development that is expressed in malignancy (Cho and Morré 2009). ENOX2 proteins catalyze both NAD(P)H or hydroquinone oxidation and carry out protein disulfide-thiol interchange. The two activities alternate creating a regular 22 min period length (Fig. 6.3).

If ENOX2 is a widespread constituent contributing to the cancer phenotype, then the protein has potential utility in cancer management either as a diagnostic aid (Chap. 12) or as a therapeutic drug or vaccine target (Chap. 11) or both. The latter possibility is enhanced by the external location of the ENOX2 protein in a position readily available to agents delivered through the circulation including antibodies and drugs conjugated to impermeant supports. Its potential as a pancancer, cancer-specific molecular marker (Morré and Reust 1997; Morré et al. 1997a; Hostetler et al. 2009; Chap. 12) adds to the overall potential utility of ENOX2 in cancer management.

8.1 ENOX2 Discovery

The discovery of ENOX2 traces its beginnings to early work from the Morré laboratories at Purdue University focused on control of plant growth and the basis for how certain growth regulatory herbicides caused plants to "grow out of control" and

D.J. Morré and D.M. Morré, *ECTO-NOX Proteins: Growth, Cancer, and Aging*,
DOI 10.1007/978-1-4614-3958-5_8,

eventually die (Chap. 10). The parallels between killing plants with such herbicides and the unregulated growth of cancer cells provided a compelling parallel with the expectation that if the target and mechanism for killing plants with growth regulatory herbicides might be discovered, important clues to understanding cancer would follow. ENOX discovery evolved out of this approach. Based on the early work with plant regulators, the plant regulatory target was shown to be localized to the plasma membrane at the cell surface. The target was deduced to have important redox properties and the activity was shown to oscillate although these very early findings were never published. It was also predicted that the target activity would be inhibited by doxorubicin, cisplatin, and other anticancer drugs since the uncontrolled growth induced by the plant regulatory herbicides was blocked specifically by these known anticancer agents (Chap. 10; Table 10.5).

A focused search for a plant regulator-responsive NAD(P)H oxidoreductase of plant plasma membranes inhibited by doxorubicin failed to reveal any candidates until, in 1986, experiments conducted at the University of Geneva in the laboratory of Claude Penel and Hubert Greppin showed that disappearance of NADH in the presence of ascorbate free radical and plasma membranes isolated from soybean was accelerated by addition of the auxin plant growth hormone, indole-3-acetic acid (Morré et al. 1986a). With further research, we found that the auxin-induced activity did not require the presence of ascorbate free radical and was, in fact, not a NAD(P)H oxido-reductase but an NADH oxidase. The activity required auxin regulators for activity which distinguished it from a constitutive NADH oxidase also present at the plasma membrane of plants and the auxin-induced activity was inhibited by doxorubicin, cisplatin, and other cancer chemotherapeutic agents (Morré 1998c; Table 10.5).

Subsequently, liver plasma membranes were shown to possess NADH oxidase activity that was responsive to mammalian hormones and growth factors (Brightman et al. 1992) but which was resistant to the antitumor drugs and agents (Chap. 3). However, when plasma membranes were isolated from rat hepatoma cells, they were found to possess an NADH oxidase activity that was unresponsive to mammalian hormones and growth factors (constitutively activated) (Bruno et al. 1992; Fig. 5.2; Table 3.1). In subsequent work, this activity was shown to be uniquely inhibited by various quinone site inhibitors all of which possessed anticancer activity (Chap. 11).

An early inhibitor of the tumor-specific NADH oxidase of hepatoma plasma membranes was capsaicin, the pungent principle of chili peppers. Capsaicin as an ENOX2 inhibitor was suggested during a flight between Chicago and the West Coast by Donald A. Lee, a native of Texas living in California with a penchant for making Texas-style chili. As it turned out Mr. Lee's chili peppers did inhibit the activity and the inhibition was quickly traced to the capsaicin present (Fig. 8.1). Inhibition by capsaicin was ultimately used as a defining criterion for identification of the tumor-specific NADH oxidase to facilitate its isolation, characterization, and eventual cloning (Chueh et al. 1997b, 2002b). The tumor-specific NADH oxidase, now ENOX2, was initially known as tNOX (a designation coined by George Todaro).

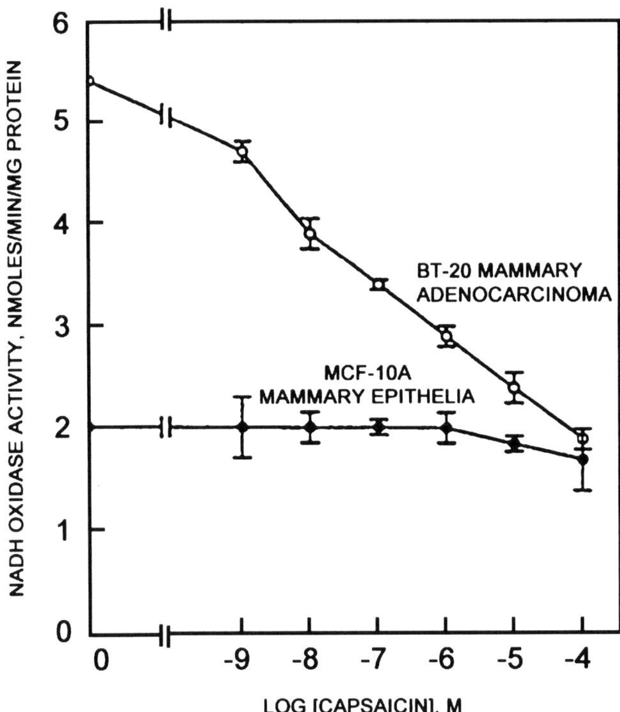

Fig. 8.1 Dose–response of ENOX activity of MCF-10A human mammary epithelial cells (*filled circles*) and BT-20 human mammary adenocarcinoma cells (*open circles*) to capsaicin. Values are averages of duplicate determinations in three separate experiments ± standard deviations. The activity of the BT-20 cells which express both ENOX1 and ENOX2 is inhibited due to ENOX2 presence whereas activity of the MCF-10A cells which express only ENOX1 was unaffected

8.2 ENOX2 Activity

The properties of ENOX2 proteins, many of which are common to other ENOX2 proteins (Chap. 1), are recapitulated in Table 8.1. ENOX2 is distinguished from ENOX1 on the basis of response to inhibitors and activators, amino acid sequence and functional motifs, period length and entrainment properties. The normal constitutive ECTO-NOX, CNOX, or ENOX1 is completely refractory to most inhibitors (an exception being the quassinoid simalikalactone D) (Morré and Grieco 1999), is stimulated by a variety of hormones and growth factors (Bruno et al. 1992), is less affected by thiol reagents ENOX2 (Morré and Morré 1995a) and oscillates with a period length of 24 min that is autoentrainable and entrainable by light and melatonin (Morré and Morré 2003a; Chap. 3).

The ENOX2 activity period is 22 min in length, 2 min shorter than the period length of ENOX1 (Wang et al. 2001, 2003b). In Fig. 6.3, NADH oxidation and protein disulfide-thiol interchange were measured simultaneously using two aliquots of

Table 8.1 Properties of ENOX2 proteins

Oxidative activity
Donor: hydroquinone/NADH/NADPH/dopamine
Acceptor: molecular oxygen → H_2O/protein disulfides
Protein disulfide-thiol interchange
Restore activity of scrambled RNase
Dithiodipyridine substrates
Two activities alternate to generate a period length of 22 min
Low turnover number: 200–500
Specific activity: 10–20 μmol/min/mg protein
Two moles of zinc and 2 mol of copper/mol of protein
Functional unit is a dimer
Propensity of purified protein to aggregate to form amyloid rods
No flavin and no cytochromes, heme or non-heme iron
Refractory to N-terminal sequencing
Protease resistant
Located at external cell surface
No GPI anchor or membrane-spanning domains
Phased by low-frequency EMF but not by melatonin
Shed from cell surface and circulates as a serum cancer marker (Chap. 12)
No ancillary proteins required for activity
Blocked by quinone site inhibitors with anticancer activity (Chap. 11)
Refractory to growth factor stimulation (constitutively activated)

the same ENOX2 preparation using two separate spectrophotometers operated in parallel showed the now characteristic 2+3 pattern (Chap. 6) and the alternation of the two activities. The pattern of hydroquinone oxidation paralleled that of NADH oxidation. The identification of maxima was facilitated by the observation that maxima ① and ② of NADH or hydroquinone oxidation are separated by about 6 min whereas maxima ③, ④, and ⑤ are separated from each other and from the ①–② diad at intervals of about 4 min ($6+(4\times4)=22$ min for recombinant ENOX2 and $6+(4\times4.5)=24$ min for ENOX1). This same pattern of asymmetry is revealed in the oscillatory patterns of structural change measured by Fourier transform infrared and circular dichroism (CD) measurements indicating changes in the proportion of β-structure and α-helix that underlie the patterns of enzymatic activity oscillations (Morré and Morré 2003a; Kim et al. 2005; Figs. 8.2 and 8.3). The known quinone-site inhibitors with anticancer activity that inhibit both the hydroquinone/NAD(P)H oxidase and the protein disulfide-thiol interchange activities of ENOX2 include doxorubicin, the antitumor sulfonylureas, the antitumor vanilloids, and capsaicinoids, e.g., capsaicin, the principal tea catechin (−)-epigallocatechin-3-gallate (EGCg), and several known differentiating agents including antitumor retinoids, sodium phenylacetate, and calcitriol (Morré 1998c; Chap. 11). These agents are largely without effect on the ENOX1 activity of normal cells (Fig. 8.4). They inhibit both ENOX2 and growth of cancer cells at potentially therapeutic dosage levels without inhibiting ENOX1 or growth of non-cancer cells. Drug inhibition of ENOX2 served as the defining ENOX1 characteristic used to guide its isolation and molecular characterization (Chueh et al. 1997b).

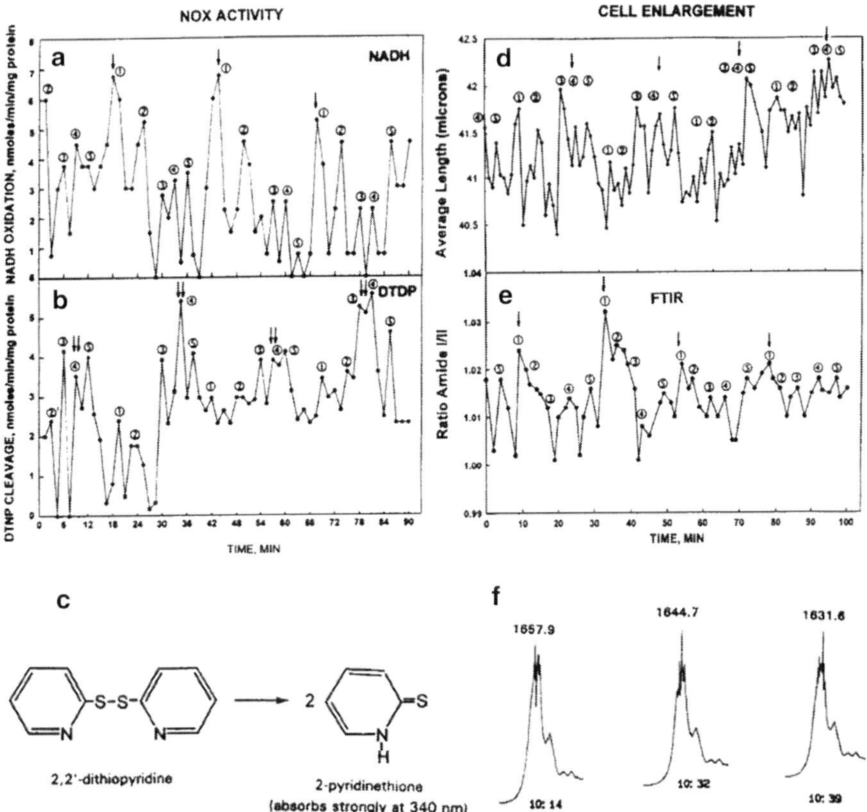

Fig. 8.2 The 3+2 pattern of NOX activity oscillations. (**a**, **b**). NADH oxidation determined by the decrease in A_{340} (*upper curve*). Maxima (*arrows*) were at 18, 42, and 66 min with secondary maxima at 24, 48, and 72 min. Three minor peaks completed each 24 min period. (**b**) Disulfide-thiol interchange activity measured simultaneously in parallel with NADH oxidation as an increase in A_{340} from the cleavage of dithiodipyridine (DTDP). Major peaks are at 6, 9, and 12 min and at 24 min intervals thereafter (*double arrows*) with minor peaks at 18 and 24 min and at 24 min intervals thereafter. The two activities, NADH oxidation (**a**) and DTDP cleavage (**b**), alternate. (**c**) DTDP substrate generating 2 mol of 340 nm-absorbing 2-pyridinethionine and cleaved as a measure of the disulfide-thiol interchange activity of the NOX protein. (**d**) Increase in length (enlargement of a single cell as determined by image enhanced light microscopy). Cell enlargement proceeds in bursts every 12 min separated by rest periods where the cells actually shrink. As with NADH oxidase activity, each 24 min period (*single arrows*) is comprised, on average, of five resolvable maxima separated by minima. Three maxima are contained within the elongation phase and correspond to the protein disulfide-thiol interchange determined in parallel (**b**). The two maxima contained within the resting period correlate with the two maxima of NADH oxidation (**a**). (**e**) Fourier transform infrared analyses of recombinant ENOX2. Sixty-one 1 min scans taken 1.5 min apart over 100 min are illustrated. The ratio of the amide I (1,645)-amide II (1,545) absorbances varied with maxima at 22 min intervals as indicated by the *arrows*. (**f**) Within the amide I region (below), peak absorbance varied between 1,658 and 1,638 indicative of alternating α-helix-β-sheet transitions. Concanavalin A, cytochrome c, or albumin when analyzed in parallel showed no such pattern of activity fluctuations. Reproduced from Morré and Morré (2003a) with permission from Taylor and Francis

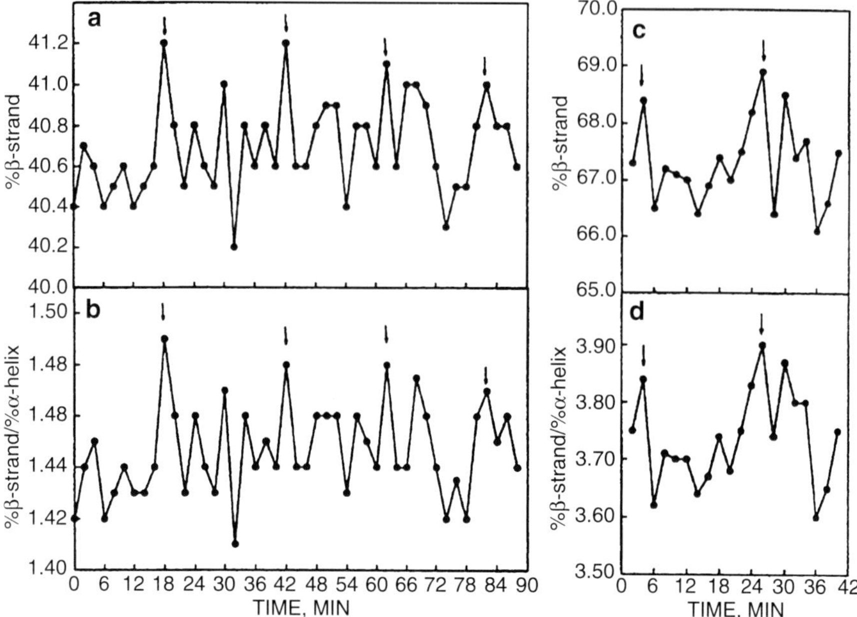

Fig. 8.3 Percent β-strand (**a**) and ratio of percent β-strand/α-helix (**b**) of soluble recombinant ENOX2 prior to acquisition of proteinase K-resistance. *Single arrows* denote maxima in β-strand structure spaced at intervals of 22 min. The double peak at 62 and 68 min (to right of *third arrow*) is a recurrent feature possibly associated with time keeping (see also (**c**) and (**d**) at 24 and 30 min). (**c**, **d**) Percent β-strand (**c**) and ratio of percent β-strand/α-helix (**d**) of recombinant ENOX2 after refolding and acquisition of proteinase K-resistance. *Single arrows* denote maxima in β-strand structure with a period length of ca. 22 min. Reproduced from Morré and Morré (2003a) with permission from Taylor and Francis

The constitutive ECTO-NOX activity, that of ENOX1, is universally present in all organisms thus far examined, plant, animal, and bacteria (Sedlak et al. 2001). In contrast, ENOX2 is universally restricted to cells, tissues, and biofluids of cancer patients. ENOX2 has not been detected in the absence of cancer. Being an ECTO protein, ENOX2 is shed into the circulation to serve as a potentially useful molecular biomarker predictive of cancer presence (Chap. 12).

ENOX2 is present in the urine of cancer patients (Yantiri et al. 1998) but not in saliva. The urine form of the activity was isolated and characterized as a 33 kDa protein with ENOX2 activity inhibited by capsaicin. The abundance of the 33 kDa ENOX2 form in urine was estimated to be between 5 and 100 μg/L depending on the particular cancer patient.

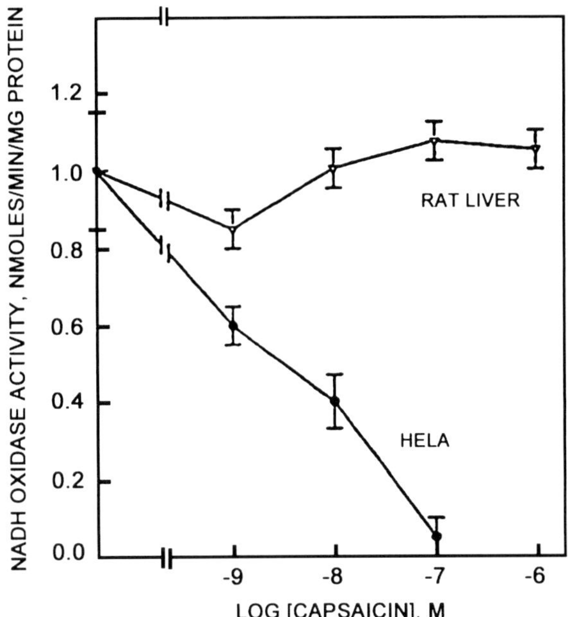

Fig. 8.4 Dose–response of NADH oxidase of HeLa cell plasma membranes (*open circles*) and of rat liver plasma membrane vesicles (*closed circles*) to capsaicin. Values are of duplicate determinations ± mean average deviation. The activity of the HeLa cell plasma membranes which contain both ENOX1 and ENOX2 proteins is inhibited due to ENOX2 presence whereas the ENOX activity liver plasma membranes which lack ENOX2 was unaffected by the capsaicin. Reproduced from Morré et al. (1995b) with permissions from PNAS

8.2.1 Biochemistry

The alternation of enzymatic activities carried out by ENOX proteins is unprecedented in the biochemical literature (Chueh et al. 2002b). For ENOX2, the specific activity of NADH oxidation by pure recombinant protein is 10–20 μmol/min/mg protein with a turnover number (the number of substrate molecules converted to product per minute with the enzyme fully saturated with substrate) of between 200 and 500 (Table 8.1). Rates of oxidation of natural hydroquinone substrates contained within the plasma membrane were similar to those for externally supplied NAD(P)H or hydroquinones.

Enzymatic assays with ENOX2 purified from the HeLa cell surface or purified recombinant ENOX2 are especially difficult due to the low specific activity along with a propensity for the protein to aggregate. The recombinant ENOX2 has been successfully assayed most often with dilute solutions (ca. 10 ng/mL). Aggregates may be disrupted by isoelectric focusing but the ENOX2 of such preparations will reaggregate upon concentration.

The natural electron donors for the oxidative activity of ENOX2 include hydroquinones (reduced coenzyme Q for animals and reduced phylloquinone for plants) and NADH or NADPH. The reduced pyridine nucleotides, NADH and NADPH, are regarded as non-physiological substrates since reduced pyridine nucleotides in the concentrations required to sustain ECTO-NOX activities are encountered rarely, if at all, at the external cell surface. Reduced quinones (coenzyme $Q_{10}H_2$) are abundant in mammalian plasma membranes and function as lipophilic *trans*-plasma membrane shuttles ferrying reducing equivalents from cytosolic NAD(P)H to molecular oxygen with the ENOX proteins functioning as terminal oxidases (Kishi et al. 1999; Chaps. 1 and 5).

As with ENOX1, molecular oxygen (Orczyk et al. 2005) and protein disulfides (Chueh et al. 1997b) both function as electron acceptors for the ENOX2-catalyzed reactions with a stoichiometry of two NAD(P)H or hydroquinone oxidized per O_2 or two disulfides reduced. Like NADH or hydroquinone oxidation, oxygen consumption is periodic with a period length of 24 min for ENOX1 and a period length of 22 min for ENOX2. The period length of 22 min is independent of temperature rather than temperature compensated per se and when water is replaced by deuterium oxide the period length increases to 26 min (Kim et al. 2005; Chap. 6). ENOX1 and ENOX2 do not cross entrain and, in contrast to ENOX1, ENOX2 is not entrained or phased by light or by melatonin, but, in common with ENOX1, is autoentrained and phased by low-frequency electromagnetic fields.

All ECTO-NOX proteins thus far examined utilize either NADH or NADPH as the pyridine nucleotide electron donor. Most ENOX1 activities utilize NADH preferentially whereas ENOX2 activity is greatest with NADPH (Morré 2002).

8.3 Sequence

The cancer-associated, drug-responsive, and constitutively activated ENOX2 form was the first ENOX family member to be cloned and expressed (GenBank Accession No. AF20788; Chueh et al. 2002b; Fig. 8.5). The ENOX2 gene consists of at least nine exons that combine to yield 1,830 bp open reading frame and 70.1 kDa protein comprised of 610 amino acids. ENOX2 protein expressed in *Escherichia coli*, COS, and MCF-10A cells exhibits the same characteristics of alternation of the two activities and drug response as the cell surface form. Identified functional motifs included a quinone binding site, an adenine nucleotide binding site, a CXXXXC cysteine motif as a potential protein disulfide-thiol interchange site and two copper binding sites (Table 8.2), one of which is conserved with superoxide dismutase (Fig. 8.5).

While there is no flavin and only one of the two C–X–X–X–X–C motifs characteristic of flavoproteins present in ENOX, the protein effectively carries out protein disulfide interchange. A CXXC motif common to protein disulfide isomerases (PDIs) (Gilberger et al. 1997) is present in full-length ENOX2 in exon 4 but is lost in exon 4 minus splice variants that retain protein disulfide-thiol interchange activity (Table 1.2).The redox active disulfide of thioredoxin reductase from the malaria

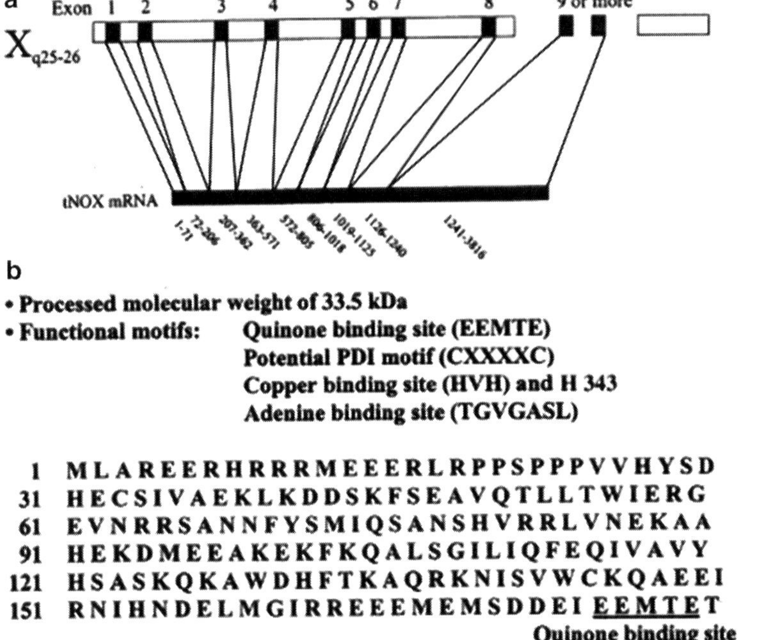

Fig. 8.5 ENOX2 sequence information (**a**) ENOX2 gene on human chromosome X. (**b**) Deduced amino acid sequence of the bacterially expressed 46 kDa functional C-terminus of ENOX2. The complete sequence data for the full-length 70.1 kDa protein are available from GenBank under Accession No. AF207881. Reproduced from Morré and Morré (2003a) with permission from Taylor and Francis

parasite *Plasmodium falciparum* contains a motif C88–X–X–X–X–C93 (Gilberger et al. 1998) similar to a second cysteine-containing motif found in ENOX2. Together with a downstream His509, the *P. falciparum* motif was shown by Gilberger et al. (1998) to be a putative proton donor/acceptor. Either the C88A or the C93A replacement resulted in complete loss of enzymatic activity. A C535–X–X–X–X–C540 motif in the same protein was shown by Kilman and Hey (1993) to be involved in substrate coordination and/or electron transfer. A C535A replacement did result in

Table 8.2 Identification of copper binding sites of ENOX2

A. Full-length amino acid sequence of the ENOX2 protein. Histidine to alanines substitutions by site-directed mutagenesis resulting in loss of NADH oxidase activity are *underlined*

1 MQRDFRWLWV YEIGYAADNS RTLNVDSTAM TLPMSDPTAW ATAMNNLGMA
51 PLGIAGQPIL PDFDPALGMM TGIPPITPMM PGLGIVPPPI PPDMPVVKEI
101 IHCKSCTLFP PNPNLPPPAT RERPPGCKTV FVGGLPENGT EQIIVEVFEQ
151 CGEIIAIRKS KKNFCHIRFA EEYMVDKALY LSGYRIRLGS STDKKDTGRL
246
201 HVDFAQARDD LYEWECKQRM LAREERHRRR MEEERLRPPS PPPVV**H**YSDH
251 ECSIVAEKLK DDSKFSEAVQ TLLTWIERGE VNRRSANNFY SMIQSANSHV
301 RRLVNEKAAH EKDMEEAKEK FKQALSGILI QFEQIVAVYH SASKQKAWDH
351 FTKAQRKNIS VWCKQAEEIR NIHNDELMGI RREEMEMSD DEIEEMTETK
401 ETEESALVSQ AEALKEENDS LRWQLDAYRN EVELLKQEQG KVHREDDPNK
451 EQQLKLLQQA LQGMQQHLLK VQEEYKKKEA ELEKLKDDKL QVEKMLENLK
546
501 EKESCASRLC ASNQDSEYPL EKTMNSSPIK SEREALLVGI ISTFL**H**VHPF
562 582
551 GASIEYICSY L**H**RLDNKICT SDVECLMGRL Q**H**TFKQEMTG VGASLEKRWK
601 FCGFEGLKLT

B. Specific activities of mutated recombinant Nus-tagged ENOX2 to identify copper binding sites

Protein	*N*	nmol/min/mg
WT	8	50 ± 10
H246A	3	6 ± 6[a]
H546A	3	8 ± 8[a]
Y556F	3	60 ± 8
Y560A	6	50 ± 13
H562A	3	8 ± 8[a]
H582A	6	10 ± 10[a]
H546A + H582A	3	5 ± 4[a]

C. Bound copper comparing Nus-tagged tENOX2 and mutant tENOX2 proteins ($n = 4$)

	Protein (μM)	Copper (μM)	Ratio
WT	8.7	15 ± 3	1.7 ± 0.3
H246A	22.4	24 ± 6	1.1 ± 0.3
H546A	11.4	10 ± 1	0.9 ± 0.1
Y556F	8.0	14 ± 2	1.8 ± 0.2
Y560A	14.0	31 ± 5	2.2 ± 0.3
H562A	2.5	2.4	0.9[b]
H582A	22.6	23 ± 2	1.0 ± 0.1
H546A + H582A	13.8	4 ± 1	0.3 ± 0.1

From Tang et al. (2010)

[a]Significantly different from wild type (WT) $p < 0.001$

[b]Determined independently by inductively coupled plasma atomic emission spectroscopy

diminution of enzymatic activity but a C540A replacement did not (Kilman and Hey 1993). Thus, either or both of the two comparable motifs present in ENOX2, C505–X–X–X–X–C510, or C569–X–X–X–X–X–C575, alone or together with downstream histidines provide potential active sites for protein disulfide-thiol

Table 8.3 Comparison of amino acid sequences within the known quinone- and sulfonylurea-binding sites of several proteins

Protein	Sequence
Q_B protein[a]	SAMHG
L/M subunit[a]	LAMHG
Acetolactate synthetase (tobacco)[b]	LGMHG
Pyruvate oxidase[a]	ATMHW
D_1 *Synechacoccus*	ETMRF
NADH (ubiquinone) dehydrogenase	GEMRE
Bovine serum albumin	ETMRE
Human serum albumin	ATLRE
Acetolactate synthetase (*Brassica*)	EDLRE

[a]Binds quinone
[b]Binds sulfonylurea
From Chueh et al. (2002b)

interchange. With ENOX2, the C505A and C569A replacements lost activity as with the C535A replacement above for the *P. falciparum* protein. The C510A and C575A replacements retained activity as did the above C540A replacement for the *P. falciparum* protein (Chueh et al. 2002a, b).

The signature ENOX2-motif is that of the potential drug binding site E394EMTE (Table 1.2). Antisera directed to this portion of the protein act as competitive inhibitors to drug binding. The sequence provides a putative quinone or sulfonylurea-binding site with four of the five amino acids in at least one other putative quinone site in the same relative positions (Table 8.3).

The various ENOX2 functional motifs of the 34 kDa processed form of ENOX2 are listed in Table 1.2. The correctness of the various assignments has, for the most part, been confirmed by site-directed mutagenesis (Chueh et al. 2002a, b; Table 6.1). While amino acid replacements that block oxidation of reduced pyridine nucleotide by ENOX2 also eliminated protein disulfide-thiol interchange and vice versa (Chueh et al. 2002b), the two activities appear to occur independently. One can be measured in the absence of the other. Also, Alzheimer's Aβ peptide carries out oscillatory oxidation of NADH (see below) but lacks the protein disulfide-thiol interchange activity. Cysteine motifs present in ENOX2 are absent from ENOX1 (Jiang et al. 2008) which carries out protein disulfide-thiol interchange comparable to that of ENOX2. ENOX proteins, in general, lack thioredoxin reductase activity (Bosneaga et al. 2009).

An ENOX2 protein isolated and characterized from the HeLa cell surface exhibited immunological cross-reactivity and amino acid sequence identity with ENOX2 cloned from a HeLa cDNA library using a monoclonal antibody to ENOX2 to provide a direct sequence link between ENOX2 of the HeLa cell surface and the cloned ENOX2 present in sera of cancer patients (Yantiri and Morré 2001).

Figure 8.6 compares the derived amino acid sequences of ENOX2 with those of ENOX1. Identity is 64 %. Similarity is 80 %. Regions forming coiled coil structures in both proteins are highlighted.

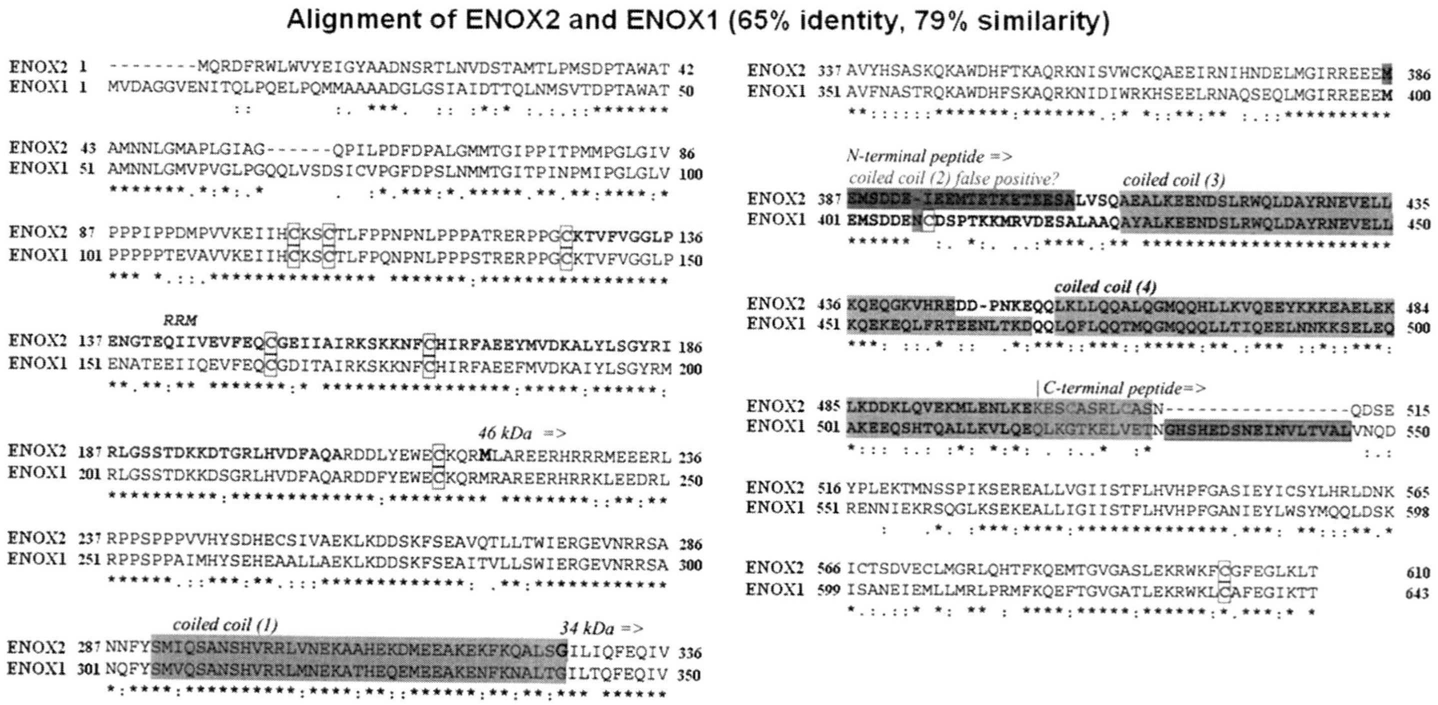

Alignment of ENOX2 and ENOX1 (65% identity, 79% similarity)

```
ENOX2 1   --------MQRDFRWLWVYEIGYAADNSRTLNVDSTAMTLPMSDPTAWAT 42
ENOX1 1   MVDAGGVENITQLPQELPQMMAAAADGLGSIAIDTTQLNMSVTDPTAWAT 50

ENOX2 43  AMNNLGMAPLGIAG------QPILPDFDPALGMMTGIPPITPMMPGLGIV 86
ENOX1 51  AMNNLGMVPVGLPGQQLVSDSICVPGFDPSLNMMTGITPINPMIPGLGLV 100

ENOX2 87  PPPIPPDMPVVKEIIHCKSCTLFPPNPNLPPPATRERPPGCKTVFVGGLP 136
ENOX1 101 PPPPPTEVAVVKEIIHCKSCTLFPQNPNLPPPSTRERPPGCKTVFVGGLP 150

                RRM
ENOX2 137 ENGTEQIIVEVFEQCGEIIAIRKSKKNFCHIRFAEEYMVDKALYLSGYRI 186
ENOX1 151 ENATEEIIQEVFEQCGDITAIRKSKKNFCHIRFAEEFMVDKAIYLSGYRM 200

                                     46 kDa =>
ENOX2 187 RLGSSTDKKDTGRLHVDFAQARDDLYEWECKQRMLAREERHRRRMEEERL 236
ENOX1 201 RLGSSTDKKDSGRLHVDFAQARDDFYEWECKQRMRAREERHRRKLEEDRL 250

ENOX2 237 RPPSPPPVVHYSDHECSIVAEKLKDDSKFSEAVQTLLTWIERGEVNRRSA 286
ENOX1 251 RPPSPPAIMHYSEHEAALLAEKLKDDSKFSEAITVLLSWIERGEVNRRSA 300

              coiled coil (1)                    34 kDa =>
ENOX2 287 NNFYSMIQSANSHVRRLVNEKAAHEKDMEEAKEKFKQALSGILIQFEQIV 336
ENOX1 301 NQFYSMVQSANSHVRRLMNEKATHEQEMEEAKENFKNALTGILTQFEQIV 350

ENOX2 337 AVYHSASKQKAWDHFTKAQRKNISVWCKQAEEIRNIHNDELMGIRREEEM 386
ENOX1 351 AVFNASTRQKAWDHFSKAQRKNIDIWRKHSEELRNAQSEQLMGIRREEEM 400

          N-terminal peptide =>
          coiled coil (2) false positive?   coiled coil (3)
ENOX2 387 EMSDDE-IEEMTETKETEESALVSQAEALKEENDSLRWQLDAYRNEVELL 435
ENOX1 401 EMSDDENCDSPTKKMRVDESALAAQAYALKEENDSLRWQLDAYRNEVELL 450

                               coiled coil (4)
ENOX2 436 KQEQGKVHREDD-PNKEQQLKLLQQALQGMQQHLLKVQEEYKKKEAELEK 484
ENOX1 451 KQEKEQLFRTEENLTKDQQLQFLQQTMQGMQQQLLTIQEELNNKKSELEQ 500

                            |C-terminal peptide=>
ENOX2 485 LKDDKLQVEKMLENLKEKESCASRLCASN-----------------QDSE 515
ENOX1 501 AKEEQSHTQALLKVLQEQLKGTKELVETNGHSHEDSNEINVLTVALVNQD 550

ENOX2 516 YPLEKTMNSSPIKSEREALLVGIISTFLHVHPFGASIEYICSYLHRLDNK 565
ENOX1 551 RENNIEKRSQGLKSEKEALLIGIISTFLHVHPFGANIEYLWSYMQQLDSK 598

ENOX2 566 ICTSDVECLMGRLQHTFKQEMTGVGASLEKRWKFCGFEGLKLT 610
ENOX1 599 ISANEIEMLLMRLPRMFKQEFTGVGATLEKRWKLCAFEGIKTT 643
```

Fig. 8.6 Alignment of ENOX2 and ENOX1 (65 % identity, 79 % similarity). *Shaded sequences* are indicative of coiled coil domains. Reproduced from Jiang et al. (2008) with permission from American Chemical Society Publications

8.4 Structural Properties

More structural information has been generated for ENOX2 than the other ENOX proteins thus far as ENOX2 was the first ENOX protein to be cloned (Chueh et al. 2002b). The functional unit is a dimer based on size exclusion criteria, ultracentrifugation and functional considerations (Chap. 1). Transmission electron microscopy of ENOX2 dimers reveal a slotted cylinder (Fig. 8.7). The cylinders open and close with an oscillatory pattern approximately equal to one half of the normal 22 min period (Fig. 8.8). A model based on these observations to explain ENOX2 cycle-dependence of drug inhibition is given in Chap. 11 (Fig. 11.42).

As determined from circular dichroism (CD) spectra (Tables 8.3, 8.4, and 8.5), the ENOX2 protein exhibits two levels of α-helix-β-structure transitions in keeping with its prion-like nature (Kelker et al. 2001). The newly synthesized recombinant protein contains ca. 30 % α-helix and about 40 % β-structure (Tables 8.4 and 8.5). It is amorphous in appearance in electron microscope preparations and is protease susceptible. When folded into the proteinase-resistance cylindrical structures, the β-structure content increases to above 60 % and the α-helix content falls to about 12 % (Table 8.6). However, this latter change in conformation is distinct from that involved in the oscillations. The α-helix-β-structure transitions that correlate with periodicity appear to be very local and were much less dramatic than those associated with protease resistance. Both the protease-resistant and the protease-susceptible forms of the protein would seem to exhibit oscillatory changes to approximately the

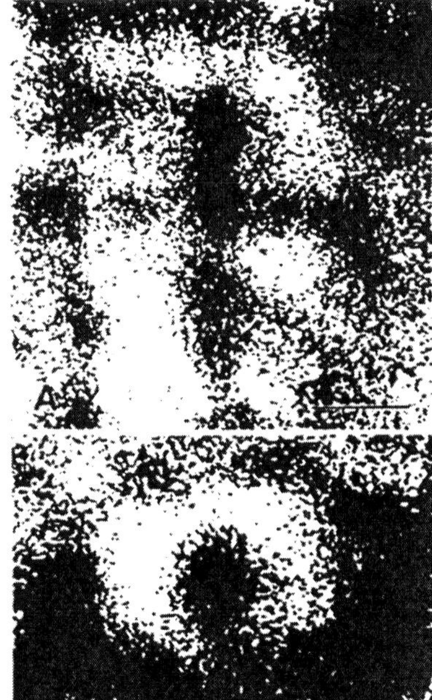

Fig. 8.7 Negative contrast electron microscopy of recombinant truncated ENOX2 (nucleotides 680–1,830) stained on a carbon-coated Formvar-covered electron microscope grid with 1 % uranyl acetate. The central channel in (**a**) is filled with electron dense uranyl acetate as is the stained-filled discontinuity in the circular profile of (**b**). The image in (**a**) is interpreted as a top view of the functional ENOX2 dimer lying flat. The image in (**b**) is interpreted as a view from the top of an ENOX2 monomer or dimer standing upright. Scale bar=0.01 μm

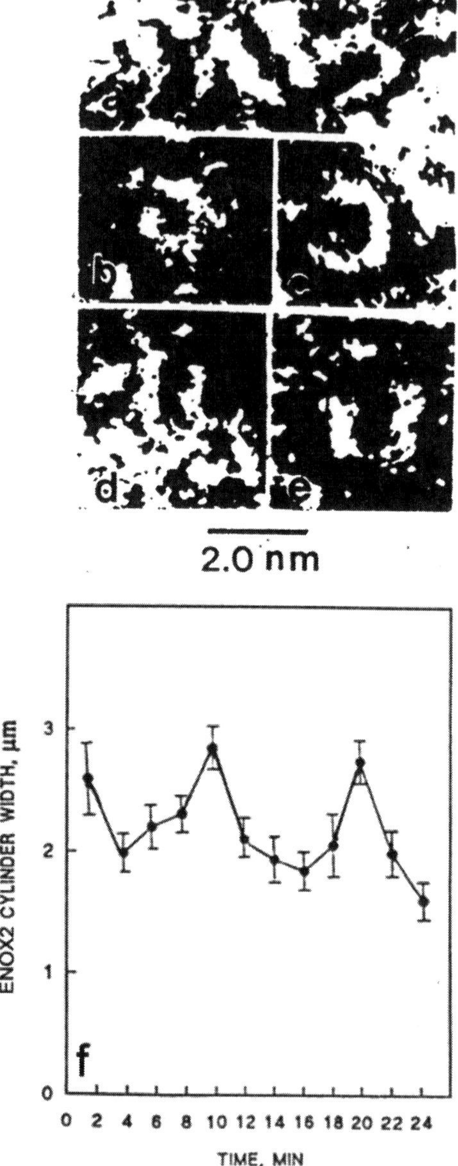

Fig. 8.8 Negatively stained images of recombinant truncated ENOX2 (nucleotides 680–1,830) in closed (*right*) and open (*left*) configurations. (**a**) The two dominant views are seen in the top consisting of an apparent hollow cylinder seen in lying flat (*double arrowheads*) or from the *top* (*single arrowheads*). (**b**, **d**) Closed configuration. (**c**, **e**) Open configuration. (**f**) Diameters of single recombinant truncated ENOX2 (nucleotides 680–1,830) particles measured from preparations stained and photographed at 2 min intervals. Diameters varied from <2 nm (**b**, **d**) to 2.8 nm (**c**, **e**)

same extent although the protease-susceptible form has a greater solubility and generates more clearly resolved CD spectra than does the protease-resistant forms. The conversion of a protease-susceptible, amorphous-recombinant CD spectrum that retains elements of α-helix is accompanied by a further increase in β-character and protease resistance. Findings show about a 10 % change in the α-helix to β-sheet

Table 8.4 Secondary structure elements calculated for recombinant ENOX2 from the CD spectrum using Neural Network Theory of Böhm et al. (1992)

	190–260 nm (%)	195–260 nm (%)	200–260 nm (%)	205–260 nm (%)	210–260 nm (%)
Helix	29.5	28.4	27.9	29.6	28.5
Antiparallel	13.6	12.8	10.4	8.8	9.7
Parallel	8.9	9.6	10.1	10.2	9.9
β-Turn	17.9	17.9	18.0	17.3	17.8
Random coil	32.1	33.5	35.4	36.1	36.6
Total sum	102.0	102.2	101.9	102.0	102.5

From Kim et al. (2005)

Table 8.5 Secondary structure elements for proteinase K-resistant ENOX2 calculated for the CD spectrum using Neural Network Theory of Böhm et al. (1992)

	190–260 nm (%)	195–260 nm (%)	200–260 nm (%)	205–260 nm (%)	210–260 nm (%)
Helix	19.4	20.7	29.2	47.3	36.7
Antiparallel	73.3	35.5	11.7	5.6	7.5
Parallel	6.8	8.6	7.4	5.9	8.0
β-Turn	24.5	21.6	19.2	15.0	16.3
Random coil	15.1	19.6	22.9	23.9	30.9
Total sum	139.1	106.1	90.4	97.7	99.4

From Kim et al. (2005)

Table 8.6 The fractions of secondary structures for proteinase K-susceptible ENOX2 from the self-consistent method of Sreerama and Woody (2000)

α-Helix 1	α-Helix 2	β-Strand 1	β-Strand 2	β-Turns	Unordered	SUM
0.146	0.124	0.114	0.085	0.218	0.281	0.967

From Kim et al. (2005)
ENOX2 (CBD), CD (190–26 nm)

ratio with a 22 min period length (the length of the activity period of ENOX2). The data were analyzed according to a neural network analysis program (Tables 8.4 and 8.5) and according to Sreerama and Woody (2000) (Table 8.6) with similar outcomes.

At least three forms of the recombinant ENOX2 protein were recognized:

1. A soluble, proteinase K-susceptible form
2. A dimeric, proteinase K-resistant form with the appearance of cylinders open along one side
3. A proteinase K-resistant form consisting of insoluble amyloid rods

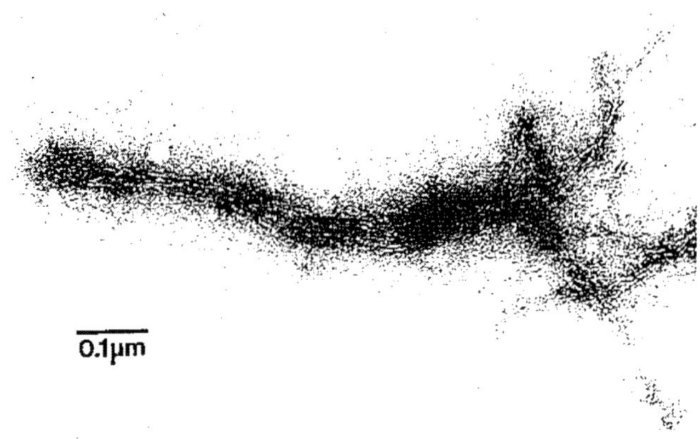

Fig. 8.9 Electron microscope picture of a concentrated preparation of purified ENOX2 incubated at 4 °C in the presence of NADH for 30 min. The precipitated proteins were collected and negatively stained with 2 % aqueous uranyl acetate. The preparation consisted almost entirely of 10 nm amyloid rods of indeterminate length. Magnification ×77,500. Reproduced from Kelker et al. (2001) with permission from American Chemical Society Publications

All three forms exhibit oscillating variations in β-strand content with maxima at intervals of 22 min coinciding with the 22 min period length of ENOX2 enzymatic activity.

When analyzed within the context of the three-state model of α-helix, β-strand, and coil, the proteinase K-susceptible form exhibits about 38 % β-strand structure (Table 8.4) whereas after refolding and acquisition of proteinase K-resistance β-strand structure, the proteinase K-resistant form was increased modestly to about 43 % (Table 8.5). Data analyzed by the neural network method agreed with data analyzed by the self-consistent method of Sreerama and Woody (2000) (Tables 8.5 and 8.6).

Structural studies designed to provide three-dimensional information were not successful. Several extensive attempts to crystallize recombinant ENOX2 have failed. Like PDI (Kemmink et al. 1995, 1999), it may not be possible to crystallize the ENOX proteins. The cause may be that ENOX proteins are likely characterized by a large number (perhaps thousands) of folding intermediates associated with the oscillatory activity. There are multiple substrate and other binding sites being exposed at intervals during each ENOX cycle as well (Fig. 11.42; Table 11.11). The ENOX proteins are likely undergoing constant changes in conformation.

The propensity of ENOX proteins to irreversibly aggregate when purified and concentrated is a second major deterrent to structural investigations (Fig. 8.9). Meaningful solution structures of ENOX2 by NMR or other physical methods have been precluded by the propensity of the ENOX proteins to dimerize, multimerize and aggregate as amyloid in concentrated suspensions. Double copper mutants of ENOX2 expressed as recombinant proteins show much reduced oscillations but are still prone to form aggregates.

In an attempt to structurally characterize just a single domain of ENOX2, the portion of the molecule containing the E_{394} EMTE drug binding site was expressed as a recombinant protein:

M386EMSDDEIE394EMTETKETEESALVSQAEALKEENDSLRWQLDAYRNE
VELLKQEQGKVHREDDPNKEQQLKLLQQALQGMQQHLLKVQEEYKKKEAELE
KLKDDKLQVEKMLENLKE502

DNA sequencing confirmed correctness of the clones. Expression was confirmed from N-terminal sequence. Mass spectrometry analysis (MALDI) confirmed the molecular weight with purification by anion exchange chromatography. Despite the fact that copper sites and substrate binding sites were excluded, the truncated protein still formed dimers and aggregated.

8.4.1 ENOX2 Protein Phosphorylation

ENOX2 is phosphorylated by protein kinase Cδ both in vitro and in vivo. Over expression of a S504A mutant with diminished capacity for phosphorylation leads as well to diminished cell proliferation and migration compared to wild type reflecting reduced stability of the unphosphorylated ENOX2 mutant protein (Zeng et al. 2012)

8.5 ENOX2 Presence and Cancer

ENOX2 is present at the cell surface of invasive and potentially invasive human cancers (Morré 1998c; Cho et al. 2002). ENOX2 is shed into the circulation and, together with the cell surface form, provides a potential drug, vaccine, and diagnostic target for cancer (Cho et al. 2002; Chaps. 11 and 12). Hydroquinone (NADH) oxidase and protein disulfide-thiol interchange activities are inhibited in parallel by the anticancer drugs that block ENOX2 activity (Morré 1998c; Table 8.7). Doxorubicin (Morré et al. 1997c) and *cis*-platinum normally bind to DNA but also occupy quinone-binding pockets such as those on ENOX2. Evidence for a cell surface site to explain the anticancer activity of doxorubicin has been in the literature since the early 1980s (Morré et al. 1997c; Chap. 11). The antitumor sulfonylureas (Morré et al. 1995g, h, i), the antitumor vanilloids (capsaicin) (Morré et al. 1995b), the principal tea catechin, EGCg (Morré et al. 2000a), as well as several known differentiating agents including antitumor retinoids, sodium phenylacetate, and calcitriol (Morré 1998c) also inhibit largely without effect on the ENOX1 activity of normal cells. They inhibit growth of cancer cells at similar therapeutic dosage levels that inhibit ENOX2 but do not inhibit growth or ENOX1 activity of non-cancer cells. Antimetabolites (e.g., methotrexate) and antimitotic (e.g., tamoxifen) agents do not inhibit recombinant ENOX2 as clear exceptions.

Table 8.7 Response of COS cells stably transfected with ENOX2 cDNA to ENOX2-targeted drugs plus the non-ENOX2-target tamoxifen and methotrexate

	EG_{50} (μM)	
Drug	Non-transfected	ENOX2 transfected
EGCg	10	0.1
Capsaicin	15	2.3
Doxorubicin	0.3	0.04
LY181984 (active)	20	3.0
LY181985 (inactive)	>100	>100
Tamoxifen	16	8
Methotrexate	1	1

From Chueh et al. (2004)

The growth involvement of ENOX2 is in cell enlargement. By blocking ENOX2 and cell enlargement, the ENOX2 inhibitors also block cell proliferation since when a cell divides, it must reach some certain minimal size to divide again (Baserga 1985; Fig. 5.18). When the activities of ENOX proteins are inhibited, cell enlargement is slowed or blocked (Morré and Morré 1995b; Morré et al. 1995b, c, f, g, h, i, 2000a). The resultant small cells are unable to enlarge, fail to undergo further divisions even though DNA and protein synthesis are not inhibited and, after a few days, undergo programmed cell death (apoptosis) (Morré and Morré 1995b, 2003a; Morré et al. 1995b, f, 2000a; Wolvetang et al. 1994; Fig. 5.18).

The presence of the ENOX2 protein was earlier demonstrated in a number of human tumor tissues and xenografts from cytochemical studies (Cho et al. 2002). However, serum analyses suggest a much broader association with cancer (Chap. 12). ENOX2 proteins are ectoproteins reversibly bound at the outer leaflet of the plasma membrane (Morré 1995a). As is characteristic of other examples of ectoproteins (sialyl and galactosyl transferases, dipeptidyl-amino peptidase IV, etc.; e.g., Hanski et al. 1986), the ENOX2 proteins are shed (Morré et al. 1997a; Wilkinson et al. 1996; Fig. 1.3). They appear in soluble form in conditioned media of cultured cells and in patient sera. The ENOX2-isoforms from sera of cancer patients exhibit the same degree and specificity of inhibition by anticancer drugs (i.e., the EC_{50} values for inhibition of activity are the same or similar) as do the membrane-associated forms (Morré and Reust 1997; Morré et al. 1997a). In contrast, no drug-responsive NOX activities have been found with sera from healthy volunteers or sera from patients with diseases other than cancer. As such, the antitumor-responsive ENOX2 activity represents the first reported cell surface change absent from non-cancer cells and potentially associated with most, if not all, forms of human cancer (Chap. 12).

ENOX proteins are released from cells into the circulation. Sera of cancer patients contain both ENOX2 and ENOX1 proteins (Cho et al. 2002; Morré and Reust 1997; Morré et al. 1997a). Sera of healthy volunteers or of patients with diseases other than cancer contain only the ENOX1 form. ENOX2 has been found in sera of patients with all major forms of cancer including leukemia and lymphomas (Cho et al. 2002; Morré and Reust 1997; Morré et al. 1997a) and serves as the basis for a diagnostic protocol under development (Chap. 12). A 22 h circadian day generated by the 22 min ENOX2 oscillations (Chap. 6) has been observed in activity patterns of some cancer patients (Levi 2000).

While ENOX2 presence provides a non-invasive approach to cancer detection based on serum analysis, until recently it offered no indication as to cancer type or location. However, analyses using a pan-ENOX2 recombinant single chain variable region (scFv) antibody carrying an S tag that cross reacted with all known ENOX2 isoforms from solid tumors of human origin and did not differentiate among different kinds of cancers (Kim 2011). When combined with two-dimensional gel electrophoretic separations, the recombinant antibody revealed a family of ENOX2 transcript variants, each with molecular weights and isoelectric points specific for a particular organ site of cancer (Hostetler and Kim 2011; Hostetler et al. 2009; Chap. 12). The family of ENOX2 transcript variants share a common antigenic determinant also recognized by a previously described pan ENOX2-specific monoclonal antibody which served as the template for creation of the recombinant antibody.

8.5.1 ENOX2 Autoantibodies Generated in Cancer Patients

ENOX2 transcription variants, while sharing a common antigenic determinant recognized by a ENOX2-specific monoclonal antibody and a corresponding scFv fragment expressed in bacteria (Cho et al. 2002), do not distinguish cancer from non-cancer in an ELISA format despite the fact that the scFv was produced from ENOX2-specific IgG rescued from the monoclonal antibody-producing hybridoma cells (Kim 2006). The basis for this anomaly only recently became understood to result from a nearly universal patient response to produce autoantibodies to the ENOX2 transcript variants that block binding of diagnostic antibodies (Chap. 12) and cross react among transcript variants to further validate the transcript variants as sharing common antigenic determinants.

8.5.2 ENOX2 Gene Present in Genome as a Single Copy

The ENOX2 (tNOX) gene is present in the human genome as a single copy, with no obvious homologs and a single constitutive ENOX1 (CNOX) ortholog (Jiang et al. 2008; Fig. 8.6). It is not a gene mutated in cancer but is universally present so there has been no reason to include the ENOX2 gene in a genomic cancer screen. The ENOX2 transcription variants all appear to be variations that include an exon 4 minus splicing event that allows for down-stream initiation and expression at the cell surface of the ENOX2 protein only in cancer cells (Tang et al. 2007, 2008). Without the exon 4 deletion, mRNA derived from the gene does not appear to be translated into protein. Thus, the exon 4 deletion is the basis for the cancer specificity of the ENOX2 isoforms. The transcript variants appear to be the result of subsequent alternative splicing events also restricted to cancer and expressed as a family of cancer-specific mRNAs encoding multiple proteins all with a common antigenic determinant (Tang et al. 2007; Sect. 8.8).

8.5.3 ENOX2 Lacks Intrinsic Membrane-Binding Motifs

The 34 kDa plasma membrane-associated form of ENOX2 is an extrinsic membrane protein that lacks intrinsic membrane-binding motifs. It contains no strongly hydrophobic regions and is not transmembrane (Morré et al. 2001c). No myristoylation or phosphatidylinositol anchor motifs were discovered. Evidence for lack of involvement of a glycosylphosphatidylinositol-linkage was derived from the inability of treatment with a phosphatidylinositol-specific phospholipase C or with nitrous acid at low pH to release the NOX protein from the surface of HeLa cells or from plasma membranes isolated from HeLa cells. Binding of NOX protein to the plasma membrane via amino acid side chain modification or by attachment of fatty acids also is unlikely based on use of specific fatty acid antisera to protein-bound fatty acids and as a result of binding to the cancer cell surface of a truncated form of recombinant ENOX2 (Morré et al. 2001c). Incubation of cells or plasma membranes with 0.1 M sodium acetate, pH 5, at 37 °C for 1 h, was sufficient to release ENOX2 from the HeLa cell surface (del Castillo-Olivares et al. 1998). Release was unaffected by protease inhibitors or divalent ions and was not accelerated by addition of cathepsin D. These findings suggest dissociable receptor binding as a possible basis for the plasma membrane association.

8.5.4 ENOX2 Has Properties of a Prion and Is Protease Resistant

Unusual characteristics exhibited by ENOX2 and shared with prions include resistance to proteases, resistance to cyanogen bromide digestion, and an ability to form amyloid filaments closely resembling those of spongiform encephalopathies (Kelker et al. 2001; Fig. 8.9). All of these are characteristics of PrP^{sc} (PrP^{res}), the presumed infective and proteinase K-resistant particle of the scrapie prion. The ENOX2 protein from the HeLa cell surface copurified with authentic glyceraldehyde-3-phosphate dehydrogenase (muscle form) (mGAPDH) (del Castillo-Olivares et al. 1998). Surprisingly, the ENOX2-associated mGAPDH also was proteinase K-resistant (Figs. 8.10 and 8.11). It was subsequently shown that combination of authentic rabbit mGAPDH with ENOX2 rendered the GAPDH resistant to proteinase K digestion. This property, that of converting the normal form of a protein into

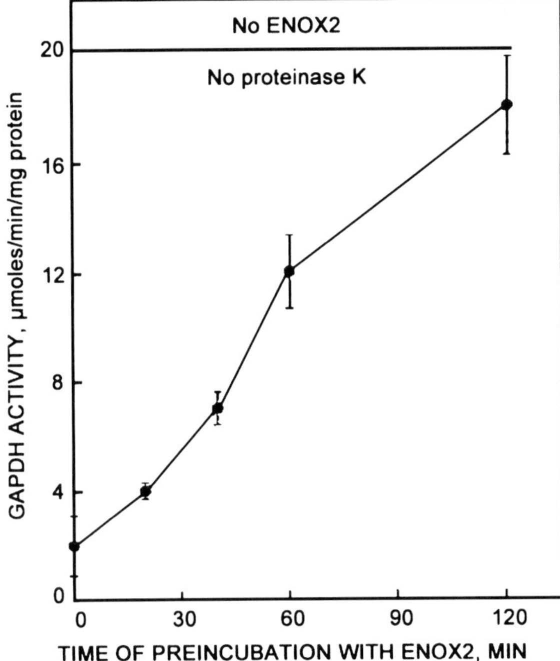

Fig. 8.10 Proteinase K resistance is imparted to muscle glyceraldehyde-3-phosphate dehydrogenase (mGAPDH) by preincubation for the times indicated with 0.2 μg of purified, recombinant ENOX2, 60 μL total volume. After the preincubation times indicated, portions of 60 μL were removed and treated for 2 h at 37 °C with 10 μg/μL proteinase K. In the absence of preincubation ($t=0$) the mGAPDH was inactivated. When incubated with recombinant ENOX2 for the times indicated, the mGAPDH gradually acquired proteinase K resistance. mGAPDH not incubated with recombinant ENOX2 was proteinase K susceptible (no incubation). Control values with no ENOX2 and no proteinase K were 20±2 μmol/min/mg of protein. Recombinant ENOX2 lacks GAPDH activity. Values are from three experiments±standard deviations. Reproduced from Kelker et al. (2001) with permission from American Chemical Society Publications

a likeness of itself, is one of the defining characteristics of the group of proteins designated as prions.

Protease resistance of ENOX2 has been established for proteinase K, trypsin, chymotrypsin, subtilisin, V-8 protease, and pronase (Chueh et al. 1997b; del Castillo-Olivares et al. 1998). If proteinase K digests of the released HeLa protein are subjected to FPLC (fast protein liquid chromatography), the 34 kDa processed (plasma membrane and serum) form of ENOX2 co-isolates in a complex with a 36 kDa peptide identified from the N-terminal amino acid sequence, amino acid composition, and enzymatic activity as muscle glyceraldehyde-3-phosphate dehydrogenase (mGAPDH) (del Castillo-Olivares et al. 1998). The mGAPDH resembled mGAPDH isolated from rabbit white muscle in every aspect except for one. The mGAPDH associated with ENOX2 was resistant both to heat (70 °C) and to proteinase K,

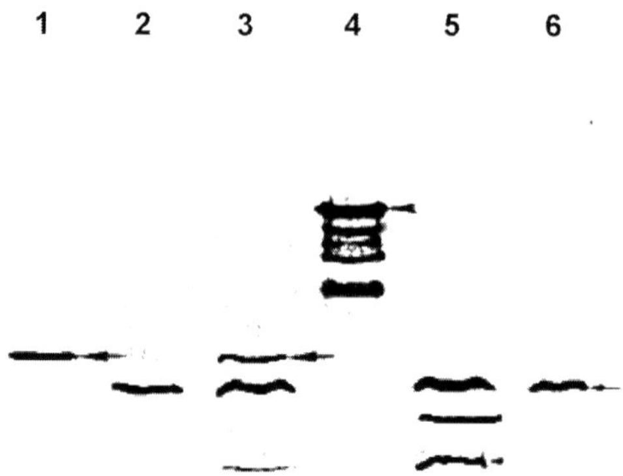

Fig. 8.11 SDS-PAGE (10 % gel) silver stained to show protection of GADPH with ENOX2. *Lane 1*: Commercial GAPDH (Table 8.1). *Lane 2*: Commercial GAPDH + proteinase K (final concentration 1 μg/mL). *Lane 3*: Commercial GAPDH preincubated for 2 h with recombinant ENOX2 prior to addition of proteinase K. A significant portion of the GAPDH was protected even with the large excess of proteinase K added and the elevated temperature of incubation. Also remaining was the proteinase K-resistant remnant of the ENOX2 protein (*small arrowhead*). *Lane 4*: Recombinant ENOX2. *Lane 5*: Recombinant ENOX2 + proteinase K (final concentration 1 μg/mL). Proteinase K-resistant remnants that retain full ENOX2 enzymatic activity are indicated by the *small arrowheads*. *Lane 6*: proteinase K (*single small arrow*) alone. Proteinase K served as a loading control for *lanes 2*, *3*, and *5*. Reproduced from Kelker et al. (2001) with permission from American Chemical Society Publications

whereas its normal cytosolic counterpart, mGAPDH, was neither heat nor proteinase K-resistant. This property of proteinase K resistance is used widely to distinguish between the disease (PrP^{sc} or PrP^{res}) and normal (PrP^{c} or PrP^{sens}) forms of transmissible spongiform encephalopathy (i.e., scrapie) prions (Baldwin et al. 1995; Griffith 1967; Prusiner et al. 1983, 1984; Stahl et al. 1990).

Both PrP^{sc} and ENOX2 are plasma membrane located (Morré 1995a; Baldwin et al. 1995; Stahl et al. 1990; Prusiner et al. 1998). Both PrP^{sc} and ENOX2 (del Castillo-Olivares et al. 1998) polymerize into insoluble aggregates and/or form characteristic rod-shaped amyloid (Fig. 8.9; del Castillo-Olivares et al. 1998; Prusiner et al. 1983). Both PrP^{sc} and ENOX2 polymerize even in the presence of detergents (Morré et al. 1998e; Prusiner et al. 1982). Both PrP^{sc} and ENOX2 appear to undergo posttranslational modification at the Golgi apparatus. Circular dichroism measurements of the recombinant ENOX2 protein suggest it to be predominantly β-sheet (Sect. 8.4). This is a further similarity between ENOX2 and PrP^{sc}. The conformational changes that convert PrP^{c} to PrP^{sc} are considered to involve largely α-helix to β-sheet transformations (Pan et al. 1993).

Both recombinant full-length mouse prion protein expressed in *E. coli* and native prion protein (PrP^{sc}) protein disulfide interchange activities are similar to those of ENOX proteins (Kim and Morré 2004). Additionally the activities of the purified prion proteins display stable and recurring patterns of oscillations but with a copper-dependent period length of 24 min (Kim and Morré 2004).

Amyloid rods of indeterminate length formed by ENOX2 and prion proteins are also formed by human amyloid peptides Aβ1-40 and Aβ1-43 which, like prion and ENOX2 proteins, bind copper. With copper bound, they exhibit an oscillating pattern of NADH oxidation also with a copper-dependent period length of 24 min (Table 1.6).

8.6 ENOX2 Has Characteristics of an Oncofetal Protein

The possibility that ENOX2 expression might be important to early stages of normal cell development was raised by experiments where the expression of a putative ENOX2 protein in chicken embryos was encountered during early developmental stages (Cho and Morré 2009; Fig. 8.12). ENOX2 cross-immunoreactive with the drug-responsive NADH oxidase of chicken hepatoma cells was used to identify ENOX2 on western blots. The putative ENOX2 protein was identified based on capsaicin-inhibited NADH oxidase activities and analyses by western blots using polyclonal antisera to a 34 kDa human serum form. The drug-responsive activity was associated with plasma membranes and sera of early chicken embryos and with chicken hepatoma plasma membranes but was absent from plasma membranes

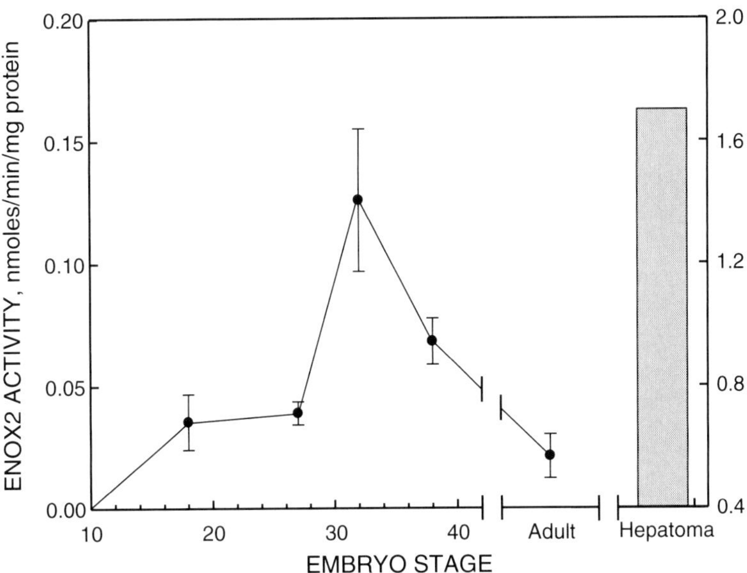

Fig. 8.12 Early developmental expression of ENOX2 activity measured with plasma membrane preparations from different stages of chicken embryos which increase to a maximum in embryo state 34 and then declines in the adult (liver) only to reappear in cancer (hepatoma). Diagram based on data of Cho and Morré (2009)

prepared from livers or from sera of normal adult chickens and from late embryo stages. The findings suggest that ENOX2 or ENOX2-like protein may fulfill some functions essential to the growth of early embryos which are lost in late embryo stages and absent from normal adult cells but which then reappear in cancer (Cho and Morré 2009).

A proposed function of ENOX2 in cancer is to support the uncontrolled cell proliferation which is characteristic of the cancer phenotype (Morré 1998c; Morré and Morré 2003a; Yagiz et al. 2006). The function of ENOX2 or an ENOX2-like protein in early embryogenesis might be similar to facilitate cell proliferation prior to developmental stages before cell proliferation comes under strict growth factor control.

Expression of ENOX2 proteins in cancer differs from other examples of cell surface proteins overexpressed in cancer which invariably are normal cell surface proteins produced in elevated amounts. MUC-1 mucin overexpressed in many adenocarcinomas is one such example (Bearz et al. 2007). Approximately 90 % of breast cancer biopsies express high levels of MUC-1 mucin and the expression is at least tenfold increased in tumors compared to normal mammary glands (Braga and Gendler 1993). Other tumor-associated antigens include human melanoma-associated glycoprotein p97 (Brown et al. 1981), the HER-2/neu protein of breast cancer patients (Disis et al. 1997), AFP (α-fetoprotein) (Mizejewski 2001), and CEA (carcinoembryonic antigen) (Lucha et al. 1997). ENOX2 is expressed only in cancer and is not an overexpression of a protein expressed by non-cancer cells.

AFP and CEA are oncofetal antigens which are found on both cancerous cells and on fetal cells. These antigens appear early in embryonic development before the immune system acquires competence. With CEA, expression in mouse embryos is developmentally regulated but expression is localized in tissues derived from mesenchyme and migratory neural crest cells (Huang et al. 1990). Human AFP expression is more general and a characteristic of fetal liver, yolk sac, and gastrointestinal tract of the human conceptus, but synthesis ceases at or near birth (Gitlin et al. 1972). ENOX2 follows an expression pattern similar to that of an oncofetal antigen. As such, if ENOX2 is expressed in early development, it is completely repressed in cancer-free adults.

8.7 Transgenic Mouse Strain Overexpressing ENOX2

A transgenic mouse line overexpressing ENOX2 was generated to determine its overall growth phenotype and susceptibility to ENOX2 inhibitors. The ENOX2 transgenic mouse line was developed using a phCMV2 vector with a hemagglutinin (HA) tag (Fig. 8.13). Transgenic mice, both males and females, exhibited both an enhanced growth rate (Figs. 8.14 and 8.15) and a response to both EGCg and phenoxodiol (Fig. 8.16) not observed with wild-type mice (Table 8.8). Both male and female transgenic mice exhibited accelerated rates of growth. Female transgenic mice grew twice as fast as wild type, and growth was reflected in an overall increased

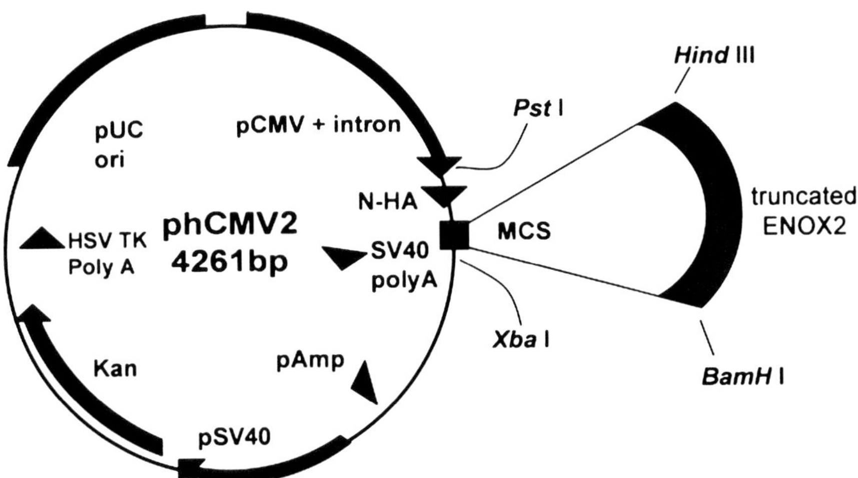

Fig. 8.13 Construct of ENOX2 plasmid. The ENOX2 cDNA was subcloned into a mammalian phCMV2-HA expression vector. The vector included the N-terminus HA tag. Reproduced from Yagiz et al. (2006) with permission from Elsevier

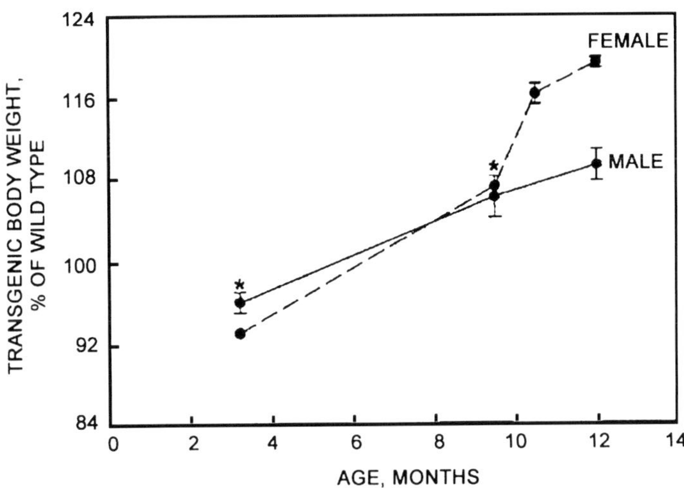

Fig. 8.14 Total body weight of transgenic male (*solid line*) and female (*dashed line*) as percent of wild type. From a study with Kadar Yagiz (unpublished)

carcass weight. Administration of EGCg in the drinking water (500 mg/kg body weight) reduced the growth rate of the transgenic mice to that of wild-type mice. The findings provide in situ validation of the hypothesis that ENOX2 represents a necessary and sufficient molecular target as the basis for the potential cancer preventive and therapeutic benefits of EGCg (Yagiz et al. 2006).

Fig. 8.15 Representative female ENOX2 transgenic mouse 9 months old compared to a female wild-type mouse of the same age. Reproduced from Yagiz et al. (2006) with permission from Elsevier

Similarly rates of enlargement of mouse embryo fibroblast (MEF) cells from transgenic animals were accelerated compared to wild type so that the cell division frequency was increased and the resultant cells were larger at confluence (Table 8.9). Cell diameters from fixed and stained tissues revealed increased cell size corresponding to increased organ weight (Table 8.10) attributed to the known phenotypic expression of ENOX2 which is to accelerate the enlargement phase of cell growth (Yagiz et al. 2008). Also increased was drug-responsive ENOX2 activity (Table 8.11). The growth of MEF cells from transgenic mice also was susceptible to inhibition by EGCg and phenoxodiol, whereas the growth of wild-type cells was not. With both EGCg and phenoxodiol, growth inhibition was followed after about 48 h by apoptosis. Growth of wild-type MEF cells from the same strain was unaffected by EGCg and phenoxodiol and neither compound induced apoptosis in wild-type MEF cells even at concentrations 100–1,000-fold higher than those that resulted in apoptotic death in the transgenic MEF cells. The findings validate earlier reports of evidence for ENOX2 presence as contributing to unregulated growth of cancer cells as well as the previous identification of the ENOX2 protein as the molecular target for the anticancer activities attributed to both EGCg and phenoxodiol. The presence of ENOX2 emerges as both necessary and sufficient to account for growth inhibitions of cancer cells by ENOX2-targeted anticancer drugs and substances such as EGCg and phenoxodiol (Yagiz et al. 2007).

The tissue expression of ENOX2 mRNA was greatest in heart, lung, and liver (Fig. 8.17) as reflected in enhanced ENOX2 activity (Table 8.12). When these tissues were analyzed for cell size, the cells from the tissues of transgenic animals

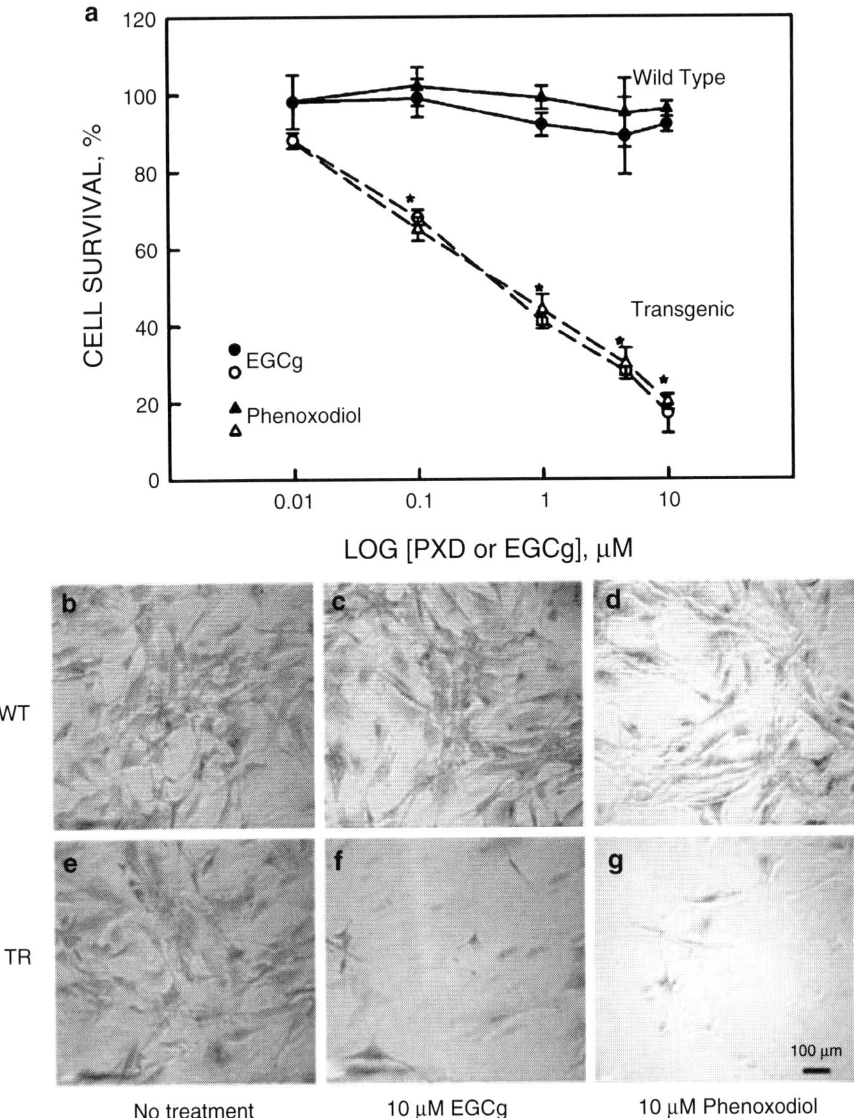

Fig. 8.16 Number of cells (growth) (**a**) and morphology (**b**–**g**) of wild-type and transgenic mouse embryo fibroblast (MEF) cells in response to (–)-epigallocatechin-3-gallate (EGCg) and phenoxodiol (PDX). Growth data are means ± standard deviations of three experiments. Inhibitions represent 48 h incubations of transgenic MEF cells with 10 μM phenoxodiol or with 10 μM EGCg. Wild-type MEF cells exhibited normal morphology with both phenoxodiol and EGCg even at the highest concentration tested of 10 μM. $\rho < 0.05$ vs. wild-type control. Bar = 100 μM. Reproduced from Yagiz et al. (2007) with permission from Wiley-SOS Press

Table 8.8 EGCg (1–1.5 mg/mL, adjusted to 500 mg/L/kg body weight) provided in drinking water

	Growth (mg/day ± SD)	
	Wild type	ENOX2 transgenic
Control	26 ± 4[a]	60 ± 20[b]
EGCg	33 ± 6[a]	33 ± 5[a]

Average consumption of 4 mL (5.5 mg/day) = 22 mg/kg body weight. Additions of EGCg to drinking water reduced growth rate of the transgenic female mice to wild-type mice rates. EGCg had no effect on growth rate of wild-type mice (which lack ENOX2). Each group within each experiment contained five mice. Averages are ±standard deviations from three experiments. Values not followed by the same letter are statistically different (a, b: $p < 0.04$). From Yagiz et al. (2006)

Table 8.9 Initial rates of cell enlargement averaged over 20 min of wild-type and transgenic MEF cells as determined by video-enhanced light microscopy

	Cell area (μm²/min)	
Treatment	Wild-type MEF cells	Transgenic MEF cells
No addition	0.73 ± 0.0	1.4 ± 0.3
0.1 μM EGCg	0.63 ± 0.09	0.41 ± 0.08[a]
0.1 μM Phenoxodiol	0.80 ± 0.2	0.43 ± 0.03[a]

From Yagiz et al. (2007)

Results are averages of three determinations ± standard deviation

[a]Significant $\rho < 0.05$

Table 8.10 Cell area of tissues of male mice corresponding to data of this table determined by Point Counting Method according to Chayes (1956)

	Cell area (μ²) ($n = 4$)		
Tissue	Wild type	Transgenic	% Increase
Heart	545 ± 14	651 ± 46[b]	19
Lung	191 ± 14	229 ± 11[b]	20
Liver	503 ± 60	616 ± 16[a]	22
Kidney	278 ± 14	286 ± 10	3
Intestine	135 ± 5	140 ± 4	4
Spleen	132 ± 4	130 ± 3	–2

From Yagiz et al. (2008)

Results were based on analyses of four tissue sections from each of four mice ± standard deviations among individual mice

[a]Transgenic significantly different from wild type $p < 0.02$

[b]Transgenic significantly different from wild type $p < 0.005$

Table 8.11 Total NADH oxidase activities of embryo fibroblast cells from wild type and transgenic mice and inhibition by EGCg and to 2281.1 ENOX2-specific antibody

		nmol/min/10^6 cells (Average ± standard deviation)	
Cell line	Addition	ENOX1	ENOX2
Wild type	None	2.3 ± 0.4	
	1 μM EGCg	2.4 ± 0.4	
Transgenic	None	2.0 ± 0.2	1.8 ± 0.3
	1 μM EGCg	1.9 ± 0.3	0.35 ± 0.2[a]
Wild type	None	2.0 ± 2.5	
	Ab 2281.1	1.9 ± 0.1	
Transgenic	None	2.1 ± 0.1	1.6 ± 0.2
	Ab 2281.1	2.2 ± 0.2	0.3 ± 0.1[a]

[a]Highly significant $p < 0.001$. From Yagiz et al. (2007)

were, on average, 20 % larger in surface area than cells from corresponding wild-type tissues (Table 8.10). Also analyzed were cells of intestine, spleen, and kidney in which ENOX2 overexpression was observed but to a lesser extent. Cell size was increased as well with intestine and kidney but less so with spleen. At the end of the study, carcass weights of the transgenic animals were greater than those of wild type. This increase in carcass weight was reflected in an increase in femur weight and thickness in both male and female transgenic mice but not in femur length. Other carcass parameters such as skin weight and body fat or body fluids were unchanged or changes were insufficient to account for the increased carcass weight. The findings are consistent with the property of ENOX2 observed in studies with cultured cells as contributing to the enlargement phase of cell growth.

The transfection experiments with MCF-10A mammary epithelial cells which lack ENOX2 at their cell surface and were refractory to growth inhibition by ENOX2 inhibitors (Morré et al. 1995b, h, 2000a) yielded similar results. Expression of ENOX2 was verified by western blot. Stable MCF-10A transfectants were sensitive to both capsaicin and EGCg (Table 8.13). Furthermore, ENOX2 transfectants of MCF-10A cells formed colonies after 1 week incubation in Matrigel and the colonies rapidly invaded the Matrigel (Fig. 8.18). The ENOX2-transfected MCF-10A cells then increased in size over a second week of incubation (Fig. 8.16c, d). With both the wild-type MCF-10A cells and the cells transfected with vector alone, many fewer colonies were formed (Fig. 8.18a, b). Additionally, HeLa cells transfected with ENOX2 antisense were no longer inhibited by capsaicin or EGCg (Chueh et al. 2004; Table 8.14). Similar findings were reported subsequently by Liu et al. (2008, 2012) and Mao et al. (2008) where RNA interference-mediated down regulation of ENOX2 expression significantly inhibited HeLa cell proliferation and migration and rendered the cells susceptible to apoptosis (Mao et al. 2008).

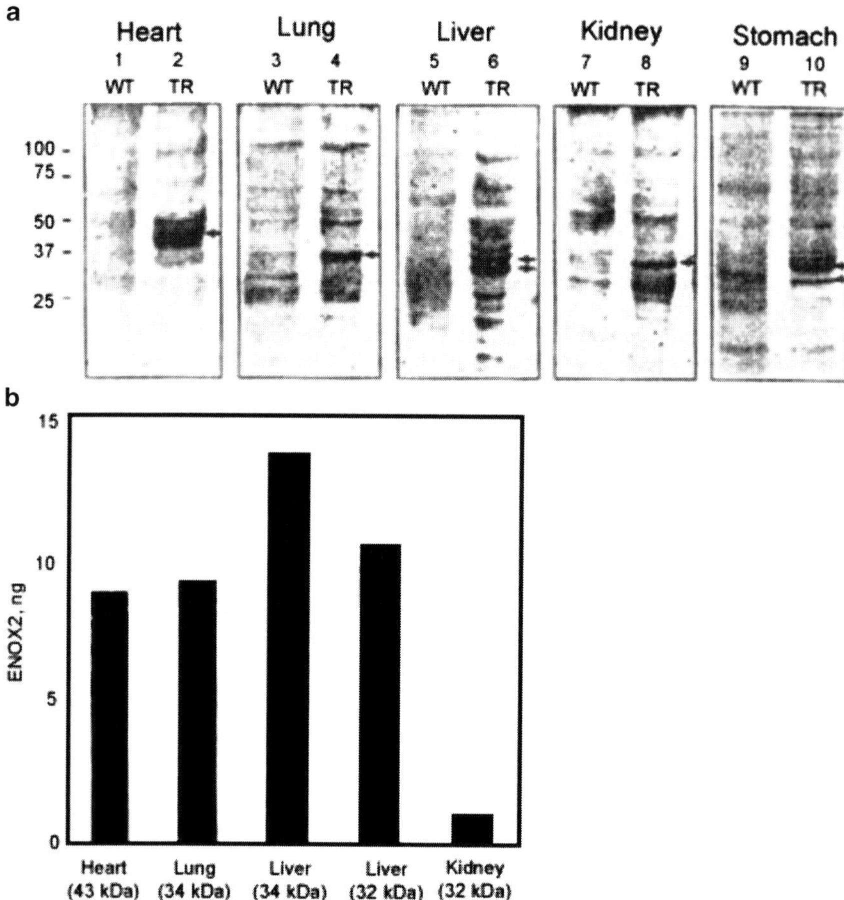

Fig. 8.17 Western blot analysis of ENOX2 presence in representative tissues comparing microsomes of wild-type and transgenic mice. Microsomes (50 μg of total protein) from tissues of transgenic animals contained protein bands reactive with anti-ENOX2 antibody that were absent from tissues of wild-type animals (*arrows*). Also analyzed were intestine and spleen (not shown). (**a**) Transgenic tissues showed specific bands corresponding to the processed molecular weights of ENOX2 of 43 kDa (heart), 34 kDa (lung, liver, and intestine), 32 kDa (liver, kidney, and stomach), or 29 kDa (stomach and spleen). (**b**) Quantitation of (**a**) by densitometry (minus background). Arbitrary units were converted to nanogram ENOX2 from a series of recombinant ENOX2 standards analyzed under comparable conditions. Reproduced from Yagiz et al. (2006) with permission from Elsevier

Table 8.12 Total plasma membrane NADH oxidase activities of liver and intestine from transgenic mice compared to wild-type mice and inhibition by EGCg

Tissue	1 μM EGCg	nmol/min/mg protein (Average ± standard deviations from three experiments) ENOX1	ENOX2
Wild-type liver	–	2.2 ± 0.4	
	+	2.0 ± 0.2	
Transgenic liver	–	1.4 ± 0.2	1.5 ± 0.3
	+	1.3 ± 0.3	0.1 ± 0.1
Wild-type intestine	–	3.1 ± 0.3	
	+	3.0 ± 0.2	
Transgenic intestine (normal assay)	–	2.4 ± 0.3	0.55 ± 0.1
	+	2.2 ± 0.2	0.4 ± 0.3
Transgenic intestine (enhanced assay)[a]	–	2.3 ± 0.6	2.1 ± 0.6
	+	2.2 ± 0.7	0.4 ± 0.4

[a]Enhanced assay conditions were in the presence of 0.2 % Triton X-100 and overnight incubation with 300 mM NADH in which ENOX2 activity was enhanced fourfold. ENOX1 was unaffected by these assay conditions. From Yagiz et al. (2006)

Table 8.13 Total cell surface NADH oxidase activities of cell lines transiently transfected with ENOX2 cDNA compared to vector alone and wild-type cells and inhibition by 1 μM capsaicin

Cell line		NADH oxidation (nmol/min/10^6 cells ± standard deviation) No addition	+1 μM Capsaicin
COS	Control	0.85 ± 0.05	0.78 ± 0.08
	Vector alone	0.80 ± 0.06	0.78 ± 0.075
	ENOX2-transfected	1.05 ± 0.07[a]	0.80 ± 0.07
MCF-10A	Control	0.80 ± 0.05	0.80 ± 0.04
	Vector alone	0.80 ± 0.04	0.83 ± 0.04
	ENOX2-transfected	1.00 ± 0.08[a]	0.8 ± 0.03
HEK-293	Control	0.80 ± 0.05	0.80 ± 0.05
	Vector alone	0.96 ± 0.09	0.84 ± 0.06
	ENOX2-transfected	1.11 ± 0.12[a]	0.87 ± 0.03

This concentration of capsaicin was sufficient to give complete inhibition of recombinant ENOX2. Averages ± standard deviations from three trials. From Chueh et al. (2004)
[a]Significantly different ($p < 0.02$)

8.8 Alternative Splicing as Basis for Specific ENOX2 Localization to the Cell Surface

The-drug responsive ENOX2 appears to arise as a splice variant from a single ENOX gene different from that encoding ENOX1 and is delivered to the cell surface as a processed 34 kDa ectoprotein (Tang et al. 2007). Full-length ENOX2 mRNA is present in both normal and tumor cells (Fig. 8.19) but appears not to be expressed

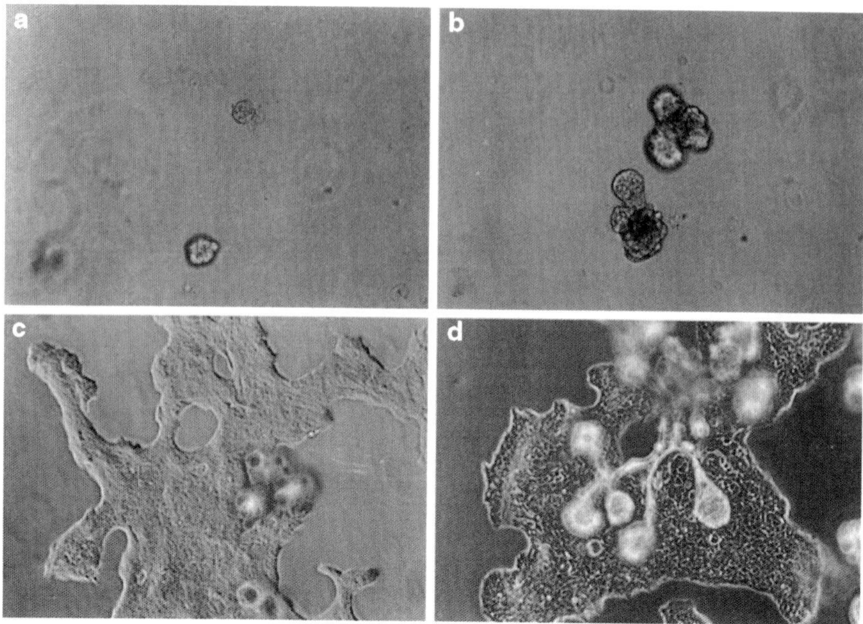

Fig. 8.18 Invasion ability of MCF-10A cells (non-cancer human mammary epithelia) was greatly enhanced by transfection with ENOX2 cDNA. Invasion was evaluated by growth in Matrigel. (**a**) Wild-type MDF-10A cells. (**b**) Empty pcDNA 3.1-transfected MCF-10A cells. Neither were able to grow in the Matrigel. (**c**, **d**) Two different clones of MCF-10A cells stably transfected with ENOX2 cDNA. The cells exhibited a transformed phenotype and formed extensive colonies in the Matrigel. Reproduced from Chueh et al. (2004) with permission from Wiley-SOS Press

Table 8.14 HeLa cells transfected with ENOX2 antisense were no longer inhibited by capsaicin or EGCg

	Percent inhibition after 72 h	
Transfection	100 μM Capsaicin	10 μM EGCg
None	76 ± 4	40 ± 8
Nonsense	60 ± 6	25 ± 6
Antisense	5 ± 2[a]	0 ± 0[a]

Determinations were in duplicate as for Table 8.12 ± mean average deviations. From Chueh et al. (2004)
[a]Significantly different ($p < 0.002$)

in either. Alternative splicing apparently is required for the cancer-specific expression of ENOX2 at the cell surface. Four splice variants have been found thus far. Two additional splice variants of ENOX2 were found. These were identified from the NCBI database (GenBank) as AK000353 (originally found from a hepatoma cell line) and AL133207 (gene located on chromosome X). Both have the same sequence as exon 2 to exon 8 of ENOX2. AK000353 has a 234 bp extra sequence before exon 2 that is different from that of full-length ENOX2. AL133207 has a 348 bp extra

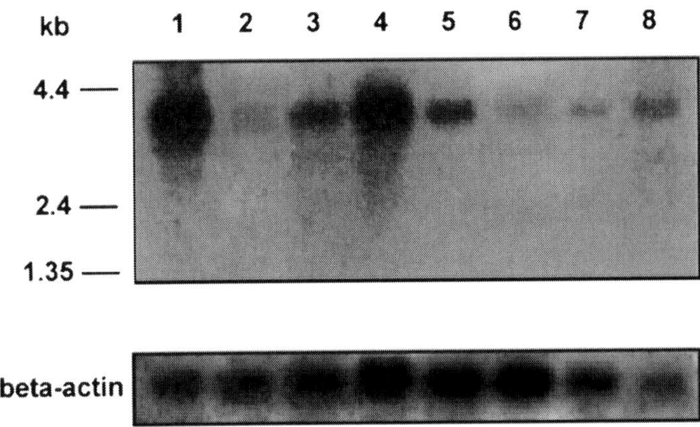

Fig. 8.19 Northern blot analysis of normal human tissues probed by truncated ENOX2 (nucleotides 680–1,830). *Lane 1* spleen; *lane 2* thymus; *lane 3* prostate; *lane 4* testis; *lane 5* ovary; *lane 6* small intestine; *lane 7* colon; *lane 8* peripheral blood leukocytes. Probes for β-actin were include to assess RNA integrity and loading. Reproduced from Tang et al. (2007) with permission from American Chemical Society Publications

sequence before exon 2 that differs from that of full-length ENOX2. Nucleotides 31–234 of AK000353 are the same as nucleotides 1–204 of AL133207. Of these, an exon 4 minus and exon 5 minus forms present in cancer cell lines and were absent in non-cancer cell lines (Fig. 8.20). The exon 5 minus cDNA yielded an open reading frame for a deduced amino acid sequence for a protein of 532 amino acids with a predicted molecular mass of 60.8 kDa.

In contrast to full-length ENOX2 cDNA (and the exon 5 minus form), transfection of COS cells with ENOX2 exon 4 minus cDNA resulted in overexpression of mature 34 kDa ENOX2 protein at the plasma membrane (Fig. 8.21). The exon 4 minus form resulted in initiation of translation at a downstream M231 initiation site distinct from that of full-length ENOX2 mRNA. With replacement of M231 by site-directed mutagenesis, no translation of exon 4 minus cDNA or cell surface expression of 34 kDa mature ENOX2 was observed (Fig. 8.22). The unprocessed molecular mass of 47 kDa of the exon 4 minus cDNA translated from methionine 231 corresponded to that of the principal native form of ENOX2 found in the endoplasmic reticulum. Taken together, the molecular basis of cancer-cell-specific expression of 34 kDa ENOX2 appears to reside in the cancer-specific expression of exon 4 minus splice-variant mRNA (Tang et al. 2007).

Studies were extended through the use of antisense oligonucleotides to different ENOX2 splice variants to further demonstrate expression of exon 4 minus ENOX2 mRNA as the basis for the cancer-specific expression of ENOX2 (Tang et al. 2008). Transfection of HeLa cells with antisense oligonucleotides and measurement of mRNA levels by real-time quantitative PCR and growth and drug response by in vitro cytotoxicity assays were combined to demonstrate that antisense to ENOX2 exon 4 mRNA blocked generation of full-length ENOX2 mRNA but not of exon

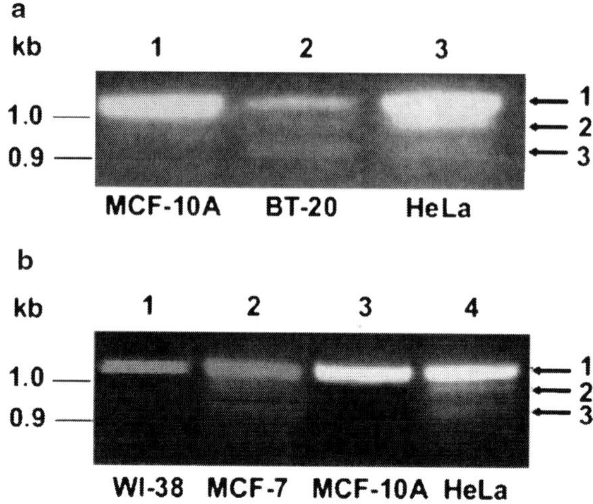

Fig. 8.20 (**a**) RT-PCR of MCF-10A, BT-20, and HeLa cells showed, in addition to ENOX2 mRNA (*band 1*), an exon 4 minus form (*band 2*) and an exon 5 minus product (*band 3*) in both cancer lines BT-20 (*lane 2*) and HeLa (*lane 3*) but not in the non-cancer MCF-10A (*lane 1*) cells. (**b**) RT-PCR of WI-38, MCF-7, MCF-10A, and HeLa cells showed an exon 4 minus form (*band 2*) and an exon 5 minus form product (*band 3*) in both MCF-7 (*lane 2*) and HeLa (*lane 4*) cancer cells but not in non-cancer WI-38 (*lane 1*) and MCF-10A (*lane 3*) cells. The ENOX2 mRNA (*band 1*) was present in all four cell lines. Primers were from the beginning of exon 2 (nucleotides 72–91) and from the end of exon 8 (nucleotides 1,212–1,238). PCR products were cloned into the T-vector and sequenced. Reproduced from Tang et al. (2007) with permission from American Chemical Society Publications

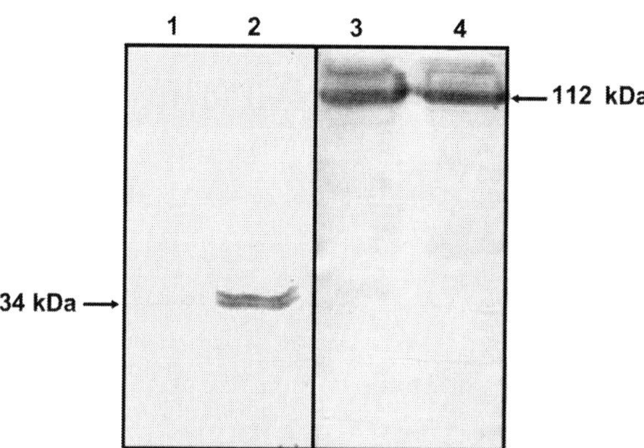

Fig. 8.21 RT-20 (human mammary adenocarcinoma) and MCF-10A (non-cancer mammary epithelia) plasma membranes (0.05 mg of protein/lane) analyzed by SDS-PAGE (10 %) and by western blot analysis with monoclonal antibody to ENOX2 with visualization using alkaline phosphatase-linked anti-mouse secondary antibody with BCIP and NBT as substrates. The 34 kDa ENOX2 (*arrow*) present in the plasma membranes of the BT-20 cells (*lane 2*) was below the level of detection from plasma membranes of MCF-10A cells (*lane 1*). Na^+, K^+-ATPase (112 kDa) detected by Na^+, K^+-ATPase monoclonal antibody was used as a loading control. *Lane 3* is for plasma membranes of MCF-10A cells. *Lane 4* is for plasma membranes of BT-20 cells. Reproduced from Tang et al. (2007) with permission from American Chemical Society Publications

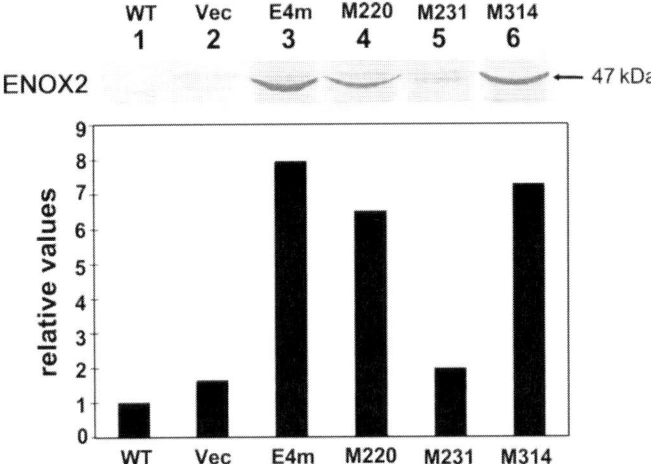

Fig. 8.22 Validation by site-directed mutagenesis of exon 4 minus mRNA with initiation at M231 as the transcriptional template for ENOX2. Replacement of Met 231 prevents the protein expression of exon 4 minus mRNA. *Lane 1* wild-type COS cells; *lane 2* COS cells transfected with pcDNA3.1 vector; *lane 3* COS cells transfected with exon 4 minus cDNA in pcDNA3.1; *lane 4* COS cells transfected with mutant M220A exon 4 minus cDNA in pcDNA3.1; *lane 5* COS cells transfected with mutant M231A exon 4 minus cDNA in pcDNA3.1; *lane 6* COS cells transfected with mutant M314A exon 4 minus cDNA in pcDNA3.1. The 47 kDa band is present in *lanes 2*, *4*, and *6*. Values of densitometry analysis were divided by the smallest value, and the division factors are plotted. The smallest value was calculated as 1. Reproduced from Tang et al. (2007) with permission from American Chemical Society Publications

4 minus mRNA (Fig. 8.23). Antisense to exon 5 mRNA inhibited the production of exon 4 minus mRNA and full-length ENOX2 mRNA. Scrambled antisense to exon 5 mRNA was without effect. Antisense to exon 5 mRNA decreased the amount of ENOX2 protein on the surface of cancer cells. As a control, antisense-mediated down regulation of exon 5 minus ENOX2 mRNA was detected using exon 4/exon 6 primers. Exon 5 antisense blocked the cell surface expression of ENOX2 whereas exon 4 antisense was without effect. In contrast to non-transfected HeLa cells, cells transfected with exon 5 antisense were not inhibited by the green tea catechin, EGCg. A relationship of ENOX2 to unregulated growth of cancer cells was provided by data where growth of HeLa cells was inhibited by transfection with the exon 5 antisense oligonucleotides. In these studies, growth inhibition was followed by apoptosis in greater than 70 % of the transfected cells (Tang et al. 2008).

Not only did the exon 5 antisense reduce exon 4 minus mRNA in the HeLa cells and result in a loss of growth inhibition due to EGCg, the ability of HeLa cells to penetrate Matrigel in a standard invasion assay was reduced by 80 % with the exon 5 antisense (Tang et al. 2008; Fig. 8.24). These results suggest that ENOX2 is critical to in vitro invasion of HeLa cells despite the complexity of the invasion process which includes cell attachment, cell detachment, secretion of proteases, cell

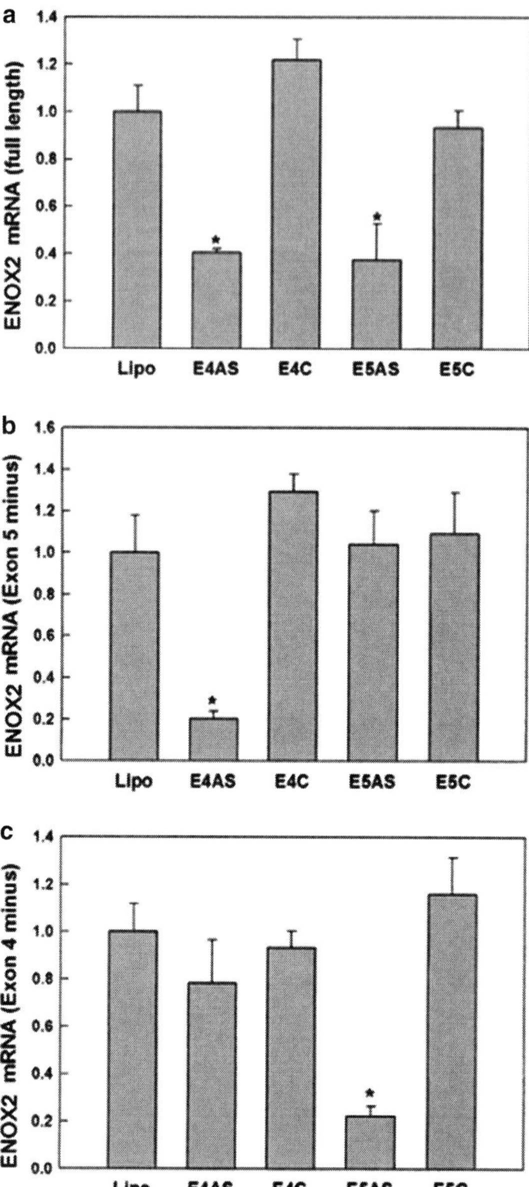

Fig. 8.23 Antisense-mediated down regulation of ENOX2 mRNA measured by real-time PCR. The ENOX2 mRNA level of each sample was calculated relative to Lipofectamine 2000 transfection control cells. (**a**) E4AS and E5AS mediated down regulation of full-length ENOX2 mRNA in HeLa cells detected by FLS and FLR primers (*E4AS and E5AS transfectants were significantly different from the remaining treatments, $\rho<0.05$). (**b**) E4AS mediated downregulation of exon 5 minus ENOX2 mRNA in HeLa cells detected by E4/6S and E4/6R primers (*the E4AS transfectant was significantly different from the other treatments, which were not different from each other, $\rho<0.05$). (**c**) E5AS mediated down regulation of exon 4 minus ENOX2 mRNA in HeLa cells detected by E3/5S and E3/5R primers (*the E5AS transfectant was significantly different from the other treatments, which were not different from each other, $\rho<0.05$). In this and subsequent figures, E4AS and E5AS are antisense oligonucleotides toward part of ENOX2 exon 4 and exon 5 sequences, respectively. E4C and E5C are scrambled controls of E4AS and E5AS, respectively. Reproduced from Tang et al. (2008) with permission from Cognizant Communication Corporation

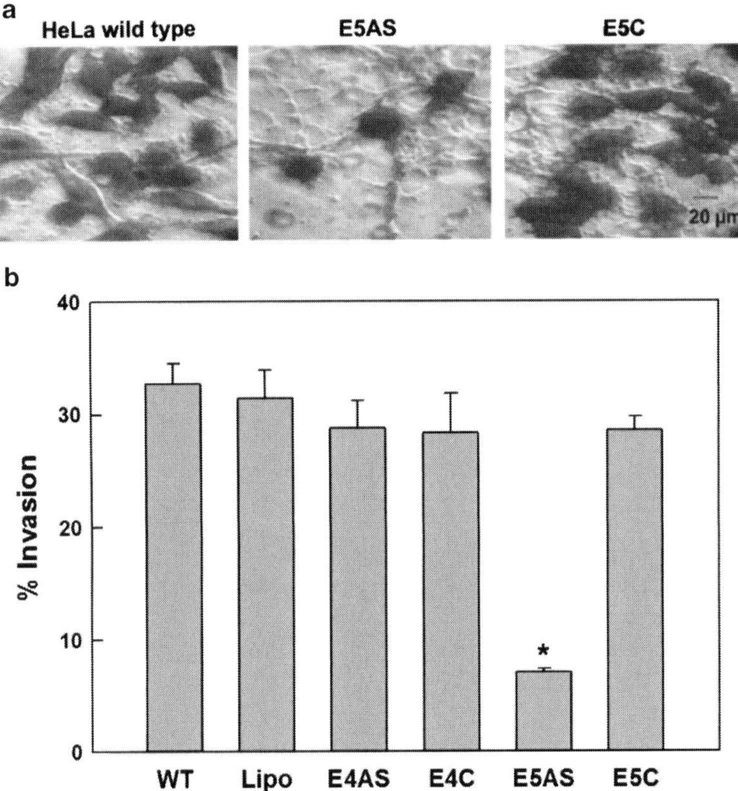

Fig. 8.24 Invasion of HeLa and transfected cells through Matrigel. (**a**) Cells (1×10^6) were placed in the upper wells of individual insects containing 8 μm pore PET membrane precoated with Matrigel. Cells were allowed to invade for 40 h at 37 °C, and then invaded cells were fixed and stained with 100 % methanol and 1 % toluidine blue, respectively. Cells on the upper surface were removed with a cotton swab, and the cells that migrated to the lower side of the membrane were mounted onto a microscope slide and photographed at ×400 magnification. (**b**) The percentage of invading cells was determined. Data shown are the mean ± SD values from three separate experiments (*the E5AS transfectant was significantly different from all other measurements, $p < 0.001$). Other treatments as described in Fig. 8.20 were not significantly different. Reproduced from Tang et al. (2008) with permission from Cognizant Communication Corporation

migration, and exchanging signals with other cells (Stetler-Stevenson et al. 1993; Bosserhoff and Buettner 2002).

A standard measure of the transformed phenotype is the ability of cells to form colonies on soft agar (Carney et al. 1980). This phenotypic characteristic of HeLa cells also was lost with the exon 5 antisense-transfected HeLa cells (Fig. 8.25; Chueh et al. 2004; Tang et al. 2008).

The presence of two additional splice variants was established by RT-PCR (Fig. 8.20). These were identified from the NCBI database (GenBank) as AK000353

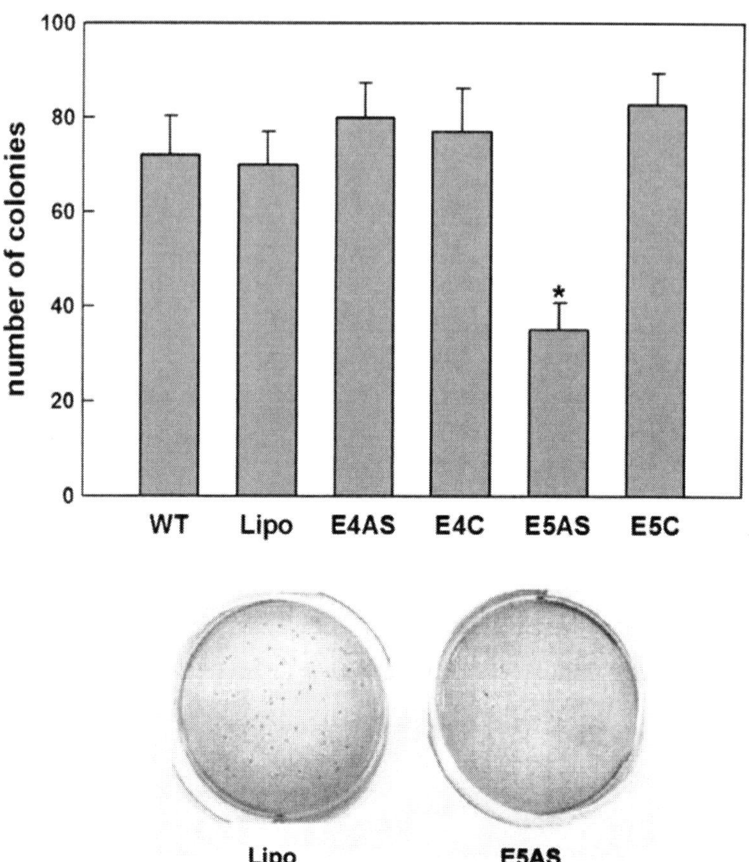

Fig. 8.25 Antisense decreases anchorage-independent growth of HeLa cells. E5AS transfection results in decreased colony formation in soft agar compared with the Lipfectamine (Lipo)-transfected cells. The total number of colonies was counted manually in three independent experiments with treatments blinded. E5AS transfection resulted in a 50 % reduction in the number of colonies formed compared with controls (*the E5AS transfectant was significantly different compared to the other treatments which were not significantly different from each other, $\rho<0.01$). Treatments are as in Fig. 8.23. Reproduced from Tang et al. (2008) with permission from Cognizant Communication Corporation

(band 1) and AL133207 (band 2). No cancer specificity was observed as both variants were present in non-cancer MCF-10A, MCF-12A, and WI-38 cells as well as in the cancer lines. The exon 5 minus form, like full-length ENOX2 cDNA (Sect. 8.8.2) appears not to be translated nor does exon 5 minus antisense affect the cancer phenotype (Tang et al. 2008).

8.8.1 Full-Length ENOX2 MRNA Identical to That of Cancer Cells Exists in Human Non-cancer Cells and Tissues

HeLa cells express the full-length mRNA and the functional 34 kDa ENOX2 protein. To investigate whether the full-length ENOX2 mRNA was cancer specific, human normal cell lines and tissues were examined separately by RT-PCR and Northern blot. In the Northern blot, commercially available human normal tissue mRNAs were probed with the ENOX2 cDNA sequence. Of the eight different human normal tissues examined, all exhibited detectable 3.8 kb full-length ENOX2 mRNA (Tang et al. 2007), although the levels of transcription varied. Probes for β-actin were used to assess RNA integrity and loading. Sequencing results of RT-PCR products showed comparable full-length ENOX2 cDNA sequences in comparing both cancer and non-cancer cell lines (Chueh et al. 2002b).

8.8.2 Full-Length 71 kDa ENOX2 Protein Not Translated

The ENOX2 cDNA, open reading frame of 1.83 kb, translates into 610 amino acids. There was no naturally expressed 71 kDa ENOX2 (corresponding to the 610 amino acid open reading frame) observed in either cancer or non-cancer cells (Sect. 8.3). No naturally expressed 71 kDa ENOX2 protein was observed in cells of either cancer or non-cancer cell lines nor did expression of full-length ENOX2 cDNA in non-cancer cells result in expression of 34 kDa processed ENOX2 protein. Whole cell extracts of COS cells transiently transfected with DNA encoding full-length ENOX2 fused with Myc tag exhibited bands corresponding to putative full-length ENOX2 plus the Myc tag. However, no bands at lower molecular mass were detected. These results further confirm that the full-length ENOX2 mRNA was unable to generate the 34 kDa ENOX2 protein.

The ENOX2 cDNA sequence has an unusually short 22 nt 5′-UTR. This may explain why translation may not start from the first ATG. On the other hand, it raises the question of whether there is additional sequence in the 5′-UTR and whether there was another ATG beyond the 5′-UTR end in cancer cells (Tang et al. 2007). 5′-RACE analysis showed that the previously determined ENOX2 cDNA sequence was complete at the 5′ end with no additional sequence in the 5′-UTR. Results with mRNA of BT-20 human mammary cancer cells were similar.

8.8.3 Cancer-Specific Expression of ENOX2

Plasma membranes purified from MCF-10A and BT-20 cells were compared by SDS-PAGE and western blot analysis with monoclonal antibody to ENOX2. The 34 kDa ENOX2 was present as a doublet (possibly due to small differences in the extent of glycosylation) in the plasma membranes of BT-20 cancer cells but not in

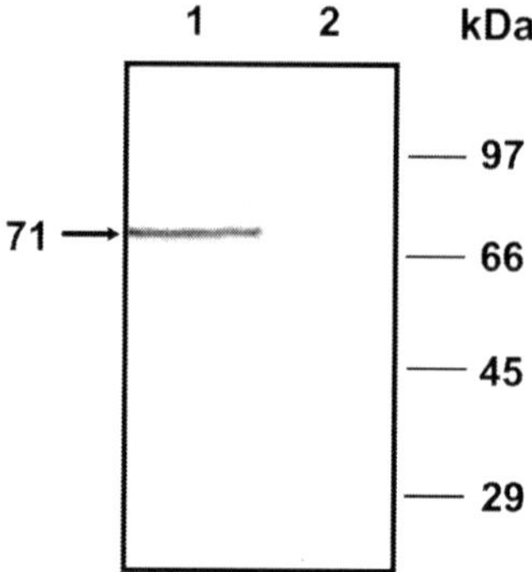

Fig. 8.26 Full-length ENOX2 cDNA expressed a ca. 71 kDa protein under the control of a vector promoter in COS cells. *Lane 1*, full-length ENOX2 cDNA in pcDNA3.1; *lane 2*, pcDNA3.1. All are COS cell transfectants. Antibody PU04, a peptide antibody to the quinine binding region on ENOX2 was utilized. Reproduced from Tang et al. (2007) with permission from American Chemical Society Publications

plasma membranes of non-cancer MCF-10A cells (Fig. 8.21). The plasma membrane marker, Na^+, K^+-ATPase (112 kDa) was used as a loading control. Other matched pair of normal and transformed pairs showing this specificity examined include rat hepatocytes and hepatoma, human melanocytes and melanoma, and SV-40 transformed and non-transformed 3T3 cells.

8.8.4 Splice Variants of ENOX2 Were Found in Cancer Cells

In all cancer lines examined, RT-PCR revealed mRNAs of reduced molecular masses which were below the limits of detection in non-cancer cells. That these mRNAs arose by degradation was ruled out by direct sequencing. Rather, band 2 yielded a sequence corresponding to an exon 4 minus form, and band 3 yielded a sequence corresponding to an exon 5 minus form. Both splice variant forms were found in BT-20, MCF-7, and HeLa cells (cancer) but not in the non-cancer MCF-10A (Fig. 8.20a) or WI-38 (Fig. 8.20b) cells.

Additional evidence for a cancer-cell-specific expression of exon 4 minus mRNA was provided using exon 4 minus-specific probes generated to the exon 4 minus-specific sequence at the splice juncture between exon 3 and exon 5. RNA preparations from HeLa human cervical carcinoma and BT-20 human mammary carcinoma cells clearly contained the expected 70 bp cancer-specific product indicative of exon 4 minus presence (Fig. 8.26). MCF-10A human non-cancer mammary epithelia and buffy coats (leukocytes and platelets from a normal volunteer) lacked the PCR product.

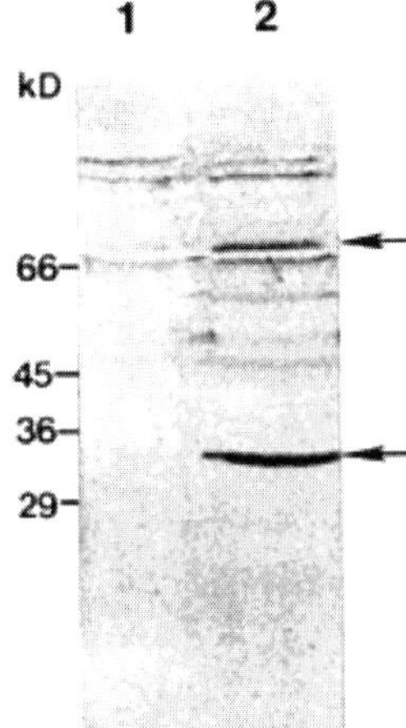

Fig. 8.27 Western blot analysis of OVCAR-3 cells using antisera raised from rabbit immunized with expressed ENOX. *Lane 1* membrane pellet after octylglucoside solubilization. *Lane 2* supernatant after octyl glucoside solubilization. *Arrows* indicate immunoreactive unprocessed ENOX2 (70 kDa) and processed ENOX (34 kDa). The regions of the gel corresponding to APK1 (29 kDa) and mesothelin (40 kDa) lacked immunoreactive material. The identities of the minor bands in the higher molecular weigh regions of the gel are unknown. Reproduced from Chueh et al. (2002b) with permission from American Chemical Society Publications

8.8.5 Expression of Exon 4 Minus and Exon 5 Minus Forms of ENOX2 in COS Cells

The exon 4 minus ENOX2 transfectants exhibited a 62 kDa band (corresponding to E4m ENOX2 expressed from M1 under the control of the CMV promoter) and 47 and 34 kDa bands corresponding to proteins translated from M231 and fully processed ENOX2, respectively, plus a 43 kDa band representing a possible processing intermediate (Chueh et al. 2002b; Fig. 8.27). Thus, with the exon 4 minus ENOX2 transfectants, a 34 kDa protein that reacted with the ENOX2 antibody was obtained from the overexpression of a naturally existing mRNA of cancer cells. This result showed that the exon 4 minus splice variant mRNA was capable of generating the 34 kDa ENOX2 protein.

8.8.6 Delivery of 34 kDa ENOX2 Protein to the Plasma Membrane

In addition to carrying out protein disulfide-thiol interchange, ECTO-NOX proteins function as terminal oxidases for plasma membrane electron transport (Morré 1998c; Morré and Morré 2003a). If the 34 kDa protein expressed from the exon 4 minus splice variant is to function as a terminal oxidase for plasma membrane electron transport, it must reach the outer surface of the plasma membrane. To investigate the subcellular localization of the 34 kDa ENOX2 protein expressed in exon 4 minus COS transfectants, plasma membranes and internal membranes of the transfectants prepared by aqueous two-phase partition were resolved on SDS-PAGE followed by western blot analysis. The western blots of plasma membranes and internal membranes of the exon 4 minus transfectants along with whole cell preparations were probed by a peptide antibody toward the quinone-binding motif of the ENOX2 protein. The internal membrane preparation exhibited the 47 and 43 kDa bands, plus a small but detectable amount of the 34 kDa protein presumably as a result of delivery of the 34 kDa processed form of the 47 kDa protein to the plasma membrane. The plasma membrane preparation exhibited the 43 and 34 kDa bands but not the 47 kDa band. The whole cell preparation exhibited all three (47, 43, and 34 kDa) bands, whereas none were present with vector alone.

8.8.6.1 Evidence for ENOX2 in Golgi Apparatus

A presence of ENOX2 in Golgi apparatus was first indicated from morphological studies where monensin-induced cisternal swelling of *trans* Golgi apparatus was blocked by the anticancer sulfonylurea LY181984 in LY181984-susceptible cells but not in LY181984-resistant cells (Morré et al. 1994f). The morphological response is related to reduced acidification in LY181984-susceptible cell lines presumably as a result of the presence of ENOX1 in *trans* Golgi apparatus enroute to the plasma membrane.

Sulfonylurea-sensitive NADH oxidase (ENOX2) activity of K-562 cells was subsequently localized to plasma membranes as well as a cytosolic membrane fraction enriched in Golgi apparatus cisternae (Moya-Camarena et al. 1995).

The enzymes of detergent-solubilized Golgi apparatus from HeLa cells efficiently process the 47 kDa ENOX2 to its mature molecular mass of 34 kDa. Processing differences may contribute but do not appear sufficient to explain the cancer-specific exon 4 minus-dependent expression of 34 kDa ENOX2 at the plasma membrane.

8.8.7 Mutation of Met 231 Blocked Expression of the Exon 4 Minus Splice Variant

Expression of a 47 kDa protein would be consistent with utilization of a downstream Met as the initiation site during translation of the exon 4 minus mRNA. Possible downstream initiation sites included Met 220, Met 231, and M314. Mutagenesis experiments of these initiation sites of E4m showed that only with COS cells transfected with E4m mRNA in which Met 231 was replaced by alanine was a ca. 47 kDa band missing (Fig. 8.22). Thus, initiation at Met 231 and subsequent processing of the 47 kDa protein and delivery to the plasma membrane of the 34 kDa active form of ENOX2 serves to explain the specific localization of ENOX2 at the cancer cell surface.

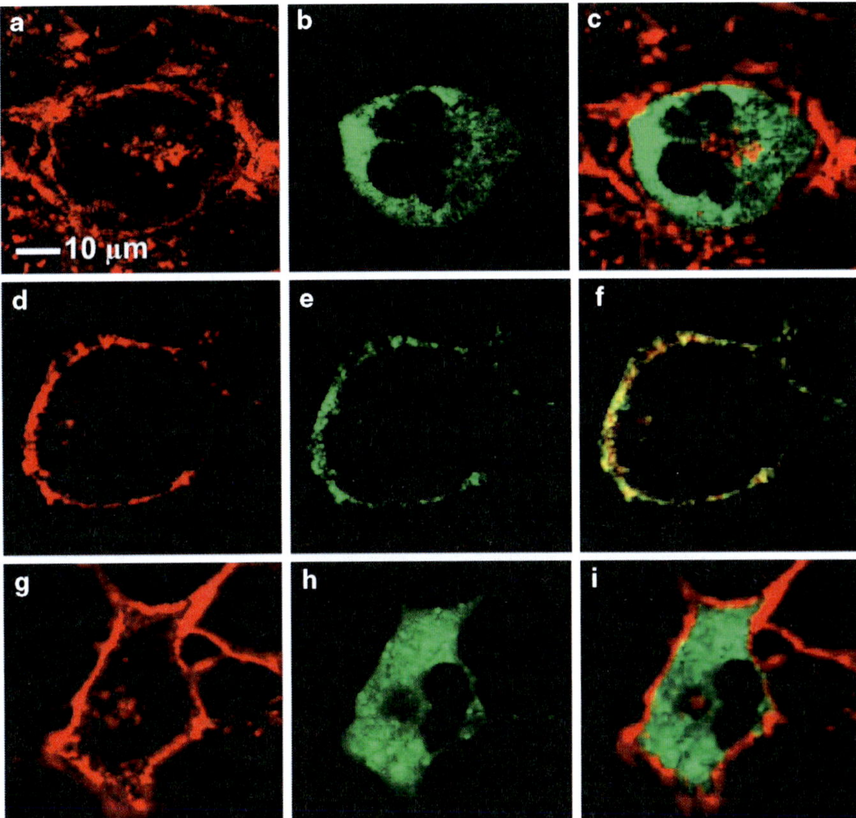

Fig. 8.28 (**a**–**c**) Confocal microscope images of full-length ENOX2-EGFP transfected COS cells. (**a**) The cell surface marker, tetramethylrhodamine concanavalin A, is in *red*. (**b**) The EGFP fusion protein is in *green*. (**c**) Colocalization of the two. (**d**–**f**) Confocal images of pE4m-EGFP transfected COS cells. (**d**) The cell surface marker, tetramethylrhodamine concanavalin A, is in *red*. (**e**) The EGF-fusion protein is in *green*. (**f**) Colocalization of the two. The transfected COS cells. (**g**) The cell surface marker, tetramethylrhodamine concanavalin A, is in *red*. (**h**) The EGFP fusion protein is in *green*. (**i**) Colocalization of the two. Reproduced from Tang et al. (2007) with permission from American Chemical Society Publications

8.8.8 Subcellular Localization of E4m ENOX2-EGFP and Full-Length ENOX2-EGFP Fusion Proteins

To test further the hypothesis that cell surface ENOX2 is the result of expression of E4m mRNA, confocal microscopy was used to determine the subcellular localization of the E4m ENOX2-EGFP and full-length ENOX2-EGFP when expressed in COS cells (Fig. 8.28). The constructs tagged with EGFP at the C terminus were expressed under the control of the cytomegalovirus promoter. Fluorescence microscopy revealed that the E4m-EGFP fusion protein was localized to the plasma membrane. In contrast, a full-length ENOX2-EGFP fusion protein was retained in internal membranes. COS cells transfected with pE4m-EGFP in which Met231 was replaced by alanine (M231A-E4m-EGFP) failed to exhibit the EGFP fusion protein at the cell surface consistent with M231 as the cancer-specific E4m mRNA initiation site.

The presence of exon 4 minus mRNA was verified by Northern blot analysis. mRNA bands spanning the correct range of 3.5–3.9 kb were obtained (full length of 3.8 kb's and exon 4 minus of 3.6 kb's) as an incompletely resolved doublet.

Taken together, the findings suggest that an exon 4 minus splice variant of ENOX2 provides the basis for the appearance of the 34 kDa processed form of ENOX2 at the cell surface specifically in cancer cells. Exon 4 minus ENOX2 mRNA was detected in all of the cancer cell lines thus far investigated but was below the limits of detection in non-cancer cells. Site-directed mutagenesis and other evidence indicate that the generation of the 34 kDa cell surface ENOX2 is the result of downstream initiation at methionine 231 (encoded by nucleotides ATG in exon 5, Fig. 8.22). Overexpression of the exon 4 minus transcript in non-cancer cells generated the expected full-length protein of molecular mass 47 kDa, plus a 43 kDa processing intermediate and the fully processed 34 kDa form at the plasma membrane, all of which reacted with anti-ENOX2 antibodies.

As stated above, there is no evidence of physiological translation of the full-length ENOX2 mRNA despite its widespread transcription. Therefore, the significance of differences in full-length mRNA abundance (Yagiz et al. 2006) may not be functionally relevant. The cDNA sequence of ENOX2 revealed that the first ATG started at nt 23. There were only 22 nt's to form a 5′-UTR. This short 5′-UTR makes the first ATG an unlikely site for initiation of translation due to the potential restrictions on ribosome binding based on the current scanning model for translation initiation of translation in eukaryotes (Kozak 1978). Normally, the initiation complex forms around 21–24 nt's after recognition of the cap of mRNA and then begins to scan for the first Met codon. This "first AUG" rule holds for about 90 % of the eukaryotic mRNAs that have been analyzed (Kozak 1989). However, the first Met typically is about 50–150 nt's downstream from the cap (Pestova et al. 1998). Thus, with ENOX2, the first Met at nt 23 may be too near to the cap to serve as a functional initiation sequence for translation especially as sequence elements in the 5′-UTR are often required as well for regulation of translation (Van der Velden and Thomas 1999). The first methionine codon does not always function as the initiator codon (Kozak 1978; Suzuki et al. 2000) and if the first ATG of full-length ENOX2 mRNA functioned as the initiator codon, the 5′-UTR of ENOX2 would be only 22 nt, shorter than most functional 5′-UTRs (Kozak 1987). Thus, it is logical to expect the initiator codon further downstream. Downstream Kozak sequences (A/G)XXAUGG (Kozak 1989) are found at nt 167 (exon 2), nt 713 (exon 5), and nt 1,163 (exon 8). A 5′-UTR length is usually smaller than 200 bp's because of the limitation of the scanning capacity of the ribosome (Kozak 1989). By deleting an upstream portion of the message through alternative splicing, it may be possible for the scanning ribosome to reach a downstream methionine codon to begin initiation to ultimately generate the cell surface 34 kDa form of ENOX2. Alternatively, downstream initiation as a result of the deletion of exon 4 might occur through disruption of the secondary structure of ENOX2 mRNA in a region where secondary structure would normally hinder ribosome binding or block translation. Involvement of cancer-specific RNA-binding proteins seems unlikely since transfection with exon 4 minus cDNA is sufficient to generate functional cell surface ENOX2 in non-cancer cell lines which would lack the binding proteins.

As the processed ENOX2 is located on the extracellular side of the plasma membrane (Morré and Morré 2003a; Morré 1995a), a functional translation initiation site would require a downstream membrane insertion sequence. The exon 4 minus splice variant results in a 207 bp deletion. There is no frame shift after the deletion of exon 4. The result is a 68 amino acid deletion plus one amino acid change. The net result is the deletion of one methionine codon at nt 542, and all downstream methionine codons are brought closer to the 5′ end. By facilitating downstream initiation at M231, the exon 4 minus alternatively spliced mRNA favors cancer-specific cell surface expression of the ENOX2 protein. Initiation at M231 has the advantage of there being both putative signal sequence and signal sequence cleavage sites. Initiation at M231 where a Kozak sequence occurs would result in a peptide sequence of 380 amino acids and a calculated molecular mass of 44.5 kDa. What is observed is a protein band with an apparent molecular mass on SDS-PAGE of 47 kDa, which after removal of the signal sequence and further processing would be expected to generate the functional 34 kDa ENOX2 protein that is present at the surface of cancer cells. This conclusion was supported by methionine to alanine replacements, which confirmed the 47 kDa protein as the expressed form of the exon 4 minus transcript with initiation at M231. The 43 kDa band likely represents the 47 kDa species after removal of a 29 amino acid sequence with characteristics of a signal peptide (Chueh et al. 2002b). The 34 kDa protein corresponds to the mature processed form of ENOX2 and appeared at the plasma membranes but was absent from internal membranes consistent with its translation as a 47 kDa peptide and subsequent processing. The 47 and 43 kDa proteins plus a small amount of the 34 kDa protein were found in the cell fractions containing internal membranes, whereas the plasma membranes contained only the 34 kDa processed form. Appearance of ENOX2 at the plasma membrane of cancer cells seems dependent upon the presence of exon 4 minus mRNA and thus would be expected to vary independently of the levels of full-length ENOX2 message. However, processing does not seem to be an obligatory consideration for delivery to the plasma membrane. The full-length exon 4 minus-EGFP construct, for example, was translated and subsequently delivered to the plasma membrane as the intact 74 kDa fusion protein without evidence of further processing as determined by fluorescence microscopy.

More than 100 genes exhibit altered pre-mRNA splicing in cancer (Kalnina et al. 2005). Several studies using bioinformatics methods have found potentially cancer-specific or cancer-associated splice variants (Wang et al. 2003a; Xu and Lee 2003;

Hui et al. 2004; Okumura et al. 2005). Several studies show specific alterations in the expression of splicing factors in cancer. Thus, expression of ENOX2 exon 4 minus mRNA may be reasonably expected to derive from cancer-specific alterations in splicing factors (Sect. 8.9). Splice variant-specific ENOX2 primers that spanned the boundary of exon 3 and exon 5 were designed to differentiate ENOX2 exon 4 minus mRNA from full-length ENOX2 as a cancer diagnostic aid (Chap. 12; Sect. 12.5).

The expression of ENOX2 is insufficient to cause cancer (Chueh et al. 2004) but ENOX2 may be important to the unregulated cell growth that typifies cancer (Morré and Morré 2003a). Unlike constitutive CNOX proteins, ENOX2 is responsive to a variety of drugs and substances, all with anticancer activity that inhibit both ENOX2-catalyzed hydroquinone and NADH oxidation (Morré 1998c). As emphasized elsewhere (Chap. 5; Fig. 5.18), when ENOX2 activity is inhibited, cancer cells, once having divided, fail to enlarge to a size sufficient to divide again and, instead, undergo apoptosis (Morré and Morré 2003a; Fig. 5.18).

A role for ENOX2 in maintaining growth of cancer cells in culture has been recently reported in experiments based on RNAi-mediated gene silencing (Liu et al. 2005). ENOX2 as the molecular target at the surface of cancer cells to explain the activity of anti-tumor sulfonylureas (Morré et al. 1995g, h) has been independently confirmed by Alonso et al. (2001).

8.8.9 Regulation of ENOX2 Expression

There have been few studies related to regulation of ENOX2 expression. Wang et al. (2009) reported that not only did capsaicin block ENOX2 activity and growth of human SCM1 stomach cancer cells leading to apoptosis but down regulation of ENOX2 protein expression also occurred concurrently with the induction of apoptosis.

Based on assays of drug-responsive NADH oxidase activities, the ENOX2 expression on tumor cells seems at least 25 times greater than that on normal cells (Cho and Morré 2009). However, the mechanism of regulation of these quantitative levels of expression remains unknown.

In a study with A-549 human non-small cell lung carcinoma cells in culture, Y.-C. Su, Y.-H. Lin, Z.-M. Zeng, K.-N. Shao and P. J. Chueh (Biom Biophys Acta, in press) have correlated ENOX2 expression levels with cell proliferation and migration under a variety of conditions including ENOX2 knock down by RNA interference and inhibition of ENOX2 activity by chemotherapeutic agents. These findings, together with ENOX2 overexpression in gain of function studies with human ENOX2-transfected NIH3T3 mouse fibroblast cells provide evidence for an essential role for ENOX2 in cell migration. These studies employed a real-time cell monitoring system where the presence of migratory cells on an electrode-bearing surface produced an elevation in electrode impedance whose magnitude was determined by cell number and degree of cell adhesion.

Unexpectedly both short term and low dose exposure to doxorubicin and tamoxifen treatment transiently up-regulated ENOX2 expression in the human lung carcinoma A-549 cells resulting in enhanced cell migration and drug response along with down-regulated epithelial markers and up-regulation of mesenchymal markers (Y.-C. Su, Y.-H. Li, Z.-M. Zeng, K.-N. Shao and P. J. Chueh, unpublished). Epithelial to mesenchymal transition has been implicated in both metastasis and cancer progression (Hugo et al. 2007). The possibility was raised that increased ENOX2 expression contributes to a defensive response to cytotoxic chemotherapeutic drugs that is important for survival of transformed cells (see also Mao et al. (2008); Wang et al. (2009, 2011) and Liu et al. (2008, 2012) for related observations with capsaicin treatment). As transient up regulation of ENOX2 was observed only at low drug concentrations, the resultant biphasic regulation of ENOX2 protein levels is reminiscent of the hormetic response of cell growth to cytotoxic agents including anticancer drugs such as doxorubicin (Morré 2000).

8.9 hnRNP F Splicing Factor Directs Formation of the Exon 4 Minus Variant of ENOX2

ENOX2 exon 4 has many different characteristics that could bind factors to direct the spliceosome to silence the exon from the main RNA product. ENOX2 is alternatively spliced (Tang et al. 2007, 2008). Furthermore, the splicing reaction is inducible via transcriptional alteration by SV-40 (Tang et al. 2011a). Within 5 days, the exon 4 minus variant of ENOX2 was produced in virus-infected HUVEC or mouse 3T3 cells. Alternative splicing may be involved, as well, in the early developmental regulation of ENOX2 expression (Cho and Morré 2009).

A minigene consisting of three exons, $EGFP_1$, either ENOX2 exon 4 or albumin exon 2, and $EGFP_2$, the link between production of the exon 4 minus variant and cancer cells was used to identify the specific transcription factor involved (Tang et al. 2011). That the two halves of the EGFP were spliced together and that the ENOX2 or albumin exon was silenced was indicated by fluorescence.

Removal of exon 4 from the processed RNA of the GFP minigene construct occurred with HeLa and to a lesser extent with BT-20 but not in non-cancer MCF-10A cells. Eight exonic splicing silencers (ESSs) for hnRNP binding in the exon 4 sequence were identified using the Splicing Rainbow Program. Each was altered by site-directed mutagenesis to determine which were responsible for the splicing skip. Mutation of MutG75 ESS changed the GFP expression to indicate splicing silence, while other mutations did not. The findings indicate that hnRNP F directs formation of the Exon 4 Minus Variant of ENOX2 (Figs. 8.29 and 8.30).

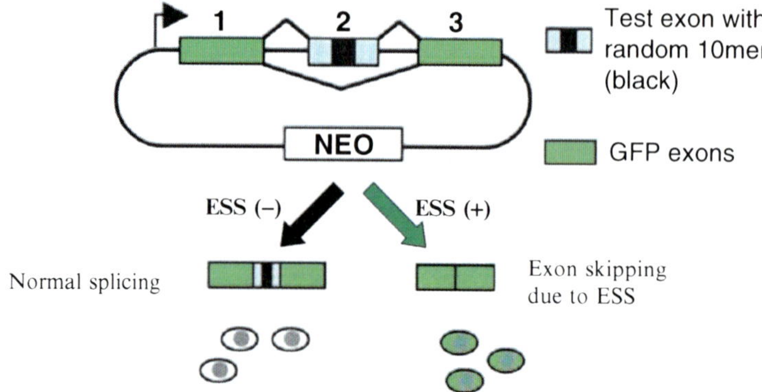

Fig. 8.29 Diagram of strategy used to screen for exonic splicing silencer (ESS). Modified from Wang et al. (2004). Using a minigene consisting of three exons, $EGFP_1$, either ENOX2 exon 4 or albumin exon 2, and $EGFP_2$, the link between production of the exon 4 minus variant and *cis*-acting regulatory elements bound by *trans*-acting factors of cancer cells was explored. Fluorescence indicated that the two halves of the EGFP were spliced together and that the ENOX2 or albumin exon was silenced. Reproduced from Tang et al. (2011) with permission from Springer-International

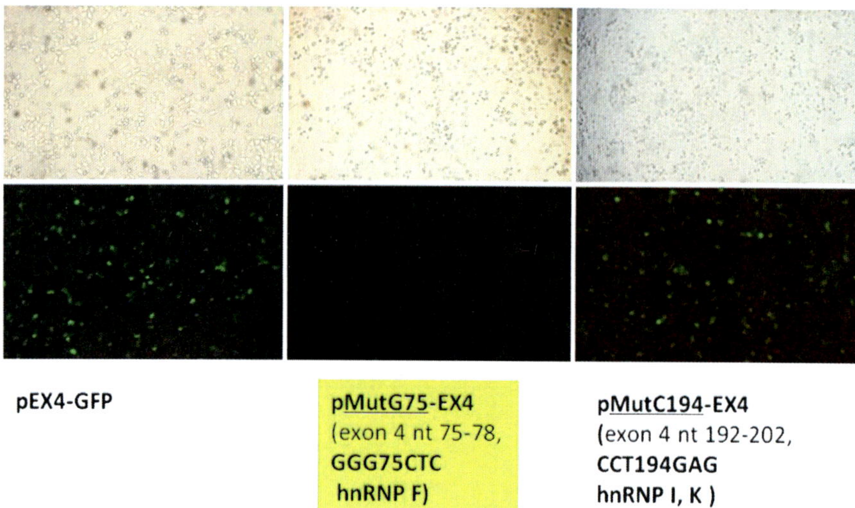

Fig. 8.30 Microscopic images of transfected HeLa cells. *Top*, phase contrast images. *Bottom*, GFP fluorescence. Results showed MutG75 ESS mutation changed the GFP expression which is a sign of splicing silence, while other mutations did not. As MutG75 changed the ESS binding site for hnRNP F, this result suggest that nhRNP F directs formation of the exon 4 minus variant of ENOX2. Reproduced from Tang et al. (2011) with permission from Springer-International

8.10 Summary

ENOX2 or tNOX (for tumor-associated NOX) proteins are ENOX proteins specific to cancer. The presence of the ENOX2 protein has been demonstrated in several human tumor tissues. However, serum analyses of ENOX2 protein shed into sera of cancer patients suggest a much broader, perhaps universal, association with human cancer.

ENOX2 proteins are cell surface-located but distinguished from ENOX1 proteins on the basis of the response of their enzymatic activities to inhibitors, activators, and drugs, amino acid sequence and functional motifs, period length of oscillatory activities (22 min) and entrainment properties. A unique characteristic that distinguishes ENOX2 family proteins from other ENOX proteins is their sensitivity to inhibition by quinone site inhibitors with anticancer activity. The NAD(P)H oxidase activity of ENOX2 is inhibited by doxorubicin, the antitumor sulfonylureas, capsaicin, the catechin EGCg, and the synthetic anticancer isoflavene phenoxodiol, as examples. The shed form of ENOX2 from cancer patients exhibits the same degree of drug responsiveness as the cell membrane-associated form.

The ENOX2 proteins are unresponsive to growth factors and are constitutively activated. They are correlated with the unregulated growth which is a hallmark of the cancer phenotype. They are not the result of oncogenic mutations. They resemble fetal NAD(P)H oxidase forms important to maintenance of unregulated cell enlargement in early development that may be re-expressed in malignancy. The phenotype of accelerated growth has been recapitulated in a transgenic mouse strain overexpressing ENOX2. The first ENOX protein to be cloned and expressed (GenBank Accession No. AF20788), the ENOX2 gene consists of at least nine exons that combine to yield a 70.1 kDa protein comprised of 610 amino acids. Despite protein disulfide-thiol interchange activity, there is no flavin and only one –C–X–X–X–X–C– motif characteristic of classical PDIs. The drug binding site of ENOX2 contains a conserved five amino acid (EMTEE) motif absent from ENOX1. ENOX2 proteins are extrinsic membrane proteins that lack intrinsic membrane-binding motifs. The protease resistance along with other characteristics shared with prions have been well studied with ENOX2. A standard measure of the transformed phenotype is the ability of cells to form colonies on soft agar. This phenotypic characteristic of HeLa cells was lost for HeLa cells transfected with exon 5 antisense to suppress ENOX2 expression.

The expressed form of ENOX2 arises as a splice variant from the single ENOX gene and in HeLa and other cultured cancer cells is delivered to the cell surface as a processed 34 kDa protein. Full-length ENOX2 mRNA is present in both normal and tumor cells but appears not to be expressed in either. Exonic Splicing Factor hnRNP F has been shown to direct formation of the Exon 4 Minus Variant of ENOX2.

Two-dimensional gel electrophoresis/western blot analyses have revealed a family of more than ten ENOX2 transcript variants. Each variant is specific to the site of origin of the cancer. Detection involved two-dimensional gel electrophoresis and western blot analysis using a recombinant scFv antibody. ELISA detection of ENOX2 transcription variants in a standard ELISA protocol is complicated by the presence in patient sera of interfering ENOX2 autoantibodies. The potential of ENOX2 as a diagnostic and therapeutic target for cancer is the subject of Chaps. 11 and 12.

Chapter 9
Age-Related ENOX Proteins (arNOX)

The age-related NADH oxidases (arNOX/ENOX3) are so named since they are absent or present at levels below the limit of detection for cells and sera of young individuals (≤30 years). They then increase with increasing age to ca. age 60–70+ years (Morré et al. 2003a). arNOX proteins have been identified in yeast and in humans. They constitute a family of TM-9 transmembrane proteins consisting of five family members encoded on different chromosomes. The classification of arNOX proteins as ENOX proteins is based on functional similarities of arNOX to ENOX1 and ENOX2 in that hydroquinones or semiquinones are electron donors for the cell surface form of the protein and molecular oxygen is an electron acceptor. arNOX proteins differ by generating superoxide during a portion of the activity cycle. The activities are periodic (period length of 26 min) and are localized in the cell at the exterior surface of the plasma membrane. Enzymatically active, truncated forms are shed into the blood and other body fluids (saliva, urine, perspiration, and interstitial fluids that percolate through the basement membranes). Protein thiols and tyrosines serve as electron donors for the shed forms of arNOX. The proteins, like ENOX1 and ENOX2, are resistant to proteolytic digestion and N-terminal sequencing and, when highly purified, form aggregates, presumably amyloid rods, devoid of enzymatic activity. Unlike the ENOX1 and ENOX2 proteins which exclusively carry out four-electron transfers to molecular oxygen to form water during the oxidative phase of the activity cycle, the arNOX proteins result in the additional generation of superoxide but only in one phase of the activity cycle (Fig. 9.1). Superoxide is not a reaction product of ENOX1 or ENOX2 as ferricytochrome c is not reduced by these proteins. There are no apparent sequence similarities to ENOX1 or ENOX2 except for the presence of required functional motifs (Fig. 9.2) which include an adenine nucleotide binding site, a thiol interchange motif, and two putative copper binding sites per monomer. Amino acid sequences within the catalytic C-termini of arNOX family members were sufficiently dissimilar to permit generation of family member-specific peptide sequences suitable for antibody production (Table 9.1). The oscillatory pattern is that of the standard 2 + 3 pattern of 5 maxima exhibited by ENOX1 and ENOX2 with the exception that maximum ③ generates as well a burst of superoxide (Fig. 9.3). Sustained production of superoxide in body

D.J. Morré and D.M. Morré, *ECTO-NOX Proteins: Growth, Cancer, and Aging*,
DOI 10.1007/978-1-4614-3958-5_9, © Springer Science+Business Media New York 2013

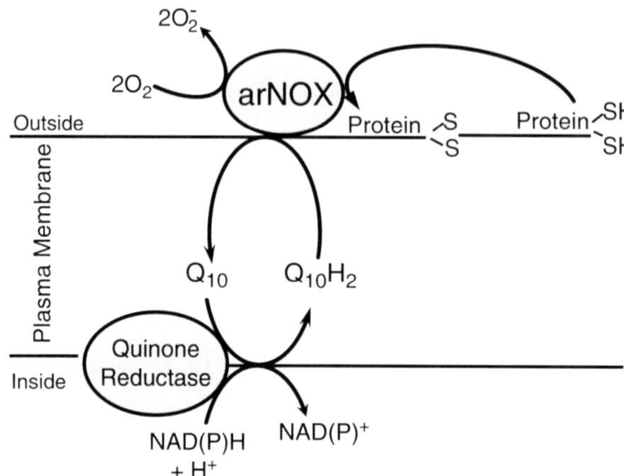

Fig. 9.1 Generation of superoxide at the cell surface from the action of plasma membrane-situated arNOX activity. Illustrated are the spatial relationships between quinones and the external arNOX protein to donate electrons from cytosolic NAD(P)H to molecular oxygen to form superoxide. The arNOX protein also utilizes protein thiols as a source of electrons as depicted on the right. Reproduced from Morré et al. (2010b) with permission from Mary Ann Liebert, Inc

```
  1   MATAMDWLPWSLLLFSLMCETSAFYVPGVAPINFHQNDPVEIKAVKLTSSRTQLPYEYYS
 61   LPFCQPSKITYKAENLGEVLRGDRIVNTPFQVLMNSEKKCEVLCSQSNKPVTLTVEQSRL
121   VAERITEDYYVHLIADNLPVATRLELYSNRDSDDKKKEKDVQFEHGYRLGFTDVNKIYLH
181   NHLSFILYYHREDMEEDQEHTYRVVRFEVIPQSIRLEDLKADEKSSCTLPEGTNSSPQEI
241   DPTKENQLVFTYSVHWEE
```

Adenine nucleotide binding site (GXVXXG)
Putative copper sites (YVH, HGY)
Putative protein disulfide interchange site (CXXXC)
Conserved CQ/CE

Fig. 9.2 Functional motifs of the ca. 30 kDa arNOX form TM9SF4

Table 9.1 Peptide antibodies were generated in rabbits to the N-terminal sequences of the exfoliated proteins

TM9SF1a and 1b	(aa 72–87)	I R H K S K S L G E V L D G D R
TM9SF2	(aa 89–104)	G K E P S E N L G Q V L F G E R
TM9SF3	(aa 70–88)	K K S I S H Y H E T L G E A L Q G V E
TM9SF4	(aa 69–84)	I T Y I A E N O G E V O R G D R

A cysteine residue was added to the N-terminus of each peptide to facilitate coupling to the carrier protein KLH

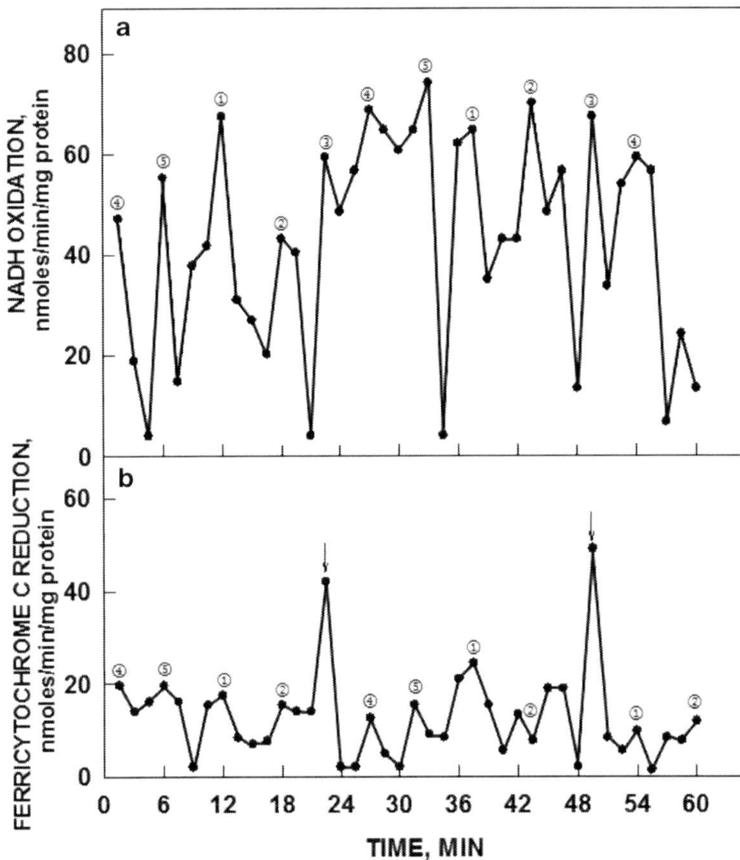

Fig. 9.3 Comparison of (**a**) NADH oxidation and (**b**) ferricytochrome c reduction of recombinant TM9 SF2 N-terminal 18 kDa peptide as a measure of superoxide production. Illustrated is the oscillatory pattern of five maxima. The maxima labeled ① and ② are separated by 6 min. The three maxima labeled ③, ④, and ⑤ that follow are separated from the major maxima and each other by 4.8 min creating the 25-min period [6 + (5 × 4) = 26]. The maximum of superoxide production arrows coincides with maximum ③ for NADH oxidation

fluids, for example, is attained by the presence of multiple family members that appear to not cross entrain (Fig. 9.4).

The reactive oxygen species generated from the superoxide such as hydrogen peroxide through dismutation can become accessible to lipoproteins in the circulation resulting in lipid oxidation and increased atherogenic risk as well as resulting in damage to adjacent cells and extracellular supporting matrices important to skin health. Alternatively, formation of superoxide and hydrogen peroxide formation may serve essential normal cell functions and major roles in subcellular redox state modulation (Linnane and Eastwood 2006).

Both cell-bound and shed arNOX are blocked by unique inhibitors not shared with ENOX1 or ENOX2 (Fig. 9.5) including coenzyme Q_{10} (CoQ_{10}) (Morré and Morré

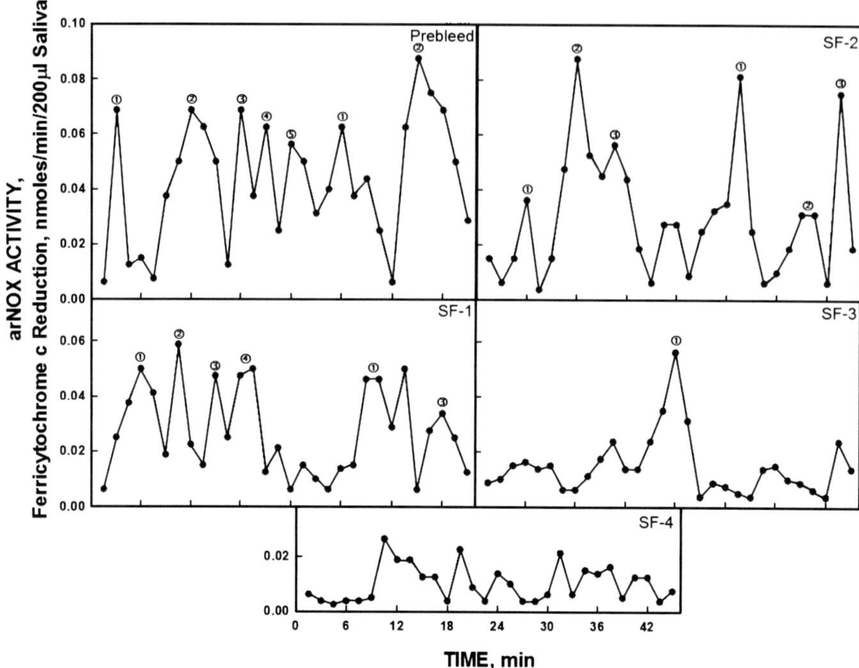

Fig. 9.4 Sequential addition of arNOX peptide antibodies to sera leads to sequential elimination of individual activity maxima. Results obtained show association of the five specific maxima unaffected by addition of the prebleed. After addition of TM9SF1-specific antisera, two maxima corresponding to SF1a and SF1b were lost. After addition of antisera specific to TM9SF2, only two maxima remained. After addition of TM9SF3-specific antisera, one maximum remained and after addition of TM9SF4-specific antisera, no maxima remained

2006a; Morré et al. 2003a; Sect. 9.11.1). The inhibition by CoQ_{10} appears to be at a site unique to arNOX and unrelated to the quinone site involved in binding and oxidation of reduced coenzyme Q_{10} ($CoQH_2$) by ENOX proteins at the plasma membrane.

9.1 arNOX Discovery

Gorman et al. (1997) reported that UV irradiation of HL-60 cells resulted in an enhanced ability to generate superoxide. Based on inhibition by diphenyliodonium, a putative specific inhibitor of NAD(P)H oxidases (Morré 2002), it was further suggested that the UV target was a cell surface NAD(P)H oxidase. Hepatic plasma membranes were reported early to generate hydrogen peroxide upon oxidation of NADH (Ramasarma et al. 1981). ENOX2 normally does not generate hydrogen peroxide suggesting the involvement of some other protein. These findings were subsequently confirmed (Morré et al. 1999c) in experiments where superoxide dismutase (SOD)-sensitive reduction of external ferricytochrome c (Butler et al. 1982)

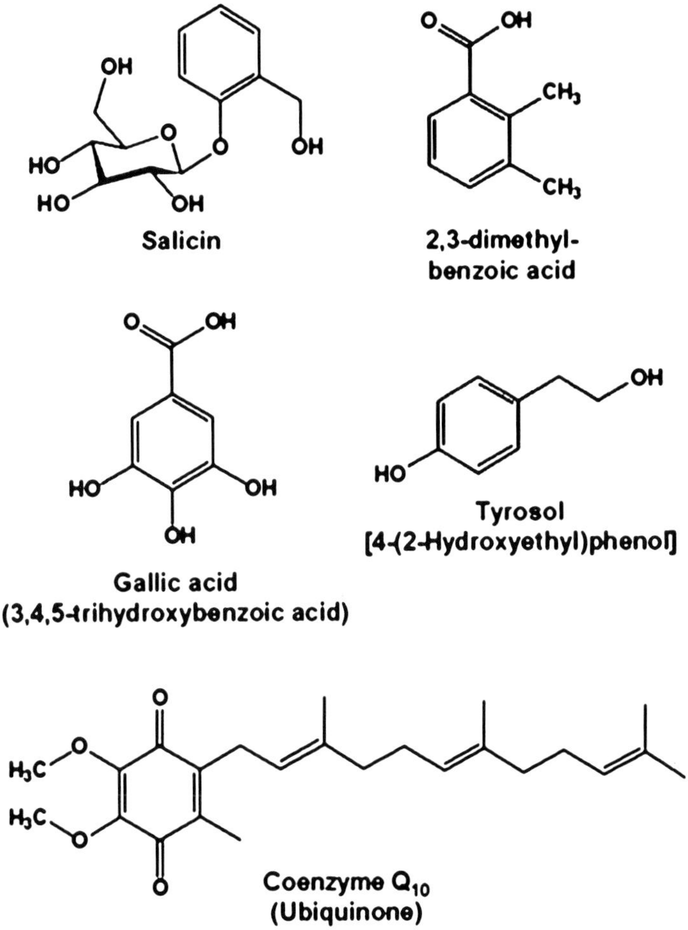

Salicin

2,3-dimethyl-benzoic acid

Gallic acid (3,4,5-trihydroxybenzoic acid)

Tyrosol [4-(2-Hydroxyethyl)phenol]

Coenzyme Q_{10} (Ubiquinone)

Fig. 9.5 Inhibitors of arNOX activity

was used as an assay for superoxide generation and subsequent dismutation to form hydrogen peroxide.

A number of features early identified the aging-related superoxide-generating oxidase (arNOX) as an ECTO-NOX. The activity was resistant to protease digestion (proteinase K) and resistant to heating to temperatures between 70 and 80 °C (Morré et al. 2003a). The superoxide-generating activity was not steady state but exhibited a pattern of oscillations with a characteristic period length as a defining characteristic of ECTO-NOX proteins (Fig. 9.3; Chap. 6). However, the period length of the arNOX was about 26 min (Morré et al. 1999c, 2003a) rather than 24 min as is characteristic of ENOX1 or 22 min as is characteristic of ENOX2. The superoxide-generating activity was demonstrated subsequently in aged cell cultures as well as associated with plasma membranes prepared from aged plant tissues. The arNOX activity was regarded as unique in that it was not inhibited by capsaicin,

(−)-epigallocatechin gallate (EGCg) or other inhibitors of the cancer-associated ENOX2 (Morré et al. 2003a) or by simalikalactone D, an inhibitor of ENOX1. arNOX-specific inhibitors, salicin, ubiquinone, gallic acid, 2,3-dimethyl benzoic acid, and tyrosol were subsequently identified (Fig. 9.5).

9.2 Measurement of Superoxide Formation by arNOX

A standard assay for arNOX activity involves measurements of superoxide production based on the reduction of ferricytochrome c by superoxide monitored from the increase in absorbance at 550 nm with reference at 540 nm (Butler et al. 1982; Mayo and Curnutte 1990; Fig. 9.6; Chap. 2). As a further check for superoxide generation, SOD normally is added near the end of the assay to ascertain that the rate of ferricytochrome c reduction returns to base line. The assay consists of 150 μL (2 mg/mL) oxidized ferricytochrome c solution and 200 μL of biofluid or tissue added to 2.5 mL phosphate-buffered saline glucose (PBSG) assay buffer (8.06 g NaCl, 0.2 g KCl, 0.18 g Na_2HPO_4, 0.13 g $CaCl_2$, 0.1 g $MgCl_2$, and 1.35 g glucose dissolved in 1,000 mL deionized water, adjusted to pH 7.4, and stored at 4 °C). Rates are determined using a SLM Aminco DW-2000 spectrophotometer in the dual wavelength mode with continuous measurements (over 1 min every 1.5 min). After 45 min, 60 μL (containing 60 units) SOD are added and the assay is continued for an additional 45 min to ascertain the level of SOD inhibition (Fig. 9.7).

Because collection is noninvasive, saliva is a very attractive biofluid for arNOX measurement and for response testing to inhibitors. Saliva is readily available and seldom in short supply. On a daily basis the average subject produces about 1–1.5 L of saliva constantly from the salivary glands (Hu et al. 2007). As saliva seems to mirror the circulation as a filtrate of our blood, one has in saliva a surrogate arNOX response marker that can be sampled repeatedly and is amenable to real-time kinetic analyses. Salivary arNOX levels are relatively constant throughout the day, so it is unimportant that specimens be collected at the same time of day for comparative purposes (Fig. 9.8).

arNOX also may be assayed using the property of the generated superoxide to reduce tetrazolium salts such as XTT (Na 3′-[(phenylamino)-carbonyl]-3,4-tetrazolim]-bis(4-methoxy-6-nitro)benzene sulfonic acid) leading to colored formazan formation. Other NOX proteins lack this activity. The reduction of XTT may be measured continuously at 470 nm at 35 °C. To monitor periodic activities, rates of XTT reduction also may be determined over 1 min at intervals every 1.5 min (Fig. 9.9). The molar concentrations of reduced XTT are determined using a molar extinction coefficient of 21.6 cm (Sutherland and Learmonth 1997).

arNOX activity is also present in human perspiration (Table 9.2). Perspiration may be conveniently collected using the Osteopatch™ Sweat Collection Device (Palacios et al. 2003), which consists of an absorbent patch placed next to the skin that collects the perspiration that when eluted with assay buffer provides arNOX activity in amounts sufficient to assay. Perspiration offers the opportunity to isolate

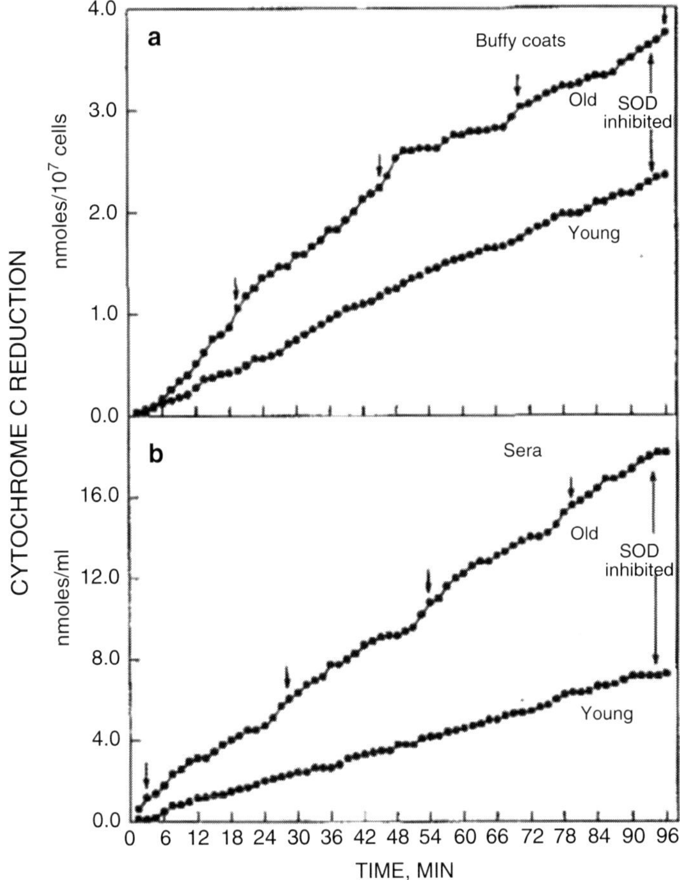

Fig. 9.6 Kinetics of ferricytochrome c reduction comparing buffy coats of old (70–100 years) or young (20–40 years) individuals (**a**) or sera pooled from old (70–100 years) or young (20–40 years) individuals (**b**). The ferricytochrome c reduction of buffy coats (**a**) or sera (**b**) of aged individuals was inhibited by superoxide dismutase (SOD) with that from buffy coats or sera of younger individuals was not. With both buffy coats and sera of the aged individuals, the kinetics were non-linear and exhibited maxima (*arrows*) at regularly spaced intervals of about 26 min. Reproduced from Morré et al. (2003a) with permission from Springer-International

arNOX and compare different isoforms from specific skin regions and compare topically treated and untreated skin from the same individual. If a particular preparation is effective as an antiaging strategy, a response of arNOX in the perspiration would be predicted to serve as a means to monitor efficacy, dosing, etc., or to identify responsive and unresponsive individuals or individuals most needful of the preparations. ENOX2 is present in urine, ENOX1 and ENOX2 are present in sera only, and arNOX is present in saliva and perspiration as well.

Oxidation of NADH and hydroquinones as well as protein disulfide interchange of arNOX proteins are assayed as for ENOX1 and ENOX2 (Chap. 2).

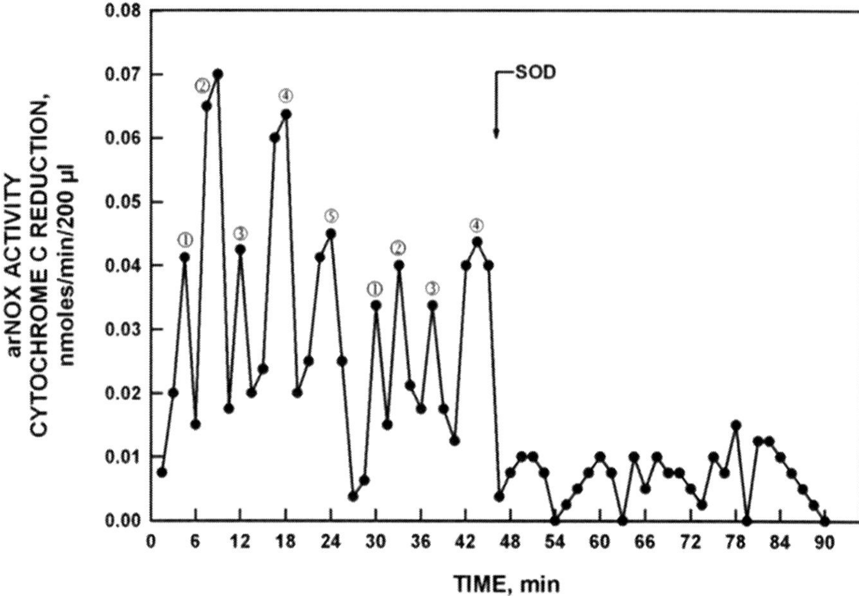

Fig. 9.7 arNOX activity oscillates with a 26-min period length. Activity measured by the rate of ferricytochrome c reduction is inhibited by addition of SOD

9.3 Characteristics

arNOX proteins are members of the ENOX family of oxidases lacking flavin and capable of oxidizing external NAD(P)H as an artificial substrate. A unifying characteristic of the ENOX proteins is that the activities oscillate. The superoxide-generating ECTO-NOX has a unique period length of 26 min rather than the usual 24 min for ENOX1 (CNOX) or 22 min for ENOX2 (tNOX). Within the oscillatory pattern, a single burst of superoxide production with a 26 min period length together with its inhibition by SOD serves as the basis for the activity assay based on superoxide production. The presence of all five isoforms in skin explants, sera, saliva, or perspiration, especially in aged individuals, gives rise to 5 maxima of superoxide production (labeled 1–5) within each 26 min period for each arNOX source (Fig. 9.4).

arNOX proteins are not modified forms of a constitutive ENOX. Rather, they are induced as part of the aging process. arNOX activities appears as individuals age in plasma membranes of red blood cells, at the cell surface of skin explants and cells of the buffy coat fraction of blood and purified lymphocytes and in serum, saliva, perspiration, and urine (Table 9.2) where the activities have been assayed. They first appears in sera at about age 30 to a near maximum at age 55 for women. Those surviving beyond age 55 have reduced arNOX activities compared to age 55. The arNOX activity of males seems to increase beyond age 55 to a maximum around age 65–70 and then may decline with increasing age (Fig. 9.10).

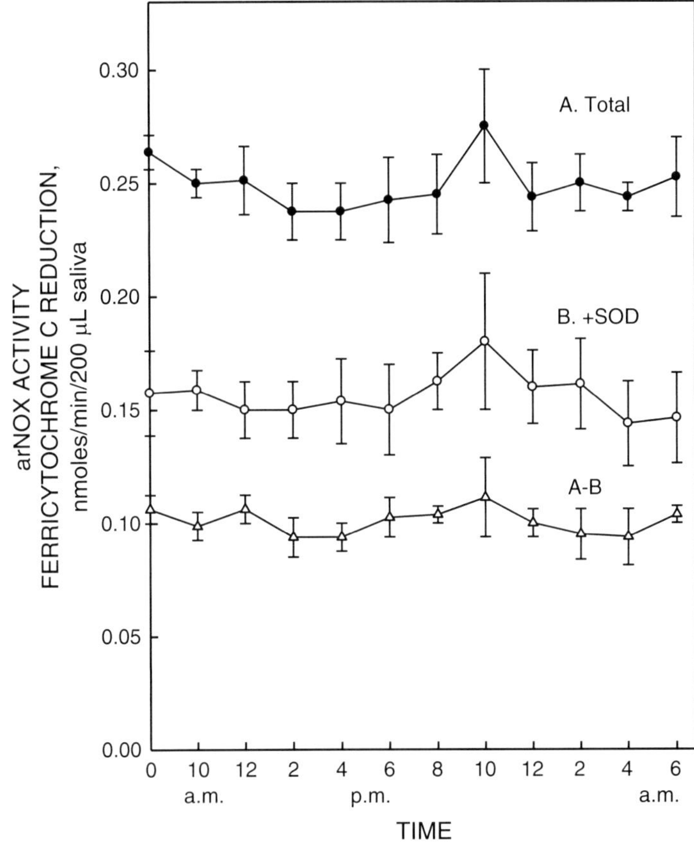

Fig. 9.8 Time course of ferricytochrome c reduction by saliva determined over 3 consecutive days and averaged. Seventy-two year male

Erythrocyte ghosts, buffy coat preparations, purified lymphocytes, and skin provide examples of the cell surface forms of the activity comparing old and young individuals. The ability to reduce ferricytochrome c for older individuals shows complete loss of activity upon the addition of SOD due to dismutation of superoxide. The correlation between patient age and arNOX of skin explants—epidermis and dermal punches—extrapolates to zero reduction of ferricytochrome c at about age 30 (Fig. 9.11). This activity was inhibited by SOD to verify superoxide production. Superoxide production in skin continued to increase at least until beyond age 70 years (Fig. 9.11) but in sera was maximal at about age 50 for females and at about age 65 for males.

The ECTO-NOX proteins have been postulated to link the accumulation of lesions in mitochondrial DNA to cell surface accumulations of reactive oxygen species as one consequence of their role as terminal oxidases in a plasma membrane electron transport (PMET) chain (Morré et al. 2000b; de Grey 1999;

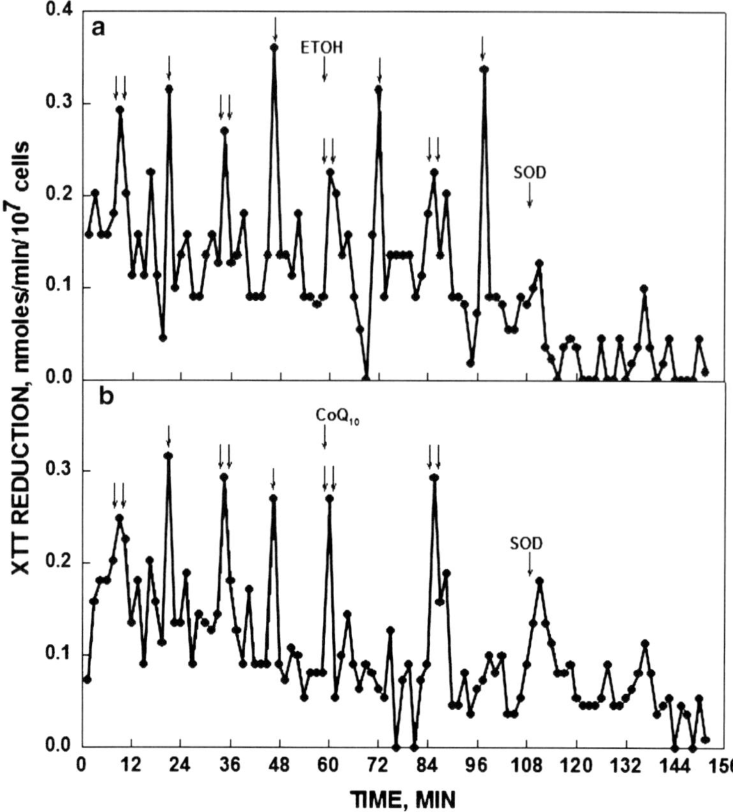

Fig. 9.9 As for Fig. 9.7 except rate of XTT reduction. Ethanol (30 μL) (**a**) or coenzyme Q_{10} (CoQ_{10}) (450 μg in 30 μL ethanol) (**b**) were added at 60 min. SOD (60 units in 60 μL) was added at 102 min (**a**, **b**) for a total of 150 min. Activity maxima separated by 26 min are indicated by the *arrows*. The activity maxima shown at *single arrows* were eliminated by CoQ_{10} whereas both activity maxima (*double* and *single arrows*) were eliminated by SOD. Following SOD addition, a minor set of oscillations with a 24-min period length remained

Fig. 9.12). Cells with functionally deficient mitochondria become characterized by an anaerobic metabolism. NADH accumulates from the glycolytic production of ATP such that an elevated PMET activity that regenerates NAD^+ is able to maintain the NAD^+/NADH homeostasis essential for survival. Hyperactivity of the plasma membrane arNOX not only restores NAD^+ but also generates superoxide at the cell surface (Fig. 9.1) (Morré et al. 1999c). The generated superoxide then serves to both propagate the aging cascade to adjacent cells and to oxidize circulating lipoproteins (Fig. 9.12). However, the overproduced arNOX activity released from cells into blood and other body fluids would extend the potential damage through action-at-a-distance including the oxidation of circulating serum lipoproteins as well as collagen and other proteins of the skin matrix.

Table 9.2 arNOX activities and inhibition by SOD

	Ferricytochrome c reductions[a]	
Human	20-40 years	50–70 years
	nmol/min/10^7 cells	
Buffy coats	0.02	0.06
+ SOD	0.02	0.02
	nmol/min/mL	
Sera	0.1	0.5
+SOD	0.1	0.1
Saliva	0.15	0.5
+SOD	0.1	0.1
	nmol/min/Osteopatch	
Perspiration	0.03	0.3
+SOD	0.03	0.03
	nmol/min/mg	
Skin	0.02	0.15
+SOD	0.01	0.01

[a]Values represent consensus means and standard deviations collected over several years of investigation. N=>20 subjects in each category

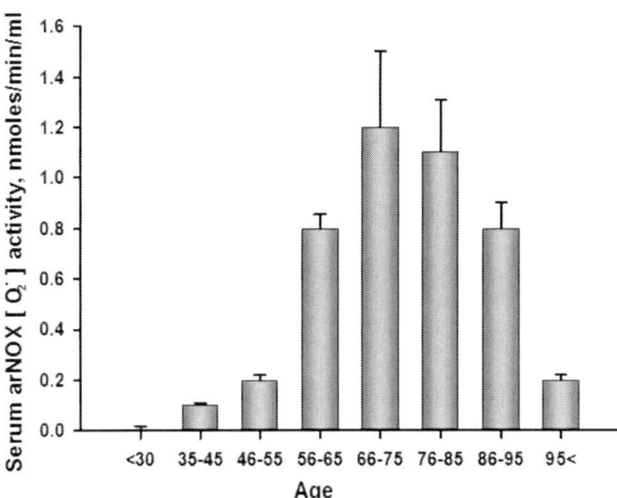

Fig. 9.10 Serum arNOX correlates with age to a maximum between 66 and 75 years. Individuals reaching age 85 and beyond have reduced levels of serum arNOX

Additional key identifying characteristics of the ECTO-NOX proteins as a family of functionally related cell surface proteins exhibited by the circulating N-terminal arNOX fragments are protease resistance (e.g., resistance to proteinase K digestions; Table 9.3), resistance to heating to temperatures between 70 and 80 °C, and an oscillating activity with a period length of 26 min independent of temperature

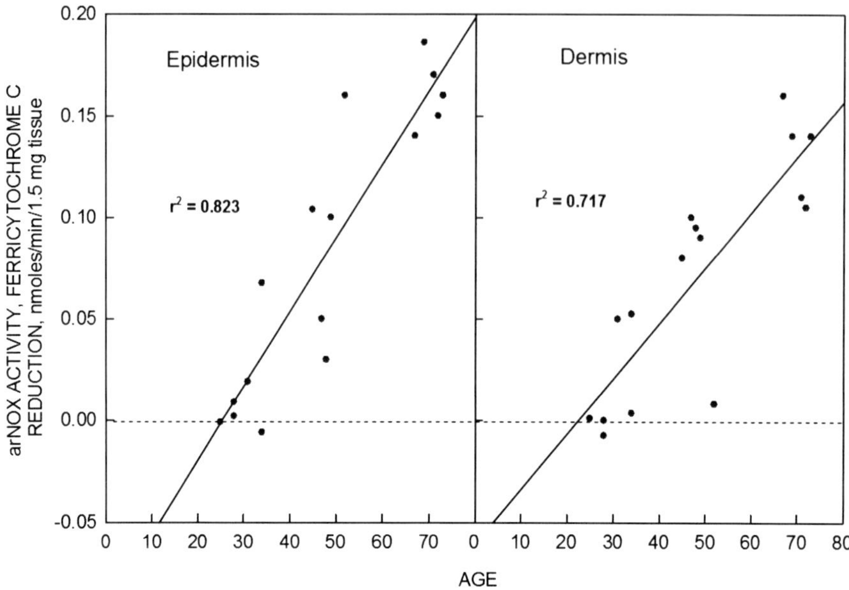

Fig. 9.11 arNOX activity of epidermal and dermal explants of skin and subject age are correlated. arNOX activity increases with age beginning about age 30. The subjects were 16 females. Values of activities susceptible to inhibition by SOD did not become measurable until after age 30. Redrawn from Morré et al. (2009c)

(Morré et al. 2003a). Age-related NOX activities have been purified and the enzymatic activity of the purified proteins also exhibits these characteristics.

One source of electrons for the circulating arNOX recognized early was protein thiols .(Morré et al. 2003a), Not only did the shed arNOX proteins generate superoxide but also oxidized proteins directly. Another important arNOX target for oxidation of proteins appears to be tyrosine residues (Morré et al. 2009a, b) to form tyrosyl radicals (van der Vlies et al. 2001). Protein tyrosyl radicals further result in intramolecular of intermolecular O,O′-dityrosine bonds (Aeschbach et al. 1976), which increase with aging (Leeuwenburgh et al. 1997; Wells-Knecht et al. 1993).

9.4 arNOX Cloning

Since arNOX on cell surfaces and in body fluids occurred together with ENOX1, discernment of characteristics attributable exclusively to arNOX required availability of recombinant arNOX proteins.

The age-related NADH oxidase of cells and body fluids was difficult to clone in order to determine its genetic origins. In retrospect, the fact that human arNOX is not

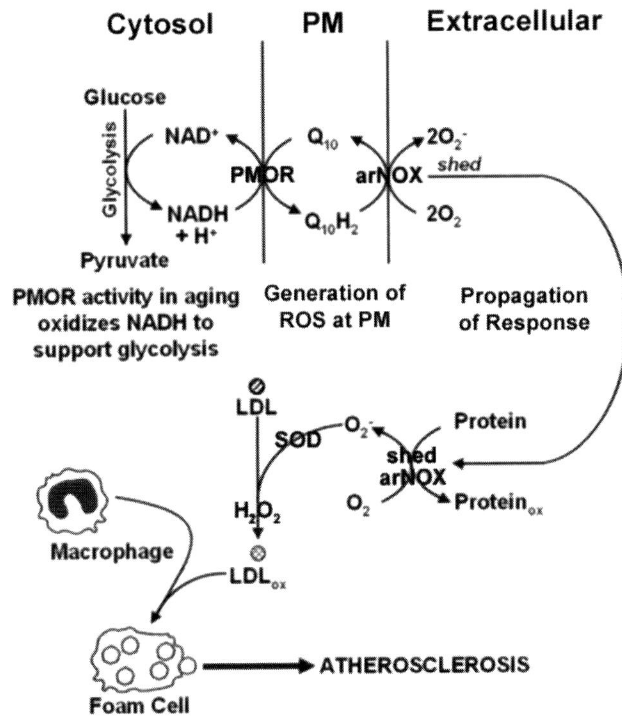

Fig. 9.12 Hypothesis to explain the mechanism whereby anaerobiosis resulting from mitochondrial lesions, the resultant stimulation of glycolysis, and the enhancement of the plasma membrane oxidoreductase (PMOR) system result in the formation of reactive oxygen species (ROS) at the cell surface that can be propagated and affect both adjacent cells and circulating blood components. *LDL* low-density lipoprotein. Based on Morré et al. (2000b)

Table 9.3 Response of rate of reduction of ferricytochrome c of pooled serum samples to proteinase K digestion

	Rate of reduction of ferricytochrome c (nmol/m/mL serum)		
Group	*N*	No addition	+ Proteinase K
<35 years female	12	0.24±0.14	0.26±0.17
35–45 males	9	0.21±0.15	0.27±0.16
75–85 males	10	0.7±0.35	0.72±0.14
75–85 females	8	1.1±0.20	1.0±0.18
>90 females	10	0.8±0.11	0.94±0.24

N=Number of subjects represented in each pooled sample

a single protein but a complex of at least five related proteins with diverse N-terminal sequences contributed to the difficulty along with their derivation from a membrane-associated precursor having nine transmembrane regions (Fig. 9.13). As is characteristic of ENOX generally, isolated proteins with arNOX activity tend to form aggregates when concentrated and are blocked to N-terminal sequencing.

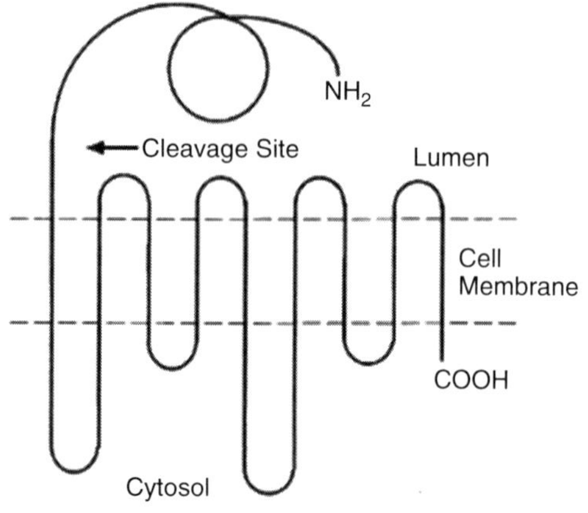

Fig. 9.13 Diagrammatic representation of the membrane association of the TM9 superfamily members. An N-terminal fragment is cleaved and released into the extracellular milieu surrounding the cells

As a cloning strategy, a soluble preparation of arNOX proteins based on activity measurements and SDS-PAGE (Fig. 9.14) was isolated to apparent homogeneity from human urine (Fig. 9.15a). When resolved by two-dimensional gel electrophoresis, the material concentrated from urine revealed only a small number of proteins of similar molecular weights (30–32 kDa) but differing acidic isoelectric points in the range of 4.2–4.8 (Fig. 9.15b). The arNOX activity based on reduction of ferricytochrome c of unfractionated urine exhibited a pattern of five different maxima, each recurring every 26 min (Fig. 9.14). The reduction of ferricytochrome c was inhibited by addition of SOD. When subjected to analysis by N-terminal sequencing, no sequence was obtained suggesting that the fragments were N-terminally blocked as is characteristic of ENOX proteins in general. Both polyclonal and monoclonal antisera were raised to the fragments isolated from urine. Both antisera sources, which reacted strongly with the arNOX fragments used as immunogen, failed in repeated attempts to result in expression cloning of the arNOX gene either from a library of arNOX expressing WI-38 cells or from lymphocytes of a 72-year male volunteer. When applied to western blots of arNOX-producing wild-type yeast (*Saccharomyces cerevisiae*), a number of proteins reacted with the antisera to human arNOX (Fig. 9.16). One of the spots corresponding to a protein of apparent molecular weight 24 kDa, isoelectric point pH 4.9, generated N-terminal sequence HN/AV/LVLPAYI/TR corresponding to yeast p24 (Singer-Krüger et al. 1993) which is derived from a ca 70-kDa precursor shown previously to be a member of the TM9 superfamily of transmembrane proteins (Chluba-de Tapia et al. 1997; Schimmöller et al. 1998).

The protein family has a relatively divergent, hydrophilic N-terminal domain and a well-conserved, highly hydrophobic C-terminal domain which contains nine potential membrane-spanning domains. The identification of the immunoreactive yeast protein permitted the cloning and sequencing of the corresponding human cDNA (X. Tang, C. Meadows, C.Phung, D. M. Morré and D. J. Morré, unpublished). The human arNOX cDNAs encode a family of proteins designated "TM9SF" (transmembrane protein 9 super family) by the Human Gene Nomenclature Committee. The lead member of the TM9SF family is the *S. cerevisiae* EMP70 gene

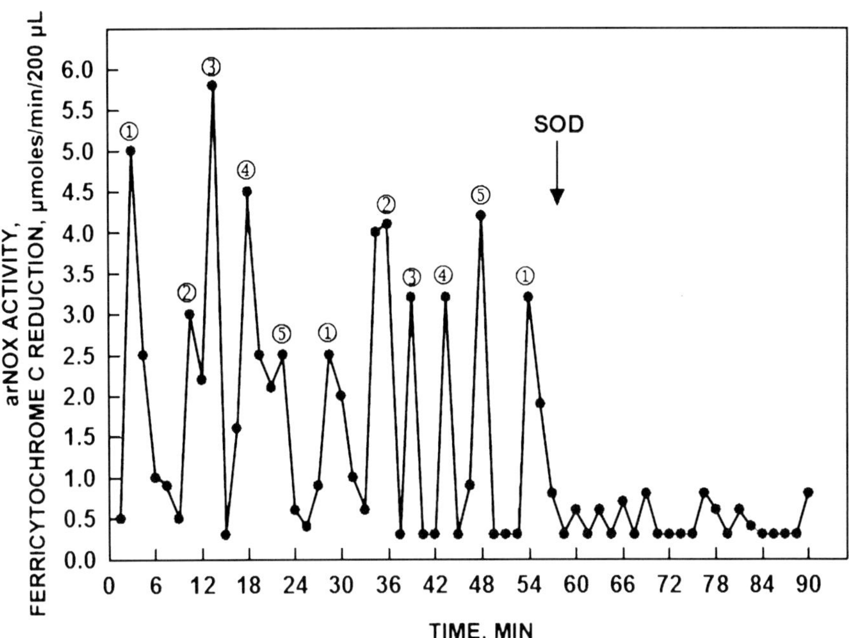

Fig. 9.14 arNOX activity of human urine utilized in the arNOX purification resulting in the data of Fig. 9.15. The repeating pattern is that of five oscillating maxima that recur at intervals of 26 min. SOD = 60 units of SOD added after 60 min

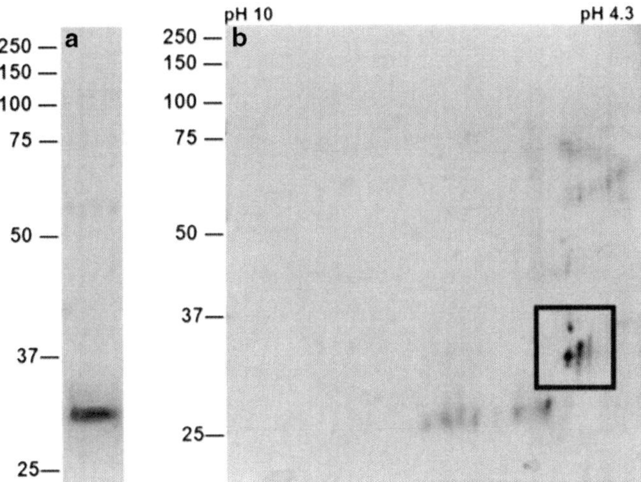

Fig. 9.15 SDS-PAGE of arNOX purified from human urine (72 year male). (**a**) The protein was purified to apparent homogeneity by SDS-PAGE. (**b**) Two-dimensional gel separation of preparation of (**a**) silver stained. Two sets of proteins are evident, one at about 24 kDa and one at about 30 kDa but with different isoelectric points

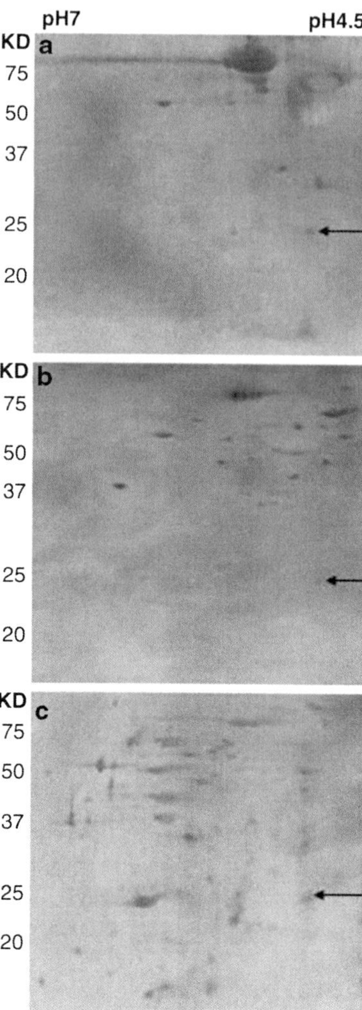

Fig. 9.16 Western blots (**a**, **b**) and coomassie blue (**c**) of two-dimensional gels of wild-type yeast homogenate soluble fraction with polyclonal rabbit antisera generated to the arNOX proteins purified from human urine or a monoclonal antibody generated from spleen cells of the same rabbit (**b**). (**a**) Reactive protein at 24 kDa, isoelectric point pH 4.9 (*arrow*) generated N-terminal sequence. Other reactive proteins reactive with the antisera including the major component at ca.72 kDa, average isoelectric point pH 5.4 failed to yield N-terminal sequence

product, a 70-kDa precursor that is processed into a 24-kDa protein (p24a) located in the endosomes (Singer-Krüger et al. 1993). To date, five subtypes of human TM9SF proteins have been distinguished, i.e., TM9SF-1 (hMP70; Chluba-de Tapia et al. 1997), TM9SF-1b (He et al. 2009), TM9SF-2 (p76; Schimmöller et al. 1998), TM9SF-3, and D87444, which exhibit 30–40 % amino acid sequence identity to each other and with the yeast p24a precursor (Sugasawa et al. 2001).

Based on homology with the yeast arNOX proteins, we then identified the five circulating human arNOX isoforms as exfoliated N-termini of the five members of the TM9 family of transmembrane proteins (GenBank NM-006405, NM-001014841, NM-004800, NM-020123, NM-014742) (X. Tang, D. Parisi, D. M. Morré and D. J. Morré, unpublished). The identification was based on cloned TM-9SF4 and TM9SF2, two of the family members and their expression in bacteria. The three additional superfamily members were identified based on the inhibition by specific peptide antisera (Fig. 9.4).

The human homolog of the *S. cerevisiae* EMP70 gene product, a precursor protein whose 24 kDa cleavage product (p24a) was found in yeast endosome-enriched membrane fractions (Singer-Krüger et al. 1993) is a 76-kDa membrane protein also found in endosomes (Schimmöller et al. 1998). Northern blot analysis indicated that p76 mRNA could be detected in all tissues examined and is conserved throughout evolution.

The EMP70 gene was cloned initially based on the N-terminal sequence information obtained by microsequencing the 24-kDa protein (Singer-Krüger et al. 1993; Genembl database entry X67316). Sequencing of the *S. cerevisiae* genome revealed that the EMP70 gene is located on chromosome XII (GenBank accession number U53880). The p76 cDNA encodes a protein of 663 amino acids with a predicted mass of 76 kDa (GenBank accession number U81006).

P76 adopts a type-I topology within the membrane, with its hydrophilic N-terminus facing the lumen of cytoplasmic membranes (Schimmöller et al. 1998). Proteolytic cleavage seems to depend on a Kex2p protease (Singer-Krüger et al. 1993) that recognizes dibasic residues (Julius et al. 1984; Fuller et al. 1989). A KR motif in the N-terminal domain of the Emp70p precursor at amino acid 249/250 (PVSIKRSSP) was suggested to represent an appropriate cleavage site to generate p24a. The TM9 proteins are unique and have no homology with G-protein-coupled receptors or with other families of transmembrane proteins (Diaz et al. 1997). Cell-associated arNOX is membrane anchored with its catalytic terminus directed toward the cells' exterior. A ca 30-kDa fragment normally internalized into endosomes (Schimmöller et al. 1998) also is shed and enters the blood and other body fluids. The soluble fragments generate superoxide and carry out all of the oxidative functions associated with the cell surface form.

Hydropathy analysis of p76 (Kyte and Doolittle 1982) and its close relatives revealed that they share a unique membrane-binding domain (Schimmöller et al. 1998; Fig. 9.17). They also contain a short N-terminal hydrophobic extension characteristic of a signal sequence, followed by a mostly hydrophilic, amino terminal portion that extends up to amino acid residue 300 in certain family members. The remaining portions of these proteins are extremely hydrophobic and contain nine potential transmembrane domains to make them integral membrane proteins that adopt a type I topology. Polypeptide translocation would be initiated via their N-terminal hydrophobic signal sequence and they would ultimately be anchored in the membrane via stop-transfer sequences.

At the protein level, p76 and the p24a protein precursor (Emp70) share 35 % identity in amino acid sequence. Strikingly, the highest level of sequence identity is localized to the C-terminal ~60 % of these proteins; in contrast, the N-terminal domains show much greater diversity. Another human homolog (GenBank accession D87444) referred to as human EMP70p to distinguish it from p76 had a predicted mass of 72 kDa.

Yeast (*S. cerevisiae*) was found to express arNOX activity and provided a second opportunity to identify a gene product associated with arNOX activity. The arNOX

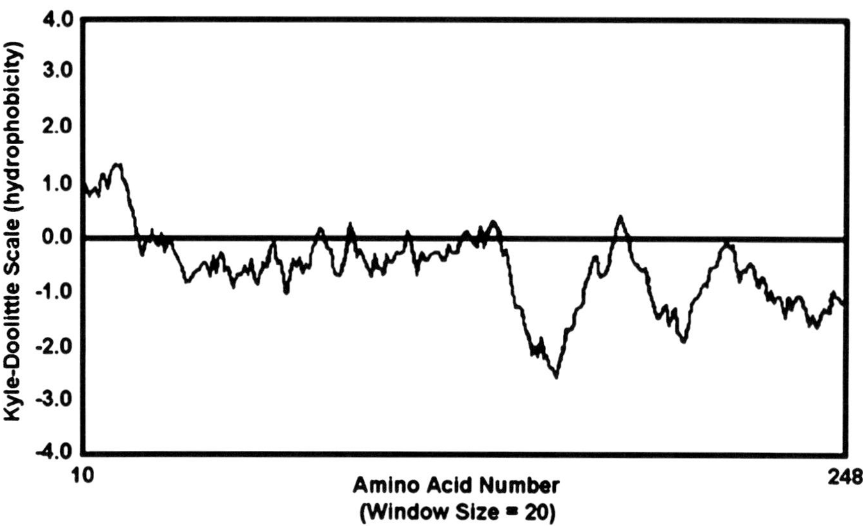

Fig. 9.17 Protein hydrophobicity plot (1–258aa) of arNOX according to the method of Kyte and Doolittle (1982)

Table 9.4 Properties of yeast arNOX

Periodic generation of superoxide based on reduction of ferricytochrome c
Period length of 26 min
Single burst corresponding to maximum ③ of the standard ENOX cycle
Inhibited by addition of SOD
Activity resistant to ENOX1 inhibitor simalikalactone D and ENOX2 inhibitors capsaicin and epigallocatechin gallate
Activity inhibited by arNOX inhibitor containing dormin + *Schizandra* + salicin (Table 9.7)
Oscillatory oxidation of NADH and reduced CoQ_{10}
26-min period
Asymmetric 2 + 3 pattern of five maxima, two of which are separated by 6 min and 3 of which are separated from each other and from the first two by 4.5 min ($6+4\times4.5=26$ min)
Oscillatory protein disulfide-thiol interchange activity measured by cleavage of dithiodipyridine also with a 26-min period length

gene was identified from yeast based on screening of a series of yeast deletion mutations (S. Dick, D. M. Morré and D. J. Morré, unpublished). The respective deletion was traced to gene YER113C and the corresponding protein was then characterized from a yeast overexpression library. The properties of the overexpressed YER113C presented in Table 9.4 were those consistent with that of an arNOX protein including a period length of 26 min, inhibition by SOD to confirm superoxide as the reaction product and inhibition by established arNOX inhibitors.

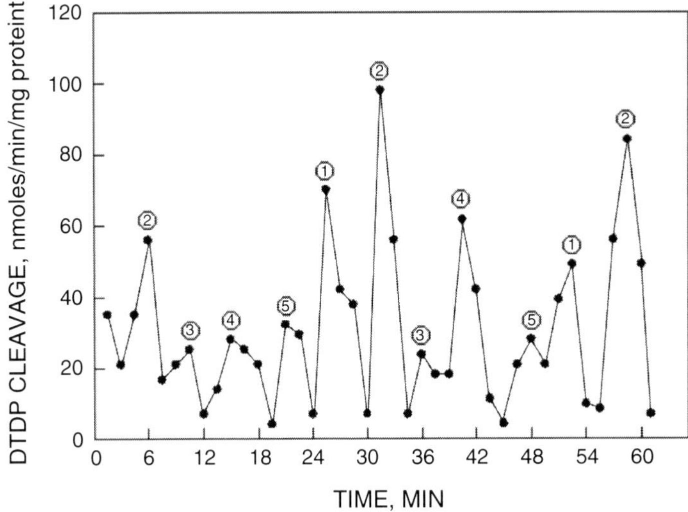

Fig. 9.18 Protein disulfide-thiol interchange activity of TM9SF2 peptide of scrambled and inactive RNase to cleave cCMP determined by cleavage of a dithiodipyridine (DTDP) substrate. The activity exhibited an oscillatory pattern

9.5 Characterization of Recombinant arNOX Proteins

To determine if the human arNOX proteins might similarly derive from TM9 superfamily members, recombinant human TM9SF4 was generated from cDNA corresponding to full length TM9SF4. TM9SF4 is the human homolog to the 70-kDa precursor to p24 from yeast. We were unable to express full-length TM9SF4 in *E. coli*. In fact, the bacteria transfected with cDNA to full-length TM9SF4 grew at only 20–30 % of the rate of nontransfected bacteria. Therefore, as we were interested primarily in the exfoliated form of the protein, new cDNAs corresponding to both ca 30 and 15 kDa N-termini of TM9SF4 and TM9SF2 were prepared and expressed in bacteria. All four of the truncated forms exhibited a marked oscillating activity characteristic of other ENOX proteins family members (Figs. 9.4–9.9).

The oxidation of NADH or hydroquinone (ubiquinol) by the recombinant arNOX proteins in contrast to that of ferricytochrome c reduction exhibited a typical ENOX five peak pattern of oscillations (Figs. 9.3 and 9.4). Two of the maxima were separated by 6 min and the remaining maxima were separated by about 5 min to generate the 26-min period [6 min + (4 × 5) min = 26 min]. The second activity exhibited by ENOX proteins, that of protein disulfide-thiol interchange, was also given by recombinant arNOX and with the same 26-min period length as for NADH or ubiquinol oxidation (Fig. 9.18).

The reduction of ferricytochrome c by sera could not be attributed to NADH-cytochrome c reductase. The activity was inhibited by thiol reagents such as ρ-chloromercuribenzoate (Table 9.1) making protein thiols a likely source of reducing equivalents for the soluble fragments equivalent to the exfoliated forms.

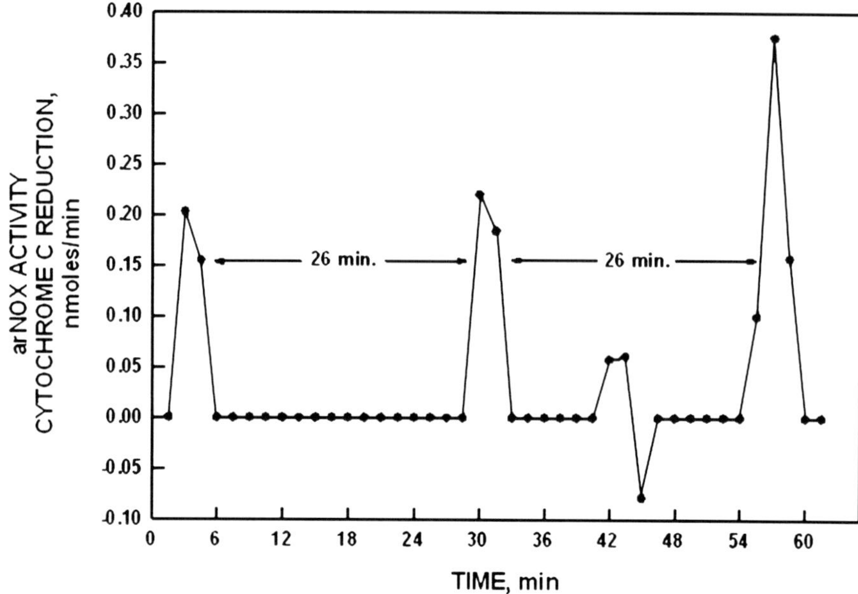

Fig. 9.19 arNOX activity of the soluble fraction of a crude bacterial lysate containing recombinant TM9SF4 showed only bursts of superoxide generation as measured by reduction of ferricytochrome c. The bursts were separated by intervals of 26 min

Alignment of the amino acid sequence of the N-terminal fragment human TM9SF4 and that of the yeast p24 revealed 50 % similarity between D222 and H255 apparently sufficient to allow for antibody cross reactivity.

Crude lysates of bacteria expressing truncated TM9SF4 exhibited single maxima of arNOX activity based on reduction of ferricytochrome c separated at intervals of approximately 26 min (Fig. 9.19) which is characteristic of the arNOX activities of the proteins purified from urine except that, with urine, there were at least five such maxima repeating every 26 min with a burst of superoxide suggestive of not one but as many of five arNOX proteins being present, each contributing to a single superoxide burst every 26 min.

The activity pattern exhibited by the expressed truncated TM9SF4 proteins purified by isoelectric focusing exhibited five activity maxima within a time span of 26 min based on NADH oxidation where two of the maxima were separated by about 6 min and the remaining three maxima were separated from each other and from the two maxima separated by 6 min by intervals of about 5 min to generate the pattern shown in Fig. 9.20a. What is most characteristic of the arNOX activity pattern is that the activity maximum labeled ③ exhibited a burst of superoxide production two to four times that of the other maxima (Fig. 9.20b) and is the only evidence of activity observed in diluted preparations of the protein (Fig. 9.19). The recombinant TM9SF4 oxidized reduced coenzyme Q in a standard assay (Fig. 9.21) with activity measured either at A_{410} (Fig. 9.21a) or at A_{290} (Fig. 9.21b).

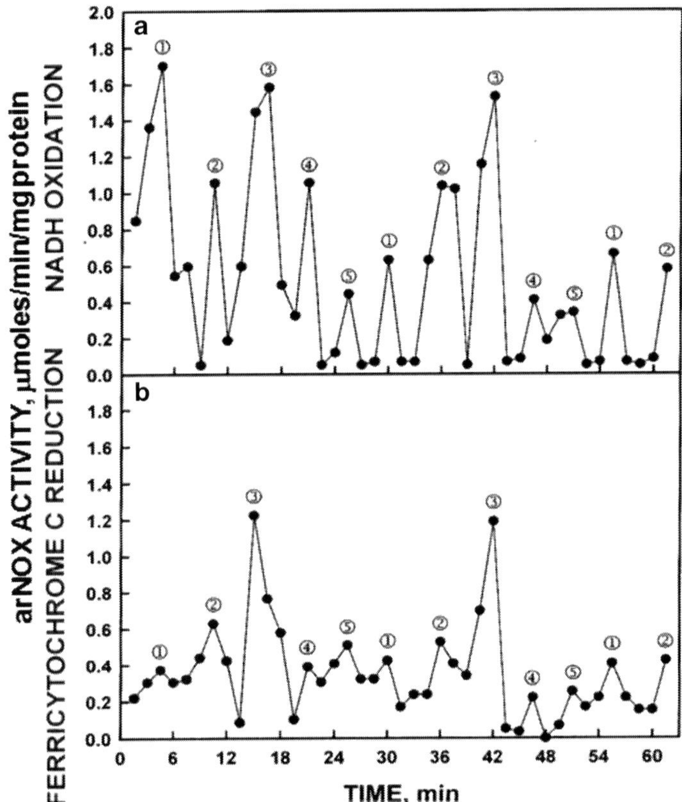

Fig. 9.20 arNOX activity of recombinant TM9SF4 purified by isoelectric focusing showing the typical 5-peak pattern of activity characteristic of ENOX proteins in general. (**a**) NADH oxidation. (**b**) Superoxide generation. Superoxide generation is intensified with maximum 3 of the 5 maxima oscillatory pattern in keeping with results of Fig. 9.19

Results with recombinant TM9SF2 protein expressed as N-terminal (ca. 15 or 30 kDa) fragments were nearly identical to those with TM9SF4 (X. Tang, C. Meadows, D. M. Morré and D. J. Morré, unpublished).

Peptide antibodies to each of the isoforms along with corresponding DNA probes to each of the isoforms were developed for the soluble forms of each of the five isoforms (Fig. 9.4). The antibodies were used to systematically identify the five isoforms in human sera and saliva and to verify that they corresponded to the known sequences of the TM9 superfamily members. Additionally, DNA sequence information was used to generate RT-PCR probes for each of the isoforms and demonstrate their expression in both human lymphocytes and human skin explants (not shown). These data confirm the TM9 superfamily of proteins as the genetic origins of the five known arNOX isoforms of human sera, plasma, saliva, and other body fluids.

Full-length members of the TM9 protein superfamily are all characterized as cell surface proteins having a characteristic series of nine membrane-spanning hydrophobic

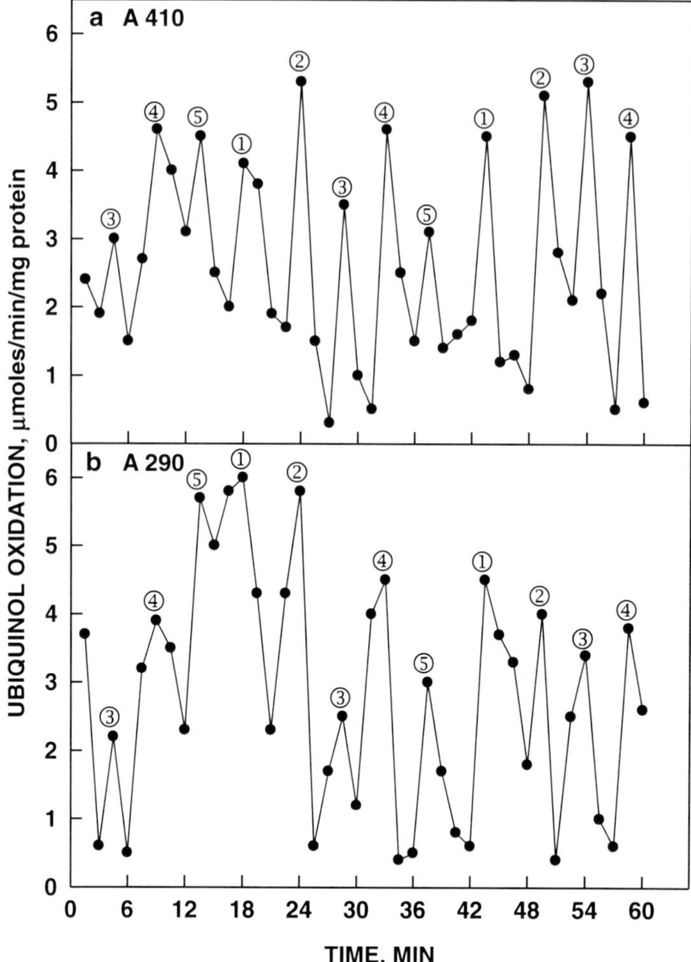

Fig. 9.21 Ability of recombinant TM9SF4 purified by isoelectric focusing to oxidize hydroquinone (reduced CoQ_{10}) either by an increase in A_{410} (**a**) or by a decrease in A_{290} (**b**). As with NADH oxidation of Fig. 9.20**a**, the activity oscillates with prominent maxima separated by 6 min plus three additional maxima ③, ④, and ⑤ separated by 5 min to create a 26-min period created by the five recurring maxima

helices that criss-cross the plasma membrane and also are present on endosomes (5). The transmembrane regions are highly conserved and similar or identical in each of the five forms. There are five such family members known (1 with two transcript variants, 2, 3, and 4). The two transcript variants of family member 1 are very similar with the exception that member 1a transcript variant contains additional C-terminal residues absent from transcription variant 1b. The different family members (1–4) are encoded by different genes and are therefore not splice variants from a single gene as for the different transcript variants of ENOX2 (Tang et al. 2007).

Table 9.5 Putative functional motifs of human TM9 superfamily members

Motif	SF1a/SF1b	SF2	SF3	SF4
Adenine nucleotide binding site	G28XGXXG	G97XVXXG	G81XAXXG	G77XVXXG
Putative protein disulfide interchange site	C107XXXL	C120XXXC	C108XXXL	C100XXXC
Putative copper sites	H60YY	Y150QH	H33TY	Y130H
	H153SH	H242TH	H75YH	H165GY
Conserved CQ/CE	C12Q	C84Q	C108E	C64Q

Table 9.6 Confirmation of functional motifs of arNOX TM9SF2 by site-directed mutagenesis

Mutation	NADH oxidation[a]	Superoxide generation[a]	Disulfide-thiol interchange[a]	Period length (min)
Wild type	2.56	0.995	2.92	26
C84A	2.46	1.1	2.5	22
Q85A	2.89	1.21	2.77	26
G97V	–	1.65	2.96	26
C120A	1.8	0.816	1.02	26
H152A	1.59	0.62	3.62	26
H242A	–	–	1.96	n/a

[a] n mole/min/mg protein

Each of the shed forms contains putative functional motifs required of an ENOX protein (Fig. 9.2, Table 9.5). The functional motifs are located in the shed forms of the TM9 proteins but despite the presence of recognizable functional motifs in each of the superfamily members, sequence identity among the shed N-terminal fragments with arNOX activity of different family members was minimal (Table 9.5). The correctness of motif assignments has been confirmed for TM9SF2 by site-directed mutagenesis (Table 9.6).

When protein disulfide-thiol interchange activity was measured, the second enzymatic activity normally associated with ENOX proteins was observed consisting of a five-maxima pattern similar to that generated for NADH oxidation (Fig. 9.22). All three activities were resistant to a specific inhibitor of ENOX1, simalikalactone D, and inhibitors of ENOX2, phenoxodiol, and capsaicin. Nor were the arNOX activity patterns phased by addition of melatonin as is characteristic of ENOX1 (Jiang et al. 2008). However, the period length was increased from 26 min to about 30 min by assay in D_2O in place of water (Fig. 9.23), a property typical of ENOX proteins generally. Both a specific arNOX inhibitor mixture of dormin + *Schizandra* + salicin to 2.5 mL of assay volume were added 60 μL of an aqueous mixture of 4 mg/mL *Schizandra* (*Schizandra chinensis*) extract, 9 % schizandrins, Draco, San Jose, CA) plus 1 mg/mL salicin (Sigma, St. Louis, MO) and 20 μL of IBR Dormin (Israli Biotechnology Research, Ramat-Gan, Israel) = AgeLoc (NuSkin Enterprises, Provo, UT) and an equal mixture of peptide antibodies representing each of the five arNOX superfamily members at a titer of 1:10,000 inhibited the activity of TM9SF4 and TM9SF2 by >90 %. Also inhibitory were gallic acid (EC_{50} = 2 μM), tyrosol (EC_{50} = 1 μM), and CoQ_{10} (EC_{50} = 80 nM). Melatonin

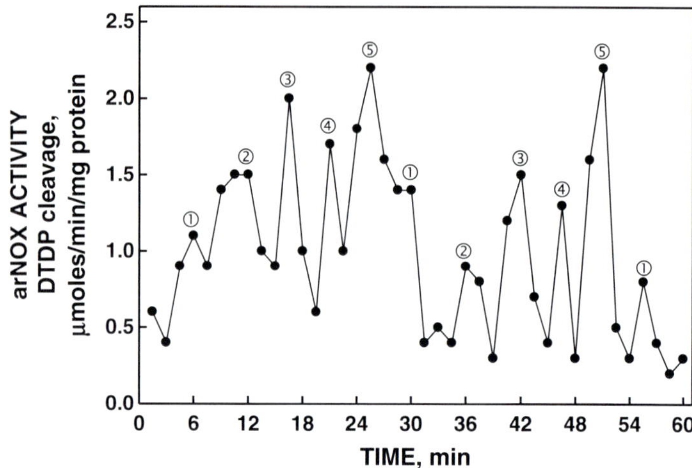

Fig. 9.22 Protein disulfide-thiol interchange activity of recombinant TM9SF4 measured from the cleavage of DTDP substrate

(100 µM) did not shift the phase of the period (not shown). However, after 1.5 periods, 39 min after melatonin addition, a maximum was missing.

To determine if the TM9SF forms required copper for activity, recombinant TM9SF4 was assayed in the presence of trifluoroacetic acid (to unfold the protein) with or without the copper chelator, bathocuproine, to remove the copper. The activity, measured either based on NADH oxidation or superoxide production, was diminished when bathocuproine was present. TFA alone did not reduce activity and activity could be restored to TFA and bathocuproine treated preparations by additions of copper.

Prior to our analysis of the expressed recombinant arNOX, the pattern of oscillations in ferricytochrome c reduction by human samples of arNOX was poorly understood. Patterns observed consisted of 4–7 maxima with the average being about five. The origins of the maxima, spaced at regular intervals of 26 min, based on the data of Fig. 9.3, correctly attributed to separate TM9SF family members each with one burst of superoxide production per 26-min activity cycle. The conclusion that ferricytochrome c reduction by body fluids of aged individuals was the combined result of at least five distinct proteins oscillating independently of each other was unexpected.

9.6 arNOX as a Biomarker of Aging

Since arNOX activity affects physiologic functions related to the pathobiology of aging, it emerges as a relevant biomarker of aging. As a potentially important parameter of antioxidant defenses and oxidative stress, measurements of arNOX may be used to determine whether, for example, an intervention achieves its intended biochemical or physiological end point or whether the enrolled subjects in a clinical trial present with the biomarker elevated at the beginning of the trial. Thus, a reliable

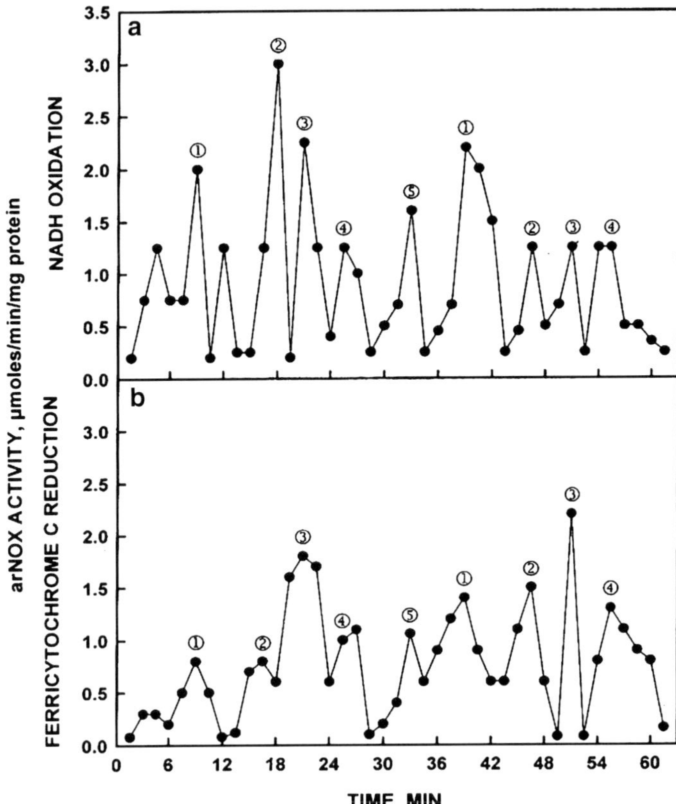

Fig. 9.23 As in Fig. 9.20a except with water replaced by D_2O in the assay. As with other ENOX proteins, assay in the presence of heavy water increased the period length by about 25 % from 26 to 30 min

biomarker correlated with aging such as arNOX has the potential to give guidance to the design and outcome measures of clinical trials with age-related disorders. To be truly useful the biomarker must have some degree of predictive validity and utility to determine which individuals are most likely to respond to a particular therapeutic or preventive intervention (Blumberg 2004). Also the biomarker should reflect both long- and short-term exposure to a particular intervention. arNOX is such a biomarker. In contrast, while DNA, lipid, and protein oxidation products provide an extensive array of alternative biomarkers for oxidative stress, most macromolecular changes of this nature appear to be secondary to the actual sources of oxidative damage particularly in aging and provide limited opportunities for intervention.

A number of observations point to arNOX as a primary generator of ROS in the body, especially with regard to oxidation of circulating lipoprotein particles. Mitochondria appear not to be a major source of cellular superoxide anion or hydrogen peroxide. Uninhibited respiring mitochondria produce very little of either (Nohl et al. 2001, 2005; St. Pierre et al. 2002). The actual amount of H_2O_2 produced by mitochondria appears to be in the order of 0.1 nmol formed/min/mg mitochondrial protein. In contrast, the bulk of the superoxide anion and H_2O_2 appears to be produced by the plasma membrane or exterior to the cells themselves.

arNOX proteins generate superoxide anion as measured by SOD-inhibited reduction of ferricytochrome c, a standard measure of superoxide anion production. At peak production which occurs on average at about age 65, approximately 0.06 nmol/min of superoxide is produced/10^7 cells (human buffy coat or epidermal epithelia). The value in sera is approximately 0.3 nmol/mL/min which results in several millimoles of superoxide being produced on a daily basis in the proximity of circulating lipoproteins just from the circulating form.

9.7 Role in Skin Aging

Accumulation of oxidative damage is considered a major contributor to age-related skin deterioration (Hensley et al. 1998; Smith et al. 1991, 1992; Stadtman et al. 1992; Leeuwenburgh et al. 1998). Although aging leads to the accumulation of mitochondrial DNA lesions, the sources of this oxidative damage and the manner by which it is directed to specific targets have been inadequately addressed since mitochondria appear not to be a major generator (Nohl et al. 2001, 2005; St. Pierre et al. 2002; Linnane et al. 2007). The bulk of the ROS actually may be produced at the cell surface or shed arNOX proteins capable of ROS generation exterior to the cells themselves (Morré et al. 2000b).

By oxidizing proteins of the supporting matrices that are important to skin health, elevated arNOX activity has been demonstrated to be a major cause of skin aging (Rehmus et al. 2008; Morré et al. 2010b; Table 9.4). To correlate mean error in estimating age with arNOX, studies were conducted with female subjects. Independent graders reviewed photographs of close-ups of the subjects faces taken at the baseline visit and estimated each subject's age. The graders estimated age and scored skin according to overall skin health, fine wrinkling, deep wrinkling, skin color, skin laxity, pore size, and evenness. Estimates of age were averaged and compared to the subject's actual age. Age differences and other skin assessment scores were compared to arNOX levels determined from collections of serum, saliva, and perspiration. Subjects with high arNOX activity had skin characteristics that made them appear on average 7 years older than their chronological age whereas subjects with low arNOX activity at the same age had on average 7 years younger appearing skin than their actual age (Fig. 9.24).

Correlative evidence that links arNOX activity levels to oxidative changes during aging includes advanced glycation endproduct (AGE) readings estimated using a Diagnoptics (San Diego, CA) fluorescence AGE reader (Reznik and Packer 1994; Terada et al. 2007). A light source excites fluorescent moieties in the tissue which then emit light of a different wavelength. In the wavelength band used, the major contribution to fluorescence comes from fluorescent AGEs linked to proteins, mostly to collagen.

Fig. 9.24 (a, b) Serum arNOX activity is correlated with mean error in estimating age. Sera were collected from 25 female subjects and compared to ages estimated from photographs taken at the same time

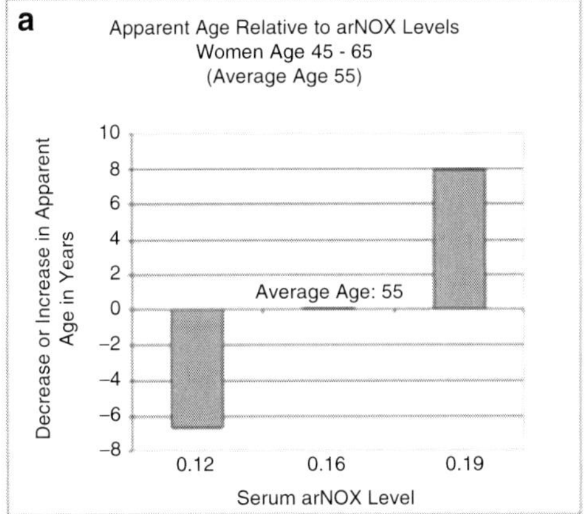

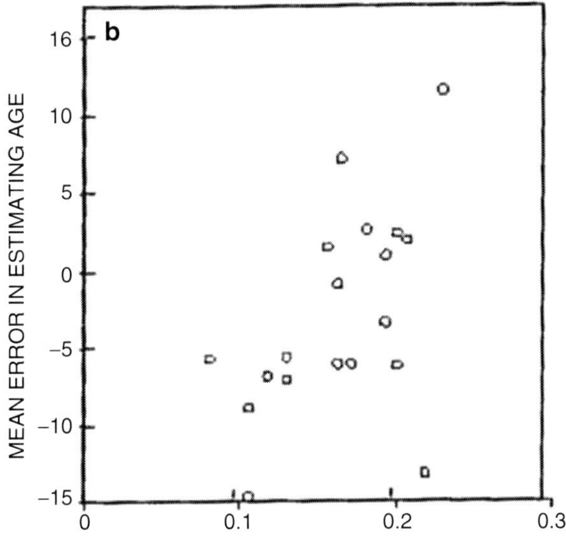

AGEs normally accumulate with aging. Figure 9.25 shows that AGEs of skin also correlate with arNOX activity. Formation of protein carbonyls in human collagen by arNOX purified from human urine was determined directly (Morré et al. 2008c) by enzyme immunoassay in a 96-well plate format following conversion to DNP hydrazone and probed with an anti-ANP antibody followed by HrP-conjugated second antibody (Cell Biolabs, San Diego, CA, OxiSelect Protein Carbonyl ELISA kit) (Reznik and Packer 1994).

arNOX-catalyzed oxidation of skin proteins based on oxidation of the amino acid tyrosine was demonstrated using the 96-well plate assay with Type I collagen, elastin, and bovine serum albumin (Table 9.7). The tyrosines, once oxidized, would normally

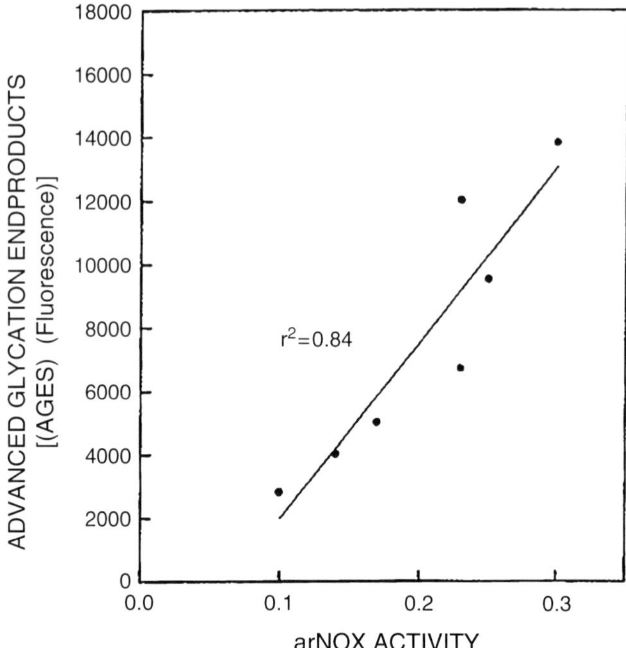

Fig. 9.25 Advanced glycation oxidation endproducts rise as arNOX activity of sera increases with increasing age

form dimers to cross link the proteins. In the assay, fluorescent tyramine formed fluorescent dimers with the oxidized tyrosines to provide a measure of tyrosine oxidation (van der Vlies et al. 2001). Fluorescent tyramine conjugation with tyrosyl radical was blocked by a cocktail of arNOX-specific inhibitors (footnote to Table 9.7) and to a mixture of peptide antibodies raised in rabbits to individual arNOX isoforms to demonstrate that the tyrosine oxidation was arNOX catalyzed. The arNOX source was concentrated from human urine and had a specific activity comparable to that measured in aged skin. Fluorescence increased with time for both collagen and elastin and was proportional to protein amount.

To demonstrate arNOX-catalyzed oxidation of skin proteins, experiments were conducted with both cultured primary keratinocytes and fibroblasts and frozen sections of human punch biopsies (Table 9.8). Epidermal keratinocytes (Fig. 9.26a) and fibroblasts (not shown) when reacted overnight with fluorescent tyramine to detect tyrosyl radical formation showed strong fluorescence that was blocked by a mixture of arNOX inhibitors (Fig. 9.26d; footnote to Table 9.7). This mixture of inhibitors provides the basis for NuSkin's ageLOC technology designed to achieve healthier skin by controlling reactive oxygen species at their source, i.e., superoxide produced by arNOX in body fluids and interstitial spaces (Kern et al. 2010). Neonatal keratinocytes were much less reactive and exhibited lower arNOX activities when assayed in parallel (Fig. 9.26b). Results with keratinocytes and fibroblasts were similar (not shown).

Table 9.7 In vitro oxidation of proteins by arNOX based on formation of fluorescent dimers by fluorescent tyramine with the oxidized proteins

	Treatment				
Target protein	arNOX	SOD	ageLoc ingredients[c]	Anti-arNOX Antibodies[d]	Relative fluorescence
None	–	–	–	–	150 ± 28
	+	+	–	–	50 ± 56
BSA	–	–	–	–	320 ± 34
	+	–	–	–	310 ± 43
Collagen[a]	–	–	–	–	664 ± 42
	+	–	–	–	2,208 ± 483
	+	+	–	–	1,321 ± 531
	+	–	+	–	330 ± 203
	+	–	–	+	413 ± 30
Elastin[b]	–	–	–	–	30 ± 24
	+	–	–	–	2,425 ± 591
	+	+	–	–	1,413 ± 1,138
	+	–	+	–	727 ± 378
	+	–	–	+	725 ± 90

[a]Source of collagen was Calbiochem Cat. No. 234138 (collagen type-human skin)
[b]Source of elastin was Sigma-Aldrich Cat. No. E7402-2MG, Lot 028 K1161 (elastin from human skin)
[c]To 2.5 mL of assay volume were added 60 μL of an aqueous mixture of 4 mg/mL *Schizandra* (*Shizandra chinensis*) extract, 9 % Schizandrins (Draco, San Jose, CA) plus 1 mg/mL salicin (Sigma, St. Louis, MO) and 20 μL of IBR Dormin (Israeli Biotechnology Research, Ramat-Gan, Israel)
[d]In equal mixture of peptide antibodies raised in rabbits to each of the five arNOX isoforms

Table 9.8 arNOX activity of frozen sections of skin punch biopsies from 16 subjects, 4 males and 4 females in each of two age categories, young and older compared to AGE Index and arNOX activities of saliva and sera of the same subjects

Chronological		Age	arNOX Activity[b]		
Age (years)	*N*	Index[a]	Saliva	Serum	Tissue
24 ± 3 (20–30)	8	1.4 ± 0.4	0.003 ± 0.013	0.003 ± 0.009	0.004 ± 0.014
55 ± 3 (51–59)	8	2.2 ± 0.8	0.122 ± 0.079[c]	0.124 ± 0.085[c]	0.118 ± 0.030[c]

[a]Arbitrary units (see text)
[b]Units of specific activity are nmol/min/three frozen sections for tissue and nmol/min/200 μL for saliva and sera
[c]Statistically significant ($p < 0.001$)

A group of 16 subjects, 8 females and 8 males, next were evaluated for fluorescent tyramine conjugation and arNOX activity with frozen sections of skin biopsy material and for arNOX activity of saliva and serum as well as for epidermal AGE readings from the same subjects (Table 9.8). The subjects were in two age groups, young, age 24 ± 3 (20–30 years) and older, age 55 ± 3 (51–59 years). The biopsy specimens were snap frozen and histological sections of the frozen tissue were examined for

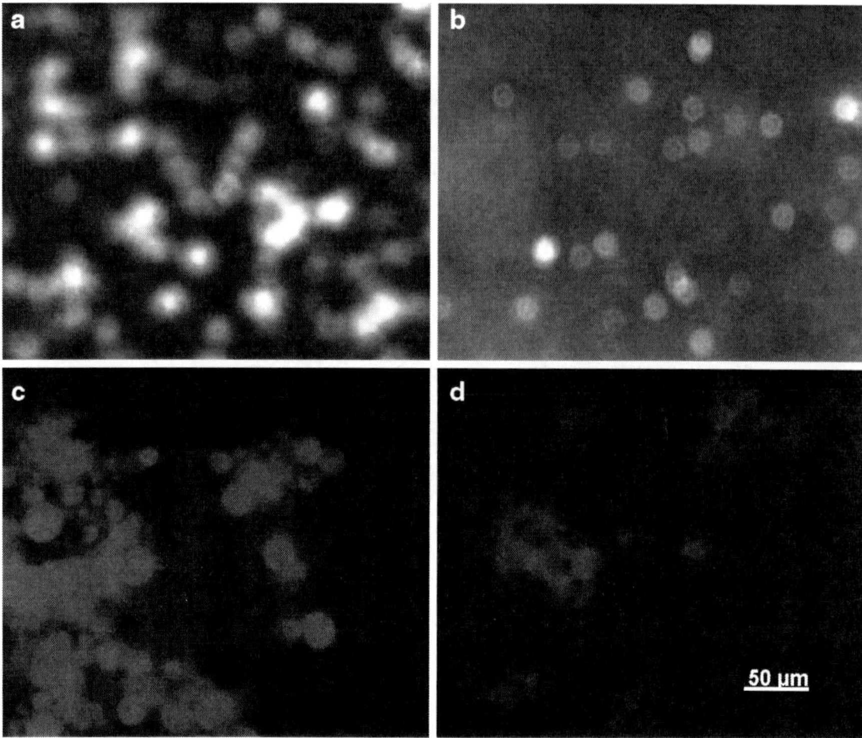

Fig. 9.26 Epidermal keratinocytes reacted for 12 h with fluorescent tyramine to detect tyrosyl radical formation as an index of arNOX activity. (**a**) 59 years face (arNOX activity 0.21 ± 0.03 nmol/min/10^6 cells). (**b**) As in A except in the presence of SOD (see text). (**c**) As in A except in the presence of the age LOC mixture of inhibitors. (**d**) Neonatal epidermal keratinocytes. Scale bar = 50 μm. Unpublished results of C. Meadows, D. J. Morré, D. M. Morré, Z. D. Draelos, and D. Kern

arNOX activity both by direct biochemical assay (Table 9.8) and by fluorescent tyramine conjugation (Figs.9.27 and 9.28). For all subjects, arNOX activity of sera and saliva was closely correlated ($r^2 = 0.84$). The AGE Index values of the younger subjects were 1.4 ± 0.4 with only one, that of the 30 year old, being different from the others with an index of 2. This subject also exhibited the highest arNOX specific activity of the younger group of 0.04 nmol/min/three frozen sections. For the older group, the AGE Index was proportional to tissue arNOX activity ($r^2 = 0.64$) with the increase in AGE score beginning at an arNOX activity of 0.1 nmol/min/three frozen sections. The arNOX activity of the younger group was on average 3 % that of the older group. In half of the young group, arNOX activity could not be detected. Average arNOX activities of females and males of either age group were not significantly different.

The fluorescent tyramine assay applied to keratinocytes showed a marked difference between aged and neonatal with the aged being much more reactive (Fig. 9.28a, b). Reaction was inhibited by the inhibitor combination of Table 9.7 (Fig. 9.28d) but less so with SOD (Fig. 9.28c). With the frozen sections, activity measured by

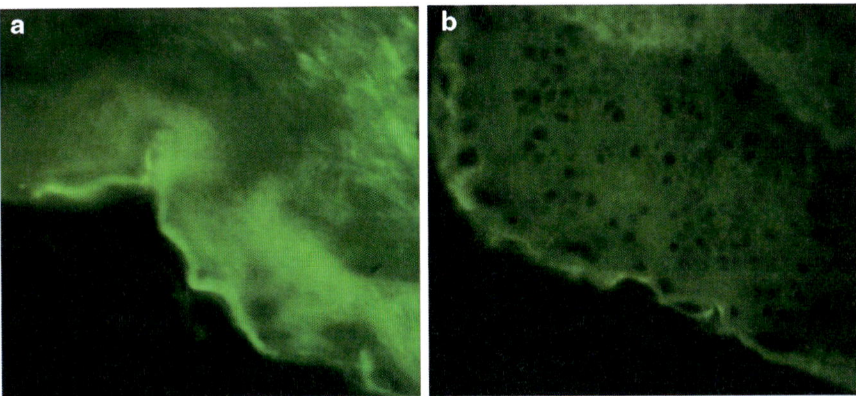

Fig. 9.27 Frozen sections reacted 12 h with fluorescent tyramine to detect tyrosyl radical formation. (**a**) 55 year female. (**b**) 22 year female. ×400. Unpublished results of C. Meadows, D. J. Morré, D. M. Morré, Z. D. Draelos and D. Kern

fluorescent tyramine conjugation was greater on average for the older group of subjects (Fig. 9.27a) compared to the younger group (Fig. 9.27b).

The reduced level of inhibition by SOD in the fluorescence assay compared to the standard ferricytochrome c reduction was traced at least in part to an inactivation of SOD by arNOX in the presence of tyramine either fluorescent or unconjugated. Apparently, in the presence of tyramine, tyrosines of SOD are oxidized by arNOX and dimerized with the tyramine to inactivate the SOD. Similar results were obtained in the biochemical assay where SOD + tyramine also were relatively ineffective in inhibiting the fluorescence labeling (Table 9.9). SODs contain a tyrosine in their active site (Sorkin et al. 1997) and derivatization (e.g., nitration) of active site tyrosines of SOD leads to enzyme inactivation (MacMillan-Crow et al. 1998; Xu et al. 2006) comparable to that observed with arNOX-catalyzed tyrosine oxidation and reaction with tyramine observed in the present study.

Unfortunately, arNOX presence using the tyramine fluorescent assay could not be detected in paraffin sections prepared from these same subjects either before or after application of a standard antigen retrieval protocol described by Shi et al. (1991).

9.8 Role in Oxidation of Serum Lipoproteins

Age and oxidative stress are major risk factors for heart disease (Schmuck et al. 1995). Despite an overwhelming mass of evidence, alterations in mtDNA per se and other forms of cellular and tissue changes related to aging have been difficult to link. Chief among these is the oxidation of low-density lipoproteins (LDLs) implied as causal to atherogenesis (Steinberg 1997).

A large body of evidence supports the notion that reactive oxygen species provide a causal link in the appearance of oxidized circulating lipoproteins such as oxidized LDLs and their subsequent clearance by macrophages and delivery to the arterial

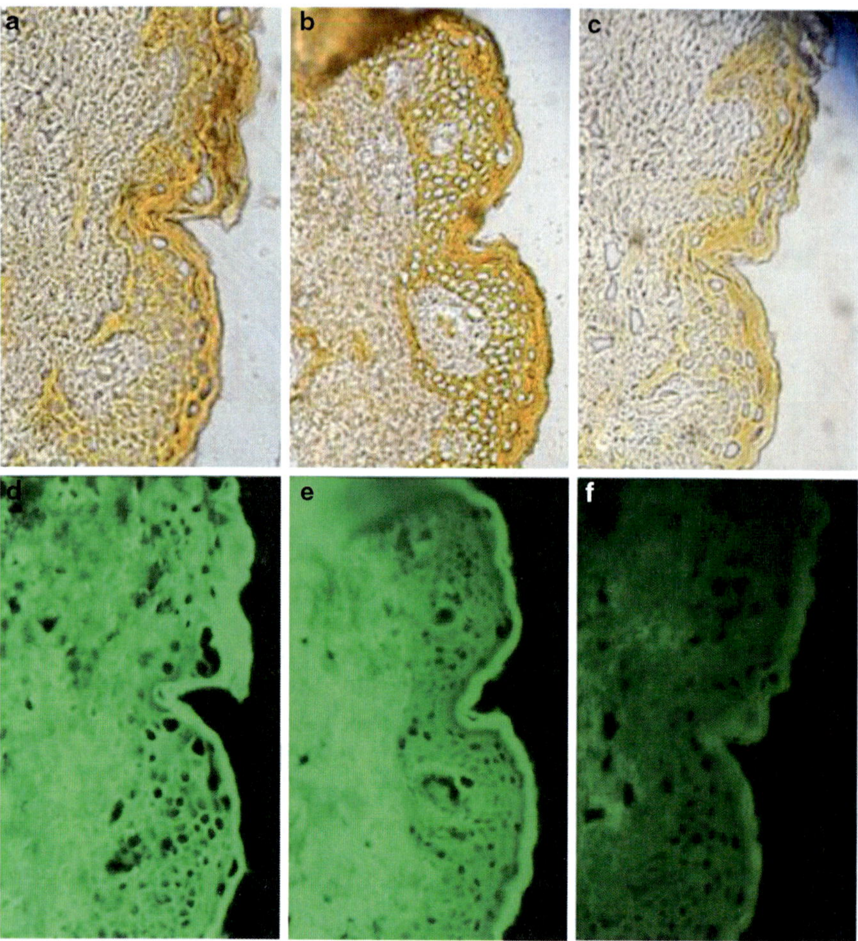

Fig. 9.28 Frozen sections of a 52 year male reacted for 12 h with fluorescent tyramine to detect tyrosyl radical formation as an index of arNOX activity. (**a–c**). Dark field illumination. (**d–f**) Fluorescence imaging. As seen by both dark field (development of orange color due to cross linking of colored tyramine-fluorescence conjugate) or fluorescence, reaction was blocked by the ageLOC mixture of inhibitors (**c**, **f**) but less so with SOD (**b**, **e**). The reduced level of inhibition of arNOX by SOD in the presence of the fluorescence-tyramine conjugate compared to the nearly complete inhibition of arNOX by SOD in biochemical assays was traced to inhibition of SOD by tyramine in the presence of arNOX (see also Fig. 9.27). ×200. Unpublished results of C. Meadows, D. J. Morré, D. M. Morré, Z. D. Draelos and D. Kern

Table 9.9 Inactivation of SOD by 5 min incubation with TM9SF2 or TM9SF4

	Amount of SOD required to dismutate 2.5 μmol O_2^- by 50%
No addition	0.175 pmol
+TM9SF2	1.11 nmol
+TM9SF4	1.75 nmol

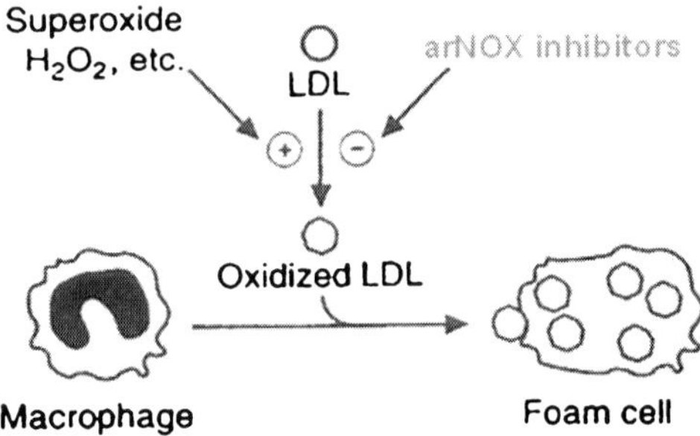

Fig. 9.29 Foam cell formation by macrophages scavenging oxidized low density lipoproteins (LDL). Foam cells deliver the oxidized LDLs to the arterial wall resulting in the formation of atherosclerotic plaques

wall. Oxidized LDL is taken up more rapidly by macrophages than native LDL (Steinberg et al. 1989). It now appears likely that oxidized LDL is a major contributor to progressive atherogenesis by enhancing endothelial injury, by inducing foam cell (lipoprotein engorged macrophages) generation and associated smooth muscle proliferation (Holvoet 1999). Macrophages clear the circulation of oxidized lipoprotein particles by internalizing them and in so doing are transformed into foam cells (Fig. 9.29). The foam cells deliver their cargo of oxidized fats and cholesterol where they are deposited beneath the arterial wall. Such progressive delivery of oxidatively damaged lipoprotein particles eventually leads to atherosclerotic plaques and advanced heart disease, arterial blockage, stroke, and death.

However, until now there has not been a clear cause of LDL oxidation in the blood. Levels of common antioxidants including α-tocopherol, β-carotene, and ascorbate decline with age in the elderly but there is no apparent correlation between ingestion of these common antioxidants and amelioration of the aging process or decreased mortality (Bjelakovic et al. 2007). The implication is that the oxidative damage leading to aging and increased atherogenic risk is the result of a much more specific causation. Why does LDL oxidation increase in the elderly and why is it greater in some individuals than in others? LDL oxidation in the elderly and in individuals at high risk for heart disease correlates with levels of circulating arNOX. Of those who die of a heart attack, 85 % are age 65 or older (American Heart Association Statistical Update 2008) when the levels of arNOX shed into the blood reach a maximum. Women surviving beyond age 65 usually have diminished arNOX levels compared to men and a lower risk of cardiovascular disease compared to men (Levine and Kannel 2003) further suggesting some causal relationship between arNOX levels and atherogenic risk.

arNOX, both at the cell surface and as a circulating shed form, affords the possibility to generate superoxide and reactive oxygen species not only in skin (Kern et al. 2008) but also in direct contact with circulating lipoproteins as an electron source (Morré and Morré 2006a, Figs. 9.30 and 9.31). Through specific

Table 9.10 arNOX-catalyzed tyrosine oxidation and crosslinking, overnight incubation

	Recombinant arNOX (TM9SF2)		
Substrate	–	+	Fold
Bovine serum albumin	14±1	22±3	1.6
Collagen	70±13	308±56	4.4
apoB	45±11	71±27	1.6

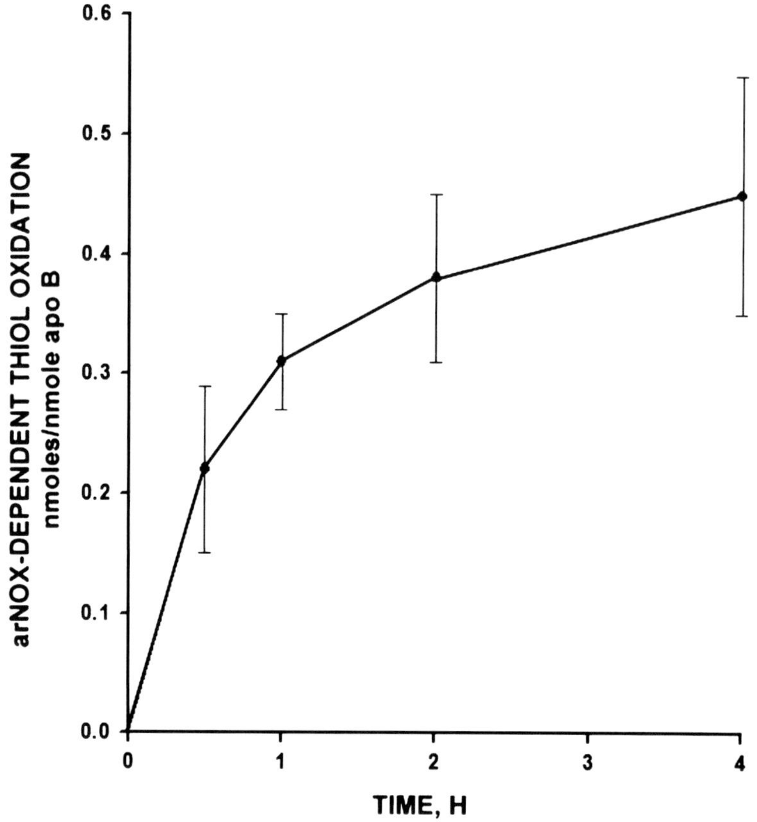

Fig. 9.30 arNOX-dependent oxidation of thiols in purified fully reduced with dithiothreitol (DTT) human apoB determined by reaction with [5,5′-dithiobis-(2-nitrobenzoic acid)] (DTNB) (Ellman's (1959) reagent)

oxidation of LDLs, arNOX proteins emerge as major contributors to cardiovascular disease. If LDL oxidation could be prevented or reduced, so might atherogenesis be reduced or prevented. The main destructive action of arNOX is to directly oxidize proteins (Table 9.10; Fig. 9.30) such that superoxide generation and conversion to H_2O_2 may be less important but the amounts of H_2O_2 generated are still substantial and may contribute to lipid oxidation.

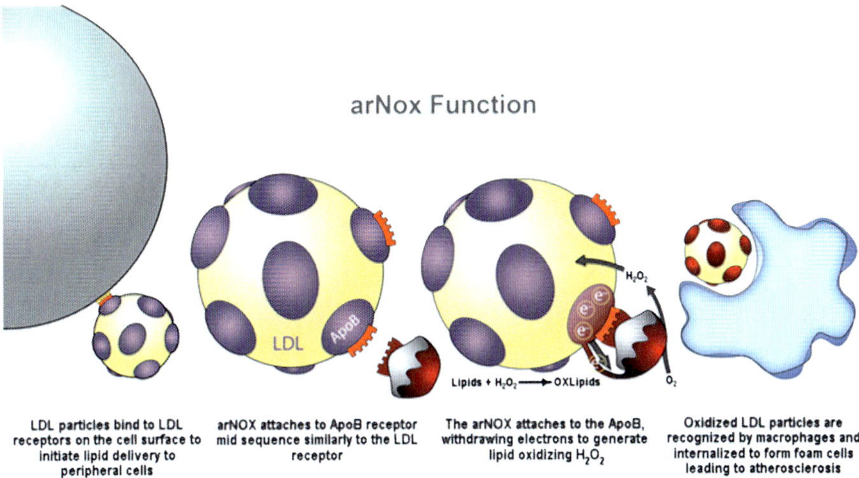

Fig. 9.31 Cartoon illustrating the role of arNOX in lipoprotein oxidation. Eric Chalko unpublished)

Table 9.11 Lipoprotein-associated arNOX-mediated oxidation of apoprotein B of serum lipoproteins determined from rate of ferricytochrome c reduction with lipoprotein particles isolated by flotation centrifugation of sera and plasma of human subjects with low vs. elevated LDL amounts

	arNOX activity (nmol/min/mL)		Lipoprotein bound, % of total activity	
	Elevated LDL	Low LDL	Elevated LDL	Low LDL
Serum	1.0	0.6	32	20
Plasma	1.2	1.0	43	27

arNOX in the blood is structured as an integral component of the LDL particle (Fig. 9.31). For either sera or plasma, an impressive 30–40 % of the arNOX is associated with the lipoprotein particles. The LDL-associated arNOX floats up through saline or distilled water, cannot be washed off, and withstands overnight flotation centrifugation (Table 9.11).

The arNOX isoform that binds LDL has a 27 amino acid sequence with 40 % similarity to the LDL receptor putative binding surface. So not only does arNOX of sera target the apoB of LDL particles as an electron source for superoxide formation, the arNOX emerges as being literally complexed with its apoB target. Thus, much of the oxidative damage is not necessarily due to ROS generation but through direct oxidation of protein thiols and tyrosines also potentially involving SODs. Extracellular SOD (exSOD) is a particular SOD that plays not only an important role in protecting the endothelium from oxidative stress such as that derived from rising levels of arNOX at the endothelial surface but also in the plasma (Littarru and Tiano 2007). A role in maintaining cardiovascular health is indicated by observations that vascular exSOD is substantially reduced in patients with coronary heart disease (Tiano et al. 2007).

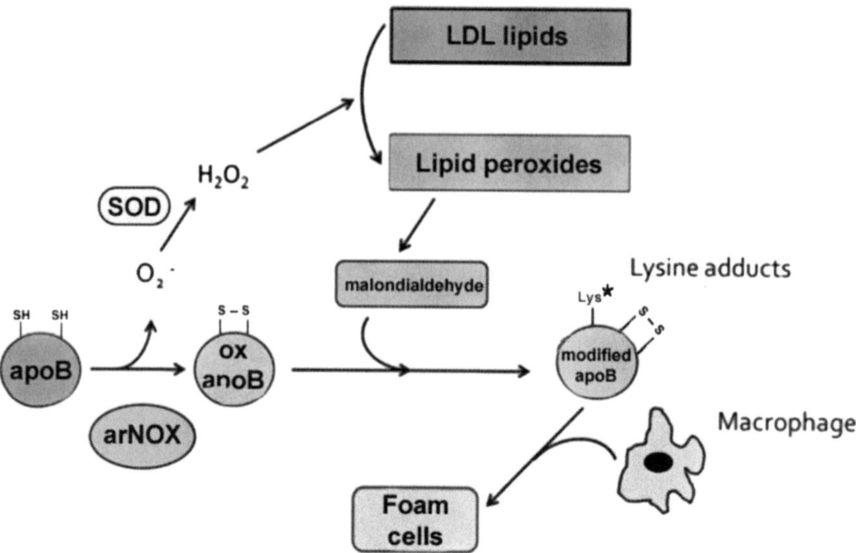

Fig. 9.32 Summary of the (apoB and lipid) cascade catalyzed by LDL-bound arNOX leading to LDL (apoB and lipid oxidation, macrophage recognition, and foam cell formation)

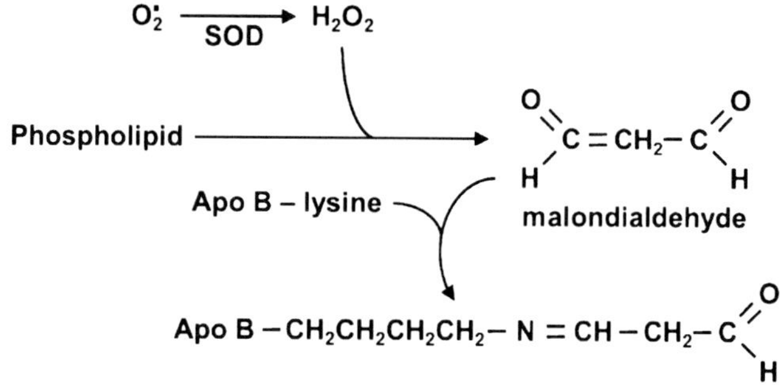

Fig. 9.33 Formation of lysine-malondialdehyde adducts

The pathway of the role for arNOX in apoB oxidation and subsequent pathology (Fig. 9.32) involves malondialdehyde formation (Fig. 9.33) as an obligatory step (Gillotte et al. 2000).

Formation of malondialdehyde-like products in human serum lipoproteins was correlated with arNOX levels in a cell free co-incubation system (Morré and Morré 2006d).

Table 9.12 arNOX specific activity in cell-free coelomic fluid of sea urchins

	N	Diameter (mm)	Ferricytochrome c reduction (nmol/min/50 μL) Total (A)	+SOD (B)	A–B	nmol/min/mg/protein
L. variegatus						
Small	34	21–31	0.10 ± 0.02	0.08 ± 0.02	0.02 ± 0.01	1.7 ± 0.9
Large	14	61–73	0.16 ± 0.03	0.07 ± 0.02	0.09 ± 0.1	8.8 ± 3.2[a]
S. purpuratus						
Small	10	46–59	0.14 ± 0.02	0.12. ± 0.02	0.02 ± 0.05	0.6 ± 0.2[b]
Large	8	75–84	0.13 ± 0.015	0.125 ± 0.02	0.005 ± 0.003	0.15 ± 0.1[c]
S. franciscanus						
Small	9	38–50	0.12 ± 0.03	0.03 ± 0.01	0.09 ± 0.01	1.7 ± 0.4[NS]
Large	6	158–165	0.09 ± 0.01	0.07 ± 0.002	0.02 ± 0.005	0.5 ± 0.3[d]

Differences were significant ([a]$p < 0.02$) or highly significant; ([b]$p < 0.0002$), ([c]$p < 0.0001$), ([d]$p < 0.005$) or not significant (NS)

SOD-inhibited ferricytochrome c reduction as a measure of superoxide formation (arNOX activity) is given by A–B where A is total ferricytochrome c reduction and B is ferricyanide reduction following addition of SOD. arNOX specific activity is given in nmol/min/mg/protein. Results are averages ± standard deviations

9.9 arNOX Activity Correlates with Life Span in Sea Urchins

Sea urchins provide an interesting model for the process of aging. Different species of sea urchins have very different natural life spans and some species display extreme longevity and negligible senescence (Bodnar 2009). The red sea urchin (*Strongylocentrotus franciscanus*) is one of the earth's longest living animals, living in excess of 100 years with no age-related increase in mortality rate or decline in reproductive capacity (Ebert and Southon 2003). In contrast, *Lytechinus variegates* has an estimated life expectancy of only 4 years, while the most widely studied species of sea urchin, *Strongylocentrotus purpuratus*, has a maximum life expectancy of more than 50 years (Moore et al. 1963). Coelomic fluid of the long-lived sea urchin species *Strongylocentrotus purpuratus* and *S. franciscanus* exhibited much longer levels of arNOX than the short-lived urchin species *L. variegates* (Table 9.12). In *L. variegates*, arNOX activity was positively correlated with increase in animal size whereas with *S. purpuratus* and *S. franciscanus*, arNOX activity and animal size were inversely correlated (E. Talbert, A. Bodnar, D. M. Morré and D.J. Morré, results unpublished). While correlations do not prove cause and effect, the inverse correlation of arNOX with sea urchin life span and decreased levels of arNOX with age in the long-lived species are consistent with the idea that reduction of arNOX activity may help reduce the consequences of natural aging.

9.10 arNOX in Plants

arNOX activity is widely distributed in plants. Assays have been limited primarily to plasma membranes purified by aqueous two-phase partition from fresh leaves, flower petals, and other foliar parts. High levels of ferricytochrome c reduction by homogenates unrelated to superoxide production and not inhibited by SOD preclude their use.

arNOX activities of a variety of plants ranging from the shortest lived plant parts analyzed (flower petals) to giant sequoia reveal an approximately 100-fold range of arNOX specific activities with petals being the highest and long-lived trees being the lowest with annuals, biennials, and herbaceous perennials exhibiting intermediate levels (Table 9.13). Early observations also indicated that arNOX levels increased as plant tissues aged and approached senescence (D. J. Morré and D. M. Morré, unpublished). Once fully senescent with loss of chlorophyll, arNOX levels decline.

Resistance to nematode infection of roots of soybean seedlings carrying agents conferring resistance to cyst nematode disease (Cyst X gene) could not be correlated either with arNOX levels or with modulation of arNOX activity by a series of arNOX inhibitors in an extensive series of green house experiments (T. Luo, J. Faghihi, R. Vierling, D. M. Morré and D. J. Morré, Purdue University, results unpublished).

Table 9.13 arNOX activities of plasma membranes from plant foliage by aqueous two-phase partition correlated with longevity

arNOX activity, species	μmoles/min/mg protein
Very short lived (days)	
Yellow day lily petals (*Heterocallis sp.*)	4.0
Annuals (months)	
Spinach (*Spinacia oleraca*)	0.4
Radish (*Raphanus sativus*)	0.32
Biennials (3 years)	
Parsley (*Petroselinum crispum*)	0.3
Queen Ann's lace/wild carrot (*Daucus carota*)	0.22
Domestic carrot (*Daucus carota* subsp. *sativus*)	0.3
Black-eyed susan (*Rudbeckia hirta*)	0.3
Herbaceous perennials (several to many years) new growth	
Echinacea (*Echinacea purpurea*)	0.3
Johnson grass (*Sorghum halapense*)	0.29
Woody perennials (many years)	
English ivy/common ivy (over wintering leaves) (*Hedera helix*)	0.06
American linden/American basswood est. age 50 years (*Tilia americana*)	0.13
European linden/Common lime est. age 1,000 years, Zinna, Germany (*Tilia x europaea*)	0.04
Douglas fir, Washington State (*Pseudotsuga minziesii*)	0.17
Giant sequoia, Washington State (*Sequioadendron giganteum*)	0.04

Unpublished data of Christopher Goyne, Beloit College, Beloit, WI

9.11 arNOX Inhibitors

A growing number of pharmaceutical companies worldwide are announcing programs to identify genes associated with the aging process as targets for anti-aging interventions (either to modulate gene expression at the level of messenger RNA or to modulate the protein products of the genes). The arNOX gene is a premier example of an aging gene amenable to such modulation. In parallel, circulating arNOX activity serves as an aging-related indicator of cardiovascular oxidative damage as a means to access efficacy of therapeutic interventions.

arNOX generates superoxide and reactive oxygen not only in body fluids but at the cell surface as well, significantly increasing the ratio of reactive oxygen species to antioxidant defense molecules, thereby accelerating age-related changes due to lipid and/or protein oxidation. Hence, agents that reduce arNOX activity (Fig. 9.5; Table 9.14) should have substantial value for anti-aging intervention and as a molecular target to explain how CoQ_{10} and other anti-aging substances offer protection to ablate undesirable cardiovascular changes or to maintain skin vitality by preventing oxidation of supporting matrices that are important to skin health (Morré et al. 2003a).

9.11.1 Coenzyme Q

A further unique feature of both cell-bound and shed arNOX is inhibition by oxidized CoQ_{10} (Morré and Morré 2006a; Morré et al. 2003a; Fig. 9.34). As such, this inhibition affords an opportunity to pharmacologically modulate arNOX levels using Q_{10} administration. Salivary measurements, particularly, provide a noninvasive method to monitor arNOX responses to Q_{10} in intervention trials and mirror the circulation as a filtrate of the blood amenable to repeated sampling at short time intervals for real-time kinetic analyses.

When tested over the range of 10–100 μM, CoQ_8, CoQ_9, and CoQ_{10} inhibited arNOX whereas CoQ_4 and CoQ_6 were without effect (Morré and Morré 2006a; Morré et al. 2003a; Table 9.15). Maximum inhibition was achieved between 100 and 150 μg/mL CoQ_{10} in the assay (Morré et al. 2003a). The mechanism of CoQ

Table 9.14 Specific inhibitors of arNOX activity

EC_{50}	TM9SF2	TM9SF4
AgeLOC[a]	1:30	1:20
CoQ_{10}	30 nM	50 nM
Tyrosol	3 μM	1 μM
Gallic acid	2 μM	2 μM

[a] [60 μL of an aqueous mixture of 4 mg/mL *Schizandra chinensis* extract, 9 % schizandrins, (Draco, San Jose, CA) plus 1 mg/mL salicin (Sigma, St. Louis, MO), and 20 μL of IBR Dormin (Israeli Biotechnology Research, Ramat-Gan, Israel) = AgeLOC (NuSkin Enterprises, Provo, UT)] added to 2.5 μL of assay volume

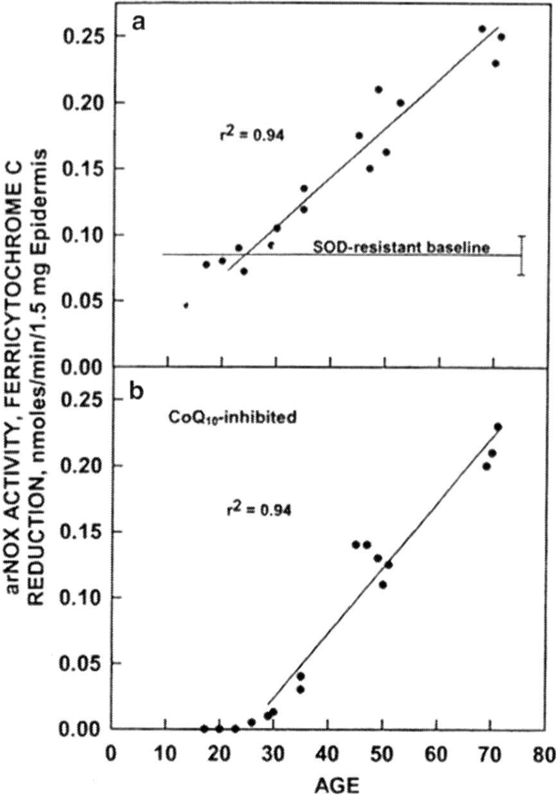

Fig. 9.34 Superoxide production of epidermal explants of both male and female subjects as (**a**) a function of chronological age and (**b**) as CoQ_{10}-inhibited activity. The CoQ_{10}-inhibited component of the activity is first observed about age 30 years

Table 9.15 Response of arNOX activities (activity maximum minus background) determined from the kinetic rates of ferricytochrome c reduction measured over 1 min at 1.5 min intervals for 96–108 min as illustrated in Fig. 9.3

	Ferricytochrome c reduction, nmol/min/10^7 cells	
	Before coenzyme Q addition	After coenzyme Q addition
Coenzyme Q_0 ($n=3$)	0.530±0.003	0.059±0.005
Coenzyme Q_2 ($n=4$)	0.050±0.008	0.056±0.008
Coenzyme Q_4 ($n=2$)	0.052±0.004	0.052±0.004
Coenzyme Q_6 ($n=3$)	0.061±0.004	0.064±0.013
Coenzyme Q_7 ($n=2$)	0.054±0.004	0.054±0.004
Coenzyme Q_8 ($n=3$)	0.050±0.004	0.007±0.011
Coenzyme Q_9 ($n=2$)	0.064±0.000	0.000±0.000
Coenzyme Q_{10} ($n=4$)	0.055±0.005	0.005±0.005

Coenzyme Q homologues having chains of different lengths at the 6 position of the benzoquinone ring were added after 48 min (450 μg in 30 μL ethanol). Values are from two to four determinations for each coenzyme Q homolog comparing buffy coat preparations from different subjects aged 65–85 years ± mean average ($n=2$) or standard ($n=3$ or 4) deviations. From Morré et al. (2003b)

Table 9.16 Effect of CoQ_{10} on the oxidation of reduced CoQ_{10}

	CoQ_{10} reduction, nmol/min/10^7 cells	
Addition	ΔA_{290}	ΔA_{410}
No CoQ_{10}	14.5 ± 2.8	16.1 ± 4.0
+CoQ_{10} (30 μL, 450 μg)	19.5 ± 5.5	17.9 ± 4.5

Table 9.17 Prenyl side chain of CoQ required for inhibition of arNOX

Compound	Inhibition, %
n-Decaprenyl	72 ± 30
Squalene	11 ± 5
Dehydroepiandrosterone sulfate	5 ± 5

inhibition of arNOX is freely reversible and not based on reduction of CoQ (Table 9.16). Reduced CoQ_{10} is not produced by arNOX nor does superoxide derived from added KO_2 appear to reduce coenzyme CoQ_{10} at physiological concentrations and pH. On the other hand, superoxide resulting from KO_2 addition to isolated lipoprotein particles does result in formation of malondialdehyde-like lipid oxidation products (Morré and Morré 2006b).

CoQ_{10} inhibition of arNOX is somehow mediated through the prenyl side chain which is chain length specific (Morré and Morré 2011). In fact, inhibition occurs with only the *N*-decaprenyl side chain and without a contribution from the benzoquinone head group (Table 9.17). The squalene side chain is too short and does not inhibit. $CoQ_{10}H_2$ is a substrate and does not inhibit. Oxidized CoQ_{10} is the product. Oxidation of NADH or protein thiols by arNOX also is inhibited by CoQ_{10} whereas CoQ_{10} is without effect on these activities of constitutive ENOX1 or the cancer-associated ENOX2.

Oral administration of CoQ_{10} to human subjects results in the reversible inhibition of arNOX in the sera (Morré et al. 2008c). In these subjects, the arNOX activities of sera and saliva correlate (Fig. 9.35) such that arNOX activity of saliva can be used as a surrogate biomarker of serum arNOX. Not only was arNOX activity reduced in sera and saliva but in perspiration as well. As with sera and saliva, arNOX activity of perspiration increased with age and was reduced comparing all subjects receiving CoQ_{10} to a level comparable to that for sera (Morré et al. 2008c).

Data for saliva demonstrated a reduction in arNOX activity by administration of CoQ_{10} in both male and female subjects, ages 52–72 years. Saliva collections were more amenable to short time sampling periods and dose–response determinations than were either serum or perspiration. With a single dose of 30 mg CoQ_{10} (Q gel, Tishcon, NY, USA), the response in saliva was rapid with a half time of <30 min and was reversible (Fig. 9.36). The arNOX activity returned to base line between 8 and 10 h after the single dose of 30 mg CoQ_{10}.

After a single dose of 120 mg CoQ_{10}, response was more pronounced and was maximal between 2 and 6 h and returned to base line at 11 h (Fig. 9.37). arNOX activity was proportional to dose over the range 0–120 mg CoQ_{10} with activity

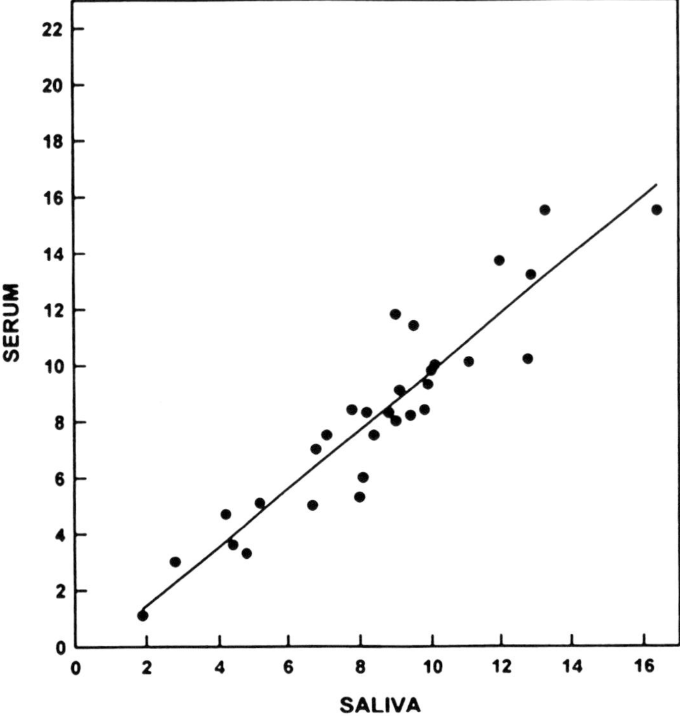

Fig. 9.35 arNOX activities of sera and saliva correlate

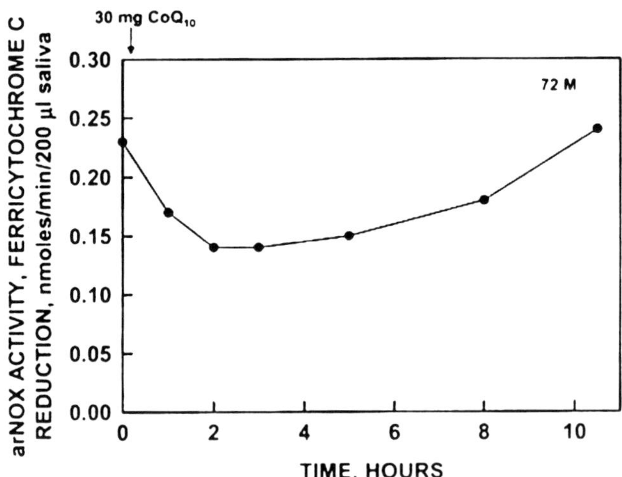

Fig. 9.36 Superoxide production of saliva of a 72 year male following ingestion of 30 mg CoQ_{10}. arNOX activity falls rapidly during the first 2 h and recovers after about 10 h

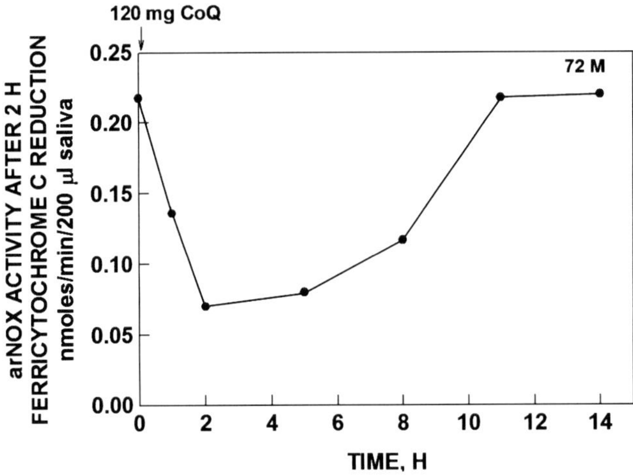

Fig. 9.37 As in Fig. 9.21 except with 120 mg CoQ_{10}. The kinetics of arNOX inhibition are similar to 30 mg CoQ_{10} except that a greater level of inhibition is achieved during the first 2 h and the return to base line occurs after 12 h

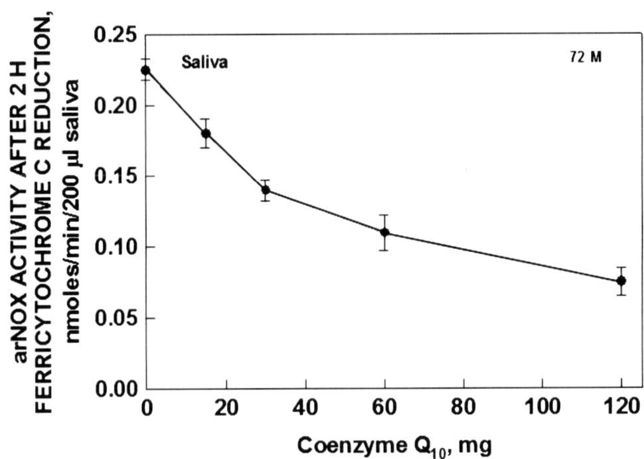

Fig. 9.38 Dose–response based on three repetitions with the same subject at each of the five doses of CoQ_{10} with superoxide production by arNOX measured 2 h after CoQ_{10} administration

measured 2 h after administration when response was greatest (Fig. 9.38). At 120 mg, inhibition of salivary arNOX activity was about 70 %. Also proportional to dose was the time of recovery which varied from 6 h for 15 mg CoQ_{10} to 11 h for 120 mg CoQ_{10}.

With the goal of maintaining arNOX levels at those of a 30 years old, a sustained release CoQ_{10} was prepared and tested. The preparations were 50 % CoQ_{10} and two sources of CoQ_{10} were compared for both saliva and sera (Fig. 9.39). Single capsules

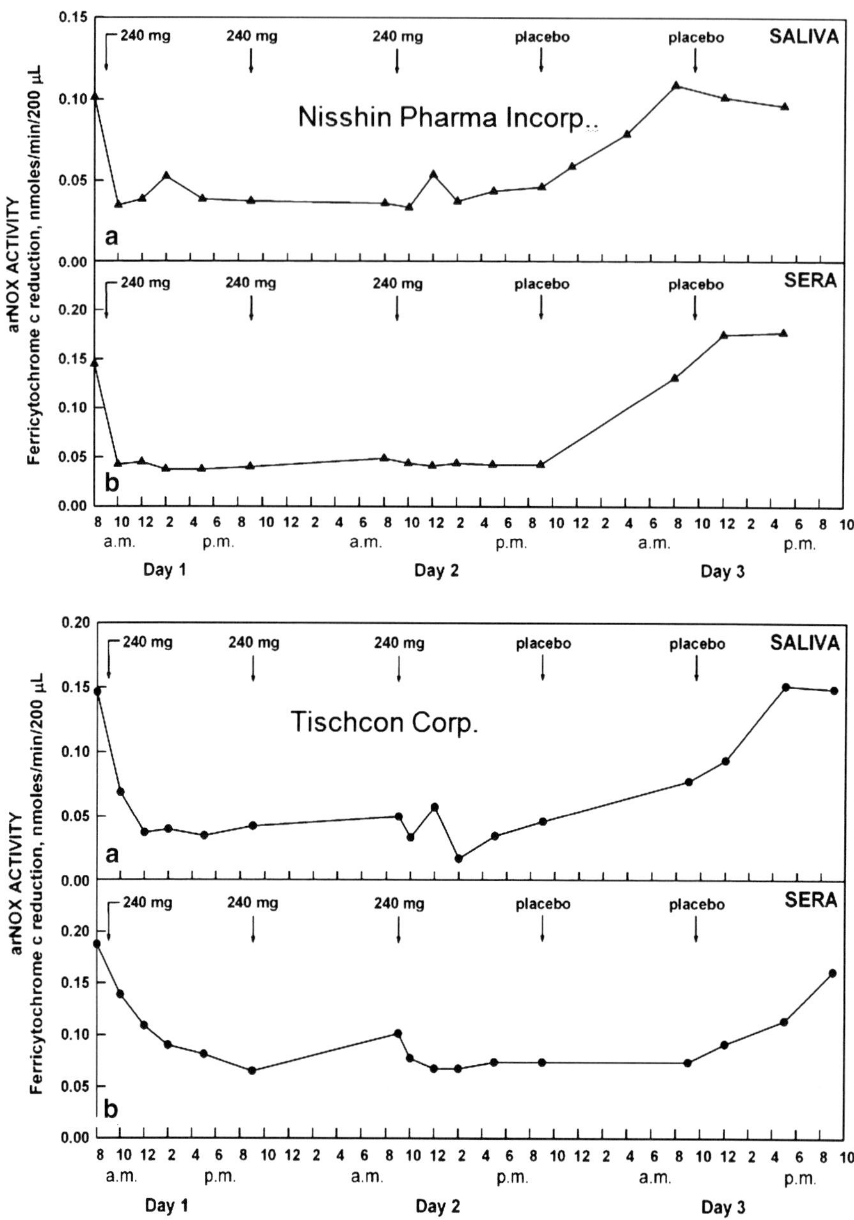

Fig. 9.39 Inhibition of arNOX-catalyzed superoxide formation by sustained release coenzyme Q_{10} clinical trial (74 M). (**a**) Nisshin Pharma. *Upper curve*, saliva. *Lower curve*, sera. (**b**) Tischcon Corp. *Upper curve*, saliva. *Lower curve*, sera. Each 240 mg capsule contained 120 mg coenzyme Q_{10}

Table 9.18 Sustained release CoQ_{10} ([a]). Preliminary clinical study of 8 subjects over 4 days ([b])

	arNOX activity, nmol/min/200 μL			
	Saliva		Sera	
CoQ_{10} Source	Start	Day 4	Start	Day 4
Nisshin	0.22	0.06	0.1	0.03
Tischcon	0.17	0.05	0.13	0.04

[a]250 mg twice daily (9:00 a.m. and 9:00 p.m.)
[b]Start and day 4 collected at 8:50 a.m.

of the 240 mg sustained release product (102 mg CoQ_{10}) were taken twice daily morning and evening. For both CoQ_{10} sources, arNOX levels were reduced to about 30 % of initial and maintained at that level for 24 h with just the two times per day administration of the sustained release material. With eight subjects given the sustained release CoQ_{10} twice daily for 4 days, both the Tishcon and the Nishin products inhibited both serum and salivary arNOX over 70 % (Table 9.18).

Skin health benefits of CoQ_{10} also are indicated (Ichihashi et al. 2007; Terada et al. 2007). In a randomized, double-blinded, placebo-controlled trial, subjects with aged skin used a skin care preparation containing 1 % CoQ_{10} or its reduced form twice daily for 5 months (Ichihashi et al. 2007). Significant reduction of wrinkle grade and an improvement of skin condition were observed as a result of the anti-aging properties of CoQ_{10}. Hoppe et al. (1999) showed that topical application of CoQ_{10} for 3 months reduced the depth and area of wrinkles around the eyes in humans. Similar findings were reported by Terada et al. (2007) following daily oral supplementation of 60 mg CoQ_{10} for 3 months.

9.11.2 *Botanical Sources of arNOX Inhibitors*

An inhibitor cocktail consisting of a blend of botanical ingredients has been investigated as a topical application which slows the deleterious effects of elevated arNOX levels in the skin (Fig. 9.40; Table 9.11). Inhibition of arNOX-generated free radical production may help to slow signs of aging potentially caused by the accelerated free radical production (Kern et al. 2010; ` 2006a; Morré et al. 2003a).

A platform for dietary intervention is provided by the inhibition of arNOX by certain dietary constituents, often referred to as “Herbes de Provence” (Morré et al. 2010c). The result is a decrease in the generation of reactive oxygen species by arNOX and provides a possible explanation for the French Paradox (Morré et al. 2010c). The paradox comes about from the reduced atherogenic risk from the French diet or lifestyle despite a cholesterol-rich diet high in cheese and butter. Previous studies attributed the reduction in risk to consumption of red wine as a natural source of the polyphenol resveratrol (Teissedre and Waterhouse 2000). However, the herbs listed in Table 9.19, which are staples of the French diet, offer a more compelling explanation.

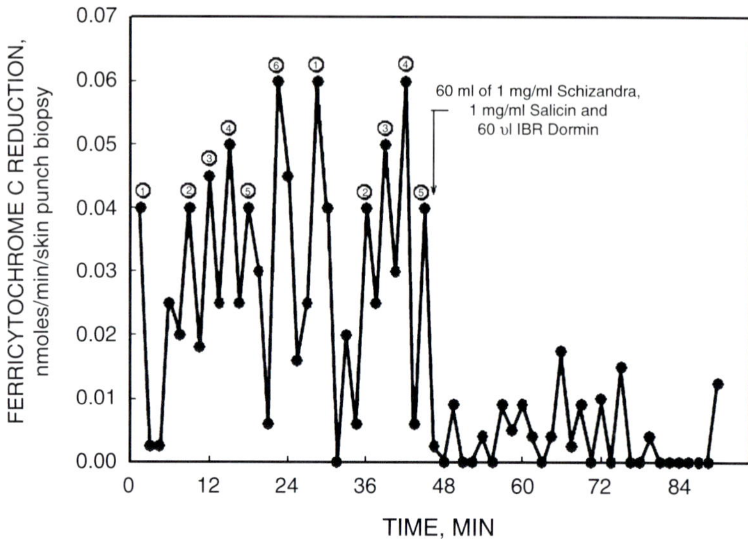

Fig. 9.40 Inhibition of superoxide production of arNOX of a homogenized skin biopsy sample by a mixture of arNOX inhibitors used in dermal preparations designed to slow skin aging (ageLOC)

Table 9.19 arNOX activity inhibition by gallic acid (±)-catechin and by herbal polyphenol sources prepared as hot water infusions at a concentration of 125 mg/mL boiling hot water

Substance or herbal infusion		Inhibition % arNOX-catalyzed arNOX activity	Lipid oxidation
Gallic acid	6 mM	83	84
(±)-Catechin	100 mM	70	80
Savory	Infusion	89	100
Marjoram leaves	Infusion	50	77
Rosemary leaves	Infusion	59	95
Basil	Infusion	82	93
Sage	Infusion	54	70
Estragon (Tarragon)	Infusion	82	93

Inhibitions by herbal infusions at a final concentration of 7.5 μg/mL in the assay varied from 50 to 90 % for savory, estragon (tarragon), basil, marjoram, rosemary, and sage along with a corresponding inhibition of arNOX-catalyzed lipid oxidation (Table 9.19). Savory and estragon (tarragon) were effective at concentrations as low as 75 ng/mL in the assay. Also effective were gallic acid at a final concentration of 0.14 μM (10 ng/mL) and (±)-catechin at 100 μM (300 ng/mL). Certain of these herbs and phenolics may have benefit as well by inhibiting platelet adhesion and aggregation (Yazdanparast and Shahriyary 2008).

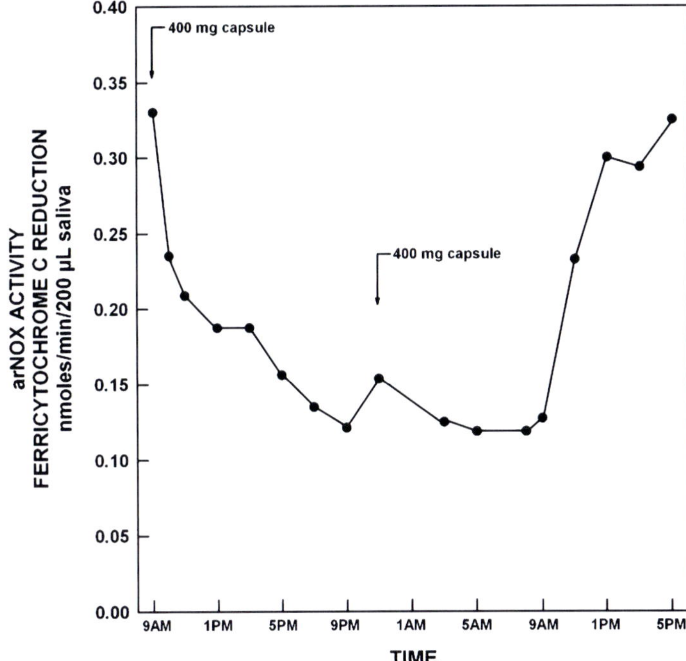

Fig. 9.41 Inhibition of arNOX-catalyzed superoxide formation of saliva by oral administration of a sustained release preparation of finely ground savory leaves (20 mg savory per 400 mg capsule). Two capsules/day, one in the morning and one in the evening reduced the salivary arNOX levels to that of a 30 year old. See Table 9.18 for similar effects on superoxide production by arNOX of sera

By incorporating the herbal preparations as sustained release formulations, 24 h protection has been attained with just two 400 mg capsules/day (morning and before bedtime) (Fig. 9.41) that would maintain arNOX levels in an aging population to nearly those of a 30 years old. It is this aspect that makes possible a therapeutic utility of the technology of importance to reducing aging-related arterial damage from oxidized circulating lipoproteins in individuals as they age beyond 30 years.

9.12 Beneficial Biological Function Associated with Superoxide Production: Physiological Roles of Superoxide

Superoxide anions are generated by single electron reduction of external oxygen using NAD(P)H as the internal electron donor. The superoxide can then dismute (spontaneously or with catalysis by SODs) to form hydrogen peroxide. Interestingly, along with having destructive properties, hydrogen peroxide may exert beneficial effects by acting as a second messenger molecule to activate specific signal transduction pathways involving phosphatases, protein kinases, and transcription factors (Forman et al. 2010). Unlike its role in serving as an age-related terminal oxidase of PMET and as a circulating generator of superoxide, a beneficial role of arNOX as a cell surface activity through generation of superoxide and transmutation to hydrogen peroxide is lacking from our information. Examples of how superoxide generation may also have benefit for diverse cellular activities at present are limited largely to NADPH oxidases of the cytosolic plasma membrane surface. The family of superoxide-generating NADPH oxidases of the inner plasma membrane surface named Nox or *phox*Nox for phagocytic oxidase is distinct from arNOX and other members of the family of ECTO-NOX or ENOX proteins and are better positioned strategically for roles in signal transduction. The phoxNoxes represent a well-characterized group of at least seven proteins (Nox1 to Nox5, Duox1 and Duox2) (Del Principe et al. 2011). Their spatial orientation as a complex facilitates a two-step electron flow beginning with electron transfer from NADPH to FAD followed by transfer from FAD to a distal heme group (Fig. 4.6). Finally, electrons move from the distal heme to molecular oxygen with the generation of superoxide anion. The most extensively studied member of the complex is Nox2, expressed in mammalian phagocytes and responsible for superoxide production during engulfment of invading microbes (Cheng et al. 2001; Sumimoto 2008). In concert with other members of the *phox*-Nox family, Nox2 has been implicated in a variety of other biological functions in addition to host defense such as signal transduction, development, angiogenesis, blood pressure regulation, and certain biosynthetic processes (Nauseef 2008). The active form of Nox2 is cytochrome b558, a heterodimer composed of two subunits, namely $p22^{phox}$ (light chain) and $gp91^{phox}$ (heavy chain) (Parkos et al. 1987).

Nox2 is usually inactive in resting cells but becomes activated during phagocytosis of invading microbes. Activation occurs as a result of translocation to the cytosolic plasma membrane surface of a ternary regulatory complex formed by $p47^{phox}$, $p67^{phox}$, and $p40^{phox}$ subunits, as well as of the small GTPase Rac (Clark et al. 1989).

Soucy-Faulkner et al. (2010) have reported that Nox2 and ROS were required for efficient detection of viral invasions through binding of RIG-1 ultimately leading to the activation of antiviral genes and Mda-5 to mitochondria-associated adaptor MAVS. Use of RNAi oligonucleotides to silence Nox2, use of Tempol, a superoxide mimetic that can enter cells, and the use of the NAD(P)H oxidase inhibitors diphenyliodonium and apocynin all effectively inhibited the antiviral response. Simple addition of hydrogen peroxide to cells did not illicit a similar response such that intracellular generation of superoxide seemed to be required. One possibility is that endosomes (phagosomes) generated in response to viral infection enter cells and that the resultant cytosol-generated superoxide is formed in the vicinity of the other cytosolic constituents whose activation is required to initiate the viral resistance cascade.

There are implications from those studies, not necessary for the potentially destructive secreted forms of arNOX isoforms but for the transplasma membrane arNOX prior to proteolytic cleavage of the active N-terminal fragments. This form of the protein is also known to be associated with endosomes. In this role the cellular form of arNOX might very well be beneficial. Thus, it might be important that the endosomal forms of arNOX be retained as active proteins and not be inhibited therapeutically below some potentially beneficial maintenance levels.

By targeting a fluorescent protein-based H_2O_2 sensor to various intracellular compartments, Enyedi et al. (2010) demonstrated a high level of H_2O_2 in the lumina of the endoplasmic reticulum of cultured HeLa cells. The enzymatic source of these high levels of H_2O_2, including the suggested thiol oxidase *Erol*-L, a flavin-containing protein that oxidizes protein disulfide isomerase, has not been determined. Complete inhibition of p22^{phox} expression by siRNA treatment was without affect indicating that the *phox*-Nox system of oxidases was an unlikely source of ROS in the endoplasmic reticulum of HeLa cells.

Regarding the regulation of cell proliferation by ROS, particularly H_2O_2, the inhibition of cell proliferation in tumor cells treated with exogenous catalase or transfected with cDNA of catalase has been observed (Onumah et al. 2009; Policastro et al. 2004). However, the source of the H_2O_2 and the nature of the receptor of the mitogenic signal have not been determined.

9.13 NQO1 (Cytoplasmic NAD(P)H: Quinone Oxidoreductases, DT-Diaphorase EC 1.6.99.2) and Plasma Membrane Electron Transport

2,3-Dimethoxy 1,4-naphthoquinone (DMNQ), which redox cycles via two-electron reduction, mediates reduction of the cell-impermeant tetrazolium dye WST-1 in kidney epithelial cells (MDCK), which express high levels of NQO1, but not in HL60 or CHO cells, which are NQO1 deficient (Tan and Berridge 2010). DMNQ-dependent WST-1 reduction by MDCK cells was strongly inhibited by low concentrations of the NQO1 inhibitor dicumarol and was also inhibited by diphenyleneiodonium, capsaicin, and SOD, but not by the uncoupler FCCP or the complex IV inhibitor cyanide. This suggests that DMNQ-dependent WST-1 reduction by MDCK cells is catalyzed by NQO1 via redox cycling and PMET, DMNQ/WST-1 reduction and extracellular H_2O_2 production were correlated as determined by cellular toxicity of DMNQ on MDCK cells and was extensively reversed by co-incubation with dicumarol or exogenous SOD, catalase, or *N*-acetylcysteine. No similar effects were observed in NQO1-deficient CHO and HL60 cells that PMET plays a significant role in DMNQ redox cycling via NQO1, leading to cellular toxicity in cells with high NQO1 levels.

Cytosolic NQO1 has been considered unlikely to be responsible for the reduction of plasma membrane coenzyme Q based on substrate specificity and its inhibition by dicumarol (Kishi et al. 2002). Overexpression of cytosolic NQO1 in pancreatic beta cells and islets of the mouse, however, increased PMET activity in the presence of 10 mM glucose based on the reduction of the impermeable dye WST-1 and ferricyanide (Gray et al. 2011), an effect abolished by dicoumarol. As PMET activities were enhanced significantly in both cells and isolated islets following exposure to aminoxyacetate, a malate-asparate shuttle inhibitor, the effect appears more related to cytosolic NADH levels than to direct reduction of plasma membrane quinones by NQO1.

9.14 Summary

Age-related ENOX proteins, arNOX (ENOX3), of the cell surface and endosomes with a period length of 26 min and shed into body fluids increases linearily with age beginning at about 30 years to a maximum at about age 65. Those surviving beyond age 65 frequently have reduced arNOX activity. Rather than only reducing molecular oxygen to water as is characteristic of ENOX1 and ENOX2, arNOX (ENOX3) proteins transfer electrons to oxygen to form superoxide during part of their functional cycle. By generating reactive oxygen species at the cell surface and in body fluids (saliva, serum, perspiration, urine), arNOX provides a mechanism to propagate reactive oxygen species generated at the cell surface to surrounding cells as occurs in skin aging and to circulating serum lipoproteins of importance to skin health and atherogenesis. arNOX is widely distributed among aged systems including late passage cultured cells and plants.

Superoxide formed by arNOX is measured by the ability of superoxide to reduce ferricytochrome c. The superoxide generated is not only active in the reduction of ferricytochrome c but also in the reduction of tetrazolium salts such as XTT (Na 3′-[(phenylamino)-carbonyl]-3,4-tetrazolium]-bis(4-methoxy-6-nitro)benzenesulfonic acid) leading to colored formazan formation. Other NOX proteins lack this activity. Activity is inhibited by SOD and CoQ_{10} as well as by a series of small molecular weight inhibitors such as salicin useful in skin care products to slow or prevent skin aging and oral supplements to reduce lipoprotein oxidation leading to coronary artery disease.

The arNOX protein family has been identified in yeast and humans and has been cloned. There are at least five family members (TM9SF1-5). The arNOX proteins are synthesized as membrane anchored proteins with their catalytic N-terminus directed toward the cell's exterior. A ca. 30-kDa fragment is shed and enters the blood and other body fluids or is internalized into endosomes. With cells and tissues, the source of electrons would be PMET. With sera, the ultimate source of electrons for the reduction of ferricytochrome c appears to be protein thiols and tyrosines. The arNOX proteins are very stable to heat, proteolysis, and chemical degradation and are released into the outer layers of the skin, for example, where they vigorously oxidize skin proteins such as collagen and elastin by converting tyrosines to tyrosyl radicals which then react to cross-link proteins through dityrosyl linkages as a major factor in altering skin texture and appearance, in wrinkle formation and other degenerative conditions related to skin aging. In the blood, arNOX is located in the near vicinity (bound to) of low density lipoprotein (LDL) particles where a particular arNOX family member TM9SF2 with sequence homology to the LDL receptor is a major contributor to LDL oxidation, a prerequisite for internalization by macrophages to form foam cells, the obligate progenitors of atherosclerotic plaques. arNOX proteins actually bind to the ApoB100 proteins of LDLs as a source of electrons for transfer to oxygen to form superoxide. The hydrogen peroxide resulting from the dismutation of the superoxide results in oxidation of the lipid core of the LDL with formation of malondialdehyde-like adducts as the

LDL coat proteins are oxidized to supply electrons to the oxidase. Protein thiols occur in sera at levels sufficient to sustain the activity of the enzyme for several months. arNOX proteins share many properties with other ENOX proteins including a cell surface location, the oscillatory pattern of activity (26-min period), resistance to proteases, N-terminal sequencing and chemical degradation, and a propensity for the purified proteins to form aggregates.

Chapter 10
The Auxin-Stimulated ENOX and Auxin Stimulation of Plant Growth

2,4-Dichlorophenoxyacetic acid (2,4-D) is a synthetic auxin which was discovered and developed during World War II as a selective herbicide for weed control in corn (maize), small grains, and other grasses. Auxins, both natural and synthetic, as a class, stimulate plant cells to increase in size (enlarge) (Fig. 10.1). A variety of mechanisms have been proposed to understand how 2,4-D and the natural auxins such as indole-3-acetic acid (IAA) promote cell enlargement (Davies 1995). Our findings implicate surface hydroquinone oxidases (ENOX proteins) as drivers of the cell enlargement process (Chap. 5). One ENOX protein in plants requires auxin for activity. When induced by 2,4-D, however, the activity persists as growth of affected tissues and plant parts becomes unregulated and cancer like. The ultimate result with 2,4-D-sensitive species is death of the plant or of the affected plant parts.

10.1 Early Evidence for Auxin-Modulated Enzymes Involved in Plant Cell Enlargement

Early indications of an enzymatic basis for auxin stimulation of plant growth came from studies by Coartney et al. (1967) where actinomycin D, now known to inhibit auxin-stimulated ENOX activities (Sect. 10.4), resulted in inhibition of both auxin-induced extensibility changes and cell elongation. Similarly, Morré and Eisinger (1968) used heat killed oat coleoptile sections to indicate that cell elongation stimulated by auxin involved enzyme-catalyzed reactions not necessarily involved with modification of cell walls. More important, especially in terms of eventual outcome, was that elongation growth of plants, especially the auxin-stimulated component, was shown to be exquisitely sensitive to thiol inhibitors (Eisinger and Morré 1968).

The first indications of a redox involvement in auxin action were studies that showed a typical auxin biphasic response with an increase in reduced cytosolic proteins at low auxin concentrations and more oxidized cytosolic proteins at high auxin concentrations resulting in measurable differences in their heat stability

D.J. Morré and D.M. Morré, *ECTO-NOX Proteins:Growth, Cancer, and Aging*, DOI 10.1007/978-1-4614-3958-5_10, © Springer Science+Business Media New York 2013

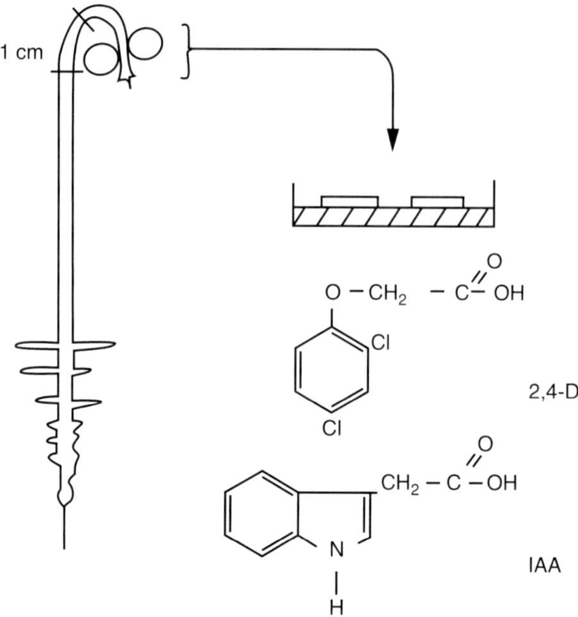

Fig. 10.1 Elongation of stem sections cut from the zone of cell elongation of dark-grown seedlings of dicotyledenous plants is stimulated by the native growth regulator indole-3-acetic acid (IAA) and the synthetic auxin herbicide 2,4-dichlorophonoxyacetic acid (2,4-D)

(Morré 1970). Reduced proteins are coagulated by heat more readily than oxidized proteins. These changes also were observed to oscillate but the observations were never published.

The first observations of a hormone-stimulated NADH oxidase activity of the external surface of the plasma membranes were with plants in response to the auxin regulators of plant cell elongation (Morré et al. 1986a, Brightman et al. 1988, Morré and Brightman 1991). With either the synthetic auxin regulators, 2,4-D, or α-naphthalene acetic acid (α-NAA), or the natural auxin regulator (IAA), auxin concentrations giving maximal stimulations of cell enlargement when applied to elongating sections of soybean hypocotyls stimulated NADH oxidase activity at these same concentrations. With 2,4-D, at least, the response was again biphasic. As optimal concentrations were exceeded, the stimulations of growth and NADH oxidase activities were reduced (Morré et al. 1988a; Morré 1994a) (Fig. 10.2). When plasma membranes were isolated from a tissue region that was no longer auxin-responsive, such as the mature region of soybean hypocotyls, the NADH oxidase was no longer stimulated by auxin (Morré et al. 1991b).

With the related but inactive auxin analogs, 2,3-dichlorophenoxyacetic acid (2,3-D) or β-naphthalene acetic acid (βNAA), neither stimulation of NADH oxidation nor stimulation of cell enlargement was observed (Morré et al. 1988b; Morré 1994a). In

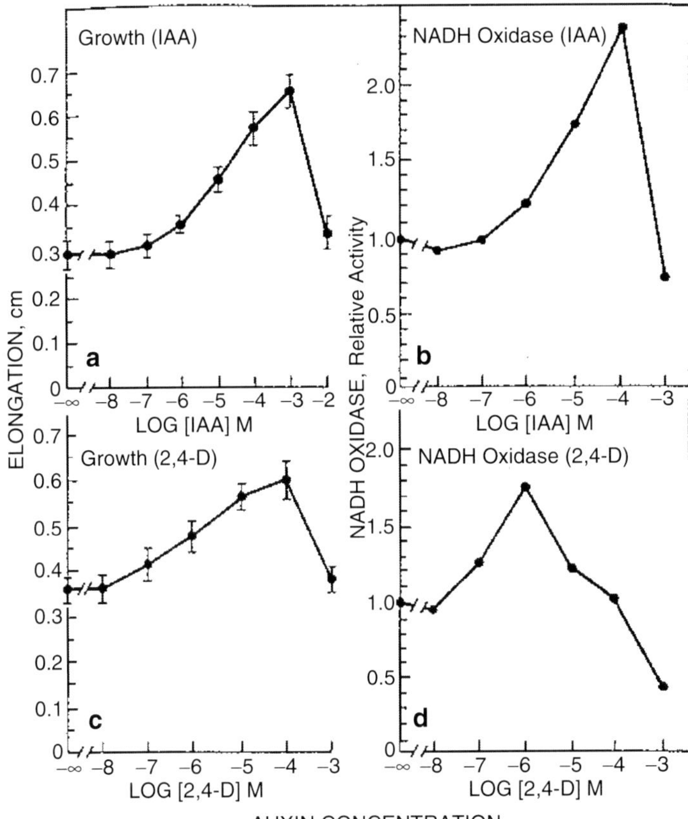

Fig. 10.2 Stimulation of growth (**a**, **c**) and NADH oxidase activity (**b**, **d**) by the natural auxin indole-3-acetic acid (**a**, **b**) and by the synthetic auxin herbicide 2,4-dichlorophenoxyacetic acid (2,4-D) (**c**, **d**) of soybean (*Glycine max*). Values are averages from eight different experiments (10 segments/experiment) ± SD for growth of 1 cm hypocotyl segments from dark-grown seedlings (Fig. 10.1) and from determinations with at least three different plasma membrane preparations also from 1 cm hypocotyls segments for NADH oxidase. The control activity (no auxin) rate of NADH oxidation was 11.8 ± 2.9 nmol NADH oxidized (mg protein)/min (Reproduced from Morré et al. 1998c with permission from Wiley-Blackwell)

these experiments, NADH oxidation was determined using isolated plasma membrane vesicles and cell enlargement was measured using 1-cm-long segments cut from the zone of cell elongation (Fig. 10.1). The plasma membrane vesicles were isolated in parallel from 1-cm-long zones of cell enlargement as represented by the tissues used for growth measurement.

Cell elongation in roots is inhibited by auxin concentrations that stimulate shoot growth (Morré and Bonner 1965). While not investigated extensively, NADH oxidation by plant roots is inhibited by auxins (Böttger and Lüthen 1986; Morré et al. 1991b).

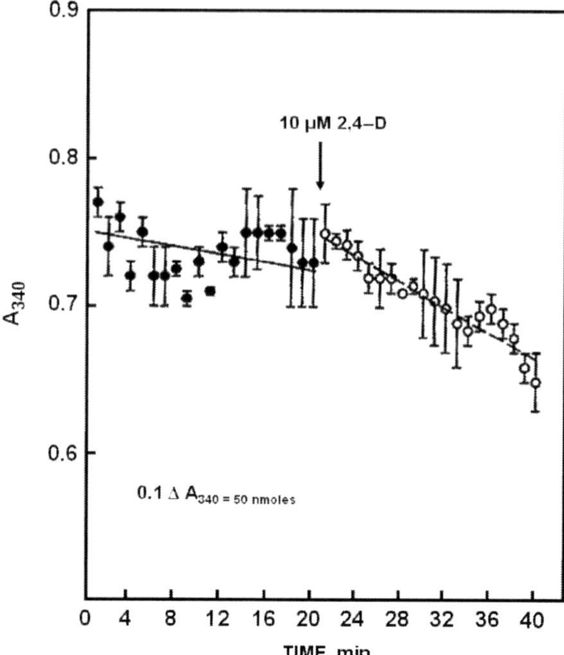

Fig. 10.3 Time course of NADH oxidation by 1 cm hypocotyl sections of dark-grown soybean seedlings. Results are averages of four replicate sets of 30 sections each. The reaction was initiated by the addition of NADH. NADH is an impermeant substrate available to the oxidase located on the external plasma membrane surface (Reproduced from Hicks and Morré 1998 with permission from Elsevier)

Oxidation of external NADH was measured directly with isolated tissue sections (Hicks and Morré 1998) (Fig. 10.3). The activity was periodic as for plasma membrane vesicles, resistant to cyanide, stimulated by 2,4-D, α-NAA, and IAA but not by the growth inactive analogs 2,3-D, β-NAA, or benzoic acid.

The marked susceptibility to inhibition by thiol reagents of both 2,4-D-induced growth and the 2,4-D-induced NADH oxidase of soybean plasma membrane suggested initially an involvement of essential active site thiols of the oxidase in the auxin growth mechanism. A similar thiol dependency was seen with the constitutively activated NADH oxidase activity (ENOX) of hepatoma and HeLa cell plasma membranes (Sect. 5.3.1).

If, indeed, the hormone- and growth factor-stimulated NADH oxidase is in reality a thiol oxidoreductase or a thiol interchange protein, then the activity should exhibit, as well, a protein disulfide isomerase-like activity. Plasma membrane vesicles were shown to exhibit protein-disulfide isomerase activity (Morré et al. 1995d; Morré 1998a). This activity with plant plasma membrane vesicles was stimulated approximately twofold by auxin (Fig. 2.16). While it is clear that long-term cell elongation extending over periods of several hours may be dependent upon the vesicular pathway of membrane addition for a source of membranes and membrane precursors, short-term auxin-induced growth, that which occurs over an initial period of several hours, seems to occur more or less independently of a vesicular mechanism (Sect. 10.6) and independently of turgor as a driving force (Sect. 10.7).

Based on the above findings, we have developed a model whereby cell enlargement is an active process involving both ENOX proteins and an ATP-requiring step amenable to evaluation in a completely cell-free system with recombinant proteins and completely synthetic membrane vesicles (Morré et al. 2006a; Chap. 5). Findings support a central role of plasma membrane ENOX proteins and associated redox constituents as essential components of the cell enlargement process applicable as well to auxin-induced cell enlargement.

10.2 The Plasma Membrane as the Subcellular Location of the Auxin-Responsive Mechanism

In 1971, Lembi et al. demonstrated that isolated plasma membranes from maize coleoptiles bound the auxin transport inhibitor *N*-1-naphthylphthalamic acid. These studies were made possible by the use of a specific electron dense stain based on phosphotungstic acid selective for plasma membrane vesicles of plants as a means to guide plasma membrane purification (Fig. 5.25; Roland et al. 1972).

Various activities associated with isolated plant plasma membrane fractions were shown to respond to auxin treatment such as a glucan synthetase (Van der Woude et al. 1972, 1974) and release of a factor that enhanced RNA polymerase activity of isolated nuclear fractions (Hardin et al. 1972) subsequently traced to an auxin-stimulated release (Clark et al. 1976; Morré and Cherry 1977) of a phosphotungstic acid-reactive non-protein small molecule (Yunghans et al. 1978) subsequently identified as a phytosterol (results unpublished).

Direct responses of plasma membranes to auxin was eventually demonstrated by a change in membrane dimensions observed in the electron microscope (Morré and Bracker 1976) from an increase in microviscosity of plasma membranes determined from fluorescence polarization measurements (Helgerson et al. 1976), from a correlation of binding of a radiolabeled auxin, IAA-2^{14}C with plasmic membrane content of isolated membrane fractions from dark-grown soybean hypocotyls (Williamson et al. 1977), and from Fourier transform infrared spectroscopic measurements demonstrating an auxin concentration-dependent broadening of amide I absorbance and changes in the ratios of amide I and amide II absorbance (Morré et al. 1987). Plasma membranes as targets for binding of auxin regulators were subsequently confirmed in a number of laboratories (Batt and Venis 1976; Batt et al. 1976; Cross et al. 1978; Dohrmann et al. 1978; Löbler and Klämbt 1985). An antibody to the auxin-binding protein of maize provided by D. Klämbt, which also was growth inhibitory, did inhibit the auxin-stimulated NADH oxidase from both maize and soybean (Morré et al. 1991b).

Soybean plasma membranes exhibit NADH oxidase activities accessible to NADH supplied to either the external or internal membrane surfaces (DeHahn et al. 1997). Activity at the external surface was demonstrated using plasma membranes of right-side-out orientation prepared by aqueous two-phase partition and using intact soybean cells grown in culture. Activity at the internal membrane surface was demonstrated using vesicles of inside-out orientation obtained by preparative free-flow electrophoresis or by aqueous two-phase partition following freeze-thaw or Brij detergent treatment to invert some of the right side-out vesicles. The NADH oxidase activity of both membrane surfaces appeared to be stimulated by the synthetic auxin 2,4-D.

10.3 Direct Effects of Auxin on Signaling Molecules Fail to Parallel Those of Mammalian Growth Factors

Isolated plant plasma membranes bind auxins (Batt and Venis 1976; Batt et al. 1976; Williamson et al. 1977; Cross et al. 1978; Dohrmann et al. 1978; Löbler and Klämbt 1985) and respond to auxins in various ways. The in vitro responses of plasma membranes to auxins include a stimulation of synthesis of hot-water soluble glucans (Van der Woude et al. 1972), release of a factor enhancing RNA polymerase activity (Hardin et al. 1972), membrane thinning (Morré and Bracker 1976), increased fluorescence polarization of a membrane-associated probe (Helgerson et al. 1976), inhibition of phosphatidic acid phosphatase (Scherer and Morré 1978a), stimulation of a membrane ATPase (Scherer and Morré 1978b) (Fig. 10.4), enhanced

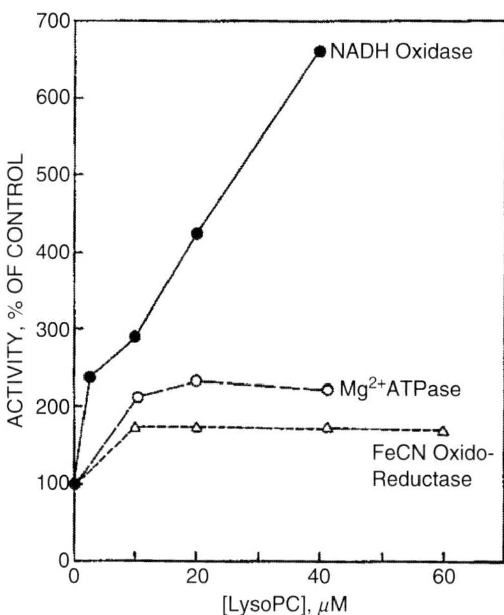

Fig. 10.4 Stimulation of NADH oxidase activity of soybean plasma membrane by Lyso PC compared with the stimulation of Mg^{2+} ATPase and NADH-ferricyanide oxidoreductases (FeCN oxidoreductases) of the same membrane preparation. Stimulation of NADH oxidase by Lyso PC is time dependent following preincubations of 10 min. (Unpublished data of Andrew Brightman, Purdue University) (Reproduced from Morré 1990a with permission from Wiley-Liss)

phospholipase C-independent turnover (Morré et al. 1984; Sandelius and Morré 1987), increased phosphorylation of membrane proteins (Varnold and Morré 1985; Varnold et al. 1986), and release of calcium ions (Buckhout et al. 1980).

Interestingly, Ca^{2+} ions reverse the auxin-induced thinning of the hypocotyl plasma membrane (Morré and Bracker 1976), inhibit auxin stimulation of plant growth (Morré and Eisinger 1968; Schnepf et al. 1979), completely block the effects of auxins on cell wall loosening (Coartney and Morré 1980a, b), and inhibit auxin-stimulated protein phosphorylation (Varnold and Morré 1986) and auxin-stimulated inositol turnover (Morré 1986; Sandelius and Morré 1987). Both the auxin-stimulated increase in membrane ATPase activity and the auxin-promoted loss of membrane phosphatidic phosphatase activity can be mimicked experimentally by removing membrane-bound calcium (Buckhout et al. 1980). One possible explanation to link these several seemingly disparate observations is that binding of Ca^{2+} to isolated plasma membranes diminishes their auxin response.

Yet, comparisons of auxin-responsive plant and hormone-responsive mammalian signal transducing systems have revealed few similarities (Morré 1990b, c; Zbell et al. 1990). Plants appear to lack adenylate cyclase (Yunghans and Morré 1977) despite occasional reports to the contrary. Plants do contain plasma membrane-associated phospholipase hydrolyzing phosphatidylcholine (Scherer and Morré 1978a; Morré 1989), phospholipase A (Morré 1990a; Brightman et al. 1991), and phosphatidylinositol-specific phospholipase C (Pfaffmann et al. 1987; Morré 1989) activities. Of these, only the plasma membrane-associated hydrolysis of phosphatidylcholine was consistently responsive to auxin as evidenced by auxin-stimulated degradation of choline-containing phospholipids of soybean plasma membranes to choline, diglyceride, and inorganic phosphorus (Morré and Drobes 1987) (Fig. 10.5). Soybean membranes contain both a choline phosphate phosphatase and a phosphatidic acid phosphatase such that the observed products could arise from either a D- or a C-type primary cleavage.

A controversial stimulatory response of a plasma membrane-associated ATPase (Scherer and Morré 1978b) has been suggested to be the ultimate target for growth-controlling substances in plants (Marré and Ballarin-Denti 1985; Hassidim et al. 1987). Palmgren et al. (1988) showed a two- to threefold activation of plasma membrane ATPase by lysophosphatidylcholine. Free fatty acids also activated the ATPase. Activation of the plasma membrane ATPase by lysophospholipids appears to depend on chain length, degree of saturation, and head group of the lysophospholipid. Maximum stimulation was given by palmitoyl 16:0 lysophosphatidylcholine and oleoyl 18:1 lysophosphatidylcholine (Palmgren et al. 1988). The ATP-dependent accumulation of protons by plant plasma membrane vesicles is stimulated by lysophospholipids as well (Palmgren and Sommarin 1989). Similar effects of lysophospholipids are seen on ENOX activities (Fig. 10.6).

A cell surface-associated electrogenic proton pump driven by ATP is considered to be responsible in part for the generation of a membrane potential and to establish a pH gradient across the plasma membrane of plant cells (Poole 1978; Spanswick 1981; Chap. 7). Although H^+ transport may be driven by a plasma membrane redox system (Crane et al. 1985; Hassidim et al. 1987), the generally accepted mechanism

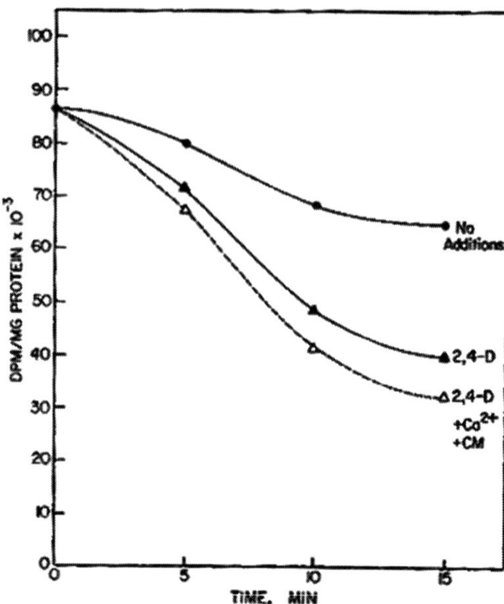

Fig. 10.5 Kinetics of loss of radioactivity from [^{14}C]choline-prelabeled plasma membranes prepared from 1 cm hypocotyl segment of dark-grown soybean seedlings as a function of time of incubation with or without 1 μM 2,4-D alone or in the presence of 1 μM Ca^{2+} + 1 μM calmodulin (Redrawn from Morré and Drobes (1987))

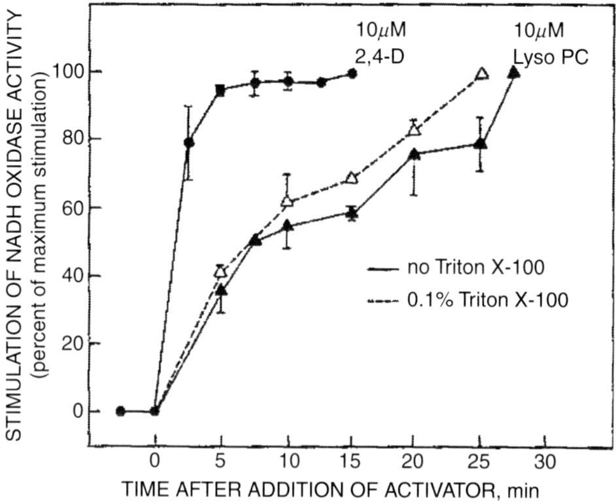

Fig. 10.6 Kinetics of stimulation of NADH oxidase of plasma membranes of dark-grown soybean stems comparing the synthetic auxin herbicide 2,4-D and lysophosphatidylcholine. Maximum stimulation of NADH oxidase by auxin was reached within 5 min of addition of the growth regulator to the enzyme reaction. In contrast, lysophosphatidylcholine activated the enzyme more slowly requiring approximately 30 min of incubation to reach a maximum stimulated rate. Incubation in the presence of 0.1 % Triton X-100 did not significantly increase the activation rate by lysophosphatidylcholine. The control rate of NADH oxidation was 0.80 ± 0.13 nmol/min/mg protein (Reproduced from Morré and Brightman 1991 with permission from Springer Scientific + Business Media)

for creating an electrochemical proton gradient is that of a plasma membrane H^+-ATPase. Proton pumping across the plasma membrane is stimulated by auxin (Rayle 1973; Cleland 1975, 1976; Felle 1987, 1988).

10.4 Evidence for a Redox-Related Plasma Membrane-Located Auxin Target

Among the first indications of a redox involvement of auxin in its interaction with plasma membranes came from observation of ascorbate-induced spectral changes (Buckhout et al. 1978).

By the mid-1980s, there were indications from several sources (Sect. 10.2) that the plant plasma membrane was the subcellular location of the auxin target for its role in cell enlargement. However, the identity of the target molecule remained elusive. In order to account for auxin-induced differentiation responses, a promoted and altered gene expression is required and does occur (Baulcombe et al. 1981) although these responses are likely downstream from the primary interaction with the ENOX protein target.

A major advance came when it was shown that cisplatin (but not transplatin), doxorubicin, and *p*-nitrophenyl acetate, agents known to inhibit both cell proliferation and plasma membrane redox activities in mammalian cells (Sun and Crane 1981, 1984, 1985; Chap. 11), inhibited both NADH-ferricyanide oxidoreductase of the isolated vesicles and elongation growth of intact hypocotyl segments. Auxin-induced growth of the isolated segments was inhibited preferentially at drug concentrations where control growth was only slightly affected (Morré et al. 1988a). These findings further indicated a connection between plasma membrane redox reactions and the control of elongation growth of plants (Fig. 10.7).

Plant plasma membranes were shown previously to contain a variety of redox-related oxidoreductases (Barr et al. 1985, 1986; Sandelius et al. 1986). None were observed to respond to auxins. However, NADH oxidation stimulated by auxin was observed in the presence of semidehydroascorbic acid (Morré et al. 1986a; Table 10.1). It was subsequently noted that the auxin-stimulated portion of the activity attributed to semidehydroascorbate reductase did not require the presence of semidehydroascorbic acid but only NADH (Morré et al. 1996a). Thus, the auxin-stimulated portion of the plasma membrane activity was simply an NADH oxidase rather than an oxidoreductase, none of which were auxin responsive (Table 10.2). A stoichometry of approximately 1 was eventually shown for NADH oxidized to 1/2 O_2 consumed for plasma membranes from hypocotyls of dark-grown soybeans (Table 10.3).

The NADH oxidase activity of soybean plasma membranes was subsequently shown to be stimulated by a variety of active auxins but not by their inactive counterparts (Table 10.4). Also the response of the plasma membrane NADH oxidase to auxins paralleled that of cell elongation (Fig. 10.2) including a log-linear dose–response to an optimum at 1 μM. However as the optimum auxin concentration was

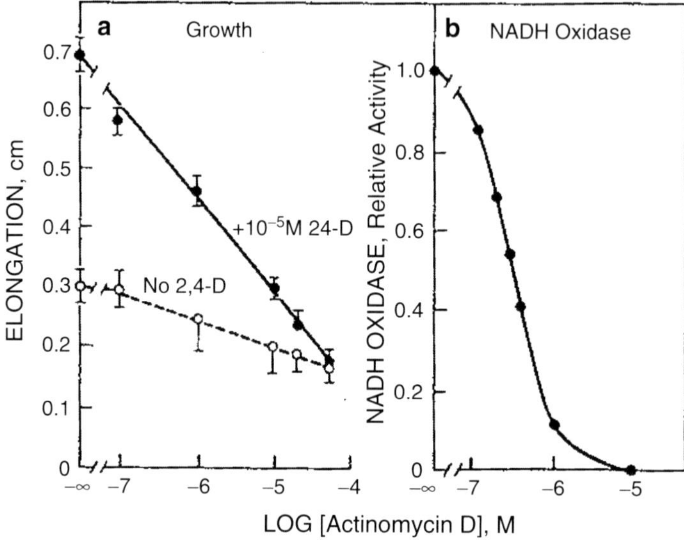

Fig. 10.7 Inhibition by actinomycin D of elongation growth (**a**) and NADH oxidase (**b**) in the presence or absence of 10 μM 2,4-dichlorophenoxyacetic acid (2,4-D). Values are averages from three different experiments (10 segments/experiment) ± SD for growth of 1 cm hypocotyl segments of dark-grown soybean seedlings and from determinations with at least two different plasma membrane preparations for NADH oxidase. For NADH oxidase, the reaction was started by addition of NADH 3 min after the addition of actinomycin D; control activity (no actinomycin D) was 18.5 ± 8.3 nmol NADH oxidized/mg/protein/min. Reproduced from Morré et al. 1988c with permission from Wiley-Blackwell

Table 10.1 NADH oxidation by soybean plasma membranes is isolated in the presence of ascorbate stimulated by auxins[a]

Additions	Concentration	nmol/min/mg/protein
None		15
2,4-D	10 μM	24
Ascorbate	1.7 mM	20
Ascorbate + 2,4-D	1.7 mM + 10 μM	31
Monodehydroascorbate	1 mM	24
Monodehydroascorbate + 2,4-D	1 mM + 10 μM	31

From Morré et al. (1986a)

[a]The reaction mixture contained in a total volume of 3 mL, 50 mM Tris-HC1, pH 7.4, 0.3 M sucrose, 85 μM NADH, and 30–50 μg plasma membrane protein. The monodehydroascorbate was generated from an equal mixture of ascorbate and dehydroascorbate adjusted to pH 7.4 with 10 mM imidazole

exceeded, the rates began to decline. The unusual response to auxin was classically explained on the basis of two-point attachment where at high auxin concentrations both of the required binding sites would be occupied by individual auxin molecules thereby unable to properly activate the target protein (McRae and Bonner 1953). A further line of evidence for the NADH oxidase being a molecular target for the

Table 10.2 Effect of auxin on NADH-dependent redox activities of soybean plasma membranes[a]

		Specific activity		
Electron acceptor	E^0, pH 7 (*V*)	Control	+Auxin	Auxin effect (%)
Oxaloacetate	−0.166	656	566	−14
Ascorbate radical	+0.058	3.1	3.3	+9
Cytochrome c	−0.240	7.9	7.8	−1
Ferricyanide	+0.360	309	273	−12
Nitrate	+0.421	0.4	0.4	0
Oxygen	+0.815	0.7	1.6	+123

From Morré and Brightman (1991)
[a]All activities were determined as the oxidation of NADH (decrease in absorbance at 340 nm) in the presence of electron acceptor. Auxin (1 μM 2,4-D) was added to the reaction after an initial rate had been established. Specific activity units are nmol/min/mg protein

Table 10.3 A comparison of NADH oxidation with O_2 uptake rates by NADH oxidase from whole soybean cells and plasma membranes from soybean hypocotyls

	nmol/mg protein/min		
Source	NADH oxidized	O_2 uptake	Ratio NADH: 1/2 O_2
Right side-out plasma membranes (two phase)	5.5±0.9	2.2±0.0	1.25
Inside-out plasma membranes (FFE)[a]	3.0±0.2	1.4±0.15	1.07
Cultured soybean cells	1.1±0.1[b]	0.75±0.7[b]	0.74

From DeHahn et al. (1997)
Results are from a minimum of three determinations±SD
[a]Free-flow electrophoresis
[b]The O_2 uptake rates and NADH oxidation rates for whole cells are given on the basis of μmol/g/dry wt/min

Table 10.4 Stimulation of NADH oxidase of isolated plasma membranes of soybean hypocotyls comparing growth-active (2,4-D, α-NAA) and inactive (2,3-D, β-NAA) synthetic auxins to indole-3-acetic (IAA) acid and benzoic acid

	Percent of control	
Compound	Growth	NADH oxidase
IAA	166 (10^{-4} M)	222 (10^{-5} M)
Benzoic acid	92 (10^{-5} M)	83 (10^{-6} M)
2,4-D	160 (10^{-5} M)	250 (10^{-6} M)
2,3-D	109 (10^{-5} M)	111 (10^{-6} M)
α-NAA	164 (10^{-6} M)	200 (10^{-7} M)
β-NAA	116 (10^{-6} M)	117 (10^{-7} M)

From Morré et al. (1988c)
Values reported are for compounds tested at the concentrations given in parenthesis. For determination of growth, 1.0 cm long hypocotyl segments were floated overnight in the dark at room temperature in aqueous solution contained in covered glass dishes. Control growth (no auxin) was 0.35±0.025 cm. For determination of effects on NADH oxidase, the compounds were incubated in the absence of detergent for 3 min at 30 °C with the membranes in the reaction buffer. NADH was then added to start the reaction. NADH oxidase control activity (no auxin) was 18±2 nmol NADH oxidized (mg protein)/min. Results are from three different preparations

Table 10.5 Effects of antiproliferative agents and inhibitors of transplasma membrane redox activities with animal cells on control and auxin (10 μM 2,4-D)-induced growth and NADH-ferricyanide oxidoreductase and NADH oxidase in purified plasma membranes of soybean hypocotyls

		Percentage inhibition			
	Inhibitor concentration	Growth		Enzymatic activity	
Inhibitor	(M)	Control	Auxin	NADH-ferricyanide oxidoreductase	NADH oxidase
Doxorubicin	10^{-4}	10	65	37	4
Cisplatin	10^{-5}	0	54	30	13
Transplatin	10^{-5}	0	0	4	0
p-Nitophenylacetate	10^{-4}	0	56	55	−18
Actinomycin D	10^{-5}	33	100	39	100

From Morré et al. (1988c)

For growth experiments, inhibitors were present at the concentration indicated during the overnight incubation. One cm long hypocotyl segments were floated in the appropriate solution in the dark at room temperature in covered glass dishes. For enzyme assays, inhibitors were incubated with the membranes in the assay buffer without detergent for 3 min at 30 °C before the addition of NADH to start the reactions. Activity was measured as described in Materials and methods. Control and auxin-induced growth in the absence of inhibitor were 0.45 ± 0.05 and 0.83 ± 0.025 cm, respectively. NADH-ferricyanide oxidoreductase specific activity was 200 nmol (mg protein)/min and NADH oxidase activity was 11.0 ± 2.7 nmol (mg protein)/min

auxin stimulation of plant cell elongation came from results with the anticancer drug inhibitors of auxin-induced growth (Morré et al. 1988a; Table 10.5). Doxorubicin, cisplatin, but not transplatin, *p*-nitrophenylacetate, and actinomycin D all blocked the activity of the auxin-stimulated plasma membrane NADH oxidase in parallel to effects on inhibition of auxin-induced (but not control) growth (Fig. 10.7).

Studies by Lin (1982a, b), Kochian and Lucas (1985), and Rubinstein et al. (1984) all demonstrated that external NADH could be oxidized by an enzyme at the outer surface of the cell, an external NADH oxidase, or ECTO-NOX = ENOX protein. Data such as that illustrated in Figures 10.12 and 10.13 suggested that there were two such activities, one unaffected by auxin and antitumor drugs and one stimulated by auxin and inhibited by antitumor drugs.

The auxin-stimulated NADH oxidase was subsequently purified from plasma membranes of hypocotyls of dark-grown soybean seedlings (Brightman et al. 1988). The most striking result was that the purified protein was completely inactive in the absence of auxin and exhibited an absolute requirement for auxin for activity (Fig. 10.8). However, it was not until 2003 (Morré et al. 2003b) that the auxin-responsive protein band on SDS-PAGE was correctly identified (Fig. 10.9). Eventually, Morré et al. (2003b) reported that plasma membranes of soybean contained two distinct NADH oxidase activities that oscillated with a period length of about 24 min. One, designated dNOX, was activated by 2,4-D. The other, a constitutive activity designated CNOX (now ENOX1), was unaffected by 2,4-D.

The auxin-stimulated NADH oxidase, like its constitutive counterpart, was resistant to N-terminal sequencing and proteolysis. A monoclonal antibody was

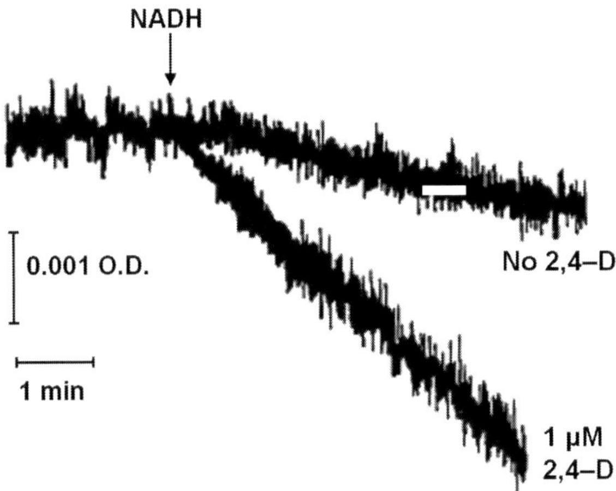

Fig. 10.8 Spectrophotomertric tracing of NADH oxidation by a preparation partially purified from plasma membranes of dark-grown soybeans. The sample was incubated with or without 1 μM 2,4-D for 3 min prior to addition of NADH. Control activity (no 2,4-D) was 9.4 nmol/min/mg protein. Reproduced from Brightman et al. 1988 with permission from Elsevier

prepared that completely blocked plant ENOX1 and cell expansion but expression cloning failed.

An auxin-responsive ENOX protein (dNOX) (ABP20)was cloned and expressed in bacteria in 2012 (L. Ades, J. Kim and D.J. Morre, unpublished). The approximately 22 kDa protein oxidizes NADH and hydroquinone and carries out protein disulfide-thiol interchange. Activity is absolutely dependent upon the presence of either a natural (indole-3-acetic acid, IAA) or a synthetic (2,4-dichlorophenoxyacetic acid, 2,4-D) in the assay. The inactive synthetic auxin analog, 2,3-dichlorophenoxyacetic acid, 2,3-D, was ineffective. The protein contains an auxin-binding motif, an adenine nucleotide (NADH) binding motif, a protein disulfide-thiol interchange motif and two putative copper binding sites as required to account for its activity spectrum. What is most significant is that the expressed recombinant protein is without activity in the absence of auxin such that the activity is, in reality, auxin-induced at the protein level as surmised from early experiments summarized in Chapters 5 and 10.

Plants contain no homologs to the cloned human ENOX1 (Jiang et al. 2008). The constitutive ENOX1 has been cloned from *Arabidopsis* based on homology with a yeast ENOX1 but does not respond to 2,4-D (X. Tang, L. Ades, D. M. Morré and D.J. Morré, unpublished).

The auxin-stimulated protein disulfide-thiol interchange activity from plasma membranes of spinach leaves was elevated in short-day plants compared to plants induced to flower by 24 h of continuous light (Morré et al. 1998d). The activity was reduced about twofold by photoinduction but remained auxin responsive. ENOX1 activity of plants is synchronized by both red (Morré et al. 1999b) and blue (Morré et al. 2002c) light. That the auxin-induced ENOX was also synchronized by light was not investigated.

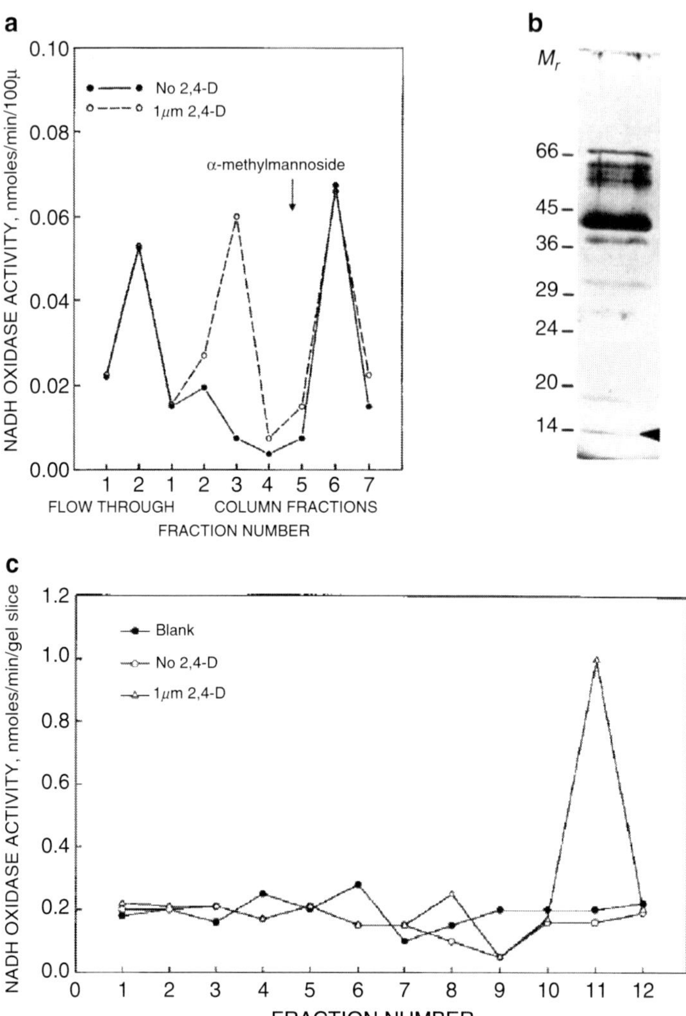

Fig. 10.9 (**a**) Separation of auxin-responsive (Fraction 3, *dotted line*) and auxin-unresponsive (Fraction 6, *solid line*) NADH oxidase activities from plasma membranes of etiolated soybean hypocotyls by differential binding to concanavalin A Sepharose. The auxin-responsive activity bound only weakly to the support and appeared in the third buffer wash. The auxin-unresponsive activity was eluted by 0.1 M α-methylmannoside. (**b**) Analysis of gel slice fractions to determine the relative molecular weight of the 2,4-D-stimulated NOX activity corresponding to fraction 3 of (**a**) in the presence of 1 μM 2,4-D (*open triangles*). An auxin-stimulated activity was found in a single gel slice of M_r range 14–17 kDa. The activity was not detected in the absence of 2,4-D (*open circles*). The auxin-unresponsive activity eluted in the 20.5–24 kDa molecular weight range, as is characteristic of ENOX1 proteins from other species (Sedlak et al. 2001). (**c**) Silver-stained SDS-PAGE (12 % acrylamide) corresponding to column fraction 3 of (**a**). Auxin-activated NADH oxidase activity in a parallel unstained gel was restricted to fraction 11 which contained only a ca. 14 kDa peptide (not shown). Reproduced from Morré et al. 2003b with permission from Springer Science+Business Media

10.4.1 Separation of Auxin-Activated and Constitutive NADH Oxidase Activities

In experiments to purify the NADH oxidase of plasma membrane vesicles from etiolated hypocotyls of soybean, the property determined previously that NOX proteins bound to a concanavalin A column (Bridge et al. 2000) were used (Fig. 10.9a). NADH oxidase activity was determined both in the presence and in the absence of 1 μM 2,4-D. Elution with 50 mM Tris-MES, pH 7, released a fraction that absolutely required 2,4-D for activity (Fig. 10.9). Elution with α-methylmannoside generated a second NADH oxidase activity that was refractory to 2,4-D.

In these experiments and in the experiments with gel slices (Fig. 10.9b), the NADH oxidase was refolded by reducing disulfide bonds with 1 μM reduced glutathione (GSH) followed by refolding in the presence of 150 μM NADH and 25 mM Tris-MES, pH 7.2, for 10 min followed by addition of 0.03 % hydrogen peroxide to reoxidize thiols to form disulfides and to initiate the reaction. Restoration of 2,4-D responsive NADH oxidation by right side-out vesicles of soybean plasma membrane also responded to activation and refolding by this method (DeHahn et al. 1997).

When further purified by SDS-PAGE (Fig. 10.9c), a fraction corresponding to a M_r of ca. 14 kDa (14–17 kDa range) exhibited 2,4-D-stimulated activity based on analyses of the 2,4-D-stimulated enzymatic activity of gel slices. The sliced gels were unstained. The slice from the gel region containing the 14 kDa band exhibited no activity in the absence of 2,4-D. The activity was observed only upon the addition of 1 μM 2,4-D to the gel slice (Fig. 10.8). A band in the M_r range of 14–17 kDa ran very near the dye front on the native gel electrophoretic separation of Brightman et al (1988). These low molecular weight bands were present but overlooked in the Brightman et al (1988) study but were near the M_r of 22 kDa found for expressed recombinant dNOX (Sect. 10.4).

Like the constitutive ENOX1 (Lin 1982a, b; Kochian and Lucas 1985; Rubinstein et al. 1984), the auxin-responsive NADH oxidase is located on the external surface of the plasma membrane (Barr et al. 2000b, c; DeHahn et al. 1997). Since NADH is an impermeant substrate, this allows measurement of auxin-induced NADH oxidation with intact cells, tissue segments (Hicks and Morré 1998; Barr et al. 2000b) (Fig. 10.3), and leaf discs (Barr et al. 2000c). It also limits auxin-induced NADH oxidase activity to right side-out plasma membrane vesicles and its absence from inside-out plasma membrane vesicles resolved, for example, by preparative free flow electrophoresis (Canut et al. 1988).

10.5 Auxin-Stimulated NADH Activity and Growth Oscillates with a Period Length of 24 min

A unifying characteristic of ENOX proteins is the periodicity of its activity. The auxin-stimulated (activated) activity is no exception (Morré and Morré 1998). Auxin-related oscillations were observed consistently in auxin-induced growth (Hicks and Morré 1998) (Fig. 10.3), in auxin-induced proton pumping (Felle 1987,

1988; Chap. 7), and in the auxin-stimulated protein disulfide-thiol interchange activity (Morré et al. 1995d). All three activities oscillated with period lengths of 24 min (Fig. 10.10). The period length was independent of temperature and entrainable (Morré and Morré 1998; Pogue et al. 2000). Patterns of oscillations in activity following addition of 2,4-D correlated with a corresponding set of 24-min oscillations in the rate of cell enlargement also determined at short time intervals.

The cell surface ENOX activity of soybean stems and of pea and cucumber tendrils responded to touch (mechanical stimulation) by a several-fold increase in activity (Barr et al. 2000a) (Fig. 10.11). However, when the ENOX activity was fully stimulated by auxin (2,4-D), the ENOX activity failed to respond further to touch. In contrast to the ENOX of stems and leaves, the ENOX of plant roots is inhibited by auxin (as is the growth response) (Unpublished) and is gravity responsive (Garcia et al. 1999; Bacon and Morré 2001) both with intact roots and with isolated plasma membrane vesicles. The naturally occurring electron donor for the auxin-stimulated ENOX is vitamin K (phylloquinone) (Bridge et al. 2000).

The renaturation of scrambled (oxidized and inactive) RNase A was catalyzed by soybean plasma membranes (Fig. 2.16). The catalysis was stimulated by both the auxin herbicide 2,4-D and by IAA (Morré et al. 1995d). The inactive auxin analog, 2,3-dichlorophenoxyacetic acid was without effect. The activity occurred in the absence of external electron acceptors or donors and appeared to be a true disulfide-thiol interchange activity between protein disulfides and thiols of RNase A and those of plasma membrane proteins. The activity was not affected by a mixture of reduced and oxidized glutathione. However, no auxin-stimulated activity was observed in the presence of either oxidized glutathione or reduced glutathione alone. Early indications of an auxin effect on the redox state of the cytoplasm were reported for soybean membranes and soluble proteins (Morré 1970) and were offered as a biochemical explanation for the earlier work of Galston and Kaur (1963) showing an increase with auxin in the heat coagulability of cytoplasmic proteins. Both the sulfhydryl content of cytoplasmic proteins and the heat coagulability showed a concentration dependency in response to auxin that paralleled the growth response (Morré 1970) and the direct effect of auxin on the plasma membrane NADH oxidase (Morré et al. 1986a, 1988a). Taken together, the results suggest the operation in the plant plasma membrane of a protein disulfide-thiol interchange activity that is stimulated by auxins. A direct measure of the protein disulfide-thiol interchange activity was provided by spectrophotometric measurement of the cleavage of dipyridyl-dithio substrates. The cleavage products absorb specifically at A_{340} (Morré et al. 1999a) (Chap. 2).

That the auxin-stimulated ENOX proteins participate in protein disulfide-thiol interchange has been reinforced by direct measurements of thiols and disulfides that show an auxin-stimulated increase in thiols of membrane proteins in the presence of NADH or cysteine with a corresponding decrease in membrane disulfides (Chueh et al. 1997a). In the absence of NADH or cysteine, a 2,4-D response of membrane thiols is also seen, but in the opposite direction, i.e., a decrease (Chueh et al. 1997a). Thus, the general reaction catalyzed by the 2,4-D-responsive protein appears to be protein disulfide-thiol interchange (Morré et al. 1995d). The response to detergent treatment suggests that the protein disulfides reduced are at the outer surface of the plasma membrane.

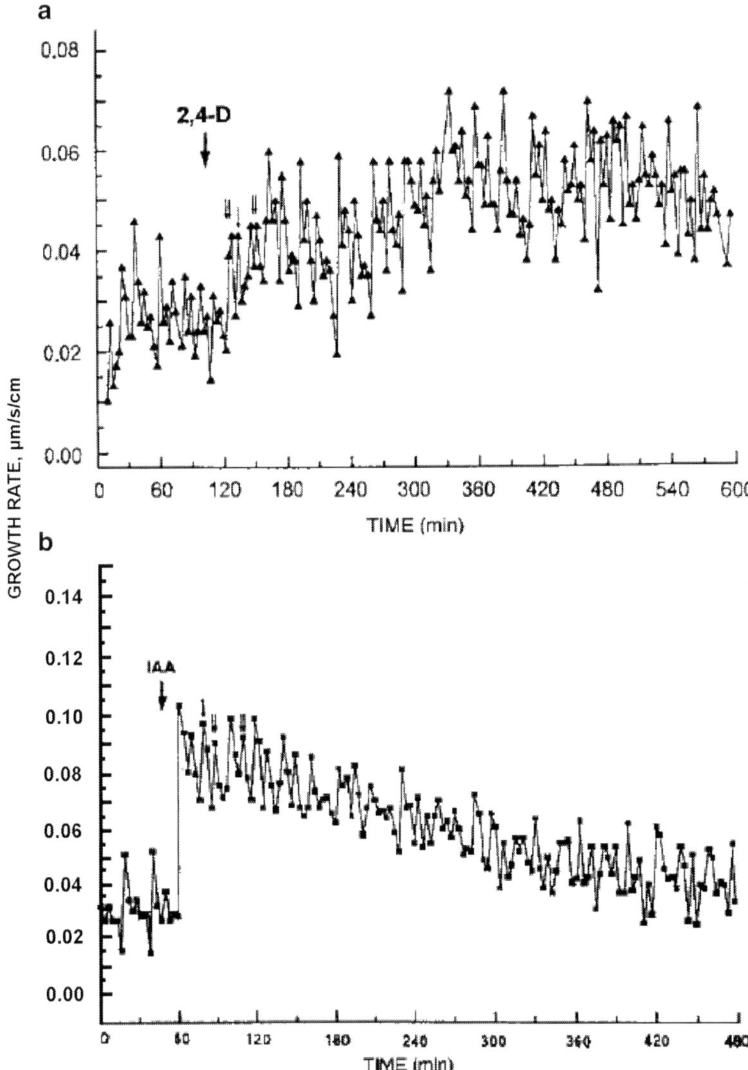

Fig. 10.10 Oscillations in the rate of elongation of four 0.5 cm sections of dark-grown seedlings of soybean measured at intervals of 3 min using a multichannel auxanometer as described by Lüthen and Böttger (1992). (**a**) After 2,4-D addition, a second set of oscillations (*double arrows*) was observed midway within the first set of oscillations. (**b**) Similar results were obtained with the natural auxin regulator indole-3-acetic acid (IAA). Normally, three of the five subdivisions of the 24-min NOX cycle contribute to a cell enlargement, with one being dominant. Between 480 and 510 min, a discontinuity was seen in the pattern where all five seem to contribute for a single cycle. Reproduced from Morré et al. 2003b with permission from Springer Science+Business Media

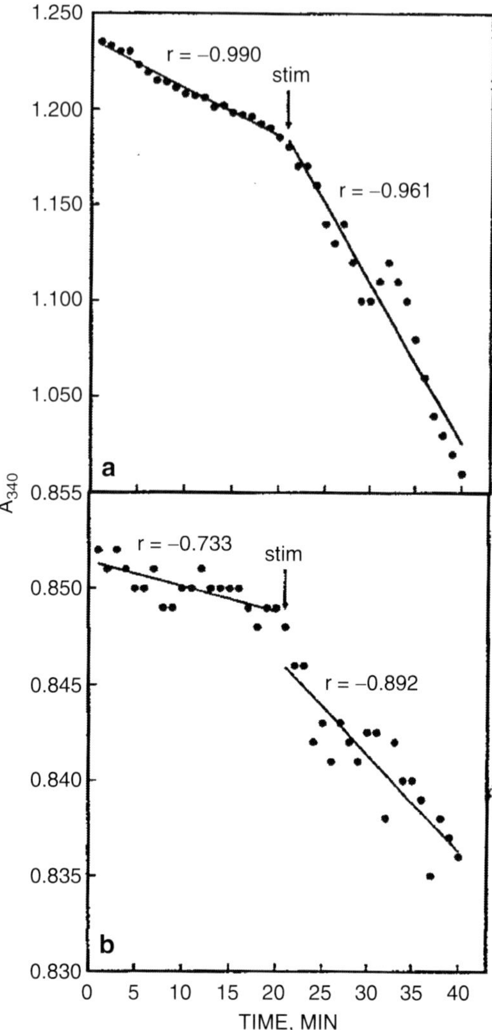

Fig. 10.11 Decrease in A_{340} as a measure of cyanide-insensitive NADH oxidation (Chap. 2) by excised tendrils in 4 mL of solution containing 50 mM Tris-MES (pH 7), 1 mM KCN, and 150 μM NADH, at 24 °C. At intervals of 1 min, the incubating solution was mixed, and a 2.5 mL portion was transferred to a second cuvette and the absorbance at 340 nm was determined. The solution was then returned to the incubation vessel. After 20 min, the tendrils were stroked six times along the concave surface (in an *upward* direction) with a wooden stick or were tapped, and the rate was redetermined for 20 min. (**a**) Four cucumber tendrils. The average rate of NADH oxidation calculated from the regression line before stimulation was 0.3 nmol/min/tendril. After stimulation, the rate was 0.8 nmol/min/tendril. (**b**) One pea tendril. The average rate of NADH oxidation calculated from the regression line before stimulation was 0.05 nmol/min/tendril. After stimulation, the rate was 0.2 nmol/min/tendril. Periods of rapid activity alternate with periods of inactivity. Two such inactive periods are clearly observed at 8 and 32 min in (**a**). Synchrony among tendrils was achieved by light exposure of plants maintained overnight in darkness. Reproduced from Barr et al. 2000a with permission from The University of Chicago Press

Indirect evidence for a thiol involvement in auxin-induced growth was provided earlier in that auxin-induced cell enlargement was inhibited by thiol reagents in parallel to inhibition of auxin-stimulated oxidation of NADH (Morré et al. 1995a). Again, an involvement of thiols at the cell surface is implicated in that auxin-stimulated cell enlargement is prohibited specifically by the impermeant thiol reagent *p*-chloromercuriphenylsulfonic acid (Chueh et al. 1997a).

10.6 Golgi Apparatus Transport Important to Sustained Cell Enlargement but not Specifically Required for Auxin-Induced Cell Enlargement

The fungal macrolide antibiotic inhibitor, brefeldin A, known to inhibit trafficking from the Golgi apparatus to the plasma membrane (Miller et al. 1992; Schindler et al. 1994) also specifically inhibits auxin-induced cell elongation in plants (Fig. 7.4). Such findings might be regarded as evidence for an obligatory role of secretory vesicles derived from the Golgi apparatus in cell enlargement at least in plants since the action of brefeldin A is normally thought to reside exclusively somewhere within the Golgi apparatus. However, several lines of evidence do not support this interpretation. At temperatures of 18 °C or less, Golgi apparatus activity of elongating segments of soybean is blocked as evidenced by extensive accumulations of membranes at the trans Golgi network (Morré and Mollenhauer 2009). Similar accumulations are seen with temperature blocks in other plant (Noguchi and Morré 1991) and animal cells (Tartakoff 1986). Yet, cell enlargement induced by auxin shows no sharp temperature transition over the entire range of 4–25 °C (Fig. 7.4) suggesting that elongation growth in plants induced by auxin may occur independent of the vesicular transport pathway.

Similar conclusions were reached earlier based on observations with monensin using the narrow setae of the moss *Pellia* (Schnepf et al. 1979). The setae elongate rapidly in response to auxin and are thin enough that the monensin is able to penetrate throughout the tissue segment as evidenced by swollen trans elements of the Golgi apparatus associated with all Golgi apparatus stacks following fixation with glutaraldehyde (Morré et al. 1986c). Despite the nearly complete inhibition of normal *trans* Golgi apparatus functioning by monensin, auxin-treated setae respond to auxin by accelerated growth for times of 4 h or more in the presence of monensin. Beyond 4 h of monensin treatment, auxin-induced elongation ceases abruptly, suggesting that essential Golgi apparatus-derived plasma membrane precursors required for sustained expansion have eventually been depleted.

10.7 Response of ENOX of Isolated Plasma Membrane Vesicles to Osmotica

The modulation of ENOX activity in response to turgor provides a parallel between this activity and elongation to accompany the responsiveness of each to both natural and synthetic auxins. Auxin-induced cell elongation exhibits a strict dependency on turgor with maximum auxin-induced elongation only with full turgor (Cleland 1959). Similar parallels to the elongation data reported for *Avena* coleoptiles (Cleland 1959) have been observed with mesocotyls (Morré and Eisinger 1968) as well as epicotyls (Coartney and Morré 1980a, b). Auxin-induced elongation reaches its maximum at full turgor and is reduced to basal levels at concentrations of osmotica that result in incipient plasmolysis (e.g., 0.2–0.25 M sucrose or mannitol) (Morré and Eisinger 1968).

In this regard it is of interest that the ENOX activity of isolated vesicles from soybean plasma membrane exhibited a sensitivity to membrane stretching that paralleled the effects of osmotica on elongation (Morré and Brightman 1997). In these studies, NADH oxidation was measured with isolated vesicles whereas growth was measured with intact segments. Elongation was measured over 20 h and ENOX was measured over 20 min. However, both ENOX activity of plasma membrane vesicles and elongation of hypocotyl segments appeared to be steady state after relatively short initial lags (Chap. 5).

The responses are considered to result from membrane stretching rather than from the solutes per se based on the following arguments. If the resuspended vesicles were mixed with sucrose, activity was greatly reduced and the 2,4-D-stimulated component of activity was eliminated. In contrast, activity was enhanced by 2,4-D when the vesicles were added to buffer lacking sucrose. Similarly, when water reentered the vesicles in concert with solute entry, NADH activity also was restored. Continuous recording of ENOX activity and vesicle dehydration (vesicle volume) determined by light scattering showed a correlation between ENOX activity and vesicle volume. A direct effect of solute also was ruled out by experiments where sucrose, sorbitol, or mannitol was without effect between 0 and 0.5 M with detergent solubilized membranes either in the presence or in the absence of auxin. A relationship between response of the oxidase and cell wall contact was ruled out since the inhibitions were observed with isolated plasma membrane vesicles.

The classical interpretation of the effects of osmotica on plant cell elongation is that cell elongation is turgor driven and as turgor is reduced, elongation is reduced in parallel. We have postulated that the ENOX activities per se are sensitive to membrane stretch and the auxin-sensitive NADH oxidase activity exhibits a greater sensitivity to increasing concentrations of osmotica than does the basal activity. Available evidence supports the concept that it is the oxidase itself that is stretch sensitive and that linkage to a stretch receptor is not required for inhibition of either growth or the oxidase.

10.8 Inhibitors of the Auxin-Stimulated NADH Oxidase of Plants

Also a molecular target for low levels of exposure to organic solvents stimulatory to plant cell elongation (Morré 1998a, b, 2000), 2,4-D induced growth was inhibited by octacosanol, an antagonist of the NADH oxidase and by the growth stimulatory

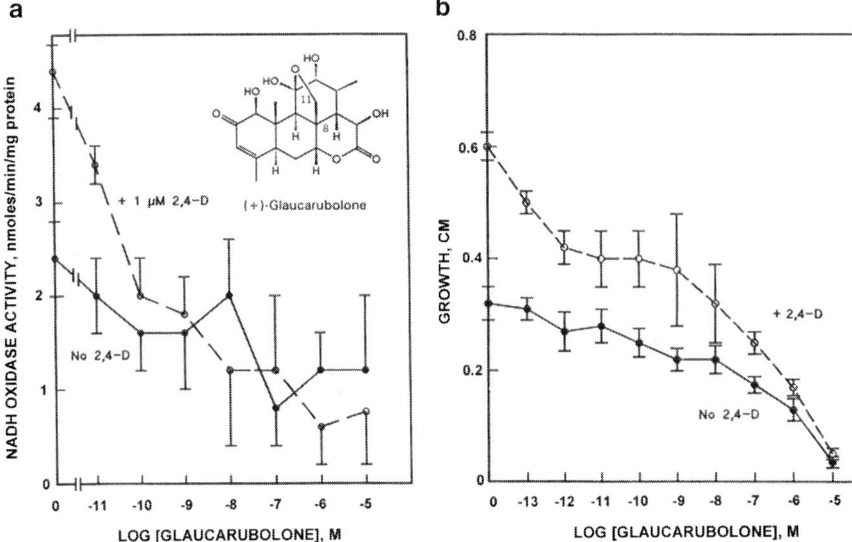

Fig. 10.12 The quassinoid anticancer drug glaucarubolone (Grieco et al. 1996, 1997) preferentially inhibits NADH oxidation and growth induced by auxin (2,4-D). (**a**) NADH oxidase activity of plasma membrane vesicles isolated from dark-grown hypocotyls of soybean to glaucarubolone in the presence (*solid symbols, solid lines*) and absence (*open symbols, broken line*) of 1 μM 2,4-D. Values are averages from three different membrane preparations ± SD. (**b**) Elongation of 1 cm segments of hypocotyls from dark-grown seedlings of soybeans over 18 h as a function of glaucarubolone concentration in the absence (*solid symbols* and *solid line*) or presence (*open symbols, dotted line*) of 1 μM 2,4-D. Results are averages of three experiments (10 sections per treatment per experiment) ± SD among experiments. Reproduced from Morré and Grieco 1999 with permission from The University of Chicago Press

straight-chain primary alcohol, triacontanol (Morré et al. 1991c). The auxin-stimulated NADH oxidase activity from soybean hypocotyls was inhibited by submicromolar concentrations of ATP (Morré 1998d; Sect. 3.5).

Distinguishing characteristics of plant ENOX1 and its 2,4-D stimulated counterpart are its response to quassinoid antitumor drugs (Morré and Grieco 1999) (Figs. 10.12 and 10.13). Glaucarubolone specifically inhibits the auxin-induced NADH oxidase and auxin-induced growth with no effect on the constitutive oxidase and growth. In contrast, simalikalactone D inhibits the constitutive oxidase and growth with no effect on its auxin-induced counterpart.

Glaucarubolone, a naturally occurring quassinoid from the root bark of *Castela polyandra*, Simaroubaceae, with a C(8), C(11) hemiacetal bridge, is a potent inhibitor of both the auxin-induced component of the plasma membrane NADH oxidase and of plant cell enlargement in soybean and other plant species (Morré and Grieco 1999). Auxin-stimulated ENOX activity of isolated plasma membrane vesicles was inhibited half maximally by ca. 0.1 nM glaucarubolone. Auxin-induced enlargement of stem segments of soybean and elongation growth of *Arabidopsis thaliana*

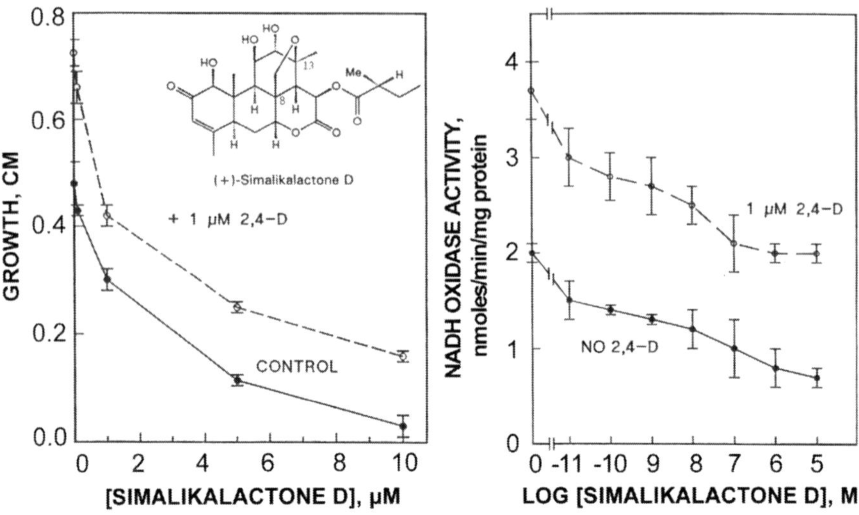

Fig. 10.13 (**a**) Elongation of 1 cm sections excised from the elongating zone of hypocotyls of dark-grown seedlings of soybeans over 18 h as a function of simalikalactone D concentration. Results are averages of three experiments of ten sections per treatment per experiment ± SD among experiments. (**b**) NAD oxidase activity of plasma membrane vesicles isolated from dark-grown hypocotyls of soybean to simalikalactone D in the absence (*solid symbols, solid line*) and presence (*open symbols, broken line*) of 1 μM 2,4-D. Values are averages from three different membrane preparations ± SD. Reproduced from Morré and Grieco 1999 with permission from The University of Chicago Press

seedlings were inhibited half maximally by ca. 10 nM glaucarubolone with significant inhibition of auxin-stimulated growth even at nannomolar glaucarubolone concentrations. Seedlings of *A. thaliana* and tomato treated with sublethal concentrations of 1–10 μM glaucarubolone remained alive but failed to elongate for periods of two to several months. Once seedlings recovered from the effects of the inhibitor, they resumed growth, flowered, and produced fruits. Treated *A. thaliana* produced viable seeds that germinated and developed into normal-appearing progeny. In contrast to glaucarubolone, which inhibited preferentially auxin-induced growth and plasma membrane ENOX, simalikalactone D, a quassinoid with a C(8), C(13) epoxymethano bridge, inhibited the constitutive ENOX activity and growth and had little or no effect on the 2,4-D responsive ENOX and growth. The findings demonstrate a functional role of both the constitutive and the auxin-stimulated plasma membrane ENOX activities in plant cell enlargement.

The sulfonylurea herbicide chlorsulfuron (2-chloro-*N*-[(4-methoxyl-6-methyl-1,3,5-triazin-2y1amino)carbonyl] benzene sulfonamide) which functions as well as a plant growth retardant (Morré and Tautvydas 1986) inhibited both auxin-induced cell enlargement in excised stem sections and the auxin-stimulated ENOX component of isolated plasma vesicles of soybean (Morré et al. 1995a). The only inhibitor with specificity for the basal ENOX and basal cell enlargement in plants is simalikalactone D (Fig. 10.13; Morré and Grieco 1999). Both the plant growth retardant mefluidide

(*N*-[2,4-dimethyl-5-[[(trifluoromethyl) sulfonyl]amino] phenyl] acetamide) (Morré and Tautvydas 1986) as well as the sulfonylurea herbicide sulfosulfuron (Trade Name: Outrider) with specificity for control of the noxious weed Johnsongrass (*Sorghum halapense*) appear to be closely linked if not directly targeted to the ENOX1 of plants as well.

The auxin-stimulated ENOX activity was purified and shown to require auxin addition for activity (Brightman et al. 1988). It was initially reasonable to assume that the effect of auxin was on the constitutive ENOX. The study of Morré and Grieco (1999) using quassinoid growth inhibitors clearly demonstrated that the auxin-stimulated cell surface ENOX was distinct from the constitutive oxidase. The constitutive NADH oxidase activity and the constitutive cell elongation were inhibited by simalikalactone D but not by glaucarubolone, whereas the 2,4-D-stimulated NADH oxidase activity and the 2,4-D-stimulated rate of cell enlargement were inhibited by glaucarubolone but not by simalikalactone D.

Glaucarubolone (Fig. 10.12) was isolated from root bark of *Castela polyandra* (as described in Grieco et al. 1995). A member of a related group of natural products derived from plants collectively known as quassinoids, the substance contains a 20-carbon picrasane skeleton and is heavily oxygenated. Simalikalactone D (Fig. 10.13), also a quassinoid, was obtained by total synthesis (Moher et al. 1992).

10.9 Cell Elongation Oscillates with a Period of 24 min and Exhibits a Second set of Oscillations in Response to 2,4-D

Rates of enlargement of 0.5-cm sections of soybean hypocotyls (Figs. 10.9 and 10.10) were markedly periodic. The rates were enhanced by 2,4-D (Fig. 10.10a) or IAA (Fig. 10.10b). While the length of the sections increased at an average steady-state rate of 0.02–0.03 μm/sec/cm and was increased to 0.04-0.07 μm/s/cm in the presence of 1 μM 2,4-D, the rate of elongation fluctuated by more than a factor of two around these means. The fluctuations were not random but exhibited a regular pattern of oscillations with a period of about 24 min (Morré et al. 2002d).

With 2,4-D-stimulated cell elongation, a new increased rate of elongation and oscillatory activity was seen within a few minutes after auxin addition (Fig. 10.10, Morré et al. 2002d). In addition, a new complex pattern of major and minor oscillations typically was observed following 2,4-D addition. However, a steady state eventually was achieved where the original 24-min period was restored but at the 2,4-D-stimulated rate.

The oscillations observed were not inherent in the instrumentation. When tissue was replaced by polystyrene tubing, a level of background variation was observed but without periodicity (Morré et al. 2002d).

The immediate response to 2,4-D was a double set of oscillations with one corresponding to the original set of constitutive oscillations in the absence of 2,4-D (Fig. 10.14; see also Fig. 5.14). A second set of oscillations corresponding to a new 24-min period length in response to the 2,4-D (indicated by double arrows) was observed as the rate of cell elongation began to accelerate following the addition of

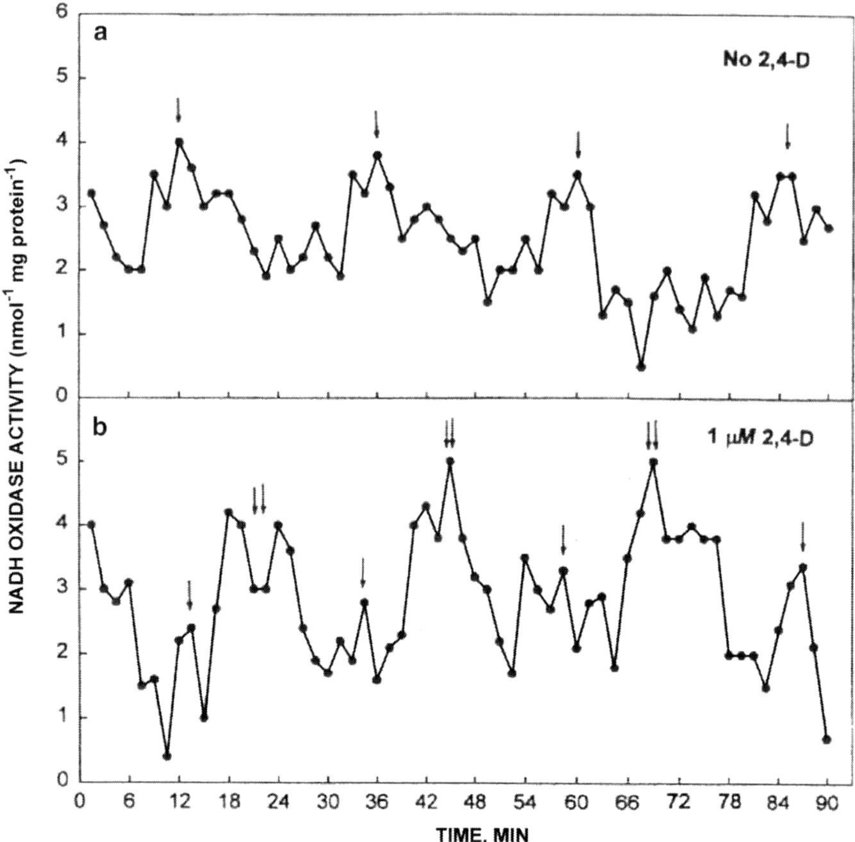

Fig. 10.14 Experiment illustrating the apparent recruitment of ENOX1 by 2,4-D-activated dNOX in a cell-free system. Shown are the oscillations in NADH activity of soybean plasma membranes prepared from hypocotyls of dark-grown soybean seedlings incubated for 3 h at 30 °C in the absence (**a**) or presence (**b**) of 1 μM 2,4-D. The assays were simultaneous and used two spectrophotometers in parallel. 2,4-D was added to one portion of the preparation and the other portion was assayed in the absence of 2,4-D. The characteristic ENOX1 period length of 24 min was seen in both preparations (*single arrows*). In addition, a 2,4-D responsive component, not seen in the absence of 2,4-D, appeared also with a period length of about 24 min (*double arrows*). The 2,4-induced increase in the rate of NADH oxidation was accompanied by a corresponding decrease in the activity ascribed to ENOX1. (**a**) Control plasma membrane preparation incubated for 3 h at 30 °C and assayed with no 2,4-D added. The oscillating NADH oxidase activity with a period length of 24 min (ENOX1) is indicated at the *single arrows*. (**b**) 2,4-D treated plasma membrane preparation after 3 h at 30 °C. The dNOX pattern of oscillating NADH oxidase activity (*double arrows*) now dominates at the expense of the ENOX1 pattern (*single arrows*). Similar results were obtained in experiments where membranes were reassayed after 48 h at 4 °C. Reproduced from Morré et al. (2003b) with permission from Springer Science+Business Media

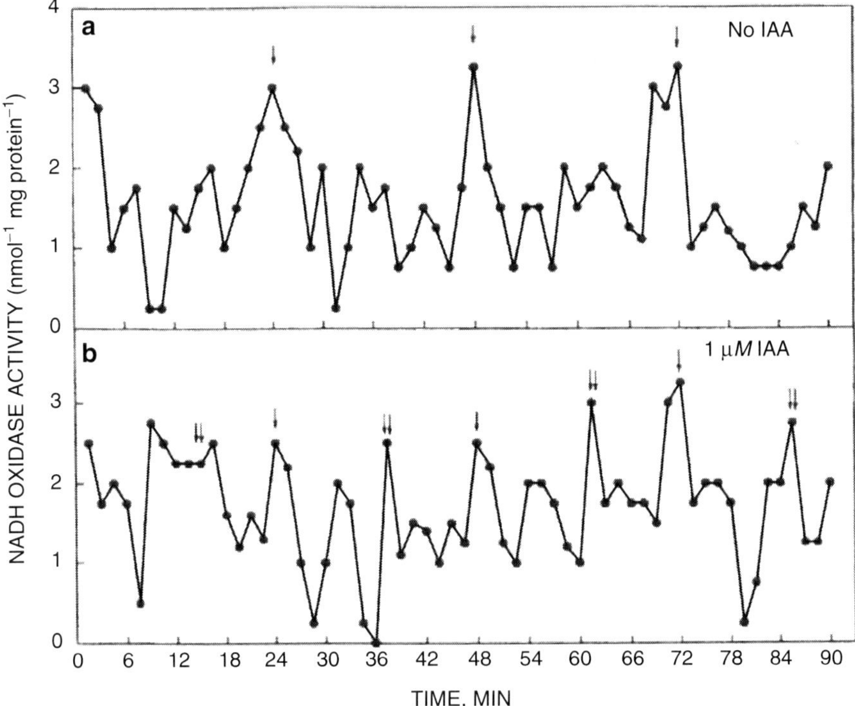

Fig. 10.15 Oscillations in NADH oxidase activity of soybean plasma membranes in the absence (**a**) and presence (**b**) of 1 μM indole-3-acetic acid (IAA). The simultaneous assays used two spectrophotometers in parallel with the same plasma membrane preparation. The characteristic ENOX1 period length of 24 min was seen in both preparations (*single arrows*). In (**b**), a second IAA-responsive component, not seen in (**a**) in the absence of IAA, appeared also with a period length of about 24 min (*double arrows*). Reproduced from Morré et al. (2003b) with permission from Springer Science + Business Media

2,4-D (Fig. 10.14). A new steady-state rate of elongation was achieved and a dominant single set of oscillations with a period length of 24 min was established after about 3 h of incubation. A similar pattern was seen with the natural growth regulator indole-3-acetic acid (IAA), where a second IAA-responsive oscillatory component was induced following IAA addition (Fig. 10.15).

10.10 The Auxin-Stimulated ENOX Has Properties of a Prion: How 2,4-D Kills Plants

The surprising finding was that what initially starts out as a double set of 24-min oscillations with 2,4-D eventually stabilizes into a single set of oscillations reflecting the 2,4-D-induced rate (Figs. 10.16 and 10.17). A progressive resolution of the two periods into a single combined period reflecting the 2,4-D rate was suggestive of

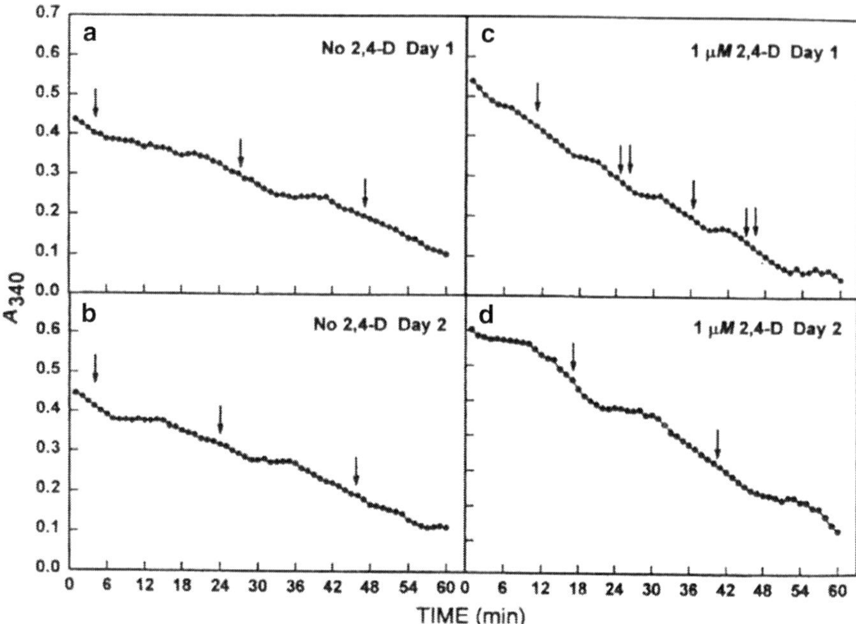

Fig. 10.16 The in vitro findings are supported by results in vivo with excised segments, 1 cm in length, cut from the elongating region of dark-grown seedlings of soybeans. In the absence of 2,4-D, the patterns of oscillations in the rate of NADH oxidation with a period length of about 24 min were similar on days 1 and 2. With 2,4-D added at day 1 and assayed immediately, the complex initial pattern of oscillations exhibited both constitutive and 2,4-D induced oscillations. By day 2, the double pattern had resolved into a single set of oscillations. Synchrony was achieved by exposure of dark-grown seedlings to light. Each experiment was repeated four times with consistent results. The ratio of NADH oxidase activity plus 2,4-D to no 2,4D on day 1 was 1.4 and 1.8 on day 2. Reproduced from Morré et al. 2003b with permission from Springer Science + Business Media

recruitment of one NOX form by the other through a process analogous to that exhibited by prions. The 3-h interval was selected, based on growth studies (Morré et al. 2002d), as the time required for apparent entrainment of the initial ENOX1 oscillations and the subsequent dNOX oscillation to unify as a single new set of oscillations.

With the preparations incubated for 3 h at 30 °C, the relative ENOX activity of the maxima responding to 2,4-D increases, whereas the area under the constitutive maxima declines (Table 10.6). The sum of the two activities remains constant. To explain these unusual findings, we postulate recruitment of ENOX1 molecules by dNOX molecules (Fig. 10.18).

The cancer form of human ENOX (ENOX2) has the properties of a prion (resistance to proteases, heat, and chemical degradation; solubility characteristics; β-sheet character; and ability to form amyloid rods) (Kelker et al. 2001). ENOX2 also has the ability to impart proteinase K resistance to proteins that are normally susceptible to proteinase K digestion. The 2,4-D-stimulated ENOX protein is proteinase

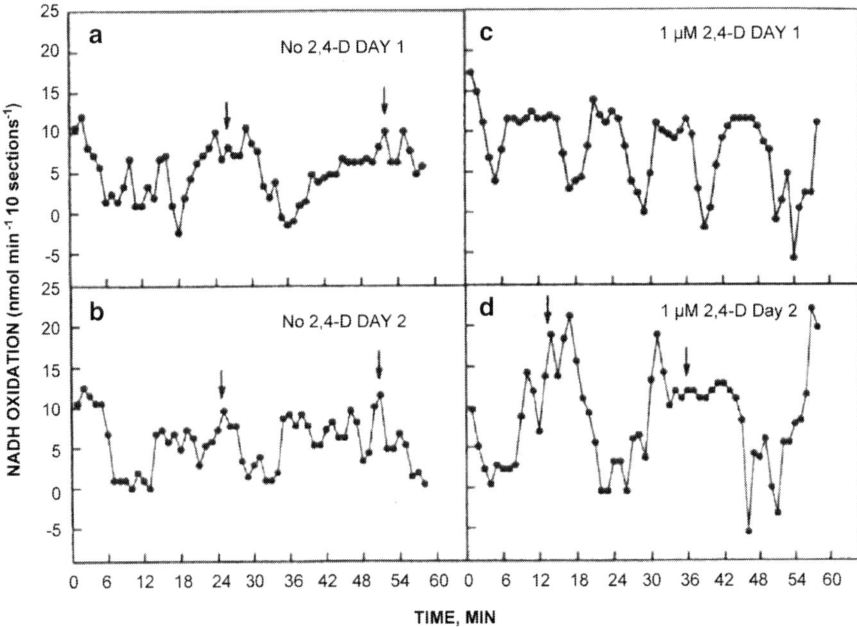

Fig. 10.17 Rates of NADH oxidation calculated from the change in A_{340} of Fig. 10.10. by numerical averaging (change in absorbance (ΔA) at time $t=(A_{t+1}-A_{(t-1)})/2$) followed by conversion to specific activity of NADH oxidation. Intervals of 24 min coinciding with maximum rates of NADH oxidation (*arrows*) were observed in the absence of 2,4-D (**a**, **b**) and at day 2 with 2,4-D (**d**). In the first h after 2,4-D addition (**c**), however, the oscillations from both the constitutive NOX plus the oscillations from the 2,4-D induced NADH oxidation were observed. Note that the overall rate of NADH oxidation by day 2 was much accelerated (**d**) and approximated the sum of the two activities (constitutive and 2,4-D induced) observed on day 1 immediately following 2,4-D addition (**c**). Reproduced from Morré et al. (2003b) with permission from Springer Science + Business Media

Table 10.6 Area under oscillatory activity maxima (nmole/min/mg protein) of plasma membrane preparations of etiolated hypocotyls of soybean incubated in the presence and absence of 1 μM 2,4-D for 3 h, stored frozen (−20 °C) for varying times (average 1 week) and reassayed

After 3 h			After 1 week (Ave)		
No 2,4-D ENOX1	+1 μM 2,4-D ENOX1	dNOX	No 2,4-D ENOX1	+1 μM 2,4-D ENOX1	dNOX
5.5	5.6	7.2	5.8	2.3[a]	10.4[a]

Modified from Morré et al. (2003b)
[a]Significantly different $p<0.001$

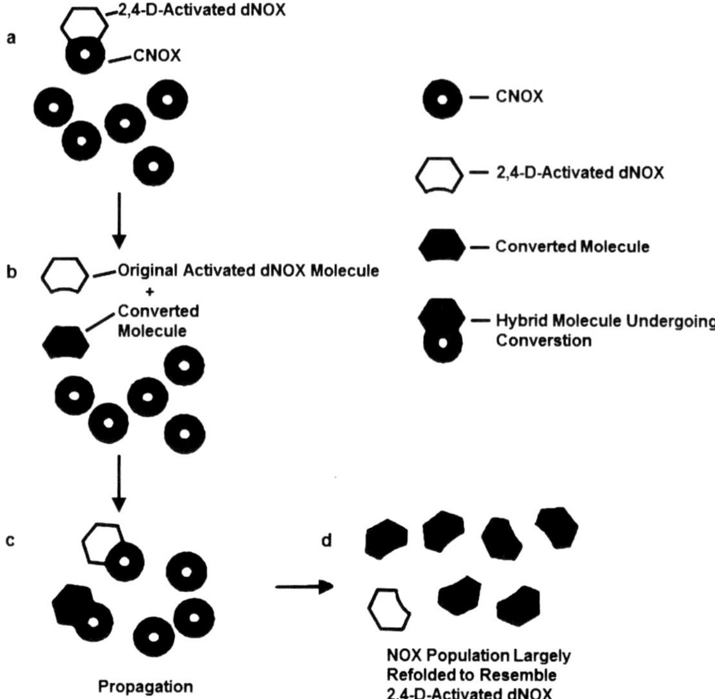

Fig. 10.18 Diagram to explain the results of Figs. 10.14–10.17 and Table 10.6. An active recruitment of ENOX1 proteins by dNOX proteins is postulated, followed by a conversion of the ENOX1 proteins into a likeness of the other (dNOX), much in the manner of the creation of infective prion proteins from non-infective prion proteins by an interaction of one with the other. (**a**) A 2,4-D-activated dNOX is postulated to recruit normal ENOX1 proteins and cause them to undergo a conformational alteration to allow their entrainment with other 2,4-D-stimulated ENOX proteins. (**b**) The converted molecules (*shaded*) are then considered to participate in the conversion of other ENOX1 proteins. (**c**) The net result is a propagation of the 2,4-D response and a net loss of unconverted ENOX1. (**d**) The end results would be an ENOX population largely refolded to resemble 2,4-D activated dNOX and entrainment to produce predominantly a single synchronous 2,4-D-stimulated oscillatory ENOX activity and growth response. Reproduced from Morré et al. (2003b) with permission from Springer Science+Business Media

K-resistant (Table 10.7) and, like ENOX2, is inhibited by anticancer drugs (Table 10.8). While other explanations may be possible, the scheme in Fig. 10.18 depicts a prion-like mechanism but for plants.

In the scheme of Fig. 10.18, the 2,4-D-responsive ENOX (dNOX) when activated by 2,4-D would recruit constitutive ENOX1 proteins and alter them to exhibit functional properties similar to those of activated dNOX. This property, that of converting a normal form of a protein into a likeness of itself, is one of the defining characteristics of the group of proteins designated as prions (Griffith 1967; Prusiner 1994).

A puzzling aspect of 2,4-D action, as well as that of other auxin herbicides such as picloran (4-amino-3,5,6-trichloropicolinic acid) (Eisinger and Morré 1971) or

Table 10.7 2,4-D stimulated NADH oxidase activity of isolated plasma membranes from etiolated soybean hypocotyls is resistant to digestion with proteinase K

	Plasma membrane NADH oxidase activity (nmol/min/mg protein)	
	Before proteinase K	After proteinase K[a]
Total	1.8 ± 0.4	0.9 ± 0.2
2,4-D stimulated	0.6 ± 0.1	0.5 ± 0.1
2,4-D insensitive	1.2 ± 0.2	0.4 ± 0.1

From Morré et al. (2003b)
Results are averages from three different plasma membrane preparations ± SD
[a]Proteinase K: 0.1 mg/mL, 37 °C, 30 min

Table 10.8 Anticancer and/or differentiation promoting substances but not the inactive transplatin inhibit specifically the elongation growth of 1 cm long excised segments of etiolated soybean (*Glycine max*) hypocotyls induced by the auxin herbicide 2,4-D (1 μM)

	EC_{50} (M)	
Substance	No 2,4-D	+2,4-D
Doxorubicin	No inhibition	10^{-4} M
Cisplatin	No inhibition	2×10^{-6} M
Transplatin	$>10^{-4}$	$>10^{-4}$
Actinomycin D	10^{-6}	$5\times>10^{-7}$
Retinol	No inhibition	10^{-4}
Retinoic acid	$>10^{-4}$	$2\times>10^{-6}$
Capsaicin	$>10^{-4}$	$5\times>10^{-5}$
p-Nitrophenylacetate	$>10^{-4}$	10^{-4}

Based on analyses with at least three repetitions with consistent results (Unpublished)

perhaps even trichlopyr [(3,5,6-trichloro-2-pyridinyl)oxy]acetic acid (Morré et al. 1998c) now explained, is that the fatal formative effects of the herbicides continue to be expressed for days, weeks, and even months following a single spray application, long after the herbicides are no longer present anywhere in the plant. The irreversibly activated auxin-stimulated ENOX together with its ability to recruit and stably activate the plants' ENOX1 proteins provides a mechanism whereby the resultant uncontrolled growth through uncontrolled cell enlargement ultimately results in the death of the affected plants or plant parts.

10.11 Summary

Elongation growth of cells is growth that results from an increase in cell size (or cell volume). Because the cell walls leave a permanent record of cell elongation, plants offer an experimental advantage over mammalian cells in that irreversible changes in cell dimensions can be readily observed and recorded. Classically, growth regulators of the auxin type have been used experimentally to modulate expansion rates of excised stem segments floated on solutions. Included among the auxins are a natural plant hormone, IAA, along with synthetic analogues such as 2,4-D and α-naphthaleneacetic acid (α-NAA). IAA and 2,4-D, at least, seem to regulate growth by similar but not identical modes of action. Focus on this chapter has been on the location of the auxin hormone target at the plasma membrane and a general lack of evidence for growth regulation by auxins involving a classic auxin-initiated signal-response cascade. Plant plasma membranes bind auxin and the target protein has been identified as a plant-specific ENOX protein that requires auxin, either natural or synthetic, for activity. In other respects, the auxin-stimulated ENOX has functional properties similar to those of plant ENOX1 including NAD(P)H oxidation, and reduction of oxygen to form water, as well as a protein disulfide-thiol interchange more directly involved in the cell enlargement process and the typical 5-maxima periodic oscillatory activity with a period length of 24 min. In contrast to plant ENOX1, auxin-stimulated ENOX is inhibited by many of the same substances that also inhibit the human cancer-associated ENOX2. The modulation of ENOX activity in plants in response to turgor provides a parallel between ENOX activity and cell enlargement and the response of both to auxin stimulation.

The hormone-stimulated and growth-related cell surface hydroquinone (NADH) oxidase activity from plants also oscillates with a period of about 24 min or 60 times per 24-h/day. Plasma membranes of hypocotyls from dark-grown soybeans contain two such NADH oxidase activities that have been resolved by purification on concanavalin A columns. One in the apparent molecular weight range of 14–17 kDa is stimulated by the auxin herbicide 2,4-dichorophenoxyacetic acid (2,4-D). The other is larger and unaffected by 2,4-D. That the auxin-stimulated cell surface NADH oxidase was distinct from the constitutive oxidase was first indicated by Morré and Grieco (1999) using quassinoid growth inhibitors. The constitutive NADH oxidase activity and the constitutive cell elongation were inhibited by simalikalactone D but not by glaucarubolone, whereas the 2,4-D stimulated NADH oxidase activity and the 2,4-D-stimulated rate of cell enlargement were inhibited by glaucarubolone but not by simalikalactone D (Morré and Grieco 1999).

The 2,4-D-stimulated activity absolutely requires 2,4-D for activity and exhibits a period length of about 24 min. Also exhibiting 24-min oscillations is the rate of cell enlargement induced by the addition of 2,4-D or the natural auxin growth regulator, IAA. Immediately following 2,4-D or IAA addition, a complex pattern of oscillations is observed (e.g., Fig. 5.14). However, after several hours, a dominant 24-min period emerges at the expense of the constitutive activity. The sum of the two activities remains constant. A recruitment process whereby an altered protein form converts a normal form of a protein into a likeness of itself analogous to that exhibited by prions (Griffith 1967; Prusiner et al. 1998) has been postulated to explain this unusual behavior to a growth-regulating herbicide (Morré et al. 2003b). 2,4-D is a synthetic auxin which was discovered and developed during World War II as a selective herbicide for general weed control in corn, small grains, and other grasses. Auxins, both natural and synthetic, as a class, stimulate plant cells to increase in size (enlarge). An auxin-stimulated ENOX activity of plasma membrane fractions of soybean was reported early (Morré et al. 1986a). The activity was purified and shown to require auxin addition for activity (Brightman et al. 1988). A variety of mechanisms have been proposed to understand how 2,4-D and the natural auxins such as IAA promote cell enlargement. Our findings implicate surface hydroquinone oxidases (NOX proteins) as drivers of the cell enlargement process, one of which requires auxin for activity. The new 2,4-D-induced period then persists as growth of affected tissues and plant parts becomes unregulated and cancer like. The ultimate result with 2,4-D-sensitive species is death of the plant or of the affected plant parts.

Chapter 11
Cancer Therapeutic Applications of ENOX2 Proteins

Tumor-associated NOX (tNOX) proteins designated ENOX2 are cancer cell-specific surface ECTO-NOX (ENOX) proteins that have many characteristics to make them the Achilles Heel of Cancer and a target for selective antitumor therapy (Fig. 11.1) that could be exploited for a broad range of patients with cancer. A unique feature is the presence of ENOX2 on the cell surface of invasive human cancers such that drugs need not enter cells to be effective (Fig. 11.2). Furthermore, there are resolute differences between the drug inhibited ENOX2 that contains a drug-binding site and the drug-resistant ENOX1 which lacks the site. As such, ENOX2 represents an attractive target for drug-, vaccine-, and immune-therapeutic strategies for cancer. Interestingly, several products currently utilized or under development for cancer treatment and/or prevention exhibit inhibition of ENOX2 as a potential, but inadequately explored, underlying mechanism. These include nonsteroidal anti-inflammatory drugs (NSAIDs), (−)-epigallocatechin gallate (EGCg), phenoxodiol (PXD), and doxorubicin hydrochloride (Adriamycin®).

ENOX proteins serve as the terminal oxidases for plasma membrane electron transport (PMET) (Chap. 4). PMET is a ubiquitous, high-capacity acute NADH redox-regulatory system responsible for maintaining a NADH/NAD^+ ratio favorable for glycolytic ATP production. With few exceptions, cancer-selective PMET inhibitors appear to target the terminal oxidase ENOX2 as a common therapeutic property.

Consistent with ENOX2 as an anticancer target is information that ENOX2 facilitates the uncontrolled growth exhibited by both cancer tissues and cancer cell lines (Morré and Morré 2003a; Chap. 8). In cancer cells, ENOX2 dominates ENOX1 coupling to the growth process (Fig. 11.2) making ENOX2-specific blocking of this pathway a rational approach to compromise the viability of rapidly proliferating cells that rely on PMET to regenerate NAD^+ from NADH (Herst and Berridge 2006). ENOX2 inhibition prevents cell enlargement and eventually cessation of growth leading, along with the elevated levels of NADH coming from blocked PMET, to apoptotic killing of the ENOX2-inhibited cancer cells (Sect. 11.2).

D.J. Morré and D.M. Morré, *ECTO-NOX Proteins: Growth, Cancer, and Aging*,
DOI 10.1007/978-1-4614-3958-5_11,

DRUG NAME	STRUCTURE	REFERENCES
(–)-Epigallocatechin gallate	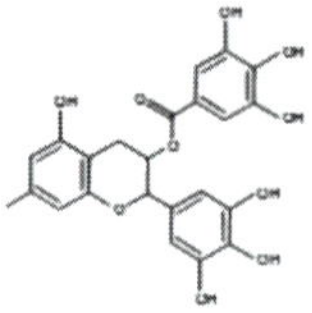	Morré et al. (2000a) Yagiz et al. (2006)
Capsaicin	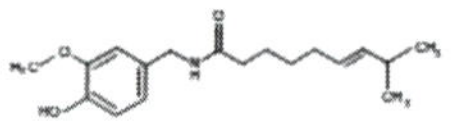	Morré et al (1995b) Wang et al. (2009)
Phenoxodiol	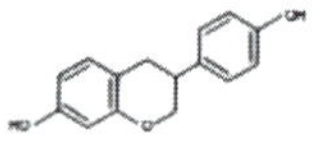	Herst et al. (2007) Morré et al. (2007a)
LY-181984I	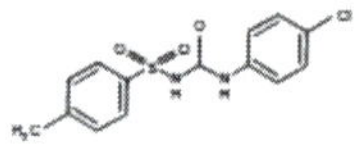	Morré et al. (1995i)
Suramin	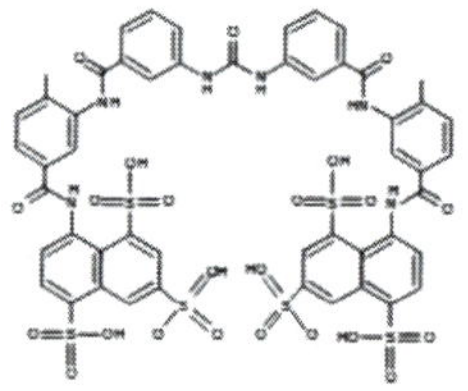	Section 11.3.10
BTS-2	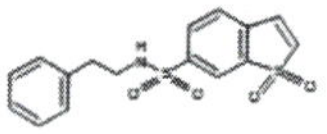	Encio et al. (2005)
Doxorubicin hydrochloride (Trade name Adriamycin®)	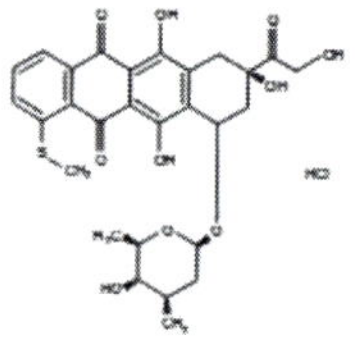	Morré et al. (1997c) Hedges et al. (2003)
Cis platinum	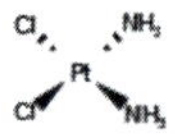	Section 11.4.3
Glaucarubolone	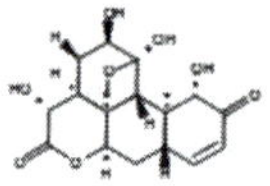	Grieco et al. (1996) Morré et al. (1998b)

Fig. 11.1 ENOX2 inhibitors

DRUG NAME	STRUCTURE	REFERENCES
Bullatacin		Morré et al. (1995c)
Sulforaphane		Section 11.3.9
Retinoic acid	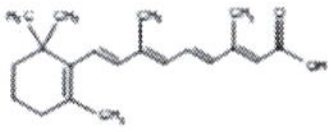	Dai et al. (1997)
Calcitriol vitamin D	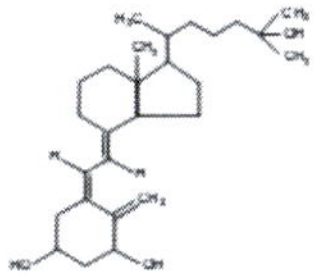	Morré et al. (1992a)
Conjugated linoleic acid		Morré et al. (2010a)
Non-steroidal antiintflammatory drugs		Morré and Morré (2006e)
Ibuprofen	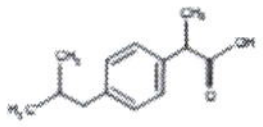	
Celecoxib	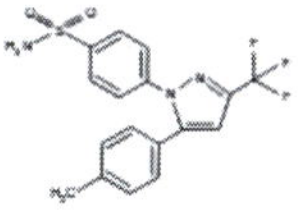	
Naproxen sodium	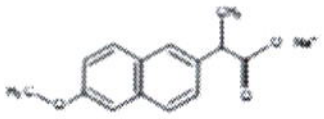	
Piroxicam	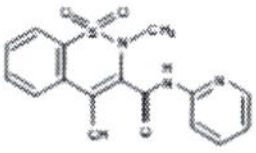	
Aspirin	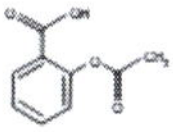	

Fig. 11.1 (continued)

Redox-directed cancer therapeutics have been the subject of a comprehensive review (Wondrak 2009). Within this broad framework, ENOX2 offers sufficient cancer cell specificity to qualify as a molecular drug target not only for new lines of research but to help explain cancer specificity of ENOX2 targeted substances and drugs under investigation or already used clinically. Emphasis in this chapter is on

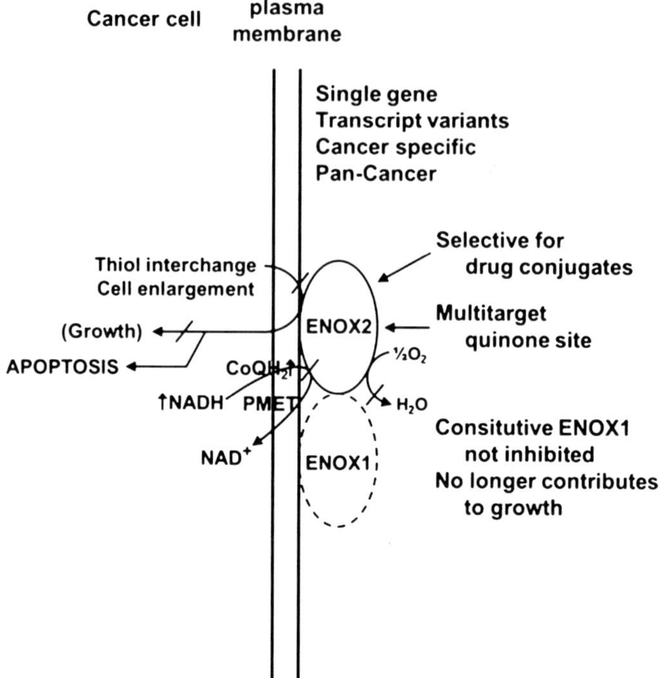

Fig. 11.2 Summary of ENOX2 properties as a cancer therapeutic target

the creation and testing of impermeant ENOX2-targeted drugs to enhance specificity and reduce dose-limiting toxicities related to cellular targets not involved in growth suppression and induction of apoptosis.

11.1 PMET as a Target for Anticancer Drug Development

Perhaps not all tumors switch to glycolytic metabolism. However, the most aggressive tumors often are also the most glycolytic (Gatenby and Gillies 2004; Yasuda et al. 2004; Kunkel et al. 2003; Schomack and Gillies 2003; Simonnet et al. 2002; Yamagata et al. 1998; Herst and Berridge 2006). The increased glycolytic rate is considered to be supported by increased PMET activity in order to regenerate NAD^+ thus assuring continued glycolytic ATP production (Fig. 4.7; Chap. 4). The possibility that some cancer cells rely on PMET for survival may offer an opportunity to use PMET as a target for anticancer drug development (Herst and Berridge 2006). Inhibition of WST-1/PMS reduction and cell surface oxygen consumption by capsaicin in HL60ρ° cells supports activity at the level of PMET. The ability of capsaicin and EGCg, as examples, to cause growth inhibition and apoptosis in cancer cells is related to inhibition by capsaicin on the cancer-specific form of the cell surface

PMET terminal oxidase, ENOX2 (tNOX) (Sects. 11.4.1). The flavin inhibitor, dicumarol, promotes superoxide formation, cell cycle arrest, and apoptosis in cancer cells, possibly through inhibition of NQO1 (Chen et al. 1993). However, recent research has questioned the involvement of NQO1 in these processes (Sect. 9.13) as dicumarol also inhibits mitochondrial electron transport at the level of respiratory complex IV, and the dicumarol-specific effects were seen even in the presence of the NQO1-specific inhibitor, ES936 (Gonzales-Aragon and Villalba 2006).

In general, compounds that affect PMET also affect cancer cell survival. One explanation for this could be that blocking electron transport through PMET by inhibiting ENOX2, for example, increases cytosolic NADH levels (De Luca et al. 2005). Increased cytosolic NADH has been shown to stimulate acid sphingomyelinase activity and inhibit sphingosine kinase activity, resulting in the conversion of sphingomyelin to ceramide and lowering of prosurvival levels of sphingosine-1-phosphate (Sect. 11.2.1). The involvement of a plasma membrane redox mechanism in the activation of acid sphingomyelinase, and the resultant formation of ceramide-enriched membrane islands, which lead to apoptosis also was postulated by Dumitru and Gulbins (2006). The advantage of the external ENOX2 target of PMET for novel anticancer drug development strategies is that drugs can be designed to specifically locate to the exterior surface of the plasma membrane without entering the cell as has been accomplished for doxorubicin (Tritton and Yee 1982; Tritton et al. 1983; Tokes et al. 1982; Rogers and Tokes 1984; Yeh and Faulk 1984; Sect. 11.4.2.3), capsaicin (Sect. 11.4.1.3), an antitumor sulfonylurea (Kim et al. 1997; Sect. 11.4.4.2), and glaucarubolone (Morré et al. 1998b; Sect. 11.4.5.1).

Tumor- or virus-transformed cells which have a defect in growth control modify the transplasma membrane redox system by changing the K_m and V_{max} for reduction of artificial electron acceptors. For diferric transferrin reduction, ferricyanide reduction or indigo disulfonate reduction, the V_{max} is depressed in tumor or transformed cells compared to nontransformed control cells (Sun et al. 1983, 1986a, b). This parallels the physiological reduction of oxygen to water which may also be decreased with more rapidly growing cancer cells (Sect. 4.9.1). The modification of transmembrane redox activity, not only in tumors but also in virally transformed cells, as indicated by an alteration of V_{max} and K_m and other criteria of these same enzymatic activities (Sun et al. 1983, 1986b; Warley and Cook 1973; Löw et al. 1991) may explain why control of growth activity is lacking in these cells and susceptibility to antitumor drugs is enhanced (Sun et al. 1983). This may also help to explain why many antiviral drugs are used clinically to treat cancers (Chap. 7), as they selectively inhibit ENOX2 and perhaps other cancer-related PMET activities.

11.1.1 Arsenicals as Unspecific Anticancer PMET Inhibitors

Trivalent arsenic [As (III)]-containing inorganic and organic compounds are one class of thiol-reactive, redox-directed therapeutic drugs (Wondrak 2009) potentially targeted to the PMET (Fig. 11.3). If NADH at the cytosolic surface of the plasma

Fig. 11.3 Trivalent arsenic [As (III)]

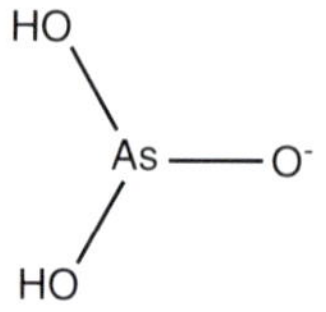

membrane is proapoptotic by enhancing ceramide and reducing sphingosine-1-phosphate (S1P), it follows that other means of acceleration of cytosolic NADH production also might be proapoptotic. In this regard, it is of interest that arsenate has been reported to induce apoptosis in pancreatic cancer cells (Li et al. 2003). Arsenate resembles inorganic phosphate in structure and reactivity and will replace phosphate in the reaction catalyzed by glyceraldehyde-3-phosphate dehydrogenase. However, the 1-arseno-3-phosphoglycerate formed is spontaneously hydrolyzed. Glycolytic production of NADH proceeds in the presence of arsenate, most likely at an accelerated rate, but the ATP normally formed in the conversion of 1,3-diphosphoglycerate into 3-phosphoglycerate is not produced.

11.2 Inhibition of PMET and Induction of Apoptosis

Although PMET activity has been associated with many critical cellular functions, such as growth control, apoptosis, and bioenergetics, its precise role(s) has (have) been difficult to resolve, possibly due to an overlapping role as a redox sensor or as a regulator of the cellular redox environment (Baker and Lawen 2000). Through the transfer of electrons from intracellular NADH to extracellular electron acceptors, including plasma membrane protein disulfides (Morré et al. 1998a), the PMOR has the ability to regulate both the intracellular (Martinus et al. 1993; Larm et al. 1994) and plasma membrane (Morré et al. 1999c) redox state and to maintain an optimized environment for redox signaling and bioenergetics, both of which play important roles in cellular decisions to live or die (Bowling and Beal 1995; Maher and Schubert 2000). The creation of ρ° Namalwa cells demonstrates that the PMOR has the capacity to maintain internal redox homeostasis even under the extraordinary condition in which mitochondrial oxidative phosphorylation has been experimentally lost (Larm et al. 1994). Similarly, the NAD(P)H-quinone oxidoreductases (NQO1; DT-diaphorase, EC 1.6.99.2) play an important role in regulating the cellular redox state (Gaikwad et al. 2001), and consequently modulates stress-activated signaling pathways and apoptosis (Cross et al. 1999; Brar et al. 2001). Moreover, overexpression of the antiapoptosis gene bcl-2 in neural cells resulted in significant decreases in the NADH/NAD ratio of two- to threefold compared with control transfectants (Ellerby et al. 1996). This contrasts with the proapoptotic response to ENOX2 inhibition by cancer therapeutic drugs where increased NADH levels favor apoptosis (De Luca et al. 2005, 2010; Wu et al. 2011). Taken together, however, these studies suggest that maintenance of local permissive redox environments, by the PMOR and other redox systems, plays a fundamental role in modulating critical cellular processes resulting in cell survival or death (Castagne et al. 1999).

NAD^+/NADH and CoQ/$CoQH_2$ ratios from PMET modulate cell death by regulating sphingolipid metabolism (De Luca et al. 2005, 2010). Either NQO1 or ENOX2 will promote oxidation of cytosolic NADH but with NQO1, not necessarily reduction of plasma membrane hydroquinone. Ubiquinone in high concentrations leads to inhibition of neutral sphingomyelinase, the enzyme responsible for hydrolysis of ceramide (Martín et al. 2003) whereas ubiquinol stimulates or is without effect. Sphingosine kinase, which generates antiapoptotic S1P in contrast is stimulated by ubiquinone and inhibited by hydroquinone and NADH (De Luca et al. 2005, 2010; Fig. 11.4). Ceramide accumulations are avoided and cells are protected from the apoptotic stimuli that lead to G_1 arrest and death (Spyridopoulos et al. 2001; Radin 2003). The inhibition of sphingomyelinase results from both a direct effect of hydroquinone on the enzyme (Martín et al. 2001, 2003; De Luca et al. 2005, 2010) or to the inhibition of lipid peroxidation, a factor indirectly involved in sphingomyelinase activation (Martín et al. 2001). Results reported by the group of Navas and coworkers showed that apoptosis induced by serum withdrawal in leukemia cells was reduced by CoQ_{10} (Navas et al. 2002) along with inhibition of an early increase of neutral-sphingomyelinase activity and caspase-3 activation (Fernández-Ayala et al. 2000). Capsaicin downregulates ENOX2 expression concomitantly with the appearance of apoptotic biomarkers in stomach cancer cells (Wang et al. 2009). Lowering ENOX2 activity slows PMET. Both reduced CoQ_{10} and NADH are elevated as are levels of plasma membrane ceramide (De Luca et al. 2005, 2010).

11.2.1 Mechanism of Induction of Apoptosis When Plasma Membrane Electron Transport Is Inhibited

The mechanism whereby apoptosis is induced by inhibitors of ENOX2 specifically and PMET generally has been studied most extensively in relationship to the mode of action of the isoflavene anticancer drug phenoxodiol (PXD) (Sect. 11.4.8). PMET is inhibited by PXD through direct binding and inhibition of ENOX2. The constitutive ENOX1 neither binds nor is inhibited by PXD (Fig. 11.5).

In epithelial ovarian cancer cells, inhibition of ENOX2 by PXD induces a caspase-3-dependent apoptotic process involving both extrinsic (death receptor) and intrinsic (mitochondrial) pathways (Alvero et al. 2008). Both the phosphorylation status and expression of Akt are downregulated so that increased degradation of X-linked inhibitor of apoptosis (XIAP) and the short version of FAS-associated death domain (FADD)-like interleukin-l-β-converting enzyme-inhibitory protein (FLIPs) are increased (Alvero et al. 2006). The activation of the extrinsic apoptotic pathway results from death receptor-mediated cleavage of procaspase-8 to its active form which enables degradation of the death receptor inhibitory FLIPs (Kamsteeg et al. 2003). Activation of the intrinsic pathway via BID cleavage and caspase-9 activation occurs through Akt down regulation-mediated activation of caspase-8 and degradation of XIAP (Panka et al. 2001; Suhara et al. 2001; Fig. 11.6). Conformation changes in Bax on the mitochondrial membrane that result in mitochondrial depolarization and truncation of BID catalyzed by caspase-2

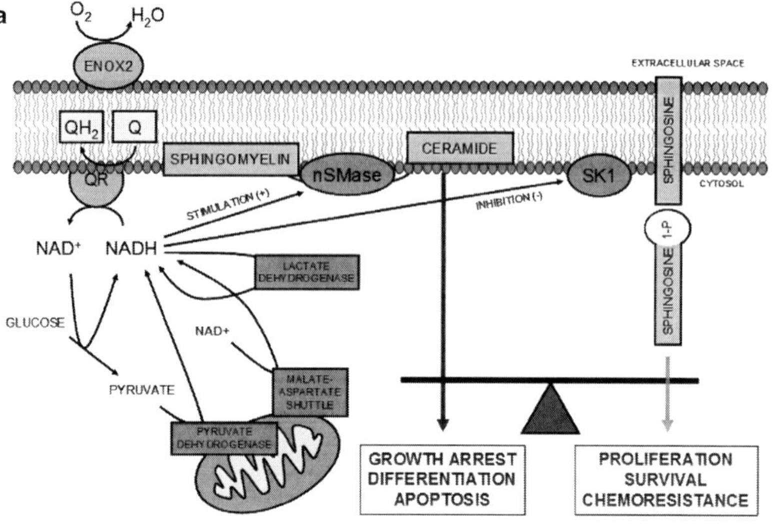

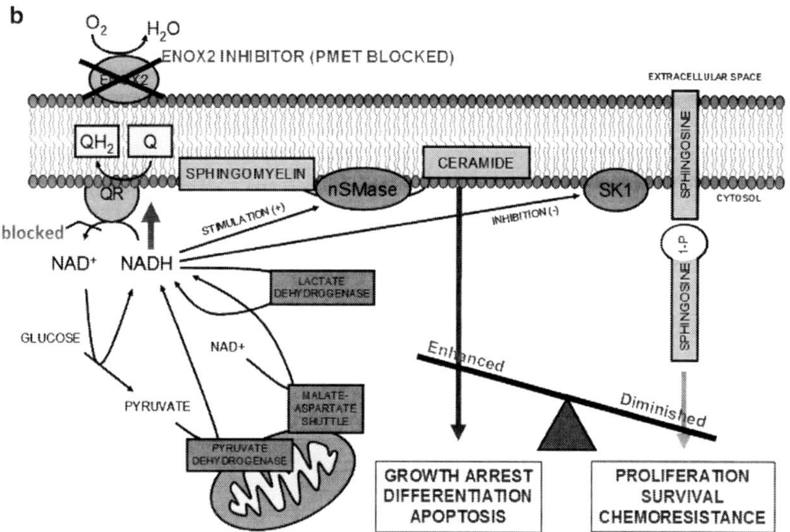

Fig. 11.4 Scheme to explain ENOX2 inhibitor-induced apoptosis. (**a**) ECTO-NOX proteins at the cell surface keep the plasma membrane pool of coenzyme Q_{10} largely oxidized through the oxidation of $CoQ_{10}H_2$. ENOX2 is a cancer-specific ECTO-NOX form absent from noncancer cells. (**b**) In cancer cells, ENOX2 inhibitors would result in accumulation of both $CoQ_{10}H_2$ in the membrane and NADH at the cytosol-membrane interface due to ECTO-NOX inhibition at the cell surface. Coenzyme Q_{10} presence in the plasma membrane blocks sphingomyelinase and prevents the accumulation of ceramide leading to G_1 arrest. Inhibition of ENOX2 leads to replacement of coenzyme Q_{10} by reduced coenzyme Q_{10}, release of sphingomyelinase from inhibition, increased ceramide, and G_1 arrest. At the same time, as sphingomyelinase is activated by the accumulation of reduced coenzyme CoQ_{10} and cytosolic NADH resulting from ENOX2 inhibition, sphingosine kinase is inhibited either by $CoQ_{10}H_2$, NADH, or both. Sphingosine kinase inhibition leads to lack of suppression of the FAS pathway as the product of Akt stimulated by sphingosine-1-phosphate (S1P), FLIP, and XIAP would no longer be available to serve as alternative substrates for caspase 8 to allow apoptosis to proceed. Courtesy of Thomas De Luca, Purdue University

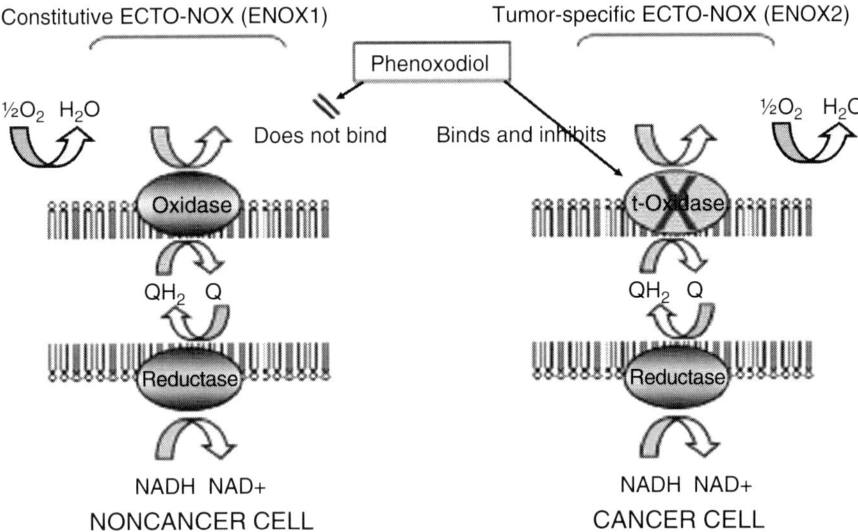

Fig. 11.5 Model to explain the resistance of ENOX1 to the synthetic ENOX2-inhibitor phenoxodiol (PXD) and its specific inhibition by ENOX2. Courtesy of Marshall Edwards, Inc. *Oncology*, used by permission

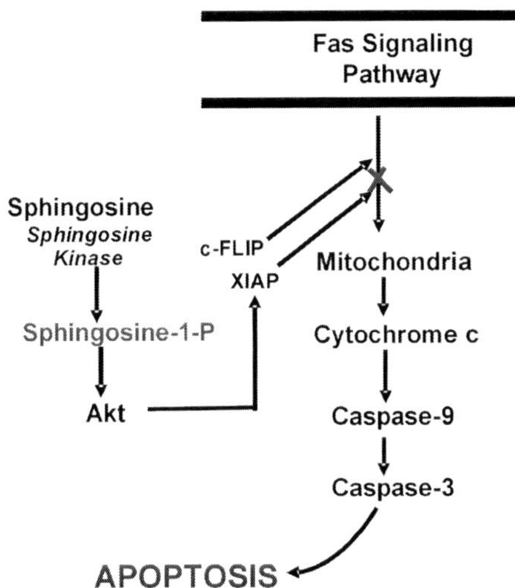

Fig. 11.6 Diagram outlining the FAS signaling pathway of apoptosis and its suppression by sphingosine-1-phosphate (S1P), the prosurvival signal. The products of Akt activation, c-FLIP and XIAP, exhibit anticaspase activities to block FAS signaling when S1P is elevated. Reproduced from De Luca et al. (2005) with permission from Wiley-SOS Press

are considered to be obligatorily involved. Accompanying the decrease in XIAP is decreased phosphorylation of Akt, a decrease in total Akt between 8 and 16 h (Alvero et al. 2006) and appearance of the p30 XIAP fragment at 16 and 24 h from cleavage by activated caspase-3, -8, and -9. The serine protease Omi/HtrA2 released from the mitochondria about 8 h post phenoxodiol treatment may contribute as well to the production of the p30 XIAP fragment since XIAP is a known substrate of the protease. As cytochrome c and Omi/HtrA2 are released from mitochondria and accumulate in the cytosol, a second mitochondria-derived activator of caspase (Smac) or "direct IAP binding protein with low pH" (Diablo) (Smac-Diablo) also appears to be released at about the same time postphenoxodiol treatment (Alvero et al. 2006). Smac-Diablo promotes apoptosis by sequestering rather than cleaving XIAP.

By whatever mechanism, the degradation of XIAP has been suggested to be responsible for the resensitization of resistant melanoma cells to carboplatin (Kluger et al. 2007). The link between the two major apoptotic pathways and apoptosis induced by inhibition of ENOX2 by PXD is provided for in cancer cells by modulation of the sphingomyelin pathway (De Luca et al. 2009). The over-production of sphingosine-1-phosphate (S1P), via ceramide either by enhanced sphingomyelinase activity and/or sphingosine kinase (SK), is known to impart positive feedback on prosurvival cell signaling in cancer cells through Akt stabilization so that apoptosis is suppressed (Cuvillier 2008; Baudhuin et al. 2002; Kim et al. 2003; Shida et al. 2008). On the other hand, production of ceramide by sphingomyelinase is known to promote G_1 cell cycle arrest and induce caspase-mediated apoptosis (Kim et al. 2000a, b; Pettus et al. 2002; Carpinteiro et al. 2008). Direct demonstration of the inhibition of production of S1P and increased ceramide accumulation has been provided by De Luca et al. (2005, 2010) for HeLa cervical cancer cells. Similarly, head and neck (KB-3-1), ovarian cancer (A2780), and doxorubicin-resistant breast cancer cells (MCF-7-Adr) exhibited an intracellular accumulation of ceramide when treated with PXD (Cabot et al. 2005). PXD may even exert control over SK activity in noncancer human endothelial cells after stimulation by TNF-α (Gamble et al. 2006). ENOX2 emerges as the putative primary molecular target of PXD (Morré et al. 2007a; Herst et al. 2007). When ENOX2 is inhibited, the plasma membrane content of ubiquinol is increased, which, in turn, results in cytosolic accumulation of NADH and the resultant decoupling of the S1P prosurvival signal transduction cascade which then appears to be linked to the inhibition of Akt, XIAP, and FLIP (De Luca et al. 2005; Morré et al. 2007a). ENOX2 also results in an accumulation of ceramide through NADH-mediated activation of plasma membrane sphingomyelinase (De Luca et al. 2005). Both events (concurrent reduction in S1P and accumulation of ceramide) appear to initiate caspase-3-dependent programmed cell death. The cancer specificity of the ENOX2 protein (Fig. 11.4) provides a mechanistic explanation for the apparent targeted toxicity of PXD for cancer cells and the lack of a PXD response with normal cells (Morré et al. 2007a) via the well-established sphingosine kinase, S1P, and apotosis link (Maceyka et al. 2002).

The role of the ENOX2 proteins is to drive cell enlargement (Morré and Morré 2003a). When inhibited, cells fail to enlarge to a size sufficient to pass the G_1 checkpoint that monitors cell size, so that they cannot divide (Fig. 5.18; Chap. 5). This may explain how PXD disrupts the cell cycle in the majority of cancer lines thus far investigated. The cell cycle in head and neck squamous cell carcinoma cell lines in G_1 is arrested by 12 h following treatment with PXD along with concomitant decrease in S-phase. Cycling of A2780 ovarian cancer cells is strongly blocked at S phase by PXD with a more moderate block at G_2M (Brown et al. 2005). The latter was elicited by inhibiting cdk2 activity through p53-independent increase $p21^{waf1}$ expression (Aguero et al. 2005).

11.3 Mechanism of Growth Arrest When Plasma Membrane Electron Transport Is Inhibited

Induction of apoptosis resulting from PMET inhibition and growth arrest appears to be related. Much of the growth arrest is due to the blockage of ENOX2-catalyzed protein disulfide-thiol interchange obligatorily required for cell enlargement (Fig. 11.2). However, links to elevation of ceramide also may contribute.

11.3.1 Elevation of Ceramide

All ENOX2 inhibitors exert similar responses such that one may postulate that the growth arrest in cancer cells when ENOX2 is inhibited is triggered by immediate products resulting from the inhibition of the ENOX2 target (i.e., elevated NADH) with a resultant elevation in ceramide accompanying or leading to growth arrest (Geilen et al. 1997; Spiegel et al. 1998).

11.3.2 Links for Elevated Ceramide and Cell Cycle Arrest

Uncontrolled cell proliferation is the most relevant feature of cancer. The reactions that link ceramide overproduction by the plasma membrane and G_1 arrest are based primarily on studies with a variety of ENOX2 inhibitors such as diphenyliodonium (Table 11.1), EGCg (Sect. 11.4.7), capsaicin (Sect. 11.4.1), and doxorubicin that indicate that the resultant G_1 block is associated with inhibition of cyclin D_1, cyclin E, and cyclin A protein expression. These inhibitions begin about 6 h after treatment plus a simultaneous upregulation of p27 (Ruvolo 2003). Additions of exogenous ceramide mimic these responses (Kim et al. 2000a, b; Ruvolo 2003; Spyridopoulos et al. 2001; Zhu et al. 2003). Expression of cyclins D_1, E, and A as well as upregulation of p27 is

Table 11.1 Inhibition of NOX activity by DPI ($C_{12}H_2Cl$)

Source of NOX activity	μmol/min/mg protein		EC_{50} for DPI inhibition	
	NADH	NADPH	NADH	NADPH
HeLa cells	1.1±0.2	1.4±0.3	Not reached	0.1 μM
Released from HeLa cells by low pH	45±10	30±5	1 μM	0.1 μM
Soybean plasma membrane	1.0±0.2	0.5±0.05	Not reached	0.1 μM
Recombinant ENOX2	450	1,350	No response to DPI	0.1 μM

From Morré (2002)

most closely related to the MAP kinase cascade which would require downregulation to account for the observed responses of down-stream effectors (Lauricella et al. 1998; Panka et al. 2001). One possible control point supported by preliminary evidence is the ceramide-sensitive activation of ras GTPase activity (or a GTP/GDP interchange protein) (Gulbins et al. 1996). The ras protein or a protein closely affecting ras activity is a logical control point to explain the ceramide response of G_1 arrest (Zhang et al. 1998). A GTP/GDP interchange protein is required for ras activation and the subsequent initiation of the MAP kinase cascade. Another possibility might be the kinase suppressor of ras (Gulbins and Grassmé 2002). A relationship between ENOX2 inhibition, redox control, and ras binding of GTP was indicated early (Wilkinson et al. 1993) but requires further study. If the interchange protein was inhibited by ceramide, the ras-bound GTP would not be released as GDP following GTP hydrolysis with a corresponding shutdown of the entire pathway. Cyclooxygenase inhibitors increase ceramide and at the same time block cell growth and inhibit the cell cycle (Kundu et al. 2002). Cyclooxygenase inhibitors also inhibit ENOX2 as an alternative target to explain the anticancer activities of NSAIDS (Morré and Morré 2006e). Cell cycle progression is tightly controlled by cyclin-dependent kinases (CDKs), which are themselves regulated in part by two classes of CDIK inhibitors (CDK1), the cip/kip family ($p21^{Cip1/Waf1}$, $p27^{Kip1}$, and $p57^{Kip2}$) and the INK family ($p16^{INK4a}$, $p15^{INKb}$, $p18^{INK4c}$, and $p19^{INK4d}$). The CDK1 $p21^{Cip1/Waf1}$ can be induced through p53-dependent and -independent mechanisms and consequently mediate arrest of cycling cells primarily in G_0/G_1, but also in G_2/M. Furthermore, induction of $p21^{Cip1/Waf1}$ is required to maintain the G_2 checkpoint following DNA damage. Protein expression of $p21^{Cip1/Waf1}$ is regulated at transcriptional and posttranslational levels. Proteasomal activity is essential in maintaining low-level basal $p21^{Cip1/Waf1}$ protein expression by regulating its degradation (Phillips et al. 2007).

11.3.3 ENOX2 Inhibitors Slow the Growth of HeLa Cells and Induce Apoptosis in Cancer But Not in Noncancer Cells

Two inhibitors, EGCg and phenoxodiol (PXD), known to target ENOX2 at low doses not effecting other targets, slow the growth of HeLa (human cervical

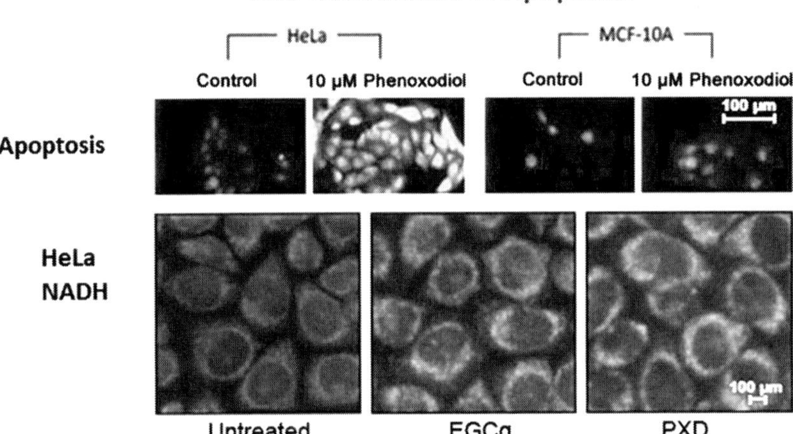

HeLa: NADH levels estimated from fluorescence measurements

Addition	NADH, pmoles/10^6 cells
None	286 ± 28
1 µM Phenoxodiol	411 ± 34[a]

[a] Values are different statistically ($p<0.02$).

Fig. 11.7 HeLa cell response to ENOX2 inhibitors. (**a**) BODIPY-duTP/TdT assay for apoptosis in HeLa and MDF-10A cells after 6 h incubation with 10 µM PXD. Scale = 100 µM. (**b**) Increased intracellular fluorescence from NADH in HeLa cells following 30-min incubation with 10 µM PXD for 30 min. Excitation wavelength: 340 nm. Emission wavelength: 460 nm. (**c**) Increase in cytosolic NADH measured at 340 nm as a result of ENOX2 inhibition by 1 µM PXD. Modified from De Luca et al. (2010)

carcinoma) cells (Sects. 11.4.7 and 11.4.8) and induce apoptosis (Fig. 11.7a) without similar effects on noncancer MCF-10A (human mammary epithelial cells). The immediate response to PXD or EGCg was a rapid cessation of cell enlargement thus creating a population of small cells, unable to divide further, which underwent apoptosis (Fig. 11.7a). With times as short as 6 h, incubation with 10 µM PXD or 100 µM EGCg, apoptosis was seen in HeLa cells using the BODIPY-duTP/TdT assay. With the noncancer MCF-10A cells, neither PXD nor EGCg resulted in apoptosis in parallel to the lack of effect of PXD on growth and ENOX activity.

Table 11.2 HeLa cells: depletion of NADH by external ferricyanide

Metabolite	Concentration (μM)	pmol/10^6 cells
Ferricyanide	1	56 ± 2[a]
	100	182 ± 29[b]

From Wu et al. (2011)
Numbers not followed by the same letter were significantly different ($p < 0.01$). The ENOX2 inhibitor PXD (1 μM) increases NADH by 125 pmol/10^6 cells. External ferricyanide depletes cytosolic NADH by an amount equivalent to the increase in response to PXD

Table 11.3 Effect of ferricyanide on growth inhibition and apoptosis of HeLa cells elicited by 1 μM PXD

		Growth inhibition (%)		Apoptotic cells (%)	
Metabolite	Concentration (μM)	Alone	+Phenoxodiol	Alone +	+Phenoxodiol
Vehicle	–	0 ± 2	56 ± 2[c]	>1	55 ± 6
Ferricyanide	1	2 ± 3	21 ± 10[a]	5 ± 1	34 ± 7[b]
	100	1 ± 8	6 ± 6[a]	3 ± 1	5 ± 2[b]

From Wu et al. (2011)
Numbers in each column significantly different from vehicle alone

11.3.4 ENOX Inhibitors Increase Cytosolic NADH Levels

ENOX2 inhibition results in an increase in the levels of cytosolic NADH especially near the inner face of the plasma membrane (Fig. 11.7a, b). On the premise that apoptosis might relate to changes in cytosolic NADH resulting from ENOX2 inhibition, relative increases in NADH estimated in HeLa cells from fluorescence measurements averaged over 30 min of inhibitor addition were 100 ± 36 pmol/10^6 cells for 50 μM EGCg and 125 ± 34 pmol/10^6 cells for 1 μM PXD representing increases of 35 % and 44 %, respectively. Addition of 100 μM ferricyanide or 100 μM pyruvate reduced the NADH levels by 182 and 113 pmol/10^6 cells, respectively, to very nearly the levels observed in the absence of inhibitor (Table 11.2) and completely blocked the ability of either 50 μM EGCg or 1 μM PXD to induce apoptosis (Table 11.3; Fig. 11.8).

HeLa cells contain 150 pmol NADH/mg protein and 55 pmol NAD^+/mg protein to give a ratio of NADH/NAD^+ of 2.7, ca. 300 pg protein/cell, 10^6 cells would contain 45 pmol NADH, with a cytosolic NADH content of 5 μM.

With both PXD and EGCg, NADH fluorescence at the borders of HeLa cells was observed by microscopy to be markedly increased in the inhibited cells to verify that cytosolic levels of NADH increased as the result of inhibition of the activity of ENOX2 by both PXD and EGCg (Fig. 11.7a).

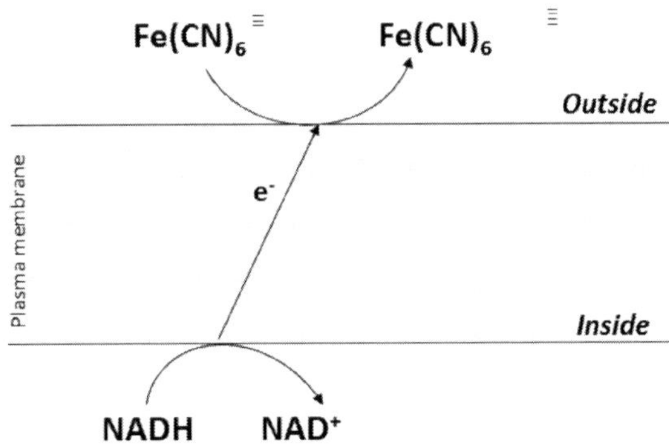

Fig. 11.8 Treatment of HeLa cells with a potent impermeant electron acceptor ferricyanide increases the rate of NADH oxidation independent of ENOX2 and prevents both EGCg- and PXD-induced apoptosis

11.3.5 Increased NADH Resulting from ENOX2 Cell Surface Inhibition Inhibits Plasma Membrane-Associated Sphingosine Kinase (SK) and Lowers Levels of Prosurvival Sphingosine-1-Phosphate

S1P decreased as a function of time for HeLa cells treated with either 10 μM PXD or 10 μM EGCg between 3 and 24 h (Fig. 11.9a). After 24 h, S1P content was reduced by about 60 % with EGCg and by nearly 9 % by PXD. Additionally, SK activity was decreased in response to 1.5 mM NADH (Fig. 11.10a). Both NADH and coenzyme Q_{10} (CoQ_{10}) inhibited SK activity for isolated HeLa cell plasma membranes. NAD^+ at the same concentration of 1.5 mM had no effect. The EC_{50} for inhibition was estimated to be about 600 μM (Fig. 11.10b, c). It was shown that the absolute level of NADH was the determining factor rather than the ratio of NADH to NAD^+.

11.3.6 Sphingomyelinase

A reciprocal relationship of sphingomyelinase (SMase) was observed in parallel to that of SK leading to accumulation of proapoptotic levels of ceramide (Fig. 11.9b).

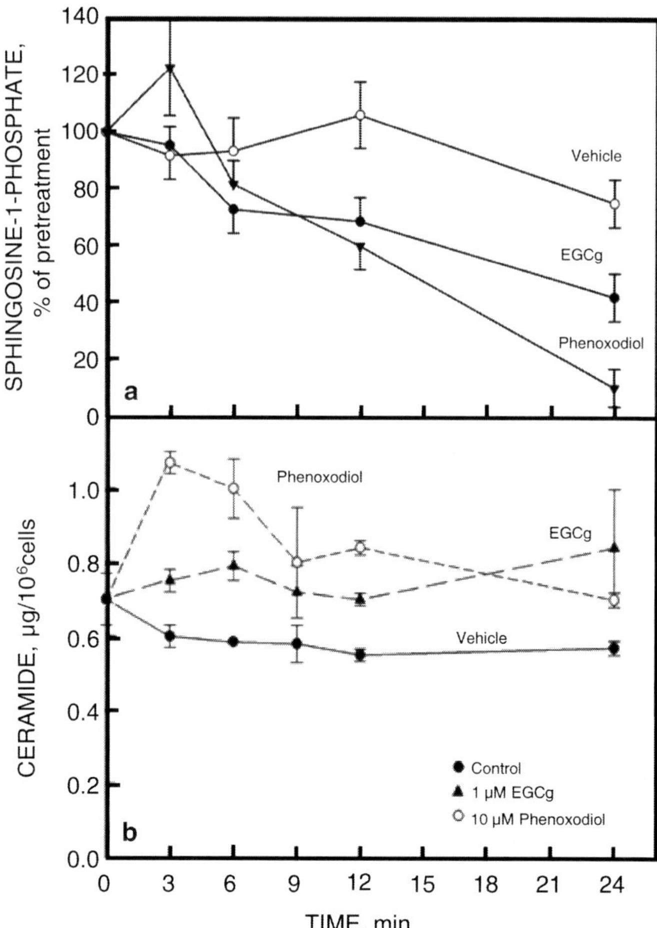

Fig. 11.9 (**a**) Sphingosine-1-phosphate content (radiolabeled) in HeLa cells treated with vehicle (DMSO) alone or 10 µM PXD or EGCg dissolved in DMSO. Sphingosine-1-phosphate was visualized on resolved silica gel plates by spraying with sodium molybdate in phosphoric acid. Results were standardized to untreated control samples. (**b**) Ceramide content of HeLa cells treated with ENOX2 inhibitors EGCg or PXD increases between 0 and 6 h and then remains elevated. Results are from three experiments ± standard deviations. The values for EGCg and PXD at 3 h ($\rho<0.01$), 6 h ($\rho<0.01$), 12 h ($\rho<0.001$), and 24 h ($\rho<0.05$) were significantly different from control values. Modified from De Luca et al. (2010)

SMase activity was stimulated by NADH in contrast to SK which was inhibited by NADH (Fig. 11.11a). This was not due to generation of a reducing environment. GSH (1 mM) inhibited whereas CoQ_{10} was largely without effect.

Purified bacterial neutral SMase was stimulated by 50 % at 1 mM NADH. In contrast, 1 mM NAD^+ resulted in an almost 60 % inhibition (Fig. 11.11a). A SMase

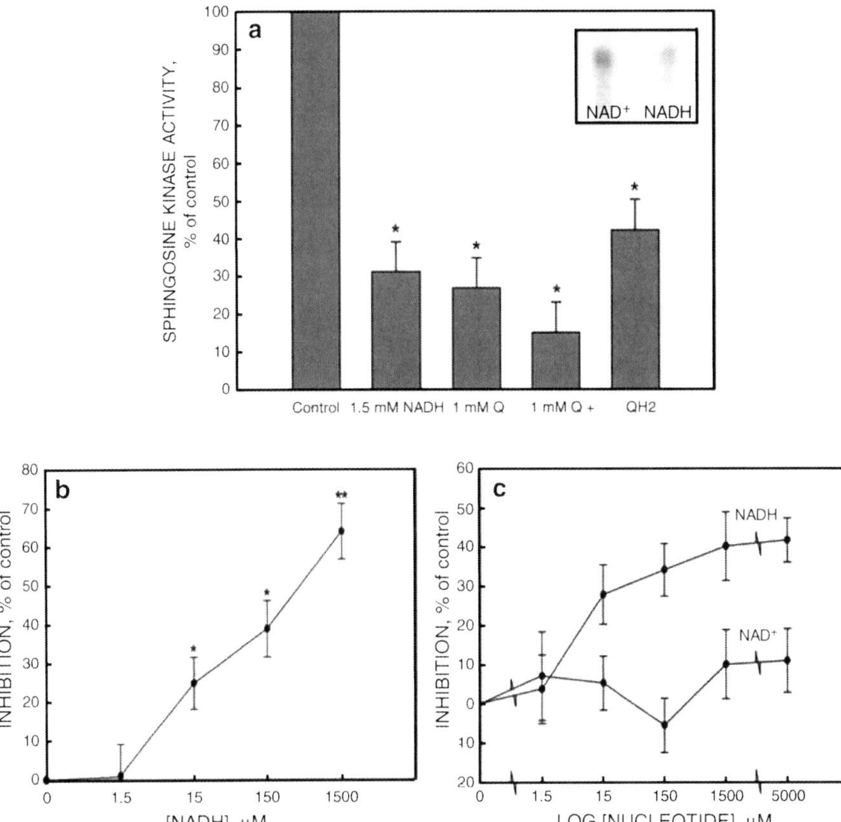

Fig. 11.10 Response of sphingosine kinase based on sphingosine-1-phosphate (S1P) formation. (**a**) Oxidized or reduced coenzyme Q. (**b**, **c**) Inhibition by NADH but not NAD⁺. A S1P formation by HeLa cell plasma membranes and inhibition by NADH, coenzyme Q_{10} (Q) and reduced coenzyme Q_{10} (QH_2). γ32-ATP donated radioactive phosphate to sphingosine in the kinase reaction. Results were standardized to untreated control samples. *Inset*: Thin layer chromatography confirming inhibition of S1P formation by NADH (1.5 mM). (**b**) HeLa-cell plasma membrane where NADH inhibited with an IC_{50} of about 150 μM ($p<0.01$). (**c**) With 0.1 μg/assay of recombinant human SK1, NAD⁺ did not inhibit. Modified from De Luca et al. (2010)

from urine also demonstrated similar responses to 150 μM NADH. The stimulation by NADH of bacterial SMase activity was linear up to 1.5 mM NADH (Fig. 11.11c).

Thus, treatments that stimulate cytosolic NADH production potentiate the antiproliferative effects of ENOX2 inhibitors whereas those that attenuate NADH production or stimulate PMET confer a survival advantage (De Luca et al. 2010).

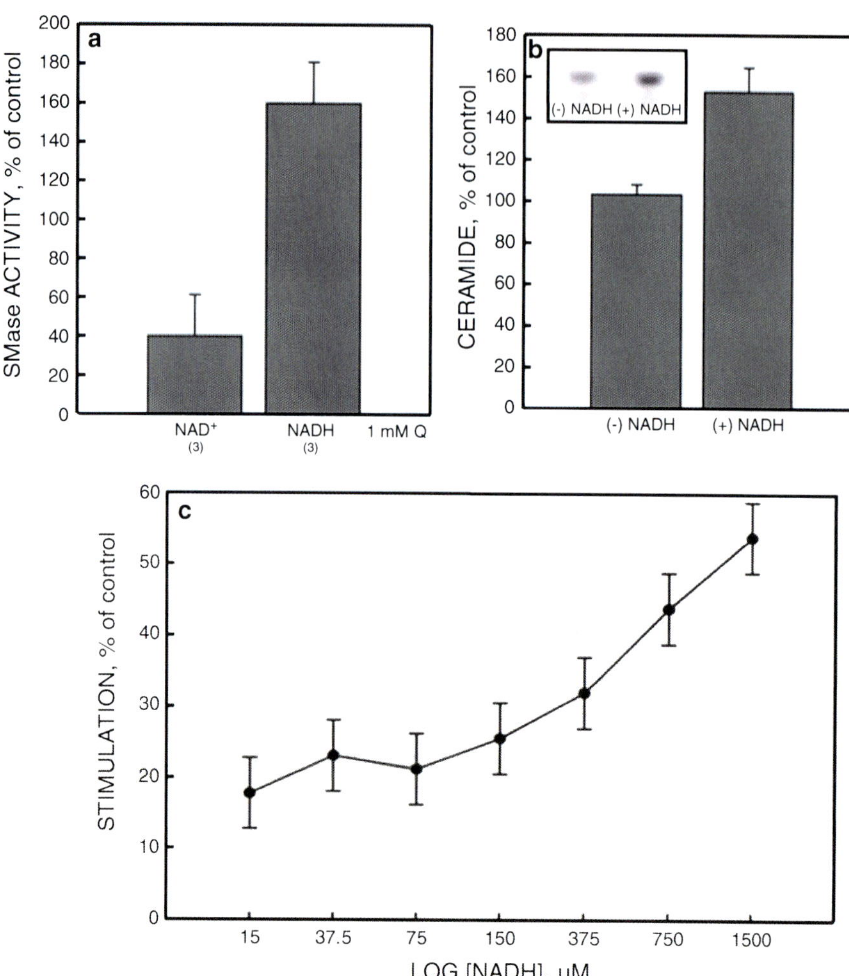

Fig. 11.11 Stimulation of sphingomyelinase by NADH but not NAD+. Averages of three determinations ± standard deviations. (**a**) Neutral sphingomyelinase from *Staphylococcus aureus* (4 mU, Sigma S-8633), values for 1 mM NAD+ or NADH over 45 min of incubation were significantly different from the control values ($p<0.05$). (**b**) Neutral sphingomyelinase (50 μg total protein) isolated from human urine stimulated by 1.5 mM NADH. *Inset*: Representative chromatogram. (**c**) Bacterial sphingomyelinase. Modified from De Luca et al. (2010)

11.4 ENOX2 Inhibitors

Interestingly, several products currently in clinical study for cancer have been shown to display ENOX2 inhibitory activity along with their other mechanisms of action (Fig. 11.1). It is important to realize that ENOX2 inhibitors do not inherently cure cancer and would do so only by apoptotic elimination of cancer cells. In the sections that follow, therapeutic application of ENOX2 proteins will be addressed first from the general view of PMET as a therapeutic target and then from the standpoint of drugs and substances specifically targeted to ENOX2.

Cancer cells expressing ENOX2 such as HeLa (human cervical carcinoma) and BT-20 (human mammary carcinoma) cells not only grow in an unregulated manner but their growth is inhibited by a variety of quinone site inhibitors with anticancer activity (Morré 1995a, 1998c; Morré et al. 1995b, c, g, h, i, 1996b, 1997c, d; Cho et al. 2002).

While some PMET inhibitors such as arsenicals may be therapeutic for cancer independently of ENOX2 (Sect. 11.1.1), the majority of PMET inhibitors, especially those with a high degree of anticancer specificity such as the anthracyclines, appear to block PMET primarily through inhibition of ENOX2 (Fig. 11.2).

Despite the fact that cancer cells express both ENOX1 and ENOX2 at their surfaces, ENOX2 appears to dominate such that complete inhibition of ENOX2 results in a complete cessation of growth.

11.4.1 Vanilloids (Capsaicinoids) as PMET Inhibitors

The vanilloids including various capsaicinoids represent a group of natural products that are vanillylamides of monocarboxylic acids of varying chain lengths from C-8 to C-11 and of varying degrees of unsaturation (Szallasi and Blumberg 1993, 1999). Capsaicinoids are found in extracts of the fruit of peppers (*Capsicum* species) with high amounts being found in the well-known chili peppers. Among the more widely studied vanilloids as inhibitors of PMET are capsaicin and resiniferatoxin (Morré et al. 1995b, 1996b; Vaillant et al. 1996; Wolvetang et al. 1996; Scarlett et al. 2005; Wang et al. 2009). Capsaicin and resiniferatoxin are potent neurotoxins for nociceptive sensory neurons that express vanilloid receptor 1 (VR1), yet they also induce apoptosis in cancer cells by inhibiting the plasma membrane ENOX2 (Morré et al. 1995b, 1996b; Wolvetang et al. 1996; Sect. 11.4.1). Inhibited chicken neuronal cells do not express functioning capsaicin receptors (Winter et al. 1990; Marin-Burgin et al. 2000). Capsaicin and resiniferitoxin also have receptor-independent effects on cells (Szallasi and Blumberg 1999), including inhibition of mitochondrial complex 1 (Shimomura et al. 1989). Capsaicin inhibition of ENOX2 leads to inhibition of cell proliferation or death in many cell types (Wang et al. 2009). Macho et al. (1999) showed that capsaicin-induced apoptosis was correlated with the cellular level of PMOR activity in different cell types. Capsaicin acts as a competitive inhibitor of ubiquinone (Satoh et al. 1996) and perturbs electron flow to oxygen via the plasma membrane NADH oxidase, which normally produces water (Morré et al. 1999c). Interestingly, inhibition of NADH oxidase by capsaicin in Jurkat and activated T-cells gives rise to higher ROS production (Macho et al. 1998, 1999, 2003), while in glioblastoma cells, capsaicin lowers the rate of endogenous ROS production (Lee et al. 2000). In both situations, perturbation of electron flow resulted in increased apoptosis.

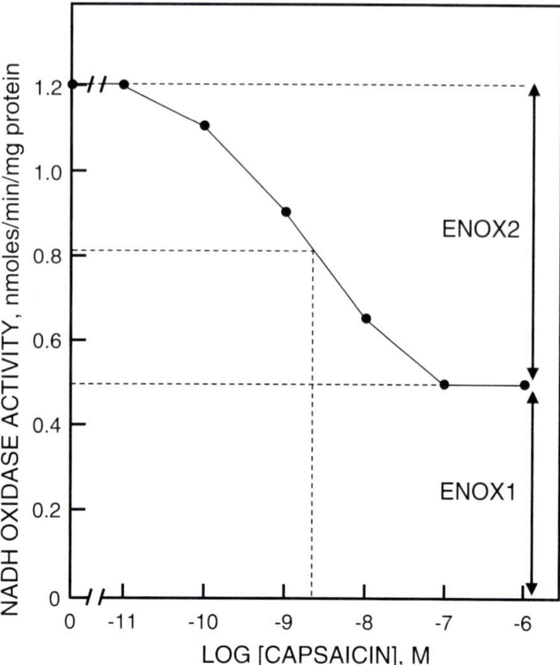

Fig. 11.12 Dose–response of NADH oxidase (NOX) activity of partially purified and solubilized ENOX2 from the surface of HeLa cells to capsaicin. The NOX activity was derived from two components. One was cancer-specific and inhibited by capsaicin (ENOX2). The other was constitutive and resistant to inhibition by capsaicin (ENOX1)

11.4.1.1 Capsaicin Specifically Inhibits ENOX2

Capsaicin was the very first ENOX2 inhibitor to be described (Morré et al. 1995). Capsaicin served in early studies of ENOX2 as a defining inhibitor to guide ENOX2 identification and purification (Chap. 8). Capsaicin inhibition of ENOX2 is both specific and complete. Cancer cells express both ENOX1 and ENOX2 activities at their surfaces. ENOX1 is refractory to capsaicin.

Capsaicin is largely without effect on the NADH oxidase of plasma membranes from rat liver (Fig. 8.4) and other noncancer sources. However, with plasma membrane from HeLa cells, the activity was inhibited by capsaicin with an EC_{50} of approximately 50 nM (Fig. 8.4). At 0.1 μM capsaicin, nearly complete inhibition was observed. Thus, the ENOX activity inhibited by 100 nM capsaicin serves as a measure of the activity of ENOX2 whereas the ENOX activity resistant to 100 nM capsaicin serves as a measure of ENOX1 activity allowing estimation of activities of both ENOX1 and ENOX2 (Fig. 11.12) with the same plasma membrane preparation. In addition to inhibition of NADH oxidase by plasma membranes, capsaicin also inhibited the growth in culture of attached HeLa cells (Morré et al. 1995b).

Noncancer MCF-10A cells were largely unaffected by capsaicin. With the BT-20 adenocarcinoma cells (Figs. 8.1 and 11.13), NADH oxidase activity of isolated plasma membrane vesicles was inhibited by capsaicin to a specific activity value approaching that of the basal activity of the MCF-10A mammary epithelial cells.

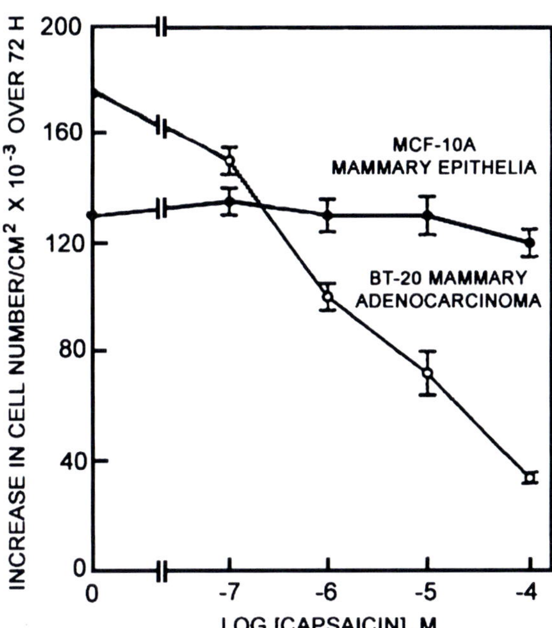

Fig. 11.13 Inhibition of growth of MCF-10A human mammary epithelial cells (*solid circle*) and BT-20 human mammary adenocarcinoma cells (*open circle*) by capsaicin. Results are growth after 72 h (initial cell number subtracted) and are averages of duplicate experiments ± mean average deviations. Reproduced from Morré et al. (1995b) with permission from PNAS

With several cell lines transiently transfected with ENOX2 cDNA, capsaicin (1 μM) completely inhibited the transiently expressed ENOX2 activity resulting from transfection (Table 8.1). Similarly, HeLa cells, which normally overexpress ENOX2, are no longer responsive to capsaicin when transfected with ENOX2 antisense (Table 8.14).

ENOX2 activity was inhibited preferentially in human A-375 melanoma cell cultures by capsaicin but not in primary melanocytes or resistant J-MEL-28 melanoma cultures. Inhibition of growth and ENOX2 activity by capsaicin could be induced by co-administration of the capsaicin with the weak oxidant *t*-butyl hydroperoxide (Morré et al. 1996b; Fig. 11.14). Growth of B16 mouse melanoma, transplanted in C57BL/6 mice also was significantly inhibited by capsaicin injected directly into tumor site when co-administrated with *t*-butyl hydroperoxide.

11.4.1.2 Vanilloids (Capsaicin) as Cancer Therapeutics

An inverse correlation exists between cancer death rates and diets traditionally high in capsaicin. Compared to the United States, cancer death rates at all sites were 157 for males and 106 for females compared to 55 and 71 for Mexico and 29 and 20 for Thailand (the lowest of any country). When capsaicin was administered to rats receiving carcinogenic agents, the incidence of certain tumors was decreased over controls (Jang et al. 1991). Capsaicinoids are important as well for their analgesic and anti-inflammatory effects.

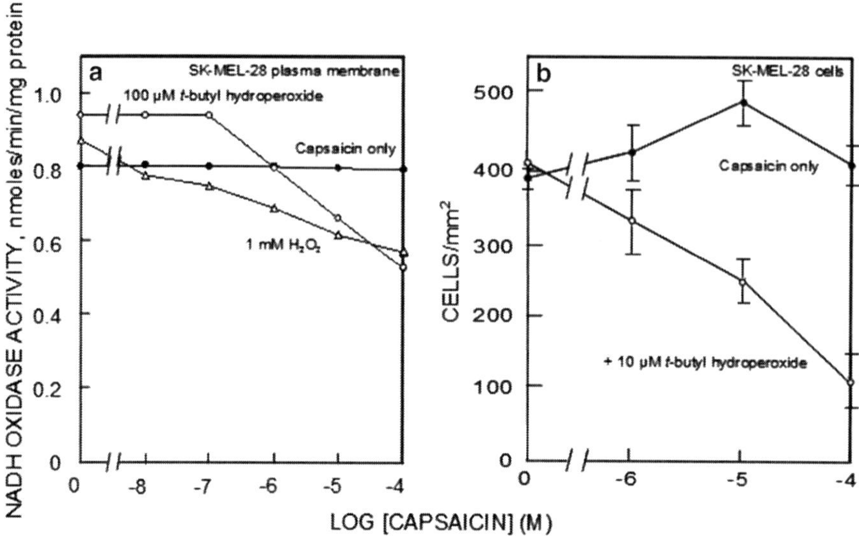

Fig. 11.14 Inhibition of (**a**) NADH oxidase activity of isolated plasma membrane (PM) vesicles and (**b**) growth in culture of SK-MEL-28 human melanoma cells in response to capsaicin alone or co-administered with oxidizing agents. Redrawn from Morré et al. (1996)

In nonneuronal cells, capsaicin induced cell death was TRPV1 receptor (transient receptor potential cation selective channels)-independent. Capsaicin causes apoptosis in transformed cells (Macho et al. 1999). Capsaicin induced intracellular accumulation of ceramide that started at 20 min and was maximal at 10 h (Sanchez et al. 2007). The source of ceramide came from sphingomyelin hydrolyses rather than de novo biosynthesis since pharmacological sphingomyelinase inhibition as well as sphingomyelinase silencing inhibited the ceramide accumulation which was not prevented by ceramide biosynthesis inhibitors (Sanchez et al. 2007).

One case of a patient with recurrent adenocarcinoma of the prostate with Gleason grade 7 experienced a slowing of his prostate-specific antigen doubling time while taking capsaicin (Jankovic et al. 2010). Inhibited by capsaicin as well were PC-3 prostate cancer xenografts in nude mice (Sanchez et al. 2006; Mori et al. 2006).

11.4.1.3 ENOX2-Targeted Capsaicin-α-Cyclodextrin Conjugates

Among the large body of accumulated evidence that ENOX2 is a low dose target for the anticancer action of capsaicin is the activity of head (H) and tail (T) conjugates of capsaicin with cyclodextrin (Fig. 11.15). The degree of inhibition of ENOX2 activity of both the conjugate (Fig. 11.16) and free capsaicin (not shown) was augmented by oxidizing conditions (GSSG>GSH). In the presence of GSSG, the EC_{50} of ENOX2 inhibition was about 10^{-6} M on a capsaicin basis provided by the conjugate (Fig. 11.16) compared to about 10^{-5} M for free capsaicin. Growth of HeLa cells

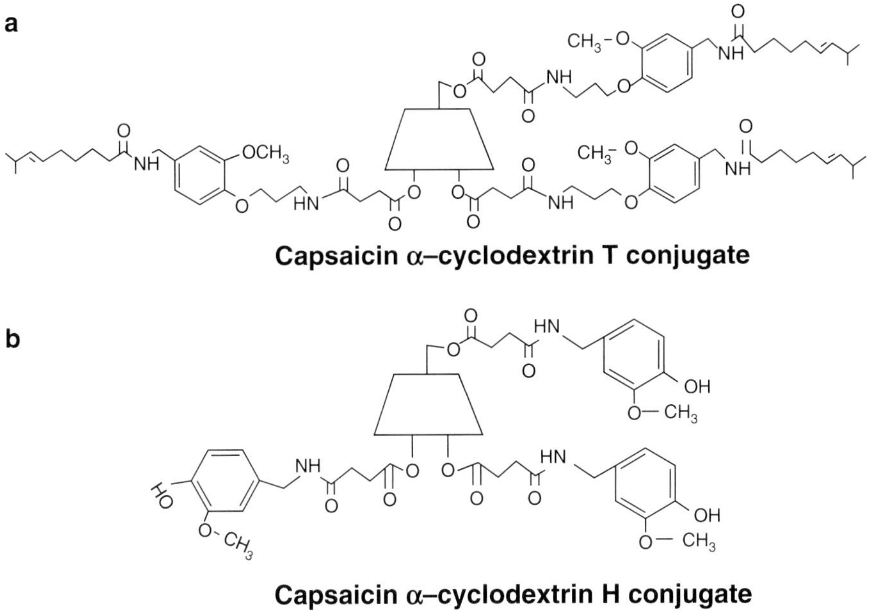

a

Capsaicin α–cyclodextrin T conjugate

b

Capsaicin α–cyclodextrin H conjugate

Fig. 11.15 Capsaicin-α-cyclodextrin conjugates. (**a**) Linked through the vanillylamine head group and (**b**) linked through the enoic acid side chain

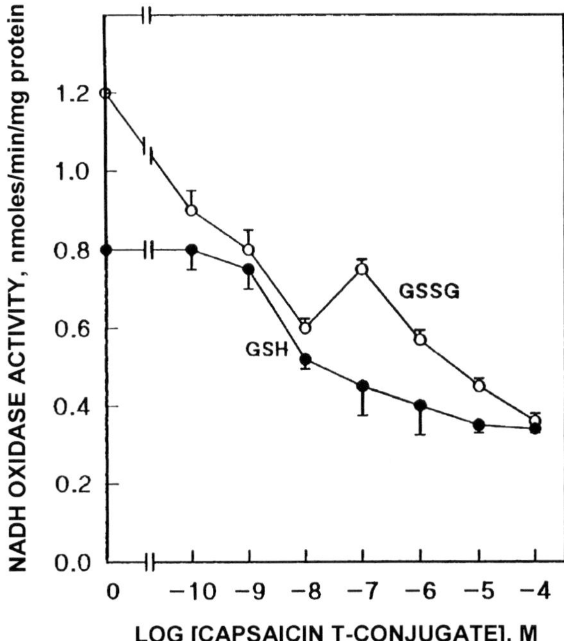

Fig. 11.16 Inhibition of NADH oxidase activity of plasma membranes of HeLa cells with the T-α-cyclodextrin-capsaicin conjugate (Fig. 11.15a) in the presence of reduced (GSH) or oxidized (GSSG) glutathione

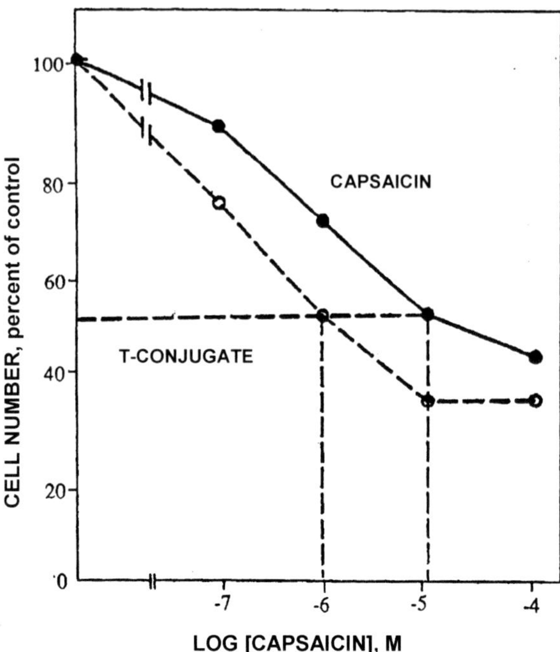

Fig. 11.17 Inhibition of growth of HeLa cells with the T-α-cyclodextrin-capsaicin conjugate (Fig. 11.15a) compared to unconjugated capsaicin

was 50 % inhibited by the T-conjugate also at 10^{-6} M compared to 10^{-5} M for unconjugated capsaicin (Fig. 11.17). Approximately tenfold enhancements of activity in inhibition of ENOX2 and growth of cancer cells by conjugated vs. unconjugated drug have also been observed for doxorubicin (Sect. 11.4.2.3) and an antitumor sulfonylurea (Sect. 11.4.4.2).

11.4.2 Anthracycline Antibiotics

Anthracycline antibiotics, doxorubicin (trade name Adriamycin®), daunomycin, and mitomycin are among the most important agents used in the treatment of human cancer (Arcamone 1985). They target PMET generally and ENOX2 specifically based on a large body of evidence not widely appreciated.

11.4.2.1 Anthracyclines Inhibit PMET of Tumor (or Transformed) Cells But Not of Noncancer Cells

Doxorubicin, the anthracycline most commonly used clinically for cancer treatment, inhibits PMET-associated ferricyanide reduction by transformed cells. Not only are the NADH-ferricyanide oxidoreductases of PMET inhibited by doxorubicin, but also they are inhibited by other anthracyclines such as bleomycin and by cisplatin and actinomycin (Sun et al. 1984b, 1992a; Sun and Crane 1984),

NADH-driven ferricyanide reduction by HeLa cells was inhibited by the above antitumor drugs at the same concentrations at which they were cytotoxic (Sun et al. 1992a; Sun and Crane 1984, 1985). Inhibition was observed at micromolar concentration of doxorubicin and the activities of rat hepatoma plasma membranes (Löw et al. 1986) as well as of plasma membranes of HeLa cells were inhibited. Moreover, the same concentration of drug gave much less inhibition of ferricyanide reduction by those cells exhibiting nontransformed phenotypes. These effects were seen with relatively short-term exposure (3 min) of the cells to doxorubicin. The sensitivity to the anthracycline antibiotics was increased by longer preincubation times. For virus-transformed cells (33 °C cultures), the drugs had strong cytotoxicity. However, primary fetal hepatocytes like RLA209-15 (40 °C cultures) and RPNA209-1 (40 °C cultures) were not sensitive to killing (Sun et al. 1983, 1986a, b). Daunomycin at 10^{-6} M inhibited ferricyanide reduction and proton release by Ehrlich ascites cells (Medina et al. 1988). Evidence using a physiological acceptor, ascorbate free radical, for a direct inhibition of transplasma membrane electron transport of HL-60 cells by doxorubicin correlated closely with doxorubicin inhibition of growth of the cells (Morré et al. 1994b, c).

11.4.2.2 Inhibition of PMET Activities by Doxorubicin (Adriamycin®)-Mediated Through a Plasma Membrane-Located Target

Doxorubicin has been shown to intercalate between the bases of DNA molecules with the result that DNA and RNA biosynthesis are inhibited (Haidle and McKinney 1986). As a result, interference with DNA and RNA synthesis has been generally regarded as the basis for the cytotoxicity of doxorubicin. Interference with DNA and RNA synthesis taken together with impairment of topoisomerase II activity (Myers et al. 1988) normally has been considered to be sufficient to account for the cytotoxicity of doxorubicin.

For various reasons, however, additional sites of action of doxorubicin were proposed as reviewed by several authors (Rogers et al. 1983; Sun et al. 1984a; Thormalley et al. 1986). These included alterations in the cell membrane (Bucher et al. 1983; Deliconstantinos 1987) which might interfere with functioning of membrane-bound enzyme systems (Rogers and Tokes 1984). Alternatively, oxygen radicals produced when doxorubicin enters the cell (Doroshow 1983) might damage the plasma membrane and/or inhibit cytoskeletal assembly (Goormaghtigh et al. 1983; Jadot 1986; Deliconstantinos 1987). Additional evidence for an alternative target for doxorubicin cytotoxicity came from studies of cytotoxic doxorubicin derivatives. One argument was that the N-substituted derivatives such as *N*-acetyl daunomycin and doxorubicin are cytotoxic under circumstances where DNA synthesis was not affected (Silvertrini et al. 1970, 1973). Neither doxorubicin conjugate nor free doxorubicin derived from the conjugate reached cell nuclei in concentrations sufficient to be cytotoxic by a mechanism based solely on DNA intercalation (Barabas et al. 1991, 1992). One compelling argument that death of the cells resulted primarily from an interruption of cellular metabolic functions rather than interference with the process of DNA replication is that most cell death occurred within less than one cell cycle.

Although suggested much earlier (Löw and Crane 1978), direct evidence for a plasma membrane target was first provided by experiments that showed that the anthracycline antibiotics need not enter tumor cells to be effective. Tritton and Yee (1982) demonstrated that doxorubicin bound to agarose beads, which cannot enter the cells, retained its cytotoxicity to sarcoma cells (Tritton and Yee 1982; Tritton et al. 1983; Tokes et al. 1982; Murphee et al. 1981). Growth inhibition was observed with doxorubicin attached to agarose beads (Tritton and Yee 1982), polyvinyl alcohol (Wingard et al. 1985), polyglutaraldehyde microspheres (Tokes et al. 1982; Rogers et al. 1983), *N*-(2-hydroxyprophyl)methacrylamide (Seymour et al. 1990), and diferric transferrin (Yeh and Faulk 1984). Likewise, the cytostatic action of doxorubicin coupled to polyglutaraldehyde microspheres was demonstrated on human leukemia cells and rat hepatocytes (Tokes et al. 1982). From these studies, it was concluded that the anthracyclines could disrupt cellular growth processes without actually entering cells.

11.4.2.3 Inhibition of PMET Activities by Doxorubicin (Adriamycin®) Mediated Through ENOX2

Doxorubicin inhibited the ENOX2 activity of both rat hepatoma plasma membrane (Fig. 11.18) and that of plasma membrane of HeLa cells (Fig. 11.19) in a dose-dependent manner (Morré et al. 1997c). In contrast, the ENOX1 activity of plasma

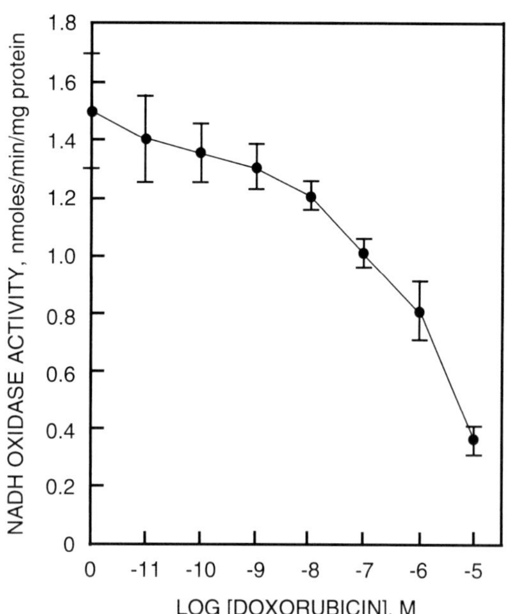

Fig. 11.18 Inhibition of NADH oxidase activity of plasma membrane of RLT-N hepatoma by doxorubicin. The EC_{50} for inhibition was ca. 1 μM. Reproduced from Morré et al. (1997c) with permission from Springer Scientific + Business Media

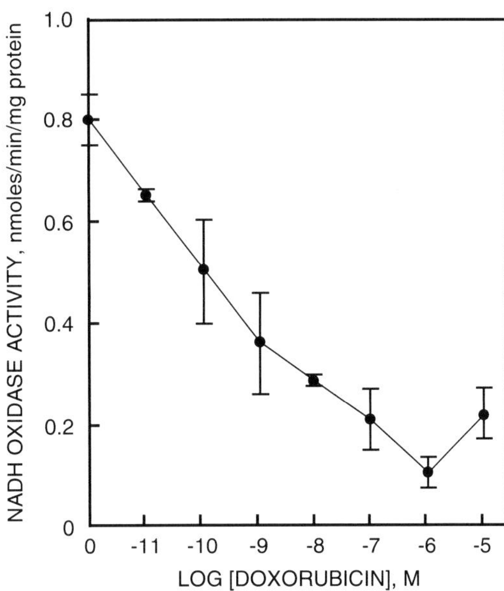

Fig. 11.19 Inhibition of NADH oxidase activity of plasma membrane of HeLa cells (human cervical carcinoma) by doxorubicin. The EC_{50} was ca. 0.7 nM. Reproduced from Morré et al. (1997c) with permission from Springer Scientific + Business Media

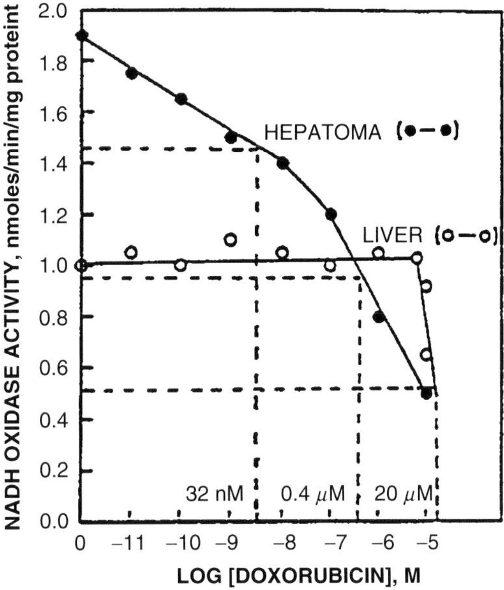

Fig. 11.20 Inhibition of NADH oxidase activity of plasma membranes of rat RLT-N hepatoma and of rat liver by doxorubicin. The activity of the hepatoma membranes is completely inhibited by about 0.1 μM doxorubicin, a concentration without effect on the NADH oxidase activity of rat liver membranes. The hepatoma plasma membranes were inhibited half-maximally at 32 nM. Total activity of the liver membrane was inhibited half-maximally at 20 μM. Redrawn from data of Morré et al. (1997c)

membrane of normal liver was inhibited only at a much higher range of doxorubicin concentrations (Figs. 11.20 and 11.21; Morré et al. 1997c; see also Sun and Crane (1981) for doxorubicin inhibition of NADH ferricyanide reductase activity in plasma membrane preparations of both mouse liver and pig erythrocytes).

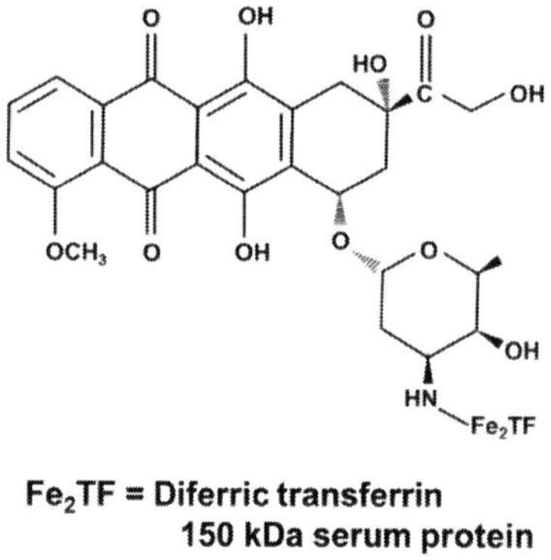

Fig. 11.21 Doxorubicin-diferric transferrin conjugate

Fe_2TF = Diferric transferrin
150 kDa serum protein

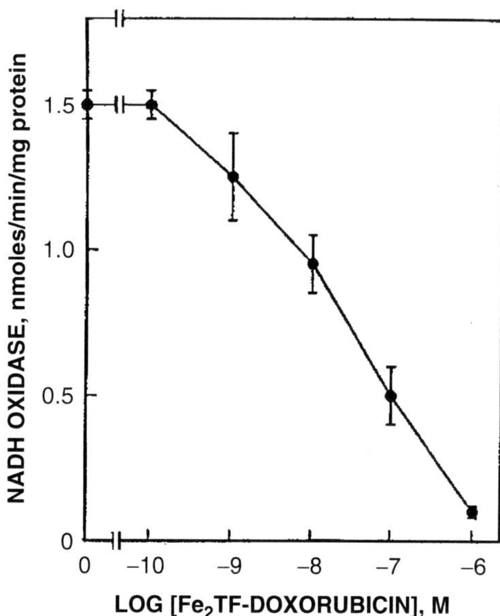

Fig. 11.22 Inhibition of NADH oxidase activity of plasma membranes of rat hepatoma RLT-N by doxorubicin conjugated to diferric transferrin. The activity was inhibited half-maximally at about 30 nM. Reproduced from Morré et al. (1997c) with permission from Springer Scientific + Business Media

Doxorubicin conjugated to diferric transferrin (Fig. 11.21) inhibited the ENOX2 activity of plasma membrane vesicles from rat hepatoma (Fig. 11.22) and HeLa cells (Fig. 11.23). The EC_{50} for inhibition of the constitutively activated NADH oxidase was about 30 nM for the conjugate and rat hepatoma plasma membranes (Fig. 11.22) compared to 1 μM for free doxorubicin (Fig. 11.18) and about 0.3 nM for the conjugate and HeLa plasma membrane (Fig. 11.21) compared to about 0.7 nM for free doxorubicin (Fig. 11.11). The doxorubicin conjugated to diferric transferrin did not inhibit the NADH oxidase activity of plasma membranes from normal rat liver even at the highest concentration tested of 100 μM (Morré et al. 1997c).

The differic transferrin conjugate of doxorubicin inhibited the plasma membrane NADH oxidase of HL-60 (human leukemia) cells as well as that of HL-60 cells normally resistant to doxorubicin (Fig. 11.24; Morré et al. 1997c). Thus, when

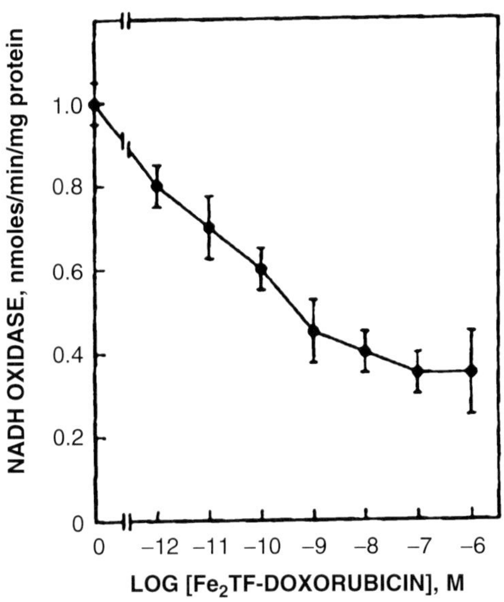

Fig. 11.23 Inhibition of NADH oxidase activity of plasma membranes of HeLa cells by doxorubicin conjugated to diferric transferrin. The activity was inhibited half-maximally at about 1 nM doxorubicin. Reproduced from Morré et al. (1997c) with permission from Springer Scientific + Business Media

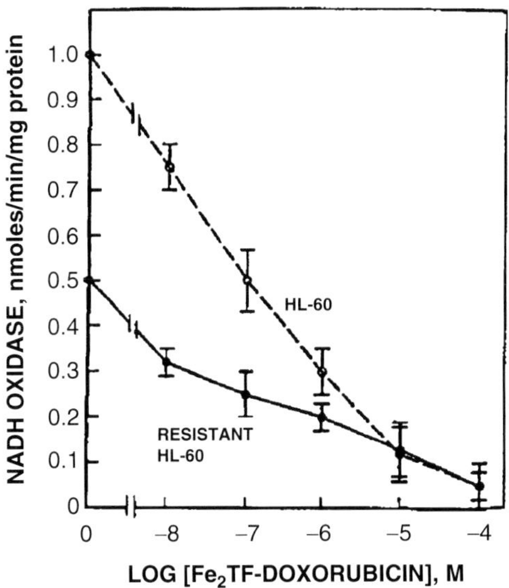

Fig. 11.24 NADH oxidase activity as a function of the concentration of doxorubicin conjugated with diferric transferrin comparing plasma membranes from HL-60 cells susceptible (*open circles*) and resistant (*solid circles*) to doxorubicin. The doxorubicin resistant cell line was provided by Drs. W. P. Faulk and K. Barabas, Methodist Center for Reproduction and Transplantation Immunology, Indianapolis, IN. The growth and electron transport characteristics of the cell line have been reported previously (Morré et al. 1994b). Reproduced from Morré et al. (1997c) with permission from Springer Scientific + Business Media

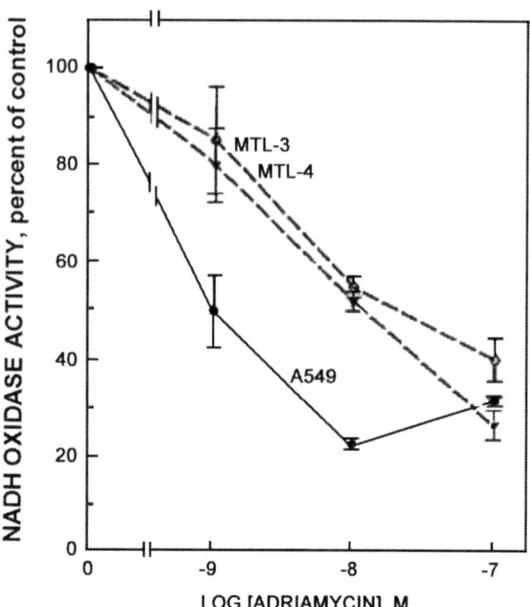

Fig. 11.25 NADH oxidase activities of plasma membranes isolated from cells resistant (MTL-3 and MTL-4) and susceptible (A549) to doxorubicin as a function of doxorubicin concentration. Results are means of three independent determinations ± standard deviations. Differences comparing A549 and MTL-3 or MTL-4 cells at 10^{-8} M ($p<0.01$) and 10^{-9} M ($p<0.001$) doxorubicin were highly significant. Reproduced from Hedges et al. (2003) with permission from Elsevier

presented to the resistant cells in a conjugated form, the resistance mechanism was passed such that NADH oxidase was inhibited, growth of the resistant cells was inhibited, and the resistant cells were killed. Thus, not only do the conjugates appear to be targeted to ENOX2, they may also provide an avenue for overcoming drug resistance in some patients with drug-resistant cancers (Morré et al. 1994c).

A correlation between the growth response to doxorubicin of doxorubicin-tolerant mesothelioma lines and the dose–response of the activity to doxorubicin of ENOX2 of the mesothelioma cell surface was reported by Hedges et al. (2003). A decreased doxorubicin responsiveness of ENOX2 correlated with drug tolerance (Fig. 11.25).

11.4.2.4 Impermeant Doxorubicin (Adriamycin®) as ENOX2-Targeted Antitumor Drugs

Not only do doxorubicin and other anthracyclines covalently linked to diferric transferrin and other impermeant supports prevent entry into cells but the conjugates were effective in inhibition of growth of tumorigenically transformed cells in culture (Fig. 11.20) where they exhibited enhanced efficacy compared to free doxorubicin (Sect. 11.4.2.3). The enhanced efficacy whereby anthracycline conjugates were more effective than was the free drug on a per mole of doxorubicin basis was an important observation implicating the conjugates as candidate antitumor drugs. With doxorubicin attached to agarose, immobilized doxorubicin was 100–1,000 times more active than was free doxorubicin (Tritton and Yee 1982). With doxorubicin bound to the polyglutaraldehyde microspheres (Tokes et al. 1982; Rogers et al. 1983; Rogers and Tokes 1984), increased effectiveness on a doxorubicin basis also

Table 11.4 Effect of doxorubicin conjugate and free doxorubicin on the component of NADH oxidase of rat liver plasma membranes stimulated by diferric transferrin[a]

	Diferric transferrin-stimulated NADH oxidation (nmol/min/mg protein)	
Inhibitor concentration	Doxorubicin	Doxorubicin conjugate
None	0.38 ± 0.09	0.38 ± 0.09
10^{-9}	0.17 ± 0.03	0.17 ± 0.02
10^{-8}	0.04 ± 0.01	0.04 ± 0.01
10^{-7}	0.03 ± 0.01	0.05 ± 0.01
10^{-6}	0.00 ± 0	0.025 ± 0.02

From Morré et al. (1997c)

[a]Assay is 0.05 M potassium phosphate buffer at pH 7.4 with 50 μM NADH, 17 μM differic transferrin, and 0.5 mg membrane in 2.8 mL. The absorbance change in response to diferric transferrin was measured at 340 nm with reference at 430 nm

was observed. The doxorubicin-transferrin conjugates introduced by Yeh and Faulk (1984) that were targeted to the transferrin receptor also proved to be disproportionately more effective than free doxorubicin (Faulk et al. 1991). Transferrin, which is the natural ligand for transferrin receptors was recognized earlier as a potent drug target due to the relative abundance of transferrin receptors on cancer cells (Faulk et al. 1980, 1990a). The transferrin-doxorubicin conjugates when tested clinically were found to be therapeutic in the treatment of leukemia (Faulk et al. 1990a; Yeh et al. 1984). Additionally, they inhibited the growth of doxorubicin-resistant cells (Yeh et al. 1984; Faulk et al. 1990b, 1991; Fritzer et al. 1992; Bérczi et al. 1993; Morré et al. 1997c; Fig. 11.24). At conjugate doses therapeutically effective in patients, cardiotoxicity was not observed (Rogers and Tokes 1984; see also Adler et al. 1995). With polymer-bound doxorubicin, the level of drug reaching the heart was reduced 100-fold compared to free drug in mice (Seymour et al. 1990). Reduction in toxicity and retention of activity has been reported as well for a doxorubicin-*N*-(2-hydroxypropyl)methylacrylamide-copolymer conjugate (Seymour et al. 1994). In another example, an inactive anthracycline analog, 4-dimethoxy-7,9-di-epi-daunorubicin, acquired significant cytostatic activity with doxorubicin-resistant and sensitive L 1210 cells when linked to polyglutaraldehyde microspheres (Rogers and Tokes 1984). Doxorubicin linked to monoclonal antibodies has proven superior to doxorubicin alone in some studies (Adler et al. 1995) but not in others (Dillman et al. 1986; Kaneko et al. 1991). Interestingly, conjugate inhibited the NADH oxidase activity of rat liver plasma membranes stimulated by diferric transferrin (Table 11.4). However, the basal ENOX1 activity of rat liver plasma membranes was not inhibited (Morré et al. 1997c).

11.4.2.5 Doxorubicin (Adriamycin®)-Inhibited NADH-Quinone Reductase of the Plasma Membrane

With hepatoma plasma membranes and, to a lesser extent, with HeLa cell plasma membranes, there appear to be at least two NADH oxidase activities of differing

Table 11.5 Rates of NADH oxidation in sera from normal vs. tumor bearing rats

	nmol/min/mg protein	
Sera	No doxorubicin	+25 μM doxorubicin
Normal	1.7±0.1	1.75±0.05
RLT-28 bearing	1.0±0.1	0.35±0.05

From Morré et al. (1997c)

sensitivities to doxorubicin. One may correspond to the activity purified from rat liver membranes that requires micromolar concentrations of doxorubicin to inhibit (Kim et al. 2002). The other, inhibited by nanomolar doxorubicin concentrations, appears to be specific to the plasma membrane of transformed cells and represented by ENOX2. ENOX2 is shed from the cell surface and appears in sera from tumor-bearing rats and cancer patients (Morré et al. 1996). In keeping with these observations, the NADH oxidase of sera of rats bearing RLT-28 hepatomas was inhibited by doxorubicin whereas that of sera of normal rats was not inhibited (Table 11.5).

The doxorubicin-inhibited NADH-quinone reductase was characterized and purified from plasma membranes of rat liver (Kim et al. 2002) First, an NADH-cytochrome b_5 reductase, which was doxorubicin-insensitive, was removed from the plasma membranes by treatment with the lysosomal protease, cathepsin D. After removal of the NADH-cytochrome b_5 reductase, the plasma membranes retained a doxorubicin-inhibited NADH-quinone reductase activity (Sect. 4.1.3.3). The enzyme, with an apparent molecular mass of 57 kDa, was purified 200-fold over the cathepsin D-treated plasma membranes. The purified enzyme had also an NADH-coenzyme Q_0 reductase (NADH: external acceptor (quinone) reductase; EC 1.6.5.) activity. Partial amino acid sequence of the enzyme showed that it was unique with no sequence homology to any known protein. Antibody against the enzyme (peptide sequence) was produced and affinity-purified. The purified antibody immunoprecipitated both the NADH-ferricyanide reductase activity and NADH-coenzyme Q_0 reductase activity of plasma membranes and cross-reacted with human chronic myelogenous leukemia K562 cells and doxorubicin-resistant human chronic myelogenous leukemia K562R cells. Localization by fluorescence microscopy showed that the reaction was with the external surface of the plasma membranes. The doxorubicin-inhibited NADH-quinone reductase remains as a candidate ferricyanide reductase for PMET (Chap. 4).

11.4.2.6 Bleomycin Control of Transplasma Membrane Redox Activity and Proton Movement in HeLa Cells

There is a significant inhibition of ferricyanide reduction rates by bleomycins (Sun and Crane 1985). Similar inhibition of proton release coupled to the redox system also was observed. Tallysomycins and bleomycin show similar effects on all these

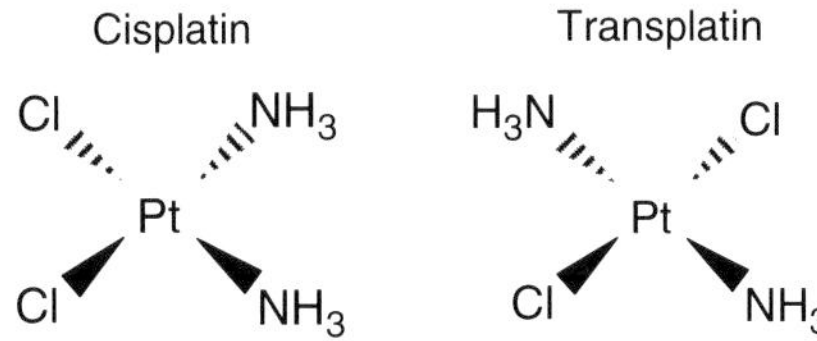

Fig. 11.26 Cisplatin [(SP-4-2)-diamminedichloroplatinum, *cis* platinum] (active) and transplatin (inactive)

reactions. Studies with purified membranes clearly showed that a target for these drugs was on the plasma membrane most likely ENOX2.

11.4.3 Cisplatin Targets ENOX2 of the PMET

Another widely used chemotherapy agent, cisplatin (*cis*-platinum; (*SP*-4-2)-diamminedichloridoplatinum), is also a potent inhibitor the PMOR of cancer cells (Chu 1994; Gonzales et al. 2001; Roos and Kaina 2006) and of ENOX2 (Morré et al. 2008a). This is despite the general consensus that its cytotoxicity is the result of cross-linking of DNA to ultimately trigger apoptosis. Cisplatin was the first member of a family of related platinum-based drugs which also includes carboplatin and oxaliplatin to be employed as a cancer therapeutic. Cisplatin was approved for use in testicular and ovarian cancer in 1978. Sometimes designated as an alkylating agent, cisplatin has no alkyl group required to carry out alkylating reactions. Transplatin, the *trans* stereoisomer of cisplatin (Fig. 11.26) is without effect on ENOX2 activity and exhibits no pharmacological effect of comparable utility in cancer chemotherapy to that of cisplatin.

Both cis- and transplatin isomers bind to DNA equally well (Hill and Grubbs 1982; Salles et al. 1983). Thus, the possibility is raised that the formation of DNA interstrand cross-links may be of little importance in determining the specific cytotoxic effect of the *cis* compound such that the involvement of other targets may be required (Sun and Crane 1990). The PMET system of HeLa cells, which transfers electrons from internal NADH to external ferricyanide, is inhibited by cisplatin at concentrations exceeding 10^{-7} M (Fig. 11.27). Incubation with transplatin under the same conditions shows very little inhibition and actually gives a stimulation of ferricyanide reduction at about 10^{-6} M (Sun and Crane 1984; Fig. 11.27). The stimulation of ferricyanide reduction by the cis- and transplatin above 5×10^{-7} M is eliminated by superoxide dismutase consistent with evidence that higher concentrations of these compounds induce superoxide production (McGinness et al. 1978). Transplatin gives much less inhibition of ferricyanide reduction than cisplatin suggesting an involvement of the transplasma membrane electron transport system as a specific site for cisplatin inhibition.

Cisplatin, but not transplatin, is a potent inhibitor of the ENOX2 of HeLa cells (Morré et al. 2008a). The molecular basis for the binding that results in ENOX2 inhibition has not been investigated.

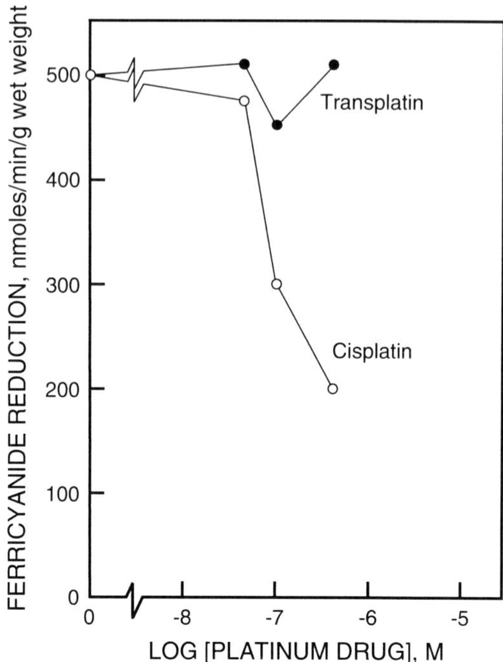

Fig. 11.27 The effect of cisplatin and transplatin on ferricyanide reduction by HeLa cells. Concentrations of 1 μM or higher were less effective (cisplatin) or stimulated (transplatin). Redrawn from Sun and Crane (1990)

11.4.4 Antitumor Sulfonylureas

Two sulfonylurea ENOX2 inhibitors with antitumor activity, LY181984 (Lilly; Morré et al. 1995i) and BTS-2 (Purdue University and Universidad Publica de Navarra; Encio et al. 2005), have been investigated extensively. The sulfonylureas represent a novel series of synthetic organic compounds identified as having activity against human solid tumors in vivo (Grindey et al. 1987; Grindey 1988; Taylor et al. 1989). They were identified as the result of a program of screening against in vivo murine solid tumors implanted subcutaneously (Grindey et al. 1987; Howbert et al. 1990). Members of the series, including the experimental *N*-(methylphenylsulfonyl)-*N′*-(4-chlorophenyl)urea (LY181984) were described with the potential of exhibiting a toxicity directed exclusively to transformed cells and tissues (Taylor et al. 1989; Howbert et al. 1990). Known collectively as diaryl sulfonylureas (sulfonylureas), their mechanism of action was apparently unrelated to previously described classes of oncolytic agents (Talbot et al. 1993). There was no evidence for cell cycle specificity of the drugs and no inhibition of DNA, RNA, or protein synthesis (Howbert et al. 1990; Houghton et al. 1990a, b, c). The sulfonylureas exhibited few, if any, mechanistic parallels to other known antitumor agents. The drugs were membrane active and weak uncouplers of mitochondrial oxidative phosphorylation (Houghton et al. 1990a, b, c; Thakar et al. 1991; Rush et al. 1992). Their mode of action was expected to be unique. One member of the series, Sulofenur, progressed in evaluation to Phase I (Taylor et al. 1989; Hainsworth et al. 1989) and Phase II (Talbot et al. 1993) clinical trials.

11.4.4.1 ENOX2 as the Target for the Anticancer Activity of the Anticancer Sulfonylureas

Interest in the sulfonylureas as potential ENOX2 inhibitors stemmed initially from the demonstration of a requirement for quinones in the proton release from HeLa cells stimulated by differic transferrin or ferricyanide (Sun et al. 1992a). Analog inhibition was used with intact cells. The coenzyme Q analogs, DCIQ or ETHOXQ, inhibited both ferricyanide- and diferric transferrin-stimulated proton release in HeLa cells, and this release could be reversed by the addition of coenzyme Q (Sun et al. 1992b). Such a relationship had been established previously for sulfonylurea interactions with quinone sites (Schloss et al. 1988; Grabau and Cronan 1986). Both the ferricyanide- and diferric transferrin-induced proton release of HeLa cells were inhibited by the active antitumor sulfonylurea LY181984 (Sun et al. 1995). Binding of diaryl sulfonylureas to quinone sites was demonstrated initially with the sulfonylurea herbicides (chlorsulfuron, sulfometuron methyl, metsulfuron methyl). These herbicides were inhibitors of acetolactate synthase, the entry enzyme into the branched chain amino acid pathway (Schloss et al. 1988).

Subsequently, a putative antitumor sulfonylurea binding protein corresponding to ENOX2 with an approximate molecular weight of 34 kDa was identified for the HeLa plasma membranes (Morré et al. 1995g, h) and found to exhibit an NADH oxidase activity (Morré et al. 1995i). Plasma membranes isolated from HeLa cells grown in culture bind the active antitumor sulfonylurea LY181984 with high affinity ($K_d = 30$ nM) (Morré et al. 1995g). NADH oxidase was inhibited by LY181984 with an EC_{50} of about 30 nM for plasma membranes of HeLa cells but not from rat liver (Morré et al. 1995i; Fig. 11.28). Also inhibited by antitumor-active but not by -inactive sulfonylureas was proton release from HeLa cells and alkalization of cytoplasm induced by diferric transferrin or ferricyanide (Sun et al. 1995; Fig. 5.17). The NADH oxidase activity of plasma membrane vesicles from HeLa cells was inhibited by the sulfonylurea only with NADH supplied to the external plasma membrane surface (Morré 1995a). These findings demonstrated that the sulfonylurea-inhibited activity was an ectoprotein of the plasma membrane (Morré 1995a). Subsequently, the activity was shown to be shed in soluble form from the cell surface of the HeLa cells (Morré et al. 1996c). The presence of a sulfonylurea-inhibited shed form of the NADH oxidase activity of media conditioned by growth of HeLa cells then prompted a search for a comparable activity in sera of tumor-bearing rats and cancer patients.

The soluble shed form of the NADH oxidase activity inhibited by LY181984 and isolated from culture media conditioned by growth of HeLa S cells was similar to that associated with the outer surface of the plasma membrane (Morré et al. 1996c). The activity was absent from media in which cells had not been grown and was present in conditioned culture media from which cells had been removed by centrifugation both for serum-containing and serum-free media. The K_m with respect to NADH and the response to thiol reagents were similar to those of the corresponding activity of the plasma membrane of HeLa cells. The conditioned HeLa culture media bound [^{3}H]LY181984 with high affinity (Morré et al. 1996c). Both antitumor sulfonylurea-inhibited (ENOX2) and -resistant (ENOX1) forms of the NADH oxidase were isolated by preparative free-flow electrophoresis. The antitumor sulfonylurea-inhibited

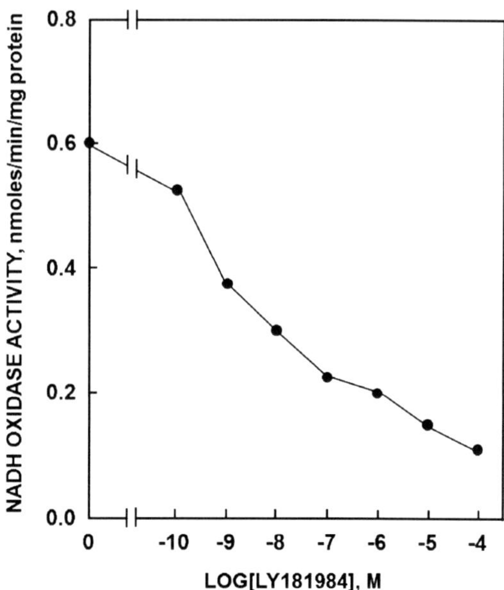

Fig. 11.28 NADH oxidase activity of plasma membrane vesicles isolated from HeLa cells and assayed in the presence of varying concentrations of LY181984 in the presence of 100 μM L-cysteine. Reproduced from Morré et al. (1998g) with permission from Elsevier

activity was purified to apparent homogeneity and was identified as ENOX2 with a molecular weight of 33.5 (34) kDa and an isoelectric point of about pH 4.5. The 33.5-kDa protein from conditioned HeLa culture medium both bound [^{3}H]LY181984 and retained an LY181984-inhibited NADH oxidase activity.

Inhibition by sulfonylureas of growth (Fig. 5.17a), NADH oxidase activity (Fig. 5.17b), diferric transferrin-stimulated proton release (Fig. 5.17c), and a protein disulfide-thiol interchange activity measured by reconstitution of inactive (scrambled) RNase (Fig. 5.17d; Morré et al. 1998a) were correlated. All responded to the active antitumor sulfonylurea LY181984 but not the inactive LY181985. Growth of COS cells which express low levels of ENOX2 was inhibited by LY181984 with an IC_{50} of 20 μM. When stably transfected with ENOX2 cDNA, the IC_{50} for inhibition of growth was reduced to 3 μM (Table 11.6). The inactive LY181985 was without effect (IC_{50} > 100 μM).

11.4.4.2 Antitumor Sulfonylureas Need Not Enter Cells to Be Effective

The antitumor sulfonylurea LY237868 (*N*-(4-aminophenyl-sulfonyl)-*N′*-(4-chlorophenyl)urea) was conjugated through the A ring to α-cyclodextrin or agarose bead material (Affigel 10) to prepare impermeant conjugates for activity measurements and affinity isolation of binding proteins from serum. When conjugated to α-cyclodextrin, the resulting LY237868 conjugate (Fig. 11.29) inhibited both NADH oxidase activity (Fig. 11.30) and growth (Fig. 11.31) of HeLa cells in

Table 11.6 Inhibition by ENOX2-targeted drugs plus LY181985, tamoxifen, and methotrexate on growth of COS cells (inhibition of the increase in cell number) as a result of stable transfection with ENOX2 cDNA

Drug[a]	IC_{50} (μM) Nontransfected	ENOX2 transfected
EGCg	10	0.1
Capsaicin	15	2.3
Adriamycin	0.3	0.04
LY181984 (active)	20	3.0
LY181985 (inactive)	>100	>100
Tamoxifen	16	8
Methotrexate	1	1

From Chueh et al. (2004)
Values were estimated from triplicate dose–response determinations
[a]*EGCg* (−)-epigallocatechin-3-gallate; Capsaicin, 8-methyl-*N*-vanillyl-6-noneamide; LY181984, *N*-(4-methylphenylsulfonyl)-*N*′-(4-chlorophenyl) urea; LY181985, *N*-(4-methylphenylsulfonyl)-*N*′-(4-phenyl)urea

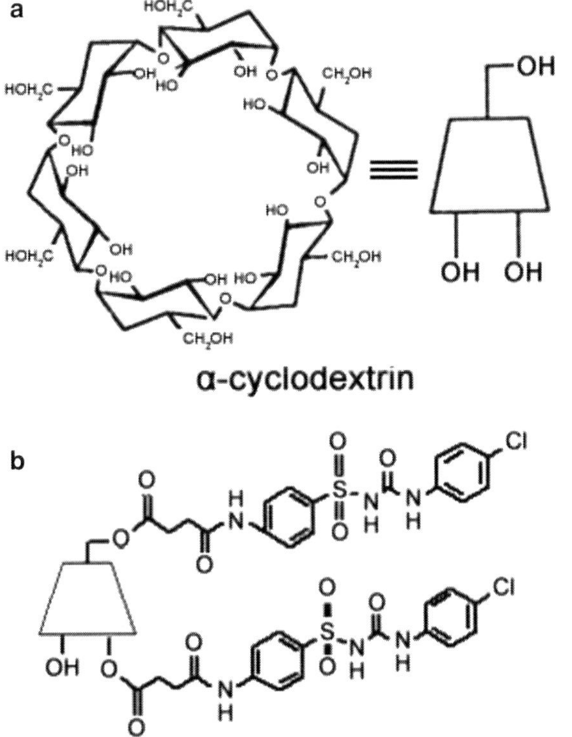

Fig. 11.29 Conjugation of antitumor sulfonylurea LY237868 with α-cyclodextrin. Reproduced from Kim et al. (1997) with permission from Elsevier

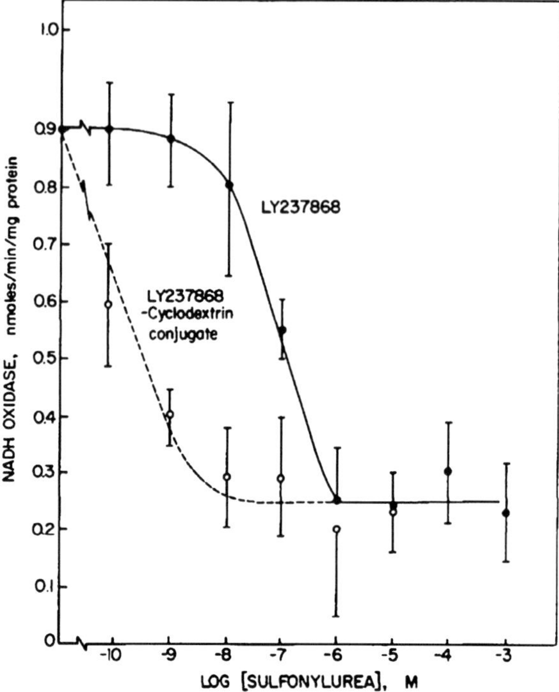

Fig. 11.30 Inhibition of NADH oxidase of sealed right-side-out vesicles of HeLa plasma membrane by the α-cyclodextrin conjugate of LY237868 as a function of concentration compared to the unconjugated LY237868. Both were inhibitory. Reproduced from Kim et al. (1997) with permission from Elsevier

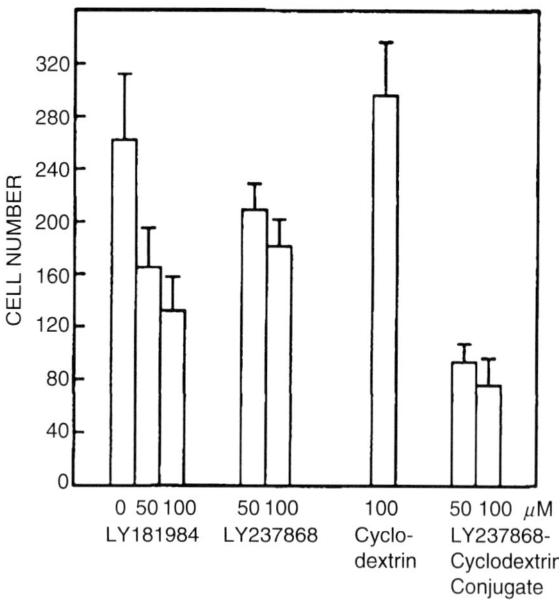

Fig. 11.31 Cytotoxicity of the α-cyclodextrin conjugate of LY237868 at two concentrations compared to α-cyclodextrin and LY181984 or LY237868 based on growth over 72 h of attached HeLa cells. The 0 is for the DMSO control. All drugs were dissolved in DMSO. Results are averages of three experiments ± standard deviations among experiments. Reproduced from Kim et al. (1997) with permission from Elsevier

culture (Kim et al. 1997). The conjugate was at least one order of magnitude more potent as an inhibitor than the parent compound.

LY237868 conjugated to agarose beads as the affinity support bound a large number of serum proteins. However, compared to serum from normal patients, the affinity support bound two proteins of M_r approximately 33.5 and 29.5 not found in sera of normal patients. The 33.5-kDa protein from human sera reacted with antisera to a 33.5-kDa protein from culture media conditioned by growth of HeLa cells that blocked and immunoprecipitated the sulfonylurea-responsive activity from HeLa cell plasma membranes. The results point to the 33.5-kDa protein from cancer patient sera that bound to the sulfonylurea affinity support as representing the circulating equivalent of the previously identified 34 kDa sulfonylurea-binding protein, with NADH oxidase activity at the external cell surface of cultured HeLa cells and a corresponding 33.5 kDa sulfonylurea-binding protein, with NADH oxidase activity at the external cell surface of cultured HeLa cells and a corresponding 33.5-kDa protein shed into culture media conditioned by growth of HeLa cells (ENOX2).

Benzo[*b*]thiophenesulphonimide1,1-dioxide (BTS) derivatives (Fig. 11.1) represent a new class of ENOX2 inhibitors which were synthesized on the basis of theoretic structure-activity studies of sulfonylureas (Martinez-Merino et al. 2000; Alonso et al. 2001) and where inhibition targeted to ENOX2 is related to redox state of the protein (Encio et al. 2005).

11.4.4.3 Response of ENOX2 Activity to Antitumor Sulfonylureas is Redox and Growth Factor Sensitive

A surprising result with the shed (serum) form of ENOX2 was that the activity could be either stimulated or inhibited by 1 μM LY181984. Whether or not the activity was inhibited or stimulated by LY181984 was explained on basis of the redox environment of the protein (Morré 1998c) If plasma membrane vesicles were first treated with dithiothreitol (DTT) or with reduced glutathione (GSH) and then assayed for NADH oxidase activity, the sulfonylurea inhibited in a concentration-dependent manner (Fig. 11.32). In contrast, if the plasma membrane vesicles were first treated with diluted hydrogen peroxide or oxidized glutathione (GSSG), and then assayed for NADH oxidase activity, the antitumor sulfonylurea stimulated the activity (Fig. 11.33).

Growth experiments were conducted in parallel (Figs. 11.34 and 11.35). LY181984 administered to HeLa cells in the presence of GSH was approximately two log orders more effective than LY181984 administered to HeLa cells in the presence of GSSG. Similar results were found in the sera of cancer patients (Table 11.7). With sera from normal individuals or with plasma membranes of rat liver, the oxidizing or reducing conditions were without effect (Table 11.8). The findings suggest that the response of the cell surface NADH oxidase of HeLa cells to the antitumor sulfonylurea LY181984 is influenced by the redox environment which may determine whether the drug will stimulate or inhibit the activity and that the degree of response may be reflected in the ability of LY181984 to inhibit HeLa cell growth.

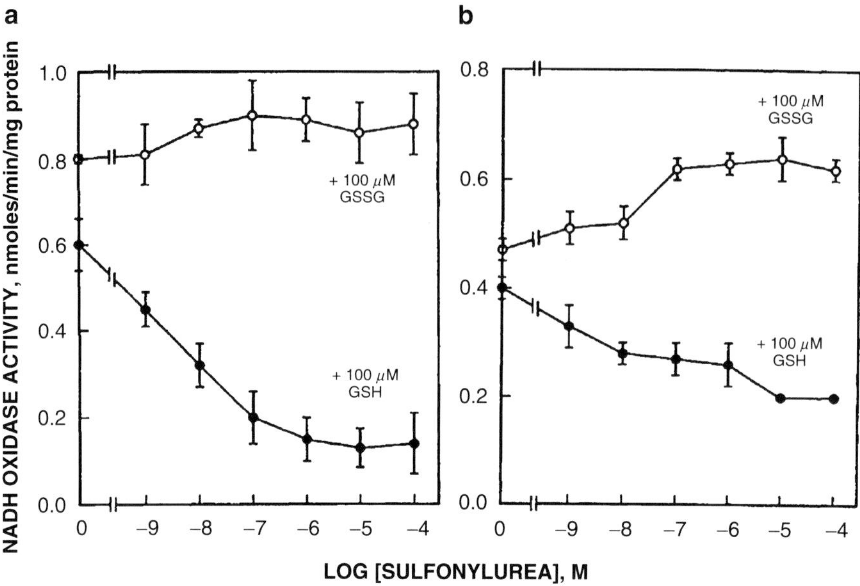

Fig. 11.32 NADH oxidase activity of isolated vesicles of plasma membranes from HeLa cells as a function of the concentration of LY181984 in the presence of either 100 μM reduced (GSH) or oxidized glutathione (GSSG). (**a**) Freshly frozen plasma membranes. (**b**) Plasma membranes stored frozen for 6 months. Results are averages from four different plasma membrane preparations ± standard deviations. Reproduced from Morré et al. (1998g) with permission from Elsevier

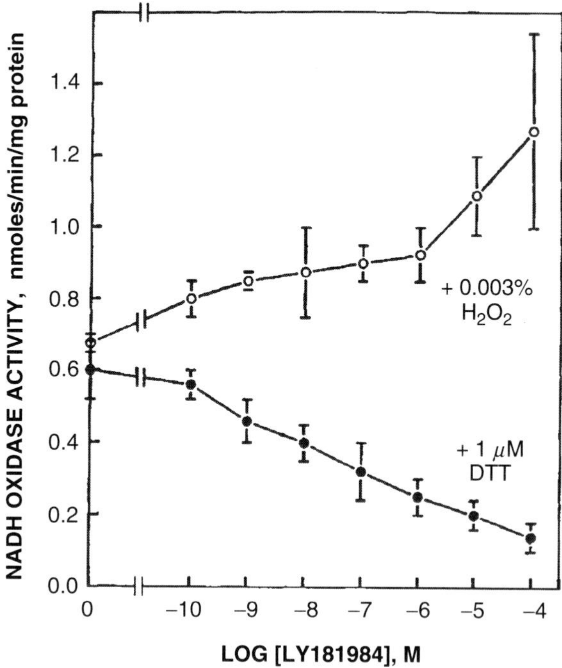

Fig. 11.33 NADH oxidase activity of isolated vesicles of plasma membranes from HeLa cells as a function of the concentration of the antitumor sulfonylurea, LY181984, in the presence of either 1 μM dithiothreitol (DTT) or 0.003 % hydrogen peroxide. Results are averages from three different plasma membrane preparations ± standard deviations. Reproduced from Morré et al. (1998g) with permission from Elsevier

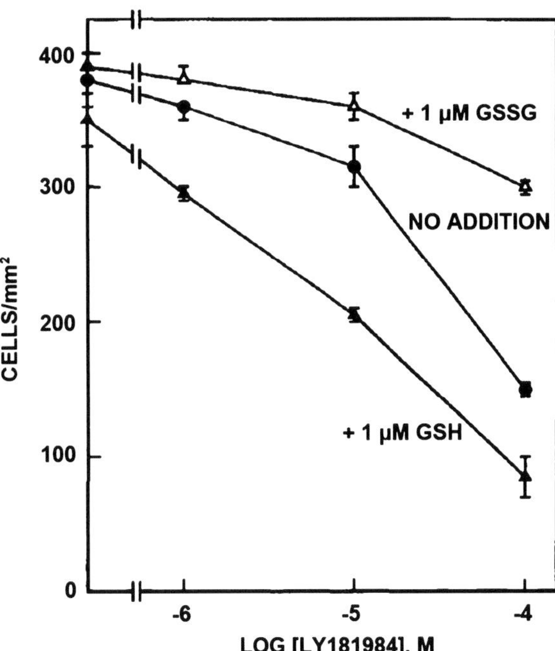

Fig. 11.34 Growth of HeLa cells after 96 h as a function of the concentration of LY181984. The *solid circles* are with no addition. The *open triangles* are with 1 μM GSSG and the *solid triangles* are with 1 μM GSH. These concentrations of GSH and GSSG were selected to be within the range of concentrations not affecting the growth of HeLa cells in the absence of LY181984. Values are from three determinations ± standard deviations. Reproduced from Morré et al. (1998g) with permission from Elsevier

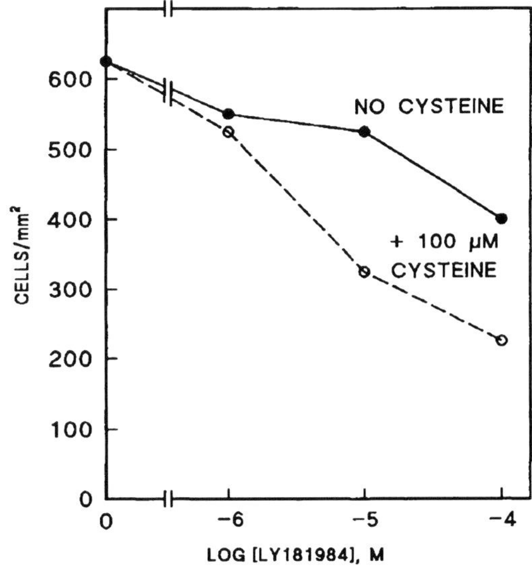

Fig. 11.35 Growth of HeLa cells after 96 h as a function of sulfonylurea concentration in the presence or absence of 100 μM L-cysteine. Reproduced from Morré et al. (1998g) with permission from Elsevier

Table 11.7 Response of NADH oxidase of patient sera (–SU) and response to 1 μM LY181984 in DMSO (+SU) comparing oxidizing (0.003 % H_2O_2) or reducing (1 μM DTT) conditions

		No addition			0.003 % H_2O_2			1 μM DTT		
Cancer	Designation	–SU	+SU	Ratio	–SU	+SU	Ratio	–SU	+SU	Ratio
Breast	SB-6	1.1	0.55	0.5	0.6	0.7	1.17	0.2	0.12	0.6
	SB-21	0.22	0.35	1.6	0.9	1.2	1.33	0.55	0.4	0.55
	SB-22	0.3	0.2	0.7	1.3	1.2	0.92	0.7	0.6	0.8
	SB-104	0.45	0.35	0.8	1.5	1.8	1.13	0.6	0.5	0.8
Prostate	SB-4	0.4	0.9	2.25	0.8	1.0	1.25	0.6	0.4	0.7
Ovarian	SB-33	0.5	0.35	0.7	0.6	0.65	1.08	0.35	0.15	0.4
Leukemia (CLL)	SB-29	0.5	0.35	0.7	0.6	0.65	1.08	0.35	0.15	0.4
Pancreatic	SB-9	0.5	0.8	1.6	0.68	0.83	1.22	0.65	0.55	0.85
Bladder	SB-69	0.4	0.35	0.9	0.35	0.5	1.43	0.3	0.1	0.3
Lung	SB-5	0.4	0.2	0.5	0.9	1.1	1.22	0.5	0.25	0.5

From Morré et al. (1998g)
Units are NADH oxidase, nmol/mL serum

Table 11.8 NADH oxidase activity of sera from healthy laboratory volunteers (−SU) and response to 1 μM antitumor sulfonylurea, LY181984 in DMSO (+SU) comparing different redox conditions during assay

		NADH oxidase nmol/mL serum		
Redox condition	*N*	−SU	+SU	Ratio, +SU/−SU
No addition	25	0.38 ± 0.08	0.38 ± 0.08	1.0 ± 0.02
0.003 % H_2O_2	23	0.78 ± 0.29	0.77 ± 0.28	0.98 ± 0.06
0.03 % H_2O_2	10	0.74 ± 0.21	0.74 ± 0.21	1.0 ± 0.01
1 μM DTT	15	0.6 ± 0.3	0.6 ± 0.3	1.0 ± 0.03
10 μM DTT	4	0.3 ± 0.1	0.3 ± 0.1	1.0 ± 0.05

From Morré et al. (1998g)

The inhibition of ENOX2 activity of right-side-out plasma membrane vesicles from HeLa cells by LY181984 was enhanced by the addition of epidermal growth factor (EGF). It was necessary, however, that the LY181984 be followed by EGF. If the EGF was administered first, the response of ENOX2 activity to LY181984 was unaffected by EGF to suggest that EGF may somehow facilitate binding of LY181984 (Morré et al. 1997).

That the plasma membrane NADH oxidase responds to EGF with vesicles from rat liver, for example, suggested that the oxidase protein may normally be coupled in the membrane to growth factor responses. This coupling, however, appeared to be lost in transformation (Morré et al. 1991a; Bruno et al. 1992). One possibility to explain the EGF-sulfonylurea interaction is that the presence of the sulfonylurea restores the coupling between growth factor response and NAD oxidase in plasma membranes of the transformed HeLa cells. The result is a response to EGF, but not one of stimulation of NADH oxidase activity and growth as in the normal cell, but one of an enhanced inhibitory response to the sulfonylurea.

11.4.5 Antitumor Quassinoids Target ENOX2

ENOX2 is inhibited by nanomolar and subnanomolar concentrations of the antitumor quassinoid, glaucarubolone (Morré et al. 1998b). Glaucarubolone is one of a group of naturally occurring and chemically modified products termed quassinoids from the family Simaroubaceae with potent anticancer activity (Grieco et al. 1993, 1996; Valeriote et al. 1998). Quassinoids derive their name from the parent compound quassin first identified as a bitter principle from plants of the family Simaroubaceae (Polonsky 1973; Sect. 10.8).

Glaucarubolone has been shown previously to have therapeutic activity in vivo against C38, a transplantable murine colon carcinoma with a % T/C of 14 and a MTTD of 151 mg/kg body weight (Valeriote et al. 1998). Glaucarubolone also was active in vitro on L1210 lymphocytic leukemia and human CX-1 colon carcinoma cells. Our findings extend activity in vitro to human cervical carcinoma (HeLa) and Kaposi's sarcoma cells.

Quassinoids have been tested extensively on M17/Adr mouse mammary carcinoma cells. Especially effective was the hydroxyl analog of glaucarubolone which demonstrated not only solid tumor selectivity to both murine and human cells but also selectivity to a resistant mammary adenocarcinoma cell line M17/Adr.

11.4.5.1 Impermeant Conjugate of Glaucarubolone

Since the drug-responsive site of the NADH oxidase is located at the cell's exterior, inhibitory quassinoids directed to ENOX2 need not enter cells to be effective. Additional specificity to glaucarubolone has been imparted by conjugates involving the C-15 hydroxyl (Fig. 11.36). Effectiveness of a C-15 acylated analog ($R{=}COCH_2NMe_2$) of glaucarubolone against MAM 16/C/RP cells (% T/C of 16) has been demonstrated. A % T/C equal to <10 constitutes a highly active agent by NCI standards and would represent >1 log cell kill.

One conjugate where preliminary data have been generated is a conjugate of glaucarubolone through the C-15 hydroxyl to amino polyethylene glycol (PEG) (Morré et al. 1998b). The amino-PEG-glaucarubolone is more effective than free glaucarubolone in inhibiting the target NADH oxidase of plasma membrane, exhibits enhanced water solubility compared to glaucarubolone, and is more effective than glaucarubolone in inhibiting the growth of HeLa cells (Fig. 11.36).

The inhibition was seen with plasma membrane vesicles of HeLa cells at two log orders less glaucarubolone than with plasma membrane vesicles of rat liver (Morré et al. 1998b). Assignments of a drug-binding site to the external surface of the HeLa cell plasma membrane were supported by findings where full activity of the glaucarubolone in the inhibition of NADH oxidase activity of isolated plasma membrane vesicles and of growth of HeLa cells was given on a molar glaucarubolone basis by an impermeant conjugate of glaucarubolone in which the glaucarubolone moiety was linked via the C-15 hydroxyl to amino PEG (ave M_r 5,000) (results unpublished).

Impermeant conjugates with potency equivalent to the acylated analogs, therefore, can be utilized as antitumor agents. These conjugates also help overcome a potential drawback of the quassinoid antitumor agents, that of a relatively high inherent cytotoxicity and narrow therapeutic ratio. By reducing or eliminating unspecific cytotoxicity and restricting the drug to the cell surface target, conjugation of quassinoids to impermeant supports greatly increases efficacy and reduces toxicity with a corresponding 10- to 100-fold potential increase in the margin of safety (tenfold increase in efficacy accompanied by a tenfold decrease in unspecific toxicity).

Glaucarubolone was selected for initial testing due to the presence of a free $C_{(15)}$ hydroxyl to allow conjugation to form an impermeant derivative to test the concept of a cell surface site of action. The plasma membrane vesicles used in these experiments were sealed and right side-out (Morré 1995a). That the glaucarubolone-amino PEG conjugate inhibited the drug-responsive NADH oxidase activity with the same efficiency as free glaucarubolone was consistent with the drug-responsive NADH site being at or near the cell surface.

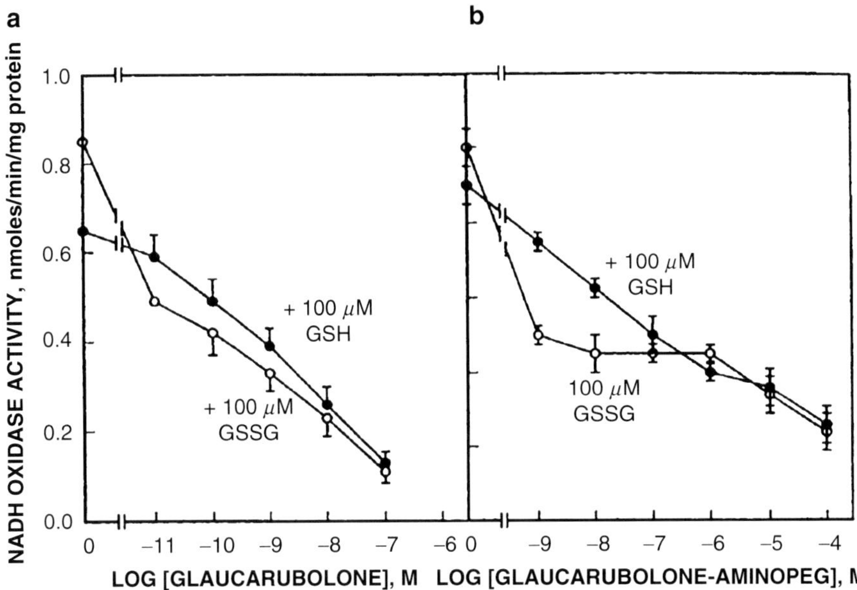

Fig. 11.36 NADH oxidase activity of isolated vesicles of plasma membrane from HeLa cells in response to varying concentrations of free glaucarubolone (**a**) or of the glaucarubolone amino PGG conjugate on a molar glaucarubolone basis. (**b**) Assays were either in the presence of 100 μM reduced (GSH) (*solid symbols*) or oxidized (GSSG) (*open symbols*) glutathione. Results are averages from three different plasma membrane preparations ± standard deviations. Reproduced from Morré et al. (1998b) with permission from Elsevier

11.4.5.2 Activity of Glaucarubolone Conjugates at the Tumor Cell Surface Is Redox Sensitive

The activity of the glaucarubolone-PEG conjugate, like that of free glaucarubolone, was modulated by the redox environment of the cells and of the plasma membrane vesicles. Activity, both in the inhibition of NADH oxidase activity and in the inhibition of growth, was enhanced by oxidizing conditions in the presence of oxidized glutathione compared to reducing conditions in the presence of reduced glutathione (Fig. 11.36).

11.4.6 Acetogenins

Various members of the plant family *Annonaceae* have yielded a group of bioactive secondary metabolites known collectively as the annonaceous acetogenins. These compounds are variously cytotoxic, pesticidal, antimalarial, antiparasitic, antimicrobial, and antineoplastic (Jolad et al. 1982; Rupprecht et al. 1990; Fang et al. 1993). A characteristic of these compounds, which may relate to cytotoxicity rather than

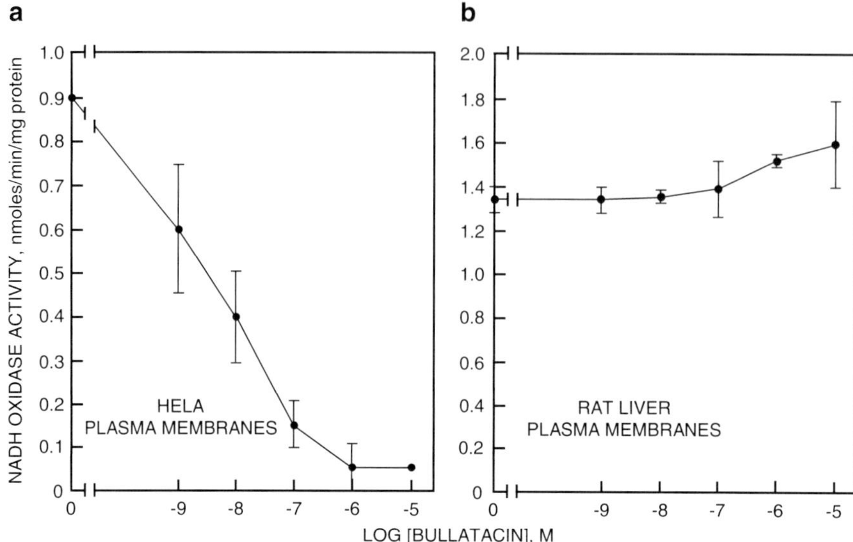

Fig. 11.37 Dose–response of isolated plasma membrane vesicles NADH oxidase to bullatacin. (**a**) Rat liver. Average of triplicate determinations ± standard deviations. (**b**) HeLa cells. Duplicate determinations ± mean average deviations. Reproduced from Morré et al. (1995c) with permission from Elsevier

cancer specificity, is activity in the inhibition of mitochondrial electron transport (Ahammadsahib et al. 1993). However, a more selective activity is necessary to explain the ability of certain acetogenins to kill cancer cells under conditions where normal cells are unharmed.

NADH oxidation by HeLa plasma membranes was reduced markedly by increasing concentrations of the acetogenin bullatacin added as DMSO solution (final DMSO concentration 0.1 %) (Morré et al. 1995c). The log-linear dose–response of HeLa plasma membranes to bullatacin of Fig. 11.37a indicated an ED_{50} of 5–10 nM with nearly complete inhibition of the activity by 1 μM. In contrast to results with HeLa cell plasma membranes, plasma membranes of rat liver were unaffected by bullatacin over the concentration range of 1 nM to 10 μM (Fig. 11.37b). NADH oxidase activity of HL-60 cells also was inhibited by bullatacin. Bullatacin and other active acetogenins (Table 5.3) inhibit growth of susceptible cell lines in culture in the nanomolar range of drug concentrations (Ahammadsahib et al. 1993).

These findings demonstrate that the inhibition of the activity of ENOX2 of the plasma membranes correlates more closely with the antitumor activity of bullatacin, and, other acetogenins, than does the inhibition of mitochondrial electron transport. Furthermore, ENOX2 as an acetogenin target helps to explain the in vitro and in vivo selectivity of the acetogens toward transformed cells.

11.4.7 EGCg

ENOX2 has been suggested as a low-dose molecular target to explain the anticancer activity of the green tea polyphenol EGCg (Morré et al. 2000a; Chueh et al. 2004). EGCg is the major catechin in green tea (Stoner and Mukhtar 1995) and uniquely has the ability to selectively block the growth of cancer cells in culture (Chen et al. 1998), in keeping with there being a cancer-specific target for EGCg (Cooper et al. 2005b), independent of its antioxidant properties (Pillai et al. 1999). Since many substances including vitamin C and catechins other than EGCg are as good or better antioxidants than EGCg, it is unlikely that the anticancer action of EGCg is mediated through its antioxidant properties (Cutter et al. 2001). Not only does EGCg selectively block the growth of tumor cells, but tumor cells also are much more sensitive to inhibition by EGCg than their normal counterparts (Morré et al. 2000a; Chen et al. 1998; Ahmad et al. 1997; Hayakawa et al. 2001).

Antisense results and the transfection and vector-forced overexpression experiments when taken together showed that cell surface expression of ENOX2 was both necessary and sufficient for the cancer-specific cell growth inhibitions attributed to EGCg (Chueh et al. 2004; Chap. 5). When ENOX2 cDNA was overexpressed in COS or MCF-10A (noncancer mammary epithelial) cells, growth was accelerated in the transfected cells, a larger cell size was reached at confluence and the susceptibility to growth inhibition by EGCg was heightened (Chueh et al. 2004; Figs. 5.20 and 5.21). Transgenic animals overexpressing ENOX2 exhibited the same level of unregulated cell enlargement and sensitivity to EGCg as cancer cells (Yagiz et al. 2006; Figs. 8.14 and 8.15). The findings with the embryo fibroblasts from ENOX2 transgenic mice are unique in that they demonstrate in a simple single cell system stable expression of a single protein, ENOX2, is both necessary and sufficient to impart drug sensitivity to EGCg. As wild-type fibroblasts are normally unresponsive, the response to EGCg of transgenic fibroblasts further confirms the hypothesis that ENOX2 represents both a necessary and sufficient molecular target for EGCg to explain both its anticancer preventative and its therapeutic potential. Similarly, mouse embryonic fibroblast (MEF) cells prepared from the ENOX2 overexpressing transgenic mice (Yagiz et al. 2006) grew at rates approximately twice those of wild-type embryonic cells (Yagiz et al. 2007; Figs. 8.12 and 8.13). EGCg does not interfere with protein synthetic activity and the implication is that ENOX2 activity is somehow directly linked to the actual process by which cells enlarge following cell division In this context, inhibition of growth by EGCg and other ENOX2 inhibitors can be understood on the basis of their direct inhibitory effects on ENOX2 activity rather than on the basis of indirect effects exerted by effects on downstream pathways (Yang and Wang 1993; Agarwal 2000; Yang et al. 2000, 2002; Bode and Dong 2003; Vayalil et al. 2004; Chueh et al. 2004).

11.4.7.1 Tea Catechins May Augment Cancer Chemo- and Radiation Therapy

Tea catechins may also be helpful to augment chemo- or radiation-therapy. Oral administration of green tea to mice bearing implanted Ehrlich ascites carcinomas increases the efficacy of doxorubicin chemotherapy (Mukhtar and Ahmad 1999). Tea components also enhanced the antitumor activity of doxorubicin against M5076 ovarian carcinoma in mice (Stoner and Mukhtar 1995). Development of mammary tumors in mice was inhibited completely by a combination of tamoxifen and green tea. Tamoxifen alone inhibited 50 % and green tea alone was without effect (Fujiki et al. 1996). EGCg reduced lung metastases in mice bearing B-16-F_{3n} melanoma cells alone and in combination with dacarbazine (Li et al. 2001). A combination of EGCg and dacarbazine was more effective than EGCg alone and reduced the number of pulmonary metastases, reduced the number of primary tumor growths, and increased the survival of the melanoma-bearing mice (Liu et al. 2001). Green tea was superior to (−)-epigallocatechin (EGC) alone for cancer prevention in humans and was synergistic with sulindac or tamoxifen in the human cancer cell line PC-9 (Suganuma et al. 1999). A favorable interaction between a tea Catechin-*Capsicum* preparation and radiation therapy has been indicated as well from compassionate intervention studies with cancer patients (Fernandez and Ganzon 2003).

In an in vivo coculture model of LNCap prostate cancer cells and human osteoblasts, pretreatment with Capsol-T followed by treatment with Taxol gave an additive response whereas cisplatin following a pretreatment with capsaicin was antagonistic (Axanova et al. 2005).

11.4.7.2 Anticancer Activity of Polyphenols Is Mediated Through the ENOX2-Specific Drug Site

Evidence from vector-forced overexpression, silencing RNA and ENOX2-overexpressing transgenic mice supports the contention that the high level of safety and pan cancer activity of the tea catechins may be related to the pan cancer distribution of the ENOX2 target (Morré and Reust 1997; Morré et al. 1997a; Cooper et al. 2005a, b) plus the virtual absence of ENOX2 from noncancer cells and tissues. In clinical studies, tea catechins have proven to be nontoxic to normal cells and tissues and exhibit both cancer preventive and cancer therapeutic activity (Cooper et al. 2005b).

The EC_{50} for inhibition of the activity of ENOX2 partially purified from the surface of HeLa cells is approximately 5 nM (Fig. 11.38b). The EGCg inhibition of NADH oxidation is ENOX2- and cancer cell-specific. NADH oxidation (Fig. 11.38b) and growth (Fig. 11.39) of noncancer MCF-10A cells which lack ENOX2 entirely were unaffected by EGCg. ENOX2 driven periodic enlargement of a single HeLa cell at 37 °C ceased within one ENOX cycle following additions of 1 μM EGCg (Fig. 11.40).

When overexpressed in COS cells or noncancer mammary epithelial cells, both the growth rate and final cell size of the cells stably transfected with ENOX2 cDNA were increased. Additionally, one to two orders of magnitude increase in susceptibility of

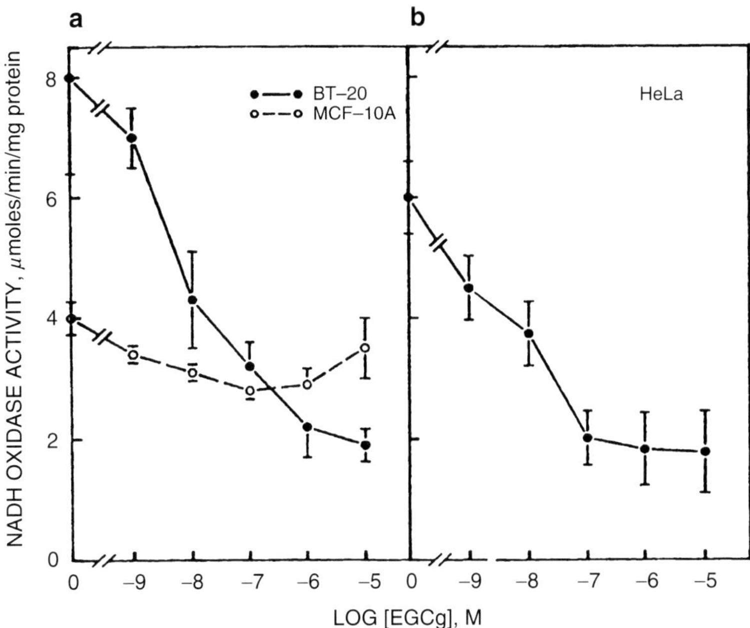

Fig. 11.38 Concentration–response curves of solubilized and partially purified NADH oxidase to EGCg. (**a**) NADH oxidase from MCF-10A and BT-20 cells. (**b**) NADH oxidase from HeLa cells. As with plasma membranes (Fig. 1 of Morré et al. 2000a), the preparations from BT-20 and HeLa cells contained NOX activities both susceptible and resistant to inhibition by EGCg, whereas the preparations from MCF-10A cells were resistant to inhibition. Results are averages of duplicate determinations in each of three separate experiments ($N=6$) ± SD among experiments ($N=3$). Reproduced from Morré et al. (2000a) with permission from Elsevier

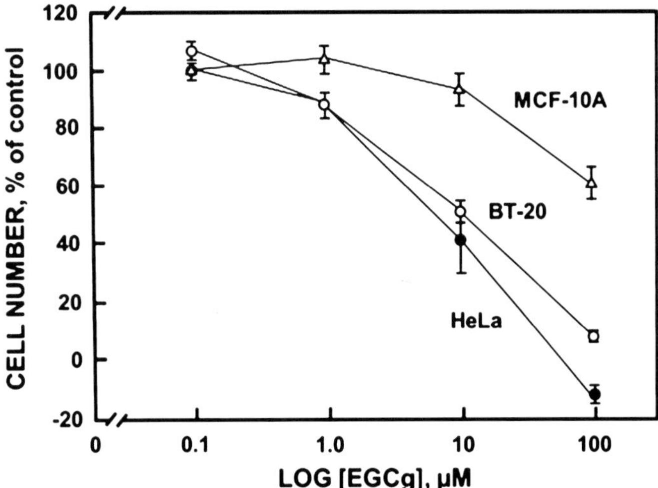

Fig. 11.39 Dose–response of cancer (HeLa or BT-20) or noncancer (MCF-10A) cells to varying concentration of EGCg measured after 72 h (unpublished results)

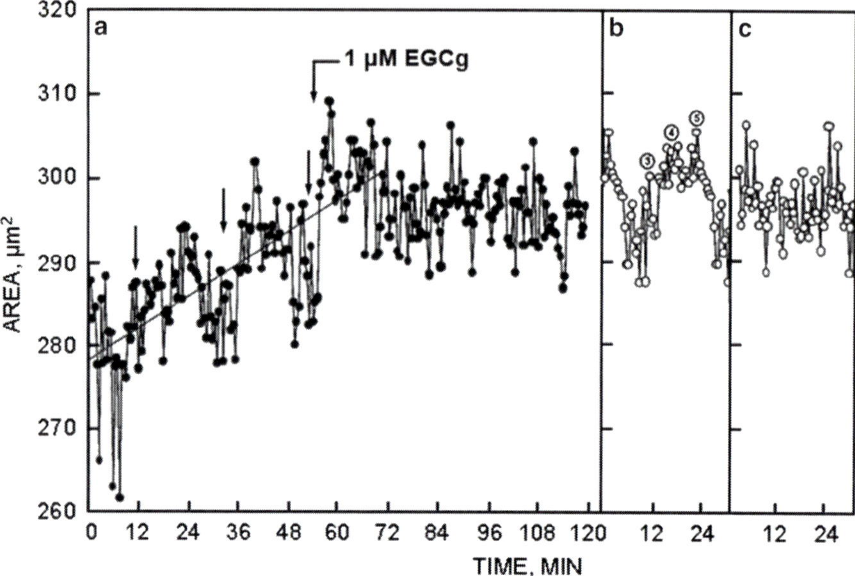

Fig. 11.40 Enlargement of a single HeLa cell at 37 °C determined using video-enhanced microscopy. An automated area determination method was used to estimate cell enlargement. (**a**) The *trend line* shows the average rate of cell enlargement prior to addition of EGCg. The pattern of oscillations between 0 and 60 min corresponds to the ③, ④, and ⑤ pattern of protein disulfide thiol interchange that is characteristic of both plant (Morré et al. 2001a) and animal (Pogue et al. 2000) cells. Within minutes after EGCg addition, net cell enlargement ceased. Decomposition fits show one full cycle (22 min) before (averaged over 0–60 min) (**b**) and after (averaged over 60–120 min) (**c**) EGCg (1 μM) inhibition. Measurements from four different cells selected at random in two separate experiments yielded consistent results (unpublished results)

growth inhibition and induction of apoptosis was induced by both catechins (EGCg) and *Capsicum* vanilloids (e.g., capsaicin) in the ENOX2 overexpressing transfectants (Fig. 11.41; Chueh et al. 2004; Figs. 5.20 and 5.21; Table 11.6).

11.4.7.3 High Dose Cellular and Biochemical Responses to Tea Catechins

There is now a vast literature on cellular, subcellular, biochemical, and molecular responses to green and black tea or to individual tea catechins. For example, more than 2,500 studies report effects of EGCg on a wide variety of molecular and biochemical reactions and processes ranging from signal transduction to telomerase shortening (see early reviews by Agarwal 2000; Manson et al. 2000; Cooper et al. 2005b). The majority of such studies report responses to EGCg in the micromolar or low millimolar ranges of EGCg concentrations which are neither cancer cell specific nor realistically attainable in tissues through dietary intervention. Usually EGCg becomes toxic to noncancer cells at about 100 μM and toxic in mice at about 200 mg/kg body weight with a single i.v. injection. Complete apoptotic killing in

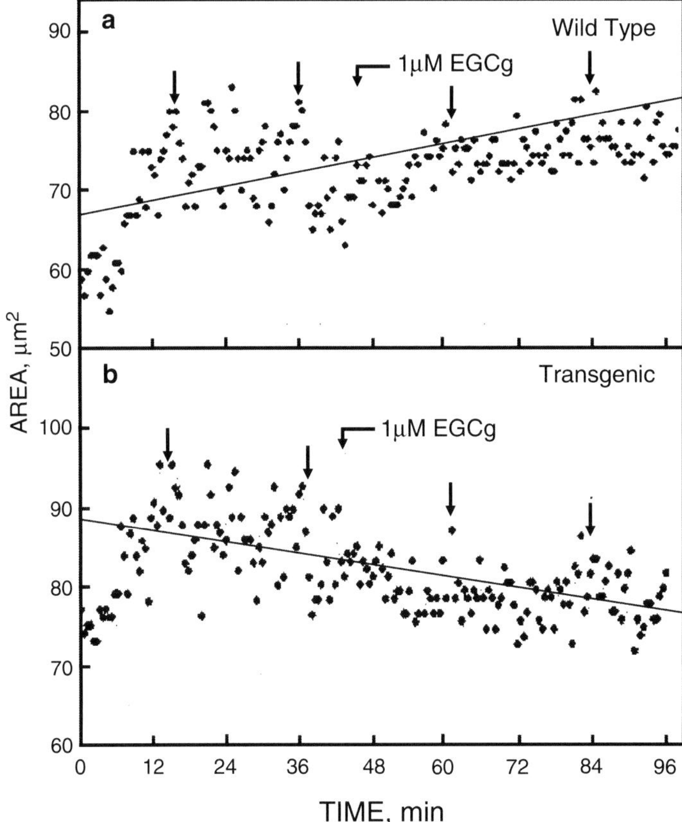

Fig. 11.41 Increase in area (enlargement growth) of small rounded MEF cells postcytokinesis comparing wild-type (**a**) and transgenic (**b**) cells as determined by image-enhanced light microscopy. The growth rates fluctuated with a complex pattern of periodicity but were approximately twofold greater with the transgenic MEF cells. The *single arrows* separated by intervals of 24 min indicate periods of rapid enlargement alternating with resting periods. (−)-Epigallocatechin-3-gallate (EGCg) (0.1 μM) was added after 60 min (*angled arrow*). Modified from Yagiz et al. (2007)

culture (BT-20 mammary carcinoma and HeLa cervical carcinoma) resulted at levels of 5 μM total polyphenol compared to 10 μM for doxorubicin when optimal catechin/vanilloid ratios were achieved. This could be reduced to between 50 and 100 nM with sustained release where catechin levels are maintained over 48–72 h (Janle et al. 2008). ENOX2 inhibition is the only biochemical parameter thus far reported where inhibitions are achieved at nanomolar concentrations of EGCg (EC_{50} of 5–10 nM) and where the dose–response of inhibition of activity and inhibition of growth of cancer cells are correlated (Morré 1998c; Chueh et al. 2004). Therefore, ENOX2 remains as a primary target to explain cancer cell-specific catechin inhibition of growth. Reports of high dose–responses may be the result of inhibition primarily of down-stream reactions in cancer cells also resulting in growth stasis and induction of apoptosis.

11.4.7.4 Tea Catechins and Cancer

Much attention to green tea catechins and health has focused on cancer (Dreosti 1996; Cooper et al. 2005a, b). Drinking green tea has been regarded traditionally in Asia as a healthful practice. A relationship between high consumption of green tea and the low incidence of breast cancer in some Asian countries was suggested (Liao et al. 1995). Anticancer effects from green tea are indicated both from animal studies in vivo (Liao et al. 1995; Ahmad et al. 1997; Chen et al. 1998; Fujiki et al. 1998) and from human epidemiological observations (Fujiki et al. 1996, 1999; Katdare et al. 1998). In general, these effects have been attributed to EGCg (Chen et al. 1998). Other antioxidant catechins and polyphenols present in green tea have less marked activities or are without anticancer properties when tested singly or even in combination. Work with cancer cell lines has related the anticancer activity of EGCg to growth inhibition, interruption of cell cycle progression, and induction of apoptosis rather than to an unspecific antioxidant function (Liao et al. 1995; Chen et al. 1998; Fujiki et al. 1999; Morré et al. 2000a; Cutter et al. 2001).

Polyphenon E, a defined, decaffeinated green tea polyphenon mixture containing (−)-epigallocatechin gallate and other catechins and vitamin C, was evaluated clinically in a phase II study at the National Cancer Institute (NCI) to prevent cancer recurrence in former smokers who had undergone surgery for bladder cancer (ClinicalTrials.gov web site 2011). Polyphenon E has also undergone evaluation for Stage 0, Stage I, or Stage II studies for lymphocytic leukemia at the Mayo Clinic (ClinicalTrials.gov web site 2011).

11.4.7.5 Catechin-Vanilloid Synergies

Of many ENOX2 inhibitors investigated (capsaicin, doxorubicin, cisplatin, the antitumor sulfonylurea LY181984, glaucarubolone, bullatacin, the synthetic anticancer isoflavone, PXD, and suramin), catechin-vanilloid combinations have the greatest margin of safety (therapeutic index). The margin of safety is even greater than that of EGCg alone. Interactions among polyphenols in cancer were investigated as a basis for evaluation of polyphenol synergies and their application to improved anticancer polyphenol compositions (Weaver et al. 2008; Morré and Morré 2006b). The most abundant vanilloid in the *Capsicum* extracts used was determined to be vanillic acid rather than capsaicin (Zhou et al. 2004).

The primary structure of ENOX2 reveals a single drug site E245EMTE located within the interior of the ENOX2 protein (sequence, Fig. 11.42). When mutated, activity is fully retained but the response to capsaicin and EGCg (Chueh et al. 2002a, b) is completely lost. For all known blockers (polyphenols, anticancer drugs, ENOX2-specific monoclonal antibody, or its scFv fragment to inhibit), a source of protons and electrons (NADH or hydroquinone, for example) is required and the protein must be renatured and enzymatically active. ENOX2 proteins are dimeric cylinders slotted along one side (Fig. 8.8) which open and close according to either a 24 (ENOX1) or a 22 (ENOX2) min period illustrated in the diagram of Fig. 11.42.

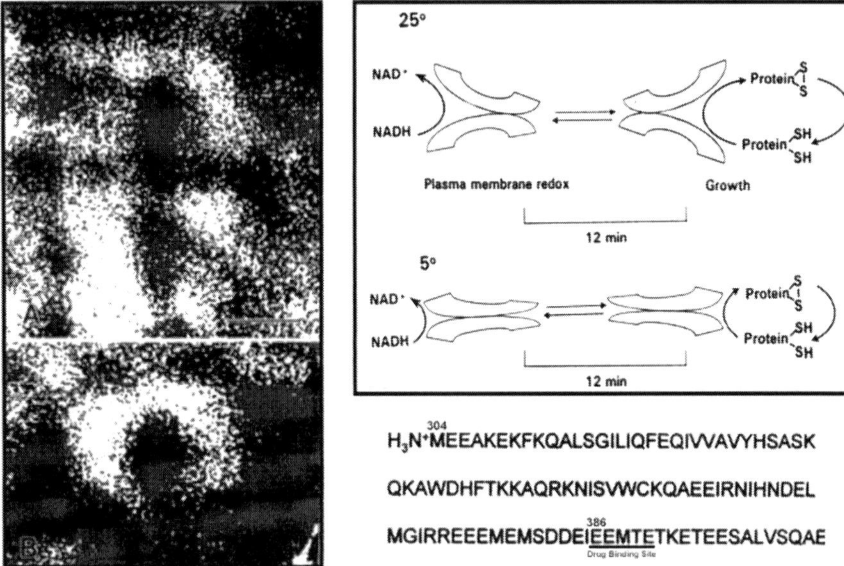

Fig. 11.42 Model to explain the temperature-independent 24 min periodic alternation of oxidative and protein disulfide-thiol interchange activities of the NOX protein. See Fig. 8.7 for interpretation of (**a**) and (**b**). Two different active sites would alternate by an opening and closing of substrate binding cavities due to reversible α-helix–β-sheet transformations. The degree of opening would increase with increasing temperatures to account for the Q_{10} of two for the catalytic activity. The rate of opening and closing would, however, be temperature dependent as in a spring. As the temperature is lowered, the spring would become more difficult to stretch. The end result depicted is that the conformational change regulating the accessibility of active sites would traverse a shorter distance at 5° but with the same time interval as at 25° before reversing. With only a single drug site, one can achieve polyphenol synergies. Some inhibitors might enter the drug site during the oxidative portion of the cycle and other inhibitors might enter the drug site during the protein disulfide-thiol interchange portion of the cycle as the basis for the synergy. The findings that EGCg binds with the protein during the oxidative portion of the cycle while capsaicin and vanilloids bind during the protein disulfide-thiol interchange portion (Table 11.9) may explain why the catechin and vanilloid inhibitors, when combined exhibit tenfold synergies

The opening and closing of the cylinder might be necessary to allow access of inhibitors to the active site.

A significant question is how, assuming a single drug site, are polyphenol synergies achieved? Part of the answer lies in the observations summarized in Table 11.9 that some inhibitors access the drug-binding site during one part of the ENOX cycle whereas other inhibitors access the drug site in another part of the ENOX cycle. This may explain why some inhibitor combinations are synergistic whereas others are antagonistic. The vanilloids combine with the protein during the oxidative portion of the cycle while the EGCg combines during the interchange portion of the cycle favored by reducing conditions. Generally two drugs that interact in the same portion of the ENOX cycle are antagonistic.

Table 11.9 Polyphenols that inhibit ENOX2 in different portions of the activity cycle or that may affect EGCg binding

Inhibit during NADH/hydroquinone oxidation	Inhibit during protein disulfide-thiol interchange	May affect EGCg binding
Capsaicin	EGCg	EC
Vanillylamine	ECG	GCG
		EGC

Table 11.10 Growth of human cervical carcinoma (HeLa) and mouse mammary cancer (4T1) cells in culture showing synergy in growth inhibition (cell killing) over 72 h by a mixture of an infusion of decaffeinated green tea (Lipton's) and Capsibiol

	Dilution to reduce cell number by 50 % after 72 h	
	HeLa	4T1
Decaffeinated green tea infusion (undiluted = 100 mM EGCg)	1:200	1:1,000
Capsibiol	Not reached	Not reached
Decaffeinated green tea extract	1:5,000	1:20,000
Decaffeinated green tea extract (50 parts) + Capsibiol (1 part)	1:20,000	1:40,000

From Morré and Morré (2003d)
The green tea infusion was standardized to provide 100 mM EGCg before dilution

Catechin-vanilloid combinations are 10–100 times more effective than either catechins or vanilloids alone (Table 11.10). A catechin-vanilloid mixture where one 350-mg capsule is equivalent to 16 cups of green tea in its ability to inhibit ENOX2 growth of cancer cells in culture (Morré and Morré 2006c). Some patients may develop resistance to catechin-*Capsicum* mixtures. Resistance may be overcome using isoflavenes, either synthetic such as PXD (Sect. 11.4.8.1) or naturally occurring as found in certain native and introduced species of *Lespedeza*. Several studies, reviewed by Miyase et al. (1999) reveal the genus *Lespedeza* to be a source rich in isoflavinoids.

11.4.7.6 Early Intervention Strategy Based on Green Tea-*Capsicum* Synergies

Some opportunity for early curative or preventive intervention is necessary if the potential benefits of early detection are to be fully realized. A decaffeinated green tea extract containing 98 % tea catechins of which 40 % are EGCg and a *Capsicum* powder in the ratio of 25 parts tea extract plus 1 part *Capsicum* powder (Capsol-T) may be sufficient to induce apoptosis in early stage cancer cells prior to development of clinical symptoms when only a small number of cells are present. Following cell division, a minimum cell size must be reached before the cell can divide again. Otherwise cell division stops and after several days, programmed cell death (apoptosis) ensues. Cancer cells with blocked ENOX2 activity being not able to enlarge are thus directed towards apoptosis. Cell cycle arrest is principally in G_1. In implanted tumors in mice, growth was inhibited in a dose-dependent manner at doses of 0.1, 0.3, and 0.5 % in the diet (Li et al. 2010).

Decaffeinated green tea when combined with anticancer substance-containing *Capsicum* preparations (Capsibiol-T) at a ratio of 25:1 resulted in a 100-fold increase in killing of cultured cancer cell lines compared to green tea alone (Morré and Morré 2003b). A current food grade *Capsicum*-green tea product (Capsol-T®) has given equivalent results. The natural combination of EGCg with other catechins, also found in green tea, is superior to EGCg alone in the mixture (Morré et al. 2003c).

One 250 mg casule of Capsol-T® every 4 h is equivalent to drinking 16 cups of green tea every 4 h based on laboratory studies with cancer cells. The dose of 1 capsule of Capsol-T® every 4 h is based on pharmacokinetic studies in rodents (Janle et al. 2008) and findings that the inhibition of ENOX2 by Capsol-T® is reversible (Morré et al. 2000a). Catechins must be present in the medium of cultured cancer cells at a level of about 100 nM and to inhibit ENOX2 continuously at that level for a period of 48–72 h to have therapeutic efficacy in selective killing of cancer cells (Morré et al. 2000a). If EGCg is removed and replaced by EGCg-free media, even after 8 h, normal rates of growth are resumed by cancer cells in vitro. Similarly, as EGCg is cleared from the culture medium and/or metabolized, normal rates of growth are resumed. In cell culture at nanomolar concentrations, the EGCg may not survive in the media for more than a few h. In vivo, the cancer cells must be inhibited from growing for at least 48 and perhaps up to 72 h for EGCg to induce apoptosis in a majority of the cancer cells present.

Pharmacokinetic results from animal studies (Janle et al. 2008) are consistent with epidemiological studies in humans and animal experiments. Cancer benefit without adverse effects has been ascribed to drinking at least 10 cups of green tea per day (Fujiki 1999; Nakachi et al. 2000). After oral administration, green tea polyphenols reach their highest plasma levels after about 1–2 h both in rats (Unno and Takeo 1995; Zhu et al. 2000; Janle et al. 2008) and in humans (Warden et al. 2001). The levels of EGCg in the rat are 12.3 nmol/mL in plasma (12.3 μM) 60 min after a single oral administration of 500 mg/kg body weight (Nakagawa 1997), more than 100 times the effective dose to stop the growth of tumor cells. The concentration of EGCg in the blood after 2–3 cups of green tea reached a maximum of about 0.6 μM (Yang 1997).

In human studies, 0.2 % of the ingested EGCg and 0.2–1.3 % of the ingested EGC were found in plasma after 90 min (Nakagawa et al. 1997). The half life for plasma levels of individuals drinking 8 cups of tea per day for 3 days to be 4.8 h for green tea and 6.9 h for black tea (Van het Hof et al. 1999). C_{max} values were observed 1.4–2.4 h after injection with a half life of 5–5.5 h (Yang et al. 1998) after ingestion of green tea by human volunteers. Studies are consistent with dosing at regular intervals of 4 h with Capsol-T. Formulated for sustained release, the expectation is that two 500 mg capsules of 50 % material per day, one in the morning and one in the evening, may eventually prove to be sufficient.

Safety and efficacy of tea catechins in combination with *Capsicum* are well documented (Cooper et al. 2005b). Safety has been the subject of a series of reports that evaluate the genotoxic, acute, dermal, subchronic short term, teratogenic, and reproductive responses of EGCg (Isbrucker et al. 2006a, b, c). Capsol-T® is free of vitamin K, pesticides, and/or heavy metal residues. EGCg, the principal catechin constituent, was affirmed to be GRAS (generally recognized as safe) by an independent review panel (Office of Dietary Supplements 2008).

Oral green tea extracts have been evaluated clinically both for safety and efficacy. In a phase 1 study, a commercially available but not decaffeinated green tea source was given 1 or three times daily for 4 weeks to 6 months (Pisters et al. 2001). Doses of 0.5–5.05 g/m^2/day and 1.0–2.2 g/m^2 three times per day tested in 49 cancer patients registered a maximum tolerated dose of 4.22 g/m^2 which was limited primarily by caffeine levels easily avoided with the use of decaffeinated green tea.

In open label sequential trials with Capsol-T®, 36 % of participants with advanced cancer reported significant prolongation of life and/or remained alive at study completion (Morré and Morré 2006c). Another 32 reported improvement while the remaining 32 % experienced a normal course of their disease. Most in the last category were diagnosed very late at a time when even compliance with 6 capsules per day, one every 4 h became difficult. For patients with severe head and neck carcinomas, a preparation containing a mixture of decaffeinated green tea and a chili pepper (*Capsicum* sp.) extract was highly efficacious in reducing tumor burden (compassionate intervention) (Fernandez and Ganzon 2003).

11.4.8 Phenoxodiol Targets the PMET Through Inhibition of ENOX2

PXD (2H-1-benzopyran-7-0,1,3-(4-hydroxy-phenyl)) (Fig. 11.10) is a synthetic anticancer isoflavene (isoflavone) chemically related to natural isoflavone phytoestrogens such as genistein (Aguero et al. 2005; Brown et al. 2005; Kamsteeg et al. 2003; Constantinou and Husbard 2002; Constantinou et al. 2003). PXD has high selectivity for tumor cells and induces cytostasis and apoptosis in many forms of human cancer (Aguero et al. 2005; Kamsteeg et al. 2003). It has no discernible effects on the biology of nontumor cells. ENOX2 has been shown to be the molecular target to explain the anticancer action of PXD through the inhibition of the PMET (Herst et al. 2007) and, more specifically, through inhibition of the protein disulfide-thiol interchange activity of ENOX2 (Morré et al. 2007a; Fig. 11.5). The protein disulfide-thiol interchange activity catalyzed by ENOX2 correlates with the enlargement phase of cell growth (Chap. 5).

Purified recombinant ENOX2 bound PXD with high affinity (K_d of 50 nM) (Morré et al. 2007a). The ENOX2 protein appeared to be both necessary and sufficient for the cancer-specific cytotoxicity of PXD. Growth inhibition of fibroblasts from embryos of mice expressing an ENOX2 transgene, but not from wild-type mice, was inhibited by PXD followed by apoptosis (Fig. 11.43). Both the oxidative and protein disulfide-thiol interchange activities that alternate to generate the complex set of oscillations with a period length of 22 min that characterize ENOX2 proteins respond to PXD. All three enzymatic activities associated with the ENOX2 protein, hydroquinone oxidation (Fig. 11.44a), NADH oxidation (Fig. 11.44b), and protein disulfide-thiol interchange (Fig. 11.44c) were blocked. Protein disulfide-thiol interchange measured classically from the restoration of

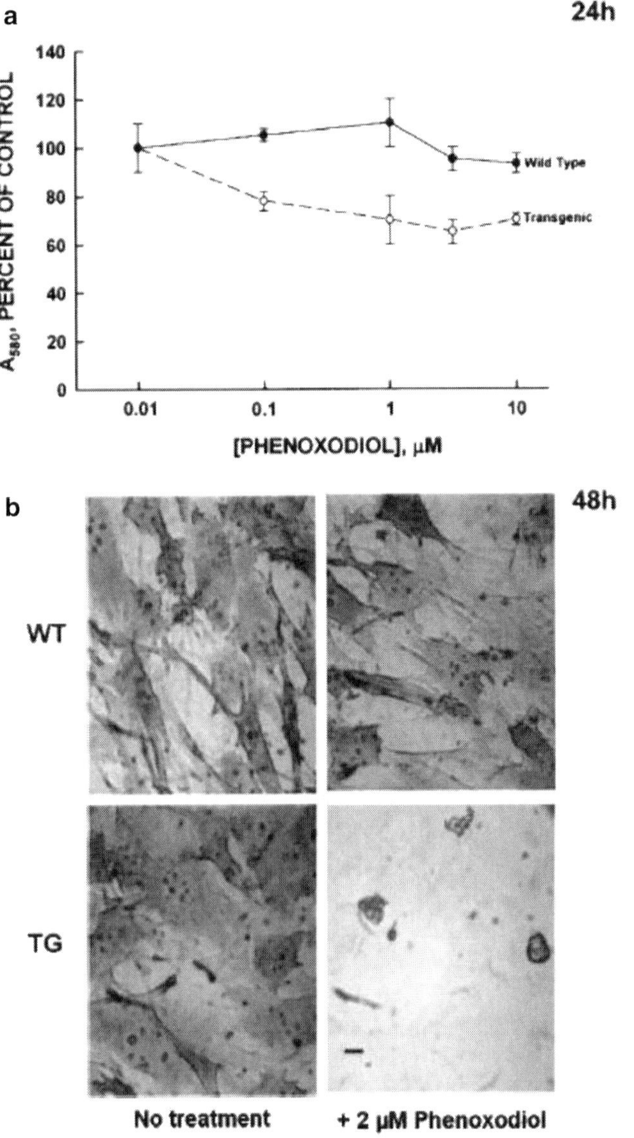

Fig. 11.43 Growth (**a**) and morphology (**b**) of wild-type (WT) mouse embryo fibroblasts and fibroblasts from embryos of ENOX2 transgenic mice (TG) (Yagiz et al. 2006). Growth was inhibited during the first 24 h of culture with the transgenic but not wild-type fibroblasts (**a**) and by 48 h the transgenic mouse fibroblasts had undergone apoptosis whereas wild-type cells exhibited normal morphology (**b**). Attached cells as percent of control were 95 ± 5 for wild type and 8 ± 3 for transgenic grown for 48 h in the presence of 2 μM PXD. Reproduced from Morré et al. (2007a) with permission from Cognizant Communication Corporation

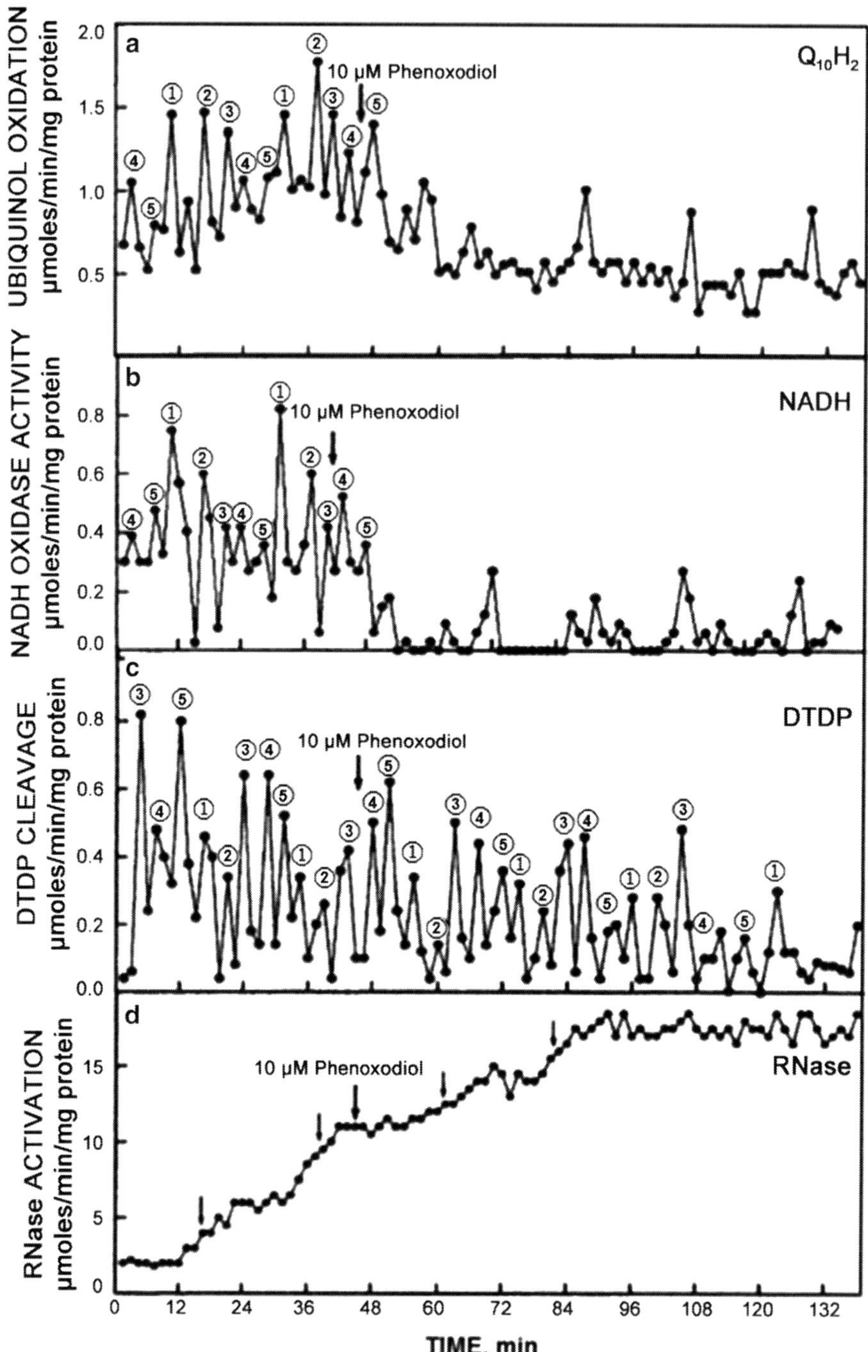

Fig. 11.44 Activity of recombinant ENOX2. Comparison of the response of recombinant ENOX2 to addition of 10 µM PXD added after 48 min (*large arrow*) comparing each of the four available activity assays on aliquots of the same ENOX2 preparation. Hydroquinone (**a**) or NADH (**b**) oxidation was inhibited rapidly by PXD. Protein disulfide-thiol interchange measured either by DTDP cleavage (**c**) or activation of scrambled ribonuclease (*arrows*) (**d**) was inhibited after about 60 min. (**b**, **c**) were determined simultaneously and were directly comparable. Oxygen consumption paralleled NADH oxidation (not shown). Reproduced from Morré et al. (2007a) with permission from Cognizant Communication Corporation

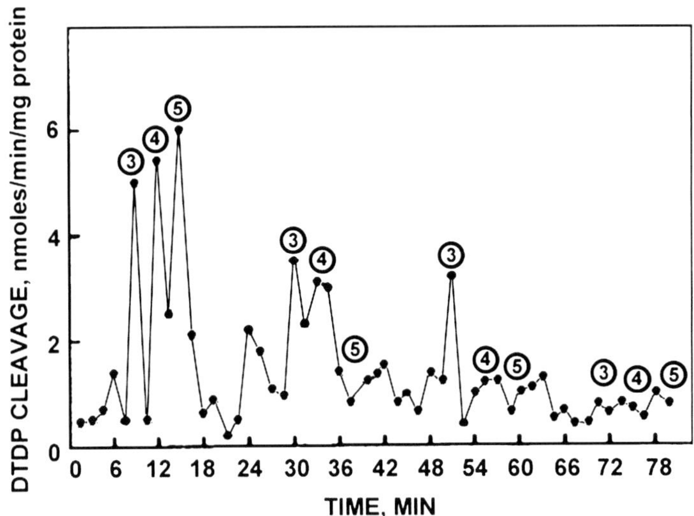

Fig. 11.45 The rate of protein disulfide-thiol interchange of recombinant ENOX2 as determined by DTDP cleavage showed that each of the three maxima in protein disulfide-thiol interchange was lost progressively by following PXD (10 μM) addition beginning with maximum ⑤ in the second period followed by a loss of maximum ④ in the third period and loss of maximum ③ in the fourth period. Reproduced from Morré et al. (2007a) with permission from Cognizant Communication Corporation

activity to denatured and inactive (scrambled) RNase which also was blocked (Fig. 11.44d). Similar results were seen using a spectrophotometric assay for protein disulfide-thiol interchange based on cleavage of dithiodipyridine (Morré et al. 2007a; Fig. 11.45).

Oxidation of NADH or reduced coenzyme Q_{10} was rapidly blocked by PXD. In contrast, the protein disulfide-thiol interchange activity measured either by the restoration of activity to scrambled and inactive RNase or from the cleavage of dithiodipyridine (EC_{50} of 50 nM) was inhibited progressively over an interval of 60 min that spanned three cycles of activity (Fig. 11.43; Table 11.11). Inhibition of the latter paralleled the inhibition of cell enlargement and the consequent inability of inhibited cells to initiate traverse of the cell cycle. Activities of constitutive ENOX1 forms of either cancer or noncancer cells were unaffected by PXD to help explain how the cytotoxic effects of PXD may be restricted to cancer cells (Morré et al. 2007a). Herst et al. (2007, 2009) report that PXD may be also immunosuppressive which suggests the presence of a target molecule on T-cells.

Ecliptic expression of ENOX2 augmented the effects of PXD in tumor cells (Table 11.12). With COS cells, which express ENOX2 at low levels, the IC_{50} for PXD was decreased from 5 μM with wild-type HeLa cells or HeLa cells transfected with vector alone to 1.5 μM in the ENOX2-transfected HeLa cells. Similarly, embryo fibroblasts from transgenic mice carrying the ENOX2 transgene were inhibited by PXD whereas wild-type mouse embryo fibroblasts lacking the ENOX2 transgene were not (Fig. 8.14).

Table 11.11 Progressive loss of maxima ③, ④, and ⑤ of the ENOX2 activity cycle of recombinant ENOX2

	Control (nmol/min/mg protein)			10 μM PXD (nmol/min/mg protein)		
	③	④	⑤	③	④	⑤
Period 1	3.8	5.4	5.4	5.0	5.4	6.0
Period 2	4.8	7.2	5.4	3.5	3.1	0.8
Period 3	6.5	4.3	5.4	3.2	1.2	0.6
Period 4	3.5	3.8	4.5	0.6	0.5	0.8

From Morré et al. (2007a)
Upon treatment with 10 μM PXD based on analyses of DTDP cleavage (Chap. 2)

Table 11.12 Summary of ENOX2 knockdown and overexpression on response to PXD

Cell line	IC_{50} (μM)
COS cells wild-type	5[a]
Vector alone	5[a]
ENOX2 transfected	1.5[b]
HeLa cells wild type	1.5[b]
RNAi C6 (control)	1.5[b]
RNAi ttNOXcl	15[c]
ENOX2-specific monoclonal antibody	Not reached

From Morré et al. (2007a)
IC_{50} values determined from complete dose–response curves (0.005, 0.05, 0.5, 5, and 50 μM) of growth of HeLa or COS (Chung Hsing University) cells measured after 24, 48, and 72 h (96-well assay) in three experiments. Values not followed by the same letter were significantly different ($p<0.001$)

With HeLa cells, siRNA knockdown increased the IC_{50} for PXD approximately tenfold from 1.5 to 15 μM (Table 11.12). Even more pronounced was the effect of an ENOX2-specific monoclonal antibody, which increased the IC_{50} of growth inhibition for PXD more than 100-fold (Table 11.12).

PXD was evaluated in early human cancer clinical trials in women with chemotherapy-resistant, advanced ovarian cancer and in men with hormone-refractory advanced prostate cancer (Kelly 2004; Wilkinson 2004). Central to the molecular mechanism for the anticancer activity of the multiple signal transduction regulator PXD is inhibition of ENOX2 (Morré et al. 2007a; Herst et al. 2004). Completed or halted clinical studies at Marshall Edwards, Inc. include: phase Ib neoadjuvant application in squamous cell carcinoma or adenocarcinoma of the cervix, vagina, or vulva (Azodi et al. 2005); phase Ib/IIa as a chemosensitizer and in combination with docetaxel in epithelial ovarian cancer or primary peritoneal cancer (Kelly et al. 2011); phase IIb as a chemosensitizer in refractory, recurrent, or persistent epithelial ovarian, fallopian, or primary peritoneal cancer (Rutherford et al. 2004); phase III as a chemosensitizer in late-stage epithelial ovarian, fallopian, or primary peritoneal cancer (OVATURE TRIAL) (http://www.marshalledwardsinc.com). PXD was also the subject of a phase I study for refractory solid tumors at the NCI, sponsored by Novogen (ClinicalTrials.gov web site 2011; Choueir et al. 2006; Davies et al. 2004) and a phase II trial with castrate and noncastrate prostatic cancer (Gibney et al. 2010). Due to a seriously underpowered patient cohort and an overall lack of efficacy apparently related to the oral route of administration, the phase III OVATURE trial was halted in 2010. Continuing clinical evaluations are anticipated, however, within the framework of the PXD platform of a related isoflavene, triphendiol, and a highly active triphendiol metabolite, MEI-143 (www.Marshalledwards.com/NEWS/2010/06/01), also an ENOX2 inhibitor (results unpublished).

The clinical trials with PXD were based on laboratory studies that identified PXD to be an effective drug at killing ovarian cancer cells, including those resistant to standard anticancer chemotherapies (Kamsteeg et al. 2003). However, the cancer-specific toxic effects of PXD were not restricted to ovarian cancer. Also inhibited was proliferation of a wide range of human cancer cell lines, including HL-60 promyelocytic and CCRF-CEM and RPMI-8226 leukemia lines, and ovarian, breast, prostate, cervical, and renal carcinomas (Brown et al. 2005; Kamsteeg et al. 2003; Constantinou and Husband 2002). In these published studies, PXD was 5–20 times more potent than genistein.

11.4.8.1 Phenoxodiol as a Chemosensitizer to Anticancer Drugs Overcoming Drug Resistance and Restoring Chemosensitivity

PXD is unusual among ENOX-targeted anticancer drugs by an ability to enhance the anticancer activity of other drugs through chemosensitization as a means to overcome drug chemoresistance by restoring chemosensitivity to resistant cancers (Kelly 2010). When combined with standard anticancer drugs such as platinum-based or taxanes such as docetaxel or paclitaxel, PXD resulted in synergy where the response to the mixture was greater than to the individual drugs tested alone. Moreover, it was not necessary for the PXD to be co-administered with the drug. It was sufficient to treat patients first with PXD and then subsequently with taxol or docetaxel to restore susceptibility in keeping with the observations that PXD restored drug sensitivity to drug-resistant cancer cells in culture.

PXD may also exhibit chemosensitization by restoring sensitivity of drug-resistant cancer cells to standard anticancer drugs such as taxanes and platinum-based drugs (Kelly 2004) used extensively in the treatment of ovarian cancer. Patients' tumors commonly become resistant to these drugs as their disease progresses. With such chemoresistant cancers, combinations of PXD and cisplatin or of PXD and paclitaxel have resulted in a therapeutic response with doses of cisplatin or paclitaxel that alone would have been ineffective. Among women with chemoresistant epithelial ovarian cancer, clinical benefit was observed in 75 % of patients receiving PXD administered with the chemotherapies against which resistance had been developed (Goss et al. 2005). The latter appears to be a true chemosensitization in that the PXD again needs not be co-administered with the taxane or platinum drugs. Sensitivity is retained many weeks after the PXD has been cleared from the system. As addressed in the next section, the response appears to be mediated through a conformational change in ENOX2, the primary drug target of PXD (Morré et al. 2008a).

11.4.8.2 The Chemosensitization Response of Phenoxodiol May Be Mediated Through ENOX2

In studies with cultured cell lines resistant to *cis*-platinum or taxanes, the chemosensitization response of Sect. 11.4.8.1 was shown to be mediated through ENOX2, the PXD target protein (Morré et al. 2008a, 2009b). With taxane and/or platinum-resistant ovarian carcinoma cells, reversal of drug resistance was attained by pretreatment with PXD followed by platinum or taxane after a wash out to remove the PXD. Thus, the phenomenon was a true chemosensitization in that the PXD did not need to be co-administered with the taxanes or platinum drugs (Morré et al. 2008a). In subsequent studies with HeLa cells which are relatively resistant to both paclitaxel and cisplatin, sequential addition of PXD (Fig. 11.46) and paclitaxel (Fig. 11.47) or PXD and cisplatin showed greater inhibition of HeLa cell ENOX2 activity and growth compared to adding the drugs simultaneously or individually (Tables 11.13 and 11.14). ENOX1 was not affected and *trans*-platinum was without effect.

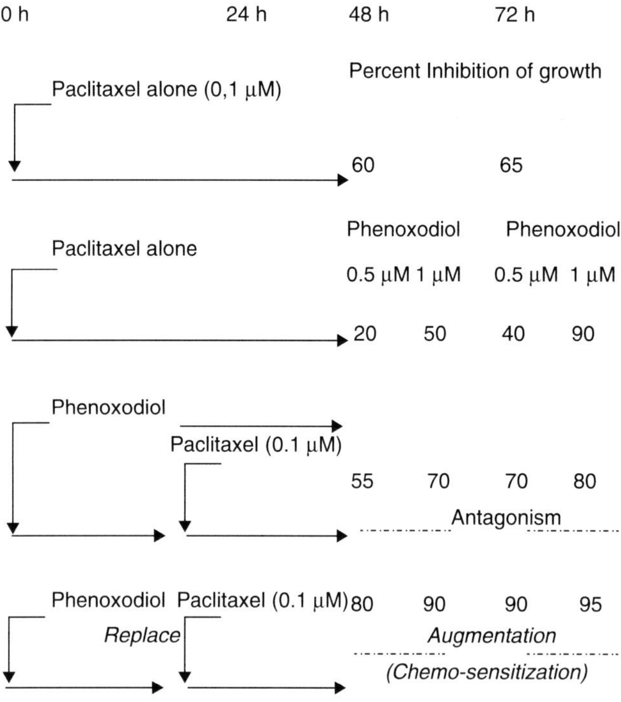

Fig. 11.46 Response of cultured HeLa cells to paclitaxel under different conditions of PXD treatment. Reproduced from Morré et al. (2009b) with permission from Springer Science+Business Media

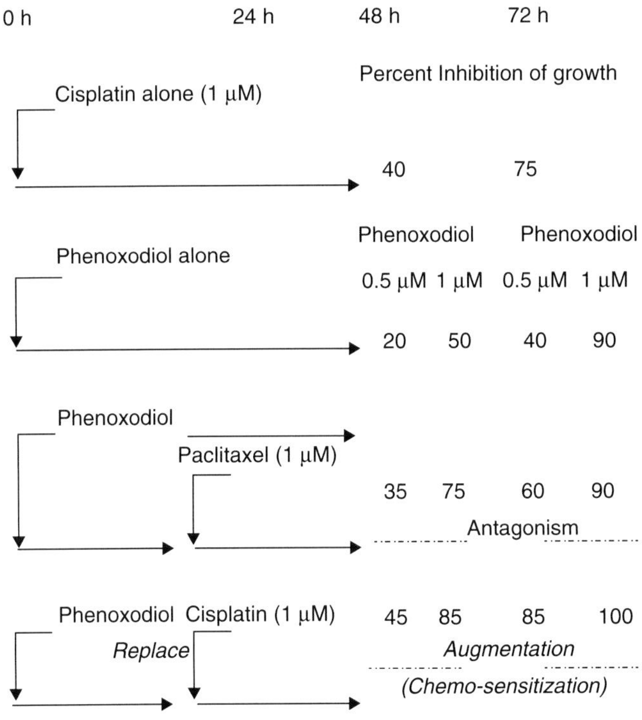

Fig. 11.47 Response of cultured HeLa cells to cisplatin under different conditions of PXD treatment. Reproduced from Morré et al. (2009b) with permission from Springer Science+Business Media

Table 11.13 Response of growth of CP-70 cells to 0.5 or 1 μM PXD and to 1 μM cisplatin alone and in combination

Treatment	(μM)	Inhibition of growth (%) 48 h	72 h
Phenoxodiol alone	0.5	22±5	21±6
	1.0	40±5	45±9
Cisplatin alone	1.0	40±6	60±6
Phenoxodiol+cisplatin	0.5+1	26±4	43±7
	1.0+1	32±7	50±4
Phenoxodiol, removed after 48 h and replaced with cisplatin	0.5+1	60±4[a]	71±2[a]
	1.0+1	74±11[a]	88±6[a]

From Morré et al. (2008a)

Chemosensitization was enhanced by removal of PXD after 48 h and replacement by cisplatin in the absence of PXD

[a]Significantly different from cisplatin alone ($\rho<0.001$)

Table 11.14 Response of ENOX2 activity of HeLa and CP-70 cells to PXD, cisplatin, and paclitaxel alone or in combination

	EC_{50}	
Drug	HeLa	CP-70
Phenoxodiol (nM)	200	20
Paclitaxel (μM)	>100	>100
Cisplatin (nM)	100	>1,000
Trans-platinum (nM)	>1,000	>1,000
Pretreat with 10 μM PXD for 60 min, then add cisplatin (no wash) (μM)	0.1–2.5	1
Pretreat with 10 μM PXD for 60 min, wash to remove drug, then add cisplatin	<50	<100
Pretreat with 10 μM PXD for 60 min, then add paclitaxel (no wash) (μM)	>100	1
Pretreat with 10 μM PXD, wash to remove drug, then add paclitaxel (nM)	<50	<50

From Morré et al. (2008a)

EC_{50} was estimated from rate measurements at five concentrations over the range 10 nM to 100 μM compared to no drug as described. Estimates of EC_{50} within this range from duplicate determinations varied less than twofold except where indicated

With spent media from PXD-treated cells, sensitivity was enhanced to both paclitaxel and cisplatin if the cells from which the spent media was derived were first pretreated with PXD (Morré et al. 2009b). Similar results were obtained with ENOX2-enriched preparations stripped from the surfaces of PXD-treated cells. One view of the mechanism based on the prion model is that the ENOX2 "remembers" the PXD and "teaches" other ENOX2 molecules to respond to paclitaxel and cisplatin as if the PXD were still present (Morré et al. 2009b). This model is further supported by observation that chemosensitization is achieved by merely adding back ENOX2 proteins that have been exposed to PXD (Morré et al. 2009b). Thus, it would appear that drug-resistant and drug-sensitive ENOX2 proteins are interconvertible idioforms that propagate through prion-like protein–protein interactions not encoded in the genome and where the resistant idioforms are converted back to the sensitive idioforms through the reversible binding of PXD (Morré et al. 2007a). See also Chap. 10 for a similar response of a plant ENOX protein to the auxin herbicide, 2,4-D.

11.4.9 Sulforaphane

Sulforaphane-induced apoptosis has been observed in many cancer cell lines and the subject of early chemoprevention trials (Antosiewicz et al. 2008; Clarke et al. 2008). Sulforaphane is yet another specific inhibitor of ENOX2 activity and growth of cancer cells with no effect on ENOX1 or growth of noncancer cells (Morré et al. 2008a). A constituent found in broccoli, sulforaphane has promise for therapeutic intervention in prostate cancer (Clarke et al. 2008).

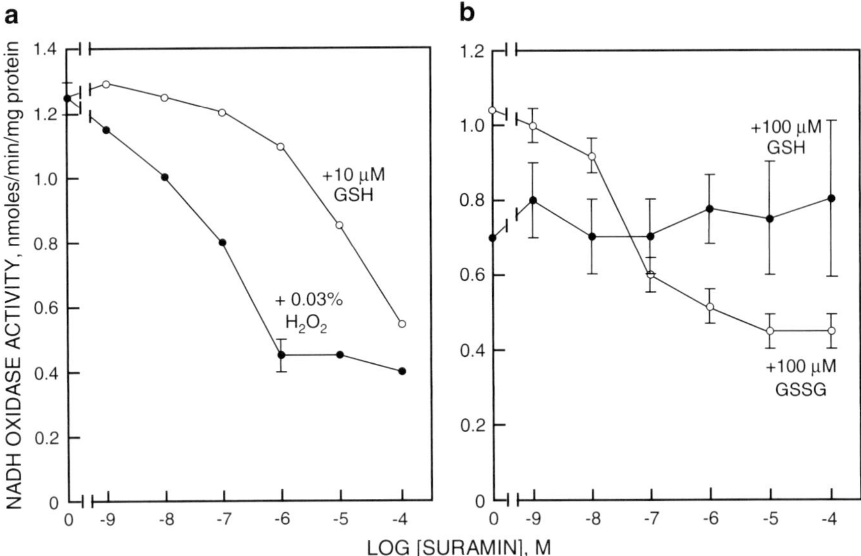

Fig. 11.48 Response of HeLa cells to suramin. (**a**) Inhibition of NADH oxidation activity of isolated plasma membrane vesicles of HeLa cells by increasing concentrations of suramin in the presence of reduced glutathione (GSH) or 0.03 % hydrogen peroxide. (**b**) Inhibition of NADH oxidase activity of isolated plasma membrane vesicles as a function of oxidizing (GSSG) and reducing (GSH) conditions

11.4.10 Suramin

Suramin is a hexanesulfonated napthylurea ENOX2 inhibitor used initially as a treatment for trypanosomiasis (human sleeping sickness caused by trypanosomes) also explored for use in the treatment of some cancers (Stein 1993). Suramin inhibited NADH oxidase activity of HeLa cell plasma membranes (Fig. 11.47) and inhibited the growth of HeLa cells in culture (Fig. 11.48). Suramin was without effect on growth and NADH oxidase of nontransformed cells. A response also was seen to redox poise (Figs. 11.48 and 11.49). Suramin activity was, like capsaicin, enhanced by oxidizing conditions.

11.4.11 Callipeltin

Callipeltin A, a cyclic depsipeptide macrocyclic lactone containing four amino acids in the L configuration is representative of a number of natural products from marine sponges that both inhibit viral entry and exhibit anticancer activity (Andavan and Lemmens-Gruber 2010). Callipeltin A exhibited ENOX2 inhibitory activity in preliminary trials (D. J. Morré, Unpublished).

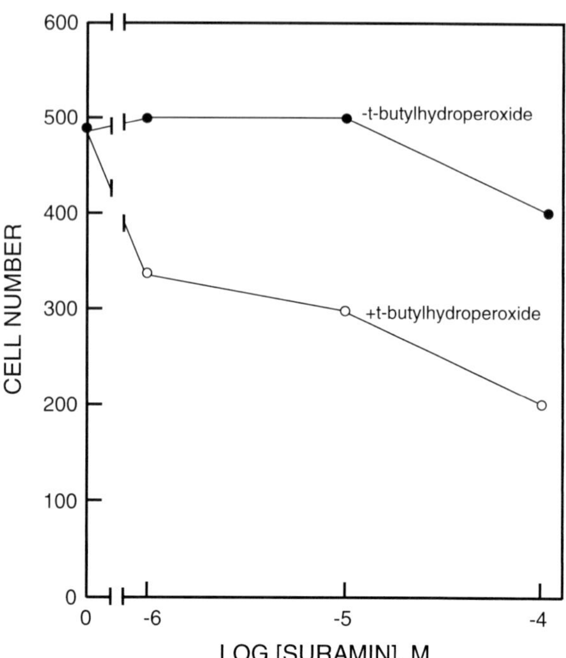

Fig. 11.49 Inhibition of growth of HeLa cells in culture by increasing concentrations of suramin in the presence or absence (10 μM) of the mild oxidizing agent *t*-butyl hydroperoxide

11.5 Nonsteroidal Anti-inflammatory Drugs

Evidence from COX-2-deficient animals indicates that the anticancer activity of NSAIDs is independent of cyclooxygenase-2 (COX-2) targeting (Liu et al. 2001; Neufang et al. 2001). In vitro studies of ENOX2 overexpressing human cervical cancer HeLa and human mammary adenocarcinoma BT-20 cells indicate that ENOX2 is an alternative drug target to COX-2 as the basis for the anticancer activities of COX-2 inhibitors (Morré and Morré 2006e). NSAIDs, piroxicam, aspirin, ibuprofen, naproxen, and celecoxib (Fig. 11.1) all inhibited ENOX2 activity of HeLa (human cervical carcinoma) and BT-20 (human mammary carcinoma) cells (IC_{50} in the nanomolar range) but did not inhibit ECTO-NOX activities of noncancer MCF-10A mammary epithelial cells. Rofecoxib was less effective with cancer cells and two NSAIDS selective for COX-1 had no effect on ENOX2 activity. The IC_{50}s for inhibition of ENOX2 activity ($r=0.96$) did not modulate NADH oxidation on noncancer cells lacking ENOX2.

Epidemiological studies had indicated a reduced risk for several malignancies suggested with NSAIDs such that cyclooxygenase-2 (COX-2) was proposed as a target to explain their anticancer activity (Moore and Simmons 2000; Dannenberg and Subbaramaiah 2003). Clinical trials were initiated to investigate the potential efficacy of selective COX-2 inhibitors in the prevention and treatment of cancers (Dannenberg and Subbaramaiah 2003) despite conflicting evidence that cancer cell lines lacking COX-2 still responded to certain selective COX-2 inhibitors (e.g., celecoxib) but not to others (e.g., rofecoxib). Inhibitors of COX-2 have afforded protection against the development of colorectal neoplasia (Thun et al. 2002; Rahme et al. 2003). However, in animals, tumor formation and growth were reduced by treatment with selective COX-2 inhibitors not only with normal animals but in animals engineered to be COX-2-deficient (Liu et al. 2001; Neufang et al. 2001) as well. These findings suggested COX-2-indepentent antitumor activities in addition to COX-2-dependent effects (Kazanov et al. 2004; Patel et al. 2005; Waskewich et al. 2002) and provide evidence for ENOX2 as a potential drug target to account for anticancer effects of NSAIDS that occur independent of COX-2 (Morré and Morré 2006b) as a rationale for the use of NSAIDS to prevent or treat human malignancies.

11.6 Retinoids and Calcitriol Agents of Differentiation

Also inhibitory to PMET and the NADH oxidase of plasma membrane vesicles from cancer cells are a group of related substances all of which are potent inducers of cell differentiation, including are retinol, retinoic acid, various synthetic retinoids, calcitriol, DMSO, and sodium phenylacetate (Table 5.4).

Differentiating agents such as retinoic acid and calcitriol are generally considered to exert their growth and antitumor response at the level of gene regulation. However, a direct effort on the cell surface NADH oxidase of cancer cells may help explain antitumor responses of both calcitriol and the retinoids and, especially, sodium phenylacetate (Adam et al. 1995). Responses of Golgi apparatus and Golgi apparatus-related membrane budding responses to retinol (Morré and Morré 1987, Morré et al. 1998f) are summarized in Sect. 10.7.1 of Morré and Mollenhauer (2009).

11.6.1 Retinoic Acid Inhibition of PMET

Retinoic acid is a very strong inhibitor of transplasma membrane ferricyanide reduction by HeLa and HL6O cells (Sun et al. 1987c). The concentrations which cause inhibition are the same as for inducing differentiation. Both retinoic acid and calcitriol inhibit NADH oxidase activity and growth of cultured human keratinocytes (Morré et al. 1992b).

11.6.2 Retinoid Inhibition of ENOX2

Retinoids with cancer therapeutic potential, both natural and synthetic, were evaluated for their ability to inhibit NADH oxidase activity of plasma membranes of cultured HeLa cells and the growth of HeLa cells in culture (Dai et al. 1997).

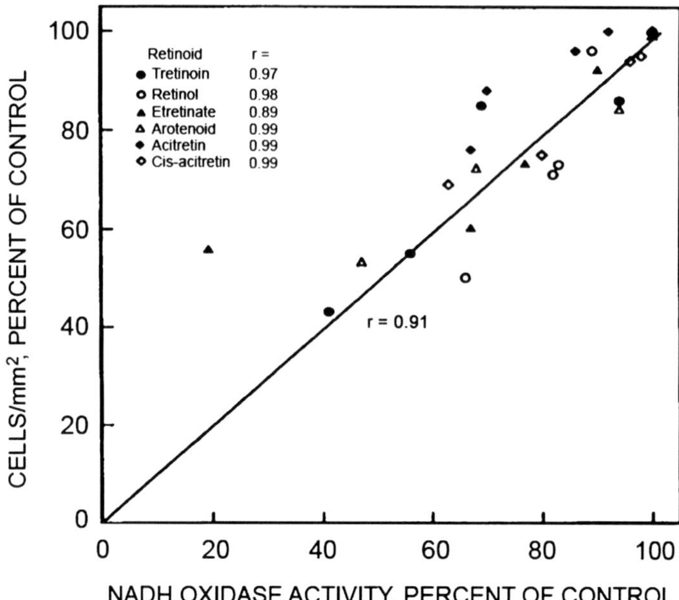

Fig. 11.50 Correlation between NADH oxidase activity and cells/min^2 as percent of control for all treatments. Correlation coefficients and symbols or individual retinoids are given in the *inset*. Reproduced from Dai et al. (1997) with permission from Springer-International

Both NADH oxidase activity and the growth of cells were inhibited by the naturally occurring retinoids, all *trans*-retinoic acid (tretinoin) and retinol as well as by the synthetic retinoids, *trans*-acitretin, 13-*cis*-acitretin, etretinate, and arotonois ethylester (Ro 13-6298). Inhibition of NADH oxidase activity and inhibition of growth were closely correlated (Fig. 11.50) suggesting that inhibition of the plasma membrane ENOX2 by retinoids may be related to their activity as anticancer agents.

Similarly, both ENOX activity and the growth of human keratinocytes immortalized by human papillomavirus type 16 DNA (HKc/HPV16) were inhibited by retinoic acid and 1,25-dihydroxy-vitamin D_3 [1,25-$(OH)_2D_3$] (Morré et al. 1992a).

11.7 ENOX2-Directed Therapeutic Antibodies

A monoclonal antibody, MAB 12.1, selectively inhibits NADH oxidase activity in sera of cancer patients (Fig. 11.51a), an activity also inhibited by the recognized ENOX2 inhibitor capsaicin. Upon selective binding of MAB 12.1 to surface membranes of human carcinoma cells and tissues, cell growth slowed followed by apoptosis (Fig. 11.52), while this monoclonal antibody did not affect the growth of noncancerous cell lines (Cho et al. 2002). Growth of BT-20 human mammary carcinoma cells was blocked by MAB 12.1 whereas growth of noncancer MCF-10A epithelial cells was unaffected (Fig. 11.51b).

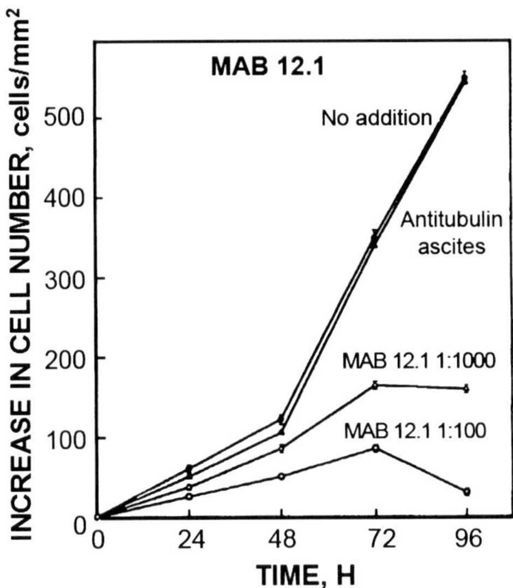

Fig. 11.51 HeLa cell growth. (**a**) mAb 12.1 at a concentration of 2 μg/mL diluted 1:100 (*open circles*) or 1:1,000 (*open triangles*) resulted in dead cells at both dilutions by 168 h. Results are the averages of two independent experiments ± SD. Growth of HeLa cells was unaffected by ascites containing mAb to α-tubulins generated in parallel (1:1,000 dilution) (*solid triangles*) compared to no addition (*solid circles*). Reproduced from Cho et al. (2002) with permission from Springer-International

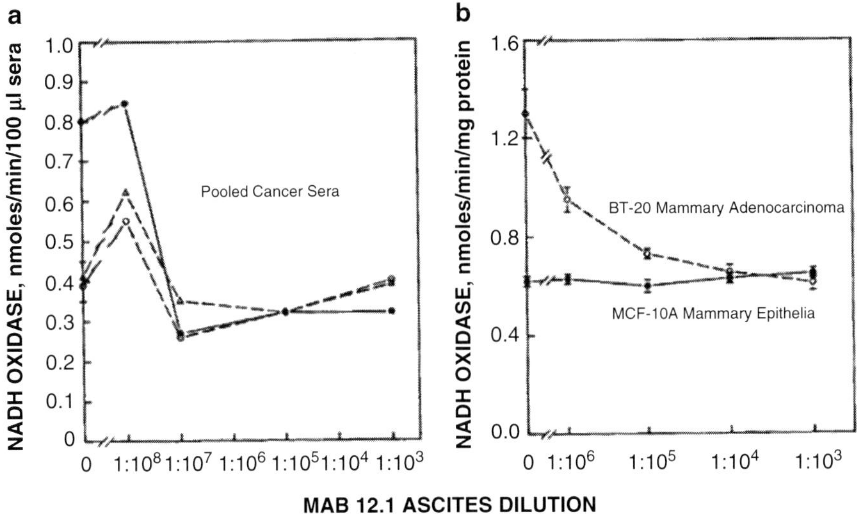

Fig. 11.52 Inhibition of NADH oxidase activity by mAb 12.1 (20 μg antibody/μL). (**a**) Pooled sera from cancer patients. *Solid symbols*: no capsaicin; *open circles*: 1 μM capsaicin; *open triangles*: 100 μM capsaicin. (**b**) Supernatants from immunoprecipitates of pooled sera from cancer patients. Three experiment ± SD. Reproduced from Cho et al. (2002) with permission from Springer-International

Survival of transformed cells is inhibited by antisera raised against bacterially expressed ENOX2 protein (Chen et al. 2006). In contrast, survival of nontransformed cells was not affected. A therapeutic antibody is made difficult by the existence of patient-generated ENOX2 autoantibodies. However, the opportunity for a vaccine remains.

mAb 12.1 was a competitive inhibitor of protein disulfide-thiol interchange both in the plasma membranes from HeLa cells and in the sera of patients with breast cancer. Ascites containing mAb to α-tubulins did not compete. With NADH, the competition was uncompetitive. The inhibition by mAb 12.1 of the ENOX2-catalyzed restoration of activity to scrambled and inactive ribonuclease was nearly complete.

Increased levels of apoptotic markers PARP cleavage and activation of capsase-9 along with activation of JNK, ROS generation, and phosphorylation of p53 were observed in HCT116 cells treated with anti-ENOX2 antibodies (Chen et al. 2006). Moreover, inhibition of ENOX2 activity by anticancer agents, such as EGCg or PXD, results in elevated cytosolic NADH to initiate the apoptotic cascade in tumor cells that results from ENOX2 inhibition (De Luca et al. 2009; see Sect. 11.2.1).

11.8 Antisense

Decreased cell growth and migration are observed when ENOX2 function was impaired through inhibition by EGCg or PXD (Yagiz et al. 2007) or by down-modulating its expression by RNA interference (Liu et al. 2008) or treatment with capsaicin (Wang et al. 2009). As the ability of tumor cells to acquire an aggressive phenotype may require ENOX2 expression, ENOX2 inhibition results in growth and induction of apoptosis in cancer cells (Wang et al. 2009).

Motility and dissemination of cancer cells may be controlled by silencing ENOX2. Targeting ENOX2 by RNA interference inhibits growth and migration of cervical cancer cells via negative modulation of membrane-associated Rac. Knockout of ENOX2 reduces Rac protein expression by more than 50 % compared to wild-type cells with a resultant decrease in both cell growth and proliferation (Liu et al. 2008). These observations are in keeping with involvement of the activation of Rho family members such as Rac, Rho, and Cdc42, important to cell motility through reorganization of the cytoskeleton (Chang et al. 2009), in cell invasion.

Tang et al. (2007) provide evidence demonstrating that ENOX2 arises as a result of alternative splicing (Chap. 8). Full-length ENOX2 mRNA was present in both normal and cancer cells but did not appear to be expressed in either. Two splice variants, an exon 4 minus and an exon 5 minus, were present in cancer cells but not in normal cells. Transfection studies of COS cells with ENOX2 exon 4 minus cDNA led to an overexpression of a 34-kDa ENOX2 protein at the plasma membrane as did exon 4 minus cDNA overexpression in transgenic mice (Yagiz et al. 2006). This 34-kDa protein corresponded to the fully processed form of ENOX2 commonly found at the cell surface (Cho and Morré 2009; Cho et al. 2002). The conclusion that the cancer cell-specific expression of 34-kDa ENOX2 was due to formation of an exon 4 minus splice variant was supported by these findings.

To determine if silencing of ENOX2 exon4 was the result of motifs located in exon 4, transfections were performed on MCF-10A (mammary noncancer), BT-20 (mammary cancer), and HeLa (cervical cancer) cells using a GFP minigene construct containing either a constitutively spliced exon (albumin exon 2) or the alternatively spliced ENOX2 exon 4 between the two GFP halves (Tang et al. 2011; Chap. 8). Removal of exon 4 from the processed RNA of the GFP minigene construct occurred with both HeLa and with BT-20 but none in noncancer MCF-10A cells. The Splicing Rainbow Program identified 8 Exonic Splicing Silencers (ESSs) for hnRNP binding in the exon 4 sequences of ENOX2. Each of these sites was mutated by site-directed mutagenesis to determine which, if any, might be responsible for the splicing skip. Results showed MutG75 ESS mutation changed the GFP expression to indicate splicing silence while other mutations did not (Tang et al., in press) to suggest that hnRNP F directs formation of the exon 4 minus variant of ENOX2. hnRNP H is upregulated 1.6 times in SV-40-transformed human keratinocytes whereas hnRNP F remained unchanged (Honoré et al. 1995).

The identification of the exon 4 minus splicing factor now may provide an approach to prevention of ENOX2 expression. Knockdown of specific splicing factors may be sufficient to reverse transformation caused by their overexpression (Karni et al. 2007). However, such a relationship between hnRNP F and the exon 4 minus splicing event with ENOX2 remains to be investigated.

11.9 ENOX Inhibitors Enhance the Response of Tumors to Radiation

Cell-based phenotypic screening identified (*Z*)-(±)-2-(1-benzenesulfonylindol-3-ylmethylene)-1-azabicyclo[2.2.2]octan-3-ol, which inhibited an ENOX1 of endothelial cells inhibited endothelial proliferation, migration, and the ability to form capillary-like structures in Matrigel at noncytotoxic concentrations. Administration of (*Z*)-(±)-2-(1-benzenesulfonylindol-3-ylmethylene)-1-azabicyclo[2.2.2]octan-3-ol prior to fractionated X-irradiation significantly diminished the number and density of tumor microvessels, as well as delayed syngeneic and xenograft tumor growth compared to results obtained with radiation alone to suggest that targeting ENOX activity represents a novel therapeutic strategy for enhancing the radiation response of tumors.

11.10 ENOX2 as a Target for Cancer Prevention Through Early Intervention

By detecting cancer in its earliest stages (Chap. 12), when perhaps only a small number of cancer cells are present, it may be possible that early intervention targeted specifically to ENOX2 will be effective in preventing development of the incipient cancer thereby resulting in what might be viewed as curative prevention.

Target molecules are provided by the ENOX2 proteins which are essential to the growth and survival of cancer cells, expressed on the surfaces of malignancies and detected as cancer-specific biomarkers in the sera of patients with cancer. The proteins provide sought-after early detection along with new opportunities for targeted preventive intervention for persons at risk for cancer well in advance of development of clinical symptoms. The combination might be expected to significantly reduce cancer mortality. Early detection based on cancer-specific ENOX2 markers in serum might also be coupled with a vaccine-based strategy targeted to the essential growth function of the ENOX2 proteins for early intervention.

11.11 Summary

As ENOX2 proteins are blocked by quinone-site inhibitors also with anticancer activity (capsaicin, EGCg, antitumor sulfonylureas, and doxorubicin as well as cisplatin), they are targets for further exploration of anticancer substances and drugs both under investigation or already used clinically.

The tNOX proteins designated ENOX2 are novel cancer cell-specific, tumor growth-related targets for selective therapy applicable to a broad range of cancer types and disease states. Especially promising is ENOX2-targeted early intervention coupled with early detection protocols based on early serum presence of ENOX2 proteins in sera as cancer-specific biomarkers (Chap. 12). The ENOX proteins, including ENOX2 proteins, serve as terminal oxidases of PMET, a ubiquitous, high capacity NADH redox regulatory system responsible for maintaining an $NADH/NAD^+$ ratio favorable to glycolytic ATP production also indicated as potential targets for anticancer intervention. While some PMET inhibitors such as arsenicals may be therapeutic for cancer independently of ENOX2, it is the cancer specificity of ENOX2 that emerges primarily as responsible for the anticancer efficacy of PMET targeted drugs.

Among the most widely studied PMET and ENOX2 inhibitors are the vanilloids, capsaicin, and resiniferatoxin. One advantage of ENOX2 and the PMET as targets for anticancer drug development strategies is that ENOX2 proteins are located on the external cell surface and do not need to enter cells to be effective. Efficacious impermeant conjugates have been prepared and tested for doxorubicin, capsaicin, an antitumor sulfonylurea, and an antitumor quassinoid, glaucarubolone.

ENOX2 and PMET inhibitors, in general, result both in growth stasis and ultimately apoptotic death of cancer cells. Growth arrest results both from blockage of ENOX2-catalyzed protein disulfide-thiol interchange required for cell enlargement and growth due, in part, to elevated ceramide. Proliferating cells unable to enlarge following division are unable to divide again. The elevated NADH at the cytosolic surface of the plasma membrane that results from inhibition of PMET reduces prosurvival levels of S1P through inhibition of sphingosine kinase as well as activation of sphingomyelinase to form ceramide. The result is caspase-3-dependent programmed cell death involving both extrinsic death receptor and intrinsic (mitochondrial) apoptotic pathways.

The antitumor sulfonylureas, PMET and ENOX2 inhibitors derived from an experimental series were identified as having activity against human solid tumors in vivo. The sulfonylureas inhibitors as well as certain other examples such as antitumor quassinoids are extremely redox sensitive and will inhibit or stimulate ENOX activity depending on the redox potential of the environment. Acetogenins from the plant family Annonaceae inhibit ENOX2 as does the principal green tea catechin EGCg. Tea catechins may augment cancer chemo- and radiation therapy and are themselves augmented by combination with *Capsicum* preparations. EGCg may affect a wide variety of cellular processes but ENOX2 inhibitions are uniquely observed in the nanomolar range of concentrations (EC_{50} of 5–10 nM). ENOX2 inhibition is mediated through an ENOX2-specific binding site containing the motif E245EMTE. PXD and related anticancer isoflavones are ENOX2 inhibitors with high selectivity for tumor cells and demonstrated clinical efficacy including chemosensitization to other anticancer agents such as cisplatin and taxanes. Sulforaphane, a constituent of broccoli, and suramin inhibit ENOX2 as do COX-2-directed NSAIDs. NSAIDS also exhibit anticancer activity along with retinoids and calcitriol. Downregulation of ENOX2 expression by RNA interference offers a further avenue for intervention as does targeting the splicing factor complexes involved in ENOX2 expression of transcription variants. Use of anti-ENOX2 antibodies as therapeutic agents is complicated by the widespread production of ENOX2 IgM class autoantibodies by cancer patients.

Chapter 12
Cancer Diagnostic Applications of ENOX2 Proteins

Cancer is the second leading disease cause of death in the United States. Estimates are that there were over 1.5 million cases in 2010 in the United States alone. Only a small fraction (less than 20 %) of cancers are diagnosed at a localized stage where curative therapy is effective. Most cancers are diagnosed only after the primary tumor has already metastasized so that chemotherapy is required for treatment. Hence, early detection is a favored opportunity to reduce mortality. By detecting cancer in its very earliest stages when perhaps only a small number of cells are present, it is possible that early intervention will be effective in preventing further development of the incipient cancer thereby resulting in what might be viewed as curative prevention.

Despite advances in early detection of major forms of human cancer (prostate, breast, lung, colon, leukemia, lymphoma), more often than not, cancers have developed to a sufficiently late stage at the time of detection to preclude most opportunities for curative therapy (Altekruse et al. 2010). The problem is exacerbated for pancreatic cancer where clinical symptoms invariably are delayed until the disease state is well advanced beyond metastatic spread. A need for early detection remains as one of the most important challenges at the forefront of cancer research, treatment, and prevention. The sooner cancer is detected, the more options an individual has for treatment and the better their chance to survive (Bertucci and Goncalves 2008).

For example, with breast cancer patients, early detection and especially early detection of recurrence or of residual breast disease postsurgery or relapse of patients with surgically resected cancer would be expected to improve chances of survival and result in a significant decrease in breast cancer mortality (Harvey et al. 2008; Rim and Chellman-Jeffers 2008). Currently only 80–85 % of breast tumors are detected by x-ray mammography often after the cancer is already advanced or has spread to distant sites (Navarrete-Montalvo et al. 2008). Generally, the tumors have progressed beyond stage I before discovery. The 15–20 % of tumors that go undetected will enlarge and become more lethal until finally detected in later screens or

D.J. Morré and D.M. Morré, *ECTO-NOX Proteins:Growth, Cancer, and Aging*,

DOI 10.1007/978-1-4614-3958-5_12,

become apparent clinically leaving chemotherapy as the only recourse for treatment (Yadav et al. 2007).

The goal, for example, is to detect more of the current 15–20 % of breast (as well as other) cancers that go undetected with mammography, tomography, or x-rays alone and prior to metastasis (Bertucci and Goncalves 2008). Application to post-surgery cancer patients is expected to help determine which patients still harbor residual disease following surgery and require chemotherapy and which patients are free of disease where chemotherapy can be delayed or averted. Additionally, such an assay would have utility as a means to monitor for remissions in patients with no evidence of disease.

12.1 Cancer Cell Surface ENOX2 Shed into Sera as Biomarkers of Cancer Presence

The concept of using ENOX2 to detect cancer is based on the discovery of the family of cell surface ECTO-NOX or ENOX proteins, one subset of which, the tumor-associated ENOX proteins, designated tNOX or ENOX2, is specific to the surface of cancer cells and is absent from noncancer cells and tissues (Chueh et al. 2002b; Morré and Morré 2003a). The presence of ENOX2 proteins in sera of cancer patients represents an origin due to shedding from the patient's cancer (Wilkinson et al. 1996). The presence of the ENOX2 protein has been demonstrated in a number of human tumor tissues and xenografts (mammary carcinoma, prostate cancer, neuroblastoma, colon carcinoma, and melanoma). However, serum analyses suggest a much broader association with cancer. ENOX2 proteins are ectoproteins reversibly bound at the outer leaflet of the plasma membrane (Morré 1995a). As is characteristic of other examples of ectoproteins (sialyl and galactosyl transferases, dipeptidylamino peptidase IV, etc.), the ENOX2 proteins are shed (Morré et al. 1996c; Wilkinson et al. 1996). They appear in soluble form in conditioned media of cultured cells and in patient sera. The ENOX2 isoforms from sera of cancer patients exhibit the same degree of drug responsiveness as does the membrane-associated form (Morré and Reust 1997; Morré et al. 1997a). With sera from more than 200 breast cancer patients, the majority (ca. 196) were found to exhibit the drug-responsive activity. In contrast, no drug-responsive NOX activities were found with sera from healthy volunteers or sera from patients with diseases other than cancer (cardiac, arthritis and other inflammatory diseases, gastric ulceration, emphysema, various nonmalignant blood disorders). As such, the antitumor-responsive ENOX2 activity represents a novel cell surface property potentially associated with most, if not all, forms of human cancer and an appropriate biomarker for serum or plasma diagnosis of cancer (Fig. 12.1).

ENOX2 proteins are robust and highly resistant to heat and protein degradation which enhances their utility as noninvasive markers for cancer detection and diagnosis (Morré and Morré 2003a).

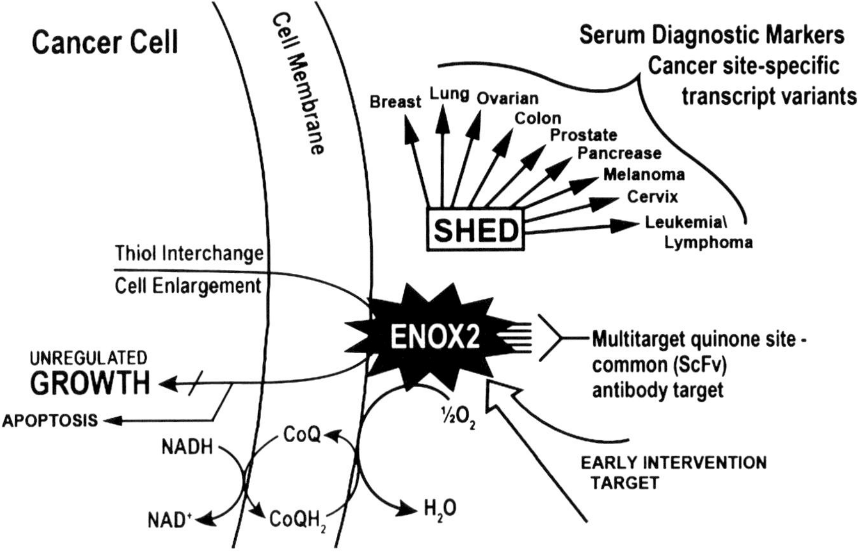

Fig. 12.1 Schematic representation of the utility of the ENOX2 family of cancer-specific cell surface proteins for early diagnosis and early intervention of cancer. Cancer site-specific transcript variants of ENOX2 are shed into the serum to permit early detection and diagnosis. The ENOX2 proteins of origin at the cell surface act as terminal oxidases of plasma membrane electron transport functions essential to the unregulated growth of cancer. When the ENOX2 proteins are inhibited, the unregulated growth ceases and the cancer cells undergo programmed cell death (apoptosis). Reproduced from Morré and Morré 2012b with permission from InTech

12.1.1 Biomarker Discovery Based on ENOX2 Activity

Circulating ENOX2 has been detected in sera of more than 500 cancer patients representing all major forms of human cancer including leukemias and lymphomas (Morré and Reust 1997; Morré et al. 1997a; Table 12.1).

12.1.2 Characteristics of ENOX2 as Cancer Biomarker Based on Activity

The ENOX proteins are responsible for the increase in cell size following cell division. After cell division, a minimum cell size must be reached or cell division stops and after several days the cells undergo programmed cell death (apoptosis). Cells with blocked ENOX activity are not able to enlarge and are thus directed towards apoptosis. The growth inhibition is due mainly to cell cycle arrest in G_1 (stage of cell division before DNA is replicated).

Table 12.1 NADH oxidase activities of sera from noncancer subjects and cancer patients with active disease ($n=70$) and response to 100 μM capsaicin

	Ratio
Cancer	100 μM capsaicin/no capsaicin
None	1.01
Breast	0.69
Prostate	0.78
Colon	0.49
Lung	0.76
Ovarian	0.87
Esophageal	0.41
Ovarian	0.65
Melanoma	0.79
Leukemia	
Granulocytic	0.52
Lymphocytic	0.62
Lymphoma (non-Hodgkin's)	0.59
Diseases other than cancer	0.99

Adapted from Morré et al. (1997a). An additional 200 cancer patient sera were analyzed by Morré and Reust (1997) for NADH oxidase activity response to the antitumor sulfonylurea LY181984

The tumor form of the NOX proteins, designated ENOX2, is blocked by quinone site inhibitors with anticancer activity [antitumor sulfonylureas (Morré et al. 1995i), antitumor vanilloids, e.g., capsaicin (Morré et al. 1995b), antitumor quassinoids (Morré et al. 1995b), antitumor catechins, e.g., (−)-epigallocatechin-3-gallate which is the active principle of green tea (Morré et al. 2000a), doxorubicin (Morré et al. 1997c), and cisplatin (11.4.3), and other substances with anticancer activity (Fig. 11.1)]. The predominant ENOX of noncancer cells (the constitutive ENOX, CNOX, designated ENOX1) is drug resistant.

That ENOX2 presence is restricted to cancer cells is indicated from lack of a drug-responsive NOX activity in normal cells and tissues (Morré 1998c), lack of a serum form of ENOX2 in healthy individuals and patients with disorders other than cancer (Morré and Reust 1997; Morré et al. 1997a; Hostetler et al. 2009; Morré et al. 2009a), and from indications with ENOX2-specific antisera (Fig. 12.2). Levels are absent or below the limits of detection (less than 10 pmol/mL of serum) based on findings determined independently from western blot analyses of two-dimensional gel separations using ENOX2-specific antisera (Cho et al. 2002) of volunteers not diagnosed as having cancer (Sect. 12.2) and from sera of patients with disorders other than cancer (Morré and Reust 1997; Morré et al. 1997a). ENOX2 has been cloned (GenBank Accession No. AF207881; Chueh et al. 2002b) using a monoclonal antibody that recognizes a common ENOX2 epitope that

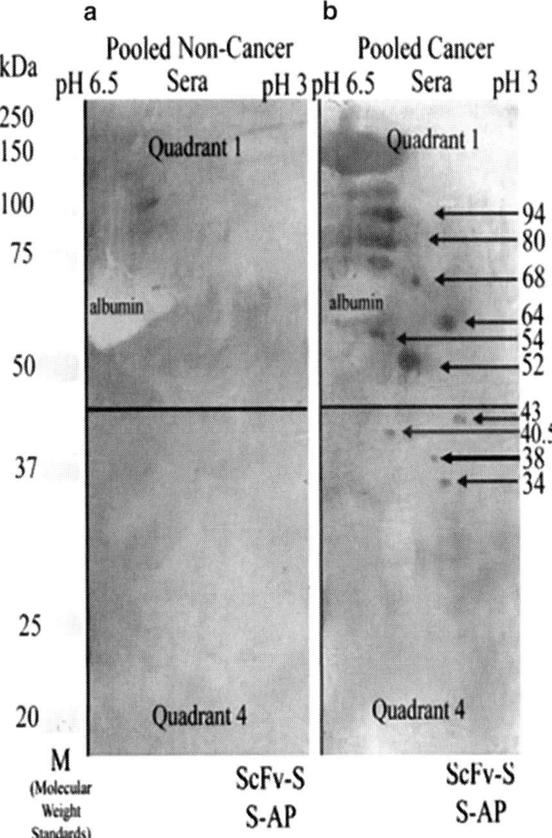

Fig. 12.2 Two-dimensional gel/western blot of ENOX2 transcript variants comparing quadrants 1 and 4 of pooled noncancer (**a**) and pooled cancer representing major carcinomas plus leukemias and lymphomas (**b**) patient sera. The approximate location of unreactive (at background) albumin is labeled for comparison. ENOX2 reactive proteins are restricted to quadrants I and IV. Detection uses recombinant scFv-S (S-tag peptide: His-Glu-Ala-Ala-Lys=Phe-Gln-Arg-Glu-His) antibody with alkaline phosphatase-linked antiS protein. The approximately 10 ENOX2 transcript variants of the pooled cancer sera are absent from noncancer (**a**) and are cancer site-specific as indicated in Fig. 12.3. Reproduced from Hostetler et al. (2009) with permission from Springer Science + Business Media

includes the cancer drug-binding site (Cho et al. 2002) and from which a pan ENOX2 isoform-specific single chain variable region (scFv) recombinant antibody was derived (Kim 2011; Sect 12.3). This binding site contains a conserved five amino acid (EMTEE) motif and appears, based on biochemical (drug inhibition of activity) and immunological evidence, to be present in all ENOX2 isoforms recognized by the pan-cancer antibody on western blots. The EMTEE drug-binding motif in exon 5 is absent from the amino acid sequence of the constitutive ENOX1 proteins of noncancer cells.

Table 12.2 Sera from patients with different cancers exhibit distinct patterns of ENOX2 isoforms with characteristic molecular weights and isoelectric points (pH)

Cancer	Sera analyzed	Molecular weight (KDa)	Isoelectric point, pH
Cervical	18	94	5.4
Ovarian	41	80 and 40.5	4.2 and 4.1
Prostate	70	75	6.3
Breast/uterine	55	64–68	4.5
Nonsmall cell lung	83	54	5.1
Small cell lung	22	52	4.3
Pancreatic	24	50	4.3
Colon	55	52 and 34	4.3 and 3.9
Lymphoma, leukemia	16	45	3.9
Melanoma	12	38	5.1

Updated from Hostetler et al. (2009)

12.1.3 Transcript Variants Detected by Two-Dimensional Gel-Western Blot Analysis Are Cancer Site-Specific Biomarkers

The opportunity to simultaneously determine both cancer presence and cancer site emerged as a result of two-dimensional gel electrophoretic separations where western blots with a pan ENOX2 recombinant scFv antibody carrying an S tag (Fig. 12.2) were employed for detection (Hostetler et al. 2009). The antibody cross reacted with all known ENOX2 forms from hematological and solid tumors of human origin but, of itself, did not differentiate among different kinds of cancers. Analyses using this antibody, when combined with two-dimensional gel electrophoretic separation, revealed specific ENOX2 species subsequently identified as transcript variants, each with a characteristic molecular weight and isoelectric point indicative of a particular form of cancer (Hostetler et al. 2009; Table 12.2).

ENOX2 transcript variants of specific molecular weights and isoelectric points are produced uniquely by patients with cancer. The proteins are shed into the circulation and have the potential to serve as definitive, noninvasive, and sensitive serum markers for early detection of both primary and recurrent cancer in at-risk populations with a low incidence of false positives, as they are molecular signature molecules produced specifically by cancer cells and absent from noncancer cells.

The ENOX2 transcript variants specific for cancer thus far appear to be either absent or too low to be detectable in sera of subjects without cancer, but are readily detectable in sera of patients with cancer. Unlike most published cancer markers, the method does not simply measure elevated levels of a serum constituent present in lesser amounts in the absence of cancer but are molecules produced specifically by cancer cells. The cancer-specific ENOX2 transcription variants result from cancer-specific expression of splice variant mRNAs (Tang et al. 2007). Neither the splice variant mRNAs nor the ENOX2 transcription variants are present in detectable levels in noncancer cells or in sera of subjects without cancer.

12.1.4 Transcript Variants of ENOX2

The scientific rationale of the proposal for early detection based on two-dimensional gel-western blot analyses is a result of the subsequent discovery of transcript variants of ENOX2 of specific molecular weights and isoelectric points that are produced uniquely by patients with cancer (Hostetler et al. 2009; Morré et al. 2009a). The transcription variants are shed into the circulation and have the potential to serve as definitive, noninvasive, and sensitive serum markers for early detection of both primary and recurrent cancer in at-risk populations when combined with the pan-cancer ENOX2 recombinant single chain variable regions (scFv) recombinant antibody carrying an S tag for detection (Morré et al. 2009a; Hostetler and Kim 2011; Kim 2011).

The antitumor drug-responsive ENOX2 activity is the first cell surface change reported to be associated with most, if not all, forms of human cancer. However, while ENOX2 presence provides a noninvasive approach to cancer detection, without methodology to identify specific cancer site-specific isoforms, it offered no indication as to cancer type or location. The ENOX2 recombinant scFv antibody carrying an S tag that cross reacted with all known ENOX2 isoforms from hematological and solid tumors of human origin per se did not differentiate among different kinds of cancer. Yet, when combined with two-dimensional gel electrophoretic separations, analyses using this antibody revealed specific ENOX2 isoforms, each characteristic of a particular form of cancer (Hostetler et al. 2009). This discovery made possible a proteomics-based detection protocol for cancer diagnosis as well as detection. ENOX2 proteins are robust and highly resistant to heat and protein degradation which enhances their utility as noninvasive markers for cancer detection and diagnosis (Morré and Morré 2003a).

Unlike most published cancer markers, cancer-specific ENOX2 variants are not simply present as elevated levels of a serum constituent present in lesser amounts in the absence of cancer. The cancer-specific ENOX2 transcript variants result from cancer-specific expression of alternatively spliced mRNAs (Tang et al. 2007; 2008). Neither the splice variant mRNAs nor the ENOX2 isoform proteins are present in detectable levels in noncancer cells or in sera of subjects without cancer.

12.2 Two-Dimensional Gel-Western Blot Cancer Detection System

Preliminary results indicate an opportunity for the serum analysis of particular isoforms of ENOX2 derived from two-dimensional gel electrophoresis (2DGE) and immunoblotting with detection using the scFv recombinant antibody produced in bacteria (Hostetler et al. 2009).

Analytical 2DGE of ENOX-enriched serum proteins from a mixed population of cancer patients (cervical, breast, ovarian, prostate, lung and colon carcinoma, leukemia, lymphoma) concentrated by nickel agarose precipitation, followed by immunoblot analysis revealed multiple species of acidic proteins of molecular weight between

Table 12.3 Protein sequence similarity between ENOX2 and the two reference proteins α1-antitrypsin inhibitor and serrotransferrin reactive with the pan ENOX2 scFv recombinant antibody

ENOX2	EEMTETK400ETEESA406LVS
Alpha-1 antitrypsin inhibitor	GTDCVAK211EATEAA216KCN
Serrotransferrin	CLDGTRK589PVEEYA595NCH

Regions of similarity are restricted to a seven amino acid sequence (*underlined*) adjacent in ENOX2 to the E394EMTE398 quinone inhibitor-binding site

34 and 100 kDa in quadrants I and IV (Fig. 12.2b), none of which were present in noncancer sera (Fig. 12.2a) (Hostetler et al. 2009). Separation in the first dimension was by isoelectric focusing over the pH range of 3–10 and separation in the second dimension was by 10 % SDS-PAGE. Isoelectric points of the ENOX2 isoforms were in the range of pH 3.9–6.3. The only reactive proteins other than the ENOX2 isoforms are a 53-kDa isoelectric point pH 4.1, α1-antitrypsin inhibitor (labeled "R" in Fig. 12.2) which served as a convenient loading control and isoelectric point reference, and 79–85 kDa, isoelectric point pH 6.8 serotransferrin which serves as a second point of reference for determination of molecular weight and isoelectric point (Table 12.3). The two cross reactive reference proteins are present in a majority of sera and plasma of both cancer and noncancer subjects. Albumin and other serum proteins do not react. On some blots, the recombinant scFv was weakly cross-reactive with heavy (ca. 52 kDa) and light (ca. 25 kDa) immunoglobulin chains.

Sera from individual patients with the various forms of cancer were analyzed by 2DGE and immunoblotting to assign each of the ENOX2 isoforms to a cancer of a particular tissue of origin (Fig. 12.3). Sera from cervical cancer patients contained the 94 kDa ENOX2 isoform (Fig. 12.3a), sera of ovarian cancer patients contained ENOX2 isoforms of 80 kDa and 40.5 kDa (Fig. 12.3a), and sera from patients with prostate cancer contained one or more 75 kDa ENOX2 isoforms resulting from small variations in isoelectric points (Fig. 12.3c). Sera of breast cancer patients contained 64 and 68 kDa ENOX2 isoform (Fig. 12.3d), sera from patients with nonsmall cell lung cancer contained a 54 kDa ENOX2 isoform (Figs. 12.3e and 12.4), and sera from patients with small cell lung cancer contained a 52-kDa ENOX2 isoform (Figs. 12.3f and 12.4). A ENOX2 isoforms of Mr 50 and 52 kDa characterized sera of pancreatic cancer patients (Fig. 12.3g) whereas sera of colon cancer patients contained ENOX2 isoforms of 52 and 43 kDa (Fig.12.3h). Figure 12.3i shows results from sera of a patient with non-Hodgkin's lymphoma illustrating the 45 kDa ENOX2 isoform of low isoelectric point characteristic of leukemias and lymphomas. Sera of patients with malignant melanoma contained an ENOX2 isoform of Mr 38 kDa (Fig. 12.3j).

Results showing corresponding molecular weights and isoelectric points of each of the ENOX2 isoforms are summarized in Table 12.2. Whereas ENOX2 proteins are absent or reduced to below the limits of detection from sera of healthy volunteers or patients with diseases other than cancer, circulating ENOX2 has been detected in sera of more than 500 cancer patients representing all major forms of human cancer including leukemias and lymphomas. All ENOX2 isoforms appear to share a common antigenic determinant recognized both by a ENOX2-specific monoclonal antibody (Cho et al. 2002) and a corresponding scFv fragment carrying

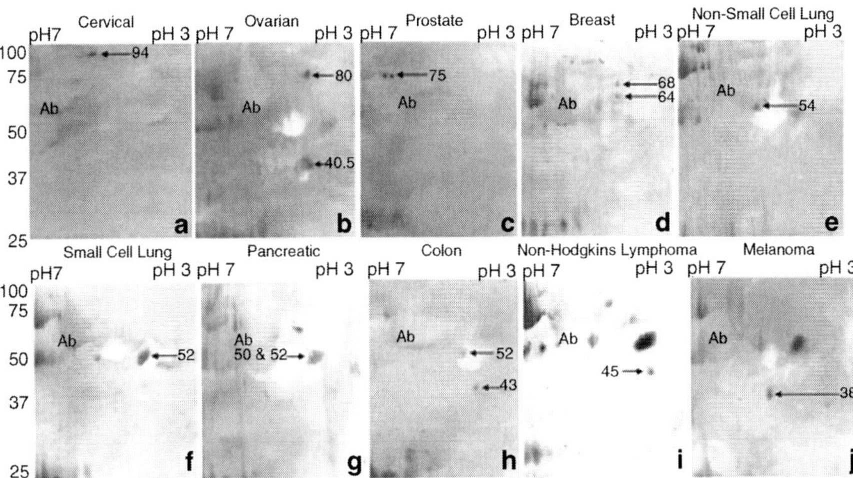

Fig 12.3 Western blots of two-dimensional gel electrophoresis/western blots of sera from cancer patients analyzed individually. Cancer sites are presented in the order of decreasing molecular weight of the major transcript variant present. (**a**) Cervical cancer. (**b**) Ovarian cancer. (**c**) Prostate cancer. (**d**) Breast cancer. (**e**) Nonsmall cell lung cancer. (**f**) Small cell lung cancer. (**g**) Pancreatic cancer. (**h**) Colon cancer. (**i**) Non-Hodgkin's lymphoma. (**j**). Melanoma. The approximate location of unreactive (at background) albumin (Ab) is labeled for comparison. Approximately 400 non-cancer patient sera were analyzed in parallel without evidence of proteins indicative of specific transcript variants. Reproduced from Hostetler et al. (2009) with permission from Springer Science + Business Media

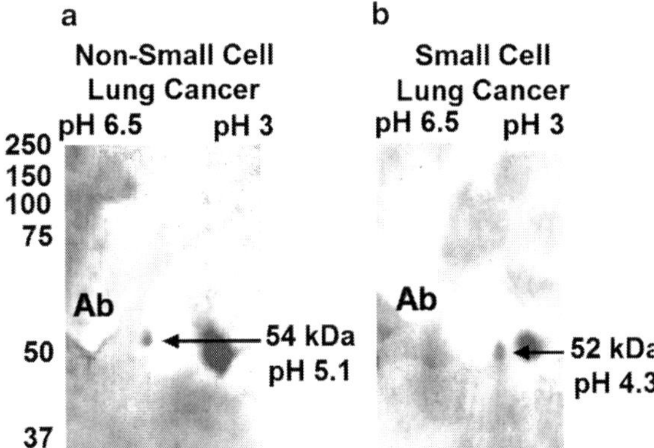

Fig. 12.4 Analytical gel electrophoresis and immunoblot of sera of nonsmall cell and small cell lung cancer. (**a**) Sera from a patient with nonsmall cell lung cancer contains the 54-kDA ENOX2 transcript variant, pH 5.1 (*arrow*). (**b**) Sera from a patient with small cell lung cancer contains the 52 kDa, isoelectric point pH 4.3 transcript variant (*arrow*). The reference spots to the right, Mr 52 kDa, isoelectric point pH 4.1, is α1-antitrypsin. Albumin and other serum proteins are unreactive. Reproduced from Morré and Morré 2012b with permission from InTech

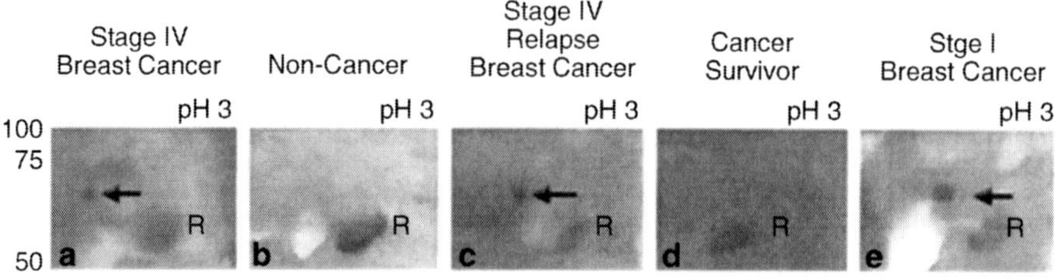

Fig. 12.5 Two-dimensional gel electrophoretic separations (**a**–**e**) and detection of ENOX2 transcript variants by western blotting of patient sera. *Arrow*=66–68 kDa breast cancer-specific transcript variant (*arrows*). *R*=52 kDa, isoelectric point pH 4.1 α1-antitrypsin inhibitor reference spot. Albumin and other serum proteins are unreactive. Reproduced from Morré and Morré 2012 with permission from InTech

an S tag for detection expressed in bacteria and derived from the ENOX2-specific IgG rescued from the monoclonal antibody-producing hybridoma cells (Hostetler et al. 2009). A 52-kDa ENOX2 transcript variant has been purified and characterized from HeLa cells (Yantiri and Morré 2001).

Particular relevance to diagnosis of breast cancer is indicated from observations where the 68-kDa ENOX2 isoform (pH 4.5) of sera correlates with disease presence in both late (Stage IV) (Fig. 12.5a) and early (Stage I) (Fig. 12.5e) disease and in Stage IV recurrence (Fig. 12.5c) but is absent from sera of noncancer (normal) volunteers (Fig. 12.5b) or in survivors free of disease for 1–5 years (Fig. 12.5d).

Additionally, the 68-kDa breast cancer-specific isoform does not apply to a subset of breast cancer patients but appears to be universally present. Analysis of sera of more than 60 patients with active disease including 20 Stage I and Stage II breast cancer patients all have tested positive.

12.2.1 Two-Dimensional Gel-Western Blot Analysis of ENOX2 Transcript Variants Provide for Very Early Detection

The diagnostic parameters are a family of cancer- and tissue-specific cell surface proteins essential to the growth of cancer cells and shed into the circulation designated as ENOX2. This strategy would provide the first ever opportunity for very early diagnosis combined with early biomarker-targeted intervention as an approach to reduce cancer mortality in at-risk populations.

The possibility that the two-dimensional gel western blot system may be able to detect cancer presence 5–7 years in advance of the appearance of clinical symptoms has been raised. This was first indicated from findings with a special lung cancer panel of sera specimens obtained through the Early Detection Research Network (EDRN) of the National Cancer Institute. The panel consisted of about 20 known lung cancer patient sera and 35 control patient sera. All 20 of the known lung cancer patient sera were successfully identified.

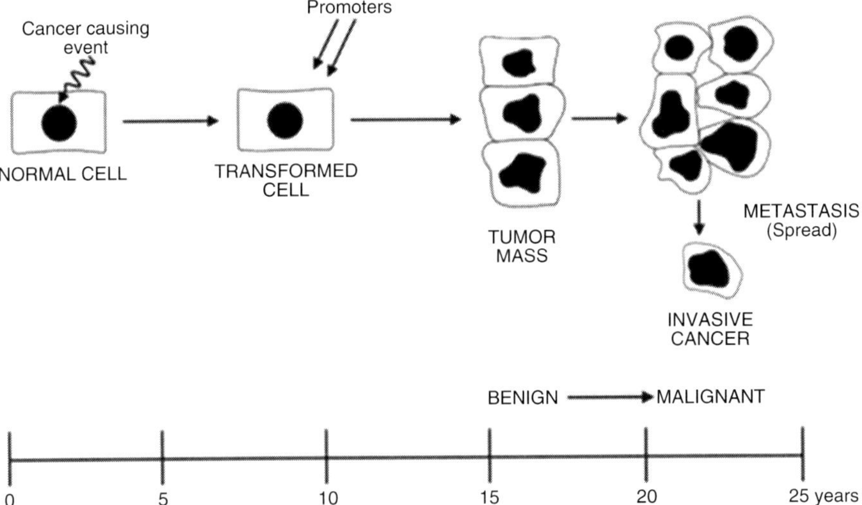

Fig. 12.6 Interpretive diagram to illustrate the various stages of cancer progression (estimated to require as long as 20 y) beginning with a cancer-causing event (initiation) through development of a clinically defined malignancy. Reproduced from Morré and Morré 2012 with permission from InTech

Unexpectedly, a high incidence of positive ENOX2 presence in sera was found with the "control" group which were derived from a community screening study. Upon further investigation, it was noted that of the 17 positive control subject samples where findings specifically indicated lung cancer (the lung cancer-specific ENOX2 marker was found) and where smoking histories were available, 16 of the subjects were smokers in the range of 15–88 pack-years (numbers of packs smoked/day multiplied by number of years smoked). However, the anticipated incidence of undetected lung cancers in such a population would be in the order of 10 % rather than nearly 50 %. Since the aberrant ENOX2 isoforms are a single molecular species produced only by lung cancer, the conclusion was reached that lung cancer was being detected much earlier than was currently possible by other methods; perhaps as early as 5–7 years before clinical symptoms, based on the estimated 20-year development time for lung cancer expression between carcinogen exposure and a clinically evident cancer (Petro et al. 2000) and diagrammed in Fig. 12.6.

Similar results were obtained with a panel of female subjects at risk for breast and ovarian cancer. An analysis of a panel of 127 sera in a Biomarker Reference Set for Cancers in Women also provided through the Early Detection Research Network of the National Cancer Institute also support indications that the two-dimensional gel-western blot system may detect cancer presence 5–7 years in advance of clinical symptoms. The panel consisted of samples pooled from 441 women in 12 different gynecologic and breast disease categories plus 115 sera from age-matched control women. Of the 127 sera samples in the panel 29 tested positive for breast cancer and another 16 tested positive for ovarian cancer. Since the aberrant transcript variants are single molecular species produced by specific cancers such as lung, breast, or ovarian, the findings suggest that cancer was detected in the control population much earlier than currently possible by other methods. As estimated for lung cancer, the indications might be detection as early as 5–7 years before clinical symptoms based on the development time estimated for a breast cancer causing event and clinically evident disease (Weinberg 2007).

12.3 Early Intervention

That the two-dimensional-western blot protocol detects cancer early, well in advance of clinical symptoms, creates the opportunity to combine early detection with early intervention as a potentially curative prevention strategy for cancer by eliminating the disease in its very earliest stages. The approach to early intervention is based on previous work in cell culture models showing that ENOX2 proteins are required to support the unregulated growth that typifies cancer cells as diagrammed in Fig. 12.1. If the growth function of ENOX2 is blocked for 48–72 h, the cancer cells cannot enlarge following division, cannot pass the checkpoint in G_1 that monitors cell size, and eventually undergo apoptosis (Morré and Morré 2003a; De Luca et al. 2010). Among the early intervention strategies under investigation are several targeted to ENOX2, production of ENOX2-directed vaccines being one promising example. Recombinant ENOX2 peptides that exhibit cancer specificity are employed as antigens. Other forms of ENOX2-directed therapeutic interventions under study include use of dietary modulators (Morré et al. 2009b). Most advanced are studies with herbal mixtures of green tea catechins and powders of ground chili peppers (*Capsicum* species) from efficacious pepper sources (e.g., guajillo or ancho) with levels of capsaicin, the pungent principle of chili peppers, sufficiently low so as to not cause discomfort (Sect. 11.4.7.6).

12.4 ENOX2 Autoantibodies May Preclude Conventional ELISA Tests for Early Appearance of ENOX2 in Serum or Plasma

ENOX2-specific mouse monoclonal antibodies have been generated to lung, breast, and ovarian transcript variants (Unpublished results, NOX Technologies, Inc. West Lafayette, IN). None have proven useful with serum or plasma Enzyme-Linked Immunoabsorption (ELISA) tests due to a relatively small margin of difference between cancer and noncancer and a high incidence of false positives. The difficulty has been traced, at least in part, to the production by cancer patients of IgM class autoantibodies to the ENOX2 transcription variants (Table 12.4; Fig. 12.7) that interfere with the ELISA protocols by already being bound to the ENOX2 proteins in the circulation. Evidence for the correctness of this supposition comes from experiments where the autoantibodies and the ENOX2 transcript variant proteins

Table 12.4 Sera from subjects with different patterns of ENOX2 transcript variants with characteristic molecular weights and isoelectric points (pH) detected by ENOX autoantibody from a breast cancer patient. From a study with Maya Koehler

	Transcript variants		Processed remnant	
Cancer	MW	pH	MW	pH
Cervical	95	5.0	33	5.1
Prostate	70	6.3–6.9	38	5.4
Breast	68	4.7	33	4.8
Lung	54	5.1	38	4.8
Colon	38 and 52	4.4 and 4.1	35	4.8
Lymph/leuk	38–45	3.9		
Ovarian	40 and 80	3.6 and 3.7	35	5.2

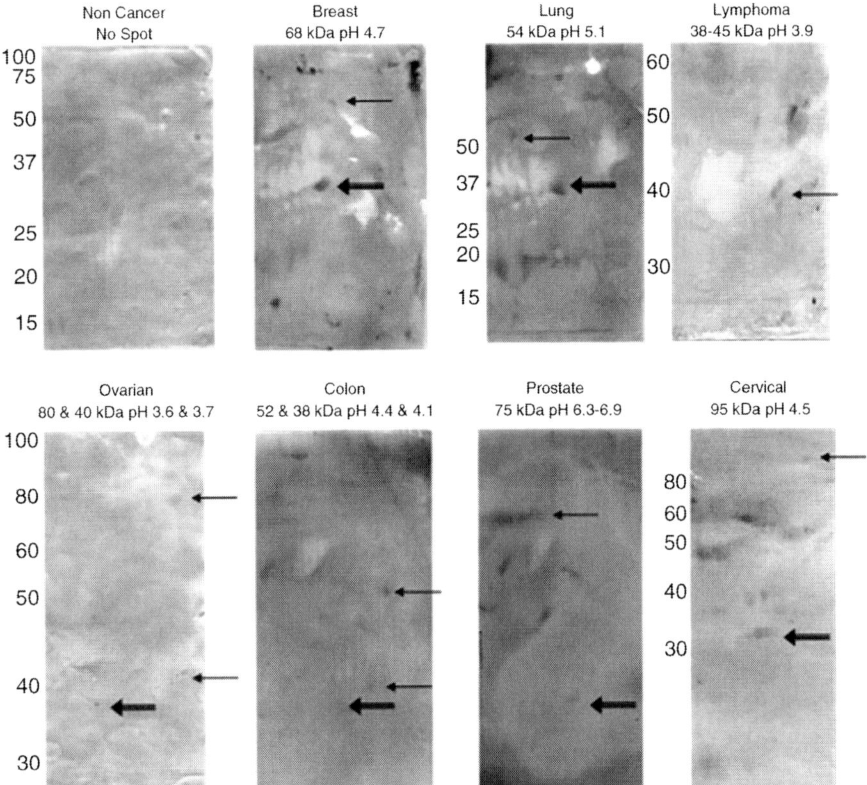

Fig. 12.7 Two-dimensional gel electrophoresis-western blot analysis demonstrating the presence of ENOX2 transcript variant-directed autoantibodies in patient sera (*thin arrows*). Sera of breast or lung cancer patients were used as the autoantibody source. Antihuman IgM coupled to alkaline phosphatase was the detection antibody. Results for breast cancer sera are summarized in Table 12.3. The thick arrows indicate the serum form of processed ENOX2 Mr 35 ± 2 kDa. From a study with Maya Koehler

are first dissociated by isoelectric focusing or separations based on the principles of free-flow electrophoresis (Morré et al. 1994a). The monoclonal antibodies then appear to perform adequately in an ELISA test. The two-dimensional gel-western blots are free of autoantibody interference due to the fact that the electrophoresis step in the first dimension displaces the autoantibody directed against ENOX2 and allows external antibodies to bind whereas prior to electrophoresis all of the ENOX2 of sera may have been occupied by binding to autoantibodies such that no additional external antibodies were able to bind. The biological role of spontaneous tumor-specific T-cell and autoantibody responses, however, are not well understood. Conflicting views include promotion of tumor growth, suppression of tumor progression, or lack of clinically relevant (Koboid et al. 2010). While not necessarily, of themselves, disruptive of tumor growth, they do appear to contribute to the ability of malignancies to avoid both detection and immune control.

12.5 Lipid-Associated Sialic Acid (LASA) Fractions from Sera of Cancer Patients Contain ENOX2 Fragments as Major Nonlipid Constituents

Beginning with the report of Skipski et al. (1975), colorimetric determinations of levels of lipid-bound sialic acid in sera were explored in several laboratories as markers of cancer presence (Kloppel et al. 1977; Portoukalian et al. 1978; Lengle 1979; Dnistrian et al. 1982; Sallay et al. 1986). Despite repeated attempts, analyses of ganglioside extracts from human sera failed to demonstrate the two- to threefold elevations in apparent sialic acid levels as reported with rodents with the monosialoganglioside G_{M3} being the dominant ganglioside in human sera. Furthermore, the commonly used procedure involving determinations with thiobarbituric acid following hydrolysis with 0.1 N acid was found to be unreliable leaving the source of the reported elevations in lipid-soluble sialic acid measured colorimetrically in humans unresolved (Sallay et al. 1986).

The LASA question was revisited in our laboratory in 1997–1998 largely through the unpublished efforts of Dagmar Sedlak. Her findings revealed both 17 kDa and 34 kDa proteinase K-resistant peptides in LASA fractions from either pooled or individual sera from cancer patients prepared by the method of Kloppel et al. (1976) corresponding to fully processed ENOX2 that were absent from pooled or individual sera from noncancer subjects. Following elution from SDS-PAGE gel slices and refolding by reducing thiols, incubation under renaturing conditions and reformation of disulfide bonds in the refolded peptides, capsaicin-inhibited ENOX2 activity (period length of 22 min) with specific activities comparable to those of recombinant ENOX2 were found. In addition, a 12 amino acid sequence from near the N-terminus of the 17-kDa peptide having 75% identity with ENOX2 of HeLa cells after subtraction of sequences identical with hemoglobin fragments was obtained by N-terminal sequencing.

12.6 RTPCR Detection of Cancer Cells in Blood Based on Presence of ENOX2 Splice Variant mRNA

Primers were developed to obtain a 150-bp RTPCR product indicative of ENOX2 splice variant mRNA in the blood of cancer patients (Morré et al. 2005; 2006b). Venous blood was collected into 7-mL vacutainers containing potassium EDTA, snap frozen and stored at −80 °C until RNA extraction. Total RNA from whole peripheral blood was extracted using a Fast Lane Cell cDNA kit (Qiagen). Purity and amounts of total RNA were assessed by electrophoresis in Tris-Acetate-EDTA buffer on 1 % agarose gels, with the DNA bands visualized by ethidium bromide staining. A total of 40 patients with histologic diagnosis of breast, lung, or ovarian cancer and 20 healthy volunteers were evaluated. Splice variant-specific ENOX2 primers that span the boundary of exon 3 and exon 5 were designed to differentiate ENOX2 exon 4 minus RNA from full length ENOX2. The primers sets used for PCR were:

E3/5 Forward	CCA AAT CCA AGT TAC CGC ATT CGC	Nucleotide Numbers 353–362, 570–583
E3/5Reverse	AGC ATA CGC TGT TTA CAC TCC CAC	Nucleotide Numbers 661–684

Each RTPCR analysis included a positive control (ENOX2-positive cell line or blood donor) and blood from a negative donor. Under the assay conditions developed, 70 % of patients with cancer were positive for ENOX2 splice variant mRNA. Samples from individuals free of disease or with disorders other than cancer were negative. The relatively high incidence of false negatives indicated that the RTPCR approach was much less sensitive than the two-dimensional gel-western blot protocol and was abandoned as having uncertain utility as an aid to cancer management.

12.7 Summary

Early cancer detection based on serum analysis offers new opportunities for cancer prevention through early curative cancer intervention. New target molecules are provided by members of a family of cancer- and tissue-specific cell surface ENOX2 proteins essential to the growth and survival of cancer cells that are expressed on the surfaces of malignancies and detectable in the sera of patients with cancer as cancer-specific biomarkers. The proteins provide sought-after early detection along with new opportunities for targeted preventive intervention for persons at risk for cancer well in advance of development of clinical symptoms. The combination would be expected to significantly reduce cancer mortality.

The concept of using specific isoforms of ENOX2 as markers to detect cancer is based on the discovery of a family of cell surface ECTO-NOX or ENOX proteins, one subset of which, the tumor-associated ENOX protein (designated ENOX2 or tNOX), is specific to the surface of cancer cells and is absent from noncancer cells and tissues. The cancer-specific ENOX2 protein isoforms are shed into the circulation and have the potential to serve as definitive, noninvasive, and sensitive serum markers for early detection of both primary and recurrent cancer in at-risk populations with a low incidence of false positives. Each transcript variant is characteristic of a particular form of cancer represented by unique patterns of ENOX2 proteins of largely nonoverlapping molecular weights and isoelectric points. Analysis is by 2DGE with detection using a panENOX2 recombinant scFv antibody. Unlike most other cancer markers which are simply present in elevated levels in sera of cancer patients, the ENOX2 transcript variants are specific molecular markers produced only by cancer cells, thus largely eliminating false positives. Not only do the ENOX2 transcript variants revealed by 2-dimensional gel electrophoresis-western blot universally detect cancer and identify the tissue of origin but appear to signal cancer presence very early, possibly 5–7 years in advance of clinical symptoms. Because ENOX2 proteins are required for uncontrolled growth and invasion of cancer cells, the ENOX2 protein also provides a rational target for early intervention when only a small number of cancer cells are present perhaps in their most vulnerable stage of development. The strategy of early detection combined with early intervention raises the possibility of a significant paradigm shift in cancer management toward early diagnosis and treatment options vastly different from those currently employed to deal primarily with advanced cancer.

Epilogue—Remaining Challenges

As an epilogue, some of the challenges relating to ENOX proteins that remain are enumerated and addressed. Of paramount importance are considerations relating to what might be required to draw wider attention to an area as potentially important as ENOX proteins than that received in the more than two decades since their discovery.

A major consideration is that ENOX proteins have never been and continue not to be investigator friendly. They are intractable to standard methods of protein analysis, irreversibly aggregate when purified, are constantly undergoing changes in conformation leading to periods of activity interspersed with little or no activity, a behavior, along with the propensity to form aggregates that precludes conventional X-ray or solution NMR structural analyses and, along with a relatively low turnover number of 200–500 and a corresponding low specific activity of 20–50 μmol/min/mg protein make routine analysis nearly impossible except for the most intrepid of investigators. There has been nothing about ENOX proteins that has followed traditional rules of chemistry or biology. Each unusual property has faced a need for an equally unusual solution. Investigations related to the infective agents of prion disease have experienced and continue to experience similar difficulties.

One hope is that greater scientific acceptance of ENOX protein existence and potential importance to cell growth and biological time keeping, for example, and/or as a target for cancer therapy and diagnosis or in the treatment of age-related diseases will follow in the wake of eventual clinical or practical agricultural applications. Might ENOX1 overexpression significantly enhance crop yields? In concert with early detection might ENOX2 inhibitors used prophylactically lower cancer risk and reduce cancer mortality? Will arNOX inhibitors impact aging or reduce coronary artery disease? Affirmative answers to any of the above questions might be expected to significantly broaden investigative interest in this unusual but potentially important family of proteins.

D.J. Morré and D.M. Morré, *ECTO-NOX Proteins: Growth, Cancer, and Aging*,
DOI 10.1007/978-1-4614-3958-5, © Springer Science+Business Media New York 2013

E.1 Need for a Facile and Sensitive Assay for ENOX Activity

One of the most important developments that would aid in ENOX advancement would be a sensitive assay for ENOX activities not requiring specialized spectrophotometers (SLM-2000) and amenable to automated kinetic analyses in a 96 well plate format. For many applications, rates of ENOX activity are low (≤1 nmol/min) and require very sensitive instrumentation capable of measuring small changes in optical density of often turbid preparations (Chap. 2).

E.2 How to Measure ENOX2 Expression

As ENOX2 proteins represent a group of splice and transcript variants from a single gene (Chap. 8), mRNA levels do not necessarily signal functional levels of ENOX2 expression. Full length ENOX2 mRNA is translated but not transcribed. It would be necessary to measure levels of exon 4 minus tNOX mRNA as this is the template for functional expression of ENOX2 at the cell surface and for the origin of cancer site-specific transcript variants (Chap. 12). RTPCR estimates of ENOX2 splice variant mRNA were attempted but technically difficult due to the small number of bases that define the splicing site (Morré et al. 2006b).

Quantitation of ENOX2 amounts present in the plasma membrane might serve as an indirect measure of ENOX2 expression. ENOX2 is not a particularly abundant plasma membrane protein and quantitation based either on measurement of drug- (capsaicin) inhibited enzymatic activity or by western blots of SDS-PAGE are approximations at best and also difficult technically.

E.3 What Triggers ENOX2 Splice Variant Expression? Can It Be Prevented?

Tang et al. (2007) provided evidence demonstrating that ENOX2 arises as a result of alternative splicing. Full-length ENOX2 mRNA was present in both normal and cancer cells but did not appear to be expressed in either (Sect. 8.8). Two splice variants, an exon 4 minus and an exon 5 minus were present in cancer cells but not in normal cells. Transfection studies of COS cells with ENOX2 exon 4 minus cDNA led to an overexpression of a 34 kDa ENOX2 protein at the plasma membrane as did exon 4 minus cDNA overexpression in transgenic mice (Yagiz et al. 2006). This 34 kDa protein corresponded to the fully processed form of ENOX2 commonly found at the cell surface (Cho and Morré 2009; Cho et al. 2002). The conclusion that the cancer-cell-specific expression of 34 kDa ENOX2 was due to formation of an exon 4 minus splice variant was supported by these findings.

To determine if silencing of ENOX2 exon 4 was the result of motifs located in exon 4, transfections were performed on MCF-10A (mammary non-cancer), BT-20 (mammary cancer), and HeLa (cervical cancer) cells using a GFP minigene construct containing either a constitutively spliced exon (albumin exon 2) or the alternatively spliced ENOX2 exon 4 between the two GFP halves (Tang et al. 2011). Removal of exon 4 from the processed RNA of the GFP minigene construct occurred with both HeLa and BT-20 but not in non-cancer MCF-10A cells. The Splicing Rainbow Program identified eight Exonic Splicing Silencers (ESSs) for hnRNP binding in the exon 4 sequences of ENOX2. Each of these sites was mutated by site-directed mutagenesis to determine which, if any, might be responsible for the splicing skip. Results showed MutG75 ESS mutation changed the GFP expression to indicate splicing silence, while other mutations did not (Tang et al. 2011) to suggest that hnRNP F directs formation of the exon 4 minus variant of ENOX2 (Sect. 8.9). hnRNP H is upregulated 1.6 times in SV-40-transformed human keratinocytes whereas hnRNP F remained unchanged (Honoré et al. 1995).

The identification of the exon 4 minus splicing factor now may provide an approach to prevention of ENOX2 expression. Knockdown of specific splicing factors may be sufficient to reverse transformation caused by their overexpression (Karni et al. 2007). However such a relationship between hnRNP F and the exon 4 minus splicing event with ENOX2 remains to be investigated.

E.4 Multiple Transcript Variants

Two dimensional gel electrophoresis (2DGE) and western blot analysis have revealed >10 different ENOX2 transcript variants (Chap. 12). They have common origins based on shared immunological determinants. The molecular basis for their origins remains unresolved and specific quantitation has proven difficult.

In contrast to ENOX2, multiple arNOX family members are all transcribed from different genes even though they exhibit functional motifs in common.

Currently, there is no evidence for multiple ENOX1 isoforms in man. Yeast appears to have at least two ENOX1 isoforms that do not cross entrain and only one of which is inhibited by the ENOX1 inhibitor simalikalactone D. A peptide antibody to ENOX1 has been produced but no simple method of quantitation exists.

E.4.1 ENOX1 Knockout Rat

Generation of 20 ENOX1 knockout rat has been proposed (Transposon Pharmaceuticals, Lexington, KY) where germ cells would be transfected by means of a piggy back transposon with 18 kb siRNA and then transferred to a sterile male.

E.4.2 Idioforms: Evidence from Chemosensitization

Mixtures of phenoxodiol, an anticancer isoflavone, targeted to ENOX2 (Morré et al. 2007a) with a variety of standard anticancer drugs such as *cis*-platinum and docetaxel increased the sensitivity of cancer cells to standard drugs (Sect. 11.4.8.1; Kelly 2010). More important phenoxodiol restored drug sensitivity to drug resistant cancer cells. It was sufficient to pretreat the patients with taxol- or docetaxel-resistant cancer with phenoxodiol to restore taxol or docetaxel susceptibility which was retained following cessation of the phenoxodiol treatment.

The chemosensitization response appears to be mediated through the primary drug target of phenoxodiol, ENOX2 (Morré et al. 2008a, 2009b). Parallel responses are observed between effects on ENOX2 activity and on cell growth.

One view of the mechanism of chemosensitization is that of learning and teaching based on the prion model where conversion of ENOX2 idioforms occurs by post-translational processes involving protein–protein interactions not encoded in the genome. For cisplatin or paclitaxel resistance, the ENOX2 learns not to respond to cisplatin or paclitaxel possibly through a conformational change and the altered ENOX2 proteins impart paclitaxel or cisplatin resistance to other ENOX2 molecules. To overcome the resistance, a drug-susceptible conformation must be restored to the ENOX population. Phenoxodiol seems to have that capability. Chemosensitization is achieved as well by merely adding back ENOX2 proteins that have been exposed to phenoxodiol (Morré et al. 2009). In keeping with the prion model, it appears as though the ENOX2 "remembers" its encounter with the phenoxodiol and "teaches" other ENOX2 molecules to respond to taxanes and to *cis*-platinum as if the phenoxodiol were still present. A similar phenomenon was observed in plants with the synthetic auxin herbicide, 2,4-D (Chap. 10).

E.5 Are Therapeutic ENOX2 Antibodies Possible?

A monoclonal antibody, MAB 12.1, selectively inhibited NADH oxidase activity in sera taken from cancer patients. The same ENOX2 activity was inhibited by the recognized ENOX2 inhibitor capsaicin. Upon selective binding of MAB 12.1 to surface membranes of human carcinoma cells and tissues, cell growth was slowed and apoptosis was induced whereas this monoclonal antibody did not affect the growth of noncancerous cell lines (Sect. 11.7; Cho et al. 2002). Subsequently, survival of transformed cells was inhibited by antisera raised against bacterially-expressed ENOX2 protein (Chen et al. 2006). Again, non-transformed cells were not affected.

Why then has there been little or no progress toward a therapeutic antibody? A major complication seems to be that patients generate circulating IgM autoantibodies to their ENOX2 transcription variants that block the action of ENOX2 antibodies deployed therapeutically. However, the opportunity for a vaccine remains (see Chap. 11).

E.6 Why Do ENOX Proteins Aggregate? What Is the Significance of ENOX Aggregation?

Purified ENOX proteins, when concentrated, aggregate as enzymatically inactive amyloid fibers and rods. (Sects. 1.10.5; 3.7; 8.5.4). Restoration of activity with guanidinium, thiocyanate, urea or other conventional renaturation protocols have been unsuccessful. Resolubilization thus far has been restricted to isoelectric focusing.

Since ENOX functions are inextricably tied to periodic behavior, all ENOX proteins must autoentrain. Autoentrainment requires that the entraining partners interact specifically and with high affinity for each other. This is exemplified by the observation that even at very low protein dilutions, the prevailing ENOX form is that of a dimer with tetramers and multimers prevailing as the concentration of protein is increased. Thus, it would seem reasonable to conjecture that the propensity for ENOX proteins to form irreversible aggregates derives in some manner from their functional need to find each other in dilute solutions mixed with other proteins in order to autoentrain.

Experimentally entrainment has been shown to be very much sequence dependent. ENOX1 and ENOX2 protein which share 60% sequence identity do not cross entrain. Nor do the five known arNOX isoforms which share common functional motifs but have very different overall amino acid sequence cross entrain.

Without entrainment, more specifically, without autoentrainment, ENOX periodicity would be futile. A complete lack of synchrony would lead to an average constant rate rather than the familiar 2+3 pattern of maxima within each 22, 24 or 26 min period. External signals such as melatonin, light or electromagnetic fields contribute to entrainment of ENOX1 but not for ENOX2 or arNOX family members. Thus, the principal means of ensuring synchrony of populations of ENOX proteins in solution may be through progressive dimer formation and exchange of monomer subunits between like ENOX proteins. Attractive interactions presumably are sufficiently strong to eventually contribute to the troublesome formation of insoluble aggregates with highly purified ENOX proteins in concentrated solutions. ENOX proteins, where copper-binding motifs have been mutated, no longer oscillate but X-ray and solution NMR are still precluded due to the propensity of even the copper mutants to aggregate.

E.7 Significance of the Periodic Oxidative Burst

Why would a single group of protein molecules as observed with the arNOX superfamily be precisely synchronized to produce a single burst of superoxide once every 26 min (Fig. 9.19) and survive through evolution as a highly conserved but potentially self destructive event (Chap. 9).

As arNOX isoforms are present in endosomes, it may be that superoxide production within endosomes provide some useful function either in terms of host defense, endocytic digestive or autophagic events.

As individuals age not only does arNOX activity increase but the number of family members expressed may also increase in number and abundance. For whatever purpose, a pulsed delivery of superoxide does provide a dose of superoxide 20–30 times more concentrated than an unsynchronized production of superoxide, produced continuously, over the same 26 min time interval.

E.8 Summary

Non-infectious, growth-related and time keeping proteins capable of replicating certain functional states in the manner of a prion seems difficult to comprehend if not heretical. Participation in a chain reaction of self perpetuating conformational changes, the ENOX proteins have proven refractory to characterization by conventional X-ray crystallography or solution NMR due to their inability to form crystals and a propensity to form insoluble aggregates upon purification and concentration. As further parallels with infective prions, ENOX proteins are insoluble in detergents, have a high β-sheet to α-helix ratio, a propensity to aggregate in the absence of stabilizing partners functioning as chaperones and a resistance to heating and to proteolysis.

Although the content of β-sheet compared to α-helix determined by CD or FTIR is much higher than that predicted from the amino acid sequence, attempts to identify a chemical change (modification) associated with the post-translational conversion from the proteinase K-susceptible to the proteinase K-resistant form of ENOX2 have been unrewarding. Attempts to solubilize the aggregates other than through isoelectric focusing have not been successful nor have efforts to restore enzymatic activity to ENOX proteins once aggregated except through disaggregation by isoelectric focusing.

The ENOX2 proteins are splice variants from a single gene that exhibit a range of transcript variants as well as idioforms with potentially important diagnostic and therapeutic applications to the field of cancer.

Appendix
Detailed Description of Two Dimensional Gel Electrophoresis-Western Blot Early Cancer Detection Protocol

A.1 Collection of Serum

Serum is prepared from 5 mL of blood collected by venipuncture (with tourniquet) in standard B and D 13 × 100 (7 mL) vacutainer clot tubes (or equivalent) with or without hemoguard closure. After approximately 30 min at room temperature to allow for clotting, the clot is pelleted by centrifugation for 5–10 min at 2,500–3,000 rpm. Clot-free serum is decanted into a clean tube, labeled and analyzed fresh or stored frozen.

A.2 Two Dimensional Gel Electrophoresis and Western Blot Analysis

For western blot analysis, enriched serum proteins are concentrated for 2DGE by nickel agarose precipitation. Nickel agarose beads (Quiagen, Hilden, Germany, Mat. No. 1018244, 50 μL) are added to 0.5 mL microcentrifuge tubes with 400 μL of sera and placed on a rotary shaker overnight at 4°C. The protein enriched beads are then collected by centrifugation at 1,000 × *g* for 30 s. The supernatants are discarded and the centrifugation step is repeated three times with pellets resuspended in distilled, deionized water. The samples are then resuspended in 150 μL of 7 M urea, 2 M thiourea, 2% (w/v) CHAPS [(3-cholamidopropyl)dimethylammonio]-1-propane-sulfonate), a nondenaturing zwitterionic detergent], 0.5% (w/v) ASB-14 (amidosulfobetaine-14, a zwitterionic detergent), 0.5% (v/v) ampholytes pH 3.10 (BioRad), 0.5% (v/v) immobilized pH gradient (IPG) buffer pH 3–10 (Amersham-Pharmacia Biotech), containing 65 mM dithiothreitol after which the samples are vortexed for 1 h at room temperature. The supernatants are recovered by removing the beads by centrifugation at 1,000 × *g* for 30 s. Four to six milligram of protein are loaded for analysis. The samples are electrophoresed in the first dimension by using a commercial flatbed electrophoresis system (Ettan IPGphor 3, Amersham-Pharmacia Biotech) with IPG dry strips (Amersham). A linear pH range of 3–10 on

D.J. Morré and D.M. Morré, *ECTO-NOX Proteins: Growth, Cancer, and Aging*, DOI 10.1007/978-1-4614-3958-5, © Springer Science+Business Media New York 2013

7 cm IPG strips is used. The IPG strips are rehydrated with the samples overnight at room temperature. The strips are then focused at 50 mA/strip and at constant voltage of 300 V for 15 min, 600 V for 30 min and 1,000 V for 1 h. Finally, the strips are focused at a constant 4,000 V for 28,000 Vh. After isoelectric focusing, the IPG strips are re-equilibrated for 30 min in 2.5% (w/v) SDS, 6 M urea, 30% (v/v) glycerol, 100 mM Tris–HCl (pH 8.8). The strips are placed onto linear SDS-PAGE gels (10% (w/v) polyacrylamide) and electrophoresed at a constant 250 V for 80 min. The samples are then transferred to nitrocellulose membranes by electroblotting using the BioRad Trans-Blot Electrophoretic Transfer Cell. The membranes are blocked using milk protein (5% low fat dry milk) at room temperature for 1 h. Detection is with recombinant anti-ENOX2 single chain variable region of antibody 12.1 (scFv) carrying an S-tag overnight at 4°C followed by alkaline phosphatase-linked anti-S (Novagen, cat. #69598-3) and, after washing, detection with Western Blue nitrotetrazolium (NBT) substrate (Promega, Madison, WI, cat. No. S3841) for 5–10 min at 4°C. Images are scanned and processed using Adobe Photoshop. Quantitation utilized an algorism developed for this purpose. Reactive proteins appear reddish blue. For interpretative purposes, the blots are divided into quadrants I–IV with unreactive serum albumin at the center (Fig. 12.2).

References

Abe M, Herzog ED, Block GD (2000) Lithium lengthens the circadian period of individual suprachiasmatic nucleus neurons. Neuroreport 11:3261–3264

Abell BA, Brown DT (1993) Sindbis virus membrane fusion is mediated by reduction of glycoprotein disulfide bridges at the cell surface. J Virol 67:5496–5501

Åberg F, Appelkvist FL, Dallner G, Ernster L (1992) Distribution and redox state of ubiquinones in rat and human tissues. Arch Biochem Biophys 295:230–234

Adam L, Crepin M, Savin C, Israel L (1995) Sodium phenylacetate induces growth inhibition and Bcl-2 down-regulation and apoptosis in MCF7ras cells in vitro and in nude mice. Cancer Res 55:5156–5160

Adler R, Hurwitz E, Wands JR, Sela M, Shoural D (1995) Specific targeting of adriamycin conjugates with monoclonal antibodies to hepatoma associated antigens to intrahepatic tumors in athymic mice. Hepatology 22:1482–1487

Aeschbach R, Amadoò R, Neukom H (1976) Formation of dityrosine cross-links in proteins by oxidation of tyrosine residues. Biochim Biophys Acta 439:292–301

Agarwal R (2000) Cell signaling and regulators of cell cycle as molecular targets for prostate cancer prevention by dietary agents. Biochem Pharmacol 60:1051–1059

Aguero MF, Facchinetti MM, Sheleg Z, Senderowica AM (2005) Phenoxodiol, a novel isoflavene, induces G_1 arrest by specific loss in cyclin-dependent kinase 2 activity by p53-independent induction of p21WAF1/CIP1. Cancer Res 65:3364–3373

Ahammadsahib KI, Hollingworth RM, McGovren JP, Hui Y-H, McLaughlin JL (1993) Mode of action of bullatacin: a potent antitumor and pesticidal annonaceous acetogenin. Life Sci 53: 1113–1120

Ahmad N, Feyes DK, Nieminen AL, Agarwal R, Mukhtar HJ (1997) Green tea constituent epigallocatechin-3-gallate and induction of apoptosis and cell cycle arrest in human carcinoma cells. J Natl Cancer Inst 89:1881–1886

Alcaín FJ, Buron MI, Villalba JM, Navas P (1991) Ascorbate is regenerated by HL-60 cells through the transplasmalemma redox system. Biochim Biophys Acta 1073:380–385

Alcaín FJ, Villalba JM, Löw H, Crane FL, Navas P (1992) Ceruloplasmin stimulates NADH oxidation of pig liver plasma membrane. Biochem Biophys Res Commun 186:951–955

Alcaín FJ, Löw H, Crane FL (1994) Iron reverses impermeable chelator inhibition of DNA synthesis in CCI 39 cells. Proc Natl Acad Sci U S A 91:7903–7906

Alessai DR, Sakamoto K, Bayascas JR (2006) Lkb1-dependent signaling pathways. Annu Rev Biochem 75:137–163

Alonso MM, Encio I, Martinez-Merion V, Gil M, Migliaccio M (2001) New cytotoxic benzo(b) thiophenilsulfonamide 1,1-dioxide derivatives inhibit a NADH oxidase located in plasma membranes of tumour cells. Br J Cancer 85:1400–1402

D.J. Morré and D.M. Morré, *ECTO-NOX Proteins:Growth, Cancer, and Aging*,
DOI 10.1007/978-1-4614-3958-5, © Springer Science+Business Media New York 2013

Altekruse SF, Kosary CL, Krapcho M, Neyman N, Aminou R, Waldron W, Ruhl J, Howlader N, Tatalovich Z, Cho H, Mariott A, Eisner MP, Lewis DR, Cronin K, Chen HS, Feuer EJ, Stinchcomb DG, Edwards BK (eds) (2010) SEER cancer statistics review, 1975–2007. National Cancer Institute, Bethesda, MD

Alvero AB, O'Malley D, Brown D, Kelly G, Garg M, Chen W, Rutherford T, Mor G (2006) Molecular mechanism of phenoxodiol-induced apoptosis in ovarian carcinoma cells. Cancer 106:599–608

Alvero AB, Kelly M, Rossi P (2008) Anti-tumor activity of phenoxodiol: from bench to clinic. Future Oncol 4:475–482

Ambrose EJ, Rose FJC (1975) Biology of cancer. Ellis Horwood, Chichester, 315pp

American Heart Association Statistical Update (2008) Heart disease and stroke statistics—2008. Circulation 117:e25–e146

Andavan GS, Lemmens-Gruber R (2010) Cyclodepsipeptides from marine sponges: natural agents for drug research. Mar Drugs 8:810–834

Andrade RP, Palmeirim I, Bajanca F (2007) Molecular clocks underlying vertebrate embryo segmentation: a 10-year-old hairy-go-round. Birth Defects Res C Embryo Today 81:65–83

Antosiewicz J, Ziolkowski W, Kar S, Powolny AA, Singh SV (2008) Role of reactive oxygen intermediates in cellular responses to dietary cancer chemopreventive agents. Planta Med 74:1570–1579

Arcamone F (1985) Properties of antitumor anthracyclines and new developments in their application: Cain memorial award lecture. Cancer Res 45:5995–5999

Asard H, Caubergs R, Renders D, DeGreef JA (1987) Duroquinone-stimulated NADH oxidase and b-type cytochromes in the plasma membrane of cauliflower inflorescences. Plant Sci 53:109–119

Asard H, Horemans N, Preger V, Trost P (1998) Plasma membrane b-type cytochromes. In: Asard H, Bérczi A, Caubergs RJ (eds) Plasma redox systems and their role in biological stress and disease. Kluwer Academic, Dordrecht, pp 1–31

Atkinson M, Kripke DF, Wolf SR (1975) Autohythmometry in manic-depressives. Chronobiologia 2:325–335

Auderset G, Morré DJ (2006) ATP- and growth substance-dependent cell-free enlargement of plasma membrane vesicles from soybean. Biofactors 28:83–90

Auderset G, Morré DJ, Williamson FA, Hess K, Greppin H (1980) Effect of light on auxin binding, cell fractionation, and ultrastructure of etiolated soybean hypocotyls. Bot Gaz 141:149–156

Axanova L, Morré DJ, Morré DM (2005) Growth of LNCaP cells in monoculture and coculture with osteoblasts and response to tNOX inhibitors. Cancer Lett 225:35–40

Azodi M, Kelly M, Rutherford T, Schwarz P, Baker L, Mor G, Kelly G (2005) Phase Ib study of oral phenoxodiol as neo-adjuvant therapy in squamous cell carcinoma of the cervix, vagina or vulva. In: Abstracts. AACR-NCI-EDRTC conference on molecular targets and cancer therapeutics, Nov 2005, p 221

Bacon E, Morré DJ (2001) Plasma membrane NADH oxidase of maize roots responds to gravity and imposed centrifugal forces. Plant Physiol Biochem 39:487–494

Baker MA, Lawen A (2000) Plasma membrane NADH-oxidoreductase system: a critical review of the structural and functional data. Antioxid Redox Signal 2:197–212

Baker MA, Ly JD, Lawen A (2004a) Characterization of VDAC1 as a plasma membrane NADH-oxidoreductase. Biofactors 21:215–221

Baker MA, Lane DJR, Ly JD, De Pinto V, Lawen A (2004b) VDAC1 is a transplasma membrane NADH-ferricyanide reductase. J Biol Chem 279:4811–4819

Balch WE, Keller DS (1986) ATP-coupled transport of vesicular stomatitis virus G protein. Functional boundaries of secretory compartments. J Biol Chem 26:14690–16496

Balch WF, Wagner KR, Keller DS (1987) Reconstitution of transport of vesicular stomatitis virus G protein from endoplasmic reticulum to the Golgi complex using a cell-free system. J Cell Biol 104:749–760

Baldwin MA, Cohen FE, Prusiner SB (1995) Prion protein isoforms, a convergence of biological and structural investigations. J Biol Chem 270:19197–19200

Barabas K, Sizensky JA, Faulk WP (1991) Evidence in support of the plasma membrane as the target for transferrin-adriamycin conjugates in K562 cells. Am J Reprod Immunol 25:120–123

Barabas K, Sizensky JA, Faulk WP (1992) Transferrin conjugates of adriamycin are cytotoxic without intercalating nuclear DNA. J Biol Chem 267:9437–9442

Barbouche R, Miqauelis R, Jones IM, Fenouillet E (2003) Protein-disulfide isomerase-mediated reduction of two disulfide bonds of HIV envelope glycoprotein 120 occurs post-CXCR4 binding and is required for fusion. J Biol Chem 278:3131–3136

Barr R, Sandelius AS, Crane FL, Morré DJ (1985) Oxidation of reduced pyridine nucleotides by plasma membranes of soybean hypocotyl. Biochem Biophys Res Commun 131:943–948

Barr R, Sandelius AS, Crane FL, Morré DJ (1986) Redox reactions of tonoplast and plasma membranes isolated from soybean hypocotyls by free-flow electrophoresis. Biochim Biophys Acta 852:254–261

Barr R, Böttger M, Crane FL, Morré DJ (1990) Electron donation to the plasma membrane redox system of cultured carrot cells stimulates proton release. Biochim Biophys Acta 1017:91–95

Barr R, Garcia C, Morré DJ (2000a) Touch-sensitive NADH oxidase activity of pea and cucumber tendrils and of soybean hypocotyl sections. Int J Plant Sci 161:387–391

Barr R, Morré DJ, Crane FL (2000b) Oxidation of NADH by hypocotyl segments of soybean is stimulated by 2,4-D. Plant Physiol Biochem 38:739–745

Barr R, Penel C, Greppin H, Morré DJ (2000c) NADH oxidase activities of intact leaf discs of spinach. Arch Sci (Geneva) 53:225–232

Baserga R (1985) The biology of cell reproduction. Harvard University Press, Cambridge

Baserga R (2007) Is cell size important? Cell Cycle 6:814–816

Bassenge E, Sommer O, Schwemmer M, Bünger R (2000) Antioxidant pyruvate inhibits cardiac formation of reactive oxygen species through changes in redox state. Am J Physiol Heart Circ Physiol 279:H2431–H2438

Batt S, Venis MA (1976) Separation and localization of two classes of auxin binding sites in corm coleoptile membranes. Planta 130:15–21

Batt S, Wilkins MB, Venis MA (1976) Auxin binding to corn coleoptile membranes: kinetics and specificity. Planta 130:7–13

Battini J-L, Danos O, Heard JM (1995) Receptor-binding domain of murine leukemia virus envelope glycoproteins. J Virol 69:713–719

Baudhuin LM, Cristina KL, Lu J, Xu Y (2002) Akt activation induced by lysophosphatidic acid and sphingosine-1-phosphate requires both mitogen-activated protein kinase kinase and p38 mitogen-activated protein kinase and is cell-line specific. Mol Pharmacol 62:660–671

Baulcombe DC, Kramer PA, Key JL (1981) Auxin and gene regulation. In: Sublenlny S, Abbott UK (eds) Levels of genetic control in development. Alan R. Liss, New York, p 83

Bearz A, Talamini R, Vaccher E, Spina M, Simonelli C, Steffan A, Berretta M, Chimienti E, Tirelli U (2007) MUC-1 (CA 15-3 antigen) as a highly reliable predictor of response to EGFR inhibitors in patients with bronchioloalveolar carcinoma: an experience on 26 patients. Int J Biol Markers 22:307–311

Beckers CJM, Keller DD, Balch WE (1987) Semi-intact cells permeable to macromolecules: use in reconstitution of protein transport from the endoplasmic reticulum to the Golgi complex. Cell 50:523–534

Beckers CJM, Plutner H, Davidson HW, Balch WE (1990) Sequential intermediates in the transport of protein between the endoplasmic reticulum and the Golgi. J Biol Chem 263:18298–18310

Bell-Pedersen D, Cassone VM, Earnest DJ, Golden SS, Hardin PE, Thomas TL, Zoran MJ (2005) Circadian rhythms from multiple oscillators: lessons from diverse organisms. Nat Rev Genet 6:544–556

Benedetti F, Barbini B, Campori E, Fulgosi MC, Pontiggia A, Colombo C (2001) Sleep phase advance and lithium to sustain the antidepressant effect of total sleep deprivation in bipolar depression: new findings supporting the internal coincidence model? J Psychiatr Res 35:323–329

Bérczi A, Sizensky J, Crane FL, Faulk WP (1991) Diferric transferrin reduction by K562 cells: a critical study. Biochim Biophys Acta 1073:562–570

Bérczi A, Barabas K, Sizensky JA, Faulk WP (1993) Adriamycin conjugates of human transferrin bind transferrin receptors and kill K562 and HL60 cells. Arch Biochem Biophys 300:356–363

Bérczi A, Van Gestelen P, Pupillo P (1998) NAD(P)H-utilizing flavor-enzymes in the plant plasma membrane. In: Asard H, Bérczi A, Caubergs RJ (eds) Plasma membrane redox systems and their role in biological stress and disease. Kluwer Academic, Dordrecht, pp 33–67

Bernstein KA, Bleichert F, Bean JM, Cross FR (2007) Ribosome biogenesis is sensed at the start cell cycle checkpoint. Mol Biol Cell 18:953–964

Berridge MV, Tan AS (2000) High-capacity redox control at the plasma membrane of mammalian cells: trans-membrane, cell surface, and serum NADH-oxidases. Antioxid Redox Signal 2:231–242

Bersuker IB (1984) Modern chemistry. Plenum Press, New York

Bertucci F, Goncalves A (2008) Clinical proteomics and breast cancer: strategies for diagnostic and therapeutic biomarker discovery. Future Oncol 4:271–287

Binhi VN (2002) Magnetobiology: underlying physical problems. Academic, San Diego

Binhi VN, Stepanov EV (2000) Tunable diode-laser spectroscopy of the para- and ortho-water vapour as a tool for investigation of metastable states of liquid water. In: Kostarakis P, Stavrolakis P (eds) Millennium international workshop on biological effects of electromagnetic fields, Heraklion, Crete, Greece, 17–20 Oct, pp 153–154

Birkmayer JG (1996) Coenzyme nicotinamide adenine dinucleotide: new therapeutic approach for improving dementia of the Alzheimer type. Ann Clin Lab Sci 26:1–9

Birkmayer JG, Vrecko C, Vole D, Birkmayer W (1993) Nicotinamide adenine dinucleotide (NADH)—a new therapeutic approach to Parkinson's disease. Comparison of oral and parenteral application. Acta Neurol Scand Suppl 146:32–35

Bishop NA, Guarente L (2007) Genetic links between diet and lifespan: shared mechanisms from yeast to humans. Nat Rev Genet 8:835–844

Bjelakovic G, Nikolova D, Gluud LL, Simonetti RG, Gluud C (2007) Mortality in randomized trials of antioxidant supplements for primary and secondary prevention. Systematic review and meta-analysis. JAMA 297:842–857

Blumberg J (2004) Use of biomarkers of oxidative stress in research studies. J Nutr 134:1188s–1189s

Bode AM, Dong Z (2003) Signal transduction pathways: targets for green and black tea polyphenols. J Biochem Mol Biol 36:66–77

Bodnar AG (2009) Marine invertebrates as models for aging research. Exp Gerontol 44:477–484

Böhm G, Muhr R, Jaenicke R (1992) Quantitative analysis of protein far UV circular dichroism spectra by neural networks. Protein Eng 5:191–195

Bohring C, Krause E, Habermann B, Krause W (2001) Isolation and identification of sperm membrane antigens recognized by antisperm antibodies, and their possible role in immunological infertility disease. Mol Hum Reprod 7:113–118

Bordone L, Guarente L (2005) Calorie restriction, SIRT1 and metabolism: understanding longevity. Nat Rev Mol Cell Biol 6:298–305

Bosneaga E, Kim C, Shen B, Watanabe T, Morré DM, Morré DJ (2009) ECTO-NOX (ENOX) proteins of the cell surface lack thioredoxin reductase activity. Biofactors 34:245–251

Bosserhoff A, Buettner R (2002) Cancer invasion-related genes. In: Heino J, Kahari V-M (eds) Cell invasion. Eurekah Press, Austin, TX

Böttger M, Lüthen H (1986) Possible linkage between NADH oxidation and proton secretion in *Zea mays* L. roots. J Exp Bot 37:666

Bowling AC, Beal MF (1995) Bioenergetic and oxidative stress in neurodegenerative diseases. Life Sci 56:1151–1171

Braga VM, Gendler SJ (1993) Modulation of Muc-1 mucin expression in the mouse uterus during the estrus cycle, early pregnancy and placentation. J Cell Sci 105:397–405

Brar SS, Kennedy TP, Whorton AR, Sturrock AB, Huecksteadt TP, Ghio AJ, Hoidal JR (2001) Reactive oxygen species from NAD(P)H:quinone oxidoreductases constitutively activate NF-kappaB in malignant melanoma cells. Am J Physiol Cell Physiol 280:C659–C676

Bridge A, Barr R, Morré DJ (2000) The plasma membrane NADH oxidase of soybean has vitamin K_1 hydroquinone oxidase activity. Biochim Biophys Acta 1463:448–458

Brightman AO, Morré DJ (1991) NADH oxidase of the plasma membrane of plants. In: Crane FL, Morré DJ, Löw H (eds) Oxidoreduction at the plasma membrane: relation to growth and transport. II. Plants. CRC Press, Boca Raton, FL, pp 85–110

Brightman AO, Barr R, Crane FL, Morré DJ (1988) Auxin-stimulated NADH oxidase purified from plasma membrane of soybean. Plant Physiol 86:1264–1269

Brightman AO, Zhu XZ, Morré DJ (1991) Activation of plasma membrane NADH oxidase activity by products of phospholipase A. Plant Physiol 96:1314–1320

Brightman AO, Wang J, Miu RK, Sun IL, Barr R, Crane FL, Morré DJ (1992) A growth factor and hormone-stimulated NADH oxidase from rat liver plasma membranes. Biochem Biophys Acta 1105:109–117

Brown FA Jr (1977) Geographic orientation, time and mudmail phototaxis. Biol Bull 152: 311–324

Brown FA Jr, Chow CK (1973) Lunar correlated variations in water uptake by bean seeds. Biol Bull 145:265–278

Brown JP, Woodbury RG, Hart CE, Hellström I, Hellström KE (1981) Quantitative analysis of melanoma-associated antigen p97 in normal and neoplastic tissues. Proc Natl Acad Sci U S A 78: 539–543

Brown DM, Kelly GE, Husband AJ (2005) Flavanoid compounds in maintenance of prostate health and prevention and treatment of cancer. Mol Biotechnol 30:253–270

Bruce VG, Pittendrigh CS (1960) Temperature independence in a unicellular "clock". J Cell Comp Physiol 56:25–31

Brummer P, Parish RW (1983) Mechanisms of auxin-induced plant cell elongation. Fed Eur Biochem Soc Lett 161:9–13

Bruno M, Brightman AO, Lawrence J, Werderitsh D, Morré DM, Morré DJ (1992) Stimulation of NADH oxidase activity from rat liver plasma membranes by growth factors and hormones is decreased or absent with hepatoma plasma membrane. Biochem J 284:625–628

Bucher J, Tien M, Morahouse L, Aust S (1983) Redox cycling and lipid peroxidation: the central role of iron chelates. Fundam Appl Toxicol 3:222–226

Buckhout TJ, Scherer GFE, Morré DJ (1978) An *in vitro*, auxin-stimulated, and ascorbate-dependent absorbance change in soybean endomembranes. In: Abdel Rahman M (ed) Proceedings of the fifth annual meeting, Plant Growth Regulator Working Group, Blacksburg, VA, 25–28 June 1978. Great Western Sugar Co., Longmont, CO, pp 266–271

Buckhout TJ, Young KA, Löw PS, Morré DJ (1980) Response of isolated plant membranes to auxins: calcium release. Bot Gaz 141:418–421

Buckhout TJ, Young KA, Löw PS, Morré DJ (1981) *In vitro* promotion by auxins of divalent ion release from soybean membranes. Plant Physiol 68:512–515

Buettner R, Papoutsoglou G, Scemes E, Spray DC, Dermietzel R (2000) Evidence for secretory pathway localization of a voltage-dependent anion channel isoform. Proc Natl Acad Sci U S A 97:3201–3206

Bulliard C, Zurbriggen R, Tornare J, Faty M, Dastoor Z, Dreyer JL (1997) Purification of a dichlorophenol-indophenol oxidoreductase from rat and bovine synaptic membranes: tight complex association of a glyceraldehyde-3-phosphate dehydrogenase isoform, TOAD64, enolase-gamma and aldolase C. Biochem J 324:555–563

Bunning E, Baltes J (1963) Zur Wirkung von schwerem Wasser au die endogene Tagersrhythmik. Naturwissenschaften 50:622

Buntkowsky G, Walaszek B, Adamczyk A, Xu Y, Limbach H-H, Chaudret B (2006) Mechanism of nuclear spin initiated para-H2 to ortho-H2 conversion. Phys Chem Chem Phys 8:1929–1935

Burgos ES, Schramm VL (2008) Weak coupling of ATP hydrolysis to the chemical equilibrium of human nicotinamide phosphoriboxyltransferase. Biochemistry 47:11086–11096

Burnstock G (2007) Physiology and pathophysiology of purinergic neurotransmission. Physiol Rev 87:659–797

Burwick NR, Wahl ML, Fang J, Zhong Z, Mosert TL, Lit B, Capaldi RA, Kenan DJ, Pizzo SV (2005) An inhibitor of the F1 subunit of ATP synthase (IF1) modulates the activity of angiostatin on the endothelial cell surface. J Biol Chem 280:1740–1745

Butler J, Koppenol WH, Margoliash E (1982) Kinetics and mechanism of the reduction of ferricytochrome c by the superoxide anion. J Biol Chem 257:10747–10750

Byus CV, Kartun K, Pieper S, Adey WR (1988) Increased ornithine decarboxylase activity in cultured cells exposed to low energy modulated microwave fields and phorbol ester tumor promoters. Cancer Res 48:4222–4226

Cabot MC, Yu JY, Kelly GE, Brown DM, Lucas KM, Tanabe K, Allen JD (2005) Phenoxodiol, a synthetic analog of genistein, generates ceramide and is equipotent in wild-type and multidrug-resistant human tumour cells. J Clin Oncol 23(16S):2075

Canelas AB, van Gulik WM, Heijnen JJ (2008) Determination of the cytosolic free NAD/NADH ratio in *Saccharomyces cerevisiae* under steady-state and highly dynamic conditions. Biotechnol Bioeng 100:734–743

Canto C, Auwerx J (2009) Caloric restriction, SIRT1 and longevity. Trends Endocrinol Metab 20:325–331

Canto C, Gerhart-Hines Z, Feige JN, Lagouge M, Noriega L, Milne JC, Elliott PJ, Puigserver P, Auwerx J (2009) AMPK regulates energy expenditure by modulating NAD^+ metabolism and SIRT1 activity. Nature 458:1056–1060

Canut H, Brightman A, Boudet AM, Morré DJ (1988) Plasma membrane vesicles of opposite sidedness from soybean hypocotyls by preparative free-flow electrophoresis. Plant Physiol 86:631–637

Cao K, Nakajima R, Meyer HH, Zheng Y (2003) The AAA-ATPase Cdc48/p97 regulates spindle disassembly at the end of mitosis. Cell 115:355–367

Cárdenas L, McKenna ST, Kunkel JG, Hepler PK (2006) NAD(P)H oscillates in pollen tubes and is correlated with tip growth. Plant Physiol 142:1460–1468

Carney DN, Gazdar AF, Minna JD (1980) Positive correlation between histological tumor involvement and generation of tumor cell colonies in agarose in specimens taken directly from patients with small-cell carcinoma of the lung. Cancer Res 40:1820–1823

Carpinteiro A, Dumitru D, Schenck M, Gulbins E (2008) Ceramide-induced cell death in malignant cells. Cancer Lett 264:1–10

Castagne V, Gautschi M, Lefevre K, Posada A, Clarke PG (1999) Relationships between neuronal death and the cellular redox status. Focus on the developing nervous system. Prog Neurobiol 59:397–423

Cavalli LK, Liang BC (1998) Mutagenesis, tumorigenicity, and apoptosis: are the mitochondria involved? Mutat Res 398:19–26

Chakkalakal DA, Mollner TJ, Bogard MR, Fritz ED, Novak JR, McGuire MH (1999) Magnetic field induced inhibition of human osteosarcoma cells treated with adriamycin. Cancer Biochem Biophys 17:89–98

Chalko CJ, Morré DM, Morré DJ (2000) Cell surface NADH oxidase activity of brine shrimp oscillates with a period of 25 min and is entrained by light. Life Sci 66:2499–2507

Chang LK, Wei TT, Chlu YF, Tung CP, Chuang JY, Hung SK, Li C, Liu ST (2003) Inhibition of Epstein-Barr virus lytic cycle by (−)-epigallocatechin gallate. Biochem Biophys Res Commun 301:1062–1068

Chang YB, Bean RR, Jakobi R (2009) Targeting RhoA/Rho kinase and p21-activated kinase signaling to prevent cancer development and progression. Recent Pat Anticancer Drug Discov 4:110–124

Chayes F (1956) Petrographic model analysis. Wiley, New York, p 113

Cheeseman IM, Desai A (2004) Cell division. AAAtacking the mitotic spindle. Curr Biol 14:R70–R72

Chen S, Hwang J, Deng PS (1993) Inhibition of NAD(P)H:quinone acceptor oxidoreductase by flavones: a structure-activity study. Arch Biochem Biophys 302:72–77

Chen W, Wieraszko A, Hogan MV, Yang HA, Komecki E, Ehriich YH (1996) Surface protein phosphorylation by ecto-protein kinase is required for the maintenance of hippocampal long-term potentiation. Proc Natl Acad Sci U S A 93:8688–8693

Chen ZP, Schell JB, Ho CT, Chen KY (1998) Green tea epigallocatechin gallate shows a pronounced growth inhibitory effect on cancerous cells but not on their normal counterparts. Cancer Lett 129:173–179

Chen J, Song H, Zhang S, Wang Y, Cui DF, Wang CC (1999) Chaperone activity of DsbC. J Biol Chem 274:19601–19605

Chen C-F, Huang S, Liu S-C, Chueh P-J (2006) Effect of polyclonal antisera to recombinant tNOX protein on the growth of transformed cells. Biofactors 28:119–133

Cheng G, Cao Z, Xu X, van Meir EG, Lambeth JD (2001) Homologs of gp91phox: cloning and tissue expression of Nox3, Nox4, and Nox5. Gene 269:131–140

Cherry JM, MacKellar W, Morré DJ, Crane FL, Jacobsen LB, Schirrmacher V (1981) Evidence for a plasma membrane redox system on intact ascites tumor cells with different metastatic capacity. Biochim Biophys Acta 634:11–18

Cheung P, Allis CD, Sassone-Corsi P (2000) Signaling to chromatin through histone modifications. Cell 103:263–271

Chi SL, Pizzo SV (2006) Angiostatin is directly cytotoxic to tumor cells at low extracellular pH: a mechanism dependent on cell surface-associated ATP synthase. Cancer Res 66:875–882

Chluba-de Tapia J, de Tapia M, Jäggin V, Eberle AN (1997) Cloning of a human multispanning membrane protein cDNA: evidence for a new protein family. Gene 197:195–204

Cho N, Morré DJ (2009) Early developmental expression of a normally tumor-associated and drug-inhibited cell surface-located NADH oxidase (ENOX2) in non-cancer cells. Cancer Immunol Immunother 58:547–552

Cho NM, Chueh P-J, Kim C, Caldwell S, Morré DM, Morré DJ (2002) Monoclonal antibody to a cancer-specific and drug-responsive hydroquinone (NADH) oxidase from the sera of cancer patients. Cancer Immunol Immunother 51:121–129

Choi SI, Jeong CS, Cho SY, Lee YS (2007) Mechanism of apoptosis induced by apigenin in HepG2 human hepatoma cells: involvement of reactive oxygen species generated by NADPH oxidase. Arch Pharm Res 30:1328–1335

Choueir TK, Mekhail T, Hutson TE, Ganapathi R, Kelly GE, Bukowski RM (2006) Phase I trial of phenoxodiol delivered by continuous intravenous infusion in patients with solid cancer. Ann Oncol 17:860–865

Choury D, Leroux A, Kaplan JC (1981) Membrane-bound cytochrome b5 reductase (methemoglobin reductase) in human erythrocytes. Study in normal and methemoglobinemic subjects. J Clin Invest 67:149–155

Chu G (1994) Cellular responses to cisplatin. The roles of DNA-binding proteins and DNA repair. J Biol Chem 269:787–790

Chu EC, Tarnawaski AS (2004) PTEN regulatory functions in tumor suppression and cell biology. Med Sci Monit 10:RA235–RA241

Chueh P-J (1997) Characterization, isolation and expression closing of a tumor-associated protein tNOX that exhibits NADH: protein disulfide reductase activity with capsaicin inhibition. Doctoral Dissertation, Purdue University

Chueh P-J (2000) Cell membrane redox systems and transformation. Antioxid Redox Signal 2:177–187

Chueh P-J, Morré DM, Penel C, DeHahn T, Morré DJ (1997a) The hormone-responsive NADH oxidase of the plant plasma membrane has properties of a NADH:protein disulfide reductase. J Biol Chem 272:11221–11227

Chueh P-J, Morré DJ, Wilkinson FE, Gibson J, Morré DM (1997b) A 33.5- kDa heat- and protease-resistant NADH oxidase inhibited by capsaicin from sera of cancer patients. Arch Biochem Biophys 342:38–47

Chueh P-J, Morré DM, Morré DJ (2002a) A site-directed mutagenesis analysis of tNOX functional domains. Biochim Biophys Acta 1594:74–83

Chueh P-J, Kim C, Cho N, Morré DM, Morré DJ (2002b) Molecular cloning and characterization of a tumor-associated, growth-related and time-keeping hydroquinone (NADH) oxidase (NOX) of the HeLa cell surface. Biochemistry 41:3732–3741

Chueh P-J, Wu L-Y, Morré DM, Morré DJ (2004) tNOX is both necessary and sufficient as a cellular target for the anticancer actions of capsaicin and the green tea catechin (−)-epigallocatechin-3-gallate. Biofactors 20:235–249

Ciriolot MR, Palamara AT, Incerpi S, Lafavia E, Buė MC, De Vito P, Garaci E, Rotilio G (1997) Loss of GSH, oxidative stress, and decrease of intracellular pH as sequential steps in viral infection. J Biol Chem 272:2700–2708

Clark JE, Morré DJ, Cherry JH, Yunghans WN (1976) Enhancement of RNA polymerase activity by non-protein components from plasma membranes of soybean hypocotyls. Plant Sci Lett 7:233–238

Clark RA, Volpp BD, Leidal KG, Nauseef WM (1989) Translocation of cytosolic components of neutrophil NADPH oxidase. Trans Assoc Am Physicians 102:224–230

Clarke CF, Williams W, Teruya JH (1991) Ubiquinone biosynthesis in *Saccharomyces cerevisiae*. Isolation and sequence of CoQ3, the 3,4-dihydroxy-5-hexaprenylbenzoate methyltransferase gene. J Biol Chem 266:16636–16644

Clarke JD, Dashwood RH, Ho E (2008) Multi-targeted prevention of cancer by sulforaphane. Cancer Lett 269:291–304

Claussen M, Lüthen H, Blatt M, Böttger M (1997) Auxin-induced growth and its linkage to potassium channels. Planta 201:227–234

Clayton DF, George JM (1998) The synucleins: a family of proteins involved in synaptic function, plasticity, neurodegeneration and disease. Trends Neurosci 21:249–254

Cleland R (1959) The effect of osmotic concentration on auxin-action and on irreversible and reversible expansion of the *Avena* coleoptile. Physiol Plant 12:809–824

Cleland RE (1975) Auxin-induced hydrogen ion excretion: correlation with growth, and control by external pH and water stress. Planta 127:233–242

Cleland RE (1976) Kinetics of hormone-induce H^+ secretion. Plant Physiol 58:210–213

Cleland RE, Rayle DL (1978) Auxin, H^+-excretion and cell elongation. Bot Mag Special Issue 1:1215–1239

Clesceri LS, Greenberg AE, Eaton AD (1998) Standard methods for examination of water and wastewater. American Health Association, Washington, DC, pp 4–132

Coartney JS, Morré DJ (1980a) Studies on the chemical basis of cell wall loosening. Bot Gaz 141:63–68

Coartney JS, Morré DJ (1980b) Studies on the role of wall extensibility in the control of cell expansion. Bot Gaz 141:56–62

Coartney JS, Morré DJ, Key JL (1967) Inhibition of RNA synthesis and auxin-induced cell wall extensibility and growth by actinomycin D. Plant Physiol 42:434–439

Cohen HY, Miller C, Bitterman KJ, Wall NR, Hekking B, Kessler B, Howitz KT, Gorospe M, de Cabo R, Sinclair DA (2004) Calorie restriction promotes mammalian cell survival by inducing the SIRT1 deacetylase. Science 16:390–392

Connett RJ, Honig CR, Gayeski TEJ, Brooks GA (1990) Defining hypoxia: a systems view of Vo_2, glycolysis, energetics, and intracellular Po_2. J Appl Physiol 68:833–842

Constantinou AI, Husband A (2002) Phenoxodiol (2H-1-benzopyran-7-0,1,3-(4-hydroxyphenyl), a novel isoflavone derivative, inhibits DNA topoisomerase II by stabilizing the cleavable complex. Anticancer Res 22:2581–2585

Constantinou AK, Mehta R, Husband A (2003) Phenoxodiol, a novel isoflavene derivative, inhibits dimethylbenz[a]anthracene (DMBA)-induced mammary carcinogenesis in female Sprague–Dawley rats. Eur J Cancer 39:1012–1018

Cooper R, Morré DJ, Morré DM (2005a) Medicinal benefits of green tea. Part I: review of noncancer health benefits. J Altern Complement Med 11:521–528

Cooper R, Morré DJ, Morré DM (2005b) Medicinal benefits of green tea: Part II. Review of anticancer properties. J Altern Complement Med 11:639–652

Crane FL, Löw H (2008) Reactive oxygen species generation at the plasma membrane for antibody control. Autoimmun Rev 7:518–522

Crane FL, Hatefi Y, Lester RI, Widmer C (1957) Isolation of a quinone from beef heart mitochondria. Biochim Biophys Acta 25:220–221

Crane FL, Sun IL, Clark MG, Grebing C, Löw H (1985) Transplasma-membrane redox systems in growth and development. Biochim Biophys Acta 811:233–264

Crane FL, Löw H, Sun IL, Isaksson M (1990a) Transmembrane electron transport and growth of transformed cells. In: Crane FL, Morré DJ, Löw HE (eds) Oxidoreduction at the plasma membrane: relation to growth and transport, I. Animals. CRC Press, Boca Raton, FL, pp 145–170

Crane FL, Morré DJ, Löw HE (eds) (1990b) Oxidoreduction at the plasma membrane: relation to growth and transport, I. Animals. CRC Press, Boca Raton, FL, 318pp

Crane FL, Morré DJ, Löw HE (eds) (1991) Oxidoreduction at the plasma membrane: relation to growth and transport, II. Plants. CRC Press, Boca Raton, FL, 292pp

Crane FL, Sun IL, Crowe RA, Alcain FJ, Löw H (1994) Coenzyme Q_{10}, plasma membrane oxidase and growth control. Mol Aspects Med 15:s1–s11

Cross AR (1987) The inhibitory effects of some iodonium compounds on the superoxide generating system of neutrophils and their failure to inhibit diaphorase activity. Biochem Pharmacol 36:489–493

Cross JW, Briggs WR, Dohrmann UC, Ray PM (1978) Auxin receptors of maize coleoptile membranes do not have ATPase activity. Plant Physiol 61:581–584

Cross JV, Deak J, Rich EA, Qian Y, Lewis M, Parrott LA, Mochida K, Gustafson D, vande Pol S, Templeton DJ (1999) Quinone reductase inhibitors block SAPK/JNK and NFkappaB pathways and potentiate apoptosis. J Biol Chem 274:31150–31154

Cunha RA (2008) Different cellular sources and different roles of adenosine: A_1 receptor-mediated inhibition through astrocytic-driven volume transmission and synapse-restricted A_{2A} receptor-mediated facilitation of plasticity. Neurochem Int 52:65–72

Curtain CC, Ali F, Volikatis I, Cherny RA, Norton RS, Beyreuther K, Barrow CJ, Masters CL, Bush AI, Barnham KJ (2001) Alzheimer's disease amyloid-beta binds copper and zinc to generate an allosterically ordered membrane-penetrating structure containing superoxide dismutase-like subunits. J Biol Chem 276:20466–20473

Cutter H, Wu L-Y, Kim C, Morré DJ, Morré DM (2001) Is the cancer protective effect of green tea (−)-epigallocatechin gallate mediated through an antioxidant mechanism? Cancer Lett 162:149–154

Cuvillier O (2008) Downregulating sphingosine kinase-1 for cancer therapy. Expert Opin Ther Targets 2:1009–1020

Dai S, Morré DJ, Geilen CC, Almond-Roesler B, Orfanos CE, Morré DM (1997) Inhibition of plasma membrane NADH oxidase activity and growth of HeLa cells by natural and synthetic retinoids. Mol Cell Biochem 166:101–109

Dalal S, Hanson PI (2001) Membrane traffic: what drives the AAA motor? Cell 104:5–8

Dannenberg AJ, Subbaramaiah K (2003) Targeting cyclooxygenase-2 in human neoplasia: rationale and promise. Cancer Cell 4:431–436

Darr D, Fridovich I (1984) Vanadate and molybdate stimulate the oxidation of NADH by superoxide radical. Arch Biochem Biophys 232:562–565

Davies PJ (1995) Plant hormones. Kluwer Academic, Dordrecht

Davies R, Tulloch A, Frydenberg M, Kelly G (2004) Final results of a phase Ib/IIa study of oral phenoxodiol in patients with late stage, hormone-refractory prostate cancer. Poster presented in the conference of AACR basic, translational and clinical advances in prostate cancer, Nov 2004

Davies JM, Tsuruta H, May AP, Weis WL (2005) Conformational changes of p97 during nucleotide hydrolysis determined by small-angle X-ray scattering. Structure 13:183–195

De Berardinis RJ, Lum JJ, Hatzivassilou G, Thompson GB (2008) The biology of cancer: metabolic programming fuels cell growth and proliferation. Cell Metab 7:11–20

de Figueiredo LF, Grossmann T, Ziegler M, Schuster S (2011) Pathway analysis of NAD^+ metabolism. Biochem J 439(2):341–348

de Grey ADNJ (1999) The mitochondrial free radical theory of aging. R. G. Landes, Austin, TX, pp 104–110

de Grey ADNJ (2003) A hypothesis for the minimal overall structure of the mammalian plasma membrane redox system. Protoplasma 221:3–9

De Luca T, Morré DM, Zhao H, Morré DJ (2005) NAD^+/NADH and/or CoQ/$CoQH_2$ ratios from plasma membrane electron transport may determine ceramide and sphingosine-1-phosphate levels accompanying G_1 arrest and apoptosis. Biofactors 25:43–60

De Luca T, Bosneaga E, Morré DM, Morré DJ (2009) Downstream targets of altered sphingolipid metabolism in response to inhibition of ENOX2 by phenoxodiol. Biofactors 34:253–260

De Luca T, Morré DM, Morré DJ (2010) Reciprocal relationship between cytosolic NADH and ENOX2 inhibition triggers sphingolipid-induced apoptosis in HeLa cells. J Cell Biochem 110:1504–1511

DeHahn T, Barr R, Morré DJ (1997) NADH oxidase activity present on both the external and internal membrane surfaces of soybean plasma membranes. Biochim Biophys Acta 1328:99–108

del Castillo-Olivares A, Yantiri F, Chueh P-J, Wang S, Sweeting M, Sedlak D, Morré DM, Burgess J, Morré DJ (1998) A drug-responsive and protease-resistant peripheral NADH oxidase complex from the surface of HeLa S cells. Arch Biochem Biophys 358:125–140

Del Giudice E, Spinetti PR, Tedeschi AL (2010) Water dynamics at the root of metamorphosis in living organisms. Water12 2:566–586

Del Principe D, Avigliano L, Savini I, Catani MV (2011) Trans-plasma membrane electron transport in mammals: functional significance in health and disease. Antioxid Redox Signal 14: 2289–2318

DeLaBarre B, Brunger AT (2003) Complete structure of p97/valosin-containing protein reveals communication between nucleotide domains. Nat Struct Biol 10:856–863

DeLaBarre B, Brunger AT (2005) Nucleotide dependent motion and mechanism of action of p97/VCP. J Mol Biol 347:437–452

Deliconstantinos G (1987) Physiological aspects of membrane lipid fluidity in malignancy. Anticancer Res 7:1011–1022

Denko N, Schindler C, Koong A, Laderoute K, Green C, Giagcia A (2000) Epigenetic regulation of gene expression in cervical cancer cells by the tumor microenvironment. Clin Cancer Res 6:480–487

Dewhurst MW, Secomb TW, Ong ET, Hsu R, Gross JF (1994) Determination of local oxygen consumption rates in tumors. Cancer Res 54:3333–3336

Diaz E, Schimmöller F, Pfeffer SR (1997) A novel Rab9 effector required for endosome-to-TGN transport. J Cell Biol 138:283–290

Dillman RO, Shawler DL, Johnson DE, Meyer DL, Koziol JA, Frincke JM (1986) Preclinical trials with combinations and conjugates of T101 monoclonal antibody and doxorubicin. Cancer Res 46:4886–4891

Disis M, Pupa SM, Gralow JR, Dittadi R, Menard S, Cheever MA (1997) High-titer HER-2/neu protein-specific antibody can be detected in patients with early-stage breast cancer. J Clin Oncol 15:3363–3367

Dnistrian AM, Schwartz MK, Katopodis N, Fracchia AA, Stock CC (1982) Serum lipid-bound sialic acid as a marker in breast cancer. Cancer 50:1815–1819

Dohrmann U, Hertel R, Kowalik H (1978) Properties of auxin binding sites in different subcellular fractions from maize coleoptiles. Planta 140:97–106

Donaldson JG, Lippincott-Schwartz J, Bloom GS, Kreis TE, Klausner RD (1990) Dissociation of a 110-kD peripheral membrane protein from the Golgi apparatus is an early event in Brefeldin A action. J Cell Biol 111:2295–2306

Döring O, Lüthje S (1996) Molecular components and biochemistry of electron transport in plant plasma membranes. Mol Membr Biol 13:127–142

Döring O, Lüthje S, Hilgendorf F, Böttger M (1990) Membrane depolarization by hexacyanoferrate (III), hexabromoiridate (IV) and hexachloroiridate (IV). J Exp Bot 43:1055–1061

Doroshow J (1983) Anthracycline antibiotic-stimulated superoxide, hydrogen peroxide, and hydroxyl radical production by NADH dehydrogenase. Cancer Res 43:4543–4551

Dowse HB, Palmer JD (1972) The chronomutagenic effect of deuterium oxide on the period and entrainment of a biological rhythm. Biol Bull 143:513–524

Dreosti IE (1996) Bioactive ingredients: antioxidants and polyphenols in tea. Nutr Rev 54:S51–S58

Dreveny I, Pye VE, Beuron F, Briggs LC, Isaacson RL, Matthews SJ, McKeown C, Yuan X, Zhang X, Freemont PS (2004) p97 and close encounters of every kind: a brief review. Biochem Soc Trans 32:715–720

Dreyer JL (1990) Plasma membrane dehydrogenases in rat brain synaptic membranes. Multiplicity and subunit composition. J Bioenerg Biomembr 22:619–633

Dumitru CA, Gulbins E (2006) TRAIL activates acid sphingomyelinase via a redox mechanism and releases ceramide to trigger apoptosis. Oncogene 24:5612–5625

Dunlap JC (1996) Genetics and molecular analysis of circadian rhythms. Annu Rev Genet 30:579–601

Dunlap JC (1999) Molecular basis for circadian clocks. Cell 26:271–286

Dunwiddie TV, Masino SA (2001) The role and regulation of adenosine in the central nervous system. Annu Rev Neurosci 24:31–55

Ebert TA, Southon JR (2003) Red sea urchins (*Strongylocentrotus franciscanus*) can live over 100 years: confirmation with A-bomb^{14} carbon. Fish Bull 101:915–922

Edmunds LN Jr (1984) Cell cycle clocks. Marcel Dekker, New York, pp 44–57

Edmunds LN Jr (1988) Cellular and molecular basis of biological clocks. New York, Berlin, 497pp

Efferth T, Sauerbrey A, Olbrich A, Gebhart E, Rauch P, Weber HO, Hengstler JG, Halatsch ME, Volm M, Tew KD, Ross DD, Funk JD (2003) Molecular modes of action of artesunate in tumor cell lines. Mol Pharmacol 64:382–394

Eisinger WR, Morré DJ (1968) The effect of sulfhydryl inhibitors on plant cell elongation. Proc Ind Acad Sci for 1967 77:136–143

Eisinger WR, Morré DJ (1971) Growth regulating properties of picloram, 4-amino-3, 5, 6-trichloropicolinic acid. Can J Bot 49:889–897

Ellem KAO, Kay GF (1983) Ferricyanide can replace pyruvate to stimulate growth and attachment of serum restricted human melanoma cells. Biochem Biophys Res Commun 112:183–190

Ellerby LM, Ellerby HM, Park SM, Holleran AL, Murphy AN, Fiskum G, Kane DJ, Testa MP, Kayalar C, Bredesen DE (1996) Shift of the cellular oxidation-reduction potential in neural cells expressing Bcl-2. J Neurochem 67:1259–1267

Ellman GL (1959) Tissue sulfhydryl groups. Arch Biochem Biophys 82:70–77

Emsley JM, Feeney J, Sutcliffe LH (1965) High resolution nuclear magnetic resonance spectroscopy, vol 1. Pergamon Press, Oxford

Encio I, Morré DJ, Villar R, Gil MJ, Martinez-Merion V (2005) Benzo[*b*]thiophenesulphonamide 1,1-dioxide derivatives inhibit tNOX activity in a redox state dependent manner. Br J Cancer 92:690–695

Engelmann W (1972) Lithium slows down the Kalanchoe clock. Z Naturforsch 27B:477

Engelmann W (1973) A slowing down of circadian rhythms by lithium ions. Z Naturforsch 28:733–736

Enright JT (1997) Heavy water slows biological timing processes. Z Vergl Physiol 72:1–16

Enyedi B, Várnai P, Geiszt M (2010) Redox state of the endoplasmic reticulum is controlled by *Ero1*L-alpha and intraluminal calcium. Antioxid Redox Signal 13:721–729

Essex DW, Li M, Miller A, Feinman RD (2001) Protein disulfide isomerase and sulfhydryl-dependent pathways in platelet activation. Biochemistry 40:6070–6075

Fang X-P, Rieser MJ, Gu ZM, Zhao G-X, McLaughlin JL (1993) Annonaceous acetogenins: an updated review. Phytochem Anal 4:27–48

Fantin VR, St. Pierre J, Leder P (2006) Attenuation of LDH-A expression uncovers a link between glycolysis, mitochondrial physiology, and tumor maintenance. Cancer Cell 9:425–434

Farquhar MG (1985) Progress in unraveling pathways of Golgi traffic. Annu Rev Cell Biol 1:447–488

Fassina G, Buffa A, Barnell R, Varnier CE, Noonan DM, Albini A (2002) Polyphenolic antioxidant (−)-epigallocatechin-3-gallate from green tea as a candidate anti-HIV agent. AIDS 16:939–941

Faulk WP, His BL, Stevens PJ (1980) Transferrin and ransferrin receptors in carcinoma of the breast. Lancet 2:390–392

Faulk WP, Harats H, Berczi A (1990a) Transferrin receptor growth control in normal and transformed cells. In: Crane FL, Morré DJ, Löw H (eds) Oxidoreduction at the plasma membrane. CRC Press, Boca Raton, pp 205–224

Faulk WP, Taylor CG, Yeh CJG, McIntyre JA (1990b) Preliminary clinical study of transferrin-adriamycin conjugate for drug delivery to acute leukemia patients. Mol Biother 2:57–60

Faulk WP, Barabas K, Sun IL, Crane FL (1991) Transferrin-adriamycin conjugates which inhibit tumor cell proliferation without interaction with DNA inhibit plasma membrane oxidoreductase and proton release in K562 cells. Biochem Int 25:815–822

Fedrowitz M, Westermann J, Loscher W (2002) Magnetic field exposure increases cell proliferation but does not affect melatonin levels in the mammary gland of female Sprague Dawley rats. Cancer Res 62:356–1363

Feiler HS, Desprez T, Santoni V, Kronenberger J, Caboche M, Traas J (1995) The higher plant *Arabidopsis thaliana* encodes a functional CDC48 homologue which is highly expressed in dividing and expanding cells. EMBO J 1422:5626–5637

Felle H (1987) Proton transport and pH control in *Sinapis alba* root hairs: a study carried out with double-barreled microelectrodes. J Exp Bot 38:340–354

Felle H (1988) Auxin causes oscillations of cytosolic free calcium and pH in *Zea mays* coleoptiles. Planta 174:495–499

Fenouillet E, Barbouche R, Courageot J, Milquelis R (2001) The catalytic activity of protein disulfide isomerase is involved in human immunodeficiency virus envelope-mediated membrane fusion after CD4 cell binding. J Infect Dis 183:744–752

Fernandez R, Ganzon DO (2003) Use of a green tea-capsicum supplement (Capsibiol-T) as adjuvant cancer treatment: case study report. Phil J Otolaryngol Head Neck Surg 18:171–177

Fernández-Ayala DJ, Martin SF, Barroso MP, Gómez-Díaz C, Rodríguez-Aguilera JM, López-Lluch G, Navas P (2000) Coenzyme Q protects cells against serum withdrawal-induced apoptosis by inhibition of ceramide release and caspase-3 activation. Antioxid Redox Signal 2:263–275

Fesenko EE, Gluvstein AY (1995) Changes in the state of water, induced by radiofrequency electromagnetic fields. FEBS Lett 367:53–55

Filipponi A, D'Angelo P, Pavel NV, Di Ciecco A (1994) Triplet correlations in the hydration shell of aquaions. Chem Phys Lett 225:150–155

Forbes-Stovall J, Morré DJ, Jacobshagen S (2008) The regular 24 min switch between the two functions of ECTO-NOX proteins might constitute the pacemaker of the circadian clock. In: Abstracts, Annual meeting of the Kentucky Academy of Science. http://www.kyacademyofscience.org/content/meeting-programs/2008-program.pdf

Forman JJ, Maiorino M, Ursini F (2010) Signaling functions of reactive oxygen species. Biochemistry 49:835–842

Forsyth LM, Preuss HG, MacDowell AL, Chiazze I Jr, Birkmayer GD, Bellanti JA (1999) Therapeutic effects of oral NADH on the symptoms of patients with chronic fatigue syndrome. Ann Allergy Asthma Immunol 82:185–191

Foster K, Anwar N, Pogue R, Morré DM, Morré DJ (2003) Decomposition analyses applied to a complex ultradian biorhythm: the oscillating NADH oxidase activity of plasma membranes having a potential time-keeping (clock) function. Nonlinearity Biol Toxicol Med 1(1):51–70

Fredholm BB, Hedqvist P (1980) Modulation of neurotransmission by purine nucleotides and nucleosides. Biochem Pharmacol 29:1635–1643

Freedman RB (1989) Protein disulfide isomerase: multiple roles in the modification of secretory proteins. Cell 57:1069–1072

Friedberg I, Kübler D (1990) The role of surface protein kinase in the ATP-induced growth inhibition in transformed mouse fibroblasts. Ann N Y Acad Sci 603:513–515

Friedman J, Kraus S, Huptman Y, Schiff Y, Seger R (2007) Mechanism of short-term ERK activation by electromagnetic fields at mobile phone frequencies. Biochem J 405:559–568

Fritzer M, Barabas K, Szüts V, Bérczi A, Szekeres T, Faulk WP, Goldenberg H (1992) Cytotoxicity of a transferrin-adriamycin conjugate to anthracycline-resistant cells. Int J Cancer 52:619–623

Froy O, Miskin R (2010) Effect of feeding regimens on circadian rhythms: implications for aging and longevity. Aging 2:7–27

Fu L, Lee CC (2003) The circadian clock: pacemaker and tumour suppressor. Nat Rev Cancer 3:350–361

Fujiki H (1999) Two stages of cancer prevention with green tea. J Cancer Res Clin Oncol 125:589–597

Fujiki H, Suganuma M, Okabe S, Komori A, Sueoka E, Sueoka N, Kosu T, Sakai Y (1996) Japanese green tea as a cancer preventative in humans. Nutr Rev 54:567–5770

Fujiki H, Suganuma M, Okabe S, Sueoka N, Komori A, Sueoka E, Kozu T, Tada Y, Suga K, Imai K, Nakachi K (1998) Cancer inhibition by green tea. Mutat Res 402:307–310

Fujiki H, Suganuma M, Okabe S, Sueoka N, Imai K, Nakachi S, Kimura S (1999) Mechanistic findings of green tea as cancer preventive for humans. Proc Soc Exp Biol Med 220:225–228

Fuller RS, Brake A, Thorner J (1989) Intracellular targeting and structural conservation of a prohormone processing endoprotease. Science 246:482–486

Fulton JL, Hoffman MM, Darab JG, Palmer BJ, Stein EA (2000) Copper (I) and copper (II) coordination structure under hydrothermal conditions at 325°C: an X-ray absorption fine structure and molecular dynamics study. J Phys Chem A 104:11651–11663

Gaikwad A, Long DJ II, Stringer JL, Jaiswal AK (2001) In vivo role of NAD(P)H:quinone oxidoreductase 1 (NQO1) in the regulation of intracellular redox state and accumulation of abdominal adipose tissue. J Biol Chem 276:22559–22564

Gallina A, Hanley TM, Mandel R, Trahey M, Broder CC, Viglianti GA, Ryser HJ-P (2002) Inhibitors of protein-disulfide isomerase prevent cleavage of disulfide bonds in receptor-bound glycoprotein 120 and prevent HIV-1 entry. J Biol Chem 277:50579–50588

Galston A, Kaur R (1963) An effect of auxins on the heat coagulability of the proteins of growing plant cells. Proc Natl Acad Sci U S A 45:1587–1590

Gamble JR, Xia P, Hahn CN, Drew JJ, Drogemuller CJ, Brown D, Vadas MA (2006) Phenoxodiol, an experimental anticancer drug, shows potent antiangiogenic properties in addition to its antitumour effects. Int J Cancer 118:2412–2420

Garber K (2006) Energy deregulation: licensing tumors to grow. Science 312:1158–1159

Garci E, Palamara AT, Di Francesco P, Favalli C, Ciriolo MR, Rotilio G (1992) Glutathione inhibits replication and expression of viral proteins in cultured cells infected with Sendai virus. Biochem Biophys Res Commun 188:1090–1096

Garcia C, Hicks C, Morré DJ (1999) Plasma membrane NADH oxidase is gravi-responsive. Plant Physiol Biochem 37:551–558

Garcia-Cañero R, Guerra M (1988) External Na dependency and variation of ferricyanide reductase in tumor cell growth. Oncologia 11:187–195

Garcia-Cañero R, Diaz-Gis JJ, Guerra MA (1987) Is transmembrane ferricyanide reductase connected with the early expression of mitogenic signals in hepatocytes. In: Ramirez J (ed) Redox function of the eukaryotic plasma membrane. Consejo Superior de Investigaciones Cientificas, Madrid, p 41

Garnett AP, Viles JH (2003) Copper binding to the octarepeats of the prion protein, affinity, specificity, folding, and cooperativity: insights from circular dichroism. J Biol Chem 278: 6795–6802

Gatenby RA, Gillies RJ (2004) Why do cancers have high aerobic glycolysis? Nat Rev Cancer 4:891–899

Gayda DP, Crane FL, Morré DJ, Löw H (1977) Hormone effects on NADH-oxidizing enzymes of plasma membranes of rat liver. Proc Indiana Acad Sci 86:385–390

Geilen CC, Wieder T, Orfanos CE (1997) Ceramide signalling: regulatory role in cell proliferation, differentiation and apoptosis in human epidermis. Arch Dermatol Res 289:559–566

Geng L, Rachakonda G, Morré DJ, Morré DM, Crooks PA, Sonar VN, Roti JL, Rogers BE, Greco S, Ye F, Salleng KJ, Sasi S, Freeman ML, Sekhar KR (2009) Indolyl-quinuclidinols inhibit ENOX activity and endothelial cell morphogenesis while enhancing radiation-mediated control of tumor vasculature. FASEB J 23:2986–2995

Giannakou ME, Partridge L (2004) The interaction between FOXO and SIRT1: tipping the balance towards survival. Trends Cell Biol 14:408–412

Gibney G, Eifiky A, Bussom S, Holmes CJ, Burns A, McDonough JA, Rowen E, Cheng YC, Kelly WK (2010) Phase II trial of phenoxodiol in patients with castrate and non-castrate prostate cancer. J Clin Oncol 28:15s

Gilberger TW, Walter RD, Müller S (1997) Identification and characterization of the functional amino acids at the active site of the large thioredoxin reductase from *Plasmodium falciparum*. J Biol Chem 272:29584–29589

Gilberger TW, Bergmann B, Walter RD, Müller S (1998) The role of the C-terminus for catalysis of the large thioredoxin reductase from *Plasmodium falciparum*. FEBS Lett 425:407–410

Gillotte KL, Hörkkö S, Witztum JL, Steinberg D (2000) Oxidized phospholipids, linked to apolipoprotein B of oxidized LDL, are ligands for macrophage scavenger receptors. J Lipid Res 41:824–833

Gitlin D, Perricelli A, Gitlin GM (1972) Synthesis of α-fetoprotein by liver, yolk sac, and gastrointestinal tract of the human conceptus. Cancer Res 32:979–982

Gomez-Diaz C, Rodriguez-Aguilera JC, Barroso MP, Villalba JM, Navarro F, Crane FL, Navas P (1997) Antioxidant ascorbate is stabilized by NADH-coenzyme Q_{10} reductase in the plasma membrane. J Bioenerg Biomembr 29:251–257

Gonzales VM, Fuertes MA, Alonso C, Perez JM (2001) Is cisplatin-induced cell death always produced by apoptosis? Mol Pharmacol 59:657–663

Gonzáles-Aragón D, Villalba JM (2006) Proceedings of the 8th international conference on membrane redox systems and their role in biological stress and disease, Szeged, p 36

Gonzalez-Gronow M, Cuchacovich M, Francos R, Cuchacovich S, del Pilar Fernandez M, Blanco A, Bowers EV, Kaczowka S, Pizzo SV (2010) Antibodies against the voltage-dependent anion channel (VDAC) and its protective ligand hexokinase-I in children with autism. J Neuroimmunol 227:153–161

Goodman R, Chizmadzhev Y, Shirley-Henderson A (1993) Electromagnetic fields and cells. J Cell Biochem 51:436–441

Goormaghtigh E, Pollakis G, Ruysschaert R (1983) Mitochondrial membrane modifications induced by adriamycin-mediated electron transport. Biochem Pharmacol 32:889–893

Gorman A, McGowan A, Cutler TG (1997) Role of peroxide and superoxide anion during tumor cell apoptosis. FEBS Lett 404:27–33

Gorospe M, Shack S, Guyton KZ, Samid D, Holbrook NJ (1996) Up-regulation and functional role of p21Waf1/Cip1 during growth arrest of human breast carcinoma MCF-7 cells by phenylacetate. Cell Growth Differ 7:1609–1615

Goss G, Quinn M, Rutherford T, Kelly GA (2005) A randomized Phase II study of phenoxodiol with platinum or taxane chemotherapy in chemoresistant epithelial ovarian cancer, fallopian tube cancer and primary peritoneal cancer. Eur J Cancer Suppl 3:261

Grabau C, Cronan JE (1986) Nucleotide sequence and deduced amino acid sequence of *Escherichia coli* pyruvate oxidase, a lipid-activated flavoprotein. Nucleic Acids Res 14:5449–5460

Graham JM, Sumner MCB, Curtis DH, Pasternak CA (1973) Sequence of events in plasma membrane assembly during the cell cycle. Nature 246:291–295

Gray JP, Eisen T, Cline GW, Smith PJS, Heart E (2011) Plasma membrane electron transport in pancreatic β-cells is mediated in part by NQO1. Am J Physiol Endocrinol Metab 301: E113–E121

Grebien F, Dolzing H, Beug H, Mullner WW (2005) Cell size control. New evidence for a general mechanism. Cell Cycle 4:418–421

Grieco PA, Collins JL, Moher ED, Fleck TJ, Gross RS (1993) Synthetic studies on quassinoids: total synthesis of (−)-chaparrinone, (−)-glaucarubolone, and (+)-glaucraubolone. J Am Chem Soc 115:6078–6093

Grieco PA, VanderRoest JM, Pineiro-Nunez MM (1995) A C_{19} quassinoid from *Castela polyandra*. Phytochemistry 38:1463–1465

Grieco PA, Morré DJ, Corbett TH, Valeriote FA (1996) Therapeutic quassinoid preparations. US Patent 08/334/735, 1996

Grieco PA, Morré DJ, Valeriote FA (1997) Therapeutic quassinoid preparations and antineoplastic antiviral and herbistatic activity. US Patent 5,639,712, 1997

Griffith JS (1967) Self-replication and scrapie. Nature 215:1043–1044
Grindey GB (1988) Identification of diarylsulfonylureas as novel anticancer drugs. Proc Am Assoc Cancer Res 29:535–536
Grindey GB, Boder GB, Grossman CS, Howbert YY, Poore GA, Shaw WH, Todd GC, Worzella JF (1987) Anticancer drugs. Proc Am Assoc Cancer Res 28:309
Grinstein S, Rothstein A (1986) Mechanisms of regulation of the Na^+/H^+ exchanger. J Membr Biol 90:1–12
Grunstein M (1997) Histone acetylation in chromatin structure and transcription. Nature 389: 349–352
Guarente L (2008) Mitochondria—a nexus for aging, calorie restriction, and sirtuins? Cell 132:171–176
Guarente L, Picard F (2005) Calorie restriction—the SIR2 connection. Cell 120:473–482
Gudkov SV, Bruskov VI, Astashev ME, Chernikov AV, Yaguzhinsky LS, Zakharov SD (2011) Oxygen-dependent auto-oscillations of water luminescence triggered by the 1264 nm radiation. J Phys Chem B 115(23):7693–7698
Gulbins E, Grassmé H (2002) Ceramide and cell death receptor clustering. Biochim Biophys Acta 1585:139–145
Gulbins E, Coggeshall KM, Brenner B, Schlottmann K, Linderkamp O, Lang F (1996) Fas-induced apoptosis is mediated by activation of a Ras and Rac protein-regulated signaling pathway. J Biol Chem 271:26389–26394
Guppy M (2002) The hypoxic core: a possible answer to the cancer paradox. Biochem Biophys Res Commun 2994:676–680
Guppy M, Leedman P, Zu X, Russell V (2002) Contribution by different fuels and metabolic pathways to the total ATP turnover of proliferating MCF-7 breast cancer cells. Biochem J 364: 309–315
Guppy M, Brunner S, Buchanan M (2005) Metabolic depression: a response of cancer cells to hypoxia? Comp Biochem Physiol B Biochem Mol Biol 140:233–239
Hager A, Menzel H, Krauss A (1971) Versuche und Hypothese zu Primärwirkung des Auxins beim Streckungswachstum. Planta 100:47–75
Haidle C, McKinney S (1986) Adriamycin-mediated introduction of a limited number of single-strand breaks into supercoiled DNA. Cancer Biochem Biophys 8:327–335
Haigis MC, Guarente LP (2006) Mammalian sirtuins-emerging roles in physiology, aging, and calorie restriction. Genes Dev 20:2913–2921
Hainsworth JD, Hande KR, Satterlee WG, Kuttesch J, Johnson DH, Grindey GB, Jackson LE, Greco FA (1989) Phase I clinical study of N-4-chlorophenylamino]carbonyl1-2,3dihydro-1-H-idene-5-sulfonamide LY186641. Cancer Res 49:5217–5220
Hanski C, Zimmer T, Gossrau R, Reutter W (1986) Increased activity of dipeptidyl peptidase IV in serum of hepatoma-bearing rats coincides with the loss of the enzyme from the hepatoma plasma membrane. Experientia 42:826–828
Hardie DG (2007) AMP-activated/SNF1 protein kinases: conserved guardians of cellular energy. Nat Rev Mol Cell Biol 8:774–785
Hardin JW, Cherry JH, Morré DJ, Lembi CA (1972) Enhancement of RNA polymerase activity by a factor released by auxin from plasma membrane. Proc Natl Acad Sci U S A 63:3146–3150
Harmer S (2010) Plant biology in the fourth dimension. Plant Physiol 154:467–470
Harvey J, Nicholson ABT, Cohén MA (2008) Finding early invasive breast cancers: a practical approach. Radiology 248:61–76
Hassidim M, Rubinstein B, Lerner HR, Reinhold H (1987) Generation of a membrane potential by electron transport in plasmalemma-enriched vesicles of cotton and radish. Plant Physiol 85:872–875
Hatefi Y (1963) Coenzyme Q (ubiquinone). Adv Enzymol 25:275–328
Hay N (2005) The Akt-mTOR tango and its relevance to cancer. Cancer Cell 8:179–183
Hayakawa S, Saeki K, Sazuka Y, Sziki Y, Shoji Y, Ohta T, Kaji K, Isemura M (2001) Apoptosis induction by epigallocatechin gallate involves its binding to Fas. Biochem Biophys Res Commun 285:1102–1106

He P, Peng Z, Luo Y, Wang L, Yu P, Deng W, An Y, Shi T, Ma D (2009) High-throughput functional screening for autophagy-related genes and identification of TM9SF1 as an autophagosome-inducing gene. Autophagy 5:52–60

Hedges KL, Morré DM, Wu L-Y, Morré DJ (2003) Adriamycin tolerance in human mesothelioma lines and cell surface NADH oxidase. Life Sci 73:1189–1198

Helgerson SL, Cramer WA, Morré DJ (1976) Evidence for an increase in microviscosity of plasma membranes from soybean hypocotyls induced by the plant hormone indole-3-acetic acid. Plant Physiol 58:548–551

Helmlinger G, Yuan F, Dellian M, Jain RK (1997) Interstitial pH and pO_2 gradients in solid tumors in vivo: high-resolution measurements reveal a lack of correlation. Nat Med 3:177–182

Henshaw DL (2002) Does our electricity distribution system pose a serious risk to public health? Med Hypotheses 59:39–51

Hensley K, Maidt ML, Yu Z, Sang H, Markesbery WR, Floyd RA (1998) Electrochemical analysis of protein nitrotyrosine and dityrosine in the Alzheimer brain indicates region-specific accumulation. J Neurosci 18:8126–8132

Herst PM, Berridge MV (2006) Plasma membrane electron transport: a new target for cancer drug development. Curr Mol Med 6:895–904

Herst PM, Tan AS, Scarlett DG, Berridge MV (2004) Cell surface oxygen consumption by mitochondrial gene knockout cells. Biochim Biophys Acta 1656:79–87

Herst PM, Petersen T, Jerram P, Baty J, Berridge MV (2007) The antiproliferative effects of phenoxodiol are associated with inhibition of plasma membrane electron transport in tumor cell lines and primary immune cells. Biochem Pharmacol 17:1587–1595

Herst PM, Davis JE, Neeson P, Berridge MV, Ritchie DS (2009) The anti-cancer drug, phenoxodiol, kills primary myeloid and lymphoid leukemic blasts and rapidly proliferating T cells. Haematologica 94:928–934

Hicks C, Morré DJ (1998) Oxidation of NADH by intact segments of soybean hypocotyls and stimulation by 2,4-D. Biochim Biophys Acta 1375:1–5

Hicks-Berger C, Morré DJ (2006) Inside-out but not right side-out plasma membrane vesicles from soybean enlarge when treated with ATP + 2,4-D as determined by electron microscopy and light scattering: evidence for involvement of a plasma membrane AAA-ATPase. Biofactors 28:91–104

Hicks-Berger C, Sokolchik I, Kim C, Morré DJ (2006) A plasma membrane-associated associated AAA-ATPase from *Glycine max*. Mol Cell Biochem 28:135–149

Hidalgo A, Garciá-Herdugo G, González-Reyes JA, Morré DJ, Navas P (1991) Ascorbate-free radical stimulated onion root growth by increasing cell elongation. Bot Gaz 152:282–288

Hill DL, Grubbs CJ (1982) The use of retinoids in combination with other chemotherapeutic agents against L1210 leukemia. Anticancer Res 2:111–124

Hillson DA, Lambert N, Freedman RB (1984) Formation and isomerization of disulfide bonds in proteins: protein disulfide isomerase. Methods Enzymol 107:281–292

Hipkiss AR (2006) Does chronic glycolysis accelerate aging? Could this explain how dietary restriction works? Ann N Y Acad Sci 1067:361–368

Hipkiss AR (2010) Aging, proteotoxicity, mitochondria, glycation, NAD and carnosine: possible inter-relationships and resolution of the oxygen paradox. Front Aging Neurosci 2:10

Hirayama J, Sassone-Corsi P (2005) Structural and functional features of transcription factors controlling the circadian clock. Curr Opin Genet Div 15:548–556

Hirayama J, Sahar S, Grimaldi B, Tamaru T, Takamatsu K, Nakahata Y, Corsi P (2007) CLOCK mediated acetylation of BMAL1 controls circadian function. Nature 450:1086–1090

Ho WZ, Douglas SD (1992) Glutathione and N-acetylcysteine suppression of human immunodeficiency virus replication in human monocyte/macrophages in vitro. AIDS Res Hum Retroviruses 8:1249–1254

Hoober JK, Cohén S (1967) Epidermal growth factor: II. Increased activity of ribosomes from chick embryo epidermis for cell-free protein synthesis. Biochim Biophys Acta 138:357–368

Hofhaus G, Johns DR, Hurko O, Attardi G, Chomyn A (1996) Respiration and growth defects in transmitochondrial cell lines carrying the I1778 mutation associated with Leber's hereditary optic neuropathy. J Biol Chem 271:12155–13161

Hofmann K, Gunderoth-Palmowski M, Wiedenmann G, Engelmann W (1978) Further evidence for period lengthening effect of Li^+ on circadian rhythms. Z Naturforsch 33C:231–234

Hogan MV, Pawlowska Z, Yang HA, Kornecki E, Ehrlick YH (1995) Surface phosphorylation by ecto-protein kinase C in brain neurons: a target for Alzheimer's beta-amyloid peptides. J Neurochem 65:2022–2030

Holvoet P, Mertens A, Verhamme P, Bogaerts K, Beyens G, Verhaeghe R, Collen D, Muls E, Van de Werf F (2001) Circulating oxidized LDL is a useful marker for identifying patients with coronary artery disease. Anterioscler Thromb Vasc Biol 21:844–848

Holvoet P (1999) Endothelial dysfunction, oxidation of low-density lipo-protein, and cardiovascular disease. Ther Apher 3:287–293

Holvoet P, Vanhaecke J, Janssens S, Van de Werf F, Collen D (1998) Oxidized LDL and malondialdehyde-modified LDL in patients with acute coronary syndromes and stable coronary artery disease. Circulation 98:1487–1494

Honoré R, Rasmusse HH, Vorum H, Dejgaard K, Liu X, Gromov P, Madsen P, Gesser B, Tommerup N, Celis JE (1995) Heterogeneous nuclear ribonucleoproteins H, H′, and F are members of a ubiquitously expressed subfamily of related but distinct proteins. J Biol Chem 270: 28780–28789

Hoppe U, Bergemann J, Dienbeck W, Ennen J, Gohla S, Harris L, Jacob J, Kielholz J, Mei W, Pollet D, Schachtschabel G, Sauermann G, Schreiner V, Staband F, Steckel F (1999) Coenzyme Q_{10}, a cutaneous antioxidant and energize. Biofactors 9:371–378

Hostetler B, Kim C (2011) Detecting neoplasia-specific tNOX isoforms. Patent GB2441860

Hostetler B, Weston N, Kim C, Morré DM, Morré DJ (2009) Cancer site-specific isoforms of ENOX2 (tNOX), a cancer-specific cell surface oxidase. Clin Proteomics 5:46–51

Houghton PJ, Bailey FC, Germain GS, Grindey GB, Howbert JJ, Houghton JA (1990a) Studies on the cellular pharmacology of *N*-(4-methylphenylsulfonylurea)-N′-(4-chlorophenyl)-urea. Biochem Pharmacol 39:1187–1192

Houghton PJ, Bailey FC, Germain GS, Grindey GB, Witt BC, Houghton JA (1990b) N-5-indanylsulfonyl)-N-(4-chlorophenyl)urea, a novel agent equally cytotoxic to nonproliferating human colon adenocarcinoma cells. Cancer Res 50:318–322

Houghton PJ, Bailey FC, Houghton JA, Murti KG, Howbert JJ, Grindey GB (1990c) Evidence for mitochondrial localization of N-(4-methylphenylsulfonylurea)-*N*′-(4-chlorophenyl)-urea in human colon adenocarcinoma cells. Cancer Res 50:664–668

Howbert JJ, Grossman CS, Crowell TA, Rieder BJ, Harper RW, Kramer KE, Tao EV, Aikins J, Poore GA, Rinzel SM, Grindey GB, Shaw WN, Todd GC (1990) Novel agents effective against solid tumors: the diarylsulfonylureas. Synthesis, activities, and analysis of quantitative structure-activity relationships. J Med Chem 33:2393–2407

Hu S, Yen Y, Ann D, Wong DT (2007) Implications of salivary proteomics in drug discovery and development: a focus on cancer drug discovery. Drug Discov Today 21–22:911–996

Huang YA, Roper SD (2010) Intracellular Ca(2^+) and TRPM5-mediated membrane depolarization produce ATP secretion from taste receptor cells. J Physiol 588:2343–2350

Huang JQ, Turbide C, Daniels E, Jothy S, Beauchemin N (1990) Spatiotemporal expression of murine carcinoembryonic antigen (CEA) gene family members during mouse embryogenesis. Development 110:573–588

Hugo H, Ackland ML, Blick T, Lawrence MG, Clement JA, Williams ED, Thompson EW (2007) Epithelial-mesenchymal and mesenchymal-epithelia transitions in carcinoma progression. J Cell Physiol 213:374–383

Hui L, Zhang X, Wu X, Lin Z, Wang Q, Li Y, Hu G (2004) Identification of alternatively spliced mRNA variants related to cancers by genome-wide ESTs alignment. Oncogene 23: 3013–3023

Huyton T, Pye VE, Briggs LC, Flynn TC, Beuron F, Kondo H, Ma J, Zhang X, Freemont PS (2003) The crystal structure of murine p97/VCP at 3.6A. J Struct Biol 144:337–348

Hynes RO (1979) Surfaces of normal and malignant cells. Wiley, New York

Hyun DH, Emerson SS, Jo DG, Mattson MP, De Cabo R (2006) Calorie restriction up-regulates the plasma membrane redox system in brain cells and suppresses oxidative stress during aging. Proc Natl Acad Sci U S A 103:19908–19912

Iancu I, Olmer A, Strous RD (2007) Caffeinism: history, clinical features, diagnosis, and treatment. In: Smith BD, Gupta U, Gupta BS (eds) Caffeine and activation theory. CRC Press, Boca Raton, pp 331–347

Ichihashi M, Ooe M, Inui M, Omura K, Fugi K (2007) Efficacy evaluation of coenzyme Q_{10} in aged human skin in vivo. In: Abstracts, Fifth conference of the international coenzyme Q_{10} association, Kobe, Japan, p 88

Imai SI (2010) "Clocks" in the NAD world: NAD as a metabolic oscillator for the regulation of metabolism and aging. Biochim Biophys Acta 1804:1584–1590

Isaacs CE, Wen GY, Xu W, Jia JH, Rohan L, Corbo C, Di Maggio V, Jenkins EC Jr, Hillier S (2008) Epigallocatechin gallate inactivates clinical isolates of herpes simplex virus. Antimicrob Agents Chemother 52:962–970

Isbrucker RA, Bausch J, Edwards JA, Wolz E (2006a) Safety studies on epigallocatechin gallate (EGCg) preparations. Part 1: genotoxicity. Food Chem Toxicol 44:626–635

Isbrucker RA, Edwards JA, Wolz E, Davidovich A, Bausch J (2006b) Safety studies on epigallocatechin gallate (EGCg) preparations. Part 2: dermal, acute and short-term toxicity studies. Food Chem Toxicol 44:636–650

Isbrucker RA, Edwards JA, Wolz E, Davidovich A, Bausch J (2006c) Safety studies on epigallocatechin gallate (EGCg) preparations. Part 3: teratogenicity and reproductive toxicity studies in rats. Food Chem Toxicol 44:651–661

Ivankina NGVA, Novak VA (1980) H^+-transport across plasmalemma: H^+-ATPase or redox-chain? In: Spanswick RM, Lucas WJ, Dainty J (eds) Plant membrane transport: current conceptual issues. Elsevier, Amsterdam, pp 503–504

Jacobs E, Morré DJ, de Cabo R, Sweeting M, Morré DM (1996) Response of a protein disulfide isomerase-like activity of transitional endoplasmic reticulum to all-trans retinol. Life Sci 59:273–284

Jadot G (1986) Anti-inflammatory activity of superoxide dismutases: inhibition of adriamycin induced edema in rats. Free Radic Res Commun 2:19–26

Jancsó G, Kiraly E, Jancsó-Gábor A (1977) Pharmacologically induced selective degeneration of chemosensitive primary sensory neurons. Nature 270:741–743

Jang JJ, Cho KJ, Lee YS, Bae JH (1991) Different modifying responses of capsaicin in a wide spectrum initiation model of F344 rat. J Korean Med Sci 6:31–36

Jankovic B, Loblaw D, Nam R (2010) Capsaicin may slow psa doubling time: case report and literature review. Can Urol Assoc J 4:E9–E11

Janle E, Morré DM, Morré DJ, Zhou Q, Chang H, Zhu Y (2008) Pharmacokinetics of green tea catechins in extract and sustained-release preparations. J Diet Suppl 5:248–263

Jentsch KD, Hinsmann G, Hartmann H, Nickel P (1987) Inhibition of human immunodeficiency virus type I reverse transcriptase by suramin-related compounds. J Gen Virol 68:2183–2192

Jiang Z, Morré DM, Morré DJ (2006) A role for copper in biological time-keeping. J Inorg Biochem 100:2140–2149

Jiang Z, Gorenstein NM, Morré DM, Morré DJ (2008) Molecular cloning and characterization of a candidate human growth-related and time-keeping constitutive cell surface hydroquinone (NADH) oxidase. Biochemistry 47:14028–14038

Jolad SD, Hoffmann JJ, Schram KH, Cole JR (1982) Uvaricin, a new antitumor agent from Uvaria accuminata (Annonaceae). J Org Chem 47:3151–3153

Jorgensen P, Tyers M (2004) How cells coordinate growth and division. Curr Biol 14:R1014–R1027

Julius D, Brake A, Kunisawa R, Thorner J (1984) Isolation of the putative structural gene for the lysine-arginine-cleaving endopeptidase required for processing of yeast prepro-alpha-factor. Cell 37:1075–1089

Kaeberiein M, McVey M, Guarente L (1999) The SIR*2/3/4* complex and SIR*2* alone promote longevity in *Saccharomyces cerevisiae* by two different mechanisms. Genes Dev 20:2570–2580

Kalén A, Norling B, Appelkvist EL, Dallner G (1987) Ubiquinone biosynthesis by the microsomal fraction from rat liver. Biochim Biophys Acta 926:70–78

Kalnina Z, Zayakin P, Silina K, Line A (2005) Alterations of pre-mRNA splicing in cancer. Genes Chromosome Cancer 42:342–357

Kamsteeg M, Rutherford T, Sapi E, Hanczaruk B, Shahabi S, Flick M, Brown D, Mor G (2003) Phenoxodiol—an isoflavene analog—induces apoptosis in chemoresistant ovarian cancer cells. Oncogene 22:2611–2620

Kaneko T, Willner D, Monkovic I, Knipe JO, Braslawsky GR, Greenfield RS, Vyas DM (1991) New hydrazone derivatives of adriamycin and their immunoconjugates—a correlation between acid stability and cytotoxicity. Bioconjug Chem 2:133–141

Karni R, de Stanchina E, Lowe SW, Sinha R, Mu D, Krainer AR (2007) The gene encoding the splicing factor SF2/ASF is a proto-oncogene. Nat Struct Mol Biol 14:185–193

Katdare M, Osborne MP, Telang NT (1998) Inhibition of aberrant proliferation and induction of apoptosis in pre-neoplastic human mammary epithelial cells by natural phytochemicals. Oncol Rep 5:311–315

Kaul N, Choi J, Forman HJ (1998) Transmembrane redox signaling activates NF-KB in macrophages. Free Radic Biol Med 24:202–207

Kazanov D, Dvory-Sobol H, Pick M, Liberman E, Strier L, Choen-Noyman E, Deutsch V, Kunik T, Arber N (2004) Celecoxib but not rofecoxib inhibits the growth of transformed cells *in vitro*. Clin Cancer Res 10:267–271

Kelker M, Kim C, Chueh P-J, Guimont R, Morré DM, Morré DJ (2001) Cancer isoform of a tumor-associated cell surface NADH oxidase (tNOX) has properties of a prion. Biochemistry 40:7351–7354

Kelly G (2004) Interim results of a phase Ib/IIa study of oral phenoxodiol in patients with late-stage, hormone-refractory prostate cancer. Proc Am Assoc Cancer Res 45(suppl 1):103–104

Kelly G (2010) A scientific odyssey: the story of phenoxodiol. http://www.Kelleymusings.com/books/

Kelly MG, Mor G, Husband A, O'Malley DM, Baker L, Azodi M, Schwartz PO, Rutherford TJ (2011) Phase II evaluation of phenoxodiol in combination with cisplatin or paclitaxel in women with platinum/taxane-refractory/resistant epithelial ovarian, fallopian tube, or primary peritoneal cancers. Int J Gynecol Cancer 21:633–639

Kemmink J, Darby NJ, Dijkstra K, Scheek RM, Creighton TE (1995) Nuclear magnetic resonance characterization of the N-terminal thioredoxin-like domain of protein disulfide isomerase. Protein Sci 4:2587–2593

Kemmink J, Dijkstra K, Mariani M, Scheek RM, Penka E, Nilges M, Darby NJ (1999) The structure in solution of the b domain of protein disulfide isomerase. J Biomol NMR 13:357–368

Kemp BE, Stapleton D, Campbell DJ, Chen ZP, Murthy S, Walter M, Gupta A, Adams JJ, Katsis F, Van Denderen B, Jennings IG, Iseli T, Michell BJ, Witters LA (2003) AMP-activated protein kinase, super metabolic regulator. Biochem Soc Trans 31:162–168

Kennett EC, Kuchel PW (2003) Redox reactions and electron transfer across the red cell membrane. IUBMB Life 55:375–385

Kern D, Draelos Z, Morré DM, Morré DJ (2008) Age-related oxidase (arNOX) activity of epidermal punch biopsies correlate with subject age and arNOX activities of serum and saliva. In: Abstracts, Society Investigative Dermatology, Kobe, Japan, May 2008

Kern DG, Draelos ZD, Meadows C, Morré DM, Morré DJ (2010) Controlling reactive oxygen species in skin at their source to reduce skin aging. Rejuvenation Res 13:165–167

Key JL (1964) Ribonucleic acid and protein synthesis as essential processes for cell elongation. Plant Physiol 39:365–370

Kilman RM, Hey J (1993) DNA sequence variation at the period locus within and among species of the *Drosophila melanogaster* complex. Genetics 133:375–387

Kim C (2011) Single chain pan ECTO-NOX variable region (ScFv) antibody, coding sequence and methods. Patent GB2442553B

Kim C, Morré DJ (2004) Prion proteins and ECTO-NOX proteins exhibit similar oscillating redox activities. Biochem Biophys Res Commun 315:1140–1146

Kim C, MacKellar WC, Cho N, Byrn SR, Morré DJ (1997) Impermeant antitumor sulfonylurea conjugates that inhibit plasma membrane NADH oxidase and growth of HeLa cells in culture.

Identification of binding proteins from sera of cancer patients. Biochim Biophys Acta 1324:171–181

Kim WH, Kang KH, Kim MY, Choi KH (2000a) Induction of p53-independent p21 during ceramide-induced G_1 arrest in human hepatocarcinoma cells. Biochem Cell Biol 78:127–138

Kim WH, Ghil KC, Lee JH, Yeo SH, Chun YJ, Choi KH, Kim DK, Kim MY (2000b) Involvement of p27 (kip1) in ceramide-mediated apoptosis in HL-60 cells. Cancer Lett 151:39–48

Kim C, Crane FL, Faulk WP, Morré DJ (2002) Purification and characterization of a doxorubicin-inhibited NADH-quinone (NADH-ferricyanide) reductase from rat liver plasma membranes. J Biol Chem 10:16441–16447

Kim DS, Hwang ES, Lee JE, Kim SY, Park KC (2003) Sphingosine-1-phosphate promotes mouse melanocyte survival via ERK and Akt activation. Cell Signal 15:919–926

Kim C, Layman S, Morré DM, Morré DJ (2005) Fourier transform infrared and circular dichroism spectroscopic analysis underlie tNOX periodic oscillations. Dose Response 3:391–413

King DP, Takahashi JS (2000) Molecular genetics of circadian rhythms in mammals. Annu Rev Neurosci 29:713–742

Kishi T, Morré DM, Morré DJ (1999) The plasma membrane NADH oxidase of Hela cells has hydroquinone reductase activity. Biochim Biophys Acta 1412:66–77

Kishi T, Takahashi T, Mizobachi S, Mori K, Okamoto T (2002) Effect of dicumarol, a NAD(P)H: quinone acceptor oxidoreductase 1 (DT-diaphorase) inhibitor on ubiquinone redox cycling in cultured rat hepatocytes. Free Radic Res 36:413–419

Kloppel TM, Keenan TW, Freeman MJ, Morré DJ (1977) Glycolipid-bound sialic acid in serum: increased levels in mice and humans bearing mammary carcinomas. Proc Natl Acad Sci U S A 74:3011–3013

Kluger HM, McCarthy MM, Alvero AB, Sznol M, Ariyan S, Camp RL, Rimm DL, Mor G (2007) The X-linked inhibitor of apoptosis protein (XIAP) is up-regulated in metastatic melanoma, and XIAP cleavage by phenoxodiol is associated with carboplatin sensitization. J Transl Med 5:6–20

Koboid S, Lütkens T, Cao Y, Bokemeyer C, Atanackovic D (2010) Autoantibodies against tumor-related antigens: incidence and biologic significance. Hum Immunol 71:643–651

Kochian LV, Lucas WJ (1985) Potassium transport in corn roots. III. Perturbation by exogenous NADH and ferricyanide. Plant Physiol 77:429–436

Korshin GV, Frenkal AI, Stern EA (1998) EXAFS study of the inner shell structure in copper (II) complexes with humic substances. Environ Sci Technol 32:2699–2705

Kozak M (1978) How do eukaryotic ribosomes select initiation regions in messenger RNA? Cell 15:1109–1123

Kozak M (1987) An analysis of 5′-noncoding sequences from 699 vertebrate messenger RNAs. Nucleic Acids Res 15:8125–8148

Kozak M (1989) The scanning model for translation: an update. J Cell Biol 108:229–241

Kripke DF, Wyborney VG (1980) Lithium slows rat circadian activity rhythms. Life Sci 26:1319–1321

Kripke DF, Mullaney DJ, Atkinson ML, Wolf S (1978) Circadian rhythm disorder in manic-depressives. Biol Psychiatry 13:335–351

Kripke DF, Judd LL, Hubbard B, Janowsky DS, Huey LF (1979) The effect of lithium carbonate on the circadian rhythm of sleep in normal human subjects. Biol Psychiatry 14:545–548

Kroesen B-J, Jacobs S, Pettus BJ, Sietsma H, Kok J-W, Hannun YA, de Leij LFMH (2003) BcR induced apoptosis involves differential regulation of C16 and C24-ceramide formation and sphingolipid-dependent activation of the proteasome. J Biol Chem 278:14723–14731

Kromkowski J, Hignite H, Morré DM, Morré DJ (2008) Response to lithium of a cell surface ECTO-NOX protein with time-keeping characteristics. Neurosci Lett 438:121–125

Kroning H, Kahne T, Ittenson A, Ansorge S (1994) Thiol-protein disulfide-oxireductase (protein disulfide isomerase), a new plasma membrane constituent of mature human B lymphocytes. Scand J Immunol 39:346–350

Kummer JT (1962) Ortho-para hydrogen conversion by metal surfaces at 21°K. J Phys Chem 66:1715

Kundu N, Smyth MJ, Samsel L, Fulton AM (2002) Cyclooxygenase inhibitors block cell growth, increase ceramide and inhibit cell cycle. Breast Cancer Res Treat 76:57–64

Kunkel M, Reichert TE, Benz P, Lehr HA, Jeong JH, Wieand S, Bartenstein P, Wagner W, Whiteside TL (2003) Overexpression of Glut-1 and increased glucose metabolism in tumors are associated with a poor prognosis in patients with oral squamous cell carcinoma. Cancer 97: 1015–1024

Kurkdjian A, Leguay JJ, Guern J (1979) Influence of fusicoccin on the control of cell division by auxins. Plant Physiol 64:1053–1057

Kutschera U, Schopfer P (1985a) Evidence against the acid-growth theory of auxin action. Planta 163:483–493

Kutschera U, Schopfer P (1985b) Evidence for the acid growth theory of fusicoccin action. Planta 163:494–499

Kyte J, Doolittle RF (1982) A simple method for displaying hydropathic character of a protein. J Mol Biol 157:105–132

L'Allemain G, Franchi A, Cragoe E Jr, Pouysségur J (1984) Blockade of the Na^+/H^+ antiport abolishes growth factor-induced DNA synthesis in fibroblasts. J Biol Chem 259:4313–4319

Lambeth JD, Cheng G, Arnold RS, Edens WA (2000) Novel homologs of gp^{91} phox. Trends Biochem Sci 25:459–461

Lange S, Richard D, Viergutz T, Kriehuber R, Weiss DG, Simko M (2002) Alterations in the cell cycle and in the protein level of cyclin D1, p21Clp1, and p16lNK4a after exposure to 50 Hz MF in human cells. Radiat Environ Biophys 41:131–137

Langmesser S, Tallone T, Bordon A, Busconi S, Albrecht U (2008) Interaction of circadian clock proteins PER2 and CRY with BMAL1 and CLOCK. BMC Mol Biol 9:41

Larm JA, Vaillant F, Linnane AW, Lawen A (1994) Up-regulation of the plasma membrane oxidoreductase as a prerequisite for the viability of human Namalwa ρ^o cells. J Biol Chem 269:30097–30100

Latini SA, Pedata F (2001) Adenosine in the central nervous system: release mechanisms and extracellular concentrations. J Neurochem 79:463–484

Lauricella M, Giuliano M, Emanuele S, Carabillo M, Vento R, Tesoriere G (1998) Increased cyclin E level in retinoblastoma cells during programmed cell death. Cell Mol Biol 44:1229–1235

Lawen A, Martinus RD, McMullen GL, Nagley P, Vaillant F, Wolvetang EJ, Linnane AW (1994) The universality of bioenergetic disease: the role of mitochondrial mutation and the putative inter-relationship between mitochondria and plasma membrane NADH oxidoreductase. Mol Aspects Med 15:s13–s27

Lawen A, Ly JD, Lane DJR, Zarschler K, Messina A, De Pinto V (2005) Voltage-dependent anion-selective channel 1 (VDAC1)—a mitochondrial protein, rediscovered as a novel enzyme in the plasma membrane. Int J Biochem Cell Biol 37:277–282

Lee YS, Nam DH, Kim JA (2000) Induction of apoptosis by capsaicin in A172 human glioblastoma cells. Cancer Lett 161:121–130

Lee C, Etchegaray JP, Cagampang FR, Loudon AS, Reppert SM (2001) Posttranslational mechanisms regulate the mammalian circadian clock. Cell 128:707–719

Lee SRKS, Yang J, Kwon C, Lee WJ, Rhee SG (2002) Reversible inactivation of the tumor suppressor PTEN by H_2O_2. J Biol Chem 277:20336–20342

Leeuwenburgh C, Wagner P, Holloszy JO, Sohal RS, Heinecke JW (1997) Caloric restriction attenuates dityrosine cross-linking of cardiac and skeletal muscle proteins in aging mice. Arch Biochem Biophys 346:74–80

Leeuwenburgh C, Hansen P, Shaish A, Holloszy JO, Heinecke JW (1998) Markers of protein oxidation by hydroxyl radical and reactive nitrogen species in tissues of aging rats. Am J Physiol 274:453–461

Lembi CA, Morré DJ, St.-Thomson K, Hertel R (1971) N-1-naphthylphthalamic acid-binding activity of a plasma membrane-rich fraction from maize coleoptiles. Planta 99:37–45

Lenaz G, Paolucci U, Fato R, D'Aurelio M, Parenti Castelli G, Sgarbi G, Biagini G, Ragni L, Salardi S, Cacciari E (2002) Enhanced activity of the plasma membrane oxidoreductase in circulating lymphocytes from insulin-dependent diabetes mellitus patients. Biochem Biophys Res Commun 290:1589–1592

Lengle EE (1979) Increased levels of lipid-bound sialic acid in thymic lymphocytes and plasma from leukemic AKR/J mice. J Natl Cancer Inst 62:1564–1579

Lesauter J, Silver R (1993) Heavy water lengthens the period of free-running rhythms in lesioned hamsters bearing SCN grafts. Physiol Behav 54:599–604

Levi F (2000) Marked 24-h rest/activity rhythms are associated with better quality of life, better response, and longer survival in patients with metastatic colorectal cancer and good performance status. Clin Cancer Res 6:3038–3045

Levi F (2002) From circadian rhythms to cancer chemotherapeutics. Chronobiol Int 25:459–461

Levi F, Schibler U (2007) Circadian rhythms: mechanisms and therapeutic implications. Annu Rev Pharmacol Toxicol 47:593–628

Levine BS, Kannel WB (2003) Coronary heart disease risk in people 65 years of age and older. Prog Cardiovasc Nurs 18:135–140

Li JJ, Chen SH, Lin CL, Tsai SH, Liang YC (2001) Inhibition of melanoma growth and metastasis by combination with (−)-epigallocatechin-3-gallate and dacarbazine in mice. J Cell Biochem 83:631–642

Li X, Ding X, Adrian TE (2003) Arsenic trioxide induces apoptosis in pancreatic cancer cells via changes in cell cycle, caspase activation, and GADD expression. Pancreas 27:174–179

Li N, Cai Y, Zuo X, Xu S, Zhang Y, Chan P, Zhang YA (2008) Suprachiasmatic nucleus slices induce molecular oscillations in fibroblasts. Biochem Biophys Res Commun 377:1179–1184

Li G-X, Chen Y-K, Hou Z, Xiao H, Jin H, Lu G, Lee M-J, Liu B, Guan F, Yang Z, Yu A, Yang CS (2010) Pro-oxidative activities and dose–response relationship of (−)-epigallocatechin-3-gallate in the inhibition of lung cancer cell growth: a comparative study *in vivo* and *in vitro*. Carcinogenesis 31:902–910

Liao S, Umekita Y, Guo J, Kokontis JM, Hiipakka RA (1995) Growth inhibition and regression of human prostate and breast tumors in athymic mice by tea epigallocatechin gallate. Cancer Lett 96:239–243

Lide DR (1992) Handbook of chemistry and physics, 73rd edn. CRC Press, Boca Raton, FL

Lim J-H, Lee Y-M, Chun Y-S, Chen J, Kim J-E, Park J-W (2010) Sirtuin 1 modulates cellular responses to hypoxia by deacetylating hypoxia-inducible factor 1α. Mol Cell 38:864–878

Lin W (1982a) Isolation of NADH oxidation system from the plasmalemma of corn root protoplast. Plant Physiol 70:326–328

Lin W (1982b) Responses of corn root protoplasts to exogenous NADH: oxygen consumption, ion uptake and membrane potential. Proc Natl Acad Sci U S A 79:3773–3776

Lin W (1984) Further characterization on the transport property of plasmalemma NADH oxidation system in isolated corn root protoplasts. Plant Physiol 74:219–222

Lin SJ, Guarente L (2003) Nicotinamide adenine dinucleotide, a metabolic regulator of transcription, longevity and disease. Curr Opin Cell Biol 15:241–246

Lin SJ, Kaeberlein M, Andalis AA, Sturtz LA, Defossez PA, Culotta V, Fink GR, Guarente L (2002) Calorie restriction extends *Saccharomyces cerevisiae* lifespan by increasing respiration. Nature 418:344–348

Linnane AW, Eastwood H (2006) Cellular redox regulation and pro-oxidant signaling systems: a new prospective on the free radical theory of aging. Ann N Y Acad Sci 1067:47–55

Linnane AW, Kios M, Vitetta I (2007) Healthy aging: regulation of the metabolome by cellular redox modulation and peroxidant signaling systems: the essential roles of superoxide anion and hydrogen peroxide. Biogerentology 8:445–467

Liochev A, Fridovich I (1968) Superoxide is responsible for the vanadate stimulation of NAD(P) H oxidation by biological membranes. Arch Biochem Biophys 263:299–304

Littarru GP, Tiano L (2007) Bioenergetic and antioxidant properties of coenzyme Q_{10}: recent developments. Mol Biotechnol 37:31–37

Liu CH, Chang SH, Narko K, Trifan OC, Wu MT, Smith E, Haudenschild C, Lane TF, Hia T (2001) Overexpression of COX-2 is sufficient to induce tumorigenesis in transgenic mice. J Biol Chem 276:18563–18569

Liu S-C, Huang S, Chueh PJ (2005) RNAi-mediated gene silencing to examine the biological function of a tumor-associated NADH oxidase (tNOX) in transformed cells in culture. Abs Am Soc Cell Biol 45:430

Liu S-C, Yang J-J, Shao K-N, Chueh PJ (2008) RNA interference targeting tNOX attenuates cell migration via a mechanism that involves membrane association of Rac. Biochem Biophys Res Commun 365:672–677

Liu N-C, Hsieh P-F, Hsieh M-K, Zeng Z-M, Cheng J-L, Liao J-W, Chueh PJ (2012) Capsaicin-mediated tNOX (ENOX2) up-regulation enhances cell proliferation and migration in vitro and in vivo. J Agric Food Chem 60:2758–2765

Lloyd D, Stupfel M (1991) The occurrence and functions of ultridian rhythms (review). Biol Rev Camb Philos Soc 66:275–299

Lloyd D, Edward SW, Fry JC (1982) Temperature-compensated oscillations in respiration and protein content in synchronous cultures of *Acanthamoeba castellanii*. Proc Natl Acad Sci U S A 79:3785–3788

Löbler M, Klämbt D (1985) Auxin-binding protein from coleoptile membranes of corn. I. Purification by immunological methods. J Biol Chem 260:9848–9853

Lobyshev VI, Shikhlinskaya RE, Ryzhikov BD (1999) Experimental evidence for intrinsic luminescence of water. J Mol Liquids 82:3–81

Lockhart JA (1965) An analysis of irreversible plant cell elongation. J Theor Biol 8:264–275

Lopez-Lazaro M (2008) The Warburg effect: why and how do cancer cells activate glycolysis in the presence of oxygen? Anticancer Agents Med Chem 8:305–312

Löw H, Crane FL (1978) Redox function in plasma membranes. Biochim Biophys Acta 515:141–161

Löw H, Werner S (1976) Effect of reducing and oxidizing agents on the adenylate cyclase activity in adipocyte plasma membranes. FEBS Lett 65:96–98

Löw H, Sun IL, Navas P, Grebin C, Crane FL, Morré DJ (1986) Transplasma membrane electron transport is part of a diferric transferrin reductase system. Biochem Biophys Res Commun 137:1117

Löw H, Crane FL, Sun IL (1990) Oxidoreduction in the plasma membrane. In: Crane FL, Morré DJ, Löw HE (eds) Oxidoreduction at the plasma membrane: relation to growth and transport, I. Animals. CRC Press, Boca Raton, FL, pp 29–65

Löw H, Crane FL, Grebing C, Isaksson M, Lindgren A, Sun IL (1991) Modification of transplasma membrane oxidoreduction by SV40 transformation of 3T3 cells. J Bioenerg Biomembr 23:903–917

Lowrey PL, Takshashi JS (2004) Mammalian circadian biology: elucidating genome-wide levels of temporal organization. Annu Rev Genomics Hum Genet 5:407–441

Lucha PA Jr, Rosen L, Olenwine JA, Reed JF III, Riether RD, Stasik JJ Jr, Khubchandani IT (1997) Value of carcinoembryonic antigen monitoring in curative surgery for recurrent colorectal carcinoma. Dis Colon Rectum 40:145–149

Lupas AN, Martin J (2002) AAA proteins. Curr Opin Struct Biol 12:746–753

Lüthen H, Böttger M (1992) A high-tech low-cost auxanometer for high-resolution determination of elongation rates in six simultaneous experimental setups. Mitt Inst Allg Bot Hamburg 24:13–22

Lüthen H, Bigdon M, Böttger M (1990) Reexamination of the acid growth theory of auxin action. Plant Physiol 93:931–939

Lüthje S, Döring OS, Heuer H, Lüthen H, Böttger M (1997) Oxidoreductases in plant plasma membranes. Biochim Biophys Acta 1331:81–102

Lüthje S, Van Gestelen P, Córdoba-Pedregosa P, González-Reyes JA, Asard H, Villalba JM, Böttger M (1998) Quinones in plant plasma membranes—a missing link. Protoplasma 25:43–51

Lyle DB, Ayotte RD, Sheppard AR, Adey WR (1988) Suppression of T-lymphocyte cytotoxicity following exposure to 60-Hz sinusoidal electric fields. Bioelectromagnetics 9:303–313

Lyles MM, Gilbert HF (1991) Catalysis of the oxidative unfolding of ribonuclease A by protein disulfide isomerase: dependence of the rate on the composition of the redox buffer. Biochemistry 30:613–619

Maceyka M, Payne SG, Milstien S, Spiegal S (2002) Sphingosine kinase, sphingosine-1-phosphate and apoptosis. Biochem Biophys Acta 1585:193–201

Macho A, Blázquez MV, Navas P, Muñoz E (1998) Induction of apoptosis by vanilloid compounds does not require de novo gene transcription and activator protein I activity. Cell Growth Differ 9:277–286

Macho A, Calzado MA, Muñoz-Blanco J, Gómez-Díaz C, Gajate C, Mollinedo F, Navas P, Muñoz E (1999) Selective induction of apoptosis by capsaicin in transformed cells: the role of reactive oxygen species and calcium. Cell Death Differ 6:155–165

Macho A, Sancho R, Minassi A, Appendino G, Lawen A, Muñoz E (2003) Involvement of reactive oxygen species in capsaicinoid-induced apoptosis in transformed cells. Free Radic Res 37:611–619

MacMillan-Crow LA, Crow JP, Thompson JA (1998) Peroxynitrite-mediated inactivation of manganese superoxide dismutase involves nitration and oxidation of critical tyrosine residues. Biochemistry 37:1613–1622

Macquet JP, Butour JL (1983) Platinum-amine compounds: importance of the labile and inert ligands for their pharmacological activities toward L1210 leukemia cells. J Natl Cancer Inst 70:899–905

Maher P, Schubert D (2000) Signaling by reactive oxygen species in the nervous system. Cell Mol Life Sci 57:1287–1305

Mailliet F, Ferry G, Vella F, Berger S, Cogé F, Chomarat P, Mallet C, Guénin SP, Guillaumet G, Viaud-Massuard M-C, Yous S, Delagrange P, Boutin JA (2005) Characterization of the melatoninergic MT3 binding site on the NRH:quinone oxidoreductase 2 enzyme. Biochem Pharmacol 71:74–88

Malik SG, Vaillant F, Lawen A (2004) Plasma membrane NADH-oxidoreductase in cells carrying mitochondrial DNA G11778A mutation and in cells devoid of mitochondrial DNA (ρ^o). Biofactors 20:189–198

Mandel R, Ryser HJ-P, Ghani F, Wu M, Peak D (1993) Inhibition of a reductive function of the plasma membrane by bacitracin and antibodies against protein disulfide-isomerase. Proc Natl Acad Sci U S A 90:4112–4416

Manson MM, Holloway KA, Howells LM, Hudson EA, Plummer SM, Squires MS, Prigent SA (2000) Modulation of signal-transduction pathways by chemopreventive agents. Biochem Soc Trans 28:7–12

Mao L-C, Wang H-M, Lin Y-Y, Chang T-K, Hsin Y-H, Chueh P-J (2008) Stress-induced down-regulation of tumor-associated NADH oxidase during apoptosis in transformed cells. FEBS Lett 582:3445–3450

Marin-Burgin A, Reppenhagen S, Klusch A, Wendland JR, Petersen M (2000) Low-threshold heat response antagonized by capsazepine in chick sensory neurons, which are capsaicin-insensitive. Eur J Neurosci 12:3560–3566

Markert C, Morré DM, Morré DJ (2004) Human amyloid peptides Aβ1-40 and Aβ1-42 exhibit NADH oxidase activity with copper-induced oscillations and a period length of 24 min. Biofactors 20:221–235

Markham RS, Frey S, Hill GJ (1963) Methods for the enhancement of image detail and accentuation of structure in electron microscopy. Virology 20:88–100

Marré E (1979) Fusicoccin: a tool in plant physiology. Annu Rev Plant Physiol 30:273–288

Marré E, Ballarin-Denti A (1985) The proton pumps of the plasmalemma and the tonoplast of higher plants. J Bioenerg Biomembr 17:1–21

Marré E, Lado P, Rasi-Caldogno F, Colombo F (1973) Correlation between cell enlargement in pea internode segments and decrease in the pH of the medium of incubation. Plant Sci Lett 1:179–184

Marré E, Lado P, Ferroni A, Ballarin-Denti A (1974) Transmembrane potential increase induced by auxin, benzyladenine and fusicoccin. Correlation with proton extrusion and cell enlargement. Plant Sci Lett 2:257–265

Marsh M, Peichen-Matthews A (2000) Endocytosis in viral replication. Traffic 1:525–532

Martín SF, Navarro F, Forthoffer N, Navas P, Villalba JM (2001) Neutral magnesium-dependent sphingomyelinase from liver plasma membrane: purification and inhibition by ubiquinol. J Bioenerg Biomembr 33:143–153

Martín SF, Gómez-Díaz C, Bello RI, Navas P, Villalba JM (2003) Inhibition of neutral Mg2+-dependent sphingomyelinase by ubiquinol-mediated plasma membrane electron transport. Protoplasma 221:109–116

Martínez-Sánchez G, Giuliana A (2007) Cellular redox status regulates hypoxia inducible factor-1 activity. Role in tumour development. J Exp Clin Cancer Res 26:39–50

Martínez-Merino V, Gil MJ, Encío I, Migliaccio M, Arteaga C (2000) Benzo(*b*)thiophene sulfonamide-1,1-dioxido derivatives and their use as antineoplastic agents. WO Patent 00/63202

Martinus RD, Linnane AW, Nagley P (1993) Growth of rho 0 human Namalwa cells lacking oxidative phosphorylation can be sustained by redox compounds potassium ferricyanide or coenzyme Q10 putatively acting through the plasma membrane oxidase. Biochem Mol Biol Int 31:997–1005

Matanoski GM (1995) Electromagnetic fields: biological interactions and mechanisms. In: Blank M (ed) Advances in chemistry, vol 250. American Chemical Society, Washington, DC, pp 157–190

Matthias LJ, Hogg PJ (2003) Redox control on the cell surface: implications for HIV-1 entry. Antioxid Redox Signal 5:133–138

Maxeiner H-G, Morré DM, Morré DJ (unpublished) (−)-Epigallocatechin gallate, the principal catechin of green tea, has anti-HIV activity

Mayo LA, Curnutte J (1990) Kinetic microplate assay for superoxide production by neutrophils and other phagocytic cells. Methods Enzymol 186:567–575

McClung CA (2007) Circadian genes, rhythms and the biology of mood disorders. Pharmacol Ther 114:222–232

McDaniel M, Sulzman FM, Hastings JW (1974) Heavy water slows the Gonyaulax clock: a test of the hypothesis that D_2O affects circadian oscillations by diminishing the apparent temperature. Proc Natl Acad Sci U S A 71:4389–4391

McGinness JE, Procter PD, Demopoulis HB, Hokanson JA, Kirkpatrick DS (1978) Superoxide production by cis and trans-Pt(II) diamine dichloride. Physiol Chem Phys 10:267–277

McKenna ST, Kunkel JG, Bosch M, Rounds CM, Vidali L, Winship LJ, Hepler PK (2009) Exocytosis precedes and predicts the increase in growth in oscillating pollen tubes. Plant Cell 21(3):026–3040

McKie AT, Barrow D, Latunde-Dada GO, Rolfs A, Sager G, Mudaly E, Mudaly M, Richardson C, Barlow D, Bomford A, Peters TJ, Raja KB, Shirali S, Hediger MA, Farzaneh F, Simpson RJ (2001) An iron-regulated ferric reductase associated with the absorption of dietary iron. Science 291:1755–1759

McRae DH, Bonner J (1953) Chemical structure and antiauxin activity. Physiol Plant 6:485–510

Medina MA (1998) Plasma membrane redox systems in tumor cells. In: Asard H, Bérczi A, Caubergs R (eds) Plasma membrane redox systems and their role in biological stress and disease. Kluwer Academic, Dordrecht, The Netherlands, pp 309–324

Medina MA, Sánchez-Jiménez F, Segura JA, Núñez de Castro I (1988) Transmembrane ferricyanide reductase activity in Ehrlick ascites tumor cells. Biochim Biophys Acta 946:1–4

Medina MA, del Castillo-Olivares A, de Castro IN (1997) Multifunctional plasma membrane redox systems. Bioessays 19:977–984

Merker M, Bongard RD, Kettenhofen NJ, Okamot Y, Dawson CA (2002) Intracellular redox status affects transplasma membrane electron transport in pulmonary arterial endothelial cells. Am J Physiol Lung Cell Physiol 282:L26–L43

Milenkl YY, Sibileva RN, Strzhemechny MA (1997) Natural ortho-para conversion rate in liquid and gaseous hydrogen. J Low Temp Phys 107:77–82

Miller S, Carnell L, Moore HH (1992) Post-Golgi membrane traffic: brefeldin A inhibits export from distal Golgi compartments to the cell surface but not recycling. J Cell Biol 118:267–283

Millet B, Badot PM (1996) The revolving movement mechanism in Phaseolus: new approaches to old questions. In: Greppin H, Degli Agosti R, Bonzon M (eds) Vistas on biorhythmicity. University of Geneva Press, Geneva, pp 77–98

Minorsky PV (2007) Solar-terrestrial effects on bean seed imbibition. Poster Abstract Book, American Society for Plant Biologists annual meeting, Chicago, 7–11 July 2007, p 186

Mitchison JM (1971) The biology of the cell cycle. Cambridge University Press, Cambridge

Miura T, Suzuki K, Kohata N, Takeuchi H (2000) Metal binding modes of Alzheimer's amyloid beta-peptide in insoluble aggregates and soluble complexes. Biochemistry 39:7024–7031

Miyase T, Sano M, Nakai H, Muraoka M, Nakazawa M, Suzuki M, Yoshino K, Nishihara Y, Tanai J (1999) Antioxidants from *Lespedeza homoloba*. (I). Phytochemistry 52:303–310

Mizejewski GJ (2001) Alpha-fetoprotein structure and function: relevance to isoforms, epitopes, and conformational variants. Exp Biol Med 226:377–408

Mizuno T, Uchino K, Toukairin T, Tanabe A, Nakashima H, Yamamoto N, Ogawara H (1992) Inhibitory effect of tannic acid sulfate and related sulfates on infectivity, cytopathic effect, and giant cell formation of human immunodeficiency virus. Planta Med 58:535–539

Moher ED, Collins JL, Grieco PA (1992) Synthetic studies on quassinoids: total synthesis of simalikalactone D and assignment of the absolute configuration of the α-methyl hydrate ester side chain. J Am Chem Soc 114:2764–2765

Møller IM, Bérczi A (1985) Oxygen consumption by purified plasmalemma vesicles from wheat roots. Stimulation by NADH and salicylhydroxamic acid. FEBS Lett 193:180–184

Møller IM, Bérczi A (1986) Salicylhydroxanic acid-stimulated NADH oxidation by purified plasmalemma vesicles from wheat roots. Physiol Plant 68:67–74

Moolenaar WH, Defize LHK, van der Saag PT, De Laat SW (1986) The generation of ionic signals by growth factor. In: Aronson PS, Bororn WF (eds) Current topics in membranes and transport, vol 26. Academic, Orlando, FL, pp 137–156

Moore BC, Simmons DL (2000) COX-2 inhibition, apoptosis, and chemoprevention by nonsteroidal anti-inflammatory drugs. Curr Med Chem 7:1131–1144

Moore HB, Jutare T, Bauer JC, Jones JA (1963) The biology of *Lytechinus variegatus*. Bull Mar Sci Gulf Caribbean 13:23–53

Moreno-Sánchez R, Rodriguez-Enriquez S, Marin-Hernández A, Saavedra E (2007) Energy metabolism in tumor cells. FEBS J 274:1393–1418

Morgan DO (1997) Cyclin-dependent kinases: engines, clocks, and microprocessors. Annu Rev Cell Dev Biol 13:261–291

Mori A, Lehmann S, O'Kelly J, Kumaggi T, Desmond JC, Pervan M, McBride WH, Kizaki M, Kaeffier HP (2006) Capsaicin, a component of red peppers, inhibits the growth of androgen-independent, p53 mutant prostate cancer cells. Cancer Res 66:3222–3229

Mormont MC, Levi F (1997) Circadian-system alterations during cancer process: a review. Int J Cancer 70:241–247

Mormont MC, Waterhouse J, Bleuzen P, Giacchetti S, Jami A, Bogdan A, Lellouch J, Misset JL, Touitou Y, Levi F (2000) Marked 24-h rest/activity rhythms are associated with better quality of life, better response, and longer survival in patients with metastatic colorectal cancer and good performance status. Clin Cancer Res 6:3038–3045

Morré DJ (1970) Auxin effects on the aggregation and heat coagulability of cytoplasmic proteins and lipoproteins. Physiol Plant 23:38–50

Morré DJ (1986) Calcium modulation of auxin-membrane interactions in plant cell elongation. In: Trewavas AJ (ed) Molecular and cellular aspects of calcium in plant development. Plenum Press, New York, pp 293–300

Morré DJ (1989) Stimulus-response coupling in auxin regulation of plant cell elongation. In: Boss WF, Morré DJ (eds) Second messengers in plant growth and development. Alan R. Liss, New York, pp 81–114

Morré DJ (1990a) Activation of phospholipase A: an alternative mechanism for signal transduction. In: Morré DJ, Boss WF, Loewus FA (eds) Inositol metabolism in plants. Wiley-Liss, New York, pp 227–257

Morré DJ (1990b) Comparison of plant and animal signal transducing systems. In: Ranjeva R, Boudet AM (eds) Signal perception and transduction in higher plants, Toulouse, France, 9–13 July 1989. NATO ASI series, vol H47. Springer, Rijeka, pp 307–322

Morré DJ (1990c) Transmembrane signal transduction and the control of plant growth. In: Randall DD, Blevins DG (eds) Current topics in plant biochemistry and physiology, vol 9. University of Missouri Press, Columbia, pp 47–65
Morré DJ (1994a) The hormone- and growth factor-stimulated NADH oxidase. J Bioenerg Biomembr 26:421–433
Morré DJ (1994b) Physical membrane displacement: reconstitution in a cell-free system and relationship to cell growth. Protoplasma 180:3–13
Morré DJ (1995a) NADH oxidase activity of HeLa plasma membranes inhibited by the antitumor sulfonylurea N-4-methylphenylsulfonyl-N′-4-chlorophenylurea LY181984 at an external site. Biochim Biophys Acta 1240:201–208
Morré DJ (1995b) The role of NADH oxidase in growth and physical membrane displacement. Protoplasma 184:14–21
Morré DJ (1998a) A protein disulfide-thiol interchange protein with NADH:protein-disulfide-reductase (NADH oxidase) activity as a molecular target for low levels of exposure to organic solvents in plant growth. Hum Exp Toxicol 17:272–277
Morré DJ (1998b) Cell-free analysis of Golgi apparatus membrane traffic in rat liver. Histochem Cell Biol 109:487–504
Morré DJ (1998c) NADH oxidase: a multifunctional ectoprotein of the eukaryotic cell surface. In: Asard H, Bérczi A, Caubergs R (eds) Plasma membrane redox systems and their role in biological stress and disease. Kluwer Academic, Dordrecht, The Netherlands, pp 121–156
Morré DJ (1998d) NADH oxidase activity of soybean plasma membranes inhibited by submicromolar concentrations of ATP. Mol Cell Biochem 187:41–46
Morré DJ (2000) Chemical hormesis in cell growth: a molecular target at the cell surface. J Appl Toxicol 20:157–163
Morré DJ (2002) Preferential inhibition of the plasma membrane NADH oxidase (NOX) activity by diphenyleneiodonium chloride with NADPH as donor. Antioxid Redox Signal 4:207–212
Morré DJ (2004) Quinone oxidoreductases of the plasma membrane. Methods Enzymol 378A:179–199
Morré DJ (2006) A CNOX-like protein disulfide-thiol interchange activity of the cell surface of mouse sperm. Acta Biol Szeged 50:71–74
Morré DJ, Bonner J (1965) A mechanical analysis of root growth. Physiol Plant 18:635–639
Morré DJ, Bracker CE (1976) Ultrastructural alteration of plant plasma membranes induced by auxin and calcium ions. Plant Physiol 58:544–547
Morré DJ, Brightman AO (1991) NADH oxidase of plasma membranes. J Bioenerg Biomembr 23:469–489
Morré DJ, Brightman AO (1997) Parallel inhibition by external osmotica of auxin-induced elongation in stem sections and auxin-stimulated NADH oxidase activity of plasma membrane vesicles from soybean hypocotyls. Plant Physiol Biochem 35:311–319
Morré DJ, Cherry JH (1977) Auxin hormone-plasma membrane interactions. In: Pilet PE (ed) Plant growth regulation. Proceedings of the ninth international conference on plant growth substances. Springer, Berlin, pp 35–43
Morré DJ, Crane FL (1990) NADH oxidase. In: Crane FL, Morré DJ, Löw H (eds) Oxidoreduction at the plasma membrane: relation to growth and transport. I. Animals. CRC Press, Boca Raton, FL, pp 67–84
Morré DJ, Drobes B (1987) A membrane-located, calcium-/calmodulin-activated phospholipase stimulated by auxin. In: Stumpf PK, Mudd JB, Ness WD (eds) The metabolism, structure and function of plant lipids. Proceedings of the seventh international symposium on methods, function and structural plant lipids, David, CA, 27 July 27 to 1 Aug 1986. Plenum, New York, pp 217–219
Morré DJ, Eisinger WR (1968) Cell wall extensibility: its control by auxin and relationship to cell elongation. In: Wrightman F, Setterfied G (eds) Biochemistry and physiology of plant growth substances. Runge Press, Ottawa, pp 625–645

Morré DJ, Grieco PA (1999) Glaucarubolone and simalikalactone D, respectively, preferentially inhibit auxin-induced and constitutive components of plant cell enlargement and the plasma membrane NADH oxidase. Int J Plant Sci 160:291–297

Morré DJ, Key JL (1967) Auxins. In: Wilt F, Wessels N (eds) Experimental techniques in developmental biology. T. Y. Crowell, New York, pp 575–593

Morré DJ, Mollenhauer HH (2009) The Golgi apparatus. The first 100 years. Springer, New York

Morré DJ, Morré DM (1987) Transition vesicles of the cis Golgi apparatus face of rat liver are increased by retinol. Cell Biol Int Rep 11:89–93

Morré DJ, Morré DM (1989) Mammalian plasma membranes by aqueous two-phase partition. Biotechniques 7:946–958

Morré DJ, Morré DM (1995a) Differential response of the NADH oxidase of plasma membranes of rat liver and hepatoma and HeLa cells to thiol reagents. J Bioenerg Biomembr 27:137–144

Morré DM, Morré DJ (1995b) Mechanism of killing of HeLa cells by the antitumor sulfonylurea N-(4-methylphenylsulfonyl)-N′-(4-chlorophenyl)urea (LY181984). Protoplasma 184:188–195

Morré DJ, Morré DM (1998) NADH oxidase activity of soybean plasma membranes oscillates with a temperature compensated period of 24 min. Plant J 16:277–284

Morré DJ, Morré DM (2000) Applications of aqueous two-phase partition to isolation of membranes from plants: a periodic NADH oxidase activity as a marker for right side-out plasma membrane vesicles. J Chromatogr B 743:377–387

Morré DJ, Morré DM (2003a) Cell surface NADH oxidases (ECTO-NOX proteins) with roles in cancer, cellular time-keeping, growth, aging and neurodegenerative disease. Free Radic Res 37:795–808

Morré DM, Morré DJ (2003b) Specificity of coenzyme Q inhibition of an aging-related cell surface NADH oxidase (ECTO-NOX) that generates superoxide. Biofactors 18:33–43

Morré DJ, Morré DM (2003c) Spectroscopic analyses of oscillations in ECTO-NOX-catalyzed oxidation of NADH. Nonlinearity Biol Toxicol Med 1:345–362

Morré DJ, Morré DM (2003d) Synergistic *Capsicum*-tea mixtures with anticancer activity. J Pharm Pharmacol 55:987–994

Morré DJ, Morré DM (2003e) The plasma membrane-associated NADH oxidase (ECTO-NOX) of mouse skin responds to blue light. J Photochem Photobiol B 70:7–12

Morré DJ, Morré DM (2006a) Aging related cell surface ECTO-NOX protein, arNOX, a preventive target to reduce atherogenic risk in the elderly. Rejuvenation Res 9:231–236

Morré DM, Morré DJ (2006b) Anticancer activity of grape and grape skin extracts alone and combined with green tea infusions. Cancer Lett 238:202–209

Morré DM, Morré DJ (2006c) Catechin-vanilloid synergies with potential clinical applications in cancer. Rejuvenation Res 9:45–55

Morré DM, Morré DJ (2006d) Coenzyme Q and lipid oxidation in aging and cardiovascular disease. In: Abstracts, 41st Annual south eastern regional lipid conference, Cashiers, NC, 1–3 Nov 2006, p 68

Morré DJ, Morré DM (2006e) tNOX, an alternative target to COX-2 to explain the anticancer activities of non-steroidal anti-inflammatory drugs (NSAIDS). Mol Cell Biochem 283:159–167

Morré DJ, Morré DM (2008a) arNOX activity of saliva as a non-invasive measure of coenzyme Q response in human trials. Biofactors 32:231–235

Morré DJ, Morré DM (2008b) ENOX proteins, copper hexahydrate-based ultradian oscillators of the cells' biological clock. In: Lloyd D, Rossi EL (eds) Ultradian rhythms from molecules to mind: a new vision of life. Springer, New York, pp 43–84

Morré DM, Morré DJ (2009) Composition and methods for treating and preventing virus infections. US Patent 7,491,413, 17 Feb 2009

Morré DJ, Morré DM (2011) Non-mitochondrial coenzyme Q. Biofactors 37:355–360

Morré DJ, Morré DM (2012a) Water in biological time keeping. In: Pepper DW, Brebbia CA (eds) Water and society. WIT Press, Southampton, Boston, pp 13–23

Morré DJ, Morré DM (2012b) Early detection: an opportunity for cancer prevention through early intervention. In: Georgakilas AG (ed) Cancer prevention. InTech, Rijeka, pp 389–402

Morré DJ, Ovtracht L (1977) Dynamics of the Golgi apparatus membrane differentiation and membrane flow. Int Rev Cytol Suppl 5:61–188

Morré DJ, Reust T (1997) A circulating form of NADH oxidase activity responsive to the antitumor sulfonylurea N-4-methylphenylsulfonyl-N′-4-chlorophenylurea LY181984 specific to sera from cancer patients. J Bioenerg Biomembr 29:281–289

Morré DJ, Tautvydas K (1986) Mefluidide-chlorsulfuron-2,4-D- surfactant combinations for roadside vegetation management. J Plant Growth Regul 4:189–201

Morré DJ, Vigil EL, Frantz C, Goldenberg H, Crane FL (1978) Cytochemical demonstration of glutaraldehyde-resistant. NADH-ferricyanide oxido-reductase activities in rat liver plasma membranes and Golgi apparatus. Eur J Cell Biol 18:213–230

Morré DJ, Gripshover B, Monroe A, Morré JT (1984) Phosphatidyl inositol turnover in isolated soybean membranes stimulated by the synthetic growth hormone, 2,4-dichlorophenoxyacetic acid. J Biol Chem 259:15364–15368

Morré DJ, Navas P, Penel C, Castillo FJ (1986a) Auxin-stimulated NADH oxidase (semidehydroascorbate reductase) of soybean plasma membrane: role in acidification of cytoplasm? Protoplasma 133:195–197

Morré DJ, Paulik M, Nowack D (1986b) Transition vesicle formation in vitro. Protoplasma 132:110–113

Morré DJ, Schnepf E, Deichgräber G (1986c) Inhibition of elongation growth in *Pellia* setae by the monovalent ionosphere monensin. Bot Gaz 147:252–257

Morré DJ, Crowe JH, Morré DM, Crowe LM (1987) Infrared spectroscopic evidence for a conformational alteration of plant plasma membranes upon exposure to the growth hormone analog, 2,4- dichlorophenoxyacetic acid. Biochem Biophys Res Commun 147:506–512

Morré DJ, Crane FL, Barr R, Penel C, Wu L-Y (1988a) Inhibition of plasma membrane redox activities and elongation growth of soybean. Physiol Plant 72:236–240

Morré DJ, Brightman AO, Wang J, Barr R, Crane FL (1988b) Role for plasma membrane redox systems in cell growth. In: Crane FL, Löw H, Morré DJ (eds) Plasma membrane oxidoreductases in control of plant and animal growth. Proceedings of the NATO advanced research workshop. Alan R. Liss, New York, pp 45–55

Morré DJ, Brightman AO, Wu L-Y, Barr R, Leak B, Crane FL (1988c) Role of plasma membrane redox activities in elongation growth in plants. Physiol Plant 73:187–193

Morré DJ, Crane FL, Eriksson LC, Löw H, Morré DM (1991a) NADH oxidase of liver plasma membrane stimulated by diferric transferrin and neoplastic transformation induced by the carcinogen 2-acetylaminofluorene. Biochem Biophys Acta 1057:140–146

Morré DJ, Brightman AO, Crane FL (1991b) Oxidoreductase activities of the plant plasma membrane and the control of plant growth. In: Penel C, Greppin H (eds) Plant signalling, plasma membrane and change of state. University of Geneva, Switzerland, pp 59–77

Morré DJ, Selldén G, Zhu XZ, Brightman AO (1991c) Triacontanol stimulates NADH oxidase of soybean hypocotyl plasma membrane. Plant Sci 79:31–36

Morré DJ, Morré DM, Paulik M, Batova A, Broome A-M, Pirisi L, Creek KE (1992a) Retinoic acid and calcitriol inhibition of growth and NADH oxidase of normal and immortalized human keratinocytes. Biochim Biophys Acta 1134:217–222

Morré DM, Spring H, Trendlenburg M, Montag M, Mollenhauer BA, Mollenhauer HH, Morré DJ (1992b) Stimulation of Golgi apparatus activity by retinol in living cells. J Nutr 122:1248–1253

Morré DJ, Brightman AO, Barr R, Davidson M, Crane FL (1993a) NADH oxidase activities of plasma membranes of soybean hypocotyls is activated by guanine nucleotides. Plant Physiol 102:595–602

Morré DJ, Davidson M, Geilen C, Lawrence J, Flesher G, Crowe R, Crane FL (1993b) NADH oxidase activity of rat liver plasma membrane activated by guanine nucleotides. Biochem J 292:647–653

Morré DJ, Lawrence J, Safranski K, Hammond T, Morré DM (1994a) Experimental basis for separation of membrane vesicles by preparative free-flow electrophoresis. J Chromatogr 668:201–214

Morré DJ, Morré DM, Wu L-Y (1994b) Adriamycin inhibits transplasma membrane electron transport of HL-60 cells. J Bioenerg Biomembr 26:137–142

Morré DJ, Morré DM, Wu L-Y (1994c) Response to adriamycin of transplasma membrane electron transport in adriamycin-resistant and non-resistant HL-60 cells. J Bioenerg Biomembr 26:137–142

Morré DJ, Navas P, Rodriguez-Aguilera JC, Morré DM, Villalba JM, de Cabo R, Lawrence J (1994d) Cyclic AMP-plus ATP-dependent modulation of the NADH oxidase activity of porcine liver plasma membranes. Biochim Biophys Acta 1244:566–574

Morré DJ, Paulik M, Lawrence JL, Morré DM (1994e) Inhibition by brefeldin A of NADH oxidation activity of rat liver Golgi apparatus accelerated by GDP. FEBS Lett 346:199–202

Morré DJ, Geilin C, Morré DM (1994f) Golgi apparatus respond to the antitumor sulfonylurea, N-4-methylphenylsulfonyl-N′-4-chlorophenylurea LY181984 in sulfonylurea-susceptible but not in resistant cell lines. Protoplasma 182:81–95

Morré DJ, Brightman AO, Hidalgo A, Navas P (1995a) Selective inhibition of auxin-stimulated NADH oxidase activity and elongation growth of soybean hypocotyls by thiol reagents. Plant Phys 107:1285–1291

Morré DJ, Chueh P-J, Morré DM (1995b) Capsaicin inhibits preferentially the NADH oxidase and growth of transformed cells in culture. Proc Natl Acad Sci U S A 92:1831–1835

Morré DJ, de Cabo R, Farley C, Oberlies NH, McLaughlin JL (1995c) Mode of action of bullatacin, a potent antitumor acetogenin: inhibition of NADH oxidase activity of HeLa and HL-60, but not liver, plasma membranes. Life Sci 56:343–348

Morré DJ, de Cabo R, Jacobs E, Morré DM (1995d) Auxin-modulated protein disulfide-thiol interchange activity from soybean plasma membranes. Plant Physiol 109:573–578

Morré DJ, Fleurimont J, Sweeting M (1995e) Chlorsulfuron blocks 2,4-D-induced cell enlargement and NADH oxidase in excised sections of soybean hypocotyls. Biochim Biophys Acta 1240:5–9

Morré DJ, Merriman R, Tanzer LR, Wu L-Y, Morré DM, MacKellar WC (1995f) Inhibition of the NADH oxidase activity of plasma membranes isolated from xenografts and cell lines by the antitumor sulfonylurea, N-4-methylphenylsulfonylurea-N′-chlorophenylurea LY181984 correlates with drug susceptibility of growth. Protoplasma 188:151–160

Morré DJ, Morré DM, Stevenson J, MacKellar W, McClure D (1995g) HeLa plasma membranes bind the antitumor sulfonylurea LY181984 with high affinity. Biochim Biophys Acta 1244:133–140

Morré DJ, Wilkinson FE, Lawrence J, Cho N, Paulik M (1995h) Identification of antitumor sulfonylurea binding proteins of HeLa plasma membranes. Biochim Biophys Acta 1236:237–243

Morré DJ, Wu L-Y, Morré DM (1995i) The antitumor sulfonylurea N-4-methylphenylsulfonylurea-N′-chlorophenylurea LY181984 inhibits NADH oxidase activity of HeLa plasma membranes. Biochim Biophys Acta 1240:11–17

Morré DM, Dai S, Wu L-Y, Morré DJ (1996a) Correlation of inhibition of growth and induction of apoptosis in HeLa cells with various synthetic retinoids. FASEB J 10:A525

Morré DJ, Sun E, Geilen C, Wu L-Y, de Cabo R, Krasagakis K, Orfanos C, Morré DM (1996b) Capsaicin inhibits plasma membrane NADH oxidase and growth of human and mouse melanoma lines. Eur J Cancer 32:1995–2003

Morré DJ, Wilkinson FE, Kim C, Cho N, Lawrence J, Morré DM, McClure D (1996c) Antitumor sulfonylurea-inhibited NADH oxidase of cultured HeLa cells shed into media. Biochim Biophys Acta 1280:197–206

Morré DJ, Caldwell S, Mayorga A, Wu L-Y, Morré DM (1997a) NADH oxidase activity from sera altered by capsaicin is widely distributed among cancer patients. Arch Biochem Biophys 342:224–230

Morré DJ, Garcia C, Mackey T, Fledderman S, Wang S, Morré DM (1997b) Is the hormone-stimulated protein disulfide-thiol oxidoreductase NADH oxidase of the plasma membrane a biological clock? Mol Biol Cell 8:146a

Morré DJ, Kim C, Paulik M, Morré DM, Faulk WP (1997c) Is the drug-responsive NADH oxidase of the cancer cell plasma membrane a molecular target for adriamycin? J Bioenerg Biomembr 29:269–280

Morré DJ, Jacobs E, Sweeting M, de Cabo R, Morré DM (1997d) A protein disulfide-thiol interchange activity of HeLa plasma membranes inhibited by the antitumor sulfonylurea N-4-

methylphenylsulfonylurea-N′-chlorophenylurea LY181984. Biochim Biophys Acta 1325:117–125

Morré DJ, Rodriguez-Aguilera JC, Navas P, Morré DM (1997e) Redox modulation of the response of NADH oxidase activity of rat liver plasma membranes to cyclic AMP plus ATP. Mol Cell Biochem 173:71–77

Morré DJ, Wu LY, Morré DM (1997f) Inhibition of NADH oxidase activity and growth of HeLa cells by the antitumor sulfonylurea, N-4-methylphenylsulfonyl-N′-4-chlorophenylurea LY181984 and response to epidermal growth factor. Biochim Biophys Acta 135:114–120

Morré DJ, Chueh P-J, Lawler J, Morré DM (1998a) The sulfonylurea-inhibited NADH oxidase activity of HeLa plasma membranes has properties of a protein disulfide-thiol oxido-reductase with protein disulfide-thiol interchange activity. J Bioenerg Biomembr 30:477–487

Morré DJ, Grieco PA, Morré DM (1998b) Mode of action of the anticancer quassinoids—inhibition of the plasma membrane NADH oxidase. Life Sci 63:595–604

Morré DJ, Morré JT, Lawrence J, Moini M (1998c) Activity of triclopyr herbicide enhanced by combination with cobalt chloride or ammonium nitrate. J Plant Growth Regul 17:125–129

Morré DJ, Morré DM, Penel C, Greppin H (1998d) Auxin modulated protein disulfide-thiol interchange activity from plasma membranes of spinach leaves responds to photoperiod and NADH. Int J Plant Sci 159:105–109

Morré DM, Sweeting M, Morré DJ (1998e) Aqueous two phase partition and detergent precipitation of a drug-responsive NADH oxidase from the HeLa cell surface. J Chromatogr B 711: 173–184

Morré DM, Wang S, Chueh P-J, Lawler J, Safranski K, Jacobs E, Morré DJ (1998f) A molecular basis for retinol stimulation of vesicle budding in vivo and in vitro. Mol Cell Biochem 187:73–83

Morré DJ, Wu L-Y, Morré DM (1998g) Response of a cell-surface NADH to the antitumor sulfonylurea N-4-methylphenylsulfonyl-N′-4-chlorophenylurea LY181984 modulated by redox. Biochim Biophys Acta 1369:185–192

Morré DJ, Zeichhardt H, Maxeiner HG, Grünert HP, Sawitzky D, Grieco P (1998h) Effect of the quassinoids glaucarubolone and simalikalactone D on growth of cells permanently infected with feline and human immunodeficiency viruses and on viral infections. Life Sci 62: 213–219

Morré DJ, Gomez-Rey ML, Schramke C, Em O, Lawler J, Hobeck J, Morré DM (1999a) Use of dipyridyl-dithio substrates to measure directly the protein disulfide-thiol interchange activity of the auxin stimulated NADH: protein disulfide reductase (NADH oxidase) of soybean plasma membranes. Mol Cell Biochem 200:7–13

Morré DJ, Morré DM, Penel C, Greppin H (1999b) NADH oxidase periodicity of spinach leaves synchronized by light. Int J Plant Sci 160:855–860

Morré DJ, Pogue R, Morré DM (1999c) A multifunctional ubiquinol oxidase of the external cell surface and sera. Biofactors 9:179–187

Morré DJ, Bridge A, Wu L-Y, Morré DM (2000a) Preferential inhibition by (−)-epigallocatechin-3-gallate of the cell surface NADH oxidase and growth of transformed cells in culture. Biochem Pharmacol 60:937–946

Morré DM, Lenaz G, Morré DJ (2000b) Surface oxidase and oxidative stress propagation in aging. J Exp Biol 203:1513–1521

Morré DJ, Pogue R, Morré DM (2001a) Soybean cell enlargement oscillates with a temperature-compensated period length of ca. 24 min. In Vitro Cell Dev Biol Plant 37:19–23

Morré DJ, Sedlak D, Tang X, Chueh P-J, Geng T, Morré DM (2001b) Cancer isoform of a tumor-associated cell surface NADH oxidase (tNOX) has properties of a prion. Biochemistry 40:7351–7354

Morré DJ, Sedlak D, Tang X, Chueh P-J, Geng T, Morré DM (2001c) Surface NADH oxidase of HeLa cells lacks intrinsic membrane binding motifs. Arch Biochem Biophys 392:251–256

Morré DJ, Chueh P-J, Pletcher J, Tang X, Wu L-Y, Morré DM (2002a) Biochemical basis for the biological clock. Biochemistry 41:11941–11945

Morré DJ, Lawler J, Wang S, Keenan TW, Morré DM (2002b) Entrainment in solution of an oscillating NADH oxidase activity from the bovine milk fat globule membrane with a temperature-compensated period length suggestive of an ultradian time-keeping (clock) function. Biochim Biophys Acta 1559:10–20

Morré DJ, Penel C, Greppin H, Morré DM (2002c) The plasma membrane-associated NADH oxidase of spinach leaves responds to blue light. Int J Plant Sci 613:543–547

Morré DJ, Ternes P, Morré DM (2002d) Cell enlargement of plant tissue explants oscillates with a temperature-compensated period length of ca. 24 min. In Vitro Cell Dev Biol Plant 38:18–28

Morré DM, Guo F, Morré DJ (2003a) An aging-related cell surface NADH oxidase (arNOX) generates superoxide and is inhibited by coenzyme Q. Mol Cell Biochem 254:101–109

Morré DJ, Morré DM, Ternes P (2003b) Auxin-activated NADH oxidase activity of soybean plasma membranes is distinct from the constitutive plasma membrane NADH oxidase and exhibits prion-like properties. In Vitro Cell Dev Biol Plant 39:368–376

Morré DJ, Morré DM, Sun H, Cooper R, Chang J, Janle EM (2003c) Tea catechin synergies in inhibition of cancer cell proliferation and of a cancer specific cell surface oxidase (ECTO-NOX). Pharmacol Toxicol 92:234–241

Morré DJ, Pennington KL, Hignite H, Orvis H, Schwartzentruber DJ, Tang X, Morré DM (2005) Detection of blood-borne cancer using polymerase chain reaction for tNOX messenger RNA. In: Abstracts, 97th annual meeting, American Association for Cancer Research, Washington, DC, 1–5 April 2006. Abstract No. 5684, pp 1336–1337

Morré DJ, Kim C, Hicks-Berger C (2006a) ATP-dependent and drug-inhibited vesicle enlargement reconstituted using synthetic lipids and recombinant proteins. Biofactors 28:105–117

Morré DJ, Tang X, Morré DM, Pennington KL, Hignite H, Orvis H, Schwartzentruber DJ (2006b) RTPCR detection of cancer cells in blood based in presence of a tNOX splice variant mRNA. In: Abstracts, Fourth Early Detection Research Network (EDRN) scientific workshop, 20–22 March 2006, Philadelphia, PA, p 60

Morré DJ, Chueh P-J, Yagiz K, Balicki A, Kim C, Morré DM (2007a) ECTO-NOX target for the anticancer isoflavene phenoxodiol. Oncol Res 16:299–312

Morré DJ, Heald S, Coleman J, Orczyk J, Jiang Z, Morré DM (2007b) Structural observations of time dependent oscillatory behavior of $Cu^{II}Cl2$ solutions measured via extended X-ray absorption fine structure. J Inorg Biochem 100:715–726

Morré DJ, Dick S, Bosneaga E, Balicki A, Wu L-Y, McClain N, Morré DM (2008a) tNOX (ENOX1) target for chemosensitization-low-dose responses in the hormetic concentration range. Am J Pharmacol Toxicol 3:16–26

Morré DJ, Jiang Z, Marjanovic M, Orczyk J, Morré DM (2008b) Response of the regulatory behavior of $copper^{II}$-containing ECTO-NOX proteins and $Cu^{II}Cl_2$ in solution to electromagnetic fields. J Inorg Biochem 102:1812–1818

Morré DM, Morré DJ, Rehmus W, Kern D (2008c) Supplementation with CoQ_{10} lowers age-related (ar) NOX levels in healthy subjects. Biofactors 32:221–230

Morré DJ, Orczyk J, Hignite H, Kim C (2008d) Regular oscillatory behavior of aqueous solutions of Cu^{II} salts related to effects on equilibrium dynamics of ortho/para hydrogen spin isomers of water. J Inorg Biochem 102:260–267

Morré DJ, Hostetler B, Weston N, Kim C, Morré DM (2009a) Cancer type-specific tNOX isoforms: a putative family of redox protein splice variants with cancer diagnostic and prognostic potential. Biofactors 34:201–207

Morré DJ, McClain N, Wu L-Y, Kelly G, Morré DM (2009b) Phenoxodiol treatment alters the subsequent response of tNOX and growth of HeLa cells to paclitaxel and cis-platin. Mol Biotechnol 42:100–109

Morré DM, Meadows C, Hostetler B, Weston N, Kern D, Draelos Z, Morré DJ (2009c) Age-related ENOX protein (arNOX) activity correlated with oxidative skin damage in the elderly. Biofactors 34:237–244

Morré DJ, Morré DM, Brightmore R (2010a) Omega-3 but not omega-6 unsaturated fatty acids inhibit the cancer-specific ENOX2 of the HeLa cell surface with no effect on the constitutive ENOX1. J Diet Suppl 7:154–158

Morré DM, Meadows C, Morré DJ (2010b) arNOX: generator of reactive oxygen species in the skin and sera of aging individuals subject to external modulation. Rejuvenation Res 13: 162–164

Morré DJ, Morré DM, Shelton TB (2010c) Aging-related nicotinamide adenine dinucleotide oxidase response to dietary supplementation: The French paradox revisited. Rejuvenation Res 13:159–161

Moya-Camarena SY, Morré DJ, Morré DM (1995) Response of NADH oxidase and growth in K-562 cells to the antitumor sulfonylurea N-4-methylphenylsulfonylurea-N′-chlorophenylurea LY181984. Protoplasma 188:151–160

Mukhtar H, Ahmad N (1999) Mechanism of cancer chemopreventative activity of green tea. Proc Soc Exp Biol Med 220:234–238

Mukoyama A, Ushijima H, Nishimura S, Koike H, Toda M, Hara Y, Shimamura T (1991) Inhibition of rotavirus and enterovirus infections by tea extracts. Jpn J Med Sci Biol 44:181–186

Multhaup G (1997) Amyloid precursor protein, copper and Alzheimer's disease. Biomed Pharmacother 51:105–111

Mumma MJ, Weaver HA, Larson HP (1987) The ortho-para ratio of water vapor in comet P/Halley. Astron Astrophys 187:419–424

Murakami M, Nakamura M, Minakami S (1986) 2,6-Dichlorophenolindophenol-reducing activity of phagocytosis-associated NADPH oxidase. J Biochem 100:1493–1497

Murphee SA, Cunningham LS, Hwang KM, Sartorelli AC (1976) Effects of adriamycin on surface properties of sarcoma 180 ascites cells. Biochem Pharmacol 25:1227–1231

Murphee SA, Tritton TR, Smith PL, Sartorelli AC (1981) Adriamycin-induced changes in the surface membrane of sarcoma 180 ascites cells. Biochim Biophys Acta 649:317–324

Myers CH, Mimmaugh EG, Yeh GC, Simha BK (1988) In: Lown JW (ed) Anthracyclines and anthracycline-based anticancer agents (chap. XIV). Elsevier, Amsterdam

Nadlinger K, Westerthaler W, Storga-Tomic D, Birkmayer JG (2002) Extracellular metabolisation of NADH by blood cells correlates with intracellular ATP levels. Biochim Biophys Acta 1573:177–182

Nagoshi E, Saini C, Bauer C, Laroche T, Naef F, Schibler U (2004) Circadian gene expression in individual fibroblasts: cell-autonomous and self-sustained oscillators pass time to daughter cells. Cell 119:693–705

Nakachi K, Matsuyama S, Miyake S, Sugaruma M, Imai K (2000) Preventive effects of drinking green tea on cancer and cardiovascular disease: epidemiological evidence for multiple targeting prevention. Biofactors 13:49–54

Nakagawa KMT (1997) Absorportion and distribution of tea catechin, (−)-epigallocatechin-3-gallate, in the rat. J Nutr Sci Vitaminol 43:679–684

Nakagawa K, Okuda S, Miyazawa T (1997) Dose-dependent incorporation of tea catechins, (−)-epigallocatechin-3-gallate and (−)-epigallocatechin, into human plasma. Biosci Biotechnol Biochem 61:1981–1985

Nakahata Y, Girmaldi B, Sahar S, Hiraymama J, Sassone-Corsi P (2007) Signaling to the circadian clock: plasticity by chromatin remodeling. Curr Opin Cell Biol 19:230–237

Nakahata Y, Kaluzova M, Grimaldi B, Sahar S, Hirayama J, Chen D, Guarente LP, Sassone-Corsi P (2008) The NAD^+-dependent deacetylase SIRT1 modulates CLOCK-mediated chromatin remodeling and circadian control. Cell 134:329–340

Nakahata Y, Sahar S, Astarita G, Kaluzova M, Sassone-Corsi P (2009) Circadian control of the NAD^+ salvage pathway by CLOCK-SIRT1. Science 324:654–657

Nakano H, Ono K (1990) Differential inhibitory effects of some Catechin derivatives on the activities of human immunodeficiency virus reverse transcriptase and cellular deoxyribonucleic and ribonucleic acid polymerases. Biochemistry 29:2841–2845

Nakayama M, Suzuki K, Toda M, Okubo S, Hara Y, Shimamura T (1993) Inhibition of the infectivity of influenza virus by tea polyphenols. Antiviral Res 21:289–299

Nauseef WM (2008) Biological roles for the NOX family NADPH oxidases. J Biol Chem 283:16961–16965

Navarrete-Montalvo D, Gonzalez MN, Montalyo VMT, Jimenez AA, Echiburu-Chau C, Calaf GM (2008) Patterns of recurrence and survival in breast cancer. Oncol Rep 20:531–535

Navarro F, Villalba JM, Crane FL, Mackellar WC, Navas P (1995) A phospholipids-dependent NADH-coenzyme Q reductase from liver plasma membrane. Biochem Biophys Res Commun 212:138–143

Navas P, Alcain FJ, Burón I, Rodriguez-Aguilera JC, Villalba JM, Morré DM, Morré DJ (1992) Growth factor-stimulated transplasma membrane electron transport in HL-60 cells. FEBS Lett 3:223–226

Navas P, Fernández-Ayala DJ, Martín SF, López-Lluch G, De Cabo R, Rodríguez-Aguilera JC, Villalba JM (2002) Ceramide dependent caspase-3 activation is prevented by coenzyme Q from plasma membrane in serum-deprived cells. Free Radic Res 36:369–374

Neufang G, Furstenberger G, Heidt M, Marks F, Muller-Decker K (2001) Abnormal differentiation of epidermis in transgenic mice constitutively expressing cyclooxygenase-2 in skin. Proc Natl Acad Sci U S A 98:7629–7634

Neuwald AF, Aravind L, Spouge JL, Koonin EV (1999) AAA, A class of chaperone-like ATPases associated with the assembly, operation and disassembly of protein complexes. Genome Res 9:27–43

Nisoli E, Tonello C, Cardile A, Cozzi V, Bracale R, Tedesco L, Carruba MO (2005) Calorie restriction promotes mitochondrial biogenesis by inducing the expression of eNOX. Science 310:314–317

Noguchi T, Morré DJ (1991) Vesicular membrane transfer between endoplasmic reticulum and the Golgi apparatus of a green alga, *Micrasterias americana*. A 16°C block and reconstitution in a cell-free system. Protoplasma 162:128–139

Noh KM, Koh JY (2000) Induction and activation by zinc of NADPH oxidase in cultured cortical neurons and astrocytes. J Neurosci 20:RC111

Nohl H, Kozlov V, Staniek K, Gille L (2001) The multiple functions of coenzyme Q. Bioorg Chem 29:1–13

Nohl H, Gille L, Staniek K (2005) Intracellular generation of reactive oxygen species by mitochondria. Biochem Pharmacol 69:719–723

Nowack DD, Morré DJ, Morré DM (1990) Retinoid modulation of cell-free membrane transfer between endoplasmic reticulum and Golgi apparatus. Biochim Biophys Acta 1051:250–258

O'Donnell VB, Smith GC, Jones OT (1994) Involvement of phenyl radicals in iodonium inhibition of flavoenzymes. Mol Pharmacol 46:778–785

Office of Dietary Supplements (2001) Botanical Fact Sheet, Green Tea

Ogura T, Wilinson AJ (2001) AAA+ superfamily ATPases: common structure—diverse function. Genes Cells 6:575–597

Ohtani H, Wakui H, Ishino T, Komatsuda A, Miura AB (1993) An isoform of protein disulfide isomerase is expressed in the developing acrosome of spermatids during rat spermiogenesis and is transported into the nucleus of mature spermatids and epididymal spermatozoa. Histochemistry 100:423–429

Okumura M, Kondo S, Ogata M, Kanemoto S, Murakami T, Yanagida K, Saito A, Imaizumi K (2005) Candidates for tumor-specific alternative splicing. Biochem Biophys Res Commun 334:23–29

Onumah OE, Jules GE, Zhao Y, Zhou L, Yang H, Guo Z (2009) Overexpression of catalase delays GO/G1- to S-phase transition during cell cycle progression in mouse aortic endothelial cells. Free Radic Biol Med 46:1658–1667

Orczyk J, Morré DM, Morré DJ (2005) Periodic fluctuations in oxygen consumption comparing HeLa (cancer) and CHO (non-cancer) cells and response to external $NAD(P)^+/NAD(P)H$. Mol Cell Biochem 273:161–167

Ou W, Silver J (2006) Role of protein disulfide isomerase and other thiol-reactive proteins in HIV-1 envelope protein-mediated fusion. Virology 350:406–417

Owens RJ, Compans RW (1989) Expression of the human immunodeficiency virus envelope glycoprotein is restricted to basolateral surfaces of polarized epithelial cells. J Virol 63:978–982

Palacios C, Wigertz K, Martin B, Weaver CM (2003) Sweat mineral loss from whole body, patch and arm bag in white and black girls. Nutr Res 23:401–411

Palk SR, Shin HJ, Lee JH, Chang CS, Kim J (1999) Copper(II)-induced self-oligomerization of alpha-synuclein. Biochem J 340:821–828

Palmer JD, Dowse HB (1969) Preliminary findings on the effect of D2O on the period of circadian activity rhythms. Abstract. Biol Bull 137:388

Palmgren MG, Sommarin M (1989) Lysophosphatidylcholine stimulates ATP-dependent proton accumulation in isolated oat root plasma membrane vesicles. Plant Physiol 90:1009–1014

Palmgren MG, Sommarin M, Ulvskov P, Jorgensen PL (1988) Modulation of plasma membrane H^+-ATPase from oat roots by lysophosphatidylcholine, free fatty acids and phospholipase A_2. Physiol Plant 74:11–19

Pan KM, Baldwin M, Nguyen J, Gasset M, Serban A, Groth D, Mehlhorn I, Huang Z, Fletterick RJ, Cohen FE, Prusiner SB (1993) Conversion of alpha-helices into beta-sheets features in the formation of the scrapie prion proteins. Proc Natl Acad Sci U S A 90:10962–10966

Panka DJ, Mano T, Suhara T, Walsh K, Mier JW (2001) Phosphatidylinositol 3-kinase/Akt activity regulates c-FLIP expression in tumor cells. J Biol Chem 276:6893–6896

Panowski SH, Wolff S, Aguilaniu H, Durieux J, Dillin A (2007) PHA-4/Foxa mediates diet-restriction-induced longevity of *C. elegans*. Nature 447:550–555

Parkos CA, Allen RA, Cochrane CG, Jesaitis AJ (1987) Purified cytochrome b from human granulocyte plasma membrane is comprised of two polypeptides with relative molecular weights of 91,000 and 22,000. J Clin Invest 80:732–742

Patel S, Latterich M (1998) The AAA team: related ATPases with diverse functions. Trends Cell Biol 8:65–71

Patel MI, Subbaranaish K, Du B, Chang M, Yang P, Newman RA, Cordon-Cardo C, Thaler HT, Dannenberg AJ (2005) Celecoxib inhibits prostate cancer growth: evidence of a cyclooxygenase-2-independent mechanism. Clin Cancer Res 11:1999–2007

Paulik M, Nowack DL, Morré DJ (1988) Isolation of a vesicular intermediate in the cell-free transfer of membrane from transitional elements of the endoplasmic reticulum to Golgi apparatus cisternae of rat liver. J Biol Chem 263:17738–17748

Paulik M, Grieco P, Kim C, Maxeiner H-G, Grünert HP, Zeichhardt H, Morré DM, Morré DJ (1999) Drug-antibody conjugates with anti-HIV activity. Biochem Pharmacol 58:1781–1790

Pedersen PL (1978) Tumor mitochondria and the bioenergetics of cancer cells. Prog Exp Tumor Res 22:190–274

Pelicano H, Xu R-H, Du M, Feng L, Sasaki R, Carew JS, Hu Y, Ramdas L, Hu L, Keating MJ, Zhang W, Plunkett W, Huang P (2006) Mitochondrial respiration defects in cancer cells cause activation of Akt survival pathway through a redox-mediated mechanism. J Cell Biol 175:913–923

Pershin SM (2005) Two-liquid water. Phys Wave Phenom 13:192–208

Pershin SM (2006) Harmonic oscillations of the concentration of H-bonds in liquid water. Laser Phys 16:114–1190

Pestova TV, Borukhov SI, Hellen C (1998) Eukaryotic ribosomes require initiation factors 1 and 1A to locate initiation codons. Nature 394:854–859

Petro R, Darby S, Deo H, Silcocks P, Whitley E, Doll R (2000) Smoking, smoking cessation, and lung cancer in the UK since 1950: combination of national statistics with two case-control studies. Br Med J 321:323–329

Pettus BJ, Chalfant CE, Hannun YA (2002) Ceramide in apoptosis: an overview and current perspectives. Biochim Biophys Acta 1585:114–125

Pfaffmann H, Hartmann E, Brightman AO, Morré DJ (1987) Phosphatidylinositol specific phospholipase C of plant stems: membrane associated activity concentrated in plasma membranes. Plant Physiol 85:1151–1155

Phillips DC, Hunt JT, Moneypenny CG, Maclean KH, McKenzie PP, Harris LC, Houghton JA (2007) Ceramide-induced G_2 arrest in rhabdomyosarcoma (RMS) cells requires $p21^{Cip1/Waf1}$ induction and is prevented by MDM2 overexpression. Cell Death Differ 14:1780–1791

Pichard GE, Zucker I (1986) Influence of deuterium oxide on circadian activity rhythms of hamsters: role of the suprachiasmatic nuclei. Brain Res 376:149–154

Pierré A, Robert-Gérco M, Tempéte C, Polonsky J (1980) Structural requirements of quassinoids for the inhibition of cell transformation. Biochem Biophys Res Commun 93:675–686

Pillai SP, Mitscher LA, Menon SR, Pillai CA, Shankel DM (1999) Antimutagenic/antioxidant activity of green tea components and related compounds. J Environ Pathol Toxicol Oncol 18:147–158

Pirotton S, Boutherin-Falson O, Robaye B, Boeynaems JM (1992) Ecto-phosphorylation on aortic endothelial cells. Exquisite sensitivity to staurosporine. Biochem J 285:585–591

Pisters KM, Newman RA, Coldman B, Shin DM, Khuri FR, Hong WK, Glisson BS, Lee JS (2001) Phase I trial of oral green tea extract in adult patients with solid tumors. J Clin Oncol 19:1830–1838

Pittendrigh CS, Caldarola PC, Cosbey ES (1973) A differential effect of heavy water on temperature-dependent and temperature-compensated aspects of circadian system of Drosophila pseudoobscura. Proc Natl Acad Sci U S A 70:2037–2041

Pogue R, Morré DM, Morré DJ (2000) CHO cell enlargement oscillates with a temperature-compensated period of 24 minutes. Biochim Biophys Acta 1498:44–51

Pogue R, Morré DM, Morré DJ (2001) Cell size continues to oscillate with a 24-min period in the absence of growth. Mol Biol Cell 12:67a

Policastro L, Molinari B, Larcher F, Blanco P, Podhajcer OL, Costa CS, Rojas P, Durán H (2004) Imbalance of antioxidant enzymes in tumor cells and inhibition of proliferation and malignant features by scavenging hydrogen peroxide. Mol Carcinog 39:103–113

Pollack GH, Clegg J (2008) Unexpected linkage between unstirred layers, exclusion zones, and water. In: Pollack GH, Chin WC (eds) Phase transitions in cell biology. Springer, Berlin, pp 143–152

Polonsky J (1973) Quassinoid bitter principles. Fortschr Chem Org Naturst 30:101–150

Poole RJ (1978) Energy coupling for membrane transport. Annu Rev Plant Physiol 29:437–460

Portoukalian J, Zwingelstein G, Abdul-Malak N, Dore JF (1978) Alteration of gangliosides in plasma and red cells of humans bearing melanoma tumors. Biochem Biophys Res Commun 85:916–920

Potekhin SA, Khusainova RS (2005) Spin-dependent absorption of water molecules. Biophys Chem 118:84–87

Prata C, Maraldi T, Fiorenstini D, Zambonin L, Hakin G, Landi L (2006) Is NOX the source of ROS involved in glut 1 activity in B1647 cells? Acta Biol Szeged 50:79–82

Prusiner SB (1994) Biology and genetics of prion diseases. Annu Rev Microbiol 48:655–686

Prusiner SB, Bolton DC, Groth DF, Bowman KA, Cochran SP, McKinley MP (1982) Further purification and characterization of scrapie prions. Biochemistry 21:6942–6950

Prusiner SB, McKinley MP, Bowman KA, Bolton DC, Bendheim PE, Groth DF, Glenner GG (1983) Scrapie prions aggregate to form amyloid-like birefringent rods. Cell 35:349–358

Prusiner SB, Groth DF, Bolton DC, Kent SB, Hood LE (1984) Purification and structural studies of a major scrapie prion protein. Cell 38:127–134

Prusiner SB, Scott MR, DeArmon MR, Cohen FE (1998) Prion protein biology. Cell 93:337–348

Pupillo P, Valenti V, DeLuca L, Hertel R (1986) Kinetic characteristics of reduced pyridine nucleotide dehydrogenases duroquinone-dependent in *Cucurbita* microsomes. Plant Physiol 80:384–389

Pye VE, Dreveny I, Briggs LC, Sands C, Beuron F, Zhang X, Freemont PS (2006) Going through the motions: the ATPase cycle of p97. J Struct Biol 156:12–28

Radin NS (2003) Killing tumors by ceramide-induced apoptosis. Critical available drugs. Biochem J 371:243–256

Rafeloff-Phail R, Ding L, Conner L, Yeh WK, McClure D, Guo H, Emerson K, Brooks H (2004) Biochemical regulation of mammalian AMP-activated protein kinase activity by NAD and NADH. J Biol Chem 279:52934–52939

Rahme E, Barkun AN, Toubouti Y, Bardou M (2003) The cyclooxygenase-2-selective inhibitors rofecoxib and celecoxib prevent colorectal neoplasia occurrence and recurrence. Gastroenterology 125:404–412

Ramasarma T, Swaroop A, MacKellar W, Crane FL (1981) Generation of hydrogen peroxide on oxidation of NADH by hepatic plasma membranes. J Bioenerg Biomembr 13:241–253

Ramsey L, Marcheva B, Kohsaka A, Bass J (2007) The clockwork of metabolism. Annu Rev Nutr 29:219–240

Ramsey KM, Yoshino J, Brace CS, Abrassart D, Kobayashi Y, Marcheva B, Hong HK, Chong JL, Buhr ED, Lee C, Takahashi JS, Imai S, Bass J (2009) Circadian clock feedback cycle through NAMPT-mediated NAD^+ biosynthesis. Science 324:651–654

Rayle DL (1973) Auxin-induced hydrogen ion secretion in *Avena* coleoptiles and its implications. Planta 114:63–73

Rehmus WE, Kern D, Janjua R, Morré DM, Morré DJ, Knaggs H (2008) Appearance of skin ageing in healthy women. Correlation with arNOX levels: a potential new mechanism in ageing? Clin Dermatol Retinoids Other Treat 24:52–56

Reiss PD, Zuurendonk PF, Beech RL (1984) Measurement of tissue purine, pyrimidine, and other nucleotides by radial compression high-performance liquid chromatography. Anal Biochem 140:162–171

Remers WA (1988) Adriamycin (p 144). In: The chemistry of antitumor antibiotics. Wiley, New York, Retinails other Treat. pp 290

Reznik AZ, Packer L (1994) Oxidative damage to proteins: spectrophotometric method for carbonyl assay. Methods Enzymol 223:357–363

Richter CP (1977) Heavy water as a tool for study of the forces that control length of period of the 24-hour clock of the hamster. Proc Natl Acad Sci U S A 74:1295–1299

Rim A, Chellman-Jeffers M (2008) Trends in breast cancer screening and diagnosis. Cleve Clin J Med 75:S2–S9

Rodriguez M, Moreau P, Paulik M, Lawrence J, Morré DJ, Morré DM (1992) NADH-activated cell-free transfer between Golgi apparatus and plasma membranes of rat liver. Biochim Biophys Acta 1107:131–138

Rogers K, Tokes Z (1984) Novel mode of cytotoxicity obtained by coupling inactive anthracycline to a polymer. Biochem Pharmacol 33:605–608

Rogers K, Carr E, Tokes Z (1983) Cell surface-mediated cytotoxicity of polymer-bound adriamycin against drug-resistant hepatocytes. Cancer Res 43:2741–2748

Roland J-C, Lembi CA, Morré DJ (1972) Phosphotungstic acid- chromic acid as a selective electron-dense stain for plasma membrane of plant cells. Stain Technol 47:195–200

Roos WP, Kaina B (2006) DNA damage-induced cell death by apoptosis. Trends Mol Med 12:440–450

Roth GS, Lesnikov V, Lesnikov M, Ingram DK, Lane MA (2001) Dietary caloric restriction prevents the age-related decline in plasma melatonin levels of Rhesus monkeys. J Clin Endocrinol Metab 86:3292–3295

Rouiller I, Butel VM, Latterich M, Milligan RA, Wilson-Kubalek EM (2000) A major conformational change in p97 AAA ATPase upon ATP binding. Mol Cell 6:1485–1490

Rouiller I, DeLaBarre B, May AP, Weis WI, Brunger AT, Milligan RA, Wilson-Kubalek EM (2002) Conformational changes of the multifunction p97 AA ATPase during its ATPase cycle. Nat Struct Biol 9:950–957

Rounds CM, Hepler PK, Fuller SJ, Winship LJ (2010) Oscillatory growth in lily pollen tubes does not require aerobic energy metabolism. Plant Physiol 152:736–746

Rozengurt E (1986) Early signals in the mitogenic response. Science 234:1612–1616

Rubinstein B, Stern AI (1986) Relationship of transplasmalemma redox activity to proton and solute transport by roots of *Zea mays*. Plant Physiol 80:805–811

Rubinstein B, Stern AI, Sout RG (1984) Redox activity at the surface of oat root cells. Plant Physiol 76:386–391

Ruggiero M, Bottaro DP, Liguri G, Gulisano M, Peruzzi B, Pacini S (2004) 0.2 T magnetic field inhibits angiogenesis in chick embryo chorioallantoic membrane. Bioelectromagnetics 25:390–396

Rupprecht JK, Hui Y-U, McLaughlin JL (1990) Annonaceous acetogenins: a review. J Nat Prod 53:237–278

Rush GF, Rinzel S, Boder G, Heim RA, Toth JE, Ponder GD (1992) Effects of diarylsulfonylurea antitumor agents on the function of mitochondria isolated from rat liver and GC3/cl cells. Biochem Pharmacol 44:2387–2394

Rutherford T, O'Malley D, Makkenchery A, Baker L, Azodi M, Schwartz P, Mor G (2004) Phenoxodiol phase Ib/II study in patients with recurrent ovarian cancer that are resistant to second line chemotherapy. In: Proceedings of the American Association for Cancer Research, vol 45, Abstract 4457

Rutter J, Reick M, Wu LC, McKnight SL (2001) Regulation of clock and NPAS2 DNA binding by the redox state of NAD cofactors. Science 293:510–514

Ruvolo PP (2003) Intracellular signal transduction pathways activated by ceramide and its metabolites. Pharm Res 47:383–392

Ryser HJ, Mandel R, Ghani F (1991) Cell surface sulfhydryls required for the cytotoxicity of diphtheria toxin but not of ricin in Chinese hamster ovary cells. J Biol Chem 266:18439–18442

Ryser HJ, Levy EM, Mandel R, Di Sciullo GJ (1994) Inhibition of human immunodeficiency virus infection by agents that interfere with thiol-disulfide interchange upon virus-receptor interaction. Proc Natl Acad Sci U S A 91:4559–4563

Sallay SL, Prosise WN, Morré DJ, Matyas GR (1986) Ganglioside and PCA sialic acid serum levels in cancer. Cancer J 1:124–129

Salles B, Butour JL, Lesca C, Macquet JP (1983) cis-Pt(NH3)2Cl2 and trans-Pt(NH3)2Cl2 inhibit DNA synthesis in cultured L1210 leukemia cells. Biochem Biophys Res Commun 112: 555–563

Sanchez AM, Sanchez MG, Malagarie-Cazenave S, Olea N, Diaz-Laviada I (2006) Induction of apoptosis in prostate tumor pc-3 cells and inhibition of xenograft prostate tumor growth by the vanilloid capsaicin. Apoptosis 11:89–99

Sanchez A, Malagarie-Cazenave S, Olea N, Vara D, Chiloeches A, Diaz-Laviada I (2007) Apoptosis induced by capsaicin in prostate pc-3 cells involves ceramide accumulation, neutral sphingomyelinase, and jnk activation. Apoptosis 12:2013–2024

Sandelius AS, Morré DJ (1986) Calcium-calmodulin requirements of phosphatidyl inositol turnover stimulated by auxin. In: Trewavas AJ (ed) Molecular and cellular aspects of calcium in plant development. Plenum Press, New York, pp 351–352

Sandelius AS, Morré DJ (1987) Characteristics of a phosphatidylinositol exchange activity of soybean microsomes. Plant Physiol 84:1022–1027

Sandelius AS, Barr R, Crane FL, Morré DJ (1986) Redox reactions of plasma membranes isolated from soybean hypocotyls by phase partition. Plant Sci 48:1–10

Sanders DA (2000) Sulfhydryl involvement in fusion mechanisms. Subcell Biochem 34:483–514

Santini MT, Rainaldi G, Ferrante A, Indovina PL, Vacchia P, Donelli G (2003) Effects of a 50 Hz sinusoidal magnetic field on cell adhesion molecule expression in two human osteosarcoma cell lines (MG-63 and Saos-2). Bioelectromagnetics 24:327–338

Saraste J, Kruismanen E (1984) Pre- and post-Golgi vacuoles operate in the transport of Semliki Forest virus membrane glycoproteins to the cell surface. Cell 38:535–549

Sarbassov DD, Guertin DA, Ali SM, Sabatini DM (2005) Phosphorylation and regulation of Akt/PKB by the rictor-mTOR complex (chapter 4). Science 307:1098–1101

Satoh T, Myoshi H, Sakamoto K, Iwamura H (1996) Comparison of the inhibitory action of synthetic capsaicin analogues with various NADH-ubiquinone oxidoreductases. Biochim Biophys Acta 1273:21–30

Sauve AA, Wolberger C, Schramm VL, Boeke JD (2006) The biochemistry of sirtuins. Annu Rev Biochem 75:435–465

Savini I, Arnone R, Rossi A, Catani MV, Del Principe D, Avigliano L (2010) Redox modulation of Ecto-NOX1 in human platelets. Mol Membr Biol 27:160–169

Savitsky A, Golay MIE (1964) Smoothing and differentiation of data by simplified least squares procedures. Anal Chem 36:1627–1639

Savitz DA (1995) Overview of occupational exposure to electric and magnetic fields and cancer: advancements in exposure assessment. Environ Health Perspect 103:69–74

Scarlett DJ, Herst PM, Tan A, Prata C, Berridge MV (2004) Mitochondrial gene-knockout (rho0) cells: a versatile model for exploring the secrets of trans-plasma membrane electron transport. Biofactors 20:199–206

Scarlett D-JG, Herst PM, Berridge MV (2005) Multiple proteins with single activities or a single protein with multiple activities: the conundrum of cell surface NADH oxidases. Biochim Biophys Acta 1708:108–119

Schemhammer ES, Laden F, Speizer FE, Willett WC, Hunter DJ, Kawachi I, Colditz GA (2001) Rotating night shifts and risk of breast cancer in women participating in the nurses' health study. J Natl Cancer Inst 93:1563–1568

Schemhammer ES, Kroenke CH, Laden F, Hankinson SE (2006) Night work and risk of breast cancer. Epidemiology 17:108–111

Scherer GFE, Morré DJ (1978a) Action and inhibition of endogenous phospholipases during isolation of plant membranes. Plant Physiol 62:933–937

Scherer GFE, Morré DJ (1978b) In vitro stimulation by 2,4-dichlorophenoxyacetic acid of an ATPase and inhibition of phosphatidate phosphatase of plant membranes. Biochem Biophys Res Commun 84:238–247

Schibler U, Naef F (2005) Cellular oscillators: rhythmic gene expression and metabolism. Curr Opin Cell Biol 17:223–229

Schimmöller F, Diaz E, Mühlbauer B, Pfeffer SR (1998) Characterization of a 76 kDa endosomal, multispanning membrane protein that is highly conserved throughout evolution. Gene 216: 311–318

Schindler T, Bergfeld R, Hohl M, Schopfer P (1994) Inhibition of Golgi-apparatus function by brefeldin A in maize coleoptiles and its consequences on auxin-mediated growth, cell-wall extensibility and secretion of cell-wall proteins. Planta 192:404–413

Schiniá ME, Carlini P, Pulticelli F, Zappacosta F, Bossa F, Calabreae L (1996) Amino acid sequence of chicken Cu, Zn-containing superoxide dismutase and identification of glutathionyl adducts at exposed cysteine residues. Eur J Biochem 237:433–439

Schloss JV, Ciskanik LM, Van Dyk DE (1988) Origin of the herbicide binding site of acetolactate synthase. Nature 331:360–362

Schmuck A, Fuller CJ, Devaraj S, Jialal I (1995) Effect of aging on susceptibility of low-density lipoproteins to oxidation. Clin Chem 41:1628–1632

Schnepf E, Herth W, Morré DJ (1979) Elongation growth of setae of *Pellia* (*Bryophyta*): effects of auxin and inhibitors. Z Pflanzenphysiol 94:211–217

Schomack PA, Gillies RJ (2003) Contributions of cell metabolism and H+ diffusion to the acidic pH of tumors. Neoplasia 5:135–145

Schwartz JP, Passonneau JV, Johnson GS, Pastan I (1974) The effect of growth conditions on NAD^+ and NADH concentrations and the NAD^+:NADH ratio in normal and transformed fibroblasts. J Biol Chem 249:4138–4143

Scott SK (1994) Oscillations, wave, and chaos in chemical kinetics. Oxford University Press, Oxford

Sebastião AM, Ribeiro JA (2000) Fine-tuning neuromodulation by adenosine. Trends Pharmacol Sci 31:341–346

Sedlak D, Morré DM, Morré DJ (2001) A drug-unresponsive and protease-resistant CNOX protein from human sera. Arch Biochem Biophys 386:106–116

Sener A, Blachier F, Malaisse WJ (1988) Crabtree effect in tumoral pancreatic islet cells. J Biol Chem 263:1904–1909

Sephton SE, Sapolsky RM, Kraemer HC, Spegal D (2000) Diurnal cortisol rhythm as a predictor of breast cancer survival. J Natl Cancer Inst 92:994–1000

Seymour LW, Ulbrich K, Strohalm J, Kopecek J, Duncan R (1990) The pharmacokinetics of polymer-bound adriamycin. Biochem Pharmacol 39:1125–1131

Seymour LW, Ulbrich K, Steyger PS, Brereton M, Subr V, Strohalm J, Duncan R (1994) Tumour tropism and anti-cancer efficacy of polymer-based doxorubicin prodrugs in the treatment of subcutaneous murine B16F10 melanoma. Br J Cancer 70:636–641

Shaw RJ (2006) Glucose metabolism and cancer. Curr Opin Cell Biol 18:598–608

Shen-Miller J, McNitt RE (1978) Quantitative assessment of ultrastructural changes in primary roots of corn (*Zea mays* L.) after geotropic stimulation. II. Curving and noncurving zones of the root proper. Plant Physiol 61:649–653

Shen-Miller J, Miller C (1972) Distribution and activation of the Golgi apparatus in geotropism. Plant Physiol 49:634–639

Shi SR, Key ME, Kalra KL (1991) Antigen retrieval in formalin-fixed, paraffin-embedded tissues: an enhancement method of immunohistochemical staining based on microwave oven heating of tissue sections. J Histochem Cytochem 39:741–748

Shida D, Takable K, Kapitonov D, Milstien S, Spiegel S (2008) Targeting SphK1 as a new strategy against cancer. Curr Drug Targets 9:662–673

Shifley ET, Cole SE (2007) The vertebrate segmentation clock and its role in skeletal birth defects. Birth Defects Res C Embryo Today 81:121–133

Shimomura Y, Kawada T, Suzuki M (1989) Capsaicin and its analogs inhibit the activity of NADH-coenzyme Q oxidoreductases of the mitochondrial respiratory chain. Arch Biochem Biophys 270:573–577

Shinohara ML, Loros JJ, Dunlap JC (1998) Glyceraldehyde-3-phosphate dehydrogenase is regulated on a daily basis by the circadian clock. J Biol Chem 273:446–452

Sijmons PC, Lanfermeijer FC, De Boer AR, Prins HB, Bienfait HF (1984) Depolarization of cell membrane potential during trans-plasma membrane electron transfer to extracellular electron acceptors in iron-deficient roots of *Phaseolus vulgaris* L. Plant Physiol 76:943–946

Silverstein SC, Steinman RM, Cohn ZA (1977) Endocytosis. Annu Rev Biochem 4:669–772

Silvertrini R, DiMarco A, Dasdia T (1970) Interference of a daunomycin with metabolic events of the cell cycle in synchronized cultures of rat fibroblasts. Cancer Res 30:966–973

Silvertrini R, Lenaz L, Gronzo CD (1973) Correlations between cytotoxicity, biochemical effects, drug effectiveness of daunomycin and adriamycin on Sarcoma 180 ascites in mice. Cancer Res 33:2954–2958

Simko M, Richard D, Kriehuber R, Weiss GG (2001) Micronucleus induction in Syrian hamster embryo cells following exposure to 50 Hz magnetic fields, benzo(a)pyrene, and TPA in vitro. Mutat Res 22:43–50

Simonnet H, Alazard N, Pfeiffer K, Gallou C, Béroud C, Demont J, Bouvier R, Schägger H, Godinot C (2002) Low mitochondrial respiratory chain content correlates with tumor aggressiveness in renal cell carcinoma. Carcinogenesis 23:759–768

Simonnet H, Dermont J, Pfeiffer K, Guenaneche L, Bouvier R, Brandt U, Schagger H, Godinot C (2003) Mitochondrial complex I is deficient in renal oncocytomas. Carcinogenesis 24: 1461–1466

Simons K, Virta H (1987) Perforated MDCK cell support intracellular transport. EMBO J 16: 2241–2247

Singer-Krüger B, Frank R, Crausaz F, Riezman H (1993) Partial purification and characterization of early and late endosomes from yeast. Identification of four novel proteins. J Biol Chem 268:14376–14386

Sisken BF, Walker J, Orgel M (1993) Prospects on clinical applications of electrical stimulation for nerve regeneration. J Cell Biochem 52:404–409

Skipski VP, Katopodis N, Prendergast JS, Stock CC (1975) Gangliosides in blood serum of normal rats and Morris hepatoma 5123tc-bearing rats. Biochem Biophys Res Commun 67:1122–1127

Slade N, Storga-Tomic D, Birkmayer GD, Pavelic K, Pavelic J (1999) Effect of extracellular NADH on human tumor cell proliferation. Anticancer Res 19:5355–5360

Smith PK, Krohn RI, Hermanson GT, Mallia AK, Gartner FH, Provenzano MD, Fujimoto EK, Goeke NM, Olson BJ, Klenk DC (1985) Measurement of protein using bicinchoninic acid. Anal Biochem 150:70–76

Smith CD, Carney JM, Starke-Reed PO, Oliver CN, Stadtman ER, Floyd RA, Markesbery WR (1991) Excess brain protein oxidation and enzyme dysfunction in normal aging and in Alzheimer disease. Proc Natl Acad Sci U S A 88:10540–10543

Smith DC, Carney JM, Tatsumo T, Stadtman ER, Floyd RA, Markesbery WR (1992) Protein oxidation in aging brain. Ann N Y Acad Sci 663:110–119

Sorkin DL, Duong DK, Miller AF (1997) Mutation of tyrosine 34 to phenylalanine eliminates the active site pK of reduced iron-containing superoxide dismutase. Biochemistry 36:8202–8208

Soucy-Faulkner A, Mukawera E, Fink K, Martel A, Jouan L, Nzengue Y, Lamarre D, Vande Velde C, Grandvaux N (2010) Requirement of NOX2 and reactive oxygen species for efficient RIG-1-mediated antiviral response through regulation of MAVS expression. PLOS Pathog 6:e1000930. doi:10.1371/journal ppat.1000930

Spanswick RM (1981) Electrogenic ion pumps. Annu Rev Plant Physiol 32:267–289

Sparla F, Tedeschi G, Trost P (1996) NAD(P)H: quinone-acceptor oxidoreductase of tobacco leaves is a flavin mononucleotide-containing flavoenzyme. Plant Physiol 112:249–258

Spiegel S, Cuvillier O, Edsall LC, Kohanna T, Menzeleev R, Olah Z, Olivera A, Pinanov G, Thomas DM, Tu Z, Van Brocklyn JR, Wang F (1998) Sphingosine-1-phosphate in cell growth and cell death. Ann N Y Acad Sci 845:11–18

Spruyt E, Verbelen JP, DeGruf JA (1987) Expression of circasepan and circammal rhythmicity in the inhibition of dry stored seeds. Plant Physiol 84:707–710

Spyridopoulos I, Mayer P, Shook KS, Axel DL, Viebahn R, Karsch KR (2001) Loss of cyclin A and G_1-cell cycle arrest are a prerequisite of ceramide-induced toxicity in human arterial endothelial cells. Cardiovasc Res 50:97–107

Sreerama N, Woody RW (2000) Estimation of protein secondary structure from circular dichroism spectra: comparison of CONTIN, SELCON, and CDSSTR methods with an expanded reference set. Anal Biochem 287:252–260

St. Pierre J, Buckingham J, Roebuck SJ, Brand MD (2002) Topology of superoxide production from different sites in the mitochondrial electron transport chain. J Biol Chem 277: 44784–44790

Stadtman ER, Starke-Reed PE, Oliver CN, Carney JM, Floyd RA (1992) Protein modification in aging. EXS 62:64–72

Stahl N, Borchelt DR, Prusiner SB (1990) Differential release of cellular and scrapie prion proteins from cellular membranes by phosphatidylinositol-specific phospholipase C. Biochemistry 29:5405–5412

Stein CA (1993) Suramin: a novel antineoplastic agent with multiple potential mechanisms of action. Cancer Res 52:2239–2248

Stein CA, La Rocca RV, Thomas R, McAtee N, Myers CE (1989) Suramin: an anticancer drug with a unique mechanism of action. J Clin Oncol 7:499–508

Steinberg D (1997) Low density lipoprotein oxidation and its pathobiological significance. J Biol Chem 272:20963–20966

Steinberg D, Carew TE, Fielding C, Fogelman AM, Mahley RW, Sniderman AD, Zilversmit DB (1989) Lipoproteins and the pathogenesis of atherosclerosis. Circulation 80:719–723

Stephens EB, Compans RW (1988) Assembly of animal viruses at cellular membranes. Annu Rev Microbiol 42:489–516

Stetler-Stevenson WG, Aznavoorian S, Liotta LA (1993) Tumor cell interactions with the extracellular matrix during invasion and metastasis. Annu Rev Cell Biol 9:541–573

Stone TW (2005) Adenosine, neurodegeneration and neuroprotection. Neurol Res 27:161–168

Stoner GD, Mukhtar H (1995) Polyphenols as cancer chemopreventive agents. J Cell Biochem Suppl 22:169–180

Strahl BD, Allis CD (2000) The language of covalent histone modifications. Nature 403:41–45

Struhl K (1998) Histone acetylation and transcriptional regulatory mechanisms. Genes Dev 12:599–606

Su D, May JM, Koury MJ, Asard H (2006) Human erythrocyte membranes contain a cytochrome b561 that may be involved in extracellular ascorbate recycling. J Biol Chem 281:39852–39859

Suganuma S, Okabe S, Kai Y, Sueoka N, Sueoka E, Fujiki H (1999) Synergistic effects of (−)-epigallocatechin gallate with (−)-epicatechins, sulindac or tamoxifen on cancer-preventive activity in the human lung cancer cell line PC-9. Cancer Res 59:44–47

Sugasawa T, Lenzen G, Simon S, Hidaka J, Cahen A, Guillaume J-L, Camoin L, Strosberg AD, Nahmias C (2001) The iodocyanopindolol and SM-11044 binding protein belongs to the TM9SF multispanning membrane protein superfamily. Gene 273:227–237

Suhara T, Mano T, Oliveira BE, Walsh K (2001) Phosphatidylinositol 3-kinase/Akt signaling controls endothelial cell sensitivity to Fas-mediated apoptosis via regulation of FLICE-inhibitory protein (FLIP). Circ Res 89:13–19

Suka N, Luo K, Grunstein M (2002) Sir2p and Sasp opposingly regulate acetylation of yeast histone H4 lysine 16 and spreading of heterochromatin. Nat Genet 32:378–383

Sumimoto H (2008) Structure, regulation and evolution of Nox-family NADPH oxidases that produce reactive oxygen species. FEBS J 175:3249–3277

Sun IL, Crane FL (1981) Transplasmalemma NADH dehydrogenase is inhibited by actinomycin D. Biochem Biophys Res Commun 101:68–75

Sun IL, Crane FL (1984) The antitumor drug, cis-platin, inhibits trans plasmalemma electron transport in HeLa cells. Biochem Int 9:299–306

Sun IL, Crane FL (1985) Bleomycin control of transplasma membrane redox activity and proton movement in HeLa cells. Biochem Pharmacol 34:617–622

Sun IL, Crane FL (1990) Interactions of antitumor drugs with plasma membranes. In: Crane FL, Morré DJ, Löw H (eds) Oxidation at the plasma membrane: relation to growth and transport, vol 1. CRC Press, Boca Raton, FL, pp 257–280

Sun IL, Crane FL, Chou JY, Löw H, Grebing C (1983) Transformed liver cells have modified transplasma membrane redox activity which is sensitive to adriamycin. Biochem Biophys Res Commun 16:210–216

Sun IL, Crane FL, Grebing C, Löw H (1984a) Properties of a transplasma membrane electron transport system in HeLa cells. J Bioenerg Biomembr 16:583–595

Sun IL, Crane FL, Löw H, Grebing C (1984b) Transplasma membrane redox stimulates HeLa cell growth. Biochem Biophys Res Commun 125:649–654

Sun IL, Crane FL, Chou JY (1986a) Modification of transmembrane electron transport activity in plasma membranes of simian virus 40 transformed pineal cells. Biochim Biophys Acta 886:327–336

Sun IL, Navas P, Crane FL, Chou JY, Löw H (1986b) Transplasmalemma electron transport is changed in simian virus 40 transformed liver cells. J Bioenerg Biomembr 18:471–485

Sun IL, Garcia-Cañero R, Lui W, Crane FL, Morré DJ, Löw H (1987a) Diferric transferrin reduction stimulates the Na^+/H^+ antiport of HeLa cells. Biochem Biophys Res Commun 145:467–473

Sun IL, Navas P, Crane FL, Morré DJ, Löw H (1987b) NADH diferric transferrin reductase in liver plasma membranes. J Biol Chem 262:15915–15921

Sun IL, Toole-Simms W, Crane FL, Golub ES, Diaz de Pagan T, Morré DJ, Löw H (1987c) Retinoic acid inhibition of transplasmalemma diferric transferrin reductase. Biochem Biophys Res Commun 146:976–982

Sun IL, Toole-Simms W, Crane FL, Morré DJ, Löw H, Chou JY (1988) Reduction of diferric transferrin by SV40 transformed pineal cells stimulates the Na^+/H^+ antiport. Biochim Biophys Acta 938:17–23

Sun IL, Sun EE, Crane FL, Morré DJ, Faulk WP (1992a) Inhibition of transplasma membrane electron transport by transferrin adriamycin conjugates. Biochim Biophys Acta 1105:84–88

Sun IL, Sun EE, Crane FL, Morré DJ, Lindgren A, Löw H (1992b) A requirement for coenzyme Q in plasma membrane electron transport. Proc Natl Acad Sci U S A 89:1126–1130

Sun E, Lawrence J, Morré DM, Sun I, Crane FL, MacKellar WC, Morré DJ (1995) Proton release from HeLa cells and alkalization of cytoplasma induced by differic transferrin or ferricyanide and its inhibition by the diarylsulfonylurea antitumor drug N-4-methylphenylsulfonylurea-N′-4-chlorophenylurea LY181984. Biochem Pharmacol 50:1461–1468

Sun P, Morré DJ, Morré DM (2000) Periodic NADH oxidase activity associated with an endoplasmic reticulum fraction from pig liver. Response to micromolar concentrations of retinal. Biochim Biophys Acta 1498:52–63

Suter RB, Rawson KS (1968) Circadian activity rhythm of the deer mouse, *Peromyscus*: effect of deuterium oxide. Science 160:1011–1014

Sutherland MW, Learmonth BA (1997) The tetrazolium dyes MTS and XTT provide new quantitative assays for superoxide and superoxide dismutase. Free Radic Res 27:283–289

Suzuki Y, Ishihara D, Sasaki M, Nakagawa H, Hata H, Tsunoda T, Watanabe M, Komatsu T, Ota T, Isogai T, Suyama A, Sugano S (2000) Statistical analysis of the 5′ untranslated region of human mRNA using "Oligo-Capped" cDNA libraries. Genomics 64:286–297

Sweiczer A, Novak B, Mitchison JM (1996) The size control of fission yeast revisited. J Cell Sci 109:2947–2957

Szallasi A, Blumberg PM (1993) Mechanisms and therapeutic potential of vanilloids (capsaicin-like molecules). Adv Pharmacol 25:123–155

Szallasi A, Blumberg PM (1999) Vanilloid (capsaicin) receptors and mechanisms. Pharmacol Rev 51:159–212

Taiz L (1984) Plant cell expansion: regulation of cell wall mechanical properties. Annu Rev Plant Physiol 35:585–657

Talbot DC, Smith IE, Nicolson MC, Powles TJ, Button D, Walling J (1993) Phase II trial of the novel sulfonylurea sulofenur in advanced breast cancer. Cancer Chemother Pharmacol 31:419–422

Tamguney T, Stokoe D (2007) New insights into PTEN. J Cell Sci 120:4071–4079

Tammariello SP, Quinn MT, Estus S (2000) NADPH oxidase contributes directly to oxidative stress and apoptosis in nerve growth factor-deprived sympathetic neurons. J Neurosci 20:RC53

Tan AS, Berridge MV (2004) Distinct trans-plasma membrane redox pathways reduce cell-impermeable dyes in HeLa cells. Redox Rep 9:302–306

Tan AS, Berridge MV (2010) Evidence for NAD(P)H: quinone oxidoreductase 1 (NQO1)-mediated quinone-dependent redox cycling via plasma membrane electron transport: a sensitive cellular assay for NQO1. Free Radic Biol Med 48:421–429

Tan DX, Manchester LC, Terron MP, Flores LJ, Tamura H, Reiter RJ (2007) Melatonin as a naturally occurring co-substrate of quinone reductase-2, the putative MT3 melatonin membrane receptor: hypothesis and significance. J Pineal Res 43:317–320

Tang X, Tian Z, Chueh P-J, Chen S, Morré DM, Morré DJ (2007) Alternative splicing as the basis for specific localization of tNOX, a unique hydroquinone (NADH) oxidase, to the cancer cell surface. Biochemistry 46:12337–12346

Tang X, Morré DJ, Morré DM (2008) Antisense experiments demonstrate an exon 4 minus splice variant mRNA as the basis for expression of tNOX, a cancer-specific cell surface protein. Oncol Res 16:557–567

Tang X, Chueh P-J, Jiang Z, Layman S, Martin B, Kim C, Morré DM, Morré DJ (2010) Essential role of copper in the activity and regular periodicity of a recombinant, tumor-associated, cell surface, growth-related and time-keeping hydroquinone (NADH) oxides with protein disulfide thiol interchange activity (ENOX2). J Bioenerg Biomembr 42:355–360

Tang X, Kane VD, Morré DM, Morré DJ (2011) hnRNP F directs formation of an exon 4 minus variant of tumorassociated NADH oxidase (ENOX2). Mol Cell Biochem 357:55–63

Tartakoff AM (1986) Temperature and energy dependence of secretory protein transport in the exocrine pancreas. EMBO J 5:1477–1482

Taylor CW, Alberts DS, Ketcham MA, Satterlee WG, Holdsworth MT, Plezia PM, Peng Y-M, McCloskey TM, Roe DJ, Hamilton M, Salmon SE (1989) Clinical pharmacology of a novel diarylsulfonylurea anticancer agent. J Clin Oncol 7:1733–1740

Teissedre PL, Waterhouse AL (2000) Inhibition of oxidation of human low density lipoproteins by phenolic substances in different essential oils varieties. J Agric Food Chem 48:3801–3805

Terada T, Takada K, Yamanishi H Ashida Y (2007) Inhibitory effects of coenzyme Q10 on skin aging. In: Abstracts, Fifth conference of the international coenzyme Q10 association, Kobe, Japan, p 156

Thakar JH, Chapin C, Berg RH, Ashmun RA, Houghton PJ (1991) Effect of antitumor diarylsulfonylureas on *in vivo* and *in vitro* mitochondrial structure and functions. Cancer Res 51: 6286–6291

Thibault A, Samid D, Cooper MR, Figg WD, Tompkins AC, Patronas N, Headlee DJ, Kohler DR, Venzon DJ, Myers CE (1995) Phase I study of phenylacetate administered twice daily to patients with cancer. Cancer 75:2932–2938

Thormalley P, Bannister W, Bannister J (1986) Reduction of oxygen by NADH/NADH dehydrogenase in the presence of adriamycin. Free Radic Res Commun 2:163–171

Thun MJ, Henley SJ, Patrono C (2002) Nonsteroidal anti-inflammatory drugs as anticancer agents: mechanistic, pharmacologic, and clinical issues. J Natl Cancer Inst 94:252–266

Thun-Battersby S, Mevissen M, Löscher W (1999) Exposure of Sprague–Dawley rats to a 50-Hertz, 100-microTesla magnetic field for 27 weeks facilitates mammary tumorigenesis in the 7,1 2-dimethylbenz[a]-anthracene model of breast cancer. Cancer Res 59:3627–3633

Tiano L, Belardinelli R, Carnevali P, Principi F, Seddalu G, Littarru GP (2007) Effect of coenzyme Q_{10} administration on endothelial function and extracellular superoxide dismutase in patients with ischaemic heart disease: a double-blind, randomized controlled study. Eur Heart J 28:2249–2255

Tichopad A, Polster J, Pecen L, Pfaffi MW (2005) Model of inhibition of *Thermos aquaticus* polymerase and Moloney murine leukemia virus reverse transcriptase by tea polyphenols (+)-epigallocatechin-3-gallate. J Ethnopharmacol 99:221–227

Tikhonov VI, Volkov AA (2002) Separation of water into its ortho and para isomers. Science 296:2363

Tokes ZA, Rogers KE, Rembaum A (1982) Synthesis of adriamycin-coupled polyglutaraldehyde microspheres and evaluation of their cytostatic activity. Proc Natl Acad Sci U S A 79: 2026–2030

Tompa P, Friedrich P (1998) Prion proteins as memory molecules: an hypothesis. Neuroscience 86:1037–1043

Tooze J, Tooze SA, Warren G (1984) Replication of coronavirus MHV-A59 in sac-cells: determination of the first site of budding of progeny virions. Eur J Cell Biol 33:281–293

Tritton TR, Yee G (1982) The anticancer agent adriamycin can be actively cytotoxic without entering cells. Science 217:248–250

Tritton TR, Yee G, Wing LB Jr (1983) Immobilized adriamycin: a tool for separating cell surface from intracellular mechanisms. Fed Proc 42:284–287

Turi JL, Wang X, McKie AT, Nozik-Grayck E, Mamo LB, Crissman K, Piantadosi CA, Ghio AJ (2006) Duodenal cytochrome b: a novel ferrireductase in airway epithelial cells. Am J Physiol Lung Cell Mol Physiol 291:L272–L280

Tzur A, Kafri R, LeBleu VS, Lahav G, Kirschner MW (2009) Cell growth and size homeostasis in proliferating animal cells. Science 325:167–171

Ueda HR, Matsumoto A, Kawamura M, Iino M, Teiichi T, Hashimoto S (2002) Genome-wide transcriptional orchestration of circadian rhythms in *Drosophila*. J Biol Chem 277: 14048–14052

Unno T, Takeo T (1995) Absorption, distribution, elimination of tea polyphenols in rats. Absorption of (−)-epigallocatechin gallate into the circulation system of rats. Biosci Biotechnol Biochem 59:1558–1559

Vaillant F, Larm JA, McMullen GL, Wolvetang EJ, Lawen A (1996) Effectors of the mammalian plasma membrane NADH-oxidoreductase system. Short-chain ubiquinone analogues as potent stimulators. J Bioenerg Biomembr 28:531–540

Vale RD (2002) AAA proteins: lords of the ring. J Cell Biol 150:13–20

Vale RD, Milligan RA (2000) The way things move: looking under the hood of molecular motor protein. Science 288:88–95

Valeriote F, Corbett T, Grieco P, Moher ED, Collins JL, Fleck TJ (1998) Anticancer activity of glaucarubinone analogues. Oncol Res 10:201–208

Van der Velden AW, Thomas AA (1999) The role of the 5′ untranslated region of an mRNA in translation regulation during development. Int J Biochem Cell Biol 31:87–106

van der Vlies D, Wirtz KWA, Pap EHW (2001) Detection of protein oxidation in rat-1 fibroblasts by fluorescently labeled tyramine. Biochemistry 40:7783–7788

Van Der Woude WJ, Lembi CA, Morré DJ (1972) Auxin (2,4-D) stimulation (in vivo and in vitro) of polysaccharide synthesis in plasma membrane fragments isolated from onion stem. Biochem Biophys Res Commun 46:245–253

Van Der Woude WJ, Lembi CA, Morré DJ, Kidinger JA, Ordin L (1974) Beta-glucan synthetases of plasma membrane and Golgi apparatus from onion stem. Plant Physiol 54:333–340

Van Gestelen P, Asard H, Caubergs RJ (1996) Partial purification of a plasma membrane flavoprotein and NADPH-oxidoreductase activity. Physiol Plant 98:389–398

van het Hof KH, Wiseman SA, Chang CS, Tijburg BM (1999) Plasma and lipoprotein levels of tea catechins following repeated tea consumption. Proc Soc Exp Biol Med 320:203–209

van Iwaarden PR, Driessen AJ, Konings WN (1992) What we can learn from the effects of thiol reagents on transport proteins. Biochim Biophys Acta 1113:161–170

Varnold RL, Morré DJ (1985) Phosphorylation of membrane-located proteins of soybean hypocotyls: inhibition by calcium in the presence of 2,4-dichlorophenoxyacetic acid. Bot Gaz 146:315–319

Varnold RL, Morré DJ, Sandelius AS (1986) Phosphorylation of membrane-located proteins of soybean: in vitro response of purified plasma membranes to auxin and calcium. In: Trewavas AJ (ed) Molecular and cellular aspects of calcium in plant development. Plenum Press, New York, pp 355–356

Vayalil PK, Mittal A, Hara Y, Elmets CA, Katiyar SK (2004) Green tea polyphenols prevent ultraviolet light-induced oxidative damage and matrix metalloproteinases expression in mouse skin. J Invest Dermatol 122:1480–1487

Vianello A, Macri F (1989) NAD(P)H oxidation elicits anion superoxide formation in radish plasmalemma vesicles. Biochim Biophys Acta 980:202–208

Vijaya S, Crane FL, Ramasarma TA (1984) A vanadate-stimulated NADH oxidase in erythrocyte membrane generates hydrogen peroxide. Mol Cell Biochem 62:175–185

Villalba JM, Navas P (2000) Plasma membrane redox system in the control of stress-induced apoptosis. Antioxid Redox Signal 2:213–230

Villalba JM, Navarro F, Cordoba F, Serrano A, Arroyo A, Crane FL, Navas P (1995) Coenzyme Q reductase from liver plasma membrane: purification and role in transplasma membrane electron transport. Proc Natl Acad Sci U S A 92:4887–4891

Villalba JM, Navarro F, Gomez-Diaz C, Arroyo A, Bello RI, Navas P (1997) Role of cytochrome b_5 reductase on the antioxidant function of coenzyme Q in the plasma membrane. Mol Aspects Med 18:S7–S13

Villalba JM, Crane FL, Navas P (1998) Antioxidative role of ubiquinone in the animal plasma membrane. In: Asard H, Bérczi A, Caubergs RJ (eds) Plasma membrane redox systems and their role in biological stress and disease. Kluwer Academic, Dordrecht, pp 247–265

Wade PA, Wolffe AP (1997) Histone acetyltransferases in control. Curr Biol 7:R82–R84

Walford RL, Harris SB, Weindruch R (1987) Dietary restriction and aging: historical phases, mechanisms and current directions. J Nutr 117:1650–1654

Walker JE, Saraste M, Runswick M, Gay NJ (1982) Distantly related sequences in the alpha- and beta-subunits of aTP synthase, myosin, kinases and other ATP-requiring enzymes and a common nucleotide binding fold. EMBO J 1:945–951

Wang CC (1998) Protein disulfide isomerase assists protein folding as both an isomerase and a chaperone. Ann N Y Acad Sci 846:9–13

Wang S, Pogue R, Morré DM, Morré DJ (2001) NADH oxidase activity (NOX) and enlargement of HeLa cells oscillate with two different temperature-compensated period lengths of 22 and 24 minutes corresponding to different NOX forms. Biochim Biophys Acta 1539:192–204

Wang Z, Lo HS, Yang H, Gere S, Hu Y, Buetow KH, Lee MP (2003a) Computational analysis and experimental validation of tumor-associated alternative RNA splicing in human cancer. Cancer Res 63:655–657

Wang S, Morré DM, Morré DJ (2003b) Sera from cancer patients contain two oscillating ECTO-NOX activities with different period lengths. Cancer Lett 190:135–141

Wang Z, Rolish M, Yeo G, Tung V, Mawson M, Burge C (2004) Systematic identification and analysis of exonic splicing silencers. Cell 119:831–845

Wang H-M, Chueh P-J, Chang S-P, Yang C-L, Shao K-N (2009) Effect of capsaicin on tNOX (ENOX2) protein expression in stomach cancer cells. Biofactors 34:209–217

Wang H-M, Chuang S-M, Su Y-C, Chueh PJ (2011) Down-regulation of tumor-associated NADH oxidase, tNOX (ENOX2), enhances capsaicin-induced inhibition of gastric cancer cell growth. Cell Biochem Biophys 61:355–366

Warburg O (1921) The physical chemistry of cell-breathing. Biochem Z 119:134

Warburg O (1929) Is the aerobic glycolysis specific for tumors? Biochem Z 2004:482–483

Warburg O (1930) The metabolism of tumors. Arnold Constable, London

Warburg O (1956) On the origin of cancer cells. Science 123:309–314

Warburg O, Posener K, Negelein E (1924) Ueber den Stoffwechsel der Tumoren. Biochem Z 152:319–344

Warden BA, Smith LA, Beecher GR, Balentine DA, Clevidence BA (2001) Catechins are bioavailable in men and women drinking black tea throughout the day. J Nutr 131:1731–1737

Warley A, Cook GMW (1973) The isolation and characterization of plasma membranes from normal and leukaemic cells of mice. Biochim Biophys Acta 323:55–68

Warren G, Woodman P, Pypaert M, Smythe E (1988) Cell-free assays and the mechanism of receptor-mediated endocytosis. Trends Biochem Sci 13:462–465

Wartenberg DA, Stapleton CP (1998) Risk of breast cancer is also increased among retired US female airline cabin attendants. Br Med J 316:1902–1916

Waskewich C, Blumenthal RD, Li H, Stein R, Goldenberg DM, Burton J (2002) Celecoxib exhibits the greatest potency amongst cyclooxygenase (COX) inhibitors for growth inhibition of COX-2-negatie hematopoietic and epithelial cell lines. Cancer Res 62:2029–2033

Wattenberg BW (1991) Analysis of protein transport through the Golgi in a reconstituted cell-free system. J Electron Microsc Tech 17:150–164

Weaver CM, Barnes S, Wyss JM, Kim H, Morré DM, Morré DJ, Simon JE, Lila MA, Janle EM, Ferruzzi MG (2008) Botanicals for age-related diseases: from field to practice. Am J Clin Nutr 97(suppl):4935–4975

Weber G (1983) Biochemical strategy of cancer cells and the design of chemotherapy. G. H. A. Clowes Memorial Lecture. Cancer Res 43:3466–3492

Wehr TA, Goodwin FK (1983) Biological rhythms in manic-depressive illness. In: Wehr TA, Goodwin EK (eds) Circadian rhythms in psychiatry. Boxwood, Pacific Grove, CA, pp 129–184

Weinberg RA (2007) The biology of cancer, Fig. 11.1, p. 400. Courtesy of Hong, W. K., compiled from SEER Cancer Statistics Review, Garland Science, Oxford

Weindruch R, Walford RL (1998) The retardation of aging and disease by dietary restriction. Charles C. Thomas, Springfield, IL, 436pp

Weindruch R, Naylor PH, Goldstein AL, Walford RL (1988) Influences of aging and dietary restriction on serum thymosin alpha 1 levels in mice. J Gerontol 43:B40–B42

Welker E, Wedemeyer WJ, Scheraga HA (2001) A role for intermolecular disulfide bonds in prion diseases? Proc Natl Acad Sci U S A 98:4334–4336

Wells MA, Jackson GS, Jones S, Hosszu LL, Craven CJ, Clarke AR, Collinge J, Waltho UP (2006) A reassessment of copper(II) binding in the full-length prion protein. Biochem J 399:435–444

Wells-Knecht MC, Huggins TG, Dyer DG, Thorpe SR, Baynes JW (1993) Oxidized amino acids in lens protein with age. Measurement of o-tyrosine and dityrosine in the aging human lens. J Biol Chem 268:12348–12352

Welsh DK, Moore-Ede MC (1990) Lithium lengthens circadian period in a diurnal primate, *Saimiri sciureus*. Biol Psychiatry 28:117–126

White TD (1988) Role of adenine compounds in autonomic neurotransmission. Pharmacol Ther 38:129–168

Whiteheart SW, Rossnagel K, Buhrow SA, Brunner M, Jaenicke R, Rothman JE (1994) N-ethylmaleimide-sensitive fusion protein: a trimeric ATPase whose hydrolysis of ATP is required for membrane fusion. J Cell Biol 126:945–954

Wigner EZ (1933) Mechanism of nuclear spin initiated para-H2 to ortho-H2 conversion. Phys Chem B 23:28

Wilkinson E (2004) Phenoxodiol offers hope for ovarian cancer. Lancet Oncol 5:201

Wilkinson FE, Paulik M, Morré DJ (1993) Modulation of guanine triphosphate nucleotide binding to p21^{ras} immunoprecipitates of rat liver plasma membranes by agents affecting redox state. Biochem Biophys Res Commun 190:229–235

Wilkinson FE, Kim C, Cho N, Chueh P-J, Leslie S, Moya-Camarena S, Wu L-Y, Morré DM, Morré DJ (1996) Isolation and identification of a protein with capsaicin-inhibited NADH oxidase activity from culture media conditioned by growth of HeLa cells. Arch Biochem Biophys 336:275–282

Williamson FA, Morré DJ, Jaffe MJ (1975) Association of phytochrome with rough-surfaced endoplasmic reticulum fractions from soybean hypocotyls. Plant Physiol 56:738–743

Williamson FA, Morré DJ, Hess K (1977) Auxin binding activities of subcellular fractions from soybean hypocotyls. Cytobiologie 16:63–71

Williamson JR, Chang K, Frangos M, Hasan KS, Idom Y, Kawamura T, Nyengaard JR, van den Enden M, Kilo C, Tilton RG (1993) Hyperglycemic pseudohypoxia and diabetic complications. Diabetes 42:801–813

Wingard LB Jr, Tritton TR, Egler AK (1985) Cell surface effects of adriamycin and carminomycin immobilized on cross-linked polyvinyl alcohol. Cancer Res 45:3529–3536

Winkler H, Carmichael SW (1982) The chromaffin granule. In: Poisner AM, Trifaro JM (eds) The secretary granule. Elsevier, Amsterdam, pp 3–79

Winter J, Dray A, Wood JN, Yeats JC, Bevan S (1990) Cellular mechanism of action of resiniferatoxin: a potent sensory neuron excitotoxin. Brain Res 520:131–140

Wolf FI, Torsello A, Tedesco B, Fasanella S, Boninsegna A, D'Ascenzo M, Grassi C, Azzena GB, Cittadini A (2005) 50-Hz extremely low frequency electromagnetic fields enhance cell proliferation and DNA damage: possible involvement of a redox mechanism. Biochim Biophys Acta 1743:120–129

Wolvetang EJ, Johnson KL, Krauer K, Ralph SJ, Linnane AW (1994) Mitochondrial respiratory chain inhibitors induce apoptosis. FEBS Lett 339:40–44

Wolvetang EJ, Larm JA, Moutsoulas P, Lawen A (1996) Apoptosis induced by inhibitors of the plasma membrane NADH-oxidase involves Bcl-2 and calcineurin. Cell Growth Differ 7:1315–1325

Wondrak GT (2009) Redox-directed cancer therapeutics: molecular mechanisms and opportunities. Antioxid Redox Signal 11:3013–3069

Woodman PG (2003) p97, a protein coping with multiple identities. J Cell Sci 116:4283–4290

Workman JL, Kingston RE (1998) Alteration of nucleosome structure as a mechanism of transcriptional regulation. Annu Rev Biochem 67:545–579 [Abstract]

Wovcehowsky KJ, Raines RT (2003) The CXC motif: a functional mimic of protein disulfide isomerase. Biochemistry 42:5387–5394

Wright MV, Kuhn TB (2002) CNS neurons express two distinct plasma membrane electron transport systems implicated in neuronal viability. J Neurochem 83:655–664

Wu L-Y, De Luca T, Watanabe T, Morré DM, Morré DJ (2011) Metabolite modulation of HeLa cell response to ENOX2 inhibitors EGCG and phenoxodiol. Biochim Biophys Acta 1810: 784–789

Wymann S, Ghielmetti M, Schaub A, Baumann MJ, Stadler BM, Bolli R, Miescher SM (2008) Monomerization of dimeric IgG of intravenous immunoglobulin (IVIg) increases the antibody reactivity against intracellular antigens. Mol Immunol 45:2621–2628

Xu Q, Lee C (2003) Discovery of novel splice forms and functional analysis of cancer-specific alternative splicing in human expressed sequences. Nucleic Acids Res 31:5635–5643

Xu S, Ying J, Jiang B, Guo W, Adachi T, Sharov V, Lazar H, Menzoian J, Knyushko TV, Bigelow D, Schöneich C, Cohen RA (2006) Detection of sequence-specific tyrosine nitration of manganese SOD and SERCA in cardiovascular disease and aging. Am J Physiol Heart Circ Physiol 290:H2220–H2227

Yadav BS, Sharma SC, Singh R, Singh G (2007) Patterns of relapse in locally advanced breast cancer treated with neoadjuvant chemotherapy followed by surgery and radiotherapy. J Cancer Res Ther 3:75–80

Yagiz K, Morré DJ, Morré DM (2006) Transgenic mouse line overexpressing the cancer-specific tNOX protein has an enhanced growth and acquired drug-response phenotype. J Nutr Biochem 17:750–759

Yagiz K, Wu L-Y, Kuntz CP, Morré DJ, Morré DM (2007) Mouse embryonic fibroblast cells from transgenic mice overexpressing tNOX express an altered growth and drug response phenotype. J Cell Biochem 101:295–306

Yagiz K, Snyder PW, Morré DJ, Morré DM (2008) Cell size increased in tissues from transgenic mice overexpressing a cell surface growth-related and cancer-specific hydroquinone oxidase, tNOX, with protein disulfide-thiol interchange activity. J Cell Biochem 105:1437–1442

Yamada K, Hara N, Shibata T, Osago H, Tsuchiya M (2006) The simultaneous measurement of nicotinamide adenine dinucleotide and related compounds by liquid chromatography/electrospray ionization tandem mass spectrometry. Anal Biochem 352:282–285

Yamagata M, Hasuda K, Stamato T, Tannock IF (1998) The contribution of lactic acid to acidification of tumours: studies of variant cells lacking lactate dehydrogenase. Br J Cancer 77:1726–1731

Yamaguchi K, Honda M, Ikigai H, Hara Y, Shimamura T (2002) Inhibitory effects of (−)-epigallocatechin gallate on the life cycle of human immunodeficiency virus type 1 (HIV-1). Antiviral Res 53:19–34

Yamamoto Y, Yamashita S (1997) Plasma ratio of ubiquinol and ubiquinone as a marker of oxidative stress. Mol Aspects Med 18:S79–S84

Yang CS (1997) Inhibition of carcinogenesis by tea. Nat Clin Proc Cardiovasc Med 389:134–135

Yang XJ, Seto E (2008) The Rpd3/Hda1 family of lysine deacetylases: from bacteria and yeast to mice and man. Nat Rev Mol Cell Biol 9:206–218

Yang CS, Wang ZY (1993) Tea and cancer. J Natl Cancer Inst 85:1038–1049

Yang CS, Chen L, Lee MJ, Balentine D, Kyo MC, Schantz SP (1998) Blood and urine levels of tea catechins after ingestion of different amounts of green tea by huam volunteers. Cancer Epidemiol Biomarkers Prev 7:679–684

Yang GY, Liao J, Li C, Chung J, Yurkow EJ, Ho CT, Yang CS (2000) Effect of black and green tea polyphenols on c-jum phosphorylation and H_2O_2 production in transformed and non-transformed human bronchial cell lines: possible mechanisms of cell growth inhibition and apoptosis induction. Carcinogenesis 21:2035–2039

Yang CS, Maliaka P, Meng X (2002) Inhibition of carcinogenesis by tea. Annu Rev Pharmacol Toxicol 42:25–54

Yang H, Yang T, Baur JA, Perez E, Matsui T, Carmona JJ, Lamming DW, Souza-Pinto NC, Bohr VA, Rosenzweig A, de Cabo R, Sauve AA, Sinclair DA (2007) Nutrient-sensitive mitochondrial NAD^+ levels dictate cell survival. Cell 130:1095–1107

Yantiri F, Morré DJ (2001) Isolation and characterization of a tumor-associated NADH oxidase (tNOX) from the HeLa cell surface. Arch Biochem Biophys 391:149–159

Yantiri F, Morré DJ, Yagiz K, Barogi S, Wang S, Chueh P-J, Cho N, Sedlak D, Morré DM (1998) Capsaicin-responsive NADH oxidase activities from urine of cancer patients. Arch Biochem Biophys 358:336–342

Yasuda S, Arii S, Mori A, Isobe N, Yang W, Oe H, Fujimoto A, Yonenaga Y, Sakashita H, Imamura M (2004) Hexokinase II and VEGF expression in liver tumors: correlation with hypoxia-inducible factor 1 alpha and its significance. J Hepatol 40:117–123

Yazdanparast R, Shahriyary I (2008) Comparative effects of *Artemisia dracumculus*, *Satureja hortensis* and *Origanum majorana* on inhibition of blood platelet adhesion, aggregation and secretion. Vascul Pharmacol 48:32–37

Yeh CJ, Faulk WP (1984) Killing of human tumor cells in culture with adriamycin conjugates of human transferrin. Clin Innunol Immunopathol 32:1–11

Yeh CJ, Taylor CG, Faulk WP (1984) Targeting of cytotoxic drug by ransferring receptors: selective killing of acute myelogenous leukemia cells. Protides Biol Fluid 32:441

Yin L, Wang J, Klein PS, Lazar MA (2006) Nuclear receptor Rev-erbá is a critical lithium-sensitive component of the circadian clock. Science 311:1002–1005

Yong V, Dreyer JL (1995) Developmental changes in the localization of the transplasma membrane NADH-dehydrogenases in the rat brain. Brain Res Dev Brain Res 89:253–263

Yoshizawa H, Tsuchiya T, Mizoe H, Ozeki H, Kanao S, Yomori H, Sakane C, Hasebe S, Motomura T, Yamakawa T, Mizuno F, Hirose H, Otaka Y (2002) No effect of extremely low-frequency magnetic field observed on cell growth or initial response of cell proliferation in human cancer cell lines. Bioelectromagnetics 23:355–368

Young MW, Kay SA (2001) Time zones: a comparative genetics of circadian clocks. Nat Rev Genet 2:702–715

Yu Q, Heikal AA (2009) Two-photon autofluorescence dynamics imaging reveals sensitivity of intracellular NADH concentration and conformation to cell physiology at the single cell-cell level. J Photochem Photobiol B 95:46–57

Yu BP, Masoro EJ, McMahan CA (1985) Nutritional influences on aging of Fischer 344 rats: 1. Physical, metabolic, and longevity characteristics. J Gerontol 40:657–670

Yunghans WN, Morré DJ (1977) Adenylate cyclase activities not found in soybean hypocotyl and onion meristem. Plant Physiol 60:144–149

Yunghans WN, Clark JE, Morré DJ, Clegg ED (1978) Nature of the phosphotungstic acid-chromic acid (PACP) stain for plasma membranes of plants and mammalian sperm. Cytobiologie 17:165–172

Zai A, Rudd MA, Scribner AW, Loscalzo J (1999) Cell-surface protein disulfide isomerase catalyses transnitrosation and regulates intracellular transfer of nitric oxide. J Clin Invest 103:393–399

Zbell B, Paulik M, Morré DJ (1990) Comparison of [^{35}S] GTP S binding to plasma membranes and endomembranes prepared from soybean and rat. Protoplasma 154:74–79

Zemková H, Teisinger T, Vyskocil F (1984) Hyperpolarization of mouse skeletal muscle plasma membrane induced by extracellular NADH. Biochim Biophys Acta 775:64–70

Zeng Z-M, Chuang S-M, Chang T-C, Hong C-W, Chou J-C, Yang J-J, Chueh PJ (2012) Phosphorylation of serine-504 of tNOX (ENOX2) modulates cell proliferation and migration in cancer cells. Exp Cell Res. http://dx/doi.org/10.1016/j.yescr.2012.04.021

Zhang L, Ashendel CL, Becker GW, Morré DJ (1994) Isolation and characterization of the principal ATPase associated with transitional endoplasmic reticulum of rat liver. J Cell Biol 127:1871–1883

Zhang Y, Yo B, Delikat S, Bayumy S, Lin XH, Basu S, McGinley M, Chan-Hui PY, Lichenstein H, Kolesnick R (1998) Kinase suppressor of Ras is ceramide-activated protein kinase. J Biol Chem 273:30419–30426

Zhang X, Shaw A, Bates PA, Newman RH, Gowen B, Oriova E, Gorman MA, Kondo H, Dokumo P, Lally J, Leonard G, Meyer H, van Heel M, Freemont PS (2000) Structure of the AAA ATPase p97. Mol Cell 6:1473–1484

Zhang Q, Wang SY, Nottke AC, Rocheleau JV, Piston DW, Goodman RH (2006) Redox sensor CtBP mediates hypoxia-induced tumor cell migration. Proc Natl Acad Sci U S A 103: 9029–9033

Zhao L-J, Subramanian T, Show Y, Chinnadurai G (2006) Acetylation by p300 regulates nuclear localization and function of the transcriptional factor CtBP2. J Biol Chem 281:4183–4189

Zhou Q, Xhu Y, Chiang H, Yagiz K, Morré DJ, Morré DM, Janle E, Kissinger PT (2004) Identification of the major vanilloid component in *Capsicum* extract by HPLC-EC and HPLC-MS. Phytochem Anal 15:117–120

Zhu M, Chen Y, Li RC (2000) Oral absorption and bioavailability of tea catechins. Planta Med 66:444–447

Zhu X, Liu ZC, Xie BF, Feng GK, Zeng YX (2003) Ceramide induces cell cycle arrest and upregulates p27kip in nasopharyngeal carcinoma cells. Cancer Lett 193:149–154

Zu XL, Guppy M (2004) Cancer metabolism: facts, fantasy, and fiction. Biochem Biophys Res Commun 313:459–465

Zurbriggen R, Dreyer JL (1994) An NADH-diaphorase is located at the cell plasma membrane in a mouse neuroblastoma cell line NB41A3. Biochim Biophys Acta 1183:513–520

Zurbriggen R, Dreyer JL (1996) The plasma membrane NADH-diaphorase is active during selective phases of the cell cycle in mouse neuroblastoma cell line NB41A3. Its relation to cell growth and differentiation. Biochim Biophys Acta 1312:215–222

Index

A

AAA-ATPase, 124–139
  ratchet model, 137
Acetogenins (annonaceous), 115, 389–390, 417
  bullatacin, 115, 390
13-cis-acetretin, 412
2-acetylaminofluorene, 98
Acrosome reaction, 202
Actinomycin D, 75, 313, 322, 324, 341, 368
Adenine receptors, cell surface, 17
Adenylate cyclase plants, 88, 319
Adriamycin®. *See* Doxorubicin
Advanced glycation endproduct (AGE) readings, 286, 289
α-fetoprotein, 234
Age-related diseases, 435
Age-related ENOX proteins (arNOX, ENOX3), 83, 185, 195, 261–311
  activities, 263
  aggregation, 273
  assay by tyramine conjugation, 288–291
  assay using ferricytochrome c reduction as measure of superoxide blocked by superoxide dismutase, 20, 266–268, 271, 274, 286, 291, 310
  as biomarker of aging, 284–286
  characteristics, 268–272
  cloning, 272–278
  copper requirement for activity, 284
  correlation with age, 271
  correlation with life span, 268, 297
    plants, 298
    sea urchins, 297
  discovery, 264–266
  electron acceptors, 261
  in endosomes, 192, 210, 277, 308
  formation of tyrosyl radicals, 272, 288, 290–292
  functional motifs, 4, 261, 283
  gene, 274
  generation of superoxide, 261, 262, 278, 308
  host defense, 308
  hydropathy analysis, 277
  identification as TM9 super family(SF) members, 274
  inhibitors, 197, 299–307
    botanical sources (savory, estragon (tarragon), basil, marjoram, rosemary and sage), 305–307
    coenzyme Q, 299–305
    coronary artery disease, 197, 310
    gallic acid, 265, 266, 283, 299, 306
    human subjects, 295, 301
    platelet adhesion and aggregation, 197, 306
    skin health, 299, 305
    sustained release, 303–305, 307
    tyrosol, 56, 265, 266, 283, 299
  as integral part of LDL particle, 8, 295
  keratinocytes/fibroblasts, 288
  Kex2ρ protease cleavage, 277
  lack of response to melatonin, 283–284
  linking functionally deficient mitochondria to cell surface accumulations of reactive oxygen species, 269
  and lipid oxidation, 195–197, 263
  malondialdehyde formation, 296
  m-DNA lesions, 269, 286
  natural electron donors, 261
  oscillations in activity, 268–284

D.J. Morré and D.M. Morré, *ECTO-NOX Proteins: Growth, Cancer, and Aging*, DOI 10.1007/978-1-4614-3958-5, © Springer Science+Business Media New York 2013

Age-related ENOX proteins (arNOX, ENOX3) (*cont.*)
- oxidation of collagen and elastin, 287
- oxidation of serum lipoproteins, 291–296
- p-chloromercuribenzoate, inhibition by, 279
- peptide antibodies to, 262, 264, 288
- in plants, 265, 298
- presence in cancer, 301
- properties of a prion, 337–341
- proteinase K resistance, 271
- protein carbonyl antibody assay, 287
- protein disulfide thiol interchange activity, 267, 279, 283, 284
- protein thiols as source of electrons, 15, 262, 294
- purification, 275
- recombinant proteins, 279–284
- role in autophagy, 192
- role in skin aging, 286–291
- saliva, 261, 266–269, 271, 281, 286, 289, 290, 301–305, 307, 310
- sequence, 261, 262, 269, 273
- serum analysis, 287, 289
- serum lipoprotein oxidation, 291–296
- shedding (shed forms), 261, 273, 283, 286, 293
- significance of oxidative burst, 439
- skin aging, role in, 286–291
- specific peptide antisera, 277
- sub types, 276
- superoxide formation, 261, 262, 266–270, 280, 286
  - beneficial roles of, 307–309
- time keeping role (lack of), 13
- transcript variants, 282
- type I topology of membrane-associated forms, 277
- tyrosine oxidation, 15, 287
- urine, 267, 268, 274–276
- yeast, 277–279
- XTT reduction, 270

Akt, 91–93, 198–200, 351–354
α-fetoprotein, 234
α-helix-β-sheet transformations, 141, 215, 223, 232, 397
Alternative splicing, 230, 241–257, 414, 436
Alzheimer's Aβ protein, 16, 61, 201
Amiloride, 78, 79, 95
Aminco SLMDW 2000, 21–25, 28, 34
Amino-PEG glaucarubolone, 388
Aminoxyacetate, 309
AMP activated protein kinase (AMPK), 88, 90, 93
AMP/ATP ratio, 93
Amyloid, 7, 14, 16–17, 22, 57, 61, 63, 201, 202, 214, 225, 226, 230, 232, 233, 261, 336, 439
Amyloid forming proteins, 16, 201, 202
- copper binding motifs, 201

Angiostatin, 76
Annonaceae, 389, 417
Annonacin, 115
Anthracycline antibiotics, 75, 368–377
Antibodies, therapeutic, 412–414, 438
Anticancer drugs, 47, 53, 84, 96, 100, 177, 211, 212, 227, 228, 236, 257, 324, 333, 340, 348–351, 396, 405, 416, 438
Antimalarials, 204, 389
Antimetabolites, 227
Antimitotics, 227
Antioxidants, 67, 76, 196, 207, 284, 293, 299, 391, 396
Antiparasitic, 389
Antiport, H^+/Na^+, 78–80, 195
Antisense, 120, 239, 243, 245–248, 259, 365, 391, 414–415
Antitrypanosamals, 204
α1-Antitrypsin inhibitor, 426, 428
Antivirals, 204, 206, 207, 209, 308, 349
- as impermeant conjugates, 209

Anxiety, 180
Aplysia, 179
ApoB, 294–296, 310
Apoptosis, 7, 57, 71, 76, 83, 91, 118, 123–124, 181, 198–200, 210, 256, 348–359, 363, 366, 377, 395–401, 408, 412, 414, 421, 430, 438
- activation of intrinsic pathway, 351
- extrinsic pathway, 351
- mechanism, 198, 351–355

Aqueous two phase partition, 125, 252, 298, 317, 318
Arabidopsis, 7, 47, 325, 333
arNOX. *See* Age-related ENOX proteins (arNOX, ENOX3)
Arotonois ethylester, 412
Arsenicals, 349–350, 363, 416
Artesunate, 204
Ascorbate, 65, 67, 72, 73, 196, 293, 321, 322
Ascorbate radical, 82, 101, 191, 212, 323, 369
Asimicin, 115
Aspirin, 410
Atherogenesis, 195, 197, 273, 291, 293, 294, 310
- inhibitors, 197

ATP, 17, 60, 76, 124–138, 191–193
generation by plasma membrane electron transport, 65, 73, 76–79
requirement for cell growth, 76, 124–138
ATPase, 80, 81, 128, 131, 136, 192, 318, 319, 321
antibody, 130
associated with different cellular cellular activities. *See* AAA-ATPase
N^+/H^+, 80, 193
Na^+/K^+, 81, 244, 250
TER, 135, 136
Autism, 201
Autoantibodies, 71, 229–230, 259, 414, 417, 430–432, 438
Autoentrainment, 169, 213, 439
Autosynchrony, 46, 146
Auxin, 101, 313–343
ascorbate-induced spectral changes, 321
auxin-stimulated ATPase, 319
auxin-stimulated NADH oxidase (*see* dNOX)
binding, 127, 136–138, 317
degradation of choline-containing phospholipids, 319
direct effect on signaling molecules, 318–321
electrogenic proton pump driven by ATP, 193, 319
enhanced RNA polymerase, 317, 318
fluorescence polarization, 317, 318
gene expression requirement, 321
heat coagulability of cytoplasmic proteins, 328
increased microviscosity of plasma membrane, 317
induced plant cell enlargement, 313–317
importance of Golgi apparatus, 331
infrared spectroscopy, 317
inhibition by thiol reagents, 100, 331
oscillations, 109, 335–337
unresponsive to monensin, 331
infrared spectroscopy, 317
interaction with calcium, 319
membrane thinning, 318
NADH oxidase as molecular target, 322
NADH oxidation in roots, 315, 333
phosphatidic acid phosphatase inhibition, 318
phosphatidylinositol specific phospholipase C, 230, 319
phospholipase A, C and D, 319
phospholipase C, 319
phosphorylation of membrane proteins, 319
plasma membrane location of response, 5
proton pumping, 321
release of calcium ions, 319
response to osmotica, 331–332
stimulated increase in membrane thiols decrease in membrane disulfides, 330
stimulation of hot water-soluble glucans, 318
stimulation of plasma membrane ATPase, 319
stimulation of protein disulfide isomerase-like activity, 316
two point attachment, 322
Avena sativa, 191
Azide, 127

B

Bacitracin, 203
Bathocuproine, 57, 58, 144, 149, 284
Bax, 351
Bcl-2, 124, 350
Benzoic acid, 316, 323
BID, 351
Biological clock, 6, 13, 18, 47, 50, 89, 141–186
and cancer, 177
links to ENOX oscillators, 62, 141, 178–184
molecular studies, 142–148
Bipolar disorder, 179
Bleomycin, 75, 368, 376–377
Blue light, 146, 169, 180
Blue light-responsive skin disorders, 180
Bodipy-duTP/TdT, 357
Brefeldin A, 124, 188, 189, 192, 204, 207–209, 331
Brij detergent, 318
Broccoli, natural source of sulforaphane, 408, 417
Bullatacin, 115, 390, 396
Bullatacinome, 115

C

Caffeine, 169, 178, 180, 185, 400
Calcitriol, 116, 214, 227, 411–412, 417
Calcium and auxin, 319
Callipeltin, 204, 409
Callipeltin A, 409
Calorie restriction, 85, 188, 197–198, 210

Cancer
- conjugates, impermeant, therapeutic drugs, 5, 209, 348, 366–369, 372–375, 380, 382, 388, 389, 416
- 2-D gel/western blot analysis, 259, 424, 425, 428, 431, 437, 441–442
- diagnosis, 420, 421, 425, 428, 433
- ENOX2 as molecular serum cancer marker, 214
- and ENOX2 presence, 213, 217, 227–233
- plasma membrane electron transport, 86–88
- progression, 257, 429
- therapeutic applications of plasma membrane redox and ENOX2 inhibitors, 345–417

Capsaicin, 115, 212, 214, 227, 241, 363–368
- as cancer therapeutic, 365–366
- as PMET inhibitor, 363
- specific inhibition of ENOX2, 364–365

Capsaicin receptor, 62, 363

Capsibiol-T, 399

Capsol-T®, 392, 398–400
- for early cancer intervention, 415, 430
- pharmacokinetics and dosing, 399
- safety and efficacy, 399–400

Carboplatin, 354, 377

Carcinoembryonic antigen, 234

Cardiovascular disease, 8, 192, 196, 293, 294

β-Carotene, 196, 293

Carrier wave, 166–168, 176, 185

Caspase-3, 199, 351, 354, 416

Caspase-8, 351

Caspase-9, 351

Castela polyandra, 333, 335
- catechin-capsicum synergies, 396–400
- chemosensitization to overcome resistance, 405
- development of resistance, 398
- early cancer intervention strategy, 433

+Catechin, 306

Catechin-vanilloid synergies, 396–398

Cathepsin D, 70, 72, 230, 376

CDC 48, 136

CD spectra. *See* Circular dichroism (CD) spectra

Celecoxib, 410, 411

Cell cycle arrest, 355–356

Cell cycle check point, 81–82

Cell enlargement
- and cell cycle control, 118–120
- cell-free, 124,138
- essential role of ENOX1, 62
- plant roots, 315
- plants, 82, 101–113
- vertebrate cells, 113–122

Cell-free analysis of endomembrane traffic, 191

Cell-free vesicle enlargement, 128, 130–132, 139

Cell-impermeable dyes, 42–43

Cell survival, 73, 198–200, 350, 359

Cellular movement, 188

Cell walls, 97, 104, 193, 313, 319, 332, 341

Ceramide, 91, 199, 349–352, 354–356, 359, 360, 366, 416

Ceramide elevation
- cell cycle arrest, 355–356
- link to ENOX2 inhibition, 199
- link to growth arrest, 355–361
- response to G_1 arrest, 199, 352, 421

Chemosensitization, 405–408, 417, 438
- *cis*platin resistance, 438
- mediated through ENOX2, 406–408
- by naturally occurring isoflavenes, 405
- by phenoxodiol, 406–408, 438
- propagation through prion-like protein-protein interactions, 408
- resistance to taxanes, 405, 406
- resistant idioforms, 408

Chili peppers, 53, 212, 363, 400, 430

Chlamydomonas reinhardii, 56

Chloroquin, 69

p-Chloromercuribenzoate, 100, 279

CHO cells, 6, 23–25, 30, 36, 37, 112–114, 116, 119, 120, 309

Chlorsulfuron, 102, 115, 127, 129, 334, 379

Cholesterol, 132, 195–197, 293, 305

Chromaffin granules, 60

Chronic fatigue syndrome, 81

Chronic myelogenous leukemia KS62 cells, 376

Chronotherapy, 177

Chymotrypsin, 231

Cip/Kip family, 356

Circadian gene transcription, 183, 184

Circular dichroism (CD) spectra, 144, 214, 223, 224, 232

13-cis-Acitretin, 412

Cisplatin, 75, 76, 212, 321, 324, 377, 406, 438
- resistance overcome by phenoxodiol, 438
- targets ENOX, 377, 378

Clock genes, 180, 181, 185

CLOCK/NPAS2:BMAL1, 89, 90

CNOX. *See* ENOX1

Cobalt chloride, 128, 192

Coenzyme Q (CoQ)
analogs, 69, 79, 379,
as arNOX inhibitor, 299–305
as inhibitor of sphingosine kinase activity, 359
role of prenyl side chain in arNOX inhibition, 301
Collagen, 178, 193, 270, 286–289, 294, 310
Colony formation on soft agar, 248
Concanavlin A column, 327, 342
Conjugates, impermeant drug, 5, 209, 348, 366, 372–383, 388
COP-II, 135
Copper, 57–60, 149–150, 201
hexaaqua ion, 152
Copper clock, 150–151, 153, 168–171, 179, 180
set by electromagnetic fields, 150, 169–174, 177
Coronary artery disease, 196, 197, 210, 310, 435
COX-1. *See* Cyclooxygenase-1 (COX-1)
COX-2. *See* Cyclooxygenase-2 (COX-2)
Crabtree effect, 87
Cryptochromes, 169, 180, 181
CtBP (C-terminal binding protein), 93
Cry 1-3, 181
Cyanide, 1, 69, 73, 309, 316,
Cyclic AMP-mediated phosphorylation, 60
Cyclin A, E and D_1, 355
Cyclin-dependent kinases (CDK), 356
Cyclodextrin conjugates, 382
α-Cyclodextrin conjugates of capsaicin, 366–368
Cyclooxygenase-1 (COX-1), 410
Cyclooxygenase-2 (COX-2), 410, 417
Cyclooxygenase inhibitors, 356
Cyst nematode resistance (Cyst X gene), 298

D

Dacarbazine augmented by green tea catechins and by EGCg, 392
Data reduction methods, 23–32
Daunomycin, N-acetyl, 369
DCIP. *See* 2,6-Dichloroindophenol (DCIP)
Decomposition fits, 11, 12, 22, 23, 29, 37, 106, 109, 112, 114, 116, 119, 143, 145, 152–157, 159–161, 166, 167, 394
Deuterium oxide, 148, 185, 218
2-D Gel-western blot detection of ENOX2 transcript variants in cancer, 423–431
Diablo, 354
Diagnosis of cancer, 420, 421, 425, 428, 433
2,6-Dichloroindophenol (DCIP), 72, 73
2,3-Dichlorophenoxyacetic acid (2,3-D), 107, 108, 126, 129, 314, 316, 323, 325, 328
2,4-Dichlorophenoxyacetic acid (2,4-D), 4, 108, 109, 125–132, 191, 313–343
how 2,4-D kills plants, 337–341
Dicoumarol, 66, 67, 309, 349
Diferric transferrin, 98
conjugates of doxorubicin, 372
1,25-Dihydroxy-vitamin D, 412
2,3-Dimethoxy 1,4-naphthoquinone (DMNQ), 309
2,3-Dimethyl benzoic acid, 266
Diode array instruments, 21, 30–32
Diphenyleneiodonium (DPI), 15, 72, 73, 84, 309, 356
Dithiodipyridine, 6, 9, 11, 19, 20, 48, 54, 117, 143, 214, 215, 278, 279, 403
assay for protein disulfide-thiol interchange, 11, 41
5, 5'-Dithiol(2-nitrobenzoic acid), 39, 100, 117, 294
dNOX, 313–343
cloning, 325
discovery, 342
Golgi apparatus transport essential to sustained enlargement, 331
inhibited by anticancer drugs, 340
inhibition by thiol reagents, 100, 316, 331
isolation of and molecular weight, 214, 327
oscillations, 131, 327–331, 335–337
properties of a prion, 337–341
protein disulfide-thiol interchange activity of, 316, 325, 328, 342, 355
vitamin K1 (phylloquinone) as electron donor, 328
Docetaxel resistance, 438
Dopamine, 81, 214
Dormin, 78, 283, 289, 299
Doxorubicin, 1, 70–72, 212, 214, 228, 345, 369–376
diferric transferrin conjugates, 372
impermeant conjugates with anticancer activity, 372–380
impermeant conjugates targeted to ENOX2, 374–375
inhibited NADH-quinone reductase of plasma membrane, 71–72, 375–376
inhibition of PMET mediated through inhibition of ENOX2, 370–374
inhibitor of PMET of cancer cells, 96
plasma membrane located target, 369–370
DPI. *See* Diphenyleneiodonium (DPI)

Drug conjugates, impermeant, 5, 209, 348, 366, 372–383, 388
DT-diaphorase. *See* NQO1
DTNB. *See* 5, 5'-Dithiol(2-nitrobenzoic acid)
Duodenal cytochrome b (Dcytb, Cybrd1), 72
Duox, 74, 308

E

Earth's magnetic field, 171
EGCg. *See* Epigallocatechin-3-gallate (EGCg)
EGF. *See* Epidermal growth factor
Elastin, 193, 287–289, 310
Electromagnetic fields
  and cancer incidence, 178
  ENOX2 phased by, 218
  and human health, 177
Electron acceptors
  oxygen as, 8, 15, 261
  protein disulfides as, 8, 15, 73, 218
Electron donors
  hydroquinones as, 14, 15, 19, 68, 261
  natural, 14–15, 19, 218
  phylloquinone (vitamin K) as, 14, 15, 32–35
  pyridine nucleotides as, 15, 218
  tyrosines as, 15–16
  vitamin K (phylloquinone) as, 14, 15
Electron transport, 80
Ellman's reagent, 30, 294
Embryonic development, 234
Endocytosis, 192–193, 204, 210
Endoplasmic reticulum, 69, 124, 135, 136, 169, 188, 190, 192, 203, 209, 243, 309
  ENOX. *See also* ENOX2; ENOX2; dNOX; Age-related ENOX proteins (arNOX, ENOX3)
  activity forms, 2
  aggregation, 7–8, 16–17, 22
    significance, 439
  alternation of oxidative and thiol interchange activities, 215
  assay, 19–46
  and cancer, 177–178
  clock, 177–178
  cloning. *See* ENOXK, ENOX2, dNOX, arNOX
  drivers of cell enlargement, 106
  endocytosis (endosomes), 192–193
  endomembrane functions, 188–192
  functional motifs, 4
  functional unit, 9
  gametogenesis, 202–203
gene regulation, 187–188
host defense, 13–14, 193
hydroquinone oxidation, 17, 20, 54, 71, 116
inhibitors increase cytosolic NADH levels, 187, 358–359
inhibitors slow growth of HeLa cells and induce apoptosis in cancer but not in non-cancer cells, 356–358
life extension, 197–198
lipid oxidation, 195–197
low specific activity, 14–17
memory, 201–202
neurodegenerative disorders, 200–201
oscillators and links to biological clock, 18, 47, 50, 141–186
  correlative experiments with heavy water, 178
  transfection experiments, 178
other potential functional roles, 187–210
participation in enlargement phase of growth, 7, 47, 62, 81
pH control, 193–195
phasing by light, 146, 180
phasing by EMF, 169–171
phosphorylation, 60
protease resistance, 7 (*see also* ENOX1, ENOX2, arNOX)
protein disulfide-thiol interchange, 1, 3, 4, 6, 9–12, 18–20, 24
protein resistance, 222
proteins
  activity measurements, 143, 274, 380
  aggregation, 16–17, 22, 39, 41
  alternation of activities, 3, 12, 215
  cell surface receptor proteins, 86
  as dicopper proteins, 8–10
  differ from phox-nox proteins of host defense, 13–14
  as ectoproteins, 2
  electron acceptors, 73
    oxygen as, 8, 15, 218
    protein disulfides as, 8, 15, 218
    electron donors
    hydroquinones, 14, 15, 19, 68
  protein thiols as, 14–15, 19, 218
  pyridine nucleotides as, 15, 218
  as terminal oxidases of plasma membrane electron transport, 2, 3, 5, 47, 65, 69
  pH control, 193–195
proton pumping, 193
response to gravity, 191
response to mechanical stimulation, 328

response to osmotica (plants), 101, 102
role as ultradian regulators of the cell's biological clock, 141–164
role in cell enlargement, 47, 62, 81–82
shed into environment, 4–6
stimulation by hormones and growth factors, 98
stimulation by lysophospholipids, 319
surface localization, 2, 17
terminal oxidases of plasma membrane electron transport, 2, 3, 5, 47, 65, 69
transcription factor function (lack of evidence for), 187
viral pathogenesis, 203–209
yield enhancement of agricultural crops, 435

ENOX1, 47–63
antibodies, 47, 50, 52, 53
aggregation and electron microscopy, 61–62
characterization, 53–57
cloning, 50–53
copper binding, 58–60
copper requirement, 57–58
functional motifs, 4, 53
gravity response, 191
homologs, 9, 50
hormone and growth factor stimulation, 48, 83, 97, 98, 213
of human platelets, 62
hydroquinone oxidation, 54
inhibitors
ATP, 60, 67
mefluidide, 334
simalika-lactone D, 47, 56, 57, 335
triacon-tanol, 101, 333
inhibitors enhance response of tumors to radiation, 37, 415
Intrinsic membrane binding motifs (lack of), 230, 259
knockout rat, 437
mechanical stimulation, 328
oscillations, 53, 54
plant roots, 328
plants, 50, 325
properties, 57
protein disulfide-thiol interchange, 39, 41, 48, 54, 62
response to nucleotides, 60
response to touch, 328
retroviral-mediated shRNA suppression, 57
site directed mutagenesis of copper sites, 58–60
terminal oxidases of plasma membrane electron transport, 47, 65, 69
transcript variants
ubiquinol reduction, 54
yeast, 47, 50, 51, 57

ENOX2
activity, 212–218
aggregation, 227
alternative splicing for cell surface localization, 241–257
antisense, 120, 241, 242, 245, 247
antisera, 8, 221, 239, 251, 264
autoantibodies, 230, 259
binding of phenoxodiol, 200
biochemistry, 217–218
biomarker of cancer, 216, 229
and cancer, 211–259
cancer biomarkers based on activity, 421–424
cancer-specific expression, 101, 242, 243, 245, 249–250, 416
as cancer therapeutic drug target, 211, 212, 227, 345–417
as cell surface biomarker of cancer presence shed into serum, 351, 416, 419–434
characteristics, 214
cloning, 212
coiled coil structures, 221
2-D gel/western blot analysis of transcript variants, 424
developmental expression (embryonic stage), 234
directed therapeutic antibodies, 412–414
discovery, 211–213
drug resistant idioforms, 422
early developmental regulation, 233, 234, 257
as ectoproteins, 228, 229, 241, 379, 420
essential role in cell migration, 256
expression levels correlated with cell proliferation, 256
formation of Exon 4 minus variant directed by hnRNPF, 257–258
full length mRNA not translated, 247
functional motifs, 4, 221
gain of function studies, 256
gene, 229–230
in Golgi apparatus, 232, 252
idioforms, 408, 438
inhibited by phenoxodiol, 198–200, 235
inhibition by calcitriol, 214, 227, 411–412
inhibition by callipeltin A, 409
inhibition by retinoic acid, 123, 411–412

ENOX2 (*cont.*)
inhibition by retinoids, 411–412
inhibitors, 412, 416, 417, 422
as therapeutic drugs, 345–417
low dose molecular target for EGCg, 391
membrane-binding motifs (lack of), 230
mRNA tissue expression, 121, 123, 236, 239
NADPH as electron donor, 218
overexpression and growth, 82, 120–123, 234, 239, 251, 256, 391, 392, 404, 414, 415
phenoxodiol target, 199, 400–408
phosphorylation, 227
as potential drug target to account for anticancer effects of NSAIDS, 396, 411
processing, 252
properties, 195, 202, 210, 214, 223–227
of a prion, 230–233
protease and heat resistant, 8, 214, 225, 226
protein disulfide-thiol interchange mediates chemosensitization response of phenoxodiol, 309
regulation of expression, 241, 256–257
response to suramin redox sensitive, 409
role in epithelial to mesenchymal transition, 257
RTPCR for detection of cancer cells in blood, 433
sequence, 218–222
serum form inhibited by anticancer drugs, 231–233
antitumor sulfonylureas, 227, 259, 378–387
catechins (EGCg), 391
cisplatin, 377–378
doxorubicin, 345, 349
vanilloids (capsaicin), 205–209
shed into circulation, 228, 420–425
site-directed mutagenesis, 221, 245, 254
specific monoclonal antibody, 8, 113, 229, 230, 396, 404, 422, 423
splice variant mRNA over expression, 122, 436
splice variants, 204, 218, 241–243, 245, 247, 250–251, 253–256, 259, 440
structural properties, 223–227
target for cancer prevention through early intervention, 415–416
target for EGCg, 391
therapeutic antibodies, 412–414
therapeutic antisense, 414–415
therapeutic applications, 345–417
transcript variants, 229, 230, 424, 425
as cancer site specific biomarkers, 424, 425
detection by 2-D gel/western blot analysis, 424, 428–430, 441–442
for early cancer detection, 433
transcript variant-specific recombinant (scFv) antibody, 229, 424
transgenic mouse over expressing, 235–241
enhanced response to antitumor sulfonylureas, 117
uncontrolled growth tissue expression, 239
urine, 216
ENOX3. *See* Age-related ENOX proteins (arNOX, ENOX3)
Entrainment, 13, 18, 146, 169, 201, 202, 213, 259, 338, 340, 439
Epidermal growth factor (EGF), 48, 98, 99, 115, 117, 253, 387
Epigallocatechin-3-gallate (EGCg)
as antioxidant, 207, 391
blocks virus infections, 205–207
combination with *Capsicum* vanilloids and other green tea catechins, 205–206 (*see also* Tea catechins)
enhancement of response to antitumor sulfonylureas, 416
response to ENOX2 overexpression, 120
tumor cell sensitivity, 391
Epigallocatechin-3-gallate sulfate, 205
Epilogue, remaining challenges, 435–440
Erk, 91–93
Erythrocytes, 269, 371
ESSs. *See* Exonic splicing silencers (ESSs)
Etretinate, 412
Euglena gracilis, 148
EXAFS investigations, 152–158
Exonic splicing silencers (ESSs), 257, 258, 415, 437

F

FAD. *See* Flavin-adenine dinucleotide (FAD)
Fas pathway, 199, 352
Fatty acid oxidation, 86, 93
Feedback regulation, 85
Feline or the human lentivirus, 207
Ferricyanide, 65, 69–73, 75, 76, 79–82, 84, 85, 91, 95, 124, 191, 195, 297, 309, 318, 323, 324, 349, 358, 359, 368, 369, 371, 376–379, 411
ferricyanide reductase, 71

Ferricyanide reduction
effect on growth and apoptosis, 358
inhibition by cisplatin, 377
inhibition by doxorubicin, 369–370
α-Fetuin. *See* α-Trypsin inhibitor
Fibroblasts, 288, 401
FIV, 207
Flavin, 8–10, 15
Favin-adenine dinucleotide (FAD), 13, 15, 171, 308
Flavoenzymes, 73, 84
Flip, 198–200, 351–354
Foam cells, 6, 195, 196, 293, 296, 310
Forkhead transcription factor (FOXO), 198
Fourier analysis, 11, 12, 22, 23, 29, 108, 110, 152, 154–156, 161, 167
Fourier transform infrared, 214, 215, 317
FOXO. *See* Forkhead transcription factor (FOXO)
Free-flow electrophoresis (preparative), 318, 327, 379, 432
French paradox, 197, 305

G

G_1 arrest, 199, 351, 352, 355, 356
G_2M block, 355
Gallic acid, 266, 283, 299, 306
Gametogenesis, 202–203, 210
Ganglioside, 432
G_1 checkpoint, 187, 200, 355, 430
Glaucarubolone, 74, 85, 102, 207, 208, 333, 388–389
impermeant conjugate, 388–389, 416
redox sensitive, 389
Glucagon, 48
Gluconeogenesis, 93
Glucose uptake, 88, 198
Glyceraldehyde-3-phosphate dehydrogenase (GAPDH)
circadian pattern of activity, 144
replacement of inorganic phosphate in reaction by arsenate, 350
Glycolysis
anaerobic, 85
ATP production, 65, 73, 85, 89, 124, 345, 348, 416
of cancer cells, 86–88
and PMET, 86–88
Golgi apparatus, 18, 60, 69, 71, 105, 124, 132, 134, 188–192, 209, 232, 252, 331, 411
Gp91 phox, 74, 308
Gp97, 234
GPI linkage, 230
Gravity, ENOX response to, 191, 328
Green tea. *See* Tea catechins
Growth, 7, 76, 77, 81–82, 97–139, 168, 313–343, 355–361, 383–387
Growth-regulatory herbicides, 211, 212
GTP/GDP exchange protein, 356

H

HAT. *See* Histone acetyl transferases (HAT)
Hatchett's brown, 191
HDAC. *See* Histone deacetylases (HDAC)
Heart beat rate model, 166–168, 176
Helianthus annus, 60
Hepatoma plasma membrane (rat), 98–100, 369–372
Herbes de Provence, 197, 305
HER-2/neu protein, 234
Hexokinase II (HK-II), 87
HIF. *See* Hypoxia-inducible factor (HIF)
Histone acetyl transferases (HAT), 183, 184
Histone deacetylases (HDAC), 182, 184, 187, 198
Hitachi U3210 dual beam spectrophotometer, 26, 27
HIV-1. *See* Human immunodeficiency virus type 1 (HIV-1)
HK-II. *See* Hexokinase II (HK-II)
HL-60 cells, 103, 264, 309, 369, 372, 373, 390
hnRNP F splicing factor, 257–258
Hormesis, 257
Host defence, 1, 13–14, 75, 193, 210, 308, 440
Human immunodeficiency virus type 1 (HIV-1), 204, 207, 208
Human melanoma-associated glycoprotein, 234
HUVEC cells, 257
Hydroquinone, 1, 2, 12, 14, 15, 17–20, 46, 54, 62, 68–69, 118, 214, 217, 218, 261, 267, 279, 282, 325, 351, 396, 398, 402
Hydroquinone oxidase activity, 32–35
measurement, 32–35
Hypoxia, 72, 88, 90, 93
Hypoxia-inducible factor (HIF), 88, 90

I

IAA. *See* Indole-3-acetic acid (IAA)
Ibuprofen, 410
Idioforms, 408, 438, 440
Idiotypic antibodies, 86
Indole-3-acetic acid (IAA), 4, 60, 104, 108, 109, 129, 189, 194, 212, 313–317, 323, 325, 328–330, 335, 337–339, 342, 343

Indolyl-quinuclidinol, 56, 57
Influenza virus, 207
Infradian rhythm, 171
INK family, 356
Insulin, 48, 98, 198
Insulin resistance, 198
Ionic radius, 158–162
Iron, 8–10, 14, 50, 57, 72, 83, 214
Iron-sulfur clusters, 8, 50
Isobrucine, 209
Isoelectric focusing to stabilize ENOX1, 55, 62
Isoflavenes (isoflavones) as chemo sensitizers to overcome drug resistance
  MEI-143 triphendiol metabolite, 405
  naturally occurring, 74
  restoration of chemo-sensitivity, 405
  synthetic, 400
  triphendiol, 405

J

Jet lag, 177

K

Kalanchoe, 179
Keratinocytes, 288, 290, 411, 412, 415, 437
Kex2p protease, 277
Kozak sequences, 254, 255

L

Lactate, 73, 74, 87
Lactate dehydrogenase, 87
Lactoferrin, 48, 98
LASA. *See* Lipid-associated sialic acid (LASA) of sera
LDL. *See* Low density lipoproteins (LDL)
LDL oxidation, 6, 196, 197, 293, 294, 310
  causal to atherogenesis, 291
Lentivirus, 207
Lespedeza, 398
Life extension, 197–198, 210
Light scattering, 11, 23, 30, 127, 130, 332
Limit oscillators, 161, 167
Lipid-associated sialic acid (LASA) of sera, 432
Lipid bilayers, 68, 130, 131
Lithium, 168, 178, 179, 185
Low density lipoproteins (LDL), 8, 192, 195, 196, 210, 273, 291, 293–296, 310, 311
LY181984, LY181985. *See* Sulfonylureas, diaryl

M

Macrophages, 195, 196, 291, 293, 296, 310
Mad cow disease, 202
Malic enzyme, 87
Malondialdehyde, 196, 296, 301, 310
Mannitol, 332
MAP kinase, 356
Mass spectroscopy, 227
Matrigel, 57, 239, 242, 245, 247, 415
Mechanical stimulation, ENOX response to, 328
MDCK cells, 309
MEF. *See* Mouse embryo fibroblasts (MEF)
Mefluidide, 334
MEI-143 triphendiol metabolite, 405
MEK, 91–93
Melatonin, 2, 13, 30, 31, 46, 52, 53, 55–57, 63, 147, 169, 171, 180, 185, 213, 214, 218, 283, 284, 439
  receptors, 171
Membrane budding, 132, 137, 188–190, 411
Membrane displacements, 125, 139, 188–192, 209, 210
Membrane thinning, 130, 318
Memory, 201–202
Menadione, 67
Methotrexate, 227, 228, 381
α-Methylmannoside, 326, 327
MOLT-4 cells, 207, 208
Monensin, 105, 190, 252, 331
Mouse embryo fibroblasts (MEF), 121, 236–238, 391, 395, 401, 403
mPMS (1-methoxy-5-methylphenazinium methyl sulfate), 42–45
mTOR, 91
mTORC, 93
MUC-1, 234
Myc tag, 249
Myristoyla tion, 230

N

α-NAA. *See* α-Naphthalene acetic acid (α-NAA)
NAD^+ cellular concentration, 359
NAD^+ homeostasis, 94, 95, 270
NAD^+:NADH ratios
  cancer cells, 87
  normal cells, 87
NADH CoQ reductases, 65
NADH ferricyanide reductase, 66, 70–72, 76, 85, 201, 371, 376
NADH given orally, 81
NADH oxidases (NOX), 1–3, 74, 308
NADH-transferrin reductase and assay, 83

NAD(P)H-quinone reductase, 15, 67, 68
Na^{+}/H^{+} ATPase, 80
Na^{+}/H^{+} exchange, 80
Namalwa rho° cells. *See* Rho° cells
Nampt, 95, 182, 186
α-Naphthalene acetic acid (α-NAA), 314, 316, 323, 339, 342
β-Naphthalene acetic acid (β-NAA), 316, 323
Naphthyl sulfonylurea, 204
Naproxen, 410
N-ethylmaleimide (NEM), 100, 104, 128, 130, 136, 192
N-ethylmaleimide-sensitive factor (NSF), 137, 192
Neurodegenerative disorders, 200–201, 210
Neurological disorders, 6, 71, 201
NF-κB, 91–93
Nicontinamide mononucleotide synthesis, 95
Nitrate, 68, 128, 192, 323
p-Nitrophenylacetate, sodium, 116, 342
NMR, 16, 226, 227, 435, 439, 440
N-1-naphthylphthalamic acid, 317
Non-inceptive sensory neurons, 363
Non-steroidal anti-inflammatory drugs (NSAIDS), 345, 410–411
  anticancer activity, 356, 410
  independent of cyclooxygenase-2 (COX-2), 410, 411, 417
p-Nitrophenyl acetate, 341
Northern blot analysis, 243, 254, 277
NOX. *See* NADH oxidases (NOX)
NPAS2, 89, 180, 181
NQO1, DT-diaphorase, 66, 67, 309, 349–351
NQO1-specific inhibitor, ES936, 349
NSF. *See* *N*-ethylmaleimide-sensitive factor (NSF)
NSAIDS. *See* Non-steroidal anti-inflammatory drugs (NSAIDS)

O
Octacosanol, 332
Omi/HtrA2, 354
Oncofetal antigen, 234
Ortho-para spin orientations of water or D2O, 156–165
  spectral evidence for disequilibrium, 161–164
Oscillations, 11–13, 19, 141–186, 335–337
  and activity measurements, 24
  of arNOX proteins, 279
  in auxin-induced cell elongation, 189, 190
  in auxin-induced proton pumping, 327
  in auxin-stimulated protein disulfide-thiol interchange, 328
  CHO cell growth, 112, 119, 121
  in dNOX (auxin-stimulated) NADH oxidase activity, 336
  in ENOX2 activity, 213
  entrainment
    dNOX, 336
    light, 218, 439
    low frequency EMF, 174, 175
  EXAFS investigations, 152–154
  generation of a carrier wave, 166–168
  of Golgi apparatus activities, 191
  in growth of elongating pollen tubes, 168
  heart beat rate model, 166–168
  inherent in structure of water, 154–158
  period length determined by ionic radius of liganded cation, 158–161
  phasing
    autoentrainment, 169
    with blue light, 169
    with caffeine, 169
    link to biological clock, 169
    with lithium, 169
    with low frequency electromagnetic fields, 174
    with melatonin, 169
  in proton pumping activity, 327
  in redox potential, 150–152
  role of copper, 149–150
  temperature independence, 106, 111, 146, 150, 185
Osteopatch, 266, 271
Ouabain, 80, 81, 128, 130, 192
Outrider. *See* Sulfosulfuron
OVCAR-3 cells, 251
Oxaliplatin, 377
Oxidative stress, 72, 76, 123, 124, 195, 197, 284, 285, 291, 295
Oxygen, electron acceptor, 8, 19, 35, 73, 218, 261
  dissolved oxygen measurement, 35

P
p21, 355, 356
p27, 355, 356
p53, 355, 356, 414
p57, 356
p97, 124–138, 192, 234
Paclitaxel, chemosensitization to, 405–408
Pannexin channels, 17
Parkinson's disease, 17, 81, 201
Pasteur effect, 86
PCMB. *See* Perichloromer-curibenzoate
PDI. *See* Protein disulfide isomerase (PDI)
Pellia (moss), 105, 190, 331

PER–3, Per 1–3, 181
Piericidin, 69
Peroxidases, 9, 69
pH control, 193–195, 210
Phenoxodiol, 1, 56, 91, 123, 198–200, 235, 237–239, 259, 283, 345, 354, 356, 358, 400–408, 438
  binding to ENOX2, 220
  chemosensitizer to overcome drug resistance and restore chemosensitivity, 405
  clinical trials, 405
  effects augmented by ecliptic expression of ENOX2, 403
  inhibits oxidation of NADH and reduced coenzyme Q, 403
  inhibits protein disulfide-thiol interchange activity of ENOX2, 400
  siRNA knockdown of ENOX2, 404
  targets the PMET through inhibition of ENOX2, 400–405
Phenylacetate, sodium, 101, 116, 214, 227, 411
Phosphofructokinase, 86, 87
Phosphotungstic acid stain to identify plant plasma membrane vesicles, 125, 128, 317, 319
Photoreceptors, 86, 95, 146, 169
Phox-NOX, 73, 75, 193, 308, 309
  role in viral resistance cascade, 308
Phylloquinone
  as electron donor (plants), 54, 66, 67
  oxidase, measurement of, 33–35
Physical membrane displacements, 124,188
  energy requirements of, 132, 190–192, 209
Phytosterol, 317
PI3-AKT-mTOR-NF-κB, 91–93
Picloram, 340
Piroxicam, 410
Plant growth, 103, 211, 313–343
Plasma membrane electron transport (PMET), 65–96
  acrosome reaction, 202
  alkalinization of cytoplasm, 76, 78–80
  and arNOX proteins, 72
  and cancer, 86–88
  as cancer therapeutic drug target, 345–417
  capacitation, 202
  composition, 65–72
  electron donors and acceptors, 73
  energetics of, 76–78
  function in electron import, 80–81
  and glycolysis, 86–88
  and growth, 71, 81–82
  inhibition of and induction of apoptosis, 350–355
  link to generation of membrane potential and pH gradient, 193
  link to major signaling pathways, 88–89
  link to proton pump, 195
  NOX2 as terminal oxidase, 69–70
  outward proton pumping driven by, 78–80
  portion of ENOX cycle, 81
  proton efflux, 78–80
  rates of, 73–76
  regulation of, 82–88
  role of NQ01, 309
  and signaling pathways, 88–94
  of sperm, 202
  target for anticancer drug development, 348, 349
  terminal oxidases, 2, 3, 5, 47, 65, 97, 138, 187, 252, 269, 345, 421
  vanilloids (capsaicin) as inhibitors, 363
  in viral infections, 203
Plasma membrane vesicles, 5, 26, 27, 35, 77, 98, 101, 116, 117, 331–332
  inside out, 5, 125–127, 129–132, 136, 138–139, 191, 327
  right side out, 5, 77, 129, 327, 382, 387
Plasmodium falciparum, 219, 221
Platelets, 62, 197, 250, 306
  aggregation, 197, 306
Platinum drugs. *See* Cisplatin
PMET. *See* Plasma membrane electron transport (PMET)
Pollen tubes, growth oscillations, 168
Polyphenols, 387, 392–394. *See also* Tea catechins
  anticancer activity mediated through ENOX2-specific drug site, 392–394
  synergies, 396, 397
Polyphenon E, 396
Porin isoform 1, 65, 71, 95, 201
Preneoplastic liver nodules, 98
Prion, 7, 14, 16, 22, 57, 61, 62, 200–202, 210, 223, 231–233, 259, 337–342, 408, 435, 438, 440
Pronase, 231
Proteinase K resistance, 216, 223, 231, 232, 338
Protein disulfide isomerase (PDI), 6, 9, 10, 17, 38, 48, 54, 62, 71, 202–204, 207, 218, 226, 309, 316
Protein disulfide-thiol interchange activity and cell enlargement restricted to protein disulfide-thiol interchange portion of NOX cycle, 113, 200–202, 400

Protein thiols as electron donors, 15–16, 261, 262
Proton efflux, 78–80
Proton gradient, 76, 80, 193, 321
Proton pumping, 78–80, 95, 127, 128, 136, 193, 195, 321, 327
Proton release inhibited by antitumor sulfonylureas, 379
PTEN, modulation by NADH, 91–93
Purigenic receptors, cell surface, 17

Q

Quassinoids (antitumor), 85, 204, 207–209, 335, 387–389, 417, 422. *See also* Glaucarubolone; Simalikalactone D
  activity of conjugates redox sensitive, 389
  impermeant conjugates, 388
  target ENOX2, 387–389
Quassinoids (antiviral), 204, 207. *See also* Glaucarubolone

R

Rac GTPase, 308
Radiation enhancement of tumor response, 57
Raman spectroscopy, 162
Ras binding of GTP, 356
Ras-Raf-Mek-Erk, 91–93
Reactive oxygen species (ROS), 13, 62, 87, 177, 195–197, 263, 269, 273, 285, 286, 288, 291, 293, 295, 299, 305, 308–310, 363, 414
Red blood cells. *See* Erythrocytes
Red light, 146, 180
Redox-directed cancer therapeutics, 347
Redox potential, 16, 150–152, 155, 157, 164, 165, 172–176, 185, 417
Red wine, 197, 305
Resiniferatoxin, 84, 124, 363, 416
Resveratrol, 197, 305
Retinoic acid, 69, 116, 123, 341, 411, 412
  inhibition of PMET, 411
  inhibition of ENOX2, 411–412
  trans-retinoic acid (*see* Tretinoin)
Retinoids, 103, 214, 227, 411–412, 417
Retinol, 116, 124, 341, 411, 412
Rev-Erba, 90, 181
Rho° (ρ°) cells, 73, 75, 76, 85, 96, 124, 348, 350
Ribosome synthesis, 120
Rora, 181
Rotational analyses, 134, 135
ROS. *See* Reactive oxygen species (ROS)
Rotaviruses, 207
RT-PCR, 244, 247, 249, 250, 281, 433

S

Saccharomyces cerevisiae, 50, 57, 68, 274, 277
Salicin, 266, 283, 289, 299, 310
Saliva, 4, 195, 216, 261, 266–269, 281, 286, 289, 290, 301–305, 307, 310
Salvage pathway, 95
scFv. *See* Single chain variable region (scFv)
Schizandra chinensis, 283, 289, 299
Scrambled RNase substrate, 38–41
Scrapie, 201, 202, 230, 232
Sea urchins, 297
Segmentation clock of somitc formation, 180, 184
Semidehydroascorbate, 321
Serrotransferrin, 426. *See also* Transferrin
Serum, collection, 286, 441
Sialic acid, lipid associated of sera, 432
Simalikalactone D, 2, 7, 47, 56, 57, 74, 77, 85, 213, 266, 278, 283, 333–335, 342, 437
Simaroubaceae, 333, 387
Single chain variable region (scFv), 229, 259, 396, 423–426, 434, 442
Sir 2 deacetylase, 182
  NAD+ requirement for activity, 182
SIR 2 gene and life extension, 198
SIR proteins. *See* Sirtuins
SIRT 1, 89, 90, 93, 182, 183, 187
  role in circadian transcriptional regulation, 182, 186
Sirtuins (SIRT1-SIRT7), 89–90, 93, 187, 198
Site-directed mutagenesis, 50, 58–60, 68, 147, 221, 245, 294, 283, 415, 437
SK. *See* Sphingosine kinase (SK)
Skin aging, 6, 193, 286–291, 306, 310
Sleep cycles, 179
Smac, 354
SNARE proteins, 136
Sodium phenylacetate, 116, 214, 227, 411
Solar/lunar rotation cycle, 171
Solution NMR, 16, 435, 439, 440
Somites, 180, 184
Sorbitol, 332
Sorghum halapense, 298, 335
Soybean, 5, 26–28, 30, 31, 39, 47, 60, 77, 100–102, 104–109, 111–116, 125–132, 136, 144, 147, 171, 189–191, 212, 298, 314–329, 331–339, 341–343, 356
S1P. *See* Sphingosine-1-phosphate (S1P)

Spectrophotometric assays, 20–22, 38, 83, 131, 403
Sphingomyelinase, 91, 199, 349, 351, 352, 354, 359–362, 366, 416
Sphingosine kinase (SK), 73, 91, 199, 349, 351, 352, 354, 359–361, 416
Sphingosine-1-phosphate (S1P), 91, 199, 349–354, 359–361, 416
Sphingosine rheostat, 91, 95
Spongiform encephalopathies, 7, 230
Standard system, 67, 68
Stretch receptor, 332
Subtilisin, 231
Sulfonylurea herbicide, 127, 334, 335, 379
Sulfonylureas, diaryl, 378–387, 416
  anti-viral, 204
  binding protein, 379, 383
  BTS-2, 378
  EGF enhancement of response, 368
  ENOX2 as target, 417
  impermeant active conjugates, 380–383, 416
  LY181984, 1, 103, 113, 117, 123, 124, 193, 194, 228, 252, 378–387, 396, 422
  LY181985, 115, 117, 193, 194, 228, 380–383
  LY 237868, 380–383
  response redox and growth factor dependent, 383–387
  sulofenur, 378
  anti-tumor, 204, 214, 227, 256, 259, 349, 368, 378–383, 416, 417, 422
Sulforaphane, 408, 417
Sulfosulfuron, 335
Sulofenur. *See* Sulfonylureas, diaryl
Superoxide, 8, 15–16, 74, 196, 266–268, 307–309
  conversion to hydrogen peroxide, 8
  ferricytochrome c assay for, 266
  generation by arNOX proteins, 192
  periodic burst by arNOX, 439–440
  physiological role, 307–309
Superoxide dismutase
  extracellular, 295
  use in arNOX assay, 286, 310
Suprachiasmatic nucleus, 182
Suramin, 251, 409, 488
  as anti-viral, 204
  ENOX target for, 409
  response redox sensitive, 409
  treatment of trypanosomiasis, 409
SV-40, 250, 257, 415, 437
Syncytia formation, 206
α-Synuclein, 16, 17, 201

T

Tallysomycin, 376
Tamoxifen, 227, 228, 257, 381, 392
Taxol, 392, 405, 438
  Tea catechins, 120, 205–209, 214, 392–400, 417, 430. *See also* Polyphenols
  as antioxidants, 391, 396
  augmentation of cancer chemo and radiation therapy, 392
  and cancer, 396
  high dose cellular and biochemical responses, 394–396
  as inducers of apoptosis, 398, 399
  polyphenon E source, 396
Temperature dependence of plant growth, 323, 325, 331
Tetrazolium salts, 266, 310
Therapeutic applications, 345–417
Thiol reagents, 2, 84, 97, 100–101, 104, 132, 138, 192, 213, 279, 313, 316, 331, 379
  inhibit growth and ENOX activity, 100
Thioredoxin (NADPH/thioredoxin), 9, 62, 91, 92
Thioredoxin reductase, 9, 62, 218, 221
Thyroid hormone, T_3, 98
TM9SF. *See* Age-related NOX
tNOX. *See* ENOX2
α-Tocopherol, 196, 293
Tomato, 334
Trans-acitretin, 412
Transcriptional feedback loops, 181
Transferrin, 48, 67, 375. *See also* Serotransferrin
Transferrin receptors, 98, 375
Transgenic mice overexpressing ENOX2, 121, 123, 391, 392
Trans Golgi network, 104, 331
Transitional endoplasmic reticulum, 124, 135, 136, 188, 192
Transitional endoplasmic reticulum ATPase, 136
Transmembrane protein 9 superfamily (TM9SF). *See* Age-related ENOX proteins (arNOX, ENOX3)
Transplatin, 75, 76, 321, 324, 341, 377, 378
Tretinoin, 412
Triacontanol, 101, 333
Trichlopyr, 341
Triphendiol, 405
TRPV1, 62, 366
Trypsin, 231
α1-trypsin inhibitor, 426, 428

Turgor (pressure), 97
Turgor not driving force of all enlargement in plants, 104–105
Tyrosol, 2, 56, 266, 283, 299
Tyrosyl radicals, 272, 288, 290–292, 310

U
Ubiquinol (reduced coenzyme Q) oxidase assay, 279
Ubiquinol reduction by ENOX1, 68
Ultradian time keeping, 142

V
Vaccine based cancer intervention strategies, 416
Valerian, 180
Valosin. *See* AAA-ATPase
Vanadate, 70, 77, 127–130, 192, 193
Vanillic acid, 396
Vanilloids as PMET inhibitors, 363
  as cancer therapeutics, 365–366
VDAC1. *See* Porin isoform 1
Vertebrate cell enlargement, 113–114
  role of ENOX proteins, 114–118
Vesicle budding, 124, 128, 134, 135, 138, 188–192
Vesicle formation, 132, 188, 191
Viral pathogenesis, 203–209
Viral resistance cascade, 308
Virus-transformed cells, 349, 369
Vitamin D_3, 412
Vitamin K. *See* Phylloquinone
Vitamin K1 as electron donor (plants), 3, 15
VNOX, 4
V-8 protease, 231

W
Walker A, walker B motifs, 136
Water
  synchrony of oscillations, 172
  UV luminescence spectra, 173, 176
WST-1, 42–45, 309, 348

X
Xenografts, 228, 366, 415, 420
X-linked inhibitor of apoptosis (XIAP), 198, 199, 351–354
X-ray, 136, 150, 152, 153, 419, 420, 435, 439, 440
XTT, 43, 266, 270, 310

Y
Yeast
  arNOX, 277
  40 min metabolic cycle, 183

Z
Zinc, 9, 159, 214

Printed by Publishers' Graphics LLC